FLORA

OF

TROPICAL EAST AFRICA

PREPARED AT THE ROYAL BOTANIC GARDENS/KEW
WITH ASSISTANCE FROM THE EAST AFRICAN HERBARIUM

EDITOR

R.M.POLHILL, B.A., Ph.D., F.L.S.

RUBIACEAE (Part 2)

BY

D. BRIDSON AND
B. VERDCOURT, B.Sc., Ph.D.

CRC Press
Taylor & Francis Group
Boca Raton   London   New York

CRC Press is an imprint of the
Taylor & Francis Group, an informa business

FLORA OF TROPICAL EAST AFRICA

RUBIACEAE (Part 2)

D. Bridson & B. Verdcourt*

SYNOPSIS OF SUBFAMILIES AND TRIBES

The synopsis given in Part 1: 2–4 needs some modifications due to further work and the correction of errors, particularly in the authorities; these changes and some additions are detailed below (see also Darwin in Taxon 25: 595–610 (1976)). Publication of Robbrecht & Puff, A survey of the *Gardenieae* and related tribes (Rubiaceae), in E.J. 108: 63–137 (1986) was too late for the consideration of any adjustments to the generic arrangement but notes are included as necessary.

Subfamily **Rubioideae** (Part 1).

Tribe **Psychotrieae** *Cham. & Schlecht.* in Linnaea 4: 1, 4 (1829), as *Psychotriaceae.*
Tribe **Morindeae**. *Lasianthus* should probably be returned back to *Psychotrieae* as was universally done before Petit's transfer.
Tribe **Hamelieae** *DC.*, Prodr. 4: 342, 438 (1830).
Tribe **Hedyotideae** *DC.*, Prodr. 4: 342, 401 (1830).
Tribe **Anthospermeae** *DC.*, Prodr. 4: 343, 578 (1830); Puff in J.L.S. 84: 355–377 (1982). *Otiophora* has been excluded by Puff since the flowers are not wind-pollinated and the karyology is different; it does not, however, fit well in any other tribe and is probably best referred to a separate tribe placed near *Spermacoceae.*
Tribe **Spermacoceae** *Dumort.*, Anal. Fam. Pl.: 33 (1829), as *Spermacoceae.*

Subfamily **Cinchonoideae** *Raf.* in Ann. Gén. Sci. Phys. 6: 81 (1820), as *Cinchonaria.* Genera 43–108. I still do not feel able to accept *Ixoroideae* Raf., loc. cit.: 84 (1820) as a separate subfamily (Verdc. in B.J.B.B. 28: 251, 280 (1958)); it is difficult enough to define adequate tribes let alone subfamilies.

Tribe **Naucleeae** *Miq.*, Fl. Ned. Ind. 2: 130, 132 (1856); Ridsdale in Blumea 24: 307–366 (1978). In the light of Ridsdale's researches the description in Part 1 needs alteration as follows. *Hallea* and *Uncaria* are transferred to *Cinchoneae* and *Cephalanthus* (which does not extend into the Flora area) to a separate tribe. Delete climbers with hooked spines. Despite the removal of *Cephalanthus* the ovules can unusually be solitary in each locule. Genera 43–46.
Tribe **Cinchoneae** *DC.*; Ridsdale in Blumea 24: 43–100 (1978). Similarly the description needs alteration as follows. Subtribe *Mitragyninae* Havil. (*Hallea, Uncaria*) is particularly characterised by the placentation; placentas 2, thick, black, adnate to the septum but becoming free with age in lower ²/₃, pendulous; ovules numerous, upwardly imbricate, attached basally; sometimes climbers with hooked spines. Genera 47–50.
Tribe **Virectarieae**. Genus 51.
Tribe **Rondeletieae** *Miq.*, Fl. Ned. Ind. 2: 130, 156 (1856).
Tribe **Isertieae** *DC.*, Prodr. 4: 342, 435 (1830). Syn: *Mussaendeae* Hook. in G.P. 2: 8 (1873). Genera 52–55.
Tribe **Heinsieae**. Genera 56, 57.
Tribe **Gardenieae** *DC.*, Prodr. 4: 342, 347 (1830), as *Gardeniaceae.* Although this tribe is still not fully understood the division into two subtribes as suggested by Robbrecht & Puff (loc. cit. 1986) has been accepted.
Subtribe **Gardeniinae**. Shrubs, small trees or sometimes lianes, often with sympodial branching. Inflorescences terminal, pseudo-axillary or leaf-opposed. Flowers often large, (4–)5–12-merous; corolla contorted to left or sometimes right. Pollen grains single or in tetrads, 3-porate or 3-colporate. Ovary 1–9-locular (or incompletely so); placentas axile or parietal with numerous or rarely few ovules. Fruits mostly large, woody, leathery or sometimes fleshy. Seeds usually embedded in fleshy or pulpy placental tissue, occasionally free, mostly lenticular; endosperm entire; testa cells with thickenings mainly along the radial and inner (rarely also outer) tangential walls. Genera 58–69.
Subtribe **Diplosporinae** *Miq.*, Fl. Ned. Ind. 2: 237 (1857), (as *Diplosporeae*); Robbrecht & Puff in E.J. 108: 114 (1986). Shrubs or occasionally pyrophytic subshrubs with branching usually monopodial. Inflorescences axillary. Flowers mostly small, 4–5(–8)-merous; corolla contorted to left. Pollen grains single, 3-colporate. Ovary 2-locular; placentas axile, 1-many-ovulate. Fruit mostly small. Seeds usually free, or wholly or partly covered by arilloid placental tissue; endosperm entire or rarely

* Genera 70 and 71 by E. Robbrecht.

ruminate; testa cells isodiametric or elongated, thickenings mostly absent or ± weak, smooth along outer tangential walls; embryo with radicle inferior.

Robbrecht & Puff included *Cremasporeae* in this subtribe, but it is maintained in this work, see below. Genera 70–71.

Tribe **Hypobathreae** *(Miq.) Robbrecht* in B.J.B.B. 50: 75 (1980). Unarmed trees or shrubs. Inflorescences axillary. Corolla-lobes contorted to left. Pollen grains single, 3-colporate. Ovary 2-locular, the ovules 1-many, often biseriate, pendulous. Fruits mostly small with fleshy or rarely leathery wall. Seeds pendulous from an apical placenta; endosperm ruminate or entire; testa fibrous, mostly with very characteristic narrow elongated cells; radicle superior. Genera 72–73.

Tribe **Pavetteae** Dumort., Anal. Fam. Pl.: 33 (1829); Robbrecht in Pl. Syst. Evol. 145: 105–118 (1984); Bridson & Robbrecht in B.J.B.B. 55: 83–115 (1985). Shrubs, trees, sometimes scandent,or occasionally pyrophytic subshrubs; branching usually monopodial. Inflorescences terminal on main or lateral branches, pseudo-axillary or occasionally cauliflorous. Corolla small or occasionally quite long; lobes 4–5(–6), contorted to the left. Anthers inserted at throat, exserted, locellate in a few species. Pollen grains colporate with perforate or reticulate exine. Style exserted or very rarely included, fusiform to clavate, not or shortly lobed to distinctly lobed. Ovary 2(–7)-locular, with 1–numerous ovules on fleshy placentas attached near the middle of septum or less often basal with single ascending ovule. Fruit mostly small, less often quite large, drupes or berries or occasionally ± woody, 1–2(–7)-locular, sometimes with 1-seeded pyrenes; seeds not immersed in placental tissue; endosperm ruminate in several genera; radicle inferior; often a characteristic annular zone of thickened exotestal cells around the basal excavation. Genera 74–81.

Tribe **Coffeeae** *DC.* sensu stricto. Shrubs or small trees, monopodial or sometimes sympodial. Flowers moderately small, (4–)5–12-merous. Pollen grains single, 3–4-colporate, with reticulate exine and without endoapertures. Ovary 2-locular, with solitary ovules. Fruit small to medium-sized, drupaceous; endocarp crustaceous. Seeds with deep ventral groove; endosperm entire; embryo with radicle inferior; testa with cells crushed during the development of endosperm, comprised of thin elongate parenchymatic cells usually containing many ± isolated fibres. Genera 85, 86.

Robbrecht & Puff (loc. cit.: 122–123 (1986)) restrict *Coffeeae* to *Coffea* and *Psilanthus*, excluding the third Asiatic genus, *Nostolachma* T. Durand (*Lachnostoma* Korth.), associated with *Coffeeae* by Leroy (ASIC 9ᵉ Colloque Londres: 473–477 (1980)). Suggestions for the tribal positions of the following genera (placed here in *Coffeeae* sensu lato), made by Robbrecht and Puff are as follows:— *Galiniera*, *Feretia* and *Kraussia* in *Hypobathreae; Heinsenia* and tentatively *Belonophora* to the new tribe *Aulacocalyceae* Robbrecht & Puff and *Calycosiphonia* to *Gardenieae* subtribe *Diplosporinae*. Genera (for *Coffeeae* sensu lato) 82–84, 87–89.

Tribe **Cremasporeae** *Darwin* in Taxon 25: 601 (1976). Shrubs, small trees or lianes. Inflorescences axillary. Corolla small, the lobes contorted to left. Style filiform, exserted, hairy, bilobed only at the extreme apex; ovary 2-locular, each locule with a solitary pendulous ovule. Fruit small, indehiscent with leathery pericarp, 1–2-seeded; placenta elongate, lying along the whole side of the seed; radicle appearing inferior; testa characteristically finely transversely wrinkled-striate with strongly thickened radial walls; endosperm not ruminate. Robbrecht & Puff (loc. cit.: 115 (1986)) include this in *Gardenieae* subtribe *Diplosporinae*. Genus 90.

Tribe **Vanguerieae**. Ovary can be up to 10-locular. Pollen grains porate to colporate, the exine finely to more openly porate or sometimes reticulate. Genera 91–108.

Subfamily **Antirheoideae** *Raf.* in Ann Gén. Sci. Phys. 6: 86 (1820). Syn.: *Guettardoideae* Verdc. Tribal name remains *Guettardeae*. Limited heterostyly does actually occur. Genus 109.

REVISED KEY TO TRIBES*

1. Rhaphides present (except where replaced by thicker styloid crystals in *Urophylleae*); corolla-lobes valvate (save in *Hamelia*, cultivated); flowers mostly, but not always heterostylous, ixoroid pollination mechanism** never present; plants herbaceous or woody (subfam. Rubioideae) 2

   Rhaphides absent; corolla-lobes contorted, valvate or sometimes imbricate; ixoroid pollination mechanism frequently present, but reduced heterostyly or rarely heterostyly can occur in combination with valvate aestivation; plants woody, only occasionally herbaceous 14

2. Calyx-tubes confluent; fruits ± united into a fleshy mass (if more than 10–35-flowered see couplet 18); 1 calyx-lobe produced into a coloured lamina in one species; trees or shrubs . . . . **2. Morindeae** (sensu stricto), p. 134

* An artificial key to genera is given on p.419
** See p. 2 for explanation.

Calyx-tubes not confluent; fruits not forming a fleshy mass 3

3. Style divided into 6–8 filiform lobes at the apex; ovary with 4–10 locules, each containing 2 (rarely 3) collateral erect ovules, but only 1 developing into a seed in each locule; fruit a drupe with woody 4–10-locular putamen; shrub with terminal corymbs . . . . **3. Triainolepideae, p.149**

Style 2-lobed or if with 3–4(–5) lobes other characters not agreeing 4

4. Ovules or seeds numerous in each locule 5
   Ovule or seed solitary in each locule 6

5. Shrubs or small trees; fruit a berry . . . . **4. Urophylleae, p. 152**
   Herbs or subshrubs or if shrubs then with eccentric calyx-limb; fruit dry, mostly a dehiscent capsule . . . . **8. Hedyotideae, p. 177**

6. Leaves and leaf-like stipules in whorls of 4–8; erect, prostrate, or climbing herbs often with ± hooked hairs rendering the plant adhesive; flowers ± rotate; fruit dry or ± fleshy . . . . **11. Rubieae, p. 380**
   Leaves opposite or if in whorls of 3 or more then stipules dissimilar and with other characters quite different 7

7. Usually foetid smelling climbers; fruits flattened, the outer pericarp falling off to expose 2 compressed winged pyrenes supported by long filiform stalks . . . . **7. Paederieae, p. 174**
   Plants not evil-smelling; fruits not of this characteristic structure 8

8. Shrubs or trees with axillary (or supra-axillary) inflorescences 9
   Herbs, subshrubs or if shrubs or trees then inflorescences terminal 10

9. Locules 4–12; ovules erect from base; inflorescences sessile or pedunculate; fruit blue . . . . **2. Morindeae (*Lasianthus*), p. 134**

   Locules 2; ovules pendulous from near apex; inflorescences mostly supra-axillary, sometimes axillary, always pedunculate; fruit probably black . . . . **5. Craterispermeae, p. 161**

10. Ovules pendulous; locules and style-arms 2–5; herbs or subshrubs; flowers very often bright blue, occasionally white . . . . **6. Knoxieae, p. 164**
    Ovules not pendulous; locules and style-arms 2(–3); herbs, shrubs or trees; flowers seldom coloured 11

11. Fruits fleshy, indehiscent; trees or shrubs or in a few cases small forest floor herbs; flowers heterostylous; ovary 2-locular or if rarely 3-locular then inflorescence with a very conspicuous involucre . . . . **1. Psychotrieae, p. 25**
    Fruits dry, usually dehiscent; herbs or subshrubs; flowers heterostylous or not 12

12. Ovules attached to the septum of the ovary; flowers ♂ mostly in globose clusters at the nodes (sometimes only terminal node); pollen pluricolpate . . . . **10. Spermacoceae, p. 333**

    Ovules erect, attached to the base of the locules; flowers unisexual or ♂, axillary or in spikes or heads; pollen 3-colpate 13

13. Flowers ♂; stigmas ± smooth; corolla-tube very fine, almost thread-like in some species; calyx-lobes unequal, 1 or more enlarged . . . . **9. Anthospermeae (sensu lato, *Otiophora*), p. 315**
    Flowers unisexual, polygamous or ♂, wind-pollinated; stigmas feathery; corolla-tube not as above, rather reduced; calyx-lobes equal . . . . **9. Anthospermeae (sensu stricto), p. 315**

14. Fruit a large subglobose drupe, with ± lobed woody 4–9-locular putamen and fibrous outer layers; seeds without albumen; corolla white, salver-shaped, velvety outside, the tube ± 2.5 cm. long with 4–9 lobes (subfam. *Antirheoideae*) . . . . **23. Guettardeae** (Part 3)

Fruit not as above, if pyrenes woody then separate; seeds with albumen; corolla not exactly as above (subfam. *Cinchonoideae*) 15

15. Herbs or subshrubs; corolla-lobes valvate, very narrowly lanceolate, subequal to tube in length, suberect; both style and stamens exserted; capsule often with 1 persistent and 1 falling lobe; seeds numerous, unwinged . . . . **14. Virectarieae, p. 457**

Small shrubs to trees, climbers or pyrophytic plants; corolla-lobes and capsule not as above; if corolla-lobes valvate, then flowers heterostylous or not; seeds 1–numerous, winged or unwinged 16

16. Corolla-lobes valvate; style characteristically topped with a knob-like structure, globose, cylindrical or mitriform, often ridged, 2–several-lobed at apex when mature; pyrenes with woody or less often crustaceous walls; inflorescence always axillary . . . . **22. Vanguerieae** (Part 3)

Corolla-lobes if valvate then style not as above, occasionally heterostylous; if pyrenes present then cartilaginous to crustaceous 17

17. Fruit a dehiscent capsule; seeds numerous, winged; inflorescences terminal, perfectly spherical heads (if trees with solitary heads see couplet 18), spike-like (if corolla hairy outside see couplet 20), paniculate or corymbose . . . . **13. Cinchoneae, p. 445**

Fruit scarcely succulent to succulent, very rarely dehiscent; seeds 2–numerous, unwinged; inflorescences terminal or axillary, occasionally spherical or spike-like, often corymbose 18

18. Inflorescences terminal, perfectly spherical, with calyx-tubes joined together or free . . . . **12. Naucleeae, p. 439**

Inflorescences sometimes tightly capitate but not as above 19

19. Corolla valvate or reduplicate-valvate; ixoroid pollination mechanism never present; ovary 2–many-locular; each locule with numerous ovules; fruit indehiscent or less often capsular; shrubs, trees or frequently climbers; sometimes one calyx-lobe foliaceous, white, yellow or red . . . . **15. Isertieae, p. 460**

Corolla-lobes contorted or less often imbricate; ixoroid pollination mechanism usually but not always present; ovules solitary or few to numerous in each locule 20

20. Corolla-lobes contorted or imbricate; ovules numerous in each locule; limited heterostyly sometimes present; stipules completely split into 2 quite separate lobes if inflorescence cymose or entire if inflorescence spicate or capitate . . . . **16. Heinsieae, p. 475**

Corolla-lobes never imbricate; ovules solitary to numerous in each locule; ixoroid pollination mechanism usually present; stipules never bifid; inflorescence never spicate, sometimes capitate 21

21. Inflorescences axillary (sometimes close to apex but overtopped by inconspicuous terminal shoots); ovules (or seeds) 1–several in each locule, pendulous; style bifid at apex, usually hairy 22

If axillary inflorescences combined with 1–several ovules then placentation different and style 2-armed 23

22. Bracteoles cupular; anthers included or only tips showing; seeds with finger-print-like striations; endosperm ruminate . . . .   **18. Hypobathreae,** p.569

Bracteoles triangular-acuminate; anthers exserted; seeds finely transversely wrinkled-striate; endosperm entire   **21. Cremasporeae,** p. 732

23. Inflorescences terminal (sometimes on reduced leafless branches); corollas mostly salver-shaped or with reflexed lobes, occasionally showy; ovules 1–few or numerous in each of the 2 locules; seeds free, either with a ± circular excavation on ventral face or with a ruminate endosperm . . . .   **19. Pavetteae,** p. 581
Inflorescences axillary, lateral, or if terminal then corolla very showy, mostly campanulate or funnel-shaped and seeds numerous, held together in a matrix 24
24. Inflorescences terminal, often on reduced branches (sympodial growth), lateral or occasionally axillary; ovary 1–2(–4)-locular with numerous ovules in each locule; flowers often but not always very showy; stigma club-like, the arms seldom divergent; fruit usually but not always large with a woody or leathery pericarp; seeds mostly small, frequently held together in a matrix; pollen grains frequently but not always in tetrads . . . .   **17. Gardenieae** subtribe **Gardeniinae,** p. 485

Inflorescences mostly axillary (but if flower and fruit large see above); ovary 2-locular with 1–several ovules in each locule; style-arms divergent, exserted or included; fruit mostly smaller; seeds never held together in a matrix; pollen grains always single 25
25. Calyx-limb mostly reduced to a rim; seed 1 in each locule, grooved on ventral face (typical coffee-bean) . . .   **20. Coffeeae** sensu stricto, p. 693

Calyx-limb clearly present, usually persistent in fruit; seeds not as above 26
26. Inflorescences fasciculate 1–several-flowered; bracteoles frequently cupular; stipules sheathing and awned on either side . . . .   **17. Gardenieae** subtribe **Diplosporinae,** p. 485

Inflorescences many-flowered and laxly branched or if as above then either stipules quite different or anthers locellate and fruit with 2 hemispherical seeds, apparently lacking a testa . . . .   **20. Coffeeae** sensu lato, p. 693

ARTIFICIAL KEY TO GENERA

1. Herbs, sometimes subshrubs; corolla always valvate; rhaphides present in all save *Virectaria* . . . . .Key 1
Plants woody, or sometimes with ± herbaceous shoots from a woody rootstock (pyrophytes); corolla valvate, contorted or sometimes imbricate; rhaphides present or absent 2
2. Climbing plants or lianes . . . . Key 3 (p. 425)
Plants not climbing, but sometimes scrambling 3
3. Plants with subshrubby to shrubby or occasionally ± herbaceous shoots 0.05–2 m. tall, mostly burnt off annually, from thickly woody or sometimes rhizomatous rootstock . . . . Key 2 (p. 423)
Shrubs or trees (0.75–)2–20 m. tall . . . . Key 4 (p. 426)

Key 1

(Herbaceous genera)

1. Nodes with whorls of 4–8 leaves and similar leaf-like
   stipules; herbs and herbaceous climbers with ± rotate
   corollas; ovules solitary in each locule; fruits globose,
   indehiscent; plants often adhesive due to prickles and
   harsh hairs (*Rubieae*) 2
   Nodes with leaves opposite or if in whorls of 3–5 then
   clearly different from stipules; corolla tubular 3
2. Leaf-blades ovate to lanceolate, the petiole very well
   developed; corolla usually 5-merous . . . . . . **41. Rubia**
   Leaf-blades mostly narrow, linear or lanceolate or, if wider
   and ± elliptic then petiole very short; corolla usually
   4-merous . . . . **42. Galium**
3. Ovules (or seeds) 2 or more in each locule 4
   Ovules (or seeds) solitary in each locule 24
4. Seeds with circular wing; leaves opposite or in whorls of
   3–4; corolla coloured or sometimes white, showy; lobes
   4, erect and small or spreading and larger (cultivated) **Bouvardia (p. 25)**
   Seeds not winged; leaves opposite; corolla if coloured and
   showy then 5-lobed 5
5. Stigma capitate; flowers not heterostylous, both style and
   stamens exserted; capsule often with 1 persistent lobe
   and 1 falling lobe; corolla-lobes very narrowly
   lanceolate and about as long as or longer than the
   corolla-tube; small shrubs and herbs without
   rhaphides . . . . **51. Virectaria**
   Stigma bilobed; flowers often heterostylous, with either
   stamens or style included (except in *Pseudo-
   nesohedyotis*); capsule mostly opening at the beak;
   corolla-lobes usually not so narrow and often shorter
   than the tube; rhaphides usually easily visible 6
6. Ovary, placentas and fruit elongated, linear; forest-floor
   herb with white funnel-shaped corolla 5.5–8 mm. long
   (T3) . . . . **20. Dolichometra**
   Ovary, placentas and fruit never linear 7
7. Flowers mostly 4-merous; leaf-blades frequently narrow
   and uninerved or with lateral nerves obscure 8
   Flowers mostly 5-merous; leaf-blades usually broad and
   with very obvious lateral and tertiary venation (but not
   always) 19
8. Corolla-lobes induplicate-valvate, covered on the inside
   with clavate hairs . . . . **24. Pentanopsis**
   Corolla-lobes valvate; lobes smooth or papillate
   inside 9
9. Anthers and stigmas included, the latter always overtopped by the former; corolla-tube narrowly
   cylindrical . . . . **22. Kohautia**
   Anthers and/or stigmas exserted or if both included then
   anthers overtopped by the stigma; corolla-tube
   cylindrical or funnel-shaped 10
10. Corolla-tube cylindrical, at least 2 cm. long; anthers
    included and style exserted; flowers not heterostylous **23. Conostomium**
    Corolla-tube cylindrical or funnel-shaped, always less than
    2 cm. long, sometimes with included anthers and
    exserted style but then usually heterostylous 11
11. Flowers solitary or fascicled at numerous nodes forming
    long interrupted spike-like inflorescences; leaves
    linear-subulate, rather sparse; corolla-tube 1.5–2 mm. **24A. Manostachya***
    Not as above 12

* *M. staelioides* (K. Schum.) Bremek. has recently been recorded from S. Tanzania (see K.B. 35: 322 (1980)).

12. Capsule opening both septicidally and loculicidally 13
    Capsule opening loculicidally, sometimes tardily dehiscent 17
13. Beak of the capsule as long as or longer than the rest of the
    capsule .14
    Beak of the capsule shorter than the rest of the capsule 15
14. Fragile annual herb; capsule of very characteristic shape,
    emarginate at the base, the beak much exceeding the
    rest of the capsule; stipules triangular with 2 lobes   **25. Mitrasacmopsis**
    Robuster subshrubby herb with distinctly discolourous
    leaves; beak as long as the rest of the capsule; stipules
    3–7-fimbriate . . . .   **26. Hedythyrsus**
15. Robust shrubby herb; stipules with a single deltoid lobe,
    sometimes ± lacerated but not fimbriate; style and
    stamens long-exserted (Uluguru Mts.) . . . . .   **27. Pseudo-
                                                         nesohedyotis**
    Herbs; stipules mostly divided into distinct fimbriae,
    flowers heterostylous or isostylous but both stamens
    and style not long-exserted 16
16. Corolla-tube bearded inside; style not shortly bifid at the
    apex (apart from the division into stigma-lobes);
    stigma-lobes subglobose or ovoid . . . .   **28. Agathisanthemum**
    Corolla-tube glabrous inside; style shortly bifid at the apex
    as well as divided into ellipsoid stigma-lobes (K4)   **29. Dibrachionostylus**
17. Rush-like plant with linear or filiform leaves; stipule-
    sheath tubular, truncate or with 2 minute teeth; seeds dorsi-
    ventrally flattened . . . .   **30. Amphiasma**
    Plant not rush-like even if leaves linear; if seeds
    dorsiventrally flattened then leaves not linear 18
18. Capsule with thick woody wall and a solid beak, tardily
    dehiscent . . . .   **32. Lelya**
    Capsule with horny wall, with or without a beak but never
    solid, early dehiscent . . . .   **33. Oldenlandia**
19. Leaves uninerved, mostly fairly small; decumbent herb of
    wet places with small white or blue flowers in very lax
    elongated axillary cymes . . . .   **31. Pentodon**
    Leaves larger, pinnately nerved; flowers mostly terminal 20
20. Creeping herbs or mat-herbs of the forest floor,
    often rooting at the nodes 21
    Erect herbs or subshrubs 23
21. Calyx-lobes conspicuously spathulate at their apices;
    corolla-tube ± 3 cm. long . . . .   **18. Chamaepentas**
    Calyx-lobes not or much less spathulate; corolla-tube
    under 2 cm. long 22
22. Inflorescences ± several-flowered cymes, not sessile   **16. Pentas**
    Inflorescences few-flowered, usually only 1 or 2 corollas
    open at a time, sessile in the axils of the leaves   **21. Parapentas**
23. Flowering inflorescences capitate or lax much-branched
    complicated cymes, not elongating into simple spikes
    in fruit, although individual branches sometimes
    become spicate; fruit globose or obtriangular, less
    often ovoid-oblong . . . .   **16. Pentas**
    Flowering inflorescences capitate, later elongating into a
    long simple "spike", rarely with axillary spikes from the
    upper axils and frequently with solitary flowers at the
    lower nodes; fruits oblong . . . .   **17. Otomeria**
24. Style characteristically topped with a knob-like structure,
    globose, cylindrical or mitriform, often ridged, 2–
    several-lobed at apex when mature; flowers never
    heterostylous; fruit with 2–several woody pyrenes   go to Key 2, couplet 3
    Style lacking knob-like structure; flowers frequently
    heterostylous; fruit very occasionally with pyrenes or
    sometimes a putamen 25
25. Ovary more than 2-locular or, if 2, then ovules pendulous
    and corolla white to blue, over 6 mm. long 26

Ovary 2-locular and ovules erect or if pendulous then
    corolla white and small, 1.75–4 mm. long 27

26. Fruit blue, succulent, with cartilaginous or bony pyrenes;
    ovary 4–12-locular; ovules erect; corolla white or pink   **6. Lasianthus**
                                                       (in part)

Fruit not blue, scarcely succulent, sometimes inner walls
    rather woody; ovary 2-5-locular; ovules pendulous;
    corolla mostly bright blue, but sometimes white   **13. Pentanisia**

27. Ovules pendulous from near apex; flowers borne in small
    sessile terminal heads; fruit breaking off to leave a
    small cup, which is the persistent woody flanged pedicel   **12. Paraknoxia**
    Ovules erect from near base or attached towards middle of
    septum; lacking cup-like organ 28

28. Ovule erect from base of locule 29
    Ovule attached near middle of septum 33

29. Fruit dry, splitting into 2 cocci; not or partially
    heterostylous; herbs or subshrubs, usually in open
    situations 30
    Fruit succulent; heterostylous; creeping herb or sometimes
    shrubby herbs; always on forest floors 31

30. Calyx-lobes unequal, 1 or more enlarged; flowers ♂, not
    heterostylous, both stamens and style exerted; stigmas
    ± smooth; corolla-tube exceedingly fine . . . **34. Otiophora**
    Calyx-lobes equal, small; flowers usually unisexual or
    polygamous and partially heterostylous in some plants;
    stigmas feathery . . . . **35. Anthospermum**

31. Pyrenes not dehiscent; seeds with a red-brown or purplish
    testa; endosperm often ruminate; subshrubby herbs   **1. Psychotria**
    Pyrenes with ± well-marked dehiscence; testa of seed pale;
    endosperm not ruminate; usually small creeping
    herbs 32

32. Stipules lacking a membranous sheath within; pyrenes
    ribbed and often rugose . . . . **2. Geophila**
    Stipules with membranous sheath within; pyrenes smooth   **3. Hymenocoleus**

33. Ovary 3-locular; stigma-lobes 3; fruit with 3 cocci . . **40. Richardia**
    Ovary 2-locular; stigmas 2, or 1 capitate; fruits indehiscent,
    capsular with 2 valves, with 2 cocci or circumscissile 34

34. Fruit circumscissile about its middle, the top coming off
    like a lid; flowers minute, in globose nodal clusters;
    seeds with a ventral impressed x-like pattern . . . **39. Mitracarpus**
    Fruit indehiscent or opening by longitudinal slits or
    2-coccous 35

35. Succulent creeping plant of the seashore with imbricated
    leaves joined by quite broad sheathing stipules with
    very short processes; stems rooting at the nodes; fruits
    indehiscent; seeds not lobed; one record from T3   **34. Phylohydrax***
                       Puff (= **Hydrophylax** pro parte)
    Plant not a littoral succulent or if somewhat so (*Diodia*
    subgen. *Pleiaulax)* then leaves not imbricated, stipules
    with longer processes, stems not rooting at the nodes,
    fruits dividing into cocci and seeds lobed; common 36

36. Capsule opening from base to apex, the calyx-limb and
    joined valves falling off together, leaving the oblong
    septum . . . . **38. Spermacoce**
                                        subgen. **Arbulocarpus**
    Capsule opening from apex to base or fruit 2-coccous 37

37. Seeds very distinctly lobed in a characteristic manner (fig.
    48/7, p. 338) . . . . **37. Diodia** subgen.
                                        Pleiaulax
    Seeds not distinctly lobed 38

38. Fruit an obvious capsule . . . . **38. Spermacoce**
    Fruit with 2 indehiscent or ± indehiscent cocci   . . . **37. Diodia**

* *Phylohydrax madagascariensis* (Roem. & Schult.) Puff.

KEY 2

(Pyrophytic genera*)

1. Ovules (or seeds) solitary in each locule; flowers not
   markedly showy 2
   Ovules (or seeds) few–numerous in each locule; flowers
   showy or occasionally small 12
2. Corolla-lobes valvate, never heterostylous; stigma
   characteristically topped with a knob-like structure,
   globose, cylindrical or mitriform, often ridged, 2–
   several-lobed at apex when mature; pyrene-walls
   woody; inflorescences always axillary 3
   Corolla-lobes contorted, or if valvate either heterostylous
   or lacking characteristic stigmatic knob; pyrene-walls
   cartilaginous to crustaceous; inflorescences terminal,
   sometimes on leafless branches that can be reduced to
   give an axillary impression 9
3. Ovary 2-locular 4
   Ovary 3–8-locular, sometimes with a few 2-locular ovaries
   as well 6
4. Corolla coriaceous, drying wrinkled, glabrous; buds
   obtuse; foliage glabrous, drying the characteristic
   yellow-green of an aluminium-accumulator; pyrenes
   very thickly woody, irregularly ridged with lines of
   dehiscence apparent . . . .   **102. Multidentia**
                                             concrescens

   Corolla not coriaceous nor drying wrinkled; buds
   acuminate; foliage not drying yellow-green, hairy or
   velvety; pyrenes not so thickly woody 5
5. Subshrub 15–30 cm.; calyx-lobes 0.8–5 mm. long, ± leafy,
   narrowly oblong, elliptic or lanceolate; leaves hairy   **96. Pygmaeo-**
                                                           thamnus

   Mostly 0.6–5 m.; calyx-lobes 1.5–4 mm., linear not leafy;
   leaves velvety beneath discolorous . . . . . .   **100. Rytigynia**
                                                    subgen. **Fadogiopsis**
6. Small subshrub 5–25 cm. tall with paired leaves; corolla
   glabrous or sparsely hairy; foliage not velvety   **93. Pachystigma**
                                                     (in part)

   Mostly larger subshrubs, often with virgate stems from a
   woody rootstock 7
7. Corolla glabrous or pubescent (rarely densely hairy in
   some extra East African species); leaves glabrous to
   velvety, almost invariably in whorls of 3–6; calyx-lobes
   absent to quite well developed (up to 3.5 mm. long);
   fruits glabrous or with sparse hairs . . . . . .   **99. Fadogia**
   Corolla and foliage velvety, entirely covered with silky
   hairs; fruits velvety or ± glabrous and then leaves
   predominantly in pairs 8
8. Calyx-lobes 2–4 mm. long, linear-lanceolate; fruits with
   orange-brown velvety tomentum on drying; leaves in
   whorls of 3–4, less often paired . . . .   **97. Tapiphyllum**
                                                    (in part)

   Calyx-lobes short and triangular, 0.5(–1) mm. long; fruits
   almost glabrous; leaves mostly paired on leafy
   branchlets, in 3's on main stem . . . .   **98. Fadogiella**
9. Corolla-lobes contorted; tube distinctly longer than lobes;
   anthers and style always exserted; rhaphides absent;
   seeds with excavation on ventral face; inflorescence
   terminal on leafless lateral branches (sometimes
   reduced or even entirely suppressed) . . . . .   **79. Pavetta**
                                                    (in part)

* Genera of doubtful habit also included in Key 4.

Corolla-lobes valvate; tube subequal to lobes or a little longer, if longer then usually bright blue; rhapides present; flowers heterostylous; seeds not excavated; inflorescences terminal on leafy stems 10

10. Corolla blue, white, lilac or purple, tube longer than lobes, ovules pendulous; fruit ± dry . . . . **13. Pentanisia schweinfurthii**

Corolla white, tube subequal to lobes or a little longer; ovules erect; fruit fleshy 11

11. Inflorescence-axes not white; pyrenes without well-marked dehiscence; testa red-brown or purplish; endosperm often ruminate, stipules entire or bifid, often caducous, never becoming corky . . . . . **1. Psychotria** (in part)

Inflorescence-axes white, tinged purple; pyrenes with ± well-marked dehiscence; testa pale; endosperm not ruminate; stipules ± truncate often becoming corky **5. Chassalia umbraticola***

12. Corolla-lobes valvate; flowers heterostylous; rhaphides present; capsules dehiscent; stipules fimbriate; leaves ternate . . . . **16. Pentas** (in part)

Corolla-lobes contorted; flowers not heterostylous; rhaphides absent; fruit indehiscent; stipules not fimbriate; leaves paired or ternate 13

13. Flowers clearly axillary (in both axils); corolla not markedly showy; fruit small to moderately large (up to 2.5 cm. long) .14

Flowers terminal, lateral or exact position obscured due to reduction of branches; corolla very large and showy; fruit moderately large to large 15

14. Stipules triangular, green, 0.7–1.5 cm. long, inflorescences few-flowered, subcapitate, subsessile or with peduncle up to 1 cm. long; corolla-tube ± 2 cm. long; fruit 2–2.5 cm. long . . . . **69. Mitriostigma greenwayi**

Stipules sheathing at base, aristate above, not green, smaller; inflorescence fasciculate with cupular bracteoles; corolla-tube less than 1 cm. long; fruit smaller, ± 0.5 cm. in diameter . . . . **70. Tricalysia cacondensis**

15. Corolla-tube slightly shorter than lobes; calyx-lobes 10, unequal, the 5 large ones very variable, elliptic-spathulate, 2–4 mm. long, the 5 small ones triangular, 1 mm. long; fruit ellipsoid with 10 distinct longitudinal nerves, crowned by persistent calyx-limb . . . . **62. Catunaregam pygmaea**

Corolla-tube much longer than lobes; calyx not lobed as above; fruit not as above 16

16. Flowers 1–3, terminal on main and slender axillary shoots; corolla-lobes (4–)5(–6); calyx-lobes linear-oblong or oblong-lanceolate, 1.2–3 cm. long, persistent on the ± fleshy fruit; seeds free, with small excavation **81. Leptactina benguelensis**

Flowers solitary on reduced lateral branches; corolla-lobes 6–8; calyx-limb with long tubular part, often split, bearing 5–9 linear-oblong or linear-triangular lobes, 5–8 mm. long; fruit fibrous-woody; seeds held together by a matrix, not excavated . . . . **63. Gardenia subacaulis**

*Vollesen, in Op. Bot. 59: 68 (1980), doubts whether subsp. *geophila* deserves subspecific rank.

KEY 3

(Climbing genera)

1. Plants covered with curved prickles; leaves and stipules similar, in whorls of (2–)4(–8), cordate at base; corolla rotate to subcampanulate; fruit fleshy, 2-seeded   **41. Rubia**
   Plants not prickly but spines sometimes present; leaves usually paired, very dissimilar to stipules; corolla distinctly tubular   2
2. Stems 4-angled, with some short branchlets reduced to curved or ± hooked spines   3
   Stems occasionally 4-angled; hooked spines absent   4
3. Flowers in completely spherical heads; fruit a fusiform capsule; seeds winged   . . . .   **48. Uncaria**
   Flowers ± capitate; fruit globose, indehiscent; seeds not winged   . . . .   **76. Cladoceras**
4. Some calyx-lobes in each inflorescence dilated into a large white or coloured lamina   . . . .   **52. Mussaenda** (in part)

   No calyx-lobes dilated into large laminas   5
5. Corolla scarlet, salver-shaped with a very narrow tube; calyx-lobes unequal; ovules numerous in each locule   **17. Otomeria volubilis**

   Corolla not scarlet or if reddish then tube not narrow   6
6. Evil-smelling plants; fruits flattened, the outer pericarp falling off to expose 2 compressed winged pyrenes supported by long filiform stalks   . . . . . .   **14. Paederia**
   Plants not evil-smelling; fruit not of this characteristic structure   7
7. Corolla-lobes contorted; rhaphides always absent; flowers never heterostylous   8
   Corolla-lobes valvate; rhaphides present or absent; flowers heterostylous or not   11
8. Corolla large and showy; tube 1.5–5.5 cm. long; fruit large, at least 2.5 cm. wide; seeds very numerous, held together in a matrix   9
   Corolla smaller, not usually conspicuous; fruit smaller, less than 1 cm. wide; seeds 1–several, free   10
9. Deciduous plants; stipules scarious, persistent; flowers several–many, terminal on leafy branches; calyx-lobes green, subulate to linear-lanceolate   . . . . .   **60. Macrosphyra**
   Evergreen plants; stipules not scarious, deciduous; flowers 1–several, appearing axillary on alternate sides of successive nodes (see fig. 75/5); calyx-lobes reddish, oblong-lanceolate, contorted in bud   . . . . .   **66. Sherbournia**
10. Ovules 1–several, impressed on a placenta attached to septum; corolla with 5 reflexed lobes; seeds several per fruit, with a small cavity and entire endosperm; stipules triangular-ovate, apiculate, dark (dark green when fresh, turning black when dry); inflorescence very lax; leaf-acumen very pronounced, rounded or somewhat spathulate at tip   . . . .   **75. Tarenna fusco-flava**

    Ovule single, erect from base of each locule; corolla with 4 or 5 spreading lobes; seed one per fruit, spherical with strongly ruminate endosperm; stipules fimbriate, or shortly sheathing and aristate, not darkened; inflorescence dense, flat-topped or pyramidal; leaf-acumen less characteristic   . . . .   **77. Rutidea**
11. Ovule (or seeds) several–numerous in each locule   12
    Ovule (or seed) solitary in each locule   14

12. Fruit a capsule; seeds winged; flowers in axillary and
    terminal cymes, often running together to form fairly
    extensive panicles; leaves glabrous; rhaphides present   **15. Danais**
    Fruit ± fleshy; seeds not winged; flowers in axillary heads
    or moderately lax panicles; leaves sparsely to very
    densely pubescent or occasionally glabrous; rhaphides
    absent 13
13. Leaves variously pubescent or sometimes glabrous, but not
    velvety-felted; stigma divided into 5 lobes or arms;
    ovary 4–5-locular . . . .   **54. Sabicea**
    Leaves velvety-felted beneath; stigmatic lobes 2, laterally
    dilated; ovary 2(–3)-locular   **55. Pseudosabicea**
14. Flowers heterostylous; stigma bifid; pyrene-walls not
    woody; rhaphides present; ovules erect 15
    Flowers not heterostylous, style always much longer than
    corolla-tube; stigmatic knob cylindrical, hollow at base,
    shortly bifid at apex when mature; pyrene-walls woody
    or crustaceous; rhaphides absent; ovules pendulous
    from near top of septum 16
15. Corolla-lobes with a median winged keel, very evident in
    bud; pyrenes ultimately dehiscing; stipules ±
    semicircular, with a linear lobe on either side   **5. Chassalia
                                                                   cristata**

    Corolla-lobes not winged; pyrenes indehiscent; stipules
    ovate-triangular, bilobed at apex . . . .   **1. Psychotria
                                                                   ealaensis**
16. Leaves coriaceous, drying bright green, glabrous; stipules
    with a strongly keeled lobe; inflorescences not
    pedunculate; calyx-limb reduced to a rim, shorter than
    disc; anthers reflexed; fruit didymous; seeds with
    endosperm not streaked by granules . . . . .   **107. Psydrax
                                                                   subgen. Phallaria**

    Leaves chartaceous to coriaceous, not drying bright green;
    stipules lacking keeled lobe; inflorescence
    pedunculate; calyx-limb at least equalling disc, often
    dentate, occasionally lobed; anthers exserted but not
    reflexed; fruit ± heart-shaped; seed with endosperm
    streaked with granules, or rarely not . . . . . .   **108. Keetia**

KEY 4

(Woody genera)

1. Flowers in globose or capitate heads 2
   Flowers solitary or in lax to compact or spike-like inflorescences 11
2. Calyces ± joined together; fruits fleshy, ± united into a
   mass 3
   Calyces not joined together; fruits not forming a fleshy
   mass 5
3. Inflorescences 4–9 mm. in diameter (excluding corollas),
   5–35-flowered; flowers heterostylous; calyx with 1 lobe
   enlarged into a lamina in one species . . . . .   **7. Morinda**
   Inflorescences 3–5 cm. in diameter (excluding corollas),
   distinctly more than 35-flowered; flowers not heterostylous; calyx-lobes never enlarged into a lamina 4
4. Stipules deltoid or short, obtuse, subpersistent; placenta
   attached to the middle of the septum, somewhat
   discoidal; ovules spreading in all directions . . .   **43. Sarcocephalus**
   Stipules ovate, elliptic, or obovate, deciduous or
   subpersistent; placenta attached to the upper third of
   the septum, Y-shaped; ovules spreading in all directions
   but predominantly pendulous . . . .   **44. Nauclea**

5. Flowers orange turning red, tubular; heads supported by tight involucral bracts, the inner toothed; style-lobes flattened, adhering but easily separable; fruit large (cultivated) . . . . **Burchellia** (p. 438)
   Flowers not brightly coloured; fruit smaller 6
6. Ovary with 1 ovule in each locule; heads capitate, often surrounded by bracts; fruit succulent 7
   Ovary with many ovules in each locule; heads perfectly spherical, bracts absent or insignificant; fruit dry or ± fleshy 8
7. Corolla valvate; flowers heterostylous; inflorescence bracteate; leaves lacking bacterial nodules; rhaphides present . . . . **1. Psychotria** (in part)

   Corolla contorted; flowers with style and anthers always exserted; inflorescence with bracts small or sometimes larger and ± membranous; leaves with bacterial nodules; rhaphides absent . . . . **79. Pavetta** (in part)
8. Fruits ± succulent; corolla-lobes contorted (fig. 1/4) **57. Bertiera** (in part)
   Fruits capsular; corolla-lobes imbricate or valvate 9
9. Corolla-lobes valvate; stipules conspicuous, leafy or membranous, free or joined near base, often reddish; flowering heads 3–many . . . . **47. Hallea**
   Corolla-lobes imbricate; stipules not so markedly conspicuous; flowering heads solitary 10
10. Leaves paired, broadly elliptic; inflorescence terminal; calyx-lobes spathulate, persistent; wood a bright orange colour **45. Burttdavya**
    Leaves in whorls of 3–4, lanceolate; inflorescence lateral; calyx-lobes oblong, at length deciduous; wood not bright orange . . . . **46. Breonadia**
11. Inflorescence spike-like or racemose 12
    Inflorescence not as above 14
12. Inflorescence 20–30 cm. long; many flowers with one calyx-lobe enlarged, crimson and laminate (cultivated) **Warszewiczia** (p. 438)

    Inflorescences shorter; flowers lacking enlarged calyx-lobes 13
13. Inflorescences subtended by large leafy paired stipitate venose red bracts; corolla funnel-shaped or narrowly campanulate; fruit a dehiscent capsule; seeds conspicuously winged . . . . **49. Hymenodictyon floribundum**

    Inflorescence not subtended by conspicuous bracts; corolla ± salver-shaped; fruit a berry; seeds not winged **57. Bertiera** (in part)
14. Calyx-limb spreading, eccentric, entire or ± shallowly lobed, venose and accrescent, up to 2.8 cm. wide in fruit; fruit dry and tardily or not dehiscent; corolla-tube narrowly tubular, 1.2–3(–3.5) cm. long . . . . . **19. Carphalea**
    Calyx-limb not as above, usually with 4–5 distinct lobes or teeth 15
15. One calyx-lobe often dilated into a large often stalked white or coloured lamina mostly exceeding 2 cm. long and wide 16
    Calyx-lobes equal or slightly unequal and green or if 1–2 enlarged into a coloured lamina then not attaining 2 cm. in length or width 17
16. Fruit ± succulent or at least indehiscent; buds without filiform apical appendages . . . . **52. Mussaenda** (in part)

    Fruit dry, dehiscing at the apex; buds bearing 5 apical filiform appendages . . . . **53. Pseudo-mussaenda**

17. Anthers 5–6 cm. long, the tips just exserted; corolla white, narrowly funnel-shaped, 15–18 cm. long; flowers solitary, axillary, 1 on each side of the node; calyx-lobes leafy, 2.5 cm. long; corolla-lobes ± valvate, the edges actually very slightly imbricate (cultivated) . . . **Portlandia** (p. 438)

Anthers shorter even if corolla as long 18
18. Anthers exserted, hard and horny in texture, with the connective pubescent; buds with limb slightly asymmetric; inflorescences terminal; corolla white, the tube narrowly cylindrical, 9–15 cm. long, 2–4 mm. wide with lobes 1.2 cm. long (cultivated) . . . . . **Posoqueria** (p. 438)

Anthers if exserted then not hard and horny; bud-limbs never asymmetric 19
19. Anthers mostly included, only the tips exserted, about ¾ the length of the corolla; ovary 5-locular; flowers in terminal branched cymes; corolla ± cylindrical, orange-red, the lobes very short; style filiform, with narrowly fusiform undivided stigma (cultivated) . . **Hamelia** (p. 25)
Anthers not ¾ the length of the corolla; other characters not combined as above 20
20. Small tree or shrub growing in the littoral zone just above high tide mark, the stems covered with large leaf scars; leaves large, crowded at the ends of the branchlets; calyx truncate; corolla white, salver-shaped, velvety outside, the tube ± 2.5 cm. long with 4–9 lobes; ovary 4–9-locular; fruit globose, fibrously woody, up to 3.5 cm. in diameter . . . . **109. Guettarda**
Not growing in the littoral zone or if so then calyx, corolla, fruit and habit all quite different 21
21. Corolla valvate; flowers heterostylous, mostly white and not markedly showy; rhaphides present 22
Corolla valvate, contorted or sometimes imbricate; flowers occasionally heterostylous, but if valvate then corolla showy and coloured or with conspicuously hairy margins to lobes; rhaphides absent 29
22. Style divided into 6–8 filiform lobes; ovary with 4–10 locules, each containing 2 (rarely 3) collateral erect ovules, but only 1 develops into a seed in each locule, the undeveloped ovule often being stuck to the seed; shrub 1.8–6 m. tall; corolla-tube ± 1 cm. long, woolly tomentose outside; drupes red, with woody putamen; mainly littoral or in coastal bushland . . . . . . **8. Triainolepis**
Style 2-fid or if with several lobes then without above characters combined 23
23. Ovules numerous in each locule 24
Ovule solitary in each locule 25
24. Inflorescences compact axillary clusters or terminal and axillary; ovary 2-locular beneath, 4-locular above; stigma-lobes 2 . . . . **9. Pauridiantha**
Inflorescences lax axillary cymes ± 3 per axil; ovary 4–5-locular beneath, 8–10-locular above; stigma-lobes 4–5 **10. Rhipidantha**
25. Inflorescence axillary or supra-axillary 26
Inflorescence terminal 27
26. Leaves with numerous arching lateral nerves and closely parallel venation; locules 4–12; ovules erect; fruit blue; inflorescences axillary, sessile or pedunculate . . . **6. Lasianthus**
Leaves with venation mostly closely reticulate; locules 2; ovules pendulous; fruit probably black; inflorescences supra-axillary or less often axillary, always pedunculate **11. Craterispermum**

27. Pyrenes without well-marked dehiscence; testa red-brown
    or purplish; endosperm often ruminate; leaves
    sometimes with bacterial nodules; stipules entire or
    bifid . . . .   **1. Psychotria**

    Pyrenes with ± well-marked dehiscence; testa pale;
    endosperm not ruminate; bacterial nodules never
    present; stipules never lobed, frequently becoming
    corky 28
28. Inflorescence-axes not white; pyrenes opening by 2
    marginal slits; flower-buds never winged; shrub with
    leaves mostly developing after the flowers have
    matured; stems corky   **4. Chazaliella**

    Inflorescence-axes often white tinged purple; pyrenes
    opening by 1 dorsal slit; flower-buds and corolla-lobes
    often with longitudinal narrow wing-like keels (but not
    present in some species); flowers borne with mature
    leaves; stems not markedly corky . . . .   **5. Chassalia**
29. Branches usually armed with spines which are modified
    lateral branchlets; flowers 1–2(–3) together, apparently
    terminating opposite pairs of very much abbreviated
    leafy lateral shoots; corolla subrotate; ovary 2-locular,
    the placentas attached to the septum which divides the
    seed mass in fruit . . . .   **62. Catunaregam**

    Branches unarmed or if with spines then corolla tubular
    and other characters not combined as above 30
30. Each stipule divided into 2 quite separate lobes, rather
    small and caducous; corolla with 1 lobe imbricate;
    lobes showy, with a small horn-like apiculum showing
    as 5 tails in bud; calyx-lobes ± leafy, often somewhat
    spathulate, persistent in fruit; fruit globose, ± 1.5 cm. in
    diameter, slightly fleshy but soon dry; seeds small,
    numerous, foveolate . . . .   **56. Heinsia**

    Each stipule not divided into 2 lobes; corolla-lobes
    contorted or valvate; other characters not combined
    as above 31
31. Corolla valvate; style characteristically topped with a knob-
    like structure, globose, cylindrical or mitriform, often
    ridged, 2–several-lobed at apex when mature; flowers
    never heterostylous; locules 2–10; pyrene-walls woody
    or sometimes crustaceous; inflorescences always
    axillary 32

    Corolla contorted or valvate; style not topped with a
    characteristic knob; if corolla valvate occasionally
    heterostylous; locules mostly 2 or occasionally several;
    if pyrenes present then walls scarious to crustaceous;
    inflorescences, terminal, lateral or axillary 58
32. Spines present 33
    Spines absent (rarely in genus 103) 39
33. Calyx-lobes conspicuous leafy, 0.7–1.6 cm. long, ± 7 times
    as long as calyx-tube, much exceeding the corolla-tube,
    persistent in fruit; spines supra-axillary or present on
    trunks of saplings, can become quite robust; pyrenes
    1–5 . . . .   **92. Lagynias
                                                         pallidiflora**

    Calyx-lobes not as conspicuous as above; other characters
    not so combined 34
34. Spines present on trunks of young trees or restricted to
    young or coppice shoots in which case ternate or less
    often paired from leaf-axils; ovary 2-locular; corolla
    lacking ring of deflexed hairs inside; corolla-lobes not
    apiculate; fruit 1.3–2.4 cm. long and wide . . . . **105. Canthium
                                                          subgen. Lycioserissa**

Spines positioned above lateral branches (sometimes reduced) or from leaf-axils of more mature stems; other characters not combined as above 35

35. Spines arising above reduced or cushion-shoots 36
    Spines arising above leaves or above normal lateral shoots 38

36. Calyx-lobes linear, 1–3.5 mm. long or acute-triangular teeth 1 mm. long, persistent in fruit; corolla-lobes shortly to distinctly apiculate . . . .   **100. Rytigynia**
                                                                  (in part)

    Calyx-lobes shorter, scarcely 1 mm. long; corolla-lobes not or shortly apiculate 37

37. Ovary 4–5-locular; fruit moderately large, ± subglobose, 1.7–2 cm. in diameter; corolla-lobes with apiculum 0.5 mm. long . . . .   **104. Meyna**
    Ovary 2-locular; fruit smaller, laterally flattened, indented at apex, ± 1 cm. across; corolla-lobes scarcely apiculate
                                                                  **105. Canthium
                                                                  glaucescens**

38. Fruits large, ± 2 cm. long; pyrenes ± verrucose; leaves probably often in 3's . . . .   **101. Vangueriella**
                                                                  (inadequately known species)

    Fruits smaller, about 1.3 cm. long; pyrenes not verrucose; leaves paired . . . .   **100. Rytigynia**
                                                                  (in part)

39. Inflorescence umbellate (sometimes 1-flowered), entirely enclosed in bud by paired connate persistent bracts; stigmatic-knob not hollow, attached at base to style; flowers ⚥ or unisexual; ovary usually 2-locular (in Flora area) . . . .   **16. Pyrostria**
    Inflorescence never enclosed by paired bracts; stigmatic knob hollow where style enters, mushroom-like, rarely as above but then other characters not so combined 40

40. Corolla-lobes linear-lanceolate, 2–4 cm. long, greatly exceeding tube, erect, tomentose or pubescent outside; fruit 2–4 cm. long with calyx-remains triangular to linear; pyrenes 2 . . . .   **95. Vangueriopsis**
    Corolla-lobes not as above; fruit smaller or if large then with more than 2 pyrenes or with calyx-remains cupular 41

41. Ovary 2-locular (rarely occasional 3-locular ones as well) 42
    Ovary 3–5-locular (rarely occasional 2-locular ones as well) 51

42. Style usually at least twice as long as corolla-tube; stigmatic knob ± twice as long as wide, ± cylindric; stipules glabrous within; disc sometimes pubescent; seed with cotyledons orientated perpendicular to ventral face 43
    Style usually much less than twice as long as corolla-tube (save in *Multidentia exserta* which has stipules hairy within); stigmatic knob mostly as broad as long, often much broader; stipules hairy or glabrous within; disc glabrous; seeds with cotyledons orientated parallel to ventral face 44

43. Trees or shrubs, sometimes scandent; leaves typically subcoriaceous to coriaceous (drying light green) or occasionally chartaceous (in deciduous species); if scandent, then stipules triangular to truncate at base with strongly keeled lobe; calyx-limb a dentate to repand rim only occasionally equalling disc, usually much smaller; anthers usually reflexed; fruit not or scarcely indented at apex (except where ± biglobose); pyrene cartilaginous to woody with shallow apical crest   **107. Psydrax**

Scandent bushes with perpendicular lateral branches, often subtended by modified leaves; leaves chartaceous to subcoriaceous, rarely coriaceous; stipules lanceolate to ovate or triangular, acuminate; anthers usually erect; fruit strongly or slightly indented at apex, typically heart-shaped; pyrene woody with lid-like area surrounding crest (either positioned on ventral face or across apex) . . . . **108. Keetia**

44. Calyx-limb with tubular part cupular, well developed, repand or lobed but lobes never exceeding it; fruit large; pyrenes very thickly woody, strongly irregularly ridged with lines of dehiscence apparent; leaves with conspicuous network or tertiary nerves; corolla-tube with a ring of deflexed hairs inside; lobes never apiculate; stigmatic knob spherical to elongate-ellipsoid, ribbed . . . . **102. Multidentia**

Calyx-limb with tube obsolete or less well developed; lobes absent or, if present, usually greatly exceeding tube; pyrenes not so thickly woody or ridged as above; other characters not combined as above 45

45. Calyx-lobes ± leafy, 2.3–4 mm. long, persistent on fruit; bracts and bracteoles present, sometimes conspicuous; corolla-lobes acuminate or with short appendages   **94. Cuviera**

Calyx-lobes absent or subulate to linear; bracts and bracteoles inconspicuous 46

46. Flowers in branched mostly many-flowered pedunculate dichasial or complicated branched cymes, occasionally subumbellate by reduction; leaves restricted to new growth or not strictly so 47

Flowers solitary or in few–several-flowered fascicles, subumbellate or less often with rudimentary branches, peduncles mostly, but not always suppressed; leaves well spaced along branches or restricted to cushion shoots .48

47. Stipules seldom sheathing at base when mature, often becoming corky outside, if lobed then lobe not decurrent and often caducous; leaves strictly restricted to new growth; inflorescence with flowers usually arranged to one side of ultimate inflorescence branch; calyx-limb ± obsolete; fruit heart-shaped and strongly indented at apex or obovate . . . . **105. Canthium** subgen. **Afrocanthium** (in part)

Stipules sheathing at base, bearing a linear to subulate often decurrent lobe; leaves occasionally restricted to new growth; inflorescence not as above; calyx-limb lobed to base or almost so; fruit slightly indented at apex . . . . **105. Canthium** subgen. **Lycioserissa**

48. Leaves restricted to very short reduced branchlets or cushion shoots giving a pseudo-verticillate appearance; inflorescences always at leafless nodes . . . . . **105. Canthium** subgen. **Afrocanthium** (in part)

Leaves opposite, not restricted to short branchlets; inflorescences in axils of normal leaves 49

49. Flowers functionally unisexual, the ♀ 1-flowered, the ♂ many-flowered; stipules not pubescent inside; corolla-tube very short, 1–2 mm. long; lobes erect . . . . **105. Canthium** subgen. **Bullockia**

Flowers not functionally unisexual; stipules pubescent to villous inside; corolla-tube often, but not always, exceeding 2 mm. long 50

50. Corolla-tube glabrous or with few hairs within; foliage glabrous; fruit 1.3–2.4 cm. long and wide; buds not acuminate nor appendaged . . . . **105. Canthium** subgen. **Lycioserissa** (in part)

    Corolla-tube with a ring of deflexed hairs within or if glabrous the leaves distinctly discolorous and velvety beneath (*R. kiwuensis*); foliage glabrous to hairy; fruit 0.8–1.2 cm. diameter; buds in some species acuminate or with distinct tails . . . . **100. Rytigynia** (in part)

51. Corolla conspicuous, yellow; tube 2.5 cm. long, slightly curved, glabrous inside; flowers solitary and long-pedicellate; stigmatic knob not hollow; ovary 3–4-locular; fruit ± 1 cm. in diameter . . . . **91. Temnocalyx**

    Corolla not as above; stigmatic knob always hollow 52

52. Calyx-lobes long and ± leafy, 0.8–1.6 cm. long 53

    Calyx-lobes much shorter, mostly linear or triangular, at the most subfoliaceous 54

53. Calyx-lobes 0.7–1.6 cm., ± 7 times as long as the calyx-tube and usually equalling or ± exceeding the corolla-tube; cymes umbel-like with a ± cup-like bract; pedicels long; one species glabrous, the other with ferruginous red indumentum . . . . **92. Lagynias**

    Calyx-lobes leafy, ± 8 × 4.7 mm., about 4 times as long as calyx-tube, much exceeding the corolla-tube; cymes not umbel-like and with scattered ± leafy bracts and bracteoles . . . . **94. Cuviera** (in part)

54. Inflorescence a dichasial cyme with flowers scattered along the arms, usually lax and many-flowered but a few species with few-flowered inflorescences; fruit large and globose, 1.4–5 cm. in diameter (when dry) **103. Vangueria**

    Inflorescences varying from solitary flowers to fascicles or dichasial cymes but mostly condensed; fruits often small, not over 3 cm. in diameter (when dry) 55

55. Plant velvety-tomentose including the corolla and dense ± many-flowered inflorescences; fruit velvety-tomentose, 0.8–1.4 cm. in diameter . . . . **97. Tapiphyllum** (in part)

    Plant glabrous to hairy, usually not velvety-tomentose; corolla and fruits not velvety 56

56. Calyx-lobes 1–3 mm. long, linear, not at all foliaceous; flowers either in pedunculate 5-flowered cymes and buds not appendaged or flowers solitary and buds and corolla-lobes with long appendages . . . . . . **100. Rytigynia***

    Calyx-lobes 1–4.5 mm. long, ovate or oblong, mostly green and subfoliaceous; flowers solitary or in ± sessile fascicles 57

57. Buds acuminate or apiculate, the corolla-lobes with distinct appendages up to 2 mm. long; leaves mostly small, under 9.5 cm. long (only rarely to 16 cm.); fruits mostly ± 1–1.5 cm. in diameter; flowers borne with the mature leaves . . . . **93. Pachystigma**

    Buds ± acute, the corolla-lobes scarcely appendaged; leaves ultimately attaining 16.5 cm. in length; fruits up to 2.8 cm. in diameter; flowering when leaves are very young . . . . **103. Vangueria** sect. **Itigi**

58. Corolla-lobes reduplicate-valvate in bud (see fig. 1/1), spreading at maturity, yellow to red, usually differently coloured (by indumentum) towards throat; flowers isostylous or heterostylous; fruit fleshy, indehiscent; seeds small, numerous, unwinged . . . . . . **52. Mussaenda** (in part)

---

* A very imperfectly known species will key here which has been temporarily placed in *Vangueriella*.

Corolla-lobes if valvate not reduplicate nor coloured as above; flowers only occasionally heterostylous; fruit 2–numerous-seeded; seeds winged or unwinged 59

59. Corolla white, pink or violet; lobes valvate, very narrowly lanceolate, subequal to tube in length, suberect; both style and stamens exserted; calyx-lobes filiform, not persistent in fruit; capsule often with 1 persistent and 1 falling lobe; seeds numerous, unwinged   . . . .   **51. Virectaria**

Corolla and fruit not as above 60

60. Fruit a capsule with numerous winged seeds; corolla-lobes contorted or valvate 61

Fruit indehiscent; seeds not winged, 1-numerous; corolla-lobes contorted 65

61. Corolla-lobes valvate 62

Corolla-lobes contorted 63

62. Corolla glabrous; style always long-exserted, with a slightly clavate stigma; capsule with large lenticels, not crowned with remains of calyx-limb   . . . . . .   **49. Hymenodictyon parvifolium**

Corolla-tube tomentose outside and lobes margined with long hairs; flowers heterostylous but style never long-exserted; capsule with ± no lenticels, crowned with the remains of the calyx-limb (cultivated)   . . . . .   **Cinchona** (p. 437)

63. Corolla 2–3 cm. long, pink, fragrant; calyx-lobes lanceolate, 1–1.5 cm. long; flowers slightly heterostylous, the stigma exserted or included; seeds small, with a reduced wing at each end (cultivated)   . . . . .   **Luculia** (p. 438)

Corolla under 1.5 cm. long; flowers not heterostylous 64

64. Style long-exserted, the stigma clavate, slightly lobed; stamens exserted; calyx-lobes 0.5–1.5 mm. long; corolla white, under 1 cm. long; seeds with a very elegant fimbriated wing all round   . . . . . .   **50. Crossopteryx**

Style and stamens not or scarcely exserted; calyx-lobes 0.5–6 mm. long; corolla pink or vermilion with yellow or orange throat, 0.7–1.3 cm. long; seeds with smaller much less fimbriated wing (cultivated)   . . . . .   **Rondeletia** (p. 438)

65. Flowers apparently in dense terminal inflorescences, but actually subterminal, the uppermost axillary components overtopping the inconspicuous terminal shoot; corolla 6–7-merous; leaves subsessile, often ± cordate at base, coriaceous and very shiny above; calyx bracteolate at base; stigma fusiform and hairy, bifid at apex; ovules solitary; restricted to coastal areas   **73. Lamprothamnus**

Flowers not arranged as above, mostly 4–5-merous but sometimes 6–8-merous; other characters not combined as above 66

66. Deciduous shrubs with chaffy stipules and bracts; flowers rather small (corolla-tube 0.4–1.5 cm. long); calyx-limb with tubular part ± obsolete and lobes linear-lanceolate to lanceolate, caducous in fruit; fruit moderately small, 0.5–2 cm. in diameter; placentas with 2–10 seeds   . . . .   **83. Feretia**

Evergreen plants, or if deciduous with chaffy stipules then either flowers and fruit very much larger or ovules solitary; other characters not combined as above 67

67. Calyx-limb reduced to a rim, usually shorter than disc, but occasionally truncate and ± equalling disc; fruit 2-seeded; seeds with a well-defined groove on ventral face (typical coffee-beans); inflorescence 1–several-flowered 68

Calyx-limb at least equalling disc, truncate, dentate or lobed; fruit 2-many-seeded; seeds lacking ventral groove 69

68. Corolla-tube ± equal to lobes; anthers and style exserted;
    fruit longer than wide; evergreen or deciduous shrubs
    with monopodial or rarely sympodial growth . . .   **85. Coffea**
    Corolla-tube longer than lobes; anthers and style included;
    fruit wider than long or as above; deciduous shrubs
    with sympodial growth . . . .   **86. Psilanthus**
69. Inflorescences axillary, in both axils of a node or
    sometimes supra-axillary, pedunculate or sessile (see
    fig. 75/9,10) 70
    Inflorescences terminal on main and lateral branches
    (occasionally appearing axillary by extreme reduction
    of lateral branches) or lateral (i.e. alternate in only one
    axil at each node); sometimes with a few axillary
    inflorescences in addition to the terminal one (see
    fig. 75/1–8) 81
70. Leaves densely finely velvety buff tomentose beneath;
    flowers in sessile bracteate axillary inflorescences, the
    outer bracts ovate, up to 1.2 cm. long; corolla-tube 0.5–1
    cm. long . . . .   **54. Pseudosabicea**
    Leaves not finely velvety beneath or if slightly so then
    bracts and corolla different 71
71. Fruit large, 2–3.5 cm. long; corolla large (tube at least 2 cm.
    long); style fusiform to clavate; ovules (or seeds)
    several–numerous in each locule 72
    Fruit not as large; corolla seldom as large; style bifid or
    undivided; ovules (or seeds) several in each locule 73
72. Flowers solitary, supra-axillary; corolla trumpet-shaped,
    longitudinally ribbed; shrubs armed with straight
    spines . . . .   **67. Didymosalpinx**
    Flowers several in pedunculate heads; corolla salver-
    shaped; unarmed subshrub . . . .   **69. Mitriostigma**
                                                               greenwayi
73. Corolla subrotate; tube ± 1 mm. long; seeds (1–)2 per
    locule, held together at apex by an arilloid structure;
    inflorescence lax . . . .   **82. Galiniera**
    Corolla distinctly tubular; seeds not as above; inflorescence
    compact or less often lax 74
74. Inflorescences many-flowered, long-pedunculate, laxly
    branched; corolla funnel-shaped to campanulate   **84. Kraussia**
    Inflorescences few to several-flowered, sessile or
    pedunculate, compact or occasionally lax (then few-
    flowered); corolla-tube cylindrical to narrowly funnel-
    shaped 75
75. Stipules often blackening when dry; inflorescence
    distinctly pedunculate, laxly or trichotomously
    branched; seeds with small ± circular excavation or
    partly ruminate endosperm . . . .   **75. Tarenna**
                                                               (in part)
    Stipules not tending to blacken when dry; inflorescences
    sessile or shortly pedunculate, compact; seeds not as
    above 76
76. Stipules 1–1.8 cm. long, triangular at base, subulate-
    attenuate above, eventually caducous; anthers and
    style included . . . .   **88. Belonophora**
    Stipules much smaller, sheathing at base and aristate or, if
    triangular, then at the most acuminate or shortly
    aristate; style always exserted; anthers exserted or
    sometimes included 77

77. Stipules triangular-acuminate, sometimes shortly aristate,
       often caducous, sometimes obscured by inflorescences
       even at apical node; lateral branches sometimes
       subtended by small rounded often cordate leaves; style
       undivided, shortly bifid at apex or sometimes distinctly
       bifid, usually hairy; flowers several in sessile or
       pedunculate glomerules, very rarely pedicellate 78
   Stipules shortly sheathing at base, distinctly aristate;
       modified leaves absent; style clearly 2-armed; flowers
       few, stalked or subsessile .79
78. Anthers included or only tips showing; flowers 4-merous;
       bracts and bracteoles cupular, rather chaffy; seeds with
       ruminate endosperm and testa with finger-print-like
       striations . . . .   **72. Polysphaeria**
   Anthers exserted; flowers 5(–6)-merous; bracts and
       bracteoles triangular-acuminate, not forming cupules;
       seeds with endosperm entire and testa finely transversely
       wrinkled-striate . . . .   **90. Cremaspora**
79. Leaves with acumens very distinctive, 0.8–4 cm. long;
       anthers locellate; ovule 1 per locule; seeds 2,
       hemispherical; testa apparently absent . . . . .   **87. Calycosiphonia**
   Leaves with acumens much shorter; anthers not locellate;
       ovules 1–several per locule; seeds often more than 2,
       not hemispherical; testa clearly present 80
80. Anthers medifixed or ± medifixed, the connective not
       enlarged; anther-thecae ± contiguous and subparallel;
       bacterial nodules absent . . . .   **70. Tricalysia**
   Anthers basifixed, the connective enlarged so that the
       anther-thecae diverge; linear bacterial nodules often
       present along midrib and petiole . . . . . .   **71. Sericanthe**
81. Inflorescence corymbose to capitate, borne on leafless
       lateral branches, sometimes very short or entirely
       suppressed, usually (but not always) on both sides of
       main shoot giving an axillary impression; leaves with
       bacterial nodules; flowers 4-merous; fruit 2-seeded;
       seeds ± hemispherical, excavated on ventral face   **79. Pavetta**
                                                           (in part)

   Inflorescence if borne on reduced leafless branches then
       usually on only one side so that sympodial branching
       pattern is clearly apparent; bacterial nodules absent 82
82. Flowers orange turning red, tubular, in terminal heads
       supported by tight involucral bracts, the inner ones
       toothed; style-lobes flattened, adhering but easily
       separable; fruit large (cultivated) . . . .   **Burchellia**
                                                           (p. 438)
                                                                  83
   Flowers and inflorescence not as above 83
83. Deciduous shrubs; calyx corky, ridged, protective in dry
       season; corolla yellow-green, purple-spotted, 8–11-
       lobed; tube shorter than lobes, 1–1.4 cm. long; fruit
       3.6–7(–10) cm. diameter; flowers terminal on main and
       lateral branches . . . .   **65. Phellocalyx**
   Evergreen or deciduous shrubs; corolla and calyx not as
       above; fruit sometimes as large 84
84. Deciduous shrubs with conspicuous scarious or chaffy
       persistent stipules; flowers conspicuous, terminal on
       main or lateral branches; fruit large 85
   Evergreen shrubs or trees; stipules not as above; flowers
       conspicuous or not; inflorescences terminal or lateral;
       fruit small or large 86
85. Flowers several in clusters; climbing or scandent shrubs;
       stigmatic knob globose; corolla salver-shaped or
       shortly funnel-shaped . . . .   **60. Macrosphyra**

Flowers solitary; shrub or small tree; stigmatic knob ellipsoid; corolla funnel-shaped, rather long and slender . . . . **61. Euclinia**

86. Inflorescences terminal on main and lateral leaf-bearing branches (lateral branches sometimes reduced when habit a compact dryland shrub), see fig. 75/1, 2 87
Inflorescences lateral or terminal on reduced naked or 1-leafed lateral branches, see fig. 75/3-8 94

87. Ovule (or seed) solitary in each locule 88
Ovules (or seeds) few–numerous in each locule 90

88. Flowers 5-merous; seeds various, ruminate or sometimes as below; stipules often with central area darkened **75. Tarenna** (in part)
Flowers 4-merous; seeds ± hemispherical, with excavation on ventral face; stipules never with central area darkened 89

89. Bacterial nodules absent from leaves; inflorescence sessile or pedunculate, rhachis sometimes white or tinted, often articulated; bracteoles linear, short; style with 2 divaricate arms; seeds dull, often rusty, not rugulose **78. Ixora**
Bacterial nodules usually present in the leaves, either scattered or arranged along the midrib; inflorescence sessile, laxly corymbose to capitate, never with rhachis articulated; bracteoles stipule-like; style entire; seeds rugulose, moderately shiny . . . . **79. Pavetta**

90. Ovary 1-locular, with 2-9 parietal placentas; pollen grains in tetrads; stipules sheathing, often truncate, usually glutinous when young; flowers usually large and showy; fruit mostly large, often fibrous-woody; seeds held together in a matrix . . . . **63. Gardenia**
Ovary 2-locular, with placentas attached to septum; pollen grains single; stipules not as above; flowers sometimes showy; fruit small to moderately large, never fibrous-woody; seeds free 91

91. Calyx-lobes subfoliaceous, 0.8-3 cm. long, persistent in fruit; corolla large (tube 0.8-11 cm.), pubescent outside; fruit moderately large, 0.8-2.2 cm. in diameter; seeds numerous; stipules conspicuous, either triangular and erect or rounded and reflexed 92
Calyx-lobes not nearly as conspicuous; corolla generally small, glabrous or occasionally pubescent outside; fruit smaller; seeds 1-several; stipules less conspicuous, never rounded and reflexed 93

92. Corolla-tube shorter than or ± equalling lobes; anthers sessile but more than half exserted; anther-thecae locellate in Flora area; style exserted . . . . . **80. Dictyandra**
Corolla-tube much longer than lobes; anthers included; anther-thecae undivided in Flora area; style included or exserted . . . . **81. Leptactina**

93. Reduced dryland shrub, with leaves borne on cushion-shoots; stigma clavate, winged; corolla-tube ± half as long as the lobes; seeds ± 6 per fruit, shaped like an orange-segment, angular, notched on inner straight margin . . . . **74. Tennantia**
Small trees or shrubs, occasionally reduced as above; stigma not winged; corolla-tube ± equal to or sometimes longer than lobes; seeds 1-many per fruit, variously shaped but never as above . . . . . . **75. Tarenna**

94. Ovules 2-10, embedded on a pendulous placenta; corolla moderately small, campanulate, tubular at extreme base, (0.7-)0.8-1.2 cm. long, excluding lobes; flowers 1-many, subtended by a single leaf, clustered at ends of very reduced shoots (see fig. 75/3); fruit small, up to 1.3 cm. in diameter, 1-several-seeded (See also p. 736) **89. Heinsenia**

Ovules more numerous; corolla if as small then cylindrical
or funnel-shaped; inflorescence if as above with
corolla much larger; fruit many-seeded, small or
much larger 95
95. Inflorescences several–many-flowered cymes, one borne
alternately at every other node (see fig. 75/7, 8);
flowers small or medium-sized (corolla-tube up to 2.5
cm.long) 96
Inflorescences 1–several or many-flowered, terminal on
reduced shoots or at alternate sides of successive nodes
(see fig. 75/3–6); flowers medium-sized to large 97
96. Corolla-tube cylindrical, ± 6 mm. long, mostly glabrous
outside; stipules triangular, not conspicuous; fruit
small, less than 1 cm. in diameter; flowering nodes
anisophyllous (see fig. 75/7) . . . .   **58. Aidia**
Corolla-tube funnel-shaped, 1.6–2.5 cm. long, densely
covered with silky hairs outside; stipules elliptic, up to
1.2 cm. long, conspicuous; fruit said to be as large as an
egg; flowering nodes not anisophyllous, but one or
both leaves sometimes fallen (see fig. 75/8) . . .   **59. Porterandia**
97. Inflorescences terminal on reduced branches or
apparently so (see fig. 75/3, 4); stipules if triangular
then brown and inconspicuous, sheathing or not;
seeds held together by a matrix, surface smooth or
slightly reticulate 98
Inflorescences strictly lateral (see fig. 75/6); stipules fairly
conspicuous, green, triangular or ovate, scarcely
sheathing; seeds not held together by a matrix, with
surface resembling a finger-print 99
98. Stipules sheathing, often truncate; young parts usually
glutinous; placentas 2–9; pollen in tetrads . . . .   **63. Gardenia**
Stipules scarcely sheathing, usually triangular; young parts
not glutinous; placentas 2 (sometimes partly fused);
pollen grains simple . . . .   **64. Rothmannia**
   68. Oxyanthus
99. Corolla white; tube narrowly cylindrical . . . . . .   **69. Mitriostigma**
Corolla purple; tube funnel-shaped . . . .   **usambarense**

Subfamily **CINCHONOIDEAE**

The genera of this subfamily cultivated in East Africa are included in the artificial keys
(see p. 419) and in Part 1: 9 in sufficient detail for them to be identified.

Cinchona was once a very important crop in East Africa and at certain periods especially during the
First and Second World Wars very extensive plantations were maintained in order to provide
supplies of quinine which had been cut off from other sources; nearly all the local needs were
supplied from these plantations. Even now quinine is being employed in some areas where synthetic
drugs have not proved entirely satisfactory. Plantations were established at the very beginning of this
century by the Germans at Amani. A very large number of species of *Cinchona* have been described
but a survey of the material suggests that most of these are superfluous and are at the most cultivars of
a very few species. Standley in his accounts of the Rubiaceae of Colombia and Peru (Field Mus. Nat.
Hist. 13(6, 1): 24–33 (1936) and ditto 7(1): 10–14 (1930)) takes this course. Naturally in the days when
Cinchona bark was an extremely valuable item of commerce the correlation of morphology with
quinine alkaloids content was important. It is of course a well-known fact that exactly the same taxon
can differ markedly in chemical content according to where it is growing. The systematics of these
variants is beyond the scope of this Flora. At this juncture the whole matter is academic but at the time
when quinine was the sole cure for malaria the fact that different variants of *Cinchona* species varied
enormously in their alkaloid content was obviously a fact of crucial importance to plantation owners
when the output of alkaloid per acre could be multiplied 10 or more fold if the correct variants were
grown. Many of these races undoubtedly bred true from seed. All the material seen can be broadly
classified under the two species *C. pubescens* Vahl and *C. officinalis* L.

C. pubescens Vahl (cv. = *C. succirubra* Klotzsch). Leaf-blades more rounded, 9–50 cm. long, 7–35 cm.
wide, pubescent to glabrescent beneath. Corolla tinged pink or red with tube 1.1–1.4 cm. long.
Capsule elongate. Selected citations: Tanzania, E. Usambaras, Amani, *Greenway* 890 & 900 & 29 Sept.
1920, Coll.? No. 1! & No. 2!; W. Usambaras, Balangai, 16 Mar. 1916, *Peter* 16156!

C. officinalis L. (cv. = *C. ledgeriana* (Howard) Trimen, *C. calisaya* Wedd. var. *ledgeriana* Howard). This, the 'yellow-bark', is perhaps the most widespread taxon in cultivation and the richest in alkaloids. It is characterised by narrowly oblong-elliptic glabrous leaf-blades, 4.5–11.5(–15) cm. long, 1–4.5 cm. wide, a yellowish corolla with tube 7–8 mm. long and compressed oblong capsules 8–10(–15) mm. long, 3.5–4.5 mm. wide, which appear rounded-ellipsoid after dehiscence. Selected citations: Kenya, Nairobi Arboretum, 23 June 1952, *G.R. Williams* 465!; Tanzania, E. Usambaras, Amani, Kiumba, Feb. 1930, *Greenway* 2205!, 2207! & W. Usambaras, Lushoto, July 1955, *Semsei* 2311!

C. officinalis L. (cv. morphologically identical with *C. josephiana* (Wedd.) Wedd. but probably a form of *C. ledgeriana*). A variant grown in Entebbe, Uganda, is represented by numerous specimens at Kew sent for naming. Sprague and Sandwith reported these as *C. josephiana* (Wedd.) Wedd. (*C. calisaya* Wedd. var. *josephiana* Wedd.) and they certainly agree well with specimens so-called. The Uganda material, however, was reported to have a high alkaloid content whereas *C. josephiana* was widely recognised to be useless, so much so that it was destroyed in many plantations in order to prevent the formation of hybrids with other more valuable strains. The Entebbe variant was finally considered to be a form of *C. ledgeriana*, but undoubtedly differs from that in its rather wider leaf-blades which are narrowly elliptic, 15–20 cm. long, 5.7–6.5 cm. wide, and larger capsules 0.9–1.7 cm. long 5–6 mm. wide; the corolla-tube is similar, 5–8 mm. long. Selected citations: Uganda, Entebbe, Forestry Dept. Plantations, May 1931, *Snowden* 1950! & Jan. 1931, *Snowden* 1951! & 1952! Specimens from Amani, e.g. Feb. 1930, *Greenway* 2203! are morphologically identical.

C. officinalis L. (cv. = *C. robusta* Hort.). This has large elliptic leaf-blades 10–22 cm. long and 5.5–12 cm. wide, a corolla-tube 1.1–1.3 cm. long and capsule 1.8–2 cm. long and resembles figures of true *C. officinalis*.

Cinchona hybrids and other cultivars. Apart from those already mentioned there are very many specimens labelled *C. hybrida* or without names. Many of these, including several sheets from Lushoto, e.g. Tanzania, W. Usambaras, Lushoto, 29 July 1929, *Ngoundai* 352!, can be identified with authentically named *C. josephiana* (Wedd.) Wedd. or *C. calisaya* Wedd., probably the latter, which as we have seen are no more than minor variants or cultivars of *C. officinalis* L. Other specimens, e.g. Tanzania, E. Usambaras, Amani, Mt. Bomole, 14 Oct. 1917, *Peter* 21781! and ditto, 30 Oct. 1914, *Peter* 55504! and Amani, Maiskopf Plantation 2, 18 Feb. 1930, *Greenway* 2204! & 17 Feb. 1930, *Greenway* 2198!, may well be hybrids between *C. ledgeriana* and *C. succirubra*. They have leaf-blades broadly elliptic, up to 8–13 cm. wide, but often with capsules and flowers not much different from *C. ledgeriana* but more usually the capsules up to 2.5 cm. long.

The following references may be useful to those needing more information about this genus:— Bentley & Trimen, Medicinal Plants 2, tt. 140–143 (1880); Howard, J.E., Illustr. Nueva Quinologie of Pavon, 27 plates, London (1862); Howard, J.E., The Quinology of the East India Plantations, 116 pp., 15 plates, London (1869–76); Howard, J.E., On *Cinchona calisaya*, var. *ledgeriana* Howard, and *C. ledgeriana* (Bern. Moens), J.L.S. 20: 317–329 (1883); Kuntze, C.E.O., Monographie der Gattung *Cinchona* L., 39 pp., Leipzig (1878); Planchon, G., Des Quinquinas, 150 pp., Paris & Montpelier (1864); Triana, J., Nouvelles Études sur les Quinquinas, 80 pp., 33 plates, Paris (1872); Trimen, H., *Cinchona ledgeriana* a distinct species, J.B. 19: 321–325, t. 222, 223 (1881); Weddell, H.-A., Revue de Genre Cinchona, Ann. Sci. Nat. sér. 3, 10: 5–13 (1848); Weddell, H.-A., Histoire Naturelle des Quinquinas, 108 pp., 30 plates, Paris (1849); Weddell H.-A., Notes sur les Quinquinas, Ann. Sci. Nat. sér. 5, 11: 346–363 & sér. 5, 12: 24–29 (1869); Fl. Rwanda 3, fig. 47/1 & 2.

Luculia gratissima Sweet (Cinchoneae), a native of India, Indo-China and China is sometimes cultivated in the Kenya highlands (Trans-Nzoia District, near Kitale, 10 Oct. 1959, *Verdcourt* 2449!); it is a small tree or shrub with calyx-lobes green at apex, red beneath and pink fragrant flowers.

Several species of *Rondeletia* (Rondeletieae) have been reported from East Africa in gardens. I have seen two in parks and gardens and *R. amoena* (Planchon) Hemsley with leaves velvety beneath has been recorded from Tanzania (Amani, *Wilson* in *E.A.H.* 67/60!) and Kenya (Nairobi, *Starzenski* 62!). *R. cordata* Benth., a native of Central America, is a shrub about 2–6 m. tall with ovate-cordate leaves up to 10 × 5.5 cm. and large thyrsoid inflorescences of pink flowers with yellow throat-hairs (Kenya, Nairobi City Park, 1 Aug. 1961, *Verdcourt* 3205!, and also recorded for Zanzibar and Pemba, see U.O.P.Z.: 434 (1949)). *R. odorata* Jacq. is a small shrub with stiff oblong-ovate-cordate leaves 4.5 × 3 cm. and smaller inflorescences of more showy vermilion flowers with orange centres (Tanzania, E. Usambaras, Amani, Bustani, 26 Dec. 1956, *Verdcourt* 1734!).

Warszewiczia coccinea (Vahl) Klotzsch (Rondeletieae), a native of tropical America, a tree to 7.5 m. with spectacular narrowly thyrsoid inflorescences of scented yellow-orange flowers, many of the calyx-lobes being enlarged into bright crimson leaf-like organs, has been grown occasionally (Tanzania, E. Usambaras, Sigi, 25 Apr. 1947, *Greenway* 7940! & same locality, 16 Dec. 1940, *Greenway* 6088!).

Portlandia grandiflora L. (Rondeletieae-Condamineineae) has been grown in Nairobi (Nairobi Arboretum, 1928, *Battiscombe* 1504!).

Posoqueria latifolia (Rudge) Roem. & Schultes (tribe uncertain, possibly deserving a separate one) has been grown around Amani in NE. Tanzania (Amani Institute grounds, 10 Feb. 1971, *Furuya* 186!; Sigi Plantations, 1 May 1934, *Greenway* 3995! & 19 Apr. 1954, *Verdcourt* 1123!). It has also been reported from Zanzibar (U.O.P.Z.: 422 (1949)).

Burchellia bubalina (L.f.) Sims (Gardenieae) has been grown in Kenya (eg. Nairobi City Park, *Mwangi* in *E.A.H.* 14402).

Other cultivated species belonging to the genera which also contain native species are dealt with under the genus concerned.

Tribe 12. NAUCLEEAE*

1. Ovaries and fruitlets connate; fruit a syncarp 2
   Ovaries and fruitlets free 3
2. Stipules small, triangular, the apex obtuse or shortly
   notched ± persistent; placenta attached to the middle
   of the septum, somewhat discoidal with ovules
   spreading in all directions . . . . **43. Sarcocephalus**
   Stipules large, elliptic or obovate, deciduous or ±
   persistent; placenta attached to upper ⅓ of septum,
   Y-shaped, ovules spreading in all directions but
   predominantly pendulous . . . . **44. Nauclea**
3. Interfloral bracteoles absent; inflorescences terminal;
   placentas attached to the middle of the septum; calyx-
   lobes spathulate, persistent; leaves opposite, ovate or
   elliptic; wood bright orange . . . . **45. Burttdavya**
   Interfloral bracteoles present; inflorescences lateral;
   placentas attached to the upper third of the septum;
   calyx-lobes oblong, at length deciduous; leaves in
   whorls of 3–4, lanceolate; wood not bright orange **46. Breonadia**

43. SARCOCEPHALUS

Sabine, in Trans. Hort. Soc. 5: 442 (1824); Ridsd. in Blumea 22: 546 (1975); Ridsd. in Blumea 24: 324 (1978)

Small to medium-sized trees. Terminal vegetative buds flattened. Leaves opposite, petiolate, the nerve-axils with or without domatia; stipules deltoid, obtuse or slightly notched at apex, ± persistent. Inflorescences terminal or terminal and lateral with unbranched axes, the node with reduced leaves and stipules which are not modified into bracts surrounding the young inflorescence; true peduncle glabrous or pubescent, elongating. Flowers 4–5-merous, calyx densely pubescent inside, limb-tubes ± developed. Calyx-lobes obtuse to triangular with or without deciduous or persistent appendages. Corolla-tube funnel-shaped; lobes oblong, imbricate in the bud. Stamens inserted in the tube, not exserted; filaments short glabrous; anthers basifixed, introrse. Ovaries entirely fused to each other, 2-locular; placentas ± discoidal, attached to the middle of the septum; ovules numerous in each locule, spreading in all directions; style exserted; stigma spindle-shaped. Individual fruits indehiscent, joined to form a fleshy syncarp. Seeds ovoid or ellipsoid, not winged.

Two species restricted to tropical Africa.

S. latifolius (*Smith*) *Bruce* in K. B. 1: 31 (1947); Ridsd. in Blumea 22: 547 (1975). Type: Sierra Leone, *Smeathman* (BM, holo.!)

Shrub or small tree 2–9 m. tall; branchlets mostly thick, glabrous or minutely puberulous, drooping; bark grey or brown, very fibrous, deeply fissured; slash yellow with crimson streaks. Leaves dark green and shining above; blade broadly elliptic to rounded-ovate, (8–)10–21 cm. long, (5–)7–14 cm. wide, shortly acuminate at the apex, cuneate to rounded or subcordate at the base, glabrous save for sometimes a fine indumentum on the costa beneath; secondary nerve-axils containing very small pubescent domatia; petiole red or purplish, 0.8–2 cm. long; stipules deltoid, 3–5 mm. long, 1–2-apiculate, glabrous. Inflorescences solitary, terminal, 4–5 cm. in diameter, fragrant; peduncle short, 1.5–2.5 cm. long. Calyx-lobes triangular 0.5–1 mm. long, pubescent with deciduous clavate appendages. Corolla white or yellowish, (0.5–)0.9–1.2 cm. long; tube narrowly funnel-shaped; lobes 4–5, almost round, ± 2–2.5 mm. long, sometimes ciliolate. Ovary ± distinctly 2-locular or sometimes with the partition imperfect above; style exserted from the corolla for 5–9 mm.; stigma 2.5–4 mm. long, conical with 2 basal swellings, acuminate at the apex. Syncarps ovoid or globose, red and fleshy, edible, (2–)5–8 cm. in diameter, rugose and pitted with pentagonal scars. Seeds numerous, superposed in series, subglobose or ellipsoidal, 1–1.2 mm. long, reticulate, the testa-cells each with 5–6 large perforations. Fig. 60.

* By B. Verdcourt

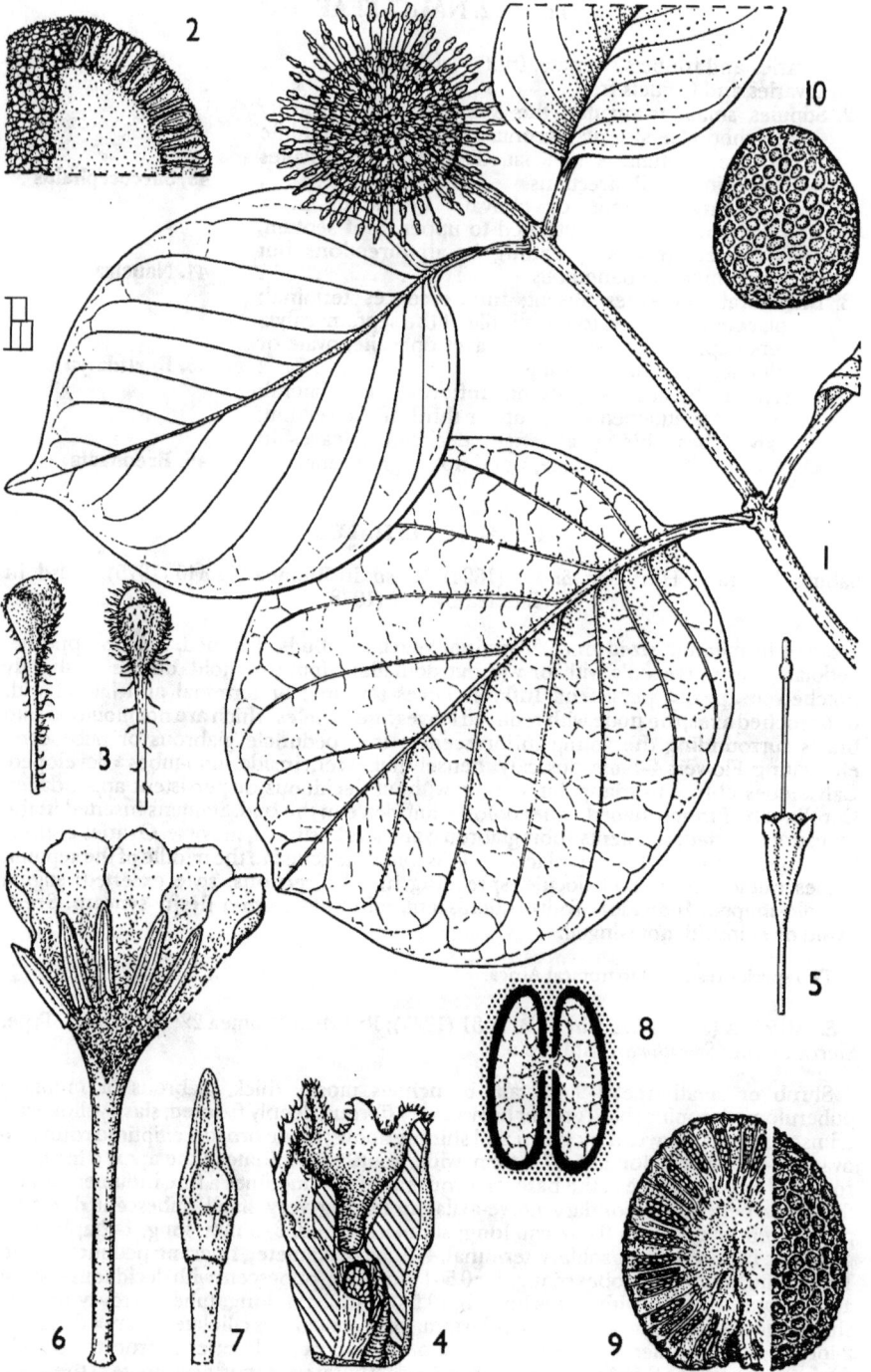

FIG. 60. *SARCOCEPHALUS LATIFOLIUS* — **1**, flowering branch, × ⅔; **2**, section of young inflorescence, × 2; **3**, calyx-lobes, 2 views, × 10; **4**, section through calyx-tube and ovary × 10; **5**, corolla with style and stigma, × 2; **6**, corolla, partly opened out, × 6; **7**, stigma, × 10; **8**, section through ovary, × 20; **9**, syncarp, exterior and section, × ⅔; **10**, seed, × 20. 1, 5–7 from *Chandler* 541; 2, 3, from *Brunt* 1412; 4, 8, from *Meikle* 1210; 9, from *Maitland* 1274; 10, from *Dawe* 836. Drawn by Diane Bridson.

UGANDA. Bunyoro District: Chiope, *Dawe* 836!; Teso District: Serere, Apr. 1932, *Chandler* 541!; Mbale District: Tororo, Dec. 1934, *Harris* 186!
KENYA. N. Kavirondo District: near Busia, 23 Jan. 1964, *Brunt* 1412!
DISTR. U 1–3; **K** 5; Senegal to Cameroun, Zaire, Cabinda, and Sudan
HAB. Scrub, moist grassland with scattered trees; 900–1500 m.

SYN. *Nauclea latifolia* Smith, Rees Cyclop. 24: 5 (1813); Milne-Redh. in K.B. 3: 459 (1948); I.T.U., ed. 2:
  352 (1952), as '*latifolius*'; Petit in B.J.B.B. 28: 8 (1958); Hepper in F.W.T.A., ed. 2, 2: 163 (1963);
  Verdc. in Kirkia 5: 276 (1966); Hallé, Fl. Gabon 12, Rubiacées, 1: 40 (1966)
Sarcocephalus esculentus Sabine in Trans. Hort. Soc. 5: 442, t. 18 (1824). Types: Sierra Leone, *Don*
  (BM, syn.!) & *Afzelius* (BM, syn.!)
S. russeggeri Schweinf., Reliq. Kotschy.: 49, t. 33 (1868); F.T.A. 3: 39 (1877). Types: Sudan,
  Fazughli, Tumat-Kassau, *Kotschy* 511 (W, syn., K, isosyn.!) & Fazughli, Tumat, Chor Adi, *von
  Würtemberg* (W, syn.)
Nauclea esculenta (Sabine) Merr. in Journ. Wash. Acad. Sci. 5: 535 (1915)

44. NAUCLEA

L., Sp. Pl., ed. 2, 1: 243 (1762); Merr. in Journ. Wash. Acad. Sci. 5: 534 (1915); Ridsd. in
Blumea 22: 547 (1975), 23: 184–186 (1976) & 24: 325 (1978)

Medium-sized to large trees. Terminal vegetative buds strongly flattened. Leaves
opposite, petiolate, the nerve-axils with or without domatia; stipules ovate, obovate or
elliptic, flattened, adpressed, deciduous or ± persistent. Inflorescences terminal or
terminal and lateral with unbranched axes; nodes with reduced leaves and stipules not
modified into bracts which surround the young inflorescence; true peduncle glabrous or
pubescent, elongating. Flowers 4–5-merous, calyx glabrous or pubescent inside; limb-
tube developed or not; lobes triangular or oblong, obtuse, with or without persistent
appendages. Corolla-tube funnel-shaped; lobes oblong, imbricate. Stamens inserted in
the throat, exserted; filaments short, glabrous; anthers basifixed, introrse. Ovaries
entirely fused to each other, 2-locular, the placentas attached to the upper third of the
septum, Y-shaped with 2 short ascending arms and long basal one; ovules numerous in
each locule, mostly pendulous, some erect or horizontal but these mostly aborting; style
exserted; stigma spindle-shaped. Fruits indehiscent, connate into a fleshy syncarp. Seeds
ovoid or ellipsoid, sometimes compressed, not winged.

Ten species, 4 tropical African, the rest Asian.

N. diderrichii (*De Wild.*) *Merr.* in Journ. Wash. Acad. Sci. 5: 535 (1915); I.T.U., ed. 2: 351
(1952); Petit in B.J.B.B. 28: 10 (1958); Hallé in Fl. Gabon 12, Rubiacées, 1: 44, t. 4/1–3, t.
5/1–6 (1966); Ridsd. in Blumea 22: 548 (1975). Type: Zaire, Mayumbe, *Diderrich* (no
material retained, notes only); Zaire, Mayumbe, Ganda Sundi, *Briey* 189 (BR, neo.!)

Tall rather slender forest tree (9–)30–40 m. tall, with bole attaining 0.9–1.5 m. in
diameter, with low or without buttresses; branchlets glabrous; bark light brown, greyish or
yellowish, shallowly and longitudinally fissured; slash fibrous, white, turning pale brown.
Leaf-blades broadly elliptic, (4–)7–18(–40) cm. long, (2.5–)4–12(–18) cm. wide (even up to
45 × 21 cm. on sterile shoots), obtuse or very slightly acuminate at the apex, rounded or
shortly cuneate at the base, glabrous; petiole 0.8–1.3(–1.8) cm. long; stipules large;
foliaceous, oblong-elliptic, 1–2.5(–7) cm. long, 0.4–1.2(–6.8) cm. wide, with a very distinct
narrow keel, glabrous. Inflorescences terminal, 3 cm. in diameter; peduncle short, 1.1–2
cm. long. Calyx-lobes prismatic or clavate, very short, 1.8–2 mm. long, the lobes unequally
joined amongst themselves, thickened and hairy at the tips. Corolla white, greenish white
or yellowish, (4–)6–8 mm. long, the tube narrowly funnel-shaped; lobes 3–4, oblong-
elliptic, 1.5–2.7 mm. long, 0.75–1.2 mm. wide. Style 1.1–1.3 cm. long, exserted 4.5–6 mm.
from the corolla; stigma 2 mm. long, acuminate. Syncarps globose, greyish brown, with
whitish flesh, 3–4 cm. in diameter, subechinate and ornamented with a disconnected
network of unequally joined calyx-lobes surrounding the floral pits. Seeds brown, ovoid, 1
mm. long, not margined, reticulate. Fig. 61.

UGANDA. Toro District: Bwamba Forest, Muntandi, Apr. 1940, *Sangster* 641! & same locality, Oct.
  1940, *Eggeling* 4069! & Bwamba, Kabango-Muntandi Reserve, *Musaka* 66!
DISTR. U 2; W. Africa, Cameroun, Central African Republic, Gabon, Zaire, Cabinda and Angola*
HAB. Evergreen forest; 840–1650 m.

SYN. *Sarcocephalus diderrichii* De Wild. in Rev. Cult. Colon 9: 7 (1901)

* The record from Mozambique in F.W.T.A., ed. 2, would appear to be erroneous.

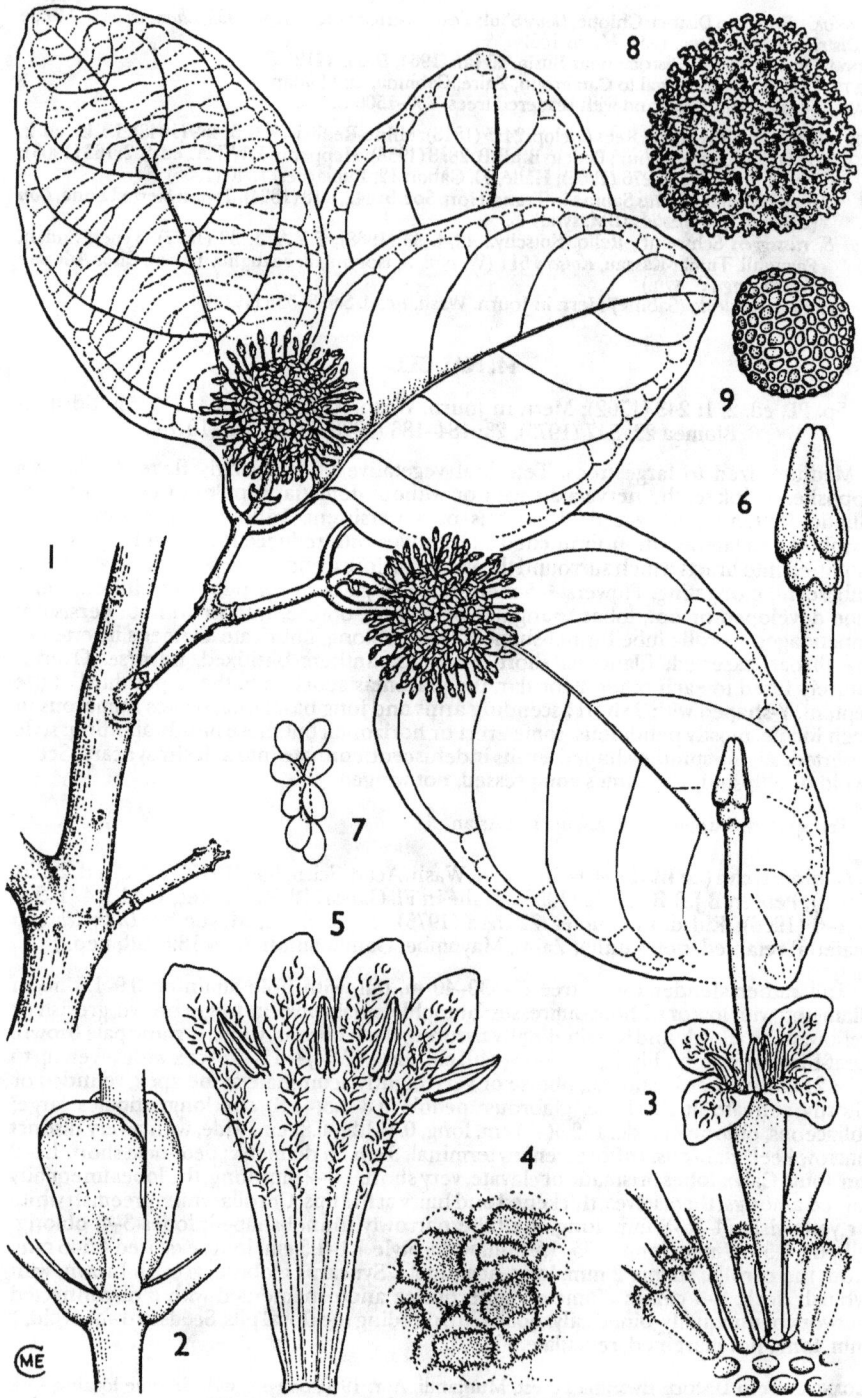

FIG. 61. *NAUCLEA DIDERRICHII* — **1**, flowering branch, × ⅓; **2**, stipules, × 1; **3**, flower with part of inflorescence, × 6; **4**, calyces, from above, × 8; **5**, corolla, opened out, × 8; **6**, stigma × 12; **7**, ovules, position of placenta indicated, × 20; **8**, syncarp, × 1; **9**, seed, × 16. 1, 3–6, from *Louis* 6125; 2, from *Eggeling* 4069; 7, from *Chipp* 212; 8, 9, from *Deighton* 3025B. Drawn by Mrs M.E. Church.

45. BURTTDAVYA

Hoyle in Hook., Ic. Pl. 34, t. 3318 (1936); Ridsd. in Blumea 22: 548 (1975).

Unarmed trees. Leaves petiolate, large; stipules large, deciduous. Inflorescences globose, solitary, terminal, with the flowers inserted on a fleshy ± pilose receptacle; interfloral bracteoles absent. Calyx 5-lobed, the lobes free, linear-spathulate, persistent. Corolla ± tubular; lobes 5, imbricate. Stamen-filaments very short; anthers 2-thecous. Ovaries closely contiguous but not fused, each 2-locular; locules with many ovules, the placentas attached to the middle of the septum; style slender, exserted; stigma red, narrowly conical with widened base and blunt tip, sulcate. Fruiting heads globose, but individual fruits not joined, red, irregularly breaking, crowned by the persistent calyx-lobes. Seeds not winged, but sometimes angular; testa reticulate; embryo straight, surrounded by albumen.

A monotypic genus confined to eastern Africa.

B. nyasica *Hoyle* in Hook., Ic. Pl. 34, t. 3318 (1936); Verdc. & Greenway in K.B. 10: 602 (1956); Ridsd. in Blumea 22: 549 (1975); Verdc. in Garcia de Orta, sér. Bot. 7(1–2): 7–8 (1985) [1987]. Type: Malawi, Nsanje (Port Herald) District, perhaps Chididi, *Townsend* 23 (K, holo.!, FHO, iso.)

Tree (6–)12–21(–40) m. tall with rounded crown and remarkable golden-orange or greenish wood; bark grey, ± smooth, apparently flaking in small pieces; branchlets striate, puberulous or glabrous. Leaf-blades ovate or broadly elliptic, 7.5–31 cm. long, 5–23 cm. wide, acute or shortly acuminate at the apex, rounded, truncate, slightly subcordate or broadly cuneate at the base, glabrous above and beneath save for sometimes a fine pubescence on the midrib beneath; petiole 2–6 cm. long, puberulous; stipules green, elliptic, oblanceolate or broadly obovate, 3–5.6 cm. long, 1.5–3.5 cm. wide, glabrous. Inflorescences strongly scented, 2–4 cm. in diameter; peduncle 1.5 cm. long. Calyx-tube 2–3 mm. long; lobes 3.5–4 mm. long, 2 mm. wide at their tips, greyish or fulvous tomentose and pilose, accrescent in fruit. Corolla yellow; tube ± 1 cm. long, glabrous; lobes oblong, 1.55–3 mm. long, 1.5 mm. wide, rounded, pilose inside in the middle. Style glabrous, ± 1.6 cm. long; stigma red with yellow base in dry state, conical with a bulbous base, 2.5 mm. long. Fruiting heads globose, 2.5–3.5(–5) cm. in diameter; fruits red, narrowly oblong-obovoid, 5–8 mm. long, 1–2 mm. wide, striate, thin-walled, very sparsely pilose. Seeds red, 20–50, ellipsoid, ± 1 mm. long. Fig. 62.

TANZANIA. Handeni District: Kideleko, 23 Mar. 1961, *Semsei* 3189! & Handeni–Kideleko–Morogoro road, 6.1 km. from Handeni, 17 June 1949, *Hoyle* 1023!; Tanga District: between Tongwe mission and Mt. Mlinga, 31 Mar. 1974, *R.B. & A.T. Faden* 74/373!; Kilosa District: Vigude, Oct. 1952, *Semsei* 944!
DISTR. T 3, 6, 8; Malawi, Mozambique
HAB. Evergreen forest, lowland ? rain-forest, often by water, on rocky outcrops or at forest margins; 200–540 m.
NOTE. The short included style described and depicted in the original description and figure refer to an immature bud. This species has been cultivated in Masasi District at Ndanda mission (*Willan* 604).

46. BREONADIA

Ridsd. in Blumea 22: 549 (1975)

Small to medium-sized trees. Terminal vegetative buds conical. Leaves in whorls of 3 or 4, petiolate, with or without domatia in the axils of the nerves; stipules narrowly triangular, deciduous. Inflorescences lateral, the axes usually unbranched; nodes with strongly cohering stipules modified into circumscissile calyptra-like bracts surrounding the young inflorescence; true peduncle elongating, glabrous or pubescent. Flowers 5-merous, subsessile on the densely pubescent receptacle, separated by numerous spathulate bracteoles. Calyx with limb-tube developed, densely hairy outside, slightly pubescent inside; lobes oblong. Corolla-tube hypocrateriform; lobes oblong, imbricate. Stamens inserted in the throat, exserted; filaments short, glabrous; anthers basifixed, introrse. Ovary 2-locular; placentas shortly obovoid, attached to the upper third of the

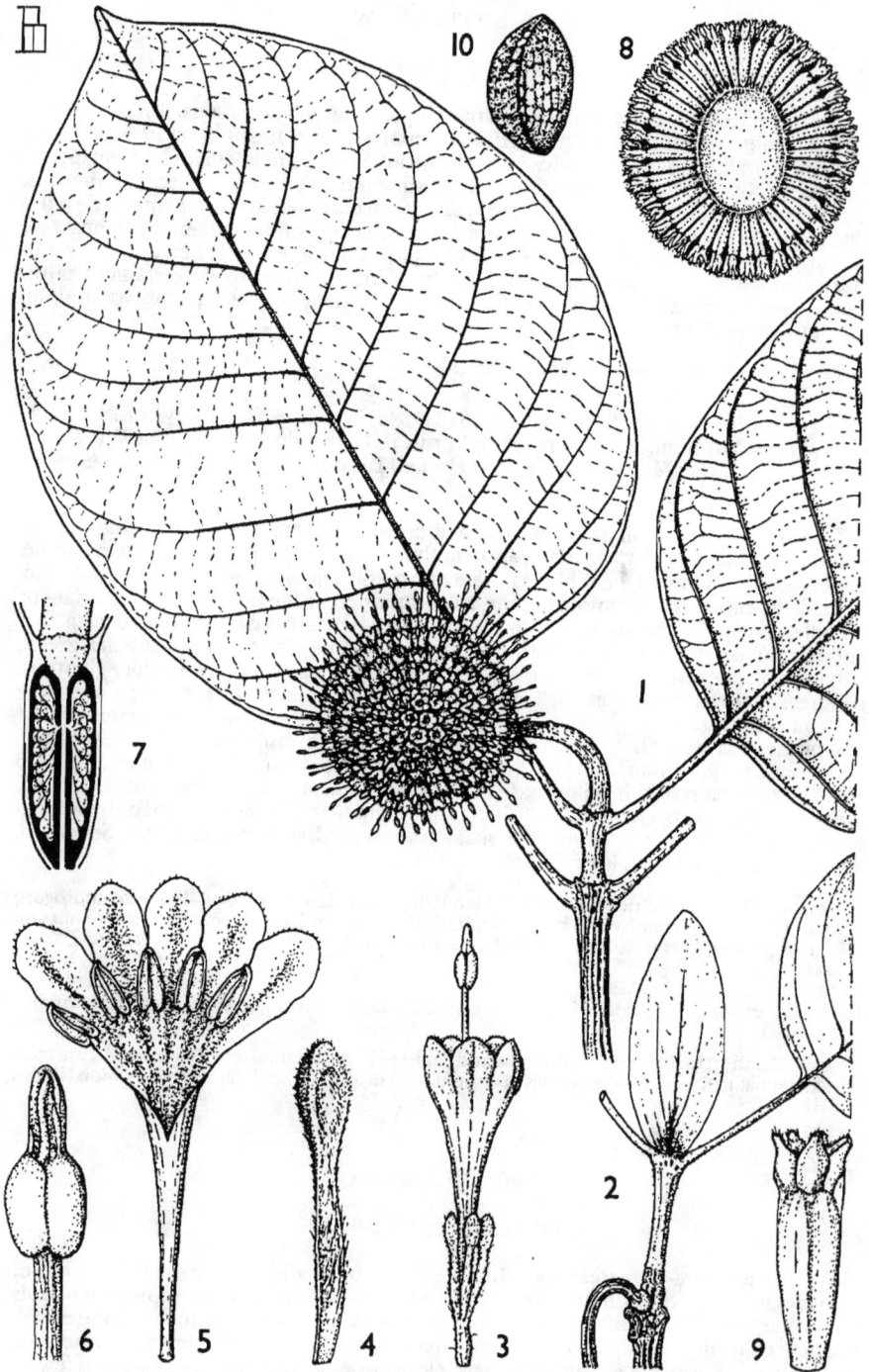

FIG. 62. *BURTTDAVYA NYASICA* — 1, flowering branch, × ⅓; 2, twig with stipule, × ⅓; 3, flower, × 3; 4, calyx-lobe, from inside, × 10; 5, corolla, partly opened out, × 6; 6, stigma, × 10; 7, section through ovary, × 20; 8, section through infructescence, × 1; 9, fruit, × 3; 10, seed, × 20. 1–7, from *Semsei* 3189; 8–10, from *Townsend* 24. Drawn by Diane Bridson.

ovary, the axis later detaching from the receptacle. Seeds ovoid, bilaterally compressed, not winged.

A single species in tropical Africa and Madagascar formerly included in the Asiatic genus *Adina* Salisb.

B. salicina *(Vahl) Hepper & Wood* in K.B. 36: 860 (1982). Types: N. Yemen, Hadie, *Forsskål* 236 & 237 (C, syn.)

Medium to large tree (1.8–)3–21 m. tall; bark grey, fissured and scaly; branches glabrous. Leaves in whorls of 3–4; leaf-blades lanceolate, 8–33 cm. long, 1.8–9 cm. wide, subacute to slightly acuminate at the apex, narrowly cuneate at the base, glabrous, at first rather shining above; petiole 0.8–2.5 cm. long; stipules ± triangular, interpetiolar, bifid, separating into a 4-toothed ring. Inflorescences solitary, the head 1.5–2.5 cm. in diameter; peduncles 2–9.5 cm. long, puberuluous with very short hairs, with 2–4 ovate membranous often connate bracts 0.4–1 cm. long just above the middle; true bracteoles filiform-spathulate, 2.8–5 mm. long. Calyx-tube 1–2 mm. long with free part narrowly tubular, ± 0.5 mm. long; lobes oblong, linear or triangular, 1–2.5(–4.5) mm. long, 0.5–0.6 mm. wide, pubescent. Corolla pinkish green or white or yellowish brown; tube 3–6.5 mm. long, densely silky grey-pubescent outside; lobes often tinged mauve, linear, ± imbricate, 1–2.5 mm. long, 0.5–0.7 mm. wide, pubescent outside. Style white, 0.9–1.45 cm. long; stigma green, ellipsoid or capitate, 0.5 mm. long. Fruit a septicidal capsule, the valves 3.5 mm. long, shining inside, pubescent outside. Seeds narrowly compressed, obovoid, 2.2–2.8 mm. long, not winged. Fig. 63.

KENYA. Fort Hall District: banks of Thika R., Fort Hall road, 24 June 1909, *Battiscombe* 31!; Teita District: Mbololo Hill, 14 Feb. 1953, *Bally* 8579!; Tana River District: Tana Falls, 4 Mar. 1934, *Sampson* 8!

TANZANIA. Lushoto/Tanga Districts: Sigi, 3 Mar. 1932, *Greenway* 2929!; Morogoro District: Nguru Mts., Turiani Falls, 25 Mar. 1933, *Drummond & Hemsley* 1795!; Rungwe District: Masoko [Massoko], 23 Mar. 1932, *St. Clair-Thompson* 1306!; Pemba I., Kaole, Wesh Road, Mile 3, 19 Jan. 1933, *Vaughan* 2050!

DISTR. K 4, 7; T 3, 6–8; Z; P; Mali to Cameroun, Central African Republic, Sudan, Ethiopia, Mozambique, Malawi, Zambia, Zimbabwe, Angola, South Africa (Transvaal), Swaziland; also in Madagascar and Yemen

HAB. Gallery forest by rivers and streams often on extreme edges with its roots in water between gravel and boulders; near sea-level to 1500 m.

SYN. *Nerium salicinum* Vahl, Symb. Bot. 2: 45 (1791)
    Nauclea microcephala Delile, Cent. Pl. Méroé: 67 (1826). Type: Sudan, Sennaar, Singue, *Cailliaud* (? MPU, holo.)
    Adina microcephala (Delile) Hiern in F.T.A. 3: 40 (1877); Havil. in J.L.S. 3: 42 (1897); T.T.C.L.: 48 (1949); K.T.S.: 424 (1961); F.F.N.R.: 398 (1962); Hepper in F.W.T.A., ed. 2, 2: 163 (1963)
    Cephalanthus spathelliferus Bak. in J.B. 20: 137 (1882). Type: Madagascar, W. Betsileo, *Baron* 86 (K, holo.!)
    Adina galpinii Oliv. in Hook., Ic. Pl. 124, t. 2386 (1895). Types: Swaziland and Transvaal, Horo Flats & Queens R., Moodies near Barberton, *Galpin* 1213 (K, syn.*!)
    A. lasiantha K. Schum. in P.O.A. C: 378 (1895). Type: S. Malawi, *Buchanan*** (B, holo.†, BM, K, iso.!)
    A. microcephala (Delile) Hiern var. *galpinii* (Oliv.) Hiern, Cat. Afr. Pl. Welw. 1: 434 (1898)
    A. lasiantha K. Schum. var. *parviflora* Hochr. in Ann. Conserv. Jard. Bot. Genève 11: 95 (1908). Type: Madagascar, Vatomandry District, *Guillot* 93 (G, holo., K, iso.!)
    Breonadia microcephala (Delile) Ridsd. in Blumea 22: 549 (1975)
    B. salicina (Vahl) Hepper & Wood var. *galpinii* (Oliv.) Hepper & Wood in K.B. 38: 85 (1983)

Tribe 13. CINCHONEAE***

1. Liane with ± hooked spines and flowers in spherical many-
    flowered heads .... **48. Uncaria**
    Trees or shrubs without spines 2
2. Flowers in collections of dense perfectly spherical heads;
    stipules large, leafy, ovate and entire .... **47. Hallea**

* One sheet with two specimens but only one label under one number mentioning both localities.
** *Buchanan* 6 (BM!) is annotated as a 'cotype'.
*** By B. Verdcourt.

FIG. 63. *BREONADIA SALICINA* — 1, flowering branch, × 1; 2, stipule, × 1; 3, section through young inflorescence, × 2; 4, flower bud surrounded by bracteoles, × 4; 5, flower, × 4; 6, corolla, opened out, × 4; 7, anther, × 8; 8, stigma, × 12; 9, transverse section of ovary, × 12; 10, fruit, × 4; 11, part of dehisced fruit, × 4; 12, seed, × 4. 1–11, from *Sampson* 8; 12, from *Drummond & Hemsley* 1795. Drawn by Miss D.R. Thompson, with 12 by Sally Dawson.

Flowers in panicles or spike-like inflorescences; stipules
    less conspicuous   3
3. Corolla glabrous outside with valvate lobes; flowers in
    spike-like inflorescences subtended in one species by
    large leafy paired stipitate venose red bracts . . .   **49. Hymenodictyon**
  Corolla pubescent outside with contorted lobes; flowers in
    panicles not subtended by bracts . . . .   **50. Crossopteryx**

47. HALLEA

J.-F. Leroy in Adansonia, sér. 2, 15: 66 (1975)

Tall trees with large petiolate leaves; small domatia present; stipules large, leafy or membranous, glabrous or pubescent, free or joined near the base. Inflorescences terminal and/or axillary of 3–many spherical heads with pubescent receptacles, the flowering branches developing monopodially; bracteoles linear-spathulate, surrounding the sessile flowers which are free from each other. Buds pubescent at apex. Calyx tubular, the limb truncate or distinctly lobed, glabrous or ciliate, an epicalyx sometimes present. Corolla-tube funnel-shaped, glabrous within and outside; lobes erect, not opening very much, pubescent outside and terminated by short subulate deciduous glabrous appendages. Anthers included. Ovaries free, 2-locular; ovules numerous, ascending and imbricate on subpeltate placentas; style distinctly exserted, the stigmatic club ovoid or mitriform. Fruits dry, dehiscing septicidally, the endocarp splitting into 2 valves each with rounded lobes. Seeds small with winged margins.

Three species in tropical Africa formerly placed in *Mitragyna*, but recognised as a distinct section by Haviland in 1897. Ridsdale (Blumea 24: 56 (1978), retains them in *Mitragyna*, but with some doubts and removes the genus to the Cinchoneae; formerly they had invariably been placed in the Naucleeae. The separation of *Hallea* on rather small but constant floral characters is supported by palynology, wood and leaf anatomy and inflorescence development; there are also some differences in the spectrum of alkaloids (see Shellard, Journ. Ethnopharmacology 8(3): 345–7 (1983)).

Calyx truncate or at the most faintly undulate; stipules thicker,
    with numerous longitudinal veins and less distinctly
    reticulate venation . . . .   *1. H. stipulosa*
Calyx distinctly toothed; stipules more membranous and with
    more obvious reticulate venation . . . .   *2. H. rubrostipulata*

1. **H. stipulosa** (*DC.*) *J.-F. Leroy* in Adansonia, sér. 2, 15: 66 (1975). Type: Gambia, R. Gambia, near Albreda, *Leprieur* (P, holo.)

Large tree 6–42 m. tall, with rounded crown and often possessing small knee-like pneumatophores; bark grey or grey-brown, with long fissures, irregularly cracked and flaking in plates; phellogen pale brown with yellowish streaks and slash pale brown to pinkish; branchlets often 4-angled, glabrous or hairy. Leaf-blades oblong, obovate or elliptic; (7–)12–45 cm. long, (3.5–)8.5–27 cm. wide, rounded at the apex, broadly cuneate to somewhat cordate at the base, glabrous above, mostly hairy beneath, particularly on the main nerves but sometimes quite glabrous; petiole 1–3.5(–5) cm. long, glabrous or hairy; stipules green, elliptic, obovate or almost round, (3–)3.5–8 cm. long, 2.5–5 cm. wide, pubescent at least at the base. Inflorescences terminal or axillary, made up of 3–10 globose heads, 1.5–2.5 cm. in diameter; peduncles (1.5–)4–20 cm. long; bracts leaf-like; secondary bracts 4, forming a caducous involucel of papery nature on the peduncle of the terminal flower head; true bracteoles spathulate, 2.5–5 mm. long, pubescent. Flowers pleasantly scented. Calyx truncate or slightly undulate, glabrous; tube 1.5–2 mm. long, free part 0.5–1 mm. long. Corolla white, greyish or yellowish; tube 1.5–3 mm. long, glabrous; lobes ovate-triangular, 2–2.5 mm. long, 1–1.5 mm. wide, densely pubescent outside in the upper part. Style 6.5–7.5 mm. long; stigma 1–1.3 mm. long, ridged. Fruit ellipsoid, 8–9 mm. long, longitudinally finely 10-ribbed, outer part bursting irregularly, endocarp dividing into 2 bilobed coriaceous valves. Seeds 2–2.5 mm. long with winged margins.

UGANDA. Bunyoro District: Budongo Forest, Jan. 1932, *Harris* 39!; Mengo District: Entebbe, Nambigirwa [Nambigiruwa], Jan. 1932, *Eggeling* 152! & Entebbe, Aug. 1937, *Chandler* 1892!

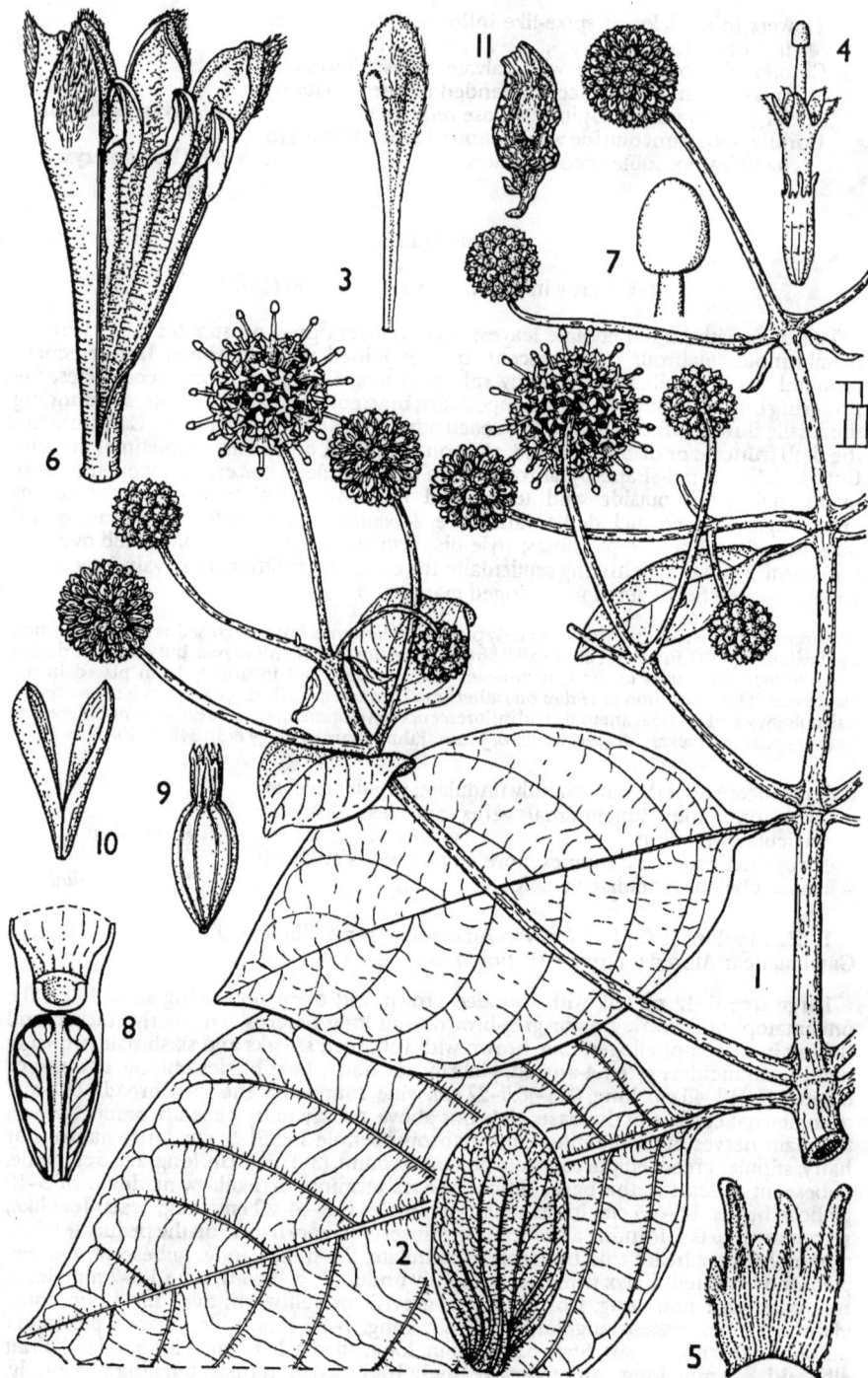

FIG. 64. *HALLEA RUBROSTIPULATA* — 1, flowering branch, × ⅔; 2, stipule, × ⅔; 3, bracteole, from outside, × 10; 4, flower, × 2; 5, calyx-limb, opened out, × 5; 6, corolla, opened out, × 6; 7, stigma, × 10; 8, section through ovary, × 10; 9, fruit, × 2; 10, valve, from inside, × 3; 11, seed, × 10. 1, 4–8, from *Styles* 191; 2, from *Procter* 1897; 3, from *Dawkins* 618; 9–11, from *Maitland* 821A. Drawn by Diane Bridson.

DISTR.   U 1, 2, 4; widespread in W. and central Africa from Senegal to Angola and Sudan to Zambia
HAB.   Swamp forest, often forming almost pure stands in East Africa, also in evergreen forest on
drier ground; 1050–1200 m.

SYN. *Nauclea stipulosa* DC., Prodr. 4: 346 (1830)
     Mitragyna macrophylla (DC.) Hiern in F.T.A. 3: 41 (1877), pro parte
     M. stipulosa (DC.) O. Kuntze, Rev. Gen. Pl. 1: 289 (1891); For. Trees & Timbers Brit. Emp. 2:
     84–90, fig. 17 & t. (1933); I.T.U., ed. 2: 349, fig. 73 (1952); F.F.N.R.: 412 (1962); Hepper in
     F.W.T.A., ed. 2, 2: 161 (1963); Hallé, Fl Gabon 12, Rubiacées, 1: 33 (1966); Ridsd. in Blumea 24:
     68 (1978)

2. H. rubrostipulata (*K. Schum.*) *J.-F. Leroy* in Adansonia, sér. 2, 15: 66 (1975). Type:
Tanzania, Kilimanjaro, Kware [Quare] steppe, Kibosho [Kiboscho], Sinas Boma, *Volkens*
1583 (B, lecto.*, BM, isolecto.!)

Tree (3–)12–18(–24) m. tall with a much-branched crown; bark greyish brown, fairly
smooth to rough, flaking off in ± quadrangular patches; phellogen yellow; slash yellow
becoming pale pink; branchlets somewhat quadrangular, glabrous or hairy. Leaf-blades
broadly elliptic, oblong-elliptic or somewhat obovate, 4.5–35 cm. long, 2.5–25 cm. wide,
shortly and obtusely acuminate at the apex, rounded to subcordate at the base, glabrous
and mostly shining above, densely yellowish-hairy or glabrous save for midrib, main
nerves and nerve-axils beneath; petiole 1–5.5 cm. long, glabrous or pubescent; stipules
often bright red, membranous, broadly elliptic to almost round, 3.5–7.5(–15) cm. long,
2.5–5(–10) cm. wide, often hairy, clearly reticulately venose. Inflorescences terminal and
axillary of up to 15 globose heads of sweetly scented flowers, ± 2.5–3 cm. in diameter;
peduncles 3–7(–11) cm. long; lower bracts like small leaves; upper bracts 0.3–2.5 mm.
long, 0.3–1.3 mm. wide; true bracteoles linear-spathulate, 5–6 mm. long, 0.8–1 mm. wide at
the apex. Calyx-tube 2–3 mm. long, ribbed, free part 2–3 mm. long, distinctly lobed; lobes
elongate-triangular, 1–2 mm. long, 1 mm. wide, glabrous, usually with an epicalyx of
alternating shorter oblong lobes less than 1 mm. long. Corolla green or yellowish white;
tube narrowly funnel-shaped, 4–6 mm. long; lobes ovate-triangular, 1.5–3 mm. long, 1 mm.
wide, outside with a diamond-shaped patch of yellowish hairs extending from the apex to
the basal midpoint of the lobe and bounded there by 2 glabrous areas; style 1–1.4 cm.
long; stigma oblong-ellipsoid, 0.85–1.5 mm. long-grooved. Fruit ellipsoid, 0.9–1.3 cm.
long, 5 mm. wide, crowned with the persistent calyx. Seeds with nucleus 1 mm. long,
winged at both ends, the wings of unequal breadth 1 mm. long. Fig. 64.

UGANDA. Ankole District: edge of main Kalinzu Forest Reserve, 3 Nov. 1962, *Styles* 191!; Kigezi
     District: Kinkizi, Rutenga, Mar. 1947, *Purseglove* 2347!; Masaka District: Malabigambo Forest, near
     Katera, 15 Aug. 1950, *Dawkins* 618!
KENYA. S. Nyeri/Embu Districts: E. Mt. Kenya, Mwea R., 29 July 1913, *Battiscombe* 690!; Meru, 18 Feb.
     1922, *R.E. & T.C.E. Fries* 1716!; Teita District: Mbololo Forest, 2 May 1985, *Faden et al.* 770!
TANZANIA. Bukoba, Apr. 1935, *Gillman* 2281!; Moshi District: Lyamungu, 25 Feb. 1954, *F.G. Smith*
     1081!; Morogoro District: Uluguru Mts., Kisaki road, 18 Dec. 1935, *E.M. Bruce* 323!
DISTR. U 1, 2, 4; K 4, ?7; T 1–4, 6, 7; E. Zaire, Rwanda, Burundi, Ethiopia (Illubabor), Malawi and
     Zambia
HAB. Swamp-forest, fringing forest by swamps and streams, wet upland forest; 900–2190 m.

SYN. *Adina rubrostipulata* K. Schum. in P.O.A. C: 378 (1895)
     Mitragyna rubrostipulata (K. Schum.) Havil. in J.L.S. 33: 73 (1897), as "*rubrostipulacea*"; T.T.C.L.:
     506 (1949); I.T.U., ed. 2: 348 (1951); K.T.S.: 450 (1961); Ridsd. in Blumea 24: 68 (1978); Fl. Pl.
     Lign. Rwanda: 586, fig. 192/1 (1982), as "*rubrostipulosa*"; Fl. Rwanda 3: 174, fig. 54/1 (1985)
     Adina rubrostipulata K. Schum. var. *discolor* Chiov., Racc. Bot. Miss. Consol. Kenya: 51 (1935).
     Type: Kenya, Meru District, Nyambeni [Giombene], *Balbo* 462 (TOM, holo.!)
NOTE. Chiovenda's varietal name refers to the form with leaves velvety beneath.

48. UNCARIA

Schreb., Gen.1: 125 (1789); Petit in B.J.B.B. 27: 442 (1957); Ridsd. in Blumea 24: 68–100
(1978), *nom. conserv.*

Woody lianes or sometimes forming scrambling bushes or thickets; flowering
branches bearing strongly curved hooked spines**. Leaves petiolate; stipules simple or

---

* Haviland (see synonymy above) does not mention *Stuhlmann* 1151 & 1566 from Tanzania,
Bukoba, and this is taken to be a lectotypification.
** Guillaumin, in C.R. Acad. Sci. 192: 1264 (1931), has shown these are the modified peduncles of
axillary inflorescences.

2-lobed, deciduous. Inflorescences globose, solitary, axillary or terminal; flowers not fused, pedicellate. Calyx-tube ellipsoid, the free part well developed, shallowly 5-undulate to distinctly 5-toothed, persistent. Corolla-tube filiform-cylindrical, funnel-shaped at the throat; lobes 5, imbricate or quincuncial. Stamens with filaments very short; anthers apiculate, 2-thecous. Ovary 2-locular; placentas peltate, entirely covered with imbricated ascending ovules; style long and narrow, well exserted from the corolla; stigma slightly thickened with 2 marks on either side. Fruits fusiform, dry, dehiscent, borne on accrescent pedicels. Seeds small, with long fine wings from opposite ends, mostly 1 from one end and 2–3 from the other.

A widely distributed genus of some 60 species, many in Asia and Indonesia, 1 in Madagascar and 2 in America. In Africa there are 3 species only one of which occurs in the Flora area. Formerly invariably placed in the Naucleeae this was transferred to the Cinchoneae by Ridsdale.

U. africana *G. Don*, Gen. Syst. 3: 471 (1834); Hook., Ic. Pl. 8, t. 781 (1848); Hiern in F.T.A. 3: 41 (1877); Havil. in J.L.S. 33: 76 (1897); Petit in B.J.B.B. 27: 445 (1957); Hepper in F.W.T.A., ed. 2, 2: 162, fig. 233 (1963); Hallé, Fl. Gabon 12, Rubiacées, 1: 30, fig. 2/1–9 (1966). Type: Sierra Leone, *Don* (BM, holo., K, iso.!)

Climbing shrub or liane (3–)5–9(–15) m. long or tall; young stems ± quadrangular, glabrous or sparsely pilose; axillary hooks opposite, equal or unequal, arranged in decussate pairs; stems in some variants with myrmecodomatia. Leaf-blades elliptic, (3.5–)5–12(–15) cm. long, (1–)2.5–6 cm. wide, acuminate at the apex, rounded to cuneate at the base, drying reddish, glabrous or pilose beneath on venation when young; petiole 0.5–1 cm. long, glabrous or pilose; stipules often red, joined at the base, puberulous or pilose near the base, persistent on unarmed shoots but very deciduous on the flowering branches. Inflorescences scented, terminal, solitary, 4–6 cm. in diameter, sometimes also with 2–4 axillary inflorescences in pairs beneath; peduncles 1–4 cm. long; pedicels obsolete or 2–3 mm. long, golden pubescent. Calyx-lobes variable, short to distinct and triangular, 0.5–3 mm. long; calyx-tube 1.5–3 mm. long the free part of the tube an additional 2–3.5 mm. tall. Corolla white to yellowish with dense golden or ochraceous adpressed bristly pubescence outside; tube 0.8–1.5 cm. long, narrow, the lower part 0.6–1 mm. wide; lobes oblong, 2.5–4 mm. long, 2 mm. wide, with an axial evanescent line of internal pubescence. Style green, 1.4–2 cm. long including the sausage-shaped stigma, up to 4.5 mm. long. Fruiting heads up to 10 cm. in diameter; capsules fusiform 1.15–2.5 cm. long, 4–6 mm. wide, 10-ribbed, pubescent, the accrescent pedicels 1–2.5 cm. long; dehiscence of capsule septicidal and ± also loculicidal into 2, 2-lobed valves, ochraceous, shining inside. Seeds brown, trigonous, the nucleus ± 0.4 mm. long with reticulate testa; wings of seed ribbon-like, 3–4 mm. long, 1 on one side, 2 on the other.

subsp. **africana**; Ridsd. in Blumea 24: 96 (1978)

Calyx with small deltoid teeth ± 0.5 mm. long; pedicels not developed in young flowering heads; leaves mostly glabrous or pilose on midrib below when young. Fig. 65/10.

TANZANIA.   Uzaramo District: Pugu Forest Reserve, 8 Mar. 1964, *Semsei* 3670! & 31 km. W. of Dar es Salaam, Mpiji R. valley near Kibaha, 24 July 1971, *Harris et al.* 5834!; Ulanga District: Ifakara, 16 July 1959, *Haerdi* 289/0!; Zanzibar I., Jozani [Josani] forest, Aug. 1972, *Robins* 102!
DISTR.   T 6; Z; Guinea Bissau to Nigeria, Cameroun, Gabon, Central African Republic, Zaire and Madagascar (fide Ridsdale)
HAB.   Not precisely recorded, apparently dry evergreen forest and associated woodland; up to ? 1000 m.

NOTE.   Haerdi (Acta Trop. Suppl. 8: 147 (1964)) gives more details of the provenance of his specimens; under the number 289/0 he included material from "Ilingera a day's journey from Ifakara in the Iringa Highlands" and "Sanje b/Kisawasawa".

subsp. **lacus-victoriae** *Verdc.* in K.B. 31: 181 (1976); Ridsd. in Blumea 24: 96 (1978). Type: Uganda, Bunyoro District, Bugoma Forest, *Dawe* 704 (K, holo.!)

Calyx with distinct narrowly triangular teeth 1.5–3 mm. long; pedicels not well-developed in young flowering heads; leaves often pilose beneath when young. Fig. 65/1–9.

UGANDA.   Kigezi District: Kinkizi, Nyamigoye, Mar. 1951, *Purseglove* 3583!; Masaka District: Malabigambo Forest, 3.2 km. SSW. of Katera, 2 Oct. 1953, *Drummond & Hemsley* 4531!; Mengo District: Kyagwe, near Mukono, Sept. 1930, *Snowden* 1748!
KENYA.   N. Kavirondo District: Kakamega Forest, Kibiri Block, S. side of Yala R., 21 Jan. 1970, *Faden et al.* 70/3! (sterile)

FIG. 65. *UNCARIA AFRICANA* subsp. *LACUS-VICTORIAE* — 1, flowering branch, × ⅓; 2, stipule, × 1; 3, flower, × 2; 4, calyx, with limb partly opened out and ovary cut through, × 6; 5, corolla, partly opened out, × 6; 6, stigma, × 8; 7, placenta with ovules, × 20; 8, fruit, × 2; 9, seed, × 10. Subsp. *AFRICANA* — 10, calyx, × 4;. 1, 3–7, from *Willan* 302; 2, from *Watkins* 533 (FH 3272); 8, 9, from *Procter* 664; 10, from *Semsei* 3670. Drawn by Diane Bridson.

TANZANIA   Bukoba District: Minziro Forest, 6 Mar. 1957, *Willan* 302! & 18 Aug. 1954, *Willan* 160! &
   Aug. 1957, *Procter* 664!
DISTR.   U 2, 4; K 5; T 1, 4; Ethiopia and probably Zaire
HAB.   Valley forest, swamp forest, wet *Croton-Celtis* evergreen forest, *Podocarpus-Baikiaea* forest on
   seasonally swampy heavy clay; 1110–1550 m.

SYN.   [*U. africana* sensu Verdc. in K.B. 14: 347 (1960), *non* G. Don sensu stricto]

DISTR.   (of species as a whole). As above with the addition of Angola for subspecies *angolensis*
   (Welw.) Ridsd.

49. HYMENODICTYON

Wall. in Roxb., Fl. Indica, ed. Carey & Wall. 2: 148 (1824), *nom. conserv.*

Kurria Hochst. & Steud. in Flora 25: 233 (1842)

   Trees, shrubs or sometimes epiphytic, rarely lianes. Leaves opposite, petiolate,
deciduous, often with domatia in the nerve-axils; stipules interpetiolar, entire or
glandular-serrate, deciduous. Inflorescences mostly elongate, spicate or capitate, often
with a pair of conspicuous petiolate leafy reticulate bracts at the base; spikes composed of
numerous 1–3-flowered cymes in the axils of minute or moderate-sized secondary bracts.
Calyx-tube ovoid or subglobose, the limb divided into 5(–6) elongate-subulate deciduous
lobes. Corolla funnel-shaped or narrowly campanulate, the tube and throat glabrous
within; lobes 5, valvate, short, erect. Stamens 5, inserted below the throat, included in the
tube; filaments short; anthers basifixed, 2-thecous. Ovary 2-locular; ovules numerous in
each locule, imbricate on cylindrical ascending placentas attached to the septum; style
slender, long-exserted; stigma fusiform or ellipsoid, obscurely 2-lobed, acting as a pollen-
carrier. Fruits capsular, oblong or subcylindric, 2-locular, the valves sulcate, dehiscing
loculicidally into 2 valves, many-seeded, the placentas thin, finally free. Seeds ascending,
imbricate; testa membranous, broadly winged, ± lacerate, the wing elongated, bifid at the
hilar end; embryo small, the cotyledons oblong or round; radicle clavate or terete.

   A genus of about 20 species, occurring in Asia, Indonesia, Madagascar and Africa where there are
4 species.

Inflorescences elongate, 6–22 cm. long, with a pair of very
   characteristic long-petiolate leafy coriaceous reticulate
   bracts 3.5–10 x 1–3.2 cm.; capsules 1–1.5 cm. long; leaf-
   blades with venation drying finely reticulate beneath   *1. H. floribundum*
Inflorescences shorter, 1–8 cm. long, without such a pair of
   bracts at the base; capsules 1.5–2.5(–4) cm. long; leaf-
   blades not drying reticulate beneath   . . . .   *2. H. parvifolium*

   1. H. floribundum (*Hochst. & Steud.*) *Robinson* in Proc. Amer. Acad. 45: 404 (1901);
T.T.C.L.: 501 (1949); I.T.U., ed.2: 346, fig.72 (1952); K.T.S.: 446 (1961); F.F.N.R.: 409, fig.72
(1962); Keay, F.W.T.A., ed.2, 2: 111 (1963); Hallé, Fl. Gabon 12, Rubiacées, 1: 58, t. 7 (1966);
Fl. Pl. Lign. Rwanda: 563, fig. 189/2 (1982); Fl. Rwanda 3, fig. 51/2 (1985). Type: Ethiopia,
Mt. Scholoda, *Schimper* 277 (B, holo. †, BM!, P, iso.)

   Shrub or small tree (1.5–)4–9 m. tall, with grey-black usually rough flaking reticulate
bark; branchlets glabrous or pubescent. Leaves turning scarlet or crimson before falling;
leaf-blades elliptic to obovate, 5–18 cm. long, 2–11.5 cm. wide, abruptly acuminate at the
apex, cuneate at the base, glabrous above and beneath save around the acarodomatia in
the axils of the main nerves or puberulous to quite densely tomentose beneath in some
forms; venation reticulate beneath; petiole 0.4–4 cm. long, glabrous to pubescent; stipules
triangular to lanceolate or strap-shaped, 0.8–1.2 cm. long, acute, often bifid, usually with
glandular margins but otherwise glabrous. Inflorescences terminal, cylindrical, dense,
raceme-like, 6–22 cm. long, 2.5–3 cm. wide; peduncle 0.5–6 cm. long, glabrous to densely
pubescent as are the rachis and pedicels; pedicels 2(–9 in fruit) mm. long; main petiolate
bracts at the base of the inflorescence yellowish green to red, elliptic-lanceolate, 3.5–10
cm. long, 1.1–3.2 cm. wide, acute, rather rigid and strongly reticulate; petiole 2–5 cm. long;
secondary bracts linear-lanceolate, 5 mm. long, 0.5 mm. wide; flowers sweet-scented.
Calyx-tube hairy, 1.5 mm. long; lobes 5–6 narrowly ovate, linear or triangular, ± 0.5–1 mm.

FIG. 66. *HYMENODICTYON FLORIBUNDUM* — **1**, flowering branch, × ⅓; **2**, stipule, × 1½; **3**, flower, × 5; **4**, fruits, × ⅓; **5**, seed × 3. *H. PARVIFOLIUM* subsp. *PARVIFOLIUM* — **6**, flowering branch, × ⅓; **7**, stipule × 5; **8**, flower, × 5; **9**, fruits, × ⅓; **10**, seed, × 2. 1, 3, from *Milne-Redhead & Taylor* 8808; 2, from *Lynes* V6; 4, 5, from *Stolz* 1596; 6, 8–10, from *Greenway & Kanuri* 12,680; 7, from *Drummond & Hemsley* 4230. Drawn by Ann Farrer, with 2, 5, 7, 10 by Sally Dawson.

long, puberulous. Corolla red, yellowish red or greenish yellow tinged purplish brown; tube 4–7 mm. long; lobes 5–6, short, narrowly ovate or triangular, 0.8 mm. long and wide. Style cream, tinged red, 0.8–1 cm. long, exserted 2.5–6 mm.; stigma greenish, 0.8–1 mm. long. Capsule brownish with pale lenticels, 1–1.5 cm. long, 4–7 mm. wide. Seeds pale brownish straw-coloured with darker nucleus, very compressed, 0.7–1 cm. long, 2.5–4 mm. wide, the nucleus 2–2.5 mm. long, 1–1.5 mm. wide, the wing, markedly bifid at the base, the testa strongly reticulate. Fig. 66/1–5.

UGANDA. Acholi District: Imatong Mts., Agoro, 8 Apr. 1945, *Greenway & Hummel* 7317!; Ankole District: Bunyaruguru, Sept. 1938, *Purseglove* 412!; Teso District: Ngora, Apr. 1949, *Philip* 354!
KENYA. E. Elgon, Cave Kopje, June 1963, *Tweedie* 2633!; N. Nandi, Kamilat Hill, Mar. 1934, *Dale* in F.D. 3177!; Kisumu-Londiani District: W. Mau, Kedowa, Feb. 1933, *R.M. Graham* 778!
TANZANIA. Ufipa District: New Sumbawanga–Mbala [Abercorn] road, 32 km. from Mbala, 25 Nov. 1960, *Richards* 13629!; Kondoa District: Kinyassi Scarp, 10 Jan. 1928, *B.D. Burtt* 973!; Songea District: Chandamara, 16 Feb. 1956, *Milne-Redhead & Taylor* 8808!
DISTR. U 1–4; K 3, 5; T 1, 2, 4–8; widespread throughout tropical Africa from Guinée, Sudan and Ethiopia to Zimbabwe and Angola
HAB. Rocky hills and pavements in bushland, *Brachystegia* woodland and wooded grassland; can colonise 25 year-old lava; 540–2250 m.
SYN. *Kurria floribunda* Steud. & Hochst. in Flora 25: 234 (1842)
     Hymenodictyon kurria Hochst. in Flora 26: 71 (1843); Oliv., F.T.A. 3: 42 (1877). Type as for *Kurria floribunda*.
NOTE. Quite a number of specimens (e.g. *Dummer* 3907!, Uganda, Mengo District Mabira Forest, Feb. 1918, and *Dale* in F.D. 3177!, Kenya, N. Nandi, Kamilat Hill, Mar. 1934) have the lower surface of the leaves densely tomentose. This variant is equivalent to *H. kurria* Hochst. var. *tomentellum* Hiern in Cat. Afr. Pl. Welw.: 437 (1898) (type: Angola, Huila, Morro de Lopollo, *Welwitsch* 3033!), a varietal epithet which could be used if considered necessary.

2. H. parvifolium *Oliv.* in Hook., Ic. Pl. 15: 69, t. 1488 (1885); T.T.C.L.: 501 (1949); K.T.S.: 446 (1961); F.F.N.R: 409 (1962). Type: Kenya, 'Mombasa', *Wakefield* (K, holo.!)

Shrub or small tree (1.2–)3.6–4.5(–10) m. tall or occasionally a liane; branchlets glabrescent or pubescent to densely scabrid pubescent with short scaly hairs; bark light grey or purplish-grey, smooth or rarely rough. Leaf-blades elliptic to oblanceolate, 1–9 cm. long, (0.4–)0.8–4.7(–5.9) cm. wide, obtuse to subacuminate at the apex, narrowly cuneate at the base, decurrent into the petiole, glabrous to densely covered with scabrid scaly hairs and more rarely thinner pubescence; petiole often crimson, 0.3–5.5 cm. long; stipules deltoid or oblong-triangular, 2–7 mm. long, 2 mm. wide, sometimes bifid, glabrous. Inflorescences terminal often with additional axillary ones, often numerous at apices of short lateral shoots, forming a thyrsoid panicle, 1–8 cm. long the axis glabrescent, pubescent or densely scabrid; flowers very sweetly (unpleasantly fide Greenway) scented; peduncle 0.3–2.5 cm. long; pedicels 1.5 mm. long; primary bracts not present; secondary bracts lanceolate, 4–6 mm. long, 0.3–1.5(–3) mm. wide, glabrescent, ciliate or scaly-pubescent, deciduous. Calyx-tube subglobose to oblong-ellipsoid, 1–1.5 mm. long, glabrous or glabrescent to scabrid-pubescent; lobes lanceolate, (0.5)1.5–2.5 mm. long with similar indumentum. Corolla white, greenish white or yellow, sometimes tinged red in bud, sometimes turning black; glabrous or puberulous; tube cylindrical below, funnel-shaped above, 2.5–5 mm. long; lobes 5–6, ovate, 1.5 mm. long, 1 mm. wide, ciliolate; style 1.4 cm. long, exserted from the corolla for ± 3–5(–7) mm.; stigma ellipsoid, 0.5 mm. long. Capsules ellipsoid, reddish brown, drying grey-brown, 1.5–2.5(–4) cm. long, 0.7–1 cm. wide, prominently lenticellate, splitting at the ± acute apex into 2 valves, each itself often splitting for a short distance. Seeds straw-coloured, broadly elliptic, 0.8–1.6(–2) cm. long, 0.6–0.9(–1.3) cm. wide, the nucleus broadly elliptic, strongly compressed, very broadly winged, the whole surface with testa strongly reticulate and slightly shining, the wing slit at the hilar end.

KEY TO INFRASPECIFIC VARIANTS

Leaves glabrous or rarely pubescent on nerves; stems mostly
     glabrescent . . . .   subsp. **parvifolium**
Leaves with at least some scaly scabrid hairs on the midrib
     beneath; stems mostly densely scabrid-pubescent when
     young (subsp. *scabrum*):
   Leaves densely scabrid-pubescent all over . . . . . .   var. **scabrum**
   Leaves sparsely scabrid-pubescent only on the main nerves
        beneath . . . .   var. **fimbriolatum**

subsp. **parvifolium**

Leaves glabrous or only with fine pubescence on main nerves beneath. Fig. 66/6–10.

KENYA. Machakos District: 1.6 km. E. of Mtito Andei on the road to Mombasa, 9 Nov. 1958, *Greenway* 9541!; Teita District: Tsavo National Park East, Voi Gate Camp Site, 7 Dec. 1966, *Greenway & Kanuri* 12680!; Kwale District: 3.2 km. E. of Mackinnon Road, 9 Sept. 1953, *Drummond & Hemsley* 4230!
TANZANIA. Pare District: Kisiwani, 2 Feb. 1936, *Greenway* 4577!; Lushoto District: Tembo, Jan. 1931, *Haarer* 1963!; Uzaramo District: Sinda I. near Dar es Salaam, 5 Jan. 1969, *Harris* 2702!
DISTR. K 1, 4, 6, 7; T 3, 5, 6, 8; Mozambique, Malawi, Zimbabwe and South Africa (Transvaal), also formerly cultivated at Kew
HAB. Mixed bushland and thicket particularly of *Acacia-Combretum* types, often on rocky outcrops; 0–1800 m.
NOTE. A few specimens from T 5 and T 8 have fine pubescence on the main nerves beneath but the hairs are not scaly nor scabrid.

subsp. **scabrum** (*Stapf*) *Verdc.* in K.B. 31: 182 (1976). Type: Sudan, Bari, *Dawe* 885 (K, holo.!)

Leaves densely scabrid-pubescent all over or at least with some scaly hairs on the main nerves beneath.

var. **scabrum**

Leaves densely scabrid-pubescent all over.

UGANDA. W. Nile District: Uleppi Escarpment ridge, 21 Mar. 1945, *Greenway & Eggeling* 7240! & E. Madi, Adzugopi [Ajugupi], *Hancock* 756!; Acholi District: Agoro, *Eggeling* 845!
TANZANIA. Mwanza District: Bukumbi [Buhumbi], Nyarwigo, 4 Feb. 1953, *Tanner* 1197!; Ufipa District: Kalambo Falls, 8 Jan. 1950, *Bullock* 2191!; Kondoa District: lower slopes of Simbo Hills, 9 Jan. 1928, *B.D. Burtt* 1883!
DISTR. U 1; T 1, 4, 5, 7; Sudan, Zambia
HAB. Open woodland, thicket, rocky hills; 600–1320 m.
SYN. *H. scabrum* Stapf in J.L.S. 37: 519 (1906); I.T.U., ed. 2: 348 (1952)
NOTE. Typical specimens of this are very distinct from the mainly coastal subsp. *parvifolium*, but var. *fimbriolatum* bridges the gap and a number of specimens from Tanzania are intermediate in many respects. *B.D. Burtt* 724 (Tanzania, Singida District) has capsules 4 cm. long and *Bjørnstad* 1075 (Tanzania, Iringa District, Ruaha National Park) slightly shorter ones, 3.2 mm. long. It is unlikely that these represent distinct taxa.

var. **fimbriolatum** (*De Wild.*) *Verdc.* in K.B. 31: 183 (1976). Type: Zaire, Shaba, *Verdick* 257 (BR, holo.)

Leaves with only sparse scabrid hairs on the main nerves beneath.

TANZANIA. Shinyanga, *Koritschoner* 1661!; Ufipa District: Rukwa, Milepa, Jan. 1949, *Burrett* 49/50!; Iringa District: Ruaha National Park, Mwagusi R. track, 21 Jan. 1966, *Richards* 21023!
DISTR. T 1, 4, 5, 7, 8; Zaire, Zambia
HAB. HAB. *Acacia-Combretum* woodland, often in rock cracks, etc.; 840–1425 m.
SYN. *H. fimbriolatum* De Wild. in Ann. Mus. Congo Bot., sér. 4, 1: 225 (1903)
NOTE. There is a tendency for this to have more capitate inflorescences and longer corolla-tubes than var. *scabra*.

50. CROSSOPTERYX

Fenzl in Nov. Stirp. Dec. Mus. Vind.: 45 (1839)

Shrubs or trees. Leaves opposite, shortly petiolate; stipules interpetiolar, triangular, subpersistent. Flowers fairly small, strongly scented, numerous in dense branched terminal corymbose panicles. Calyx-tube obovoid or globose; teeth 4–6, erect obtuse, deciduous. Corolla hypocrateriform; tube slender, not hairy at the throat; lobes 4–6, contorted, spreading. Stamens 4–6, inserted at the mouth of the tube; filaments very short. Ovary 2-locular; placentas affixed peltately to the septum; ovules few in each locule, imbricate-adpressed; style filiform, long-exserted; stigma clavate or ellipsoid, 2-lobed. Capsule ellipsoid or globose, crustaceous, loculicidally splitting into 2 persistent valves, each divisible into 2 parts along the septum, few-seeded. Seeds peltate, dorsally very compressed; nucleus ovoid; testa membranous, expanded into a broad lobulate and elegantly fimbriate wing; albumen sparse, fleshy.

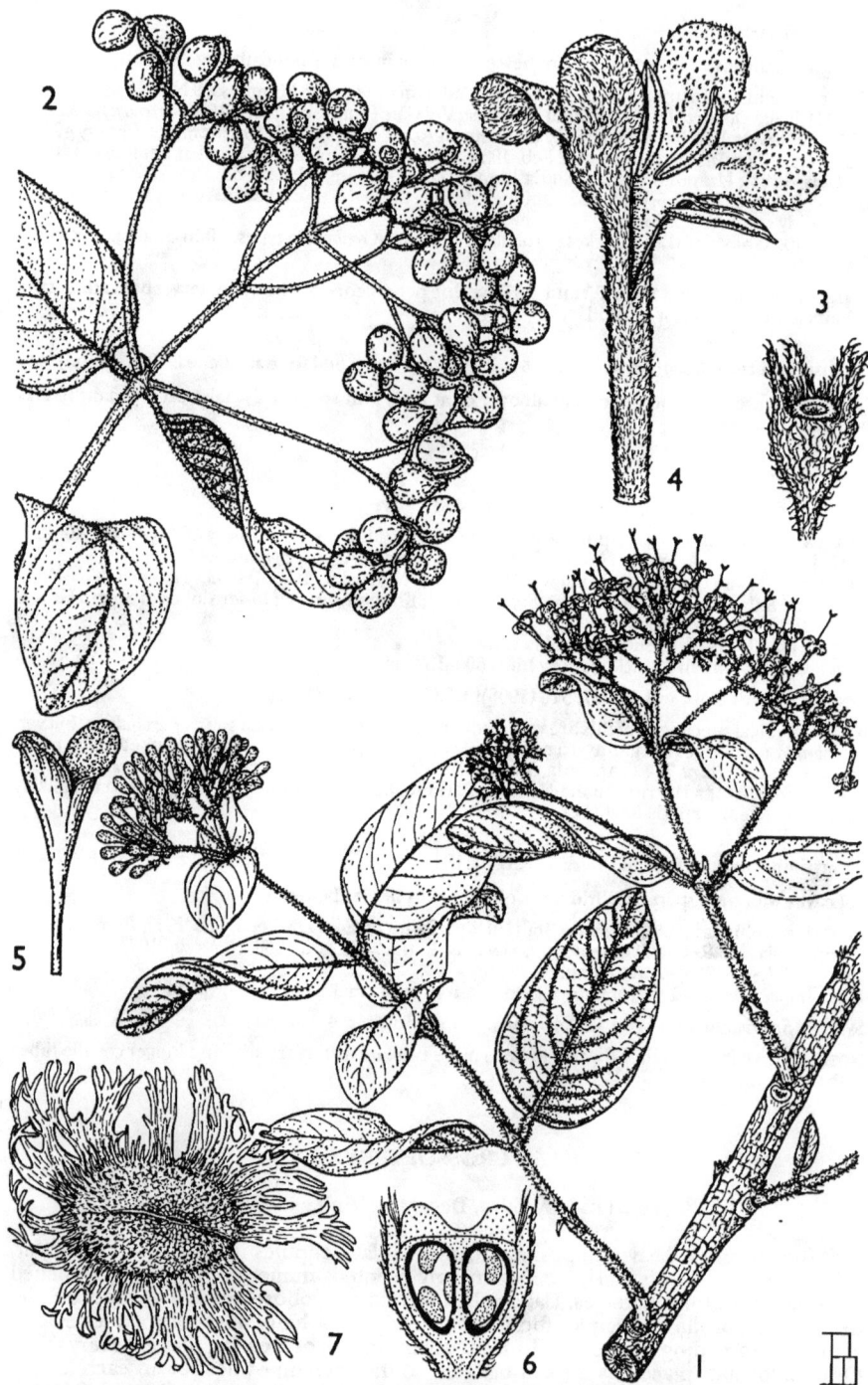

FIG. 67. *CROSSOPTERYX FEBRIFUGA* — 1, flowering branch, × ⅔; 2, fruiting branch, × ⅔; 3, calyx, with one lobe removed, × 8; 4, corolla, partly opened out, × 6; 5, stigma, × 10; 6, section through ovary, × 20; 7, seed, × 10. 1, 3–6, from *Milne-Redhead & Taylor* 7700; 2, 7, from *Tanner* 4336. Drawn by Diane Bridson.

A genus now recognised to be monotypic and confined to the savanna areas of tropical Africa. Bremekamp has suggested that it, together with *Coptosapelta* Korth., should form a tribe *Coptosapelteae.*

C. febrifuga (*G. Don*) *Benth.* in Hook., Niger Fl.: 381 (1849); Hiern in F.T.A. 3: 43 (1877); T.T.C.L.: 494 (1949); I.T.U., ed. 2: 342 (1952); K.T.S.: 437 (1961); F.F.N.R.: 406 (1962); Keay, F.W.T.A., ed. 2: 113 (1963); Fl. Pl. Lign. Rwanda: 558, fig. 187/3 (1982); Fl. Rwanda 3: 156, fig. 48/3 (1985). Type: Sierra Leone, near Freetown, *Don* (BM, holo.!, K, iso.!)

Mostly fairly small savanna tree 1.8–15 m. tall, with rounded crown and pendulous branchlets, but sometimes shrubby; bark pale grey to dark brown, scaly, finely reticulate; young stems glabrous to densely pubescent hairy or tomentose. Leaf-blades elliptic, elliptic-oblong, ovate, obovate or almost round, 1.5–13.5 cm. long, 1.2–7.5 cm. wide, rounded to shortly acuminate at the apex, broadly cuneate to rounded at the base, glabrous to densely pubescent or velvety; petioles 0.5–1.8 cm. long; stipules 2–3 mm. long, acuminate. Inflorescences dense, up to 6–10 cm. long; peduncles up to 6 cm. long; pedicels 0.5–2 mm. long; primary and secondary bracts linear, 1.5–3.5 mm. long. Calyx-tube 1 mm. long; lobes elliptic to linear, 0.5–1.5 mm. long, obtuse or acute. Corolla creamy white or pale yellow, densely pubescent outside; tube tinged pink, 5–8(–11) mm. long; lobes round, 1.5 mm. long and wide. Style exserted for 3–7.5 mm. glabrous. Capsule blackish, 0.6–1 cm. long. Seeds thin, flat, 3.2–5 mm. long, 2.5–3.5 mm. wide. Fig. 67.

UGANDA. W. Nile District: Madi Woods, Dec. 1862, *Grant* 698!; Teso District: Serere, Jan. 1932, *Chandler* 396!; Mengo District: Sungiri, Nakasongola, 11 July 1956, *Langdale-Brown* 2200!
KENYA. Kwale District: Shimba Hills, Mwele Mdogo Forest, 19.2 km. SW. of Kwale, 4 Feb. 1953, *Drummond & Hemsley* 1209! & Makadara, *R.M. Graham* in F.D. 1696! & Shimba Hills, without locality, *Donald* in F.D. 2372!
TANZANIA. Mbulu District: Derakuta [Derakata] Ranch, *Leippert* 6163!; Tabora District: Kaliua [Kaliuwa], 7 Jan. 1950, *Shabani* 89!; Tunduru District: 1.5 km. E. of R. Muhuwesi [Mawese], near Puchapucha, 18 Dec. 1955, *Milne-Redhead & Taylor* 7700!
DISTR. U 1–4; K 7; T 1–8; widespread in tropical Africa from Senegal to Sudan and Ethiopia south to South Africa (Transvaal) and northern border of Namibia
HAB. Deciduous woodland and wooded grassland, *Brachystegia* woodland, often on stony hillsides; 0–1350 m.

SYN. *Rondeletia febrifuga* G. Don, Gen. Syst. 3: 516 (1834)
    Crossopteryx kotschyana Fenzl in Nov. Stirp. Dec. Mus. Vind.: 45 (1839); Hiern in F.T.A. 3: 44 (1877). Type: Sudan, Schangul, *Kotschy* (W, holo., BM, iso.!, K, iso.!*)
    Tarenna angolensis Hiern in F.T.A. 3: 89 (1877). Type: Angola, Quiballa, *Monteiro* (K, holo.!)
    Crossopteryx africana Baillon, Hist. Pl. 7: 489 (1880); K. Schum. in P.O.A. C: 378 (1895). Type: based on *Rondeletia africana* Winterb., Acc. Sierra Leone 2: 46 (1803) mentioned without description
    Chomelia buchananii K. Schum. in P.O.A. C: 380 (1895). Type: Malawi, *Buchanan* 32 (B, holo.†, K, iso.!)

NOTE. There is a good deal of variation in the indumentum of the stems and foliage and in the length and shape of the calyx-lobes.

Tribe 14. **VIRECTARIEAE****

One genus only . . . . .51. Virectaria

51. **VIRECTARIA**

Bremek. in Verh. Nederl. Akad. Wet. Afd. Nat., ser. 2, 48: 21, adnot (1952); Verdc. in B.J.B.B. 23: 35–52 (1953) & B.J.B.B. 28: 242 (1958)

Virecta Smith in Rees, Cyclop. 37 (1817), *non* L.f. (1781)

Annual or perennial herbs or subshrubby herbs with erect or procumbent glabrescent to stiffly hairy stems from a fibrous rootstock. Leaves paired; stipules small, entire, deltoid, or fimbriated with linear setae from a short base. Flowers small, ⚥, never dimorphic, in

* The Kew sheet is *Kotschy* 532 from Fazughli but may be an isotype; the BM sheets are also *Kotschy* 532 but are labelled Schangul.
** By B. Verdcourt

few–many-flowered cymose clusters at the apices of the main shoots. Calyx-tube ovoid or urceolate, often bristly hairy; lobes 5–6, equal or unequal, spathulate, filiform or lanceolate with small glands between each pair. Corolla-tube filiform to narrowly infundibuliform; lobes 4–6(–7), deltoid to lanceolate, suberect; throat practically glabrous save for an obscure ring of hairs some distance below the orifice. Stamens 4–6, much exserted. Ovary bilocular; ovules attached to placentas which are affixed to the central partition; style much exserted, often overtopping the corolla-lobes; stigma subcapitate. Disc cylindrical or consisting of two separate cones, Capsule very characteristic, subglobose, splitting in a plane at right-angles to the central partition into 2 valves one of which usually falls away entirely but the other remains attached to the rachis by means of a woody pedicle. Seeds small, brownish, subglobose, reticulately pitted.

A small genus of 7 species restricted to tropical Africa.

This genus has long been placed in the Hedyotideae next to *Pentas* but I would agree wholeheartedly with Bremekamp that the two genera are in no way related. Hallé (Fl. Gabon 12, Rubiacées, 1: 79 (1966)), keeps it in the Hedyotideae but admits it has an anomalous position. The morphology of the capsule is totally different in the two genera, as is also the structure of the corolla, stigma and pollen grains. The absence of rhaphides in the tissues in fact removes *Virectaria* from the Rubioideae; I would not, however, agree to including it in the Ophiorrhizeae as is done by Bremekamp but would suggest it formed a separate tribe near the Rondeletieae, a course followed here.

V. major *(K. Schum.) Verdc.* in B.J.B.B. 23: 42, fig. 5/E, G, J, K, M & O (1953); Hepper in F.W.T.A., ed. 2, 2: 208 (1963); Fl. Pl. Lign. Rwanda: 612, fig. 208/1 (1982); Fl. Rwanda 3: 230, fig. 74/1 (1985). Type: Tanzania, Bukoba, *Stuhlmann* (B, holo.†)

Erect or somewhat scandent herb or subshrub 0.5–2(–3) m. tall, with ± branched stems, glabrescent or covered with adpressed to spreading hairs. Leaf-blades lanceolate to ovate-lanceolate, 2.3–9.5(–14.5) cm. long, 0.75–4.5(–5.5) cm. wide, acute to narrowly acuminate at the apex, cuneate at the base, glabrescent or with long and short hairs on both surfaces, sometimes quite velvety beneath with short hairs; petiole 0.2–1.5 cm. long; stipules consisting of a single deltoid tooth 0.1–1 cm. long or 2 subulate-lanceolate lobes 4–8 mm. long from a short base together with small colleters on either side. Flowers white, rose or violet, in dense branched terminal hairy or setose cymes 1.5–4.5(–5.5 in fruit) cm. wide. Calyx-tube urceolate, 1–2 mm. long, 1–1.5 mm. wide, densely hairy with adpressed or patent hairs; lobes 4–6 filiform to distinctly spathulate, 3.5–8(–11) mm. long, setose. Corolla-tube 0.7–1(–1.3) cm. long, 1–1.5 mm. wide at the base, 2.5–7 mm. wide at the apex, hairy with spreading setae; lobes 0.6–1.2(–1.5) cm. long, 1–2(–2.3) mm. wide, ciliate with long setae. Stamens exserted 0.6–1.6(–2) cm.; disc cylindrical 0.8 mm. tall and wide; style exserted 0.8–1.9 (–2.3) cm. long; stigma slightly capitate. Fruit chestnut-coloured, globose 2–3 mm. diameter, adpressedly pilose.

var. **major**; Verdc. in B.J.B.B. 23: 42, fig. 5/E, G, J, K, M & O (1953); Hepper in F.W.T.A., ed. 2, 2: 208 (1963)

Calyx-lobes mostly filiform, scarcely if at all spathulate. Fig. 68.

UGANDA. Ankole District: Igara, Feb. 1939, *Purseglove* 573!; Kigezi District: above Mushongero, Lake Mutanda, 30 Jan. 1939, *Loveridge* 439!; Masaka District: Nkose I., 21 Jan. 1956, *Dawkins* 866!
TANZANIA. Ngara District: Bugufi, Kirushya, 5 Apr. 1960, *Tanner* 4826!; Buha District: near Kasulu, Marera mission, 7 Apr. 1964, *Pirozynski* 640!; Iringa District: Ulevi, Mar. 1954, *Carmichael* 399!
DISTR. U 2, 4; T 1, 4, 7; Sierra Leone, Nigeria, Cameroun, Bioko [Fernando Po], Zaire, Rwanda, Burundi and Zambia
HAB. Grassland, bracken scrub, forest edges, evergreen forest, on rocky hillsides, moist places, often on lava soils; 1110–2350 m.
SYN. *Virecta major* K. Schum. in P.O.A. C: 377 (1895) & in E.J. 23: 422 (1896); Verdc. in K.B. 7: 363 (1952)
       [*V. multiflora* auctt., e.g. K. Schum. in E.J. 28: 487 (1900), *non* (Smith) Bremek.]
       V. kaessneri S. Moore in J.B. 48: 220 (1910). Type: Zaire, Lake Tanganyika, Kibanda, *Kassner* 3052 (BM, holo.!)
       Pentas fililoba K. Krause in Z.A.E. 2: 312 (1910). Type: Rwanda, Lake Mohasi, *Mildbraed* 460 (B, holo.!, BR, iso.!)
       Virectaria kaessneri (S. Moore) Bremek. in Verh. Nederl. Akad. Wet. Afd. Nat., ser. 2, 48: 21, adnot (1952)
NOTE. K. Schumann later mistakenly reduced his *V. major* to *V. multiflora*, an error which caused considerable confusion but the two species differ considerably in their indumentum and disc structure.
DISTR. (of species as a whole). As above with the addition of Sierra Leone, Liberia and Ivory Coast.

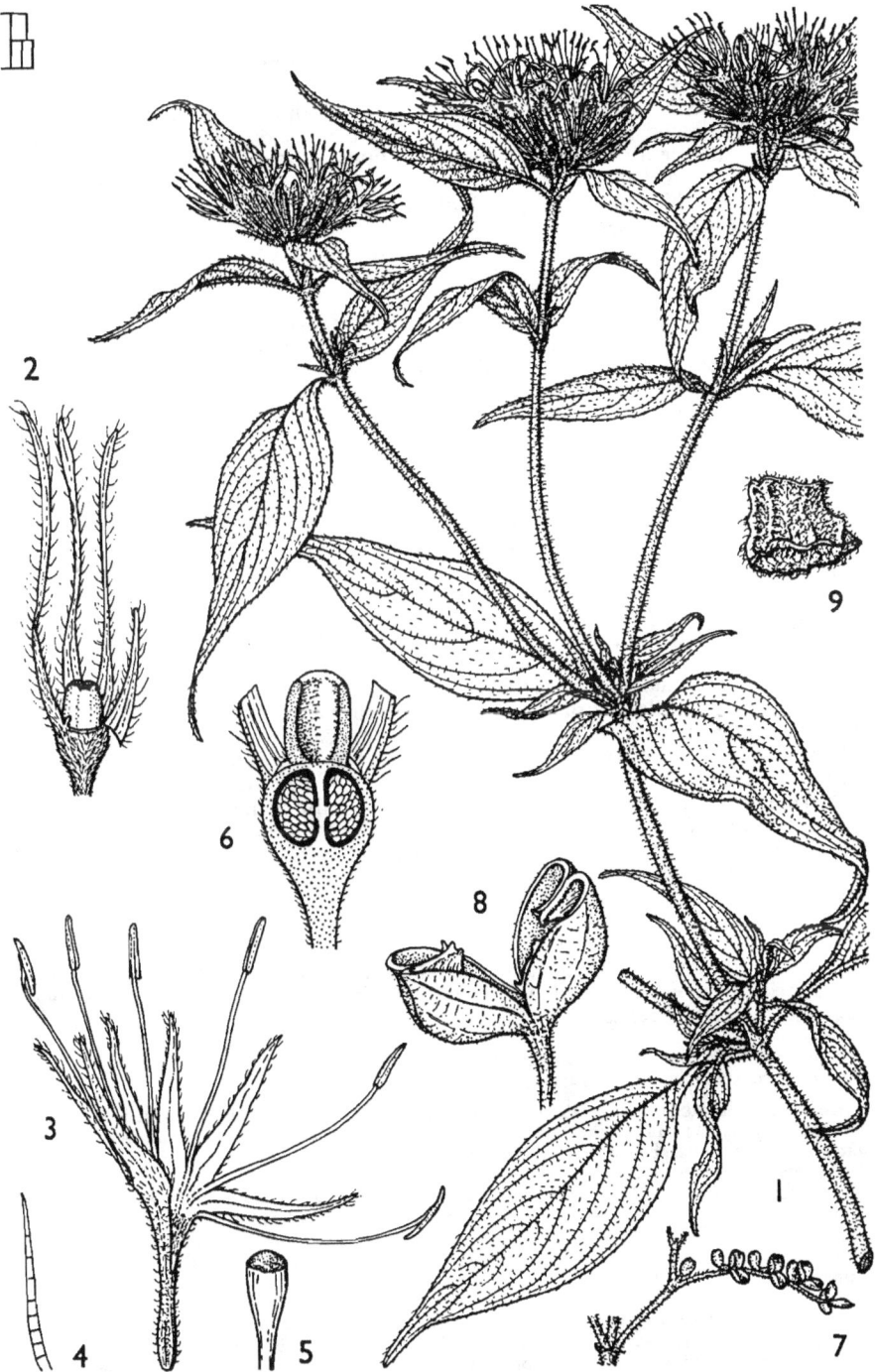

FIG. 68. *VIRECTARIA MAJOR* var. *MAJOR* — **1**, habit, × ⅓; **2**, calyx, with one lobe turned back, × 3; **3**, corolla, partly opened out, × 2; **4**, hair from inside corolla, × 20; **5** stigma, × 12; **6**, section through ovary, × 10; **7**, branch of infructescence, × 1; **8**, dehisced capsule, × 6, **9** seed, × 20. 1, from *Newbould & Harley* 4646; 2–8, from *Pirozynski* 498; 9, from *Wood* 89. Drawn by Diane Bridson.

Tribe 15. ISERTIEAE*

1. Flowers in terminal panicles; one calyx-lobe often
   enlarged, leaf-like, white or coloured 2
   Flowers in axillary heads or panicles; calyx never with 1
   lobe enlarged 3
2. Fruits ± succulent or at least indehiscent; buds without
   filiform apical appendages . . . . **52. Mussaenda**
   Fruits dry, dehiscing at the apex; buds bearing 5 apical
   filiform appendages . . . . **53. Pseudomussaenda**
3. Leaves glabrous to tomentose beneath; inflorescences
   various; stigma divided into 4–5 lobes or arms; ovary
   4–5-locular . . . . **54. Sabicea**
   Leaves velvety-felted beneath; flowers in E. African species
   in sessile axillary clusters at the nodes; stigmatic lobes
   2, laterally dilated; ovary 2(–3)-locular . . . . . **55. Pseudosabicea**

52. MUSSAENDA

L., Sp. Pl.: 177 (1753) & Gen. Pl., ed. 5: 85 (1754); Hiern in F.T.A. 3: 65 (1877); Wernham in J.B. 51: 233–240 & 274–278 (1913); Petit in B.J.B.B. 25: 149–167 (1955); F. Hallé in Adansonia sér. 2, 1: 266–298 (1961)

Shrubs, scandent shrubs or lianas. Leaves petiolate; blades usually elliptic with an acute or acuminate apex, pubescent or less often glabrous; stipules entire to completely bilobed, persistent or caducous. Flowers usually yellow to red or sometimes white, occasionally sweet-scented, isostylous or heterostylous, borne in terminal panicles. Calyx-tube oblong, turbinate or ovoid; lobes 5, usually subulate, linear, slightly spathulate or sometimes short and dentate or rounded, persistent or caducous; frequently several lobes on each inflorescence develop into a stalked white, creamy yellow or sometimes red dilated lamina. Colleters usually present between the calyx-lobes. Corolla-tube cylindrical, narrowly funnel-shaped or abruptly widened to accommodate the anthers, with flattened or fine hairs at the throat and between the anthers; limb 5-lobed, lobes reduplicate in bud, spreading at maturity, often with the centre portion connate, with short hairs and often a star of longer, coloured hairs above. Anthers 5, with tips just level with the throat or inserted up to halfway down the tube, linear, connective very shortly acuminate at the apex, attached by very short filaments near the base. Ovary 2 (rarely –3–4)-locular; ovules numerous, inserted on fleshy placentas; style slender, sometimes divided into 2 arms; stigma included or occasionally exserted; lobes 2, ± oblong to fusiform. Fruit fleshy, indehiscent, globose, ellipsoid or oblong, sometimes crowned by the persistent calyx-lobes. Seeds numerous, pale brown to blackish, small, usually ± flattened, the surface reticulate.

A genus of ± 100 species distributed throughout the tropics with the exception of the New World and northern Australia; about 35 species occur in tropical Africa.

Specimens cultivated in Kenya, Tanzania and Zimbabwe have been named *Mussaenda frondosa* L., and probably originated from the Indian sub-continent; although no exact match has been made with any of the Indian specimens at Kew, they doubtless belong to this aggregate. The epithet is used in a very general sense and includes several subspecific taxa (raised to specific rank by Hutchinson in Gamble, Fl. Madras 2: 610 (1921)); misidentifications or misapplications of the epithet in other areas further obscures the issue. The material cultivated in Africa, best named *M. frondosa* L. sens. lat., can be recognised by the long (1–1.7 cm.) linear calyx-lobes and conspicuous 3-lobed bracts (which are chaffy on dry specimens); the foliaceous calyx-lobes are cream-yellow and the flowers orange.

M. philippica L.C. Rich. var. *aurorae* Sulit (Philip. Journ. For. 2 (1): 39 (1939)) has been recorded from T6 (*Mwasumbi & Wingfield* 11230!). It is a sterile cultivar propagated from a sport found in the Philippine Is., and very distinctive, all the calyx-lobes being large, foliaceous and usually white (red forms are known but have not been recorded from Africa). *M. philippica* most closely resembles *M. microdonta* Wernham, but the calyx-tube is longer and has a somewhat coarser indumentum, the corolla-tube more slender, the lobes bright yellow, the fruit ovoid and the bark (of var. *aurorae* at least) less conspicuously lenticellate.

* Genus 52 by D.M. Bridson, 53–55 by B. Verdcourt

1. Stipules persistent (or at length deciduous) 2
   Stipules early caducous 5
2. Foliaceous calyx-lobes absent (or if present dull red and
   not fully expanded); stipules lobed to the base or
   nearly so; flowers large, corolla-lobes 1.1–3.5 cm. long   2. *M. elegans*
   Foliaceous calyx-lobes usually present; stipules entire or if
   lobed then not right to the base; flowers smaller,
   corolla-lobes 0.3–1.2 cm. long 3
3. Foliaceous calyx-lobes bright red; stipules bilobed for ⅓ of
   their length and usually reflexed; calyx-tube densely
   covered with long spreading hairs . . . . . .   *3. M. erythrophylla*
   Foliaceous calyx-lobes whitish to yellow; stipules entire or
   if lobed then not reflexed; calyx-tube sparsely covered
   with shorter hairs 4
4. Stipules ovate, reflexing; calyx-lobes not exceeding 3–4
   mm. long, persistent on the cylindrical to narrowly
   ellipsoid fruit (2–2.9 cm. long) . . . .   *5. M. monticola*
   Stipules large, bilobed for ± ½ their length (chaffy on dried
   specimens), not reflexing; calyx-lobes not less than 1
   cm. long (chaffy on dried specimens), caducous; fruit
   ovoid, ± 1 cm. long (cultivated) . . . .   *M. frondosa*
5. Foliaceous calyx-lobes absent (or if present not fully
   expanded); flowers with flattened hairs present in
   throat; fruit with the scar pale all over or only gradually
   darkened towards centre; leaves shining above (with
   bronze-blackish mottling in dried specimens)   *1. M. arcuata*
   Foliaceous calyx-lobes present (or rarely absent), cream to
   yellow; flowers with fine hairs present in the throat;
   fruit with annular whitish scar surrounding blackish
   centre; leaves dull above . . . .   *4. M. microdonta*

1. **M. arcuata** *Poir* in Lam., Encycl. Méth. Bot. 4: 392 (1797); Hiern in F.T.A. 3: 68 (1877); K. Schum. in P.O.A. C: 379 (1895); V.E. 1 (1): 311, fig.278 (1910); Z.A.E. 2: 314 (1911); Wernham in J.B. 51: 274 (1913); K. Krause in N.B.G.B. 10: 603 (1929); F.P.N.A. 2: 334 (1947); T.T.C.L.: 506 (1949); Aubrév., Fl. For. Soud.-Guin.: 478, t. 108/5–7 (1950); F.P.S. 2: 446 (1952); Petit in B.J.B.B. 25: 157 (1955); Codd in Fl. Pl. Afr. 31, 4, t. 1231 (1956); K.T.S.: 451 (1951); F.F.N.R.: 413, fig. 71H (1962); Hepper in F.W.T.A., ed. 2, 2: 615 (1963); Hallé in Fl. Gabon 12, Rubiacées, 1: 152 (1966). Type: Mauritius, *Commerson* 304 (P-LAM, holo.)

Shrub, scandent shrub or climber, 0.6–7(–14) m. tall; branches usually glabrous but often covered with shortish spreading hairs (2 specimens from Gabon have much longer reddish hairs), said to exude a watery to milky or sticky sap; lenticels not markedly apparent on young branches. Leaf-blades coriaceous, elliptic to round, 3–16.3(–20) cm. long, 1.3–8.5(–10.8) cm. wide, apex acuminate to caudate, base cuneate, acute or rounded, with 5–7 pairs of lateral nerves, medium to dark green and shiny above and pale green beneath (dried specimens have a characteristic bronze to blackish mottling between the tertiary veins) glabrous above and glabrous or sparsely pubescent with denser hairs on the nerves beneath; tertiary nerves close together ± parallel to mid-rib; petiole 0.3–2.2 cm. long; stipules with lobe entire or divided almost to the base, 3–12 mm. long, reflexing and soon caducous revealing a line of red to brown hairs above the scar. Flowers isostylous or heterostylous, sweet-scented, in dense or lax panicles, or few-flowered panicles, terminal on short lateral branches (in climbing specimens); peduncles 0.8–4 cm. long; secondary peduncles often present; pedicels 0–5.3(–7) mm. long; bracts and bracteoles entire or rarely 2–3-lobed, 0.2–2.7 cm. long and up to 2 mm. wide. Calyx-tube turbinate to ellipsoid, 2–4 mm. long, usually glabrous but occasionally sparsely hairy; lobes dentate, linear or slightly spathulate, 0.7–1.5 cm. long, 0.25–2.5 mm. wide, curving outwards when mature, usually caducous; occasionally tending to produce an expanded creamy-white limb (most developed in specimens from N. Zambia), never as large as the leaves. Corolla with greenish yellow tube and pale to bright yellow lobes with a distinctive star of orange-red hairs at the centre which become chocolate-brown at maturity; tube 1.3–2.7 cm. long, 2.5–5 mm. wide at the top, 0.75–1.5 mm. wide at the base, glabrous, hairy all over or with the hairs concentrated into 5 vertical stripes outside, with flattened yellow hairs at the throat and between the anthers, becoming glabrous towards the base within; lobes narrowly to broadly ovate, 4–19 mm. long, 2.5–11 mm. wide, acute to apiculate at apex, shortly connate for (0–)1.5–3 mm. at base, somewhat reflexed at maturity, midrib produced to form a low ridge. Distance from the throat to the top of the anthers 0–1.6 mm. in isostylous flowers (both stigma and anthers near the throat), 0–3 mm. in short-styled flowers and 4–6 mm. in long-styled flowers. Style 1.2–2.1 cm. in long-styled and isostylous flowers and 0.75–1 cm. in short-styled flowers; arms 0–(1.3–2) mm. long; stigma entirely exserted or included, each lobe 2.2–5.2 mm. long. Fruit edible, pale green to yellow, ellipsoid or subglobose, (0.8–)1.1–2.5 cm. long, (0.6–)0.8–1.8 cm. wide, with a round whitish scar at the apex. Seeds straw-coloured to pale brown, ± rectangular or triangular in outline, 1–1.3 mm. long, slightly thickened, clearly reticulate. Fig. 1/1.

UGANDA. W. Nile District: Koboko, July 1938, *Hazel* 661!; Ankole District: Kalinzu Forest, 26 Jan. 1970, *Synnott* 468!; Mengo District: 29 km. Kampala–Entebbe road, Jan. 1931, *Snowden* 1924!
KENYA. Nandi/N. Kavirondo Districts: Nandi Escarpment, 7 July 1966, *Makin* 307!; S. Kavirondo District: Kisii, Sept. 1933, *Napier* 2857!; Kisumu-Londiani District: 20 km. N. of Kisumu on Eldoret road, 8 July 1956, *Napper* 560!
TANZANIA. Lushoto District: Amani, 22 July 1929, *Greenway* 1669!; Buha District: Kasakela Reserve, 17 Nov. 1962, *Verdcourt* 3333!; Songea District: ± 32 km. E. of Songea, R. Mkurira, 19 Jan. 1956, *Milne-Redhead & Taylor* 8367!
DISTR. U 1–4; K 3, 5, ?7; T 1, 3, 4, 6–8; throughout tropical Africa (excluding Somalia) as far south as Angola and Zimbabwe; Madagascar, Mauritius and Réunion
HAB. Grassland, bushland, open or closed forest, evergreen rain-forest; 700–1830 m.

SYN. *M. arcuata* Poir. var. *parviflora* S. Moore in J.L.S. 37: 301 (1906). Type: Angola, near Malange, Quizanga, *Gossweiler* 1235 (BM, holo.!, K, iso.!)
   M. abyssinica Chiov. in Ann. Bot. Roma 9: 67 (1911); Wernham in J.B. 51: 237 (1913). Type: Ethiopia: Tzellemti, Mai Jaclit, neighbourhood of 4th cataract, *Chiovenda* 701 (FT, holo.!)
   M. arcuata Poir. var. *pubescens* Wernham in J.B. 51: 274 (1913); F.P.S. 2: 446 (1952); Petit in B.J.B.B. 25: 161 (1955). Types: Tanzania, Uluguru Mts., *Goetze* 208 (K, syn.!), Zaire, Kundelungu, *Kassner* 2585, 2608 (BM, syn.!), Sudan, Bongoland, *Schweinfurth* 2891* (K, syn.!) Angola, High Plateau, *Gossweiler* 4247 (BM, syn.!)
   M. laurifolia A. Chev., Expl. Bot. Afr. Occ. Fr.: 314 (1920), *nomen nudum*

NOTE. This species varies a great deal and much of the variation can probably be correlated with differences in habitat. Wernham's var. *pubescens* does not seem to be worth maintaining as many intermediates with the typical variety occur. All the material at Kew from Madagascar and Mauritius is ± isostylous, with the stigmas exserted to just included and the top of the anthers not more than 1.6 mm. below the throat. With the exception of a few specimens from Mozambique the majority of the African plants are distinctly heterostylous, or occasionally isostylous, with both the anthers and style further down the corolla-tube.
   M. arcuata Poir. var. *fusco-pilosa* Mildbr. has been used as a manuscript name on *Zerny* 101, but no evidence of publication has been found.

2. M. elegans *Schumach. & Thonn.*, Beskr. Guin. Pl.: 117 (1827); Hiern in F.T.A. 3: 69 (1877); Mildbr., Z.A.E. 1907–8, 2: 315 (1911); Wernham in J.B. 51: 276 (1913); A. Chev., Expl. Bot. Afr. Occ. Fr.: 314 (1920); F.P.S. 2: 449 (1952); Petit in B.J.B.B. 25: 151 (1955); F. Hallé in Adansonia, sér. 2, 1: 272 t. 31 & 292, t. 121, 2 (1961); Hepper in F.W.T.A., ed. 2, 2: 167 (1963). Type: Ghana, Aquapim Mt., *Thonning* 170 (G-DC, S, iso., K, microfiche!)

Shrub, scandent shrub or climber 1.5–10.5 m. tall; branches hispid or pubescent, becoming glabrous with age; whitish lenticels apparent. Leaf-blades oblanceolate to obovate or elliptic to round, 3–14.3(–16) cm. long, 2–7.8 cm. wide, apex acuminate, base rounded or cuneate, with 7–15 pairs of lateral nerves, sparsely pubescent above and beneath, with denser pubescence on nerves beneath, occasionally entirely glabrous; petioles 0.3–6 cm. long; stipules deeply bilobed or with 2 separate subulate or linear lobes, 4–12 mm. long, up to 1 mm. wide, hairy outside, persistent. Flowers heterostylous, in panicles with 2–3 peduncles 0–5 cm. long, each bearing 2–4 flowers; pedicels 0–1.2(–1.8) cm. long; bracts and bracteoles simple, linear, up to 7 mm. long. Calyx with short adpressed hairs or long (up to 3 mm.) patent hairs; tube cylindrical, 4–8 mm. long, 1.5–3 mm. wide; limb 0–1.5 mm. long; lobes linear or occasionally slightly spathulate, (4–)8–20 mm. long, 0.5–2(–3) mm. wide, occasionally tending to produce an expanded, dull red limb (noted in specimens from N. Nigeria, Bauchi Plateau, and Liberia). Corolla greenish yellow or yellowish orange outside with the lobes pink to orange or bright red above, salver-shaped; tube 1.6–4 cm. long, 2–7 mm. wide at top, 1–2 mm. wide at base, covered with short adpressed hairs or long patent hairs outside, with flattened yellow hairs at the throat and extending to the base of the anthers within; lobes strongly reduplicate in bud,

---

* *Schweinfurth* 1741, also cited by Wernham, is quite glabrous.

± round, 1.1–3.5(–4.3) cm. long, 0.8–2.5 cm. wide, the inner portion connate for 3.3–18 mm., apex rounded or acuminate, midrib produced to form a low ridge, glabrous or glabrescent below and papillose with a few hairs a short distance along the midrib above. Distance from the throat to the top of the anthers 4.5–12 mm. in long-styled flowers and 0–5 mm. in short-styled flowers. Style 1.9–2.7 cm. long in long-styled flowers and 0.6–1.3 cm. long in short-styled flowers; arms 0–1.6 mm. long; stigma included; lobes 2.2–5 mm. long. Fruit green with whitish to pale green flecks, ellipsoidal, 1.1–2.4 cm. long, 0.7–1 cm. wide, usually sparsely hairy but sometimes glabrous, calyx persistent or rarely caducous. Seeds yellowish brown to dark reddish brown, flattened and somewhat variable in outline, 0.6–0.7 mm. long and usually slightly narrower, clearly reticulate.

UGANDA. Bunyoro District: Unyoro, 18 Oct. 1910, *Dawe* 1064!; Toro District: Bwamba, June 1933, *Eggeling* 1255!; Mengo District: Kyagwe, Mabira Forest, Sept. 1933, *Brasnett* 1387!
DISTR. U 2, 4; K 3 cultivated; throughout W. tropical Africa, Cameroun, Cabinda, Zaire, ? Angola (see note) and Sudan
HAB. In forest undergrowth and in open forest; 760–1150 m.

SYN. *Gardenia coccinea* G. Don in Edin. Phil. Journ. 11: 343 (1824). Type: Sierra Leone, near Freetown, *G. Don* (BM, holo.!), non *Mussaenda coccinea* Poir.
Mussaenda discolor DC., Prodr. 4: 372 (1830), nom. *superfl.* Type as for *M. elegans*.
Bertiera coccinea (G. Don) G. Don, Gen. Syst. 3: 506 (1834). Type as for *Gardenia coccinea*
Mussaenda elegans Schumach. & Thonn. var. *minor* De Wild. & Th. Dur. in Ann. Mus. Congo, Bot., sér 2, 1: 27 (1899); De Wild. in B.J.B.B. 7: 278 (1921). Type: Zaire, Lemba, *Cabra* (BR, holo.)
M. elegans Schumach. & Thonn. var. *rotundifolia* Wernham in J.B. 51: 277 (1913). Type: S. Nigeria, Katagum, *Dalziel* 397 (K, holo.!)
M. elegans Schumach. & Thonn. var. *psilocarpa* Wernham in J.B. 51: 277 (1913). Type: S. Nigeria, Lagos, *MacGregor* 104 (K, holo.!)

NOTE. This species is very variable, especially in habit, leaf- and flower-size and indumentum. The material examined from Uganda, however, seems to be ± homogeneous and has large flowers with spreading hairs on the corolla and calyx-tubes and medium to large leaves.
   M. zenkeri Wernham, from Cameroun is very close to *M. elegans* Schumach. & Thonn. and is probably not worthy of specific rank. It can be distinguished by the glabrous leaves and broad (up to 6 mm.), caducous calyx-lobes. Specimens from Cabinda and Angola do not have such broad calyx-lobes, but are otherwise identical with *M. zenkeri* Wernham.

3. M. erythrophylla *Schumach. & Thonn.*, Beskr. Guin. Pl.: 116 (1827); Hiern in F.T.A. 3: 69 (1877); Watson in Bot. Mag. 134, t. 8222 (1908); Z.A.E. 2: 315 (1911); Wernham in J.B. 51: 275 (1913); A. Chev., Expl. Bot. Afr. Occ. Fr.: 314 (1920); Pellegrin, Fl. Mayombe 3: 8 (1938); F.P.N.A. 2: 335 (1947); F.P.S. 2: 446, fig. 160 (1952); Petit in B.J.B.B. 25: 154 (1955); Irvine, Woody Pl. Ghana: 690, fig. 130 (1961); F. Hallé in Adansonia sér. 2, 1: 274 t. 4/3 & 278 t. 6/1, 2 (1961); Hepper in F.W.T.A., ed. 2, 2: 165, fig. 234 (1963); N. Hallé in Fl. Gabon 12, Rubiacées, 1: 148 (1966). Type: Ghana, Aquapim Mt., *Thonning* 93 (C, holo., G-DC, iso., K, microfiche, holo. & iso.!)

Shrub, scandent shrub or climber, 1.5–8 m. tall; branches ± densely covered with patent or crisped, straw-coloured to reddish hairs; lenticels sometimes visible as small whitish streaks. Leaf-blades elliptic to round, 2.5–18 cm. long, 1.7–11 cm. wide, apex acute or more usually acuminate, base rounded, ± cordate or cuneate, with 7–10 pairs of lateral nerves, hairy on both surfaces, with denser, often red hairs on the nerves beneath; petioles 0.3–5 cm. long; stipules bilobed for ± ⅓ of their length, 5–12 mm. long, 2.5–8 mm. wide at base, hairy outside, usually reflexed, persistent. Flowers heterostylous, in usually dense, compressed panicles; peduncles 0.5–2.7(–8) cm. long; secondary peduncles present or suppressed; pedicels 0–10 mm. long; 2–6 foliaceous calyx-lobes present on each inflorescence; bracts and bracteoles 3–5-lobed, 1–1.5 cm. long, glabrous inside. Calyx-tube obovoid, 3–5 mm. long, densely covered in spreading hairs; lobes green or red, lanceolate, (0.3–)0.8–1.6(–2.4) mm. long, 1.2–3(–5) mm. wide, covered with long red hairs outside and shorter pale hairs inside, caducous; foliaceous lobes bright red, elliptic to round, 3–11 cm. long, 2–9.4 cm. wide, acute at apex, rounded at base, pubescent on both sides; stipe 0.3–1 cm. long. Corolla with tube greenish to reddish outside and lobes white to yellow or pinky above; tube 1.5–2.8 mm. long, 3–5 mm. wide at top, 1.5–3 mm. wide at base, covered with red spreading hairs outside, with yellow flattened hairs at the throat and extending below the anthers inside; lobes round, (3.5–)5–12 mm. long, 5–11 mm. wide, apex rounded or apiculate, the inner portion connate for (2–)2.5–5 mm., covered with red spreading hairs below, above ± glabrous or with short whitish hairs becoming denser towards the centre, then with a ring of long dark red hairs surrounding the flattened yellow throat hairs. Distance from the throat to the top of anthers 5.5–8.5 mm. in long-styled flowers and 1–5 mm. in short-styled flowers. Style 1.4–2.4 cm. in long-styled flowers, 0.7–1 cm. in short-styled flowers; arms 0–1.5 mm; stigma included or partly exserted; lobes 2.6–4.2 mm. long. Fruit yellowish, ellipsoid, 1–2.4 cm. long, 0.7–1.4 cm. wide, covered in spreading brown to red hairs; calyx-lobes eventually caducous. Seeds straw-coloured to blackish, ± square or oblong in outline, 0.6 mm. long, ± flattened, reticulate, sometimes with a few short, fine hairs.

UGANDA. Bunyoro District: Masindi, 15 May 1921, *Lankester;* Toro District: NW. of Fort Portal, Bwamba Forest, June 1956, *Bally* 10593!; Mengo District: Entebbe, Nambigirwa [Nambigiruwa] Forest, Jan. 1932, *Eggeling* 162!
KENYA. N. Kavirondo District: Kakamega Forest, Yala R., June 1933, *Dale* in *F.D.* 3110!
TANZANIA. Bukoba District: Minziro Forest, Jan. 1958, *Procter* 787!; Buha District: Gombe Stream Reserve, Kasakela, 18 Nov. 1962, *Verdcourt* 3357!; Mpanda District: Kasiha [Kasieha], 26 Sept. 1958, *Newbould & Jefford* 2651!
DISTR. U 2, 4; K 3 cult.; K 5; T 1, 4; throughout W. tropical Africa, Cameroun, Gabon, Zaire, Burundi, Sudan and Angola
HAB. In forest, along rivers and at the edges of clearings; 780–1220 m.

SYN. *M. fulgens* Tedlie in Bowdich Mission: 374 (1849), *nom. nud.*, based on *Tedlie* 37 (BM, spm. & drawing!)
     M. splendida Welw. in Trans. Linn. Soc. 27: 36, t. 13 (1869). Type: Angola, *Welwitsch* 1116 (LISU, holo., BM!, K!, P, iso.)

4. M. microdonta *Wernham* in J.B. 51: 239 (1913); T.T.C.L.: 507 (1949). Type: Tanzania, Usambara Mts., *Buchwald* 431 (BM, holo.!; BR, K!, iso.)

A shrub or small tree 3–9(–?27) m. tall; young branches covered with dense to sparse, short adpressed hairs or rarely red spreading hairs, becoming glabrous with age, lenticels conspicuous, numerous, whitish and slightly prominent. Leaf-blades elliptic, 7–21 cm. long, 2.9–11 cm. wide, apex acuminate, base cuneate, with 6–13 pairs of lateral nerves, glabrescent to sparsely hairy on both surfaces with dense hairs, occasionally much longer red hairs on the nerves beneath; petioles 0.4–2.3 cm. long; stipules narrowly triangular, 3–13 mm. long, with adpressed hairs outside, caducous. Flowers heterostylous or isostylous, sweet-scented, in ± dense panicles; peduncles 1.2–4.7 cm. long; secondary peduncles present; pedicels 3–5.5 mm. long; 0–8 foliaceous calyx-lobes present on each inflorescence; bracts and bracteoles inconspicuous, 1–3-lobed, up to 7 mm. long, caducous or persistent. Calyx covered with adpressed hairs; tube green, turbinate, 2.5–5 mm. long; lobes brown, subulate to narrowly triangular or linear, 2.5–12 mm. long, up to 1 mm. wide, caducous; foliaceous lobes white to cream, ovate to broadly ovate, 3.2–9.3 cm. long, 2.5–8.2 cm. wide, acute to acuminate at apex, truncate at base, glabrescent or covered with velvety hairs; stipe 1–2.1 cm. long. Corolla with tube greenish yellow and lobes lemon-yellow above; tube 2.8–4 cm. long, 3–6 mm. wide at top, 1.5–3 mm. wide at base, densely covered with buff or pink adpressed hairs outside, flattened hairs absent but with fine hairs present at the throat and extending between the anthers, becoming sparse for the remainder of the tube; lobes ovate, 5–12 mm. long, 3–7 mm. wide, acuminate at the apex with acumen 1–4 mm. long, the inner portion free or connate for up to 0.4 mm. long, with adpressed hairs below and short sparse hairs and longer yellow hairs on the midrib above. Distance from the throat to the top of the anthers 5–10 mm. in long-styled flowers and 0–4 mm. in short-styled flowers. Style 1.9–2.8 cm. in long-styled flowers and 0–4 mm. in short-styled flowers; arms absent; stigma included, lobes 4–7 mm. long. Fruit green, dotted with brown lenticels, globose, sometimes slightly bilobed, 6–10 mm. long, with annular whitish buff scar encircling the blackish remains of the disc, sparsely covered with short hairs. Seeds golden-brown to blackish, ± oblong to square in outline, 0.6–0.8 mm. long, thickened and angular, distinctly reticulate.

subsp. microdonta

Leaves 7–17 cm. long, with 6–10 pairs of lateral nerves. Flowers usually heterostylous. Calyx-lobes subulate to narrowly triangular, 2.5–3(–5) mm. long; foliaceous lobes glabrescent. Corolla-lobes with short acumen 1–1.9 mm. long. Fruit small, 6–9 mm. diameter. Fig. 69.

TANZANIA. Lushoto District: W. Usambara Mts., Shagayu Forest, Jan. 1925, *Grant* 167!; Morogoro District: Uluguru Mts., Mgeta R., just above Hululu Falls, Dec. 1952, *Eggeling* 6435! & Bunduki, Mar. 1953, *Semsei* 1089!
DISTR. T 3, 6; not known elsewhere
HAB. Evergreen forest; (1000–)1120–2130 m.

FIG. 69. *MUSSAENDA MICRODONTA* subsp. *MICRODONTA* — 1, flowering branch, × ⅔; 2, stipule, × 4; 3, flower-bud, × 3; 4, flower, × 1; 5, corolla, opened out, from short-styled flower, × 1; 6, same from long-styled flower, × 1; 7, calyx with style, from short-styled flower, × 2; 8, section through ovary, × 7; 9, part of fruiting branch, × 1; 10, fruit × 2; 11, seed, × 16. 1–5, 7, 8, from *Bruce* 173; 6, *Eggeling* 6456; 9–11, from *Harris* 1600. Drawn by Mrs M.E. Church.

SYN. *M. holstii* Wernham in J.B. 51: 275 (1913); T.T.C.L.:506 (1949). Type: Tanzania, Lushoto District, Usambara Mts., *Holst* 2470 (K, holo.!)
  M. ulugurensis Wernham in J.B. 51: 275 (1913); T.T.C.L.: 507 (1949). Type: Tanzania, Morogoro District, Uluguru Mts., *Goetze* 187* (K, holo.!)

subsp. **odorata** *(Hutch.) Bridson* in K.B. 30: 696 (1976). Type: Kenya, Embu District, streams on SE. Mt. Kenya, *Battiscombe* 708 (K, holo.!, BM, fragment of holo.!, EA, iso.)

Leaves 13–21 cm. long, with 11–13 pairs of lateral nerves. Flowers long-styled, with the stigma just exceeding the anthers or level with them (no short-styled flowers seen). Calyx-lobes narrowly triangular to linear, (3–)5–12 mm. long; foliaceous lobes with short velvety hairs near the base becoming glabrescent towards the apex. Corolla-lobes with long acumen (2–)3–4 mm. long. Fruit larger, ± 1 cm. diameter.

UGANDA. Mbale District: Bugishu, Sipi, Dec. 1938, *A.S. Thomas* 2593!
KENYA. Fort Hall District: Kimakia, Jan. 1960, *Greenway* 9680!; Meru District: Nyambeni Hills, Stone Bridge where Thangatha R. is crossed by South Circular road to Maua, Oct. 1960, *L. Verdcourt* in *Polhill & Verdcourt* 282!; Embu District: Mt. Kenya Forest, vicinity of first water crossing below Castle Forest Station, Jan. 1967, *Perdue & Kibuwa* 8357!
TANZANIA. Moshi District: Barankata, 8 Nov. 1958, *Semkiwa* 57! & Kilimanjaro, 22 Dec. 1933, *Schlieben* 4409! & *Bigger* 1058
DISTR. U 3; K 4, 5; T 2; not known elsewhere
HAB. In evergreen forest, usually near streams; 1830–2100 m.

SYN. *M. odorata* Hutch. in K.B. 1914: 247 (1914); K.T.S.: 451 (1961)
  M. keniensis K. Krause in N.B.G.B. 10: 603 (1929). Types: Meru District, Kaseri R., *R.E. & Th. C.E. Fries* 1813 & S. Nyeri District, Mukengeria R., *R.E. & Th. C.E. Fries* 2060 (both UPS, syn.!)

NOTE. The one specimen from Uganda (*A.S. Thomas* 2593) has been placed in this subspecies although the calyx-lobes and corolla acumens are shorter than those in material from Kenya (measurements given in parentheses in description) and there are no foliaceous calyx-lobes present; further collections from this area might justify the recognition of an additional subspecies. The only record from K 5, Kericho District, Changana Tea Estate, *Perdue & Kibuwa* 9237, is an immature specimen with no foliaceous calyx-lobes. *Greenway* 9680 from Thika District has much longer red hairs on the stems and leaf-venation beneath, also the stipules are not as readily caducous; as there is some variation within the gathering itself it seems doubtful if even varietal rank is merited. The material from Moshi District has the corolla-acumens shorter than the Kenya material (approaching subsp. *microdonta*, but is otherwise a good match.

5. M. monticola *K. Krause* in E.J. 48: 406 (1912); Wernham in J.B. 51: 237 (1913); T.T.C.L.: 507 (1949). Type: Tanzania, Morogoro District, Uluguru Mts., *Holtz* 1720 (B, holo. †)

Shrub or small tree, sometimes scandent, 2.4–9.6 m. tall; branches covered with spreading hairs or more rarely with very sparse adpressed hairs; lenticels whitish, small, inconspicuous. Leaf-blades elliptic to ovate, 9.8–18 cm. long, 4.5–10.9 cm. wide, acuminate at apex, acute to obtuse or sometimes rounded at base, with 10–13 pairs of lateral nerves, sparsely hairy on both surfaces or more rarely glabrescent; petioles 0.8–5.5 cm. long; stipules ovate, 6–12 mm. long, 4–9 mm. wide, acute at apex or sometimes shortly bilobed, hairy outside, glabrous within, reflexing. Flowers ? heterostylous (only short-styled flowers seen), in lax panicles; peduncles 1–8.5 cm. long; secondary peduncles present; pedicels 0.2–1.4 cm. long; up to 5 foliaceous calyx-lobes present; bracts and bracteoles inconspicuous, simple, linear, up to 7(–10) mm. long. Calyx sparsely hairy or glabrescent; tube narrowly turbinate, 5.5–6.5 mm. long; ± 1.5 mm. wide; lobes linear, 3–4 mm. long 0.75–1 mm. wide; foliaceous lobes cream to yellow, elliptic, 5.6–7 cm. long, 3.5–5.5 cm. wide, shortly acuminate at apex, acute at base, glabrescent; stipes 1–2.2 cm long. Corolla yellowish, bright orange or red; tube 1.4–2 cm. long, 2–3 mm. wide at the top, 0.75–1 mm. wide at the base, with sparse to dense adpressed hairs outside, flattened hairs absent but fine hairs present at the throat and extending between the anthers, becoming glabrous towards the base; lobes broadly ovate, 3–3.8 mm. long, 2.5–3 mm. wide, shortly acuminate, sparsely hairy below and with sparse short hairs and longer yellow to rusty-brown hairs at the centre above. Distance from the throat to the top of the anthers 3.8–5 mm. long. Styles 6–7.2 mm. long in short-styled flowers; stigma included; lobes 3 mm. long. Fruit narrowly ellipsoid to cylindrical, 2–2.9 cm. long, 0.7–0.9 cm. wide, crowned by persistent calyx-lobes, very sparsely hairy. Seeds brown to black, ± square to oblong, 0.5–0.6 mm. long, flattened, edges rounded, reticulate, with minute raised dots visible.

* This specimen was incorrectly referred to as *M. tenuiflora* Benth. by K. Schumann in E.J. 28: 488 (1900).

var. **monticola**

Stems and leaves with sparse, long, spreading hairs.

KENYA.   Kwale District: Shimba Hills, Mkongani Forest, May 1968, *Magogo & Glover* 1040! & Mwele Mdogo Forest, 8 Dec. 1972, *Spjut & Ensor* 2739; Kilifi District: Chonyi–Ribe road 1 km. NE. of Pangani, 28 July 1974, *Faden* 74/1264!
TANZANIA.   Uzaramo District: Pugu Forest Reserve, Mar. 1964, *Semsei* 3668!; Iringa District: Karenga (Gologolo) Mt., near Kidatu, Kilombero Scarp Forest Reserve, Mar. 1970, *Harris & Pócs in Harris* 4247!; Lindi District: Rondo Plateau, Jan. 1952, *Semsei* 631!
DISTR.   K 7; T 6–8; not known elsewhere
HAB.   Evergreen-forest; 45–820 m.

SYN.   *M. perlaxa* Wernham in J.B. 51: 240 (1913). Type: Tanzania, Kilosa/Morogoro District, Wamiland (not Zanzibar as cited by Wernham), *Kirk* (K, holo.!)

var. **glabrescens** *Bridson* in K.B. 30: 696, (1976). Type: Tanzania, Ulanga District, Ruaha R., ± 35 km. S. of Mahenge Station, *Schlieben* 1894 (K, holo.!, BM!, BR, iso.)

Stems and leaves with sparse to very sparse short adpressed hairs.

TANZANIA.   Kilosa District: Mvuma Hill Range, 14 June 1973, *Greenway & Kanuri* 15146!; Ulanga District: Ruaha R., ± 35 km. S. of Mahenge Station, 16 Mar. 1932, *Schlieben* 1894! & Kiberege, 6 Mar. 1958, *Haerdi* 208!
DISTR.   T 6; not known elsewhere
HAB.   Riverine associations; 400–700 m.

53. PSEUDOMUSSAENDA

Wernham in J.B. 54: 298 (1916); Troupin & Petit in B.J.B.B. 23: 227–232 (1953) & Petit in B.J.B.B. 24: 339–346 (1954); Hallé, Fl. Gabon 12, Rubiacées, 1: 136 (1966)

Erect or sometimes probably scrambling shrubs or subshrubs. Leaves opposite (or lower leaves rarely in whorls of 3 in *P. capsulifera*), petiolate; stipules with 1–2 subulate lobes. Flowers fairly large, ⚥, 5-merous, sessile, heterostylous, in several–many-flowered cymose terminal inflorescences; 1–several of the peripheral flowers have 1 of the calyx-lobes enlarged into a large coloured venose stipitate lamina. Calyx-tube elongate, ± oblong; lobes 5, ± subulate. Corolla-tube narrowly cylindrical, the upper part containing the anthers swollen; lobes (4–)5, ovate, induplicate-valvate, usually drawn out into a filiform appendage that is very evident in the buds which have 5 apical subulate projections; throat velvety hairy. Stamens always included in the swollen part of the tube. Style ± ¾ the length of the corolla-tube in short-styled flowers; stigmas just exserted in long-styled flowers; stigma bilobed, the lobes thick, linear or ovoid. Ovary 2-locular, each locule with many ovules borne on the marginal portions of very narrowly oblong curved placentas which resemble lamellae. Fruit dry, mostly capsular, with rather incomplete loculicidal dehiscence or sometimes not dehiscent. Seeds small, reticulate.

A small genus of 4–5 species closely allied to *Mussaenda* and perhaps not distinct (as maintained by Bakhuizen f.). The genus was erected by Wernham for 4 species including one which he and Delile misidentified with Forsskål's *Ophiorrhiza lanceolata*. Delile called the plant described below *Mussaenda luteola* but cited Forsskål's name in synonymy. Forsskål's plant is actually a *Pentas*. This complicated nomenclatural problem is dealt with in K.B. 6: 377 (1952). Wernham placed his new genus in the Condamineae but it is clearly very closely allied to *Mussaenda*. The very aberrant species *P. capsulifera* (Balf. f.) Wernham, with emarginate corolla-lobes and diverse capsule from Socotra may need placing in a separate genus.

P. flava *Verdc.* in K.B. 6: 378 (1952); F.P.S. 2: 459 (1952); K.T.S.: 464 (1961); E.P.A.: 1000 (1965). Type: Uganda, Toro District, Bwamba, Kabango, *A. S. Thomas* 725 (EA, holo.!, K, iso.!)

Shrub 0.9–3.6 m. tall, with branched pubescent to densely hairy stems. Leaf-blades elliptic or oblong-elliptic, 1.8–11 cm. long, 0.7–4.7(–5) cm. wide, acute to shortly or distinctly acuminate at the apex, cuneate at the base, more or less softly pubescent, often densely so beneath, particularly on the nerves; petiole 4–8 mm. long; stipules with 2 filiform fimbriae up to 4.5 mm. long. Flowers in lax rather few-flowered cymes; peduncle ± 2 cm. long; bracts filiform, 8–9 mm. long, slightly spathulate at the apex. Calyx-tube narrowly turbinate, 3 mm. long, hairy; lobes 5–6, filiform, subequal, 2–6 mm. long or 1

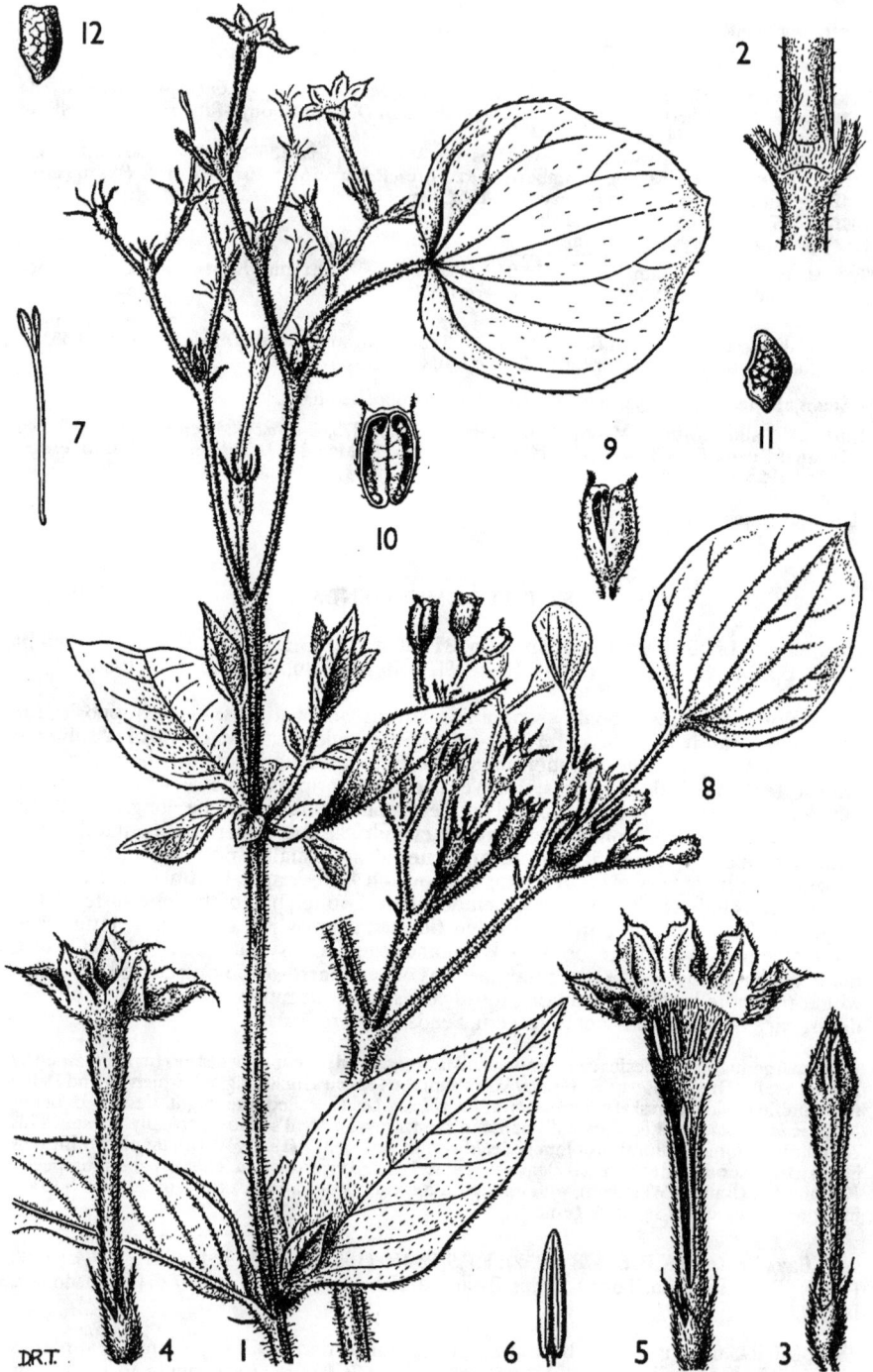

FIG. 70. *PSEUDOMUSSAENDA FLAVA* — 1, flowering branch, × 1; 2, stipule, × 4; 3 flower-bud, × 2; 4, flower, × 2; 5, flower, with corolla partly opened out, × 2; 6, stamen, × 4; 7, style and stigmatic arms, × 2; 8, portion of fruiting branch, × 1; 9, fruit, × 2; 10, one valve of dehisced fruit, × 2; 11 & 12, seeds, × 12. All drawn from *Purseglove* 1053. Drawn by Miss D.R. Thompson with 2 by Sally Dawson.

often enlarged into a stipitate yellow, white or cream pubescent lamina; blade oblong, ovate or elliptic, 1.5–5.5 cm. long, 1–4.5 cm. wide, rounded and apiculate to acute at the apex, rounded or cordate at the base; stipe 1–3.5 cm. long. Corolla yellow, hairy; tube 2.5–3.5 cm. long, greenish at the base, throat with orange-yellow hairs; lobes ovate, 4–8 mm. long, 3–5 mm. wide, caudate at the apex. Stigma green. Capsule oblong, woody, 7 mm. long, 4.5–5 mm. wide, obscurely ribbed, loculicidally 2-valved. Seeds angular, 0.5 mm. long. Fig. 70.

UGANDA. W. Nile District: Nebbi, Sept. 1940, *Purseglove* 1053!; Acholi District: Gani, ? near Koki, Dec. 1962, *Grant* 669!; Toro District: Bwamba, Oct. 1925, *Fyffe* 28!.
KENYA. Lake Turkana [Rudolf], *Wellby* (probably in Kenya, K 1); Turkana District: Songot Hills [Zingout], May 1933, *Champion* T. 180!.
DISTR. U 1–3; K ?1, 2; Nigeria, Zaire, Ethiopia, Sudan; also cultivated in Ceylon, Singapore, Nairobi and Kew
HAB. Rocky places in wooded grassland and bushland, also in evergreen forest; 630–1360 m.

SYN. *Mussaenda luteola* Delile, Cent. Pl. Afr.: 65, t. 1/1 (1826); K. Schum. in P.O.A. C: 379 (1895), *nom. illegit.*, pro parte quoad descr. et ic., excl. syn.
Vignaudia luteola (Delile) Schweinf., Beitr. Fl. Aethiop. 1: 140, 282 (1867)
Pseudomussaenda lanceolata Wernham in J.B. 54: 298 (1916), pro parte, excl. syn. Forssk. & Vahl
Mussaenda flava (Verdc.) Bakh. f. in Backer, Beknopte Fl. Java. Afl. xv, Fam. 173, 72 (1956) & in Blumea 12: 63 (1963)

NOTE. *Pole Evans and Erens* 1742 (Uganda, N. slopes of Ruwenzori Mts.) has the large lobes up to 7.5 × 5.5 cm. which are said to be pinkish and large leaf-blades but is no more than a form. A specimen *Fyffe* 130 is labelled "loc. Entebbe on Hoima roadside" and ostensibly could be from U 4 or U 2. It is unlikely to be from Entebbe; no further collections have been made from there and it has been a centre of botanical activity.

54. SABICEA

Aubl., Pl. Guian. 1: 192, tt. 75, 76 (1775); Wernham, Monograph of the Genus *Sabicea* (1914); Hallé in Adansonia sér. 2, 3: 168–177 (1963) & in Fl. Gabon 12, Rubiacées, 1: 161 (1966)

Lianes, straggling shrubs or sometimes scarcely woody climbers, usually ± hairy. Leaves opposite, petiolate; stipules entire or emarginate, often comparatively large, mostly with conspicuous colleters within. Flowers sometimes showing limited heterostyly, small or medium-sized, in sessile or pedunculate several–many-flowered paniculate or capitate axillary inflorescences or in a few cases plant cauliflorous; bracts often conspicuous, sometimes forming an involucre. Calyx-tube ellipsoid-oblong or ovoid; tubular part of limb sometimes cylindrical and elongated; lobes often well-developed. Corolla usually ± white, tube mostly narrowly cylindrical; lobes small, valvate. Stamens with very short filaments or ± sessile, the anthers ± medifixed, narrowly oblong, inserted towards the base of the tube in long-styled flowers and near the throat in shorter styled flowers but only the tips exserted if at all. Ovary 4–5-locular, with elliptic placentas usually externally covered with ovules; style slender, glabrous, divided into 5 long narrow stigmatic lobes or style with 5 arms terminating in slightly thicker stigmatic lobes. Berries globose, many-seeded, fleshy. Seeds irregularly ovoid, compressed, mostly angular, finely reticulate or striate.

A large genus of about 120 species about equally divided between tropical America and tropical Africa with 4 in Madagascar.

Peduncles obsolete, short or scarcely developed, 0–2.5(–4) cm. long; calyx-lobes lanceolate, spathulate, triangular or elliptic:
　Inflorescences ± sessile, rather laxer and without distinct tightly adpressed bracts:
　　Calyx-lobes oblong-elliptic, ± blunt, reflexed (**U 4**)　　1. *S. entebbensis*
　　Calyx-lobes lanceolate, very acute, erect (**T 3, 4, 6**)　　2. *S. orientalis*
　Inflorescences on peduncles 0.6–2.5(–4) cm. long, tightly capitate and with an involucre of adpressed bracts:
　　Bracts 0.6–1 cm. long, 4–6 mm. wide; calyx-lobes spathulate, elliptic or oblong, 1–3 mm. long, the tubular part of limb 1–2 mm. long . . . . . .　　4. *S. dinklagei*

Bracts 1–3.5 cm. long and wide; calyx-lobes triangular
or ovate-triangular, 2–4(-6) mm. long, the tubular
part of limb ± 1.2 cm. long . . . .   *5. S. dewevrei*
Peduncles well developed, 2–12 cm. long; calyx-lobes ovate   *3. S. calycina*

1. S. entebbensis *Wernham*, Monogr. *Sabicea:* 33, t. 12/15 (1914). Type: Uganda, Mengo
District, Entebbe, *E. Brown* 296 (K, holo.!)

Liane 3–3.6 m. long, with flowers mostly borne on short vertical shoots which seem not
to twine; stems rather slender, brown, ridged, covered with adpressed rather bristly hairs
when young, later the epidermis peeling off. Leaf-blades elliptic to oblong-elliptic, 4–12.5
cm. long, 2–6.5 cm. wide, acuminate at the apex, broadly cuneate or rounded at the base,
with rather sparse subscabrid adpressed hairs above, with dense subscabrid adpressed
hairs on the venation below and white tomentum on the surfaces between, the
undersurface thus appearing grey; petiole 0.4–3 cm. long, pubescent; stipules leafy,
ovate-elliptic, 5–7 mm. long, 2–4 mm. wide, ± acute, reflexed, glabrous on inner face, i.e.
that turned to the light, adpressed hairy on the outer face. Flowers in axillary paniculate
several–many-flowered ± sessile clusters; pedicels 4 mm. long, adpressed pubescent;
bracts small and not in any way forming an involucre. Calyx-tube subglobose, 1.3 mm.
long, adpressed pubescent; tubular part of limb 1 mm. long, adpressed pubescent to
glabrous; lobes ovate to oblong-elliptic, 1–2(-2.5) mm. long, 1 mm. wide, subacute or ±
obtuse, reflexed, glabrous inside and outside or with scattered hairs on the margin and
outside. Corolla dull purple flushed, pale yellow towards the base of the tube or tube
greenish with inner surface of lobes red-brown; tube 7–8.5 mm. long, the upper half or
third adpressed pubescent outside, the lower glabrous; ring of hairs at the mouth of the
tube shining white; lobes ovate-triangular 1.2–1.5 mm. long, 1 mm. wide, adpressed hairy
outside. Style 7 mm. long in long-styled flowers, stigmas and style-arms shining white,
together 2.5 mm. long and just overtopping the lobes. Fruit glistening white, globose, ±
0.85–1 cm. in diameter (alive), very soft with crimson juice inside. Fig. 71/11, 12.

UGANDA. Masaka District: Sese Is., Bugala, 3 June 1932, *A.S. Thomas* 28!; Mengo District: Kyagwe,
    Nakiza Forest, near Nansagazi, 24 Jan. 1950, *Dawkins* 702! & near Entebbe, Kitubulu Forest, Oct.
    1935, *Chandler* 1435!
DISTR. U 4; not known elsewhere
HAB. Forest edges and associated thickets; 1170–1200 m.

2. S. orientalis *Wernham*, Monogr. *Sabicea* : 34, t. 1/2–5 (1914); T.T.C.L.: 532 (1949);
Verdc. in K.B. 14: 348 (1960). Type: Tanzania, Uluguru Mts., *Goetze* 209 (K, lecto.!)

Liane, climber or straggling shrub, 1.2–4.5 m. long; stems rather slender, brown,
ridged, with adpressed hairs and ± tomentose as well, the epidermis peeling on the older
shoots. Leaf-blades elliptic or oblong-elliptic, 4–13.5 cm. long, 1.7–6.3 cm. wide, shortly to
distinctly acuminate at the apex, broadly cuneate to rounded at the base, subscabridly
pilose above, with dense adpressed silky hairs on the venation beneath and with ±
"rucked" tomentum between so that whole underside is pale and ± velvety; petiole 0.5–1.9
cm. long, densely hairy and tomentose; stipules leafy, ovate, 6–12 mm. long, 5.5–10 mm.
wide, glabrous on inner surface, hairy outside, reflexed. Flowers in axillary several–many-
flowered paniculate clusters, ± sessile; pedicels 5 mm. long, densely hairy; bracts if
present very small, not forming an involucre. Calyx-tube subglobose, 1–1.5(-2.5) mm.
long, densely hairy; tubular part of limb very short, 0.5 mm. long; lobes leafy, lanceolate,
4–6 mm. long, 1–1.5 mm. wide, glabrous inside, hairy outside and with long hairs on the
margin. Corolla white; tube 0.8–1.2 cm. long, hairy outside in upper half, glabrous
beneath or sparsely hairy all over; lobes lanceolate, 2–3 mm. long, 0.8–1 mm. wide,
glabrous inside, densely hairy outside. Style 9 mm. long in long-styled flowers, 5 mm. long
in short-styled flowers, the stigma-lobes 1.75–2 mm. long. Fruits whitish when ripe with a
pinkish juice, globose, 10 mm. in diameter, hairy, crowned with green calyx-lobes, the
fruiting pedicels up to 1.2 cm. long. Fig. 71/1–10.

TANZANIA. Lushoto District: Amani, 26 Nov. 1935, *Greenway* 4191!; Buha District: Gombe Stream
    Chimpanzee Reserve, Kasakela, 18 Nov. 1962, *Verdcourt* 3358!; Morogoro District: Uluguru Mts.,
    Kitundu, 22 Nov. 1934, *E.M. Bruce* 178!
DISTR. T 1, 3, 4, 6, 7; Cameroun, ?Zaire, Burundi
HAB. Rain-forest, streamside forest and particularly forest edges; 775–2100 m.

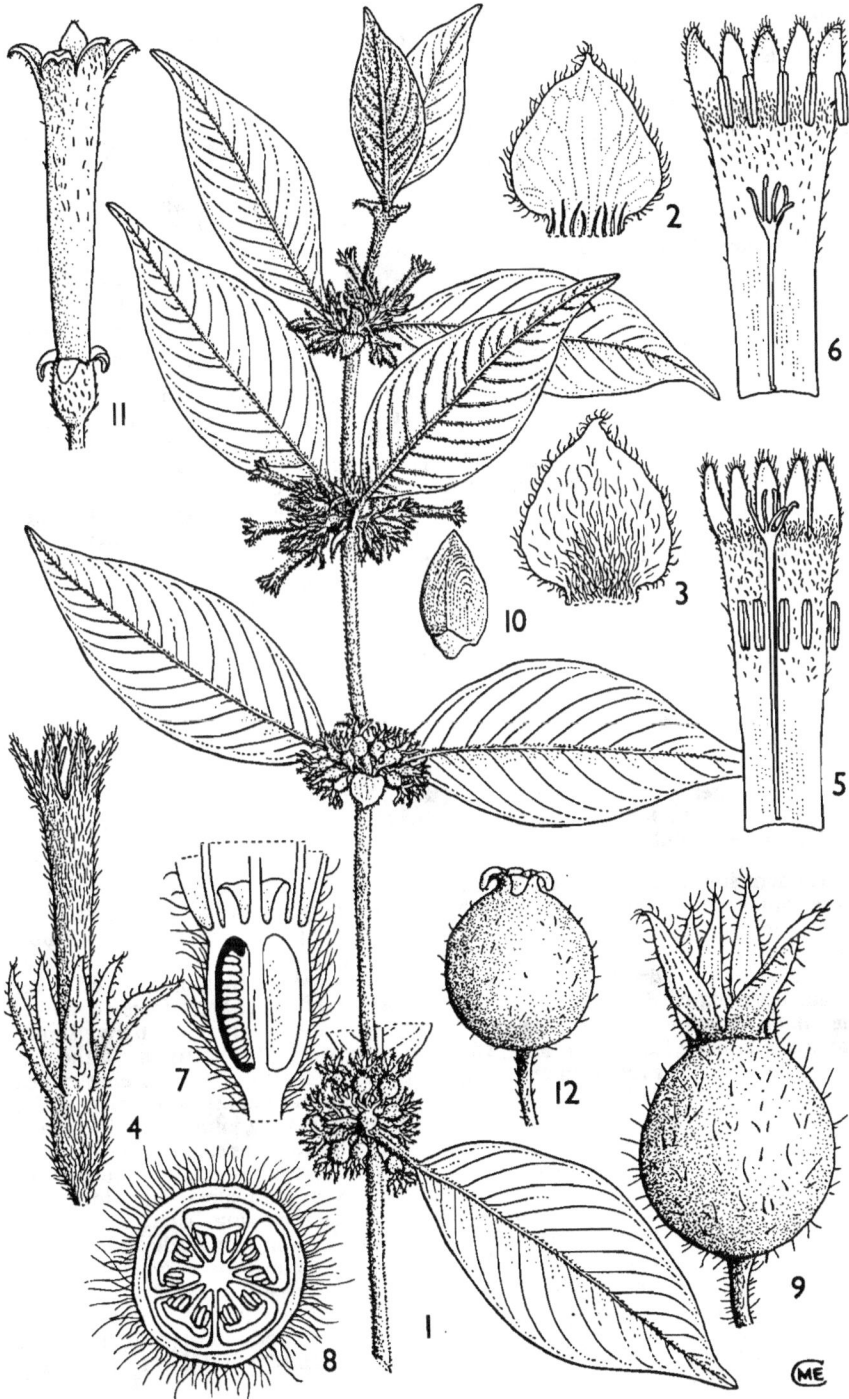

FIG. 71. *SABICEA ORIENTALIS* — 1, habit, × ⅓; 2, stipule, from inside, × 3; 3, stipule from outside, × 3; 4, flower, long-styled, × 4; 5, corolla, opened out, from long-styled flower, × 4; 6, same, from short-styled flower, × 4; 7, longitudinal section of ovary, × 12; 8, transverse section of ovary, × 20; 9, fruit, × 3; 10, seed, × 20. *S. ENTEBBENSIS* — 11, flower, × 4; 12, fruit, × 3. 1, 6, from *Greenway* 4191; 2, 3, from *Renvoize* 1632; 4, 5, 7, 8, from *Peter* 23314; 9, 10, from *Batty* 441; 11, 12, from *Dawkins* 702. Drawn by Mrs M.E. Church.

NOTE. *S. orientalis* and *S. entebbensis* may eventually both be considered as subspecies of *S. venosa* Benth., a W. African species, but all three are usually easily identified by calyx-lobe and indumentum characters.

3. S. calycina *Benth.* in Hook., Niger Fl.: 399 (1849); Hiern in F.T.A. 3: 76 (1877); F.W.T.A., ed. 2, 2: 172 (1963); Hallé in Fl. Gabon 12, Rubiacées, 1: 186, fig. 5/1-3 (1966). Type: Bioko [Fernando Po], *Vogel* 35 (K, holo.!)

Climber 1-2 m. long; stems red-brown, slender, densely covered with long adpressed and ± spreading hairs when young, very soon glabrescent and epidermis peeling off to reveal a pale surface. Leaf-blades oblong, elliptic-oblong or oblong-ovate, 3-8.5 cm. long, 1.5-2.8 cm. wide, shortly acuminate at the apex, rounded or ± cordate at the base, with sparse adpressed hairs above particularly on the venation and longer denser adpressed hairs beneath again mainly on the venation; petioles 0.5-1.7(-4.5) cm. long, hairy; stipules leafy, ovate, 5-6 mm. long, 3-4 mm. wide, ± glabrous except at margins, reflexed. Flowers in compact bracteate inflorescences; peduncles long and slender, 2-8(-12) cm. long, glabrous; pedicels 1-2 mm. long, but attaining 8-10 mm. in fruit; bracts 3, leafy, ovate, 1.3 cm. long, 0.9-1.8 cm. wide, ± acute at the apex, ± cordate at the base, glabrous except for few ciliae on the margins. Calyx-tube globose or ovoid, 2-3 mm. long, glabrous; lobes (2-)3(-4) leafy, ovate, 0.8-1.3 cm. long, 3-7 mm. wide, eventually ± scarious and persistent. Corolla white, glabrous outside; tube 1.3-2.1 cm. long; lobes short, ovate, 1.3-1.8 mm. long, 1.5-1.8 mm. wide. Anthers situated about halfway in corolla-tube in long-styled flowers, the tips just exserted in short-styled flowers. Style 1.6-1.7 cm. long in long-styled flowers, the style-arms and stigma-lobes together 2.5 mm. long, just exserted; style 6-8.5 mm. long in short-styled flowers, the arms and lobes together 3-4 mm. long. Fruit dark red-violet or blackish, globose, 8-9 mm. in diameter.

UGANDA. Kigezi District: S. Maramagambo Forest Reserve, Bitereko track, 9 May 1970, *Lock* 70/43!
DISTR. U 2; Sierra Leone to Mayombe
HAB. Undergrowth of fairly dry forest; 1080 m.
NOTE. The field label of the specimen cited above states "= *Lock* 69/349" but I have not seen that specimen.

4. S. dinklagei *K. Schum.* in E.J. 23: 428 (1897); Wernham, Monogr. *Sabicea*: 70 (1914); Hallé, Fl. Gabon 12, Rubiacées, 1: 169, t. 35/1,2 (1966). Type: Cameroun, Greet Batanga, *Dinklage* 1124 (B, holo. †)*

Scandent shrub 2.3-4.5(-9) m. long; stems fairly slender, reddish brown, the young parts with scattered longer spreading hairs, often purplish in life, and short denser adpressed pubescence, the older stems with bark tending to peel. Leaf-blades ± elliptic, 1.6-7(-11) cm. long, 0.7-3.6(-5.5) cm. wide, shortly to distinctly acuminate at the apex, cuneate at the base, drying darkish green above, paler green beneath, with reddish very reticulate venation, with scattered adpressed subscabrid hairs above, adpressed hairy on the main venation beneath; petioles 0.3-1.2 cm. long, spreading hairy; stipules transversely oblate, 2.5-5.5 mm. long, 2.8-5 mm. wide, rounded or emarginate at the apex, with bristly colleters inside. Flowers in tight capitate few-flowered inflorescences ± 1 cm. wide; peduncles 0.6-2.5(-4) cm. long; pedicels obsolete; bracts conspicuous, all reddish at the base when dry, the outer pair 6-10 mm. long, 4-6 mm. wide, joined near the extreme base to form a ± boat-shaped involucre, the inner 6 lateral ones slightly smaller, 5 mm. long, 3 mm. wide, ± hairy. Calyx-tube subglobose, pilose or pubescent, the tubular part of limb 1-2 mm. long, glabrous; lobes elliptic, oblong or subspathulate, 1-3 mm. long, 0.6-1 mm. wide, minutely pubescent on inner face, with longer marginal hairs and on midrib outside. Corolla white or pinkish; tube cylindrical, 6-8 mm. long, silky pilose above outside; lobes 4-5, narrowly ovate, 3-3.5 mm. long, 1.8-2 mm. wide, silky pilose outside. Anthers near top of corolla-tube in long-styled flowers, tips just exserted in short-styled flowers. Style 8 mm. long in long-styled flowers, the style-arms and stigma together 1.5-2.5 mm. long; style 4.5 mm. long in short-styled flowers, the arms and stigmas 3 mm. long. Fruit globose, purple or crimson when ripe with whitish bloom, ± 1 cm. in diameter, pubescent or glabrous, with a wine-coloured pulp.

UGANDA. Kigezi District: Ishasha Gorge, May 1947, *Purseglove* 2428!
DISTR. U 2; Cameroun, Rio Muni, Gabon, Central African Republic, Zambia and Angola

* Hallé mentions two syntypes, 1124 and 1284 but 1284 is not given in the original description.

HAB. Forest edge; 1350 m.

SYN. *S. laurentii* De Wild., Miss. Laurent.: 276 (1906); Wernham, Monogr. *Sabicea:* 70 (1914); F.P.N.A.
2: 335 (1947); F.F.N.R.: 420 (1962). Type: Zaire, Eala, *Laurent* 902 (BR, holo.)
S. homblei De Wild. in B.J.B.B. 5: 32 (1915). Type: Zaire, Shaba, Biano Plateau, near Katentania,
Homblé 504 (BR, holo.)

NOTE. Hallé shows a flower with the style itself exserted.

5. S. dewevrei *De Wild. & Th. Dur.* in Ann. Mus. Congo 3(1): 112 (1901); Wernham,
Monogr. *Sabicea:* 71 (1914); Hallé in Fl. Gabon 12, Rubiacées, 1 : 185, fig. 4/2–5 (1966).
Type: Zaire, Orientale, Waboundou [Wabundu, Ponthierville], *Dewèvre* 1143 (BR, holo.)

Liane 1.8–4 m. or more long; young stems glabrous to densely spreading pilose with
hairs ± 2.5 mm. long; older stems sparsely hairy, drying dark blackish purple. Leaf-blades
elliptic or oblong-elliptic, 7–15(–22.5) cm. long, 3–7.5(–12) cm. wide, narrowly and
distinctly acuminate at the apex, cuneate to rounded at the base, glabrous save for a few
sparse adpressed bristly hairs on the midrib above and main nerves beneath; reticulation
of nerves scarcely apparent beneath; petioles 0.8–2.5(–6) cm. long, with ± adpressed
bristly hairs; stipules rounded, 0.6–2 cm. long and wide, glabrous save for ciliate margins.
Inflorescence solitary, axillary, 6–28-flowered; peduncle 0.6–3.5 cm. long, glabrous;
bracts large, rounded, 1–3.5 cm. long and wide, joined at the base for 2–4 mm., glabrous;
flowers sessile, with hairs at base mixed with very small linear bracteoles. Calyx glabrous
outside; tube 1.2–2.8 cm. long; limb-tube up to 1.2(–1.8) cm. long, 5–7 mm. wide; teeth 5,
triangular, 2–4(–6) mm. long, 2–3 mm. wide, ciliate. Corolla white; tube 1.5–1.8 cm. long, 1
mm. wide, glabrous outside; lobes 5, ovate, 2 mm. long, 1 mm. wide, adpressed pilose
outside, glabrous inside; throat pilose. Stamens included. Ovary ± glabrous. Fruit red,
subglobose, ± 9 mm. diameter.

var. **dewevrei**

Leaf-blades up to 15 cm. long. Stems and petioles pilose.

UGANDA. Kigezi District: Ishasha Gorge, 5 Aug. 1971, *Katende* 1252!
DISTR. U 2; Gabon, Zaire
HAB. Forest; 1700 m.

55. PSEUDOSABICEA

Hallé in Adansonia sér. 2, 3: 168–177 (1963) & in Fl. Gabon 12, Rubiacées, 1: 199 (1966)

Lianes or robust climbers or sometimes slender creepers, more rarely shrubs. Leaves
often markedly unequal (anisophyllous) at each node but sometimes almost equal or only
showing a slight tendency to anisophylly; blades usually felted beneath; stipules elliptic,
entire. Flowers ♂, 5-merous, heterostylous*, in (few-)several–many-flowered
inflorescences, always axillary, often at leafless nodes, lax and paniculate or condensed,
sometimes sessile; bracts and bracteoles present. Calyx-tube globose, ellipsoid or
obovoid, the tubular parts of the limb very short; lobes 5, ovate, lanceolate or linear-
lanceolate, often ± recurved. Corolla whitish, becoming rusty brown with age, mostly
small; tube cylindrical, the throat pubescent with hairs shaped like strings of beads or
papillate; lobes triangular, short, ± pubescent outside towards the apex or glabrous.
Anthers medifixed. Ovary 2(–3)-locular; placentas rounded or oblong, peltate, slightly
emarginate at the apex, covered with numerous ovules; style slender, often shortly bifid at
apex so that there is a short portion of style-branch before the thick opposite laterally
dilated stigmatic lobes or stigmatic lobes sessile at end of simple style. Infructescence
often ± accrescent. Fruits fleshy, whitish or greenish, rarely red or black, the pulp usually
not coloured. Seeds angular, small, nearly smooth but with extremely fine reticulation.

A small genus of about 12 species formerly included in *Sabicea*, restricted to tropical Africa, mainly
in the west.

P. arborea *(K. Schum.) Hallé* in Adansonia sér. 2, 3: 172 (1963) & in B.J.B.B. 34: 397
(1964). Type: Tanzania, Uluguru Mts., Ngluwenu, *Stuhlmann* 8775 (B, holo. †); neotype:
Tanzania, Uluguru Mts., Morogoro, *Schlieben* 2970 (BR, holo., P, iso.)

* I am not certain what N. Hallé means by 'par pieds distincts' — perhaps different inflorescences
on the same plant can have different forms.

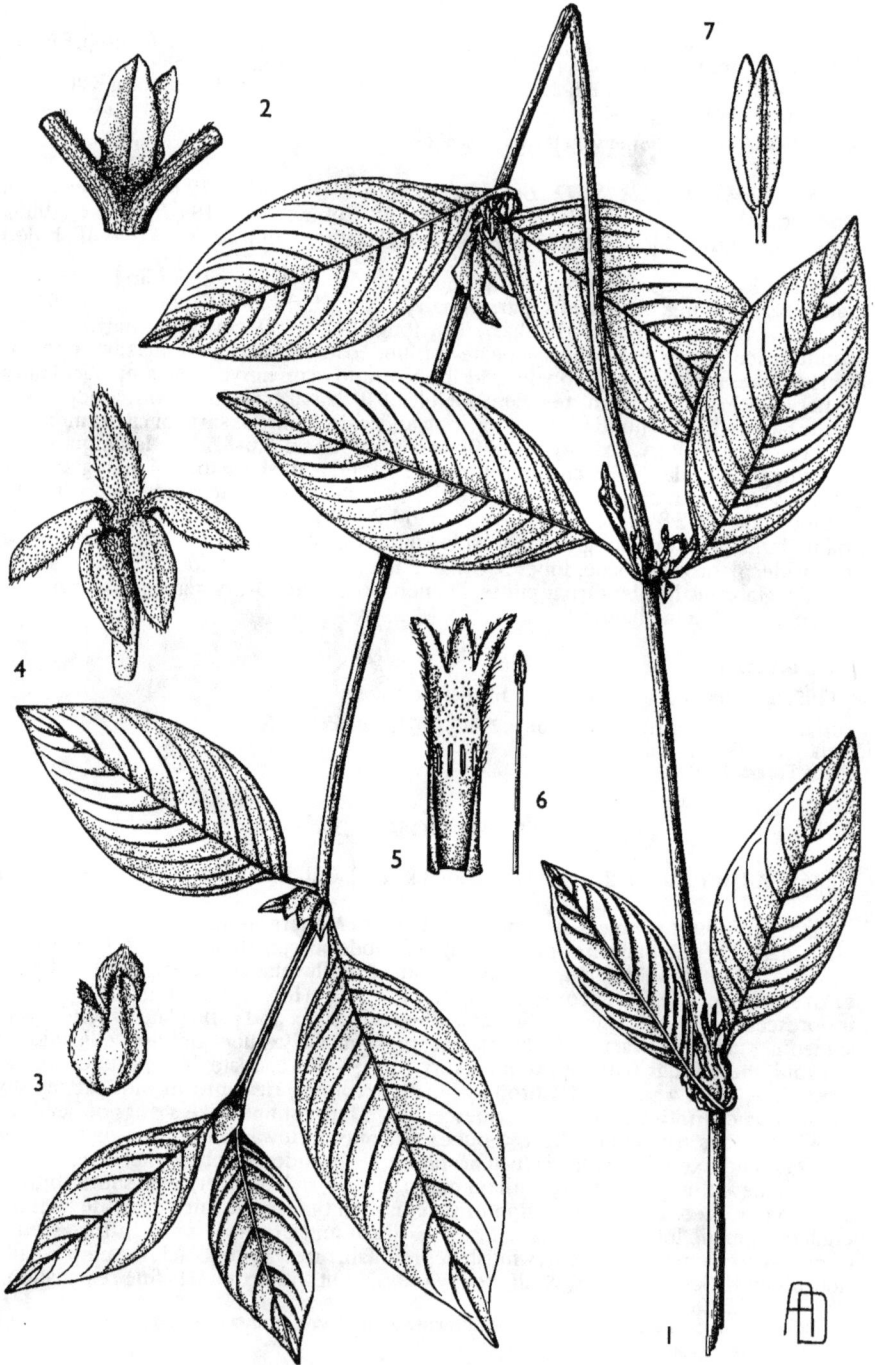

FIG. 72. *PSEUDOSABICEA ARBOREA* subsp. *BEQUAERTII* var. *BEQUAERTII* — **1**, habit, × ¼; **2**, stipule, × 2; **3**, bud surrounded by bract, × 2; **4**, corolla × 2; **5**, section through corolla, × 3; **6**, style and stigmatic arms, × 3; **7**, stigmatic arms, × 4. **1**, from *Christiansen* 1399; **2, 5, 6**, from *Purseglove* 2387; **3, 4, 7**, from *Bridson* 334. Drawn by Ann Farrer, with **3, 4**, and **7** by Sally Dawson.

Scandent shrub, climber or scrambling herb up to 4.5 m. long but also reported to be a shrub (*fide* Bequaert and Schlieben)*; stems covered with buff tomentum when very young but soon glabrous. Leaf-blades elliptic to elliptic-lanceolate, (3–)4–11.5(–20.5) cm. long, 1–4.7(–8.5) cm. wide, mostly distinctly and narrowly acuminate at the apex, cuneate at the base, covered with cottony tomentum above when very young but almost immediately glabrous, densely finely velvety persistently buff tomentose beneath; petiole 0.3–1.2 cm. long, tomentose and with longer hairs on margins of upper channel spreading on to the margins of the extreme base of the leaf; stipules elliptic-oblong, 8–9 mm. long, 4–4.5 mm. wide, ± acute, glabrescent. Flowers in sessile bracteate axillary inflorescences; outer sheathing bracts ovate, 1.2 cm. long, 9 mm. wide, ± glabrous on inner side, with long hairs on margins and outer side which is also ± tomentose particularly towards the apex, densely bristly pilose inside; bracteoles falcate, 6 mm. long, 2.5 mm. wide, keeled outside, ciliate at the margins, pilose inside. Calyx-tube obconic, 1.5 mm. long, pilose; tubular part of limb ± glabrous, 1.5 mm. long; lobes ovate, 2 mm. long, hairy at the apex outside, bristly pilose inside. Corolla white turning brown with age, ± fleshy; tube 0.5–1 cm. long, the lobes lanceolate, 2–5 mm. long, 1.5 mm. wide, both hairy outside. Style in long-styled flowers 1 cm. long; stigmatic lobes 2.2 mm. long. Fruits not seen.

subsp. **arborea**

Mostly shrubby, less often a liane. Corolla-tube shorter, ± 5–6 mm. long; lobes ± 2–3 mm. long.

TANZANIA. Morogoro District: Uluguru Mts., Bunduki, 27 Jan. 1935, *E.M. Bruce* 668! & summit of Lupanga Peak, 23 Dec. 1933, *Michelmore* 854! & 26 Dec. 1931, *B.D. Burtt* 3472!
DISTR. T 6; not known elsewhere
HAB. Evergreen forest; 1650–2106 m.

SYN. *Sabicea arborea* K. Schum. in E.J. 28: 57 (1899); Wernham, Monogr. *Sabicea:* 54 (1914)

subsp. **bequaertii** (*De Wild.*) *Verdc.* in K.B. 31: 183 (1976); Fl. Pl. Lign. Rwanda: 586, fig. 198/1 (1982); Fl. Rwanda 3: 200, fig. 63/1 (1985). Type: Zaire, Kabango, *Bequaert* 6178 (BR, holo.)

Mostly a liane, more rarely shrubby. Corolla-tube longer, 0.8–1 cm. long; lobes 3.5–5 mm. long.

var. **bequaertii**

Leaves smaller, not attaining 20 × 8.5 cm., not so shiny above and with no more than 13 lateral nerves. Fig. 72.

UGANDA. Kigezi District: Kayonza, Marambo (?Mulamba), Mar. 1947, *Purseglove* 2387! & outskirts of Kayonza Forest, Oct. 1940, *Eggeling* 4172!
DISTR. U 2; Zaire, Rwanda, Burundi
HAB. Evergreen forest; 1800 m.

SYN. *Sabicea bequaertii* De Wild., Pl. Bequaert. 2: 229 (1923)

NOTE. Hallé sinks *S. bequaertii* completely, but the difference in flower-size, coupled with the marked geographical disjunction, and perhaps also a real habit difference seems to indicate subspecific distinction. It must be admitted that very little adequate flowering material has been seen to confirm this.
  Var. *tersifolia* Hallé occurs in Zaire and its relation to the two subspecies is not clear; it has been assumed to be a variety of subspecies *bequaertii*.

Tribe 16. **HEINSIEAE****

Inflorescences lax to dense terminal and lateral cymes; calyx-
  lobes subfoliaceous, persistent in fruit; corolla showy, with
  yellow hairs around throat   . . . .   **56. Heinsia**
Inflorescences spicate or compact spherical heads; calyx-lobes
  small; corolla moderately large, but not showy   . . .   **57. Bertiera**

---

* Stuhlmann's original description as a tree must be an error.
** Genus 56 by B. Verdcourt, 57 by D.M. Bridson.

56. HEINSIA

DC., Prodr.4: 390 (1830)

Shrubs, or small trees, rarely subshrubs or climbers. Leaves opposite, shortly petiolate; stipules with 2 distinct teeth, sometimes subulate, sub-persistent or eventually falling. Flowers rather large, mostly sweetly scented, 4–6-merous, solitary or few–many in lax to dense terminal and lateral cymes. Calyx-tube turbinate or oblong; lobes ± leafy, ovate to oblong, elliptic or lanceolate. Corolla hypocrateriform; tube slender, appressed hairy outside and densely hairy at the throat; lobes variable, mostly broadly elliptic but sometimes narrow, imbricate (quincuncial) in bud, mostly finely puberulous, spreading. Stamens situated in the tube in both long- and short-styled forms but at different levels; anthers linear, affixed near their base on very short filaments. Ovary 2-locular; ovules numerous; disc small, swollen; style slender, with two linear thickened stigma-arms just exserted in long-styled flowers but only reaching to about halfway up the tube in short-styled flowers. Fruit slightly fleshy but soon dry, indehiscent, mostly globose or oblong-globose, crowned by the persistent calyx-lobes, many-seeded, the seeds adhering in 2 masses. Seeds small, strongly foveolate, the floors of the depressions pitted.

A small genus of 4–5 species restricted to tropical and subtropical Africa.

One W. African species has the internodes thickened and excavated for ants. A good deal of morphological and biological information will be found in F. Hallé in Adansonia sér. 2, 1: 266–298 (1961). He has accepted the proposal I made (B.J.B.B. 28: 249 (1958)) that *Heinsia* should be referred to a subtribe *Heinsiinae* of the *Mussaendeae*. Its previous position in the *Hamelieae* was quite untenable. Since then it has been raised to tribal rank, *Heinsieae* (Verdc.) Verdc.

Leaf-blades hairy only on the main venation beneath, mostly under 6 cm. long; stipule-lobes short, not long-attenuate at the apex; flowers solitary or in few-flowered cymes · · · · · · 1. H. crinita
subsp. *parviflora*

Leaf-blades velvety hairy beneath, up to 15 × 6 cm.; stipules attenuate into long subulate apical extensions; flowers in few–several-flowered dense cymes:
Undersurfaces of leaf-blades with straight hairs which do not hide the surface of the leaf-blade · · · · · 2. H. zanzibarica
Undersurfaces of leaf-blades with curly interwoven hairs which form a greenish white tomentum hiding a good deal of the surface of the leaf-blade · · · · · · 3. H. bussei

1. **H. crinita** (*Afzel.*) *G. Taylor* in Exell, Cat. Vasc. Pl. S. Tomé: 209 (1944); F.W.T.A., ed. 2, 2: 161, fig. 232 (1963); Hallé in Fl. Gabon 12, Rubiacées, 1: 132, t. 26 (1966). Type: Sierra Leone, *Afzelius* (BM, iso.! (one flower))

Compact shrub or small tree (0.2–)1.5–7.5(?–12) m. tall; stems glabrous to bristly pubescent. Leaf-blades narrowly to broadly elliptic to oblong or lanceolate, 1.2–14 cm. long, 0.4–7.1 cm. wide, slightly to markedly acuminate at the apex, cuneate at the base, glabrous to slightly bristly pubescent above, sparsely to densely bristly pubescent on the venation beneath; tertiary venation obscure to very evident; petiole 1–8 mm. long; stipules 1.5–5 mm. long, bifid, soon falling. Flowers sweetly scented, solitary or in lax terminal cymes; peduncles (including stalks of solitary flowers) 0.2–4 cm. long; pedicels 1–3 mm. long. Calyx-tube turbinate, 3 mm. long, appressed bristly pubescent; lobes 5–6, oblong, elliptic or lanceolate, rarely ovate, 0.35–2.1 cm. long, 0.8–8.5 mm. wide, clawed at the base, often quite leaf-like. Corolla white, the tube greenish outside, the throat hairs yellow; tube 1.5–3 cm. long, appressed pubescent outside with ± stiff hairs; lobes linear-oblong to broadly elliptic, very variable in size and shape, (1–) 1.8–2.8(–3.5) cm. long, (0.5–)1.1–1.5 cm. wide, shortly apiculate at the apex, densely puberulous. Stigmas 2–4 mm. long in long-styled flowers, exserted, the anthers 2–3 mm. below the throat; style reaching ± halfway up the tube in short-styled flowers, the stigmas 3 mm. long, the anthers not exserted but situated just below the throat. Fruit oblong-ellipsoid to subglobose, 0.8–1.8 cm. long, 0.8–1.4 cm. wide, sparsely shortly bristly pubescent, mostly crowned with the persistent calyx-lobes. Seeds blackish ± triangular in outline, ± 1.3 mm. long, strongly compressed; testa deeply pitted.

SYN. *Gardenia pulchella* G. Don in Edin. Phil. Journ. 11: 343 (1824), *nom. nud.*
  G. crinita Afzel., Stirp. Guin. Medic.Sp. Nov. 13, no. 5 (1829)
  Heinsia jasminiflora DC., Prodr. 4: 390 (1830); Hiern in F.T.A. 3: 81 (1877). Type: Sierra Leone,
    Smeathman (GE, holo.)
  H. pubescens Klotzsch, Philipp Schoenlein's bot. Nachlass auf Cap Palmas (Abh. K. Akad. Wiss.
    Berlin, 1856): 228 (1857). Type: Liberia, Cape Palmas, *Schoenlein* (B, holo. †)
  H. pulchella (G. Don) K. Schum. in E. & P. Pf. IV. 4: 84 (1891)

subsp. **parviflora** (*K. Schum. & K. Krause*) *Verdc.* in K.B. 31: 184 (1976). Types: Tanzania, Uzaramo District, Mogo Forest [Sachsenwald], *Holtz* 341 (B, syn. †) & Masasi/Newala District, Makonde Plateau, Mkomadatchi, *Busse* 1083 (B, syn. †, EA, isosyn.!)

Small shrub or rarely almost a tree, 1–4.5 m. tall. Leaf-blades mostly small, usually well under 6 × 2.5 cm., the tertiary venation not very evident beneath. Calyx-lobes mostly quite small, 3.5–6.5 mm. long, 0.8–2.5 mm. wide, the claw part widened at the base. Fruits mostly smaller, 0.8–1.2 cm. long. Fig. 1/2.

KENYA. Kwale District: Tanga road near Marere pumping station, 15 Apr. 1968, *Magogo & Glover* 865!; Mombasa, Mar. 1876, *Hildebrandt* 1983!; Kilifi District: Marafa, 22 Nov. 1961, *Polhill & Paulo* 840!
TANZANIA. Tanga District: Steinbrüch Forest, 29 May 1957, *Faulkner* 1988A!; Morogoro District: Uluguru Mts., 24 Feb. 1936, *E.M. Bruce* 851!; Mikindani, road to Rovuma and Mozambique border, 8 Mar. 1963, *Richards* 17779!
DISTR. K 7; T 3, 6, 8; Somalia, Mozambique, Malawi, Zimbabwe and South Africa (Transvaal)
HAB. Bushland, degraded coastal *Brachystegia* and particularly forest edges; 15–660 m.

SYN. [*H. jasminiflora* sensu Hiern in F.T.A. 3: 81 (1877), pro parte, *non* sensu stricto]
  H. pulchella K. Schum. in E. & P. Pf. IV. 4: 84 (1891), pro parte, *non* (G. Don) K. Schum. sensu
    stricto
  H. parviflora K. Schum. & K. Krause in E.J. 39: 530 (1907); T.T.C.L.: 501 (1949)
  [*H. crinita* sensu T.T.C.L.: 500 (1949); K.T.S.: 446 (1961) *non* (Afzel.) G. Taylor sensu stricto]

NOTE. Subsp. *crinita*, often a distinct tree, frequently has much larger leaves with the tertiary venation quite obvious beneath, almost invariably much larger calyx-lobes and bigger fruits. If it were not for some difficult intermediates in southern Africa the two could perhaps be kept as distinct species. K. Schumann & K. Krause intended their name to apply only to a small-flowered plant on the southern Tanzanian coast and considered that the W. African plant also occurred in East Africa. Subsp. *crinita* occurs from Guinée to Angola and NW. Zambia.

2. H. zanzibarica (*Bojer*) *Verdc.* in K.B. 35: 422 (1980). Type: Zanzibar, *Bojer* (P, syn.!)

Straggling to erect shrub or small tree 1.5–4.5 m. tall; young shoots densely covered with pale brown shaggy hairs ± 2 mm. long; older shoots glabrescent or glabrous, often somewhat fissured. Leaf-blades elliptic, oblong-elliptic or sometimes slightly elliptic-oblanceolate, 5–15 cm. long, 2.5–6 cm. wide, acuminate at the apex, cuneate to almost rounded at the base, hispidly hairy to glabrescent above, densely rather velvety hairy beneath but leaf-surface clearly visible under a lens; petiole 2–6 mm. long, densely hairy like the stems; stipules 0.6–1.3 cm. long, divided to near the base into 2 subulate attenuate lobes, pilose. Flowers sweetly scented, mostly numerous in dense terminal cymose inflorescences; peduncle 1–2 cm. long, secondary branches 0.5–1 cm. long, pedicels very short, all densely hairy; bracts filiform. Calyx-tube turbinate, 2 mm. long, densely hairy; lobes 5, leafy, 1.2 cm. long, 3–4.5 mm. wide, including the claw 3–4.5 mm. long, 1 mm. wide, pubescent. Corolla white with greenish tube; tube slender, 2–3 cm. long, densely pale yellowish hairy at the throat and densely velvety hairy outside; lobes elliptic to oblong-elliptic, 2.1–3.4 cm. long, 0.9–1.9 cm. wide, caudate-acuminate at the apex, the tail being 4–5 mm. long, hairy on midrib beneath and densely finely pubescent on both surfaces. Anthers situated slightly below the throat in short-styled flowers and ± 5 mm. below in long-styled flowers. Tips of stigma-arms exserted ± 2 mm. in long-styled flowers but style only ± 1 cm. long in short-styled flowers; stigma-arms 2–2.5 mm. long. Fruit dry, indehiscent ± oblong-globose or globose, 6–9 mm. long, 7.5–9 mm. wide, 7–7.5 mm. thick, compressed laterally, bristly pilose. Seeds irregularly oblong-angular ± 1.5 mm. long, strongly reticulate. Fig. 73.

KENYA. Kwale District: Shimba Hills, Mwele Mdogo Forest, 4 Feb. 1953, *Drummond & Hemsley* 1110! & Shimba Hills, 14 Oct. 1907, *Battiscombe* 72! & Shimba Hills, Shimba Settlement road, 15 May 1968, *Magogo & Glover* 1065!
TANZANIA. Lushoto District: E. Usambara Mts., Kisiwani, 30 May 1950, *Verdcourt* 249! & Amani-Muhesa road, 3.2 km. E. of site of Sigi railway station, 27 July 1953, *Drummond & Hemsley* 3491!; Uzaramo District: Kisarawe, [Kiserawe], 8 Jan. 1939, *Vaughan* 2716!; Zanzibar I., Kidichi, 18 Dec. 1919, *Faulkner* 2437! & 14 June 1960, *Faulkner* 2604!
DISTR. K 7; T 3, 6, 8; Z; Mozambique

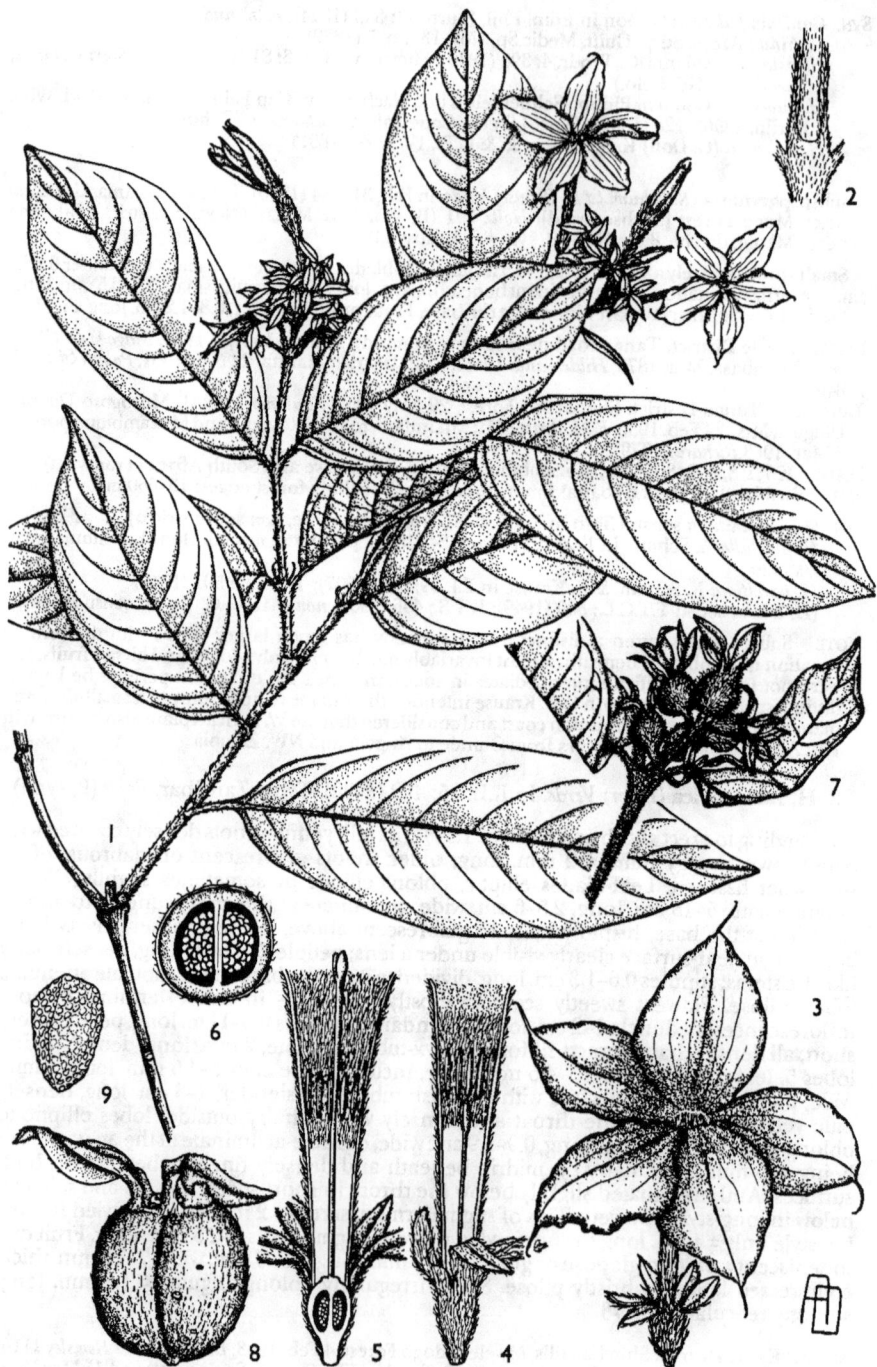

FIG. 73. *HEINSIA ZANZIBARICA* — 1, flowering branch, × ⅓; 2, stipule, × 1; 3, flower, × 1; 4, calyx with half corolla-tube, from short-styled flower, × 2; 5, same, with calyx cut through, from long-styled flower, × 2; 6, transverse section of ovary, × 5; 7, infructescence, × ⅓; 8, fruit, × 2; 9, seed, × 10. 1-3, 5-7, from *Drummond & Hemsley* 1110; 4, from *Tanner* 3614; 8, 9, from *Magogo & Glover* 39. Drawn by Ann Farrer, with 2 and 9 by Sally Dawson.

HAB.   Forest edges; 90–480(–700) m.

SYN. *Mussaenda zanzibarica* Bojer in Ann. Sci. Nat., sér. 2, 4: 264 (1835); Wernham in J.B. 51: 277
        (1913)
        M. rufa Bojer in Sept. Rapp. Ann. Trav. Soc. Hist. Nat. Maurice: 36 (1836), *non* A. Rich. (1830).
        Type: Zanzibar, *Bojer* (see note)
        Heinsia densiflora Hiern in F.T.A. 3: 81 (1877); K. Schum. in P.O.A. C: 382 (1895); T.T.C.L.: 501
        (1949); K.T.S.: 446 (1961). Type: Zanzibar, Mupanda, *Kirk* 80 (K, holo.!)

NOTE.   In his description of *M. zanzibarica* Bojer puts "*M. rufa* Boj. in herb." in synonymy. Presumably
he changed the name realising it was a homonym. The type of *M. rufa* is presumably the same
specimen.
        Brown 735 (Kenya, Kwale, 21 Dec. 1963) has the leaf-blades hairy only on the venation beneath but
is undoubtedly this species rather than a hybrid with *H. crinita*, but the possibility of such hybrids
needs investigation.

3. H. bussei *Verdc.* in K.B. 12: 353 (1957). Type: Tanzania, Masasi/Newala Districts,
Makonde Plateau, Mkomadatchi*, *Busse* 1083a (EA, holo.!, K, iso.!)

Shrub or small tree**; branches pendulous, pubescent to hairy but finally glabrescent,
covered with a blackish lenticellate bark. Leaf-blades oblong or ovate-elliptic, 2.2–8.5 cm.
long, 0.8–4.4 cm. wide, acute or ± acuminate at the apex, rounded to cuneate at the base, very
discolourous, often somewhat bullate, green above and with a few sparse hairs, grey
beneath and velutinous with interwoven hairs; petiole 3–4 mm. long; stipules black with a
median pale part, bilobed, up to 8(–11) mm. long, the lobes linear-lanceolate, 3–5 mm. long.
Flowers 1–3 in terminal inflorescences. Calyx-tube campanulate, 5 mm. long, 4 mm. wide,
densely pilose; lobes leafy, unequal, probably ± reflexed in life, 4 ovate-elliptic, 1.2–1.4 cm.
long, 6–7.5 mm. wide, acute at the apex, clawed at the base, the claw 3–5 mm. long, 2 mm.
wide, channelled and pilose, and one 3 mm. long and 2 mm. wide. Corolla white, ±
salver-shaped, adpressed pilose; tube (1.2–) 2–2.5 cm. long, 2 mm. wide at the base, 5 mm.
wide at the apex; lobes narrowly ovate-lanceolate, 2.2–3 cm. long, 0.7–1.2 cm. wide, aristate,
sericeous pilose outside, puberulous inside. Anthers inserted just below the throat in
short-styled flowers, 5–6 mm. below in long-styled flowers. Style 5.7 mm. long in short-styled
flowers, the stigma-lobes 6 mm. long; stigma-lobes 4 mm. long in long-styled flowers, the tips
just exserted at the throat. Fruit broadly ovoid, 1.2 cm. long, 9 mm. wide, densely pilose and
with short pubescence, crowned by the leafy calyx-lobes.

TANZANIA   Lindi District: Rondo Plateau, Mchinjiri, Jan. 1952, *Semsei* 619! & Mlinguru, 22 Dec. 1936,
        Schlieben 5771!; Mikindani District: Mtwara, Naliendele, 5 Jan. 1985, *Fison* 42!
DISTR.   T 8; not known elsewhere
HAB.   ? Woodland or forest; 100–810 m.

NOTE.   *Hay* 23 (Tanzania, Newala District: 53 km. from Newala, Chiumo, 18 Jan. 1959) probably
belongs here, but the indumentum is atypical.

57. BERTIERA

Aubl., Hist. Pl. Guiane Fr. 1: 180, t. 69 (1775); Wernham in J.B. 50: 110–117 & 156–164 (1912);
Hallé in Not. Syst. 16: 280–292 (1960) & in Adansonia sér. 2, 3: 294–306 (1963) & sér. 2, 4:
457–459 (1964)

Justenia Hiern, Cat. Afr. Pl. Welw. 1: 451 (1898)

Small trees or shrubs, rarely lianas. Leaves shortly petiolate; petioles normally
channelled above; blades thin or scarcely coriaceous, mid- and lateral nerves usually
prominent beneath; stipules persistent, usually connate above the nodes for a short
distance. Flowers of moderate size (rarely very small), borne in terminal, spicate thyrsoid
panicles, scorpioid cymes or spherical compact heads, very occasionally axillary (not in
Flora area). Calyx with limb produced above the tube, cup-shaped and frequently wider

---

* Assuming 1083a came from the same locality as 1083, one of the cited types of *Heinsia parviflora* K.
Schum. & K. Krause.
** Semsei's estimate of 30 ft., i.e. 9 m., seems an exaggeration; he gives the habit as a straggling forest
tree; Schlieben says 'strauch' which is probably all this species ever becomes.

than the tube, truncate or 5-toothed or -lobed. Corolla with hairy throat, often with the walls of tube toughened or thick; tube exceeding the calyx-limb; lobes (4–)5, usually acuminate or apiculate, contorted in bud. Anthers (4–)5, subsessile, attached by very short filaments near the base, included, linear, sagittate or bilobed at the base, the connective extended into an apiculate apex. Ovary 2-locular with the numerous ovules inserted on a thickened placenta; disc glabrous, fleshy, cup-shaped, or annular; stigma borne on style ± club-shaped, formed from 2 flattened appressed branches (seldom separating), with 10 membranous linear vanes which fit between the anthers in the bud. Fruit indehiscent, ovoid or globose, often coriaceous; calyx persistent; pedicels sometimes accrescent. Seeds red-brown to black, angular, rugulose or granulose, numerous.

About 55 species occurring in Africa, Madagascar, and America, of which 41 are African.

1. Inflorescence spike-like (often paniculate in fruit) 2
   Inflorescence capitate (often umbellate in fruit) 3
2. Calyx glabrous to sparsely hairy; fruit with disc hidden by
     the persistent calyx-limb . . . .   *1. B. racemosa*
   Calyx densely hirsute; fruit with disc prominent above the
     rim of the persistent calyx . . . .   *2. B. pauloi*
3. Stems glabrous; corolla-lobes long-caudate; fruiting
     pedicel accrescent, reaching 2.7 cm. long . . . .   *3. B. naucleoides*
   Stems covered with dense adpressed hairs; corolla-lobes
     acute; fruit sessile . . . .   *4. B. globiceps*

1. B. racemosa *(G. Don) K. Schum.* in Bol. Soc. Brot. 10: 127 (1892); Wernham in J.B. 50: 160 (1912); I.T.U., ed. 2 : 338 (1952); Hallé in Not. Syst. 16: 284, fig. 3/k, 4/m (1960); F.W.T.A., ed. 2, 2: 160 (1963); Hallé, Fl. Gabon 17, Rubiacées, 2: 54, t. 12/11–14 (1970); Fl. Pl. Lign. Rwanda: 544, fig. 182.1 (1982); Fl. Rwanda 3: 142, fig. 42/1 (1985). Type: Sierra Leone, *Don* (BM, holo.!)

Small tree 2.4–5 m. tall; branches angular, ± square in cross-section, almost glabrous or more usually covered with pale brown adpressed hairs, often with short white and brown streaks apparent on the bark. Leaf-blades discolourous, large, elliptic, 10–36 cm. long and 3.5–17.7 cm. wide, apex acute to shortly acuminate, base obtuse to cuneate or sometimes cordate, with 11–14 pairs of lateral nerves, glabrous above, sparsely pubescent and with ± densely adpressed pubescence on the nerves beneath; petioles (0–) 0.3–1.7 cm. long, usually covered with adpressed hairs beneath or sometimes glabrous; stipules ± ovate above, narrower at the connate portion, 1.4–4.4 cm. long and 0.6–1.6 cm. wide, connate for 0.2–1.1 cm. above the node, apex obtuse. Flowers sessile, in ± 18-flowered fascicles, sub-sessile on a spike 4–17 cm. long; internodes 0.5–3 cm. long; peduncles 1.5–7.3 cm. long, densely adpressed hirsute, probably pendulous; bracteoles inconspicuous, ± deltoid and seldom larger than 1.2 mm. long and wide, pubescent or with the hairs confined to the margins. Calyx almost glabrous or with dense adpressed hairs on the tube, becoming sparser on the limb; tube ± cylindrical, 1.8–3 mm. long and 1.7–3.2 mm. wide; limb 1.6–3.6 mm. long, 2–5.5 mm. wide, truncate, shallowly undulate or shortly dentate, the dentations not exceeding 0.8 mm., sparsely hairy within. Corolla white or green; tube 8.5–15 mm. long, widening 5–9.5 mm. above the base, 1.4–3.8 mm. wide at base and 3.2–5.5 mm. wide at top, the wall of the lower, narrow portion distinctly lined with a whitish fawn thickened layer, glabrous or covered with adpressed hairs outside and with white or fawn, striated coarser hairs in the throat and a short distance up the lobes; lobes narrowly ovate or ovate, 2.7–5.8 mm. long and 1.4–2.3 mm. wide, long-acuminate or acute (in var. *elephantina*), glabrous or sparsely hairy. Anthers 3.3–4.3 mm. long; apex 0.2–1.1 mm. long. Disc shortly conical. Style 4–9 mm. long; stigma 3.75–5.6 mm. long. Infructescences 4.5–17 cm. long; peduncles 4–9 cm. long; fascicles expanding to 5 cm. long; pedicels absent or accrescent and reaching 1 cm. long (in var. *elephantina*). Fruit greenish brown, ovoid, 6–11 mm. long and 6–9 mm. wide, glabrous or sparsely hairy (in var. *elephantina*). Seeds 0.9–1.1 mm. long. Fig. 1/4 (but note drawing reversed, contortion should be to the left).

var. racemosa

Corolla hairy outside; tube narrow, not exceeding 2.5 mm. at base and 4.2 mm. at top; lobes long-acuminate giving the buds a distinctly pointed appearance. Fruit glabrous and sessile.

Uganda. Masaka District: Sese Is., Bugala, Kisozi, 3 Mar. 1933, *A.S. Thomas* 946! & 48 km. on Masaka road, 25 May 1938, *Tothill* 2708!; Mengo District: Kiagwe, July 1932, *Eggeling* 810!

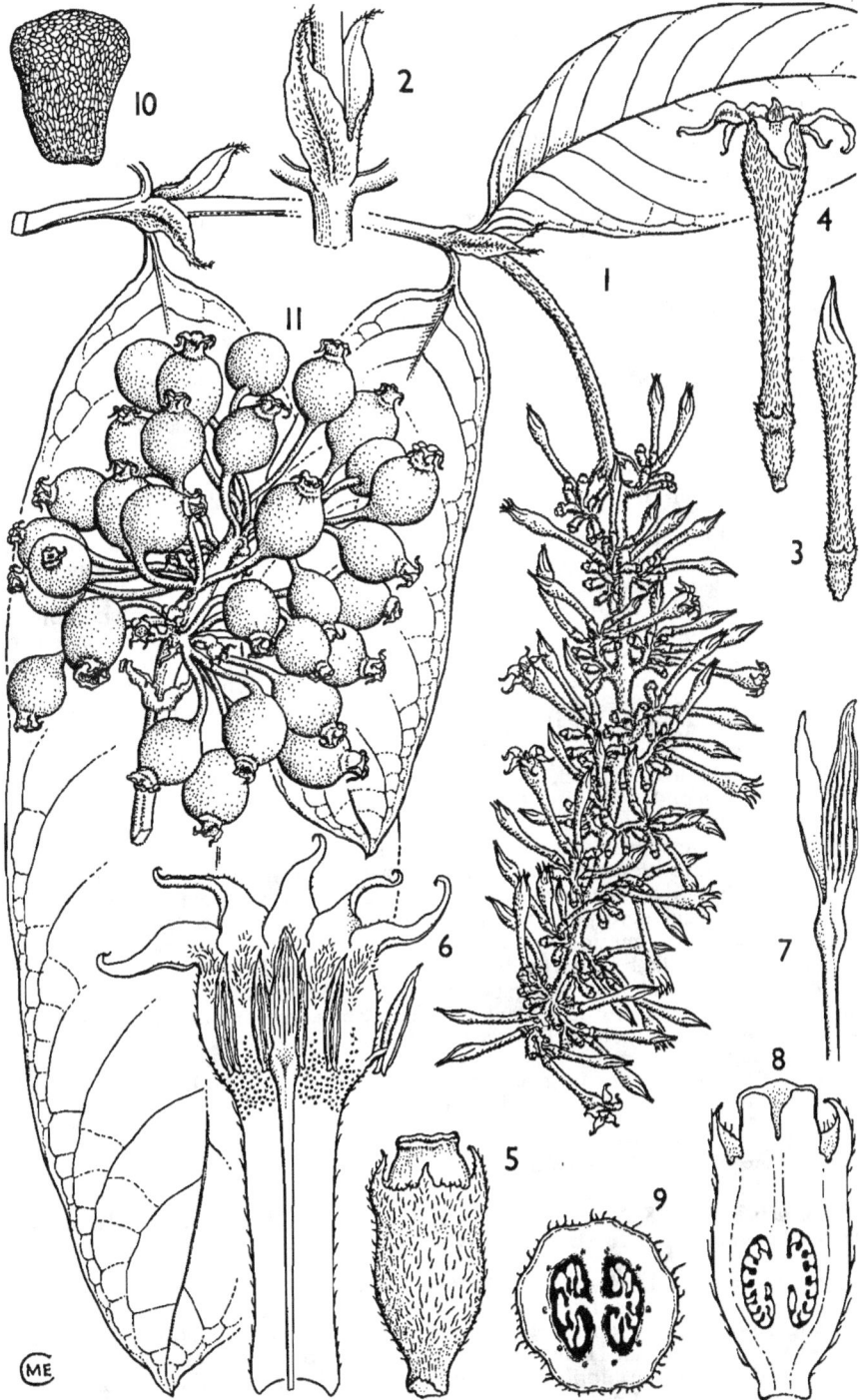

FIG. 74. *BERTIERA PAULOI* — 1, flowering branch, × ⅔; 2, stipules, × 1; 3, flower-bud, × 2; 4, flower, × 3; 5, calyx, × 8; 6, corolla, opened out, × 4; 7, stigma, × 6; 8, longitudinal section through ovary, × 10; 9, transverse section through ovary, × 12; 10, seed, × 14. *B. NAUCLEOIDES* — 11, infructescence, × ⅔. 1, from *Cribb et al. 11467B*, 2–10, from *Cribb et al. 11021*; 11, from *Maitland 803*. Drawn by Mrs M.E. Church.

TANZANIA. Bukoba District: Buyango, 14 Sept. 1934, *Gillman* 153! & Sept.–Oct. 1935, *Gillman* 427!
DISTR. U 4; T 1; W. Africa from Guinée to Cameroun, Gabon, Cabinda, Congo, Zaire and Rwanda
HAB. Alluvial-swamp, lake-shore forest or at forest edges on moist ground; 1140–1190 m.

SYN. *Wendlandia racemosa* G. Don, Gen. Syst. 3: 519 (1834)
    Bertiera macrocarpa Benth. in Hook., Niger. Fl.: 394 (1849); Hiern in F.T.A. 3: 84 (1877), *nom.*
    superfl. Type as for *Wendlandia racemosa*
    B. montana Hiern in F.T.A. 3: 83 (1877); Wernham in J.B. 50: 161 (1912); F.W.T.A. 2: 97 (1931);
    Hallé in Not. Syst. 16: 284, fig. 3/j & 4/n (1960). Type: Bioko [Fernando Po], *Mann* 292 (K,
    holo.!)
    B. dewevrei De Wild. & Th. Dur., Pl. Gillet. Congol.: 27 (1900), *nomen;* Ann. Mus. Congo 3, 1: 113
    (1901). Type: Zaire, *Dewèvre*, 1167 bis, (BR, holo.)
    B. racemosa (G. Don) K. Schum. var. *dewevrei* (De Wild. & Th. Dur.) Hallé in Fl. Gabon 17,
    Rubiacées, 2: 56 (1970)

NOTE. The material from the Flora area apparently shows slight variation in the length of
corolla-tube (10–11 mm.), lobes (4–4.1 mm.) and in the apex of the anthers (0.6 mm.) which are
shorter than those of the West African specimens, but not so distinct from the Zaire material. The
shortage of mature flowers on specimens from the Flora area makes the consistency of this hard to
verify. Hallé (Fl. Gabon.) recognises two additional varieties, var. *dewevrei* (De. Wild. & Th. Dur.)
Hallé from Zaire and var. *elephantina* Hallé from Sierra Leone to E. Zaire and Gabon.

2. B. pauloi *Verdc.* in K.B. 12: 352 (1957); Hallé in Not. Syst. 16: 285 (1960). Type:
Tanzania, Morogoro District, Mt. Ruhamba [Koruhamba], *Paulo* 224 (EA, holo!, K, iso.!)

Small tree; branches slightly angular, covered with dense adpressed straw-coloured
hairs. Leaf-blades oblong-elliptic, 15–22.5 cm. long and 5–7.6 cm. wide, apex acute, base
cuneate, rounded or subcordate, with 9–14 pairs of lateral nerves, glabrous except for the
margins above and beneath and the nerves beneath; petioles 2–6 mm. long, glabrous
above, covered with adpressed hairs beneath; stipules ovate-lanceolate to triangular,
12–18 mm. long and 4–8.5 mm. wide at base, connate for up to 3 mm. above the node,
apex acute, covered with adpressed hairs outside. Flowers sessile in 3–12-flowered cymes
subsessile in dense spikes 3–4.5(–11.5) cm. long; internodes ± 6–10 mm. long; peduncle
2–3.5(–10) cm. long, densely adpressed hirsute; bracteoles deltoid, 0.7–1 mm. long and
0.4–0.8 mm. wide, acuminate, adpressed hirsute outside. Calyx densely adpressed hirsute;
tube suborbicular to cylindrical, 1.75–2.25 mm. long and 1.25–2 mm. wide; limb 0.5–2.25
mm. long, only slightly wider than the tube, with 5 irregular crenulations ± 0.3 mm. long
and 1 mm. wide, apex obtuse or with ± subulate lobes from a triangular base, 0.75 mm.
long. Corolla white; tube 1.1–1.3 cm. long, widening 9–10 mm. above the base, ± 1.5 mm.
wide at base, 3 mm. wide at top, covered with yellowish to greyish adpressed hairs outside
and inside with short hairs restricted to the throat; lobes narrowly ovate, 4.75–5 mm. long,
± 1.5 mm. wide; acumen very long with inturned edges, glabrous or with a few adpressed
hairs towards the base outside. Anthers 3.25 mm. long, apex 1 mm. long. Disc annular,
exceeding the calyx-limb. Style 1 cm. long; stigma 6 mm. long. Infructescences up to 9–13
cm. long; peduncle up to 6 cm. long; pedicels thick, 2–3 mm. long. Fruit ovoid, 6–7 mm.
long and 5.5 mm. wide; disc accrescent, conical, 1–2 mm. wide and high and protruding
above the persistent calyx-limb. Seeds ± 1.2 mm. long. Fig. 74/1–10.

TANZANIA. Morogoro District: Uluguru Mts., Kipogoro, Kiemba, 12 Feb. 1933, *Schlieben* 3425!;
Ulanga District: Uzungwa Mts., Sanje, 14 June 1984, *Lovett* 292!; Iringa District: Mufindi, Lulanda
Forest Reserve, 16 Feb. 1979, *Cribb et al.* 11467B!
DISTR. T 6, 7, not known elsewhere
HAB. Evergreen forest; ± 1200–1700 m.

3. B. naucleoides *(S. Moore) Bridson* in K.B. 33: 554 (1979). Type: Uganda, Mengo
District, Entebbe, *Bagshawe* 771 (BM, holo.!)

Small tree or shrub 1.8–4(–5) m. tall; branches square in cross section, glabrous except
for a short line of adpressed hairs below the nodes. Leaf-blades discolorous, elliptic, 9–28
cm. long, 4.4–12 cm. wide, apex acute or shortly acuminate, base cuneate or truncate and
then shortly cuneate, with 11–14 pairs of lateral nerves, both surfaces sparsely covered
with hairs, denser on the midrib above and dense and adpressed on the mid- and lateral
nerves beneath; petiole 1–2.3 cm. long, sparsely covered with adpressed hairs or glabrous
except for the crests of the channels; stipules ovate, narrowing at the connate portion,
then widening at the base, 1–2 cm. long and (0.4–)0.7–1.2 cm. wide, connate for ± 1.5 mm.
above the node, apex usually obtuse, or sometimes rounded or acute, glabrous except for
a central line of adpressed hairs outside. Flowers closely packed in sessile spherical heads

1.3–3.5 cm. diameter; pedicels short and thick, 0.8–1.5 mm. long; bracteoles narrowly triangular, sometimes with 2 filiform lobes on either side, 5.3–11 mm. long and 1.5–3 mm. wide, with sparse adpressed hairs outside. Calyx adpressed hirsute, dense at the base and becoming sparse towards the lobes; tube cylindrical, 1.4–3 mm. long and 1.5–2.5 mm. wide; limb 3.5–4.4 mm. long and 3.25–3.5 mm. wide, bearing narrowly triangular lobes 2–4.8 mm. long and 0.5–1.2 mm. wide, apex caudate. Corolla white; tube 6–6.8 mm. long, widening 3.2–3.6 mm. above the base, occasionally distinctly widened at the base, 4–4.5 mm. wide at top, 2–2.2 mm. wide below, hairy on widened portion outside and with dense yellow hairs at throat and between anthers within; lobes ovate, 6–6.6 mm. long and 2–3 mm. wide, apex long-caudate, glabrous or with a few hairs near the margins outside. Anthers 1.9–2.6(–3) mm. long, with acuminate apex 0.2–0.5 mm. long. Disc annular, 1.3–2 mm. diameter and 0.7–1 mm. deep. Style 3–3.5 mm. long; stigma 3.6–5.3 mm. long. Infructescences ± globose, 5.5–8.8 cm. diameter, the accrescent pedicels reaching 1–2.7 cm. long, glabrous or sparsely hairy. Fruit green (? red), ± globose, 1–1.5 cm. diameter, glabrous or with a very few hairs. Seeds 1.5–1.8 mm. long. Fig. 74/11.

UGANDA. Masaka District: Sese Is., Bubembe I., June 1925, *Maitland* 803!; Mengo District: Busiro county, Kyiwaga [Kyewaga] Forest, E. side of Entebbe bay ± 3.2 km. from the town, 6 Sept. 1949, *Dawkins* 358!
TANZANIA. Bukoba District: Musira I. [Busira], 1935, *Gillman* 294! & Nov. 1958, *Procter* 1070! & 4.8 km. S. of Bukoba, by Lake Victoria, Mar. 1952, *Jervis!*
DISTR. U 4; T 1; Cameroun, Gabon, Zaire
HAB. Evergreen forest, in the open parts; 1110–1160 m.

SYN. *Randia naucleoides* S. Moore in J.B. 44: 83 (1906)
Bertiera capitata De. Wild. in Ann. Mus. Congo, Bot., sér. 5, 2: 169 (1907); Wernham in J.B. 50: 163 (1912); Verdc. in K.B. 7: 359 (1952); Hallé in Not. Syst. 16: 285, fig. 3/a (1961) & in Fl. Gabon 17, Rubiacées 2: 64, t. 14 (1970). Type: Zaire, Equatoria, Eala, *Laurent* 131 (BR, lecto.)

4. B. globiceps *K. Schum.* in E.J. 23: 451 (1897); Wernham in J.B. 50: 163 (1912); Verdc. in K.B. 7: 359 (1952); Hallé in Not. Syst. 16: 285, fig. 3/b, 4/k (1960) & Fl. Gabon 17, Rubiacées, 2: 66, t. 15/1–7 (1970). Type: Cameroun, Lolodorf, *Staudt* 128 (B, holo. †, BM!, K!, P, iso.)

Evergreen shrub up to 7.6 m. high; branches round in cross-section, covered with dense pale to golden-brown adpressed hairs. Leaf-blades discolorous elliptic, (6–)13.5–18.5 cm. long and (2–)4–6 cm. wide, acuminate at apex, cuneate or attenuate at base, with 7–11 pairs of lateral nerves, glabrous above and with sparse adpressed hairs and dense adpressed hairs on nerves beneath; petiole (1.5–)3–5(–7) mm. long, with adpressed hairs beneath; stipules ovate, 1.1–3.6 cm. long and 0.5–0.6 cm. wide (the Uganda material being the longest), apex usually long-caudate or sometimes obtuse, connate for up to 6 mm. above the node, adpressed hairs on midvein and near base outside. Flowers sessile in closely packed globose or slightly elongated sessile heads, 1–2 cm. long and 1.2–1.7 cm. wide (excluding the corolla length); bracteoles triangular, up to 8 mm. long and 3.5 mm. wide, with adpressed hairs outside. Calyx densely adpressed hirsute; tube suborbicular to cylindrical, 1.1 mm. long and 1.8 mm. wide; limb 2–3.5 mm. long (including lobes), slightly wider than the tube; lobes 0.8–1.4 mm. long and wide, obtuse or narrowly triangular, sparsely covered with adpressed hairs within. Corolla densely adpressed hirsute on the outside with the exception of the base of the tube; tube 1 cm. long, widening 6.5 mm. above the base, 1.5–2.5 mm. wide at the base and 2.8–4 mm. wide at the top, with a few hairs at throat and between the anthers; lobes ovate, 3.5 mm. long and 2 mm. wide, apex acute, glabrous within. Anthers 2.5–2.8 mm. long, with acuminate apex, 0.5 mm. long. Disc annular, 1 mm. diameter and 0.3 mm. deep. Style 5.8 mm. long; stigma up to 4.2 mm. long. Infructescences ± globose, 2.8 cm. long and 2.3 cm. wide, sessile or with a peduncle up to 2.8 cm. long. Fruit ± globose, 0.7–1 cm. diameter, indistinctly rugose and covered with sparse adpressed hairs. Seeds ± 1.5 mm. long.

UGANDA. Kigezi District: Kayonza, Ishasha Gorge, 10 Feb. 1945, *Greenway & Eggeling* 7100!
HAB. Evergreen forest, on upper slopes of rocky gorge; 1524 m.
DISTR. U 2; Cameroun, Gabon, Cabinda, Congo (Brazzaville)

NOTE. This species is only represented in the Flora area by one gathering, lacking mature flowers and fruit. It differs from the West African material in having larger stipules and a greater number of lateral nerves in the leaf; more material might show that the East African material is worthy of recognition.

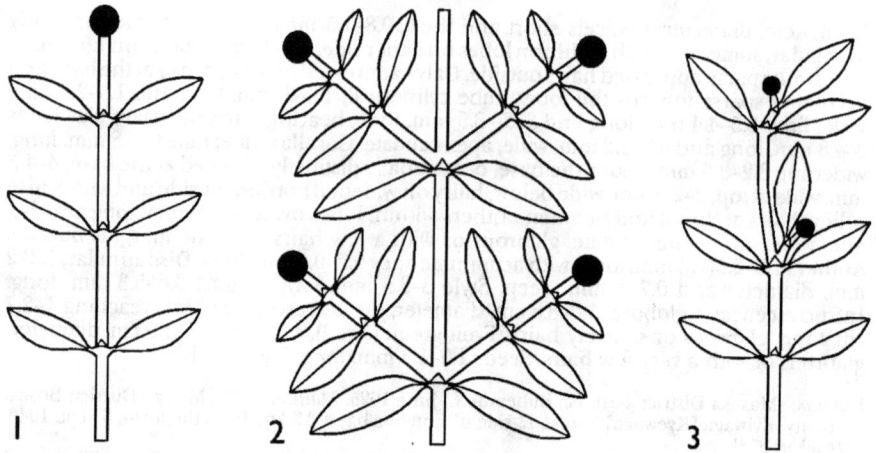

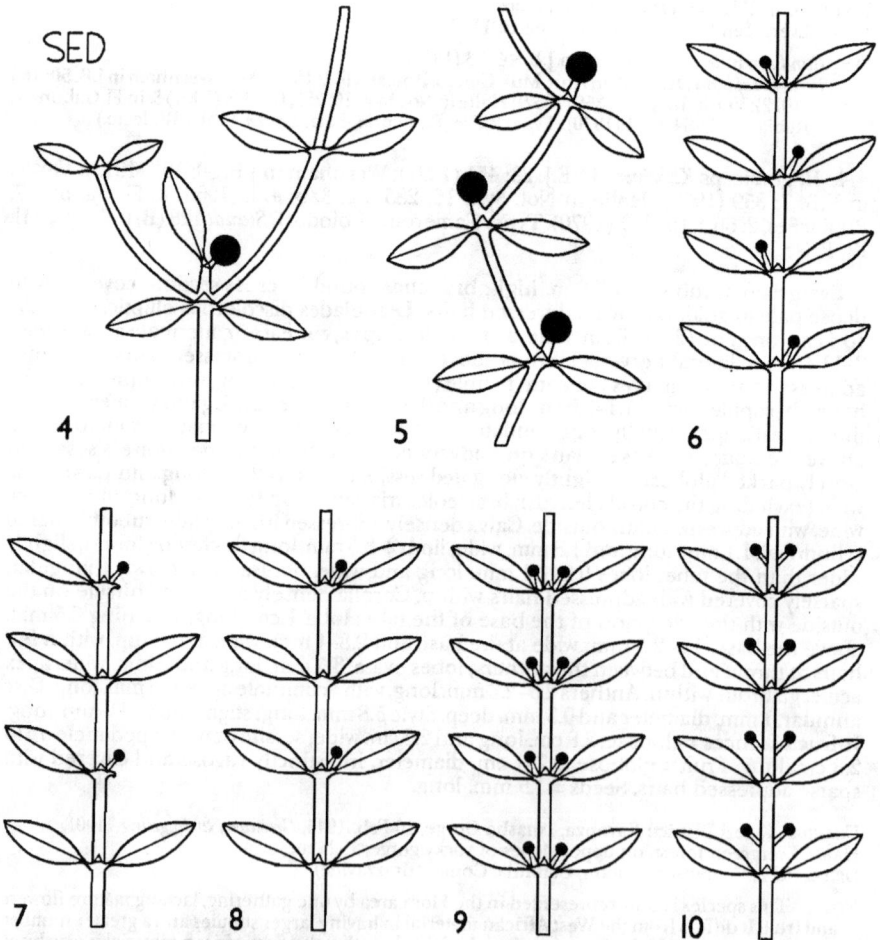

FIG. 75. *GARDENIEAE*, diagramatic representation of inflorescence-types, — 1, *Macrosphyra*-type; 2, *Gardenia*-type; 3, 4, *Rothmannia*-type; 5, *Sherbournia*-type; 6, *Oxyanthus*-type; 7, *Aidia*-type; 8, *Porterandia*-type; 9, *Tricalysia*-type; 10, *Didymosalpinx*-type. Drawn by Sally Dawson.

Tribe 17. GARDENIEAE*

Flowers orange turning red, tubular, in terminal heads
    supported by tight involucral bracts, the inner ones
    toothed; style-lobes flattened, adhering but easily
    separable (cultivated) . . . .   **Burchellia**
                               (see p. 438)

Flowers and inflorescence not as above:
  Inflorescences axillary, 2 per node, fig. 75/9, 10, if on short
    lateral spurs see below; stem monopodial:
  Style fusiform to clavate; corolla large (tube at least 2 cm.
    long); fruit large, 2–3.5 cm. long:
    Flowers solitary, often supra-axillary, fig. 75/9; corolla
      trumpet-shaped, longitudinally ribbed; shrubs
      armed with straight spines . . . .   **67. Didymosalpinx**
    Flowers several in pedunculate heads; corolla salver-
      shaped; unarmed rhizomatose shrub . . . .   **69. Mitriostigma**
                                  greenwayi

  Style 2-armed; corolla mostly smaller; fruit mostly smaller:
    Anthers medifixed or ± medifixed, the connective not
      enlarged; anther-thecae ± contiguous and
      subparallel; bacterial nodules absent . . . .   **70. Tricalysia**
    Anthers basifixed, the connective enlarged so that the
      anther-thecae diverge; linear bacterial nodules
      often present along midrib and petiole . . .   **71. Sericanthe**
Inflorescences terminal, often on very reduced branches, or
  lateral one per node, fig. 75/1–8; stems sympodial or
  monopodial:
  Spines usually present above reduced branches; corolla
    subrotate . . . .   **62. Catunaregam**
  Spines absent; branches seldom reduced; corolla with
    well-developed tube:
    Deciduous plants with conspicuous scarious or chaffy
      persistent stipules; inflorescence terminal on main
      or lateral branches, fig. 75/1:
      Flowers several in clusters; climbing or scandent
        shrubs; stigmatic knob globose; corolla salver-
        shaped or shortly funnel-shaped . . . . .   **60. Macrosphyra**
      Flowers solitary; shrub or small tree; stigmatic knob
        ellipsoid; corolla funnel-shaped, rather long and
        slender . . . .   **61. Euclinia**
    Evergreen plants or if deciduous then stipules not as
      above; inflorescence lateral, terminal on reduced
      leafless branches or sometimes as above:
    Deciduous shrubs; flowers terminal on main and
      lateral branches; calyx with corky ridges,
      protecting bud in dry season; corolla yellow-
      green, purple spotted, 8–11-lobed; tube shorter
      than lobes, 1–1.4 cm long . . . .   **65. Phellocalyx**
    Evergreen shrubs or climbers; other characters not
      combined as above:
      Ovules 2–10, embedded on a pendulous placenta;
        corolla small, campanulate, tubular at extreme
        base, (0.7–)0.9–1.2 cm. long excluding lobes;
        flowers 1–many subtended by a single leaf,
        clustered at ends of very reduced shoots, fig.
        75/3; fruit small, up to 1.3 cm. diameter,
        1–several-seeded . . . .   **89. Heinsenia**

---

*By B. Verdcourt (genera 58–63, 66, 67, 69 in part), D.M. Bridson (genera 64, 65, 68, 69 in part) and
E. Robbrecht (genera 70, 71). See also p. 736 for additional genus.

Ovules more numerous; corolla if as small
then cylindrical or funnel-shaped;
inflorescence if as above then corolla much
larger; fruit many-seeded, small or much
larger:

Inflorescences several–many-flowered cymes,
one borne at every other node, fig. 75/7, 8;
flowers small or medium-sized (corolla-tube
up to 2.5 cm. long):

Corolla-tube cylindrical, ± 6 mm. long, usually
glabrous outside; stipules triangular, not
conspicuous; fruit small, less than 1 cm.
in diameter; flowering nodes anisophyllous, fig.75/7 . . . .   **58. Aidia**

Corolla-tube funnel-shaped, 1.6–2.5 cm. long,
densely covered with silky hairs outside;
stipules elliptic, up to 1.2 cm. long,
conspicuous; fruit said to be as large as an
egg; flowering nodes not anisophyllous,
but one or both leaves sometimes fallen,
fig. 75/8 . . . .   **59. Porterandia**

Inflorescences 1–several- or many-flowered,
terminal on reduced shoots or on alternate
sides of successive nodes, fig. 75/2–6;
flowers medium-sized to large:

Lianes or scandent shrubs; calyx-lobes
reddish, contorted in bud, oblong or
lanceolate, persistent in fruit; stipules
rather large oblong or elliptic, deciduous;
inflorescence apparently lateral,
fig. 75/5 . . . .   **66. Sherbournia**

Shrubs or small trees; calyx-lobes and stipules
not as above:

Stipules green, fairly conspicuous,
triangular or ovate, scarcely sheathing;
inflorescences strictly lateral, fig. 75/6;
seeds not held together by a matrix,
surface with a finger-print-like pattern:

Corolla white; tube narrowly cylindrical   **68. Oxyanthus**
Corolla purple; tube funnel-shaped   **69. Mitriostigma usambarense**

Stipules truncate or, if triangular, then
brown and inconspicuous, sheathing
or not; inflorescence terminal on
reduced branches, fig. 75/2–4; seeds
held together by a matrix, surface
smooth or slightly reticulate:

Stipules sheathing, often truncate; young
parts usually glutinous; flowers
mostly terminal on short monopodial shoots, fig. 75/2; placentas
parietal, 2–9; pollen-grains in
tetrads . . . .   **63. Gardenia**

Stipules scarcely sheathing, usually
triangular; young parts not
glutinous; branches sympodial,
flowers usually above a single leaf,
fig. 75/3, 4; placentas parietal, 2
(sometimes partly fused); pollen
grains simple . . . .   **64. Rothmannia**

58. AIDIA

Lour., Fl. Cochinch.: 143 (1790); G. Taylor in Exell, Cat. Vasc. Pl. S. Tomé: 197 (1944); Keay in B.J.B.B. 28: 22 (1958); Hallé in Fl. Gabon 17, Rubiacées, 2: 164 (1970); Tirvengadum in Bull. Mus. natn. Hist. nat., 4e sér., 8, sect. B, Adansonia 3: 257–275 (1986)

Small trees or rarely (1 species) epiphytes. Leaves opposite, petiolate, in equal pairs or at flower-bearing nodes unequal or even one reduced to a deciduous scale; domatia sometimes present; stipules basally ± triangular, erect, acute to subulate, often deciduous. Flowers pedicellate, ♂, 5-merous, in unilateral apparently axillary inflorescences at every other node; inflorescences lax or condensed cymes up to 7 cm. long, few–many-flowered, subsessile or pedunculate; bracts small, often deciduous. Calyx-tube ovoid or turbinate; tubular part of the limb well-developed; lobes reduced or at least not over 3 mm. long. Corolla white, yellow or green, the central part sometimes red; tube cylindrical, with throat hairy; lobes contorted, overlapping to the left, acute. Stamens exserted, the filaments inserted in the upper part of the throat; anthers sagittate; pollen grains simple. Ovary 2(rarely–3)-locular, each locule containing a peltate hemispherical many-ovuled placenta; style exserted, developed into an oblong striate club of 2 adhering lobes. Fruit globose, small, marked with a circular scar left by the deciduous calyx-limb; seeds numerous.

A genus of the Old World containing ± 24 species reaching Australia and the Pacific islands. There are 8 species in Africa, 3 of which are endemic to the island of S. Tomé; one is well known from the Flora area, but a possible second species occurs in Tanzania known only from very imperfect material.

1. A. micrantha *(K. Schum.) F. White*, F.F.N.R.: 455 (1962); Petit in B.J.B.B. 32: 174 (1962) & in B.J.B.B. 34: 529 (1964); Hallé in Fl. Gabon 17, Rubiacées, 2: 165 (1970). Type: Gabon, Sibang, *Soyaux* 185 (B, holo. †, P, iso.)

Shrub or much-branched tree 1.8–9(–?12) m. high, with glabrous branches. Leaf-blades glabrous, drying olive or grey-brown, elliptic, 10–16(–18) cm. long, 3–7 cm. wide, acuminate at the apex, ± cuneate at the base; petiole 4–8 (–10) mm. long; stipules 3–5(–9) mm. long, 2–3(–6) mm. wide at the base, attenuate-subulate at the apex, glabrous. Inflorescences with one normal leaf at the node opposed by a filiform deciduous scale, 0.6–1.5 cm. long, few–many-flowered, 1.5–2 cm. long; peduncle very short, the branches with dense short deltoid bracteoles 1.5 mm. long; pedicels 3–9(–16) mm. long, usually glabrous. Calyx glabrous, puberulous or pilose; tube 2 mm. long; tubular part of limb somewhat funnel-shaped, 3–4 mm. long, the margin minutely denticulate. Flowers sweet-scented; corolla-tube apricot, purple or white, ± 6 mm. long, densely barbate with white hairs in the throat; lobes apricot, white, yellow or green, ovate, 5–8 mm. long, 3–4.5 mm. wide, acuminate at the apex, the part exposed in the bud densely very finely white pubescent or glabrous. Ovary with 50–100 ovules; style purple, 1–1.2 cm. long; stigma ellipsoid, 3–4 mm. long, 2 mm. wide, grooved. Fruit green turning red, globose, 6–8(–10) mm. in diameter, mostly glabrous. Seeds yellowish, angular, compressed, ± 1.5 mm. long.

Syn.* *Randia micrantha* K. Schum. in E.J. 23: 438 (1896)
     R. micrantha K. Schum. var. *poggeana* K. Schum. in E.J. 23: 438 (1896). Types: Zaire, between Kingenge and Kassai, *Pogge* 977 (B, syn. †) & Mukenge, *Pogge* 1051 (B, syn. †)
     R. lucidula Hiern, Cat. Afr. Pl. Welw. 1: 457 (1898). Types: Angola, Golungo Alto, Serra de Alto Queta, *Welwitsch* 3093 (LISU, syn., BM, isosyn.!) & Sobato de Quilombo, Quiacatubia, *Welwitsch* 3093b (LISU, syn.) & Cazengo, Serra de Muxâulo, *Welwitsch* 3092 (LISU, syn.)

var. **msonju** *(K. Krause) Petit* in B.J.B.B. 32: 181 (1962). Type: Tanzania, Morogoro District, Manyangu [Manjangu], *Bittkau* in Herb. *Amani* 2667 (B, holo. †)

Flower buds acute to acuminate; calyx glabrous; corolla 1.5–2 cm. long, with lobes glabrous save for ciliate margins or ± densely adpressed silky pubescent on the parts exposed in bud. Fig. 76.

UGANDA. Ankole District: Kashoya Forest, Aug. 1936, *Eggeling* 3211!; Kigezi District: Ishasha Gorge, Kayonza Forest Reserve, 5 Aug. 1960, *Paulo* 662!; Masaka District: Buddu, R. Mijusi Ravine, 25 Sept. 1957, *Osmaston* 4203!

* These all refer to var. *micrantha;* other synonyms, e.g. *Randia congolana* De Wild., *R. congestiflora* K. Krause, *R. sphaerogyne* K. Schum., *R. micrantha* var. *zenkeri* S. Moore, and *R. acarophyta* De Wild. are referable to the other four varieties.

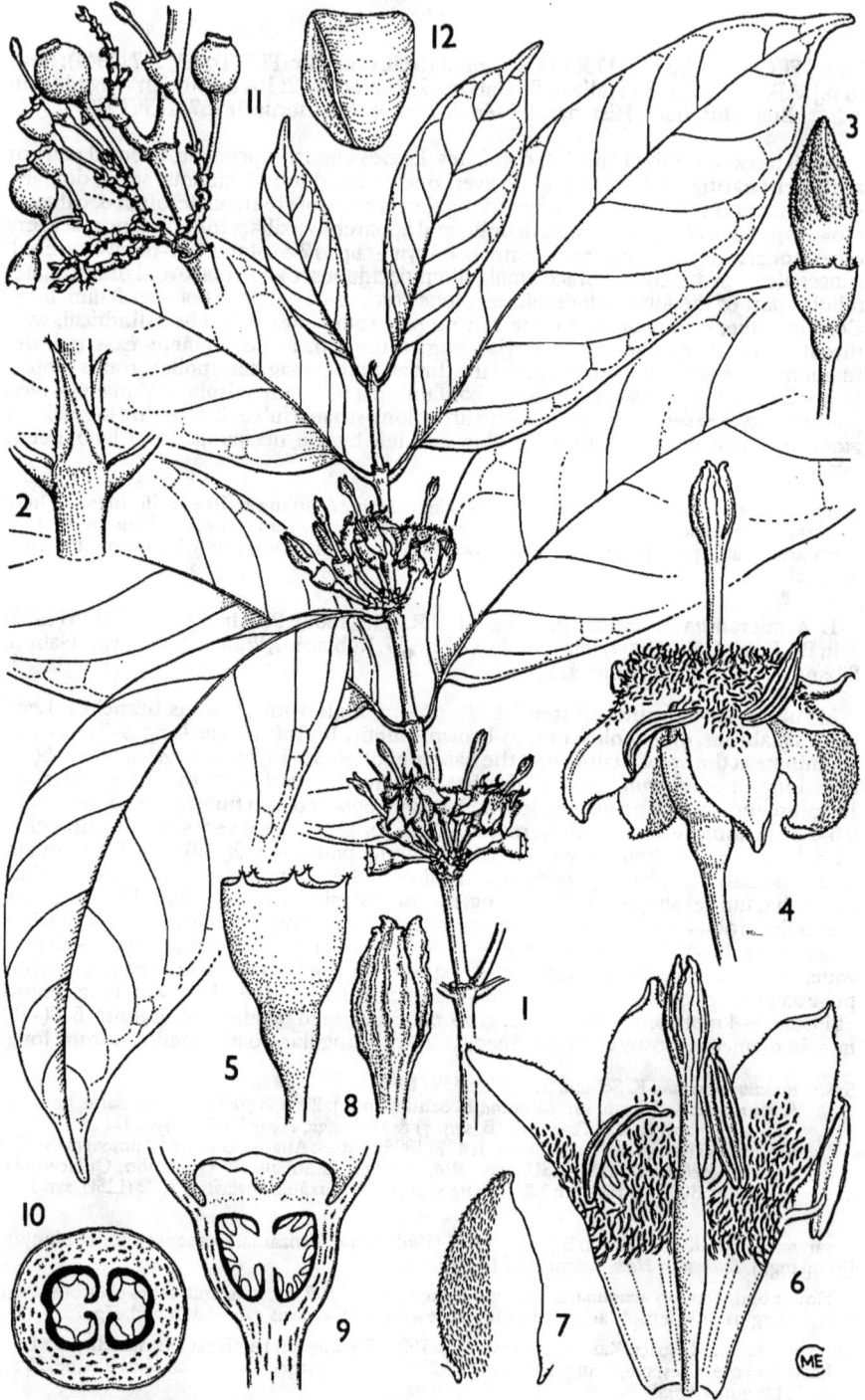

FIG. 76. *AIDIA MICRANTHA* var. *MSONJU* — 1, flowering branch, × ⅔; 2, stipules, × 4; 3, flower-bud, × 2; 4, flower, × 3; 5, calyx, × 4; 6, section through corolla, × 4; 7, corolla-lobe, from outside, × 4; 8, stigma, × 6; 9, longitudinal section of ovary, × 10; 10, transverse section of ovary, × 12; 11, portion of fruiting branch, × 1; 12, seed, × 8. 1, 3–10, from *Dawe* 41; 2, 11, from *Eggeling* 3211; 12, from *Pierlot* 2877. Drawn by Mrs M.E. Church.

TANZANIA. Bukoba District: Kiamawa, Sept. 1935, *Gillman* 409!; Morogoro District: Manyangu Forest Reserve, 19 Sept. 1960, *Paulo* 806! & NW. Uluguru Mts., 29 Sept. 1932, *Schlieben* 2748!; Rungwe District: Bundali Mts., Ngulugulu R., *Stolz* 1794!
DISTR. U 2, 4; T 1, 4, 6, 7; Zaire, Burundi, Malawi, Zambia, Mozambique and E. Zimbabwe
HAB. Evergreen forest and thicket including semi-swamp and riverine forest; 1140–1800 m.

SYN. *Randia msonju* K. Krause in E.J. 57: 30 (1920); T.T.C.L.: 527 (1949); I.T.U., ed. 2: 355 (1952)
   [*R. lucidula* sensu I.T.U., ed. 2: 355 (1952), *non* Hiern sensu stricto]
   [*Aidia micrantha* sensu F.F.N.R.: 398 (1962), *non* (K. Schum.) F. White sensu stricto]

NOTE. Four other varieties including the typical one extend the range of the species to Cameroun, Gabon, Central African Republic and Angola.

2. A. sp.

Scandent shrub; young shoots glabrous. Leaf-blades drying bronze-brown, oblong-elliptic, 3–12.5 cm. long, 1–4.2 cm. wide, sharply acuminate at the apex, cuneate at the base, rather dull to slightly glossy, glabrous with domatia present in some axils beneath; petioles 2–6 mm. long; stipules subulate, ± 3–4 mm. long. Inflorescences unilateral, apparently axillary, either leaf-opposed, the supporting leaf reduced to a slender subulate remnant 7–8 mm. long, or borne at nodes with a pair of leaves, sparsely branched, up to 3.5 cm. long; bracts triangular sheathing scales ± 1.5 mm. long, acute; youngest parts of inflorescence sparsely puberulous. Flowers known only as very young buds. Calyx narrowly turbinate, glabrous, shortly toothed. Corolla glabrous. Ovules 12–14 on fleshy placentas in a 2-locular ovary; style clavate. Fruit subglobose, wrinkled in dry state, 5–6 mm. in diameter; calyx-limb persistent, 1 × 2 mm.

TANZANIA. Lushoto District: near Kwamtili, 5 Dec. 1917, *Peter* 56098!; Tanga District: by R. Sigi below Longuza [Longusa], 18 Nov. 1917, *Peter* 56092!; Bagamoyo District: Zaraninge Plateau, Dec. 1964, *Procter* 2813!
DISTR. T 3, 6; not known elsewhere
HAB. Evergreen forest, sometimes riverine; 250–300 m.

NOTE. Mrs Bridson's suggestion that this is a species of *Aidia* close to the Sri Lankan *A. gardneri* (Thw.) Tirvengadum seems correct but the material is in extremely young bud and nothing can be done until more is available in flower and fruit. It differs from *A. micrantha* in the leaves which turn brown on drying, the laxer inflorescence and much smaller calyx-limb.

59. PORTERANDIA

Ridley in K.B. 1939: 593 (1940); Keay in B.J.B.B. 28: 23 (1958); Hallé, in Fl. Gabon 17, Rubiacées, 2: 118 (1970)

Randia L. sect. *Anisophyllea* Hook. f., G.P. 2: 89 (1873) & in Fl. Brit. India 3: 113 (1880)

Shrubs, small trees or lianes, mostly with pubescent stems and foliage. Leaves shortly petiolate, opposite, usually large; stipules ± oblong, obtuse or subacute, erect, slightly connate at the base. Flowers in several-many-flowered cymes, which although terminal actually appear unilaterally axillary; peduncles short; pedicels short, ± accrescent in fruit; bracts opposite, clasping at the base. Calyx-limb tubular with 5 rather short teeth. Corolla-tube shortly funnel-shaped, velvety outside, the narrow part of the tube closed by a ring of erect hairs; lobes 5, contorted, overlapping to the left, the covered right-hand part glabrous. Anthers ± sessile, inserted below the middle, included; pollen grains single. Ovary 2-locular; placentas 2, narrowly peltate, oblong-subcordate; ovules very numerous; style included, the stigma narrowly oblong, grooved. Fruit subglobose or ellipsoid, with persistent calyx. Seeds numerous, angular, reticulate.

A genus of 14 species, 9 occurring in Malaya and Indonesia and 5 species in tropical Africa. Robbrecht (pers. comm.) is of the opinion that the African species should perhaps be generically separated, thus requiring a new name (*Aoranthe* Somers).

P. penduliflora (*K. Schum.*) *Keay* in B.J.B.B. 28: 24 (1958). Type: Tanzania, E. Usambara Mts., Derema, *Volkens* 127 (B, holo.†, BM, K, iso.!)

Shrub or small slender tree 3–9(–?15) m. tall, with laxly branched stems; young branchlets

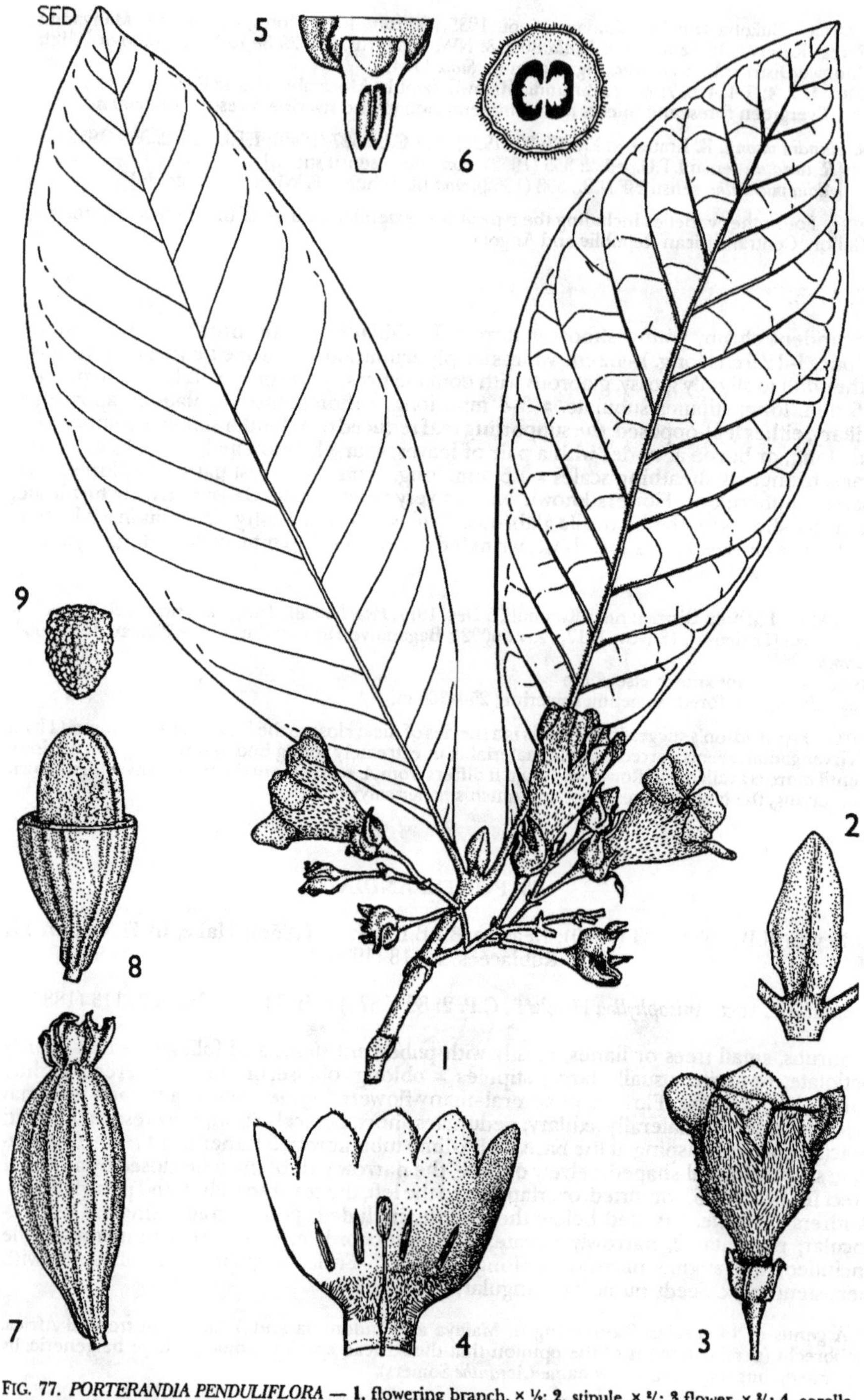

FIG. 77. *PORTERANDIA PENDULIFLORA* — **1**, flowering branch, × ⅓; **2**, stipule, × ⅔; **3** flower, × ⅔; **4**, corolla opened out with style and stigma, × ⅔; **5**, longitudinal section through ovary, × 2; **6**, transverse section through ovary, × 4; **7**, fruit, × ⅔; **8**, fruit with upper part of wall removed to show seed-mass, × ⅔; **9**, seed, × 6. **1**, from *Toms* 1; **2**, from *Bridson* s.n.; **3**, from *Verdcourt* 1752; **4–6**, from *Greenway* 7414; **7–9**, from *Rodgers et al.* 302. Drawn by Sally Dawson.

densely pubescent but later glabrous, pale and rugose. Leaf-blades elliptic or oblanceolate-elliptic, 11–30 cm. long, 5–16 cm. wide (or probably larger), acute at the apex, cuneate at the base, venation fine and reticulate beneath, pale and ± prominent in the dry state; midrib with fine bristly hairs above but otherwise glabrous; venation beneath bristly pubescent (see note for specimens with a much denser indumentum); petiole 0.8–2 cm. long; stipules elliptic, 1–1.2 cm. long, 4.5–6 mm. wide, pubescent, soon deciduous. Flowers pendulous, several in inflorescences up to about 10 cm. long; peduncles and secondary peduncles 0.8–1.7 cm. long; pedicels 0.7–1.7 cm. long; all parts ferruginous pubescent. Calyx-tube obconic, 7 mm. long, ferruginous pubescent; tubular part of limb 8 mm. long, pubescent outside, densely hairy inside; lobes ± triangular, (0.35–)0.6–1 cm. long, pubescent, the tube with winged ribs decurrent below the lobes. Corolla cream outside, sometimes tinged pink or red, deep red-crimson-brown inside, funnel-shaped above, constricted and tubular at the base, upper part 1.6–2.5 cm. long, silky pilose outside, basal part 1 cm. long, glabrous outside in the lower part; lobes rounded, 1 cm. long and wide, silky pilose save for glabrous overlapped part. Style, stigma and anthers ± yellow. Fruits ? orange, said to be as large as an egg; those seen are immature, ellipsoid to fusiform, 3–5 cm. long, 1.6–2 cm. wide, ± ribbed. Fig. 77.

TANZANIA. Lushoto District: Amani, 26 Oct. 1045, *Greenway* 7414! & Kwamkoro road, 29 Dec. 1956, *Verdcourt* 1752!; Uzaramo District: Pugu Forest Reserve, June 1954, *Semsei* 1744!
DISTR. T 3, 6, 8; not known elsewhere
HAB. Evergreen forest; 250–960 m.

SYN. *Randia penduliflora* K. Schum. in P.O.A. C: 380 (1895)
 Amaralia penduliflora (K. Schum.) Wernham in J.B. 55: 3(1917); T.T.C.L.: 482 (1949)

NOTE. The name *Randia sericantha* K. Schum. (N.B.G.B. 3: 84 (1900)) has never been properly published but has been used, eg. on *Warnecke* in Herb. Amani 394. The T 8 records are based on *Semsei* 662, a fruiting specimen from Lindi District, Rondo Plateau, Mchinjiri, Mar. 1952, and *Schlieben* 5544, from Mwera Plateau, Oct. 1934. These differ from typical material in having long hairs on both sides of the leaf-blades but particularly beneath, in fact the young leaves are quite densely hairy. Further material is required to assess the status of these plants, which are probably not even varietally distinct.

60. MACROSPHYRA

Hook. f. in G.P. 2: 86 (1873); Keay in B.J.B.B. 28: 27 (1958)

Climbing ± hairy shrubs or small erect shrubs. Leaves petiolate, opposite, mostly clustered at the apices of the shoots; stipules rather large, interpetiolar, ovate-lanceolate, thickly scarious, striate, imbricate with the leaves at the shoot-apices, at length deciduous. Flowers scented, 5–6-merous, shortly pedicellate, ♂, several–many in terminal sessile clusters. Calyx-tube ellipsoid or obconic; tubular part of limb short; lobes subulate or linear-lanceolate, erect, persistent. Corolla white, cream, yellow, pink or red, ± salver-shaped, with a narrow tube, or funnel-shaped, glabrescent to pubescent outside, glabrous inside save at base which is densely hairy or glabrous at base with rest of tube pubescent; lobes elliptic to rounded, shortly clawed, spreading, contorted. Anthers just exserted between the corolla-lobes or included; pollen in tetrads. Ovary 1-locular; ovules numerous on 2(–?4) parietal placentas; style long-exserted or included, thick; stigma large, globose, grooved. Fruit subglobose, many-seeded.

A genus of 3 species, one rather aberrant but probably best retained within the genus; all tropical African; a further species remains to be described.

Corolla salver-shaped, with narrow tube; style very long-
exserted . . . . 1. *M. longistyla*
Corolla funnel-shaped, tubular only at base; style included 2. *M. brachystylis*

1. M. longistyla (*DC.*) *Hiern* in F.T.A. 3: 106 (1877); Verdc. in K.B. 11: 452 (1957); Keay in B.J.B.B. 28: 28 (1958) & F.W.T.A., ed 2, 2: 116 (1963). Type: Gambia, *Leprieur & Perrottet* (G-DC, holo.!)

Climbing or ± erect shrub, 0.9–3.6 m. tall, with densely pubescent or velvety young stems. Leaf-blades obovate, ovate or elliptic, 3.5–20 cm. long, 2.8–12.5 cm. wide, acuminate at the apex, cuneate to subcordate at the base, densely woolly hairy on both

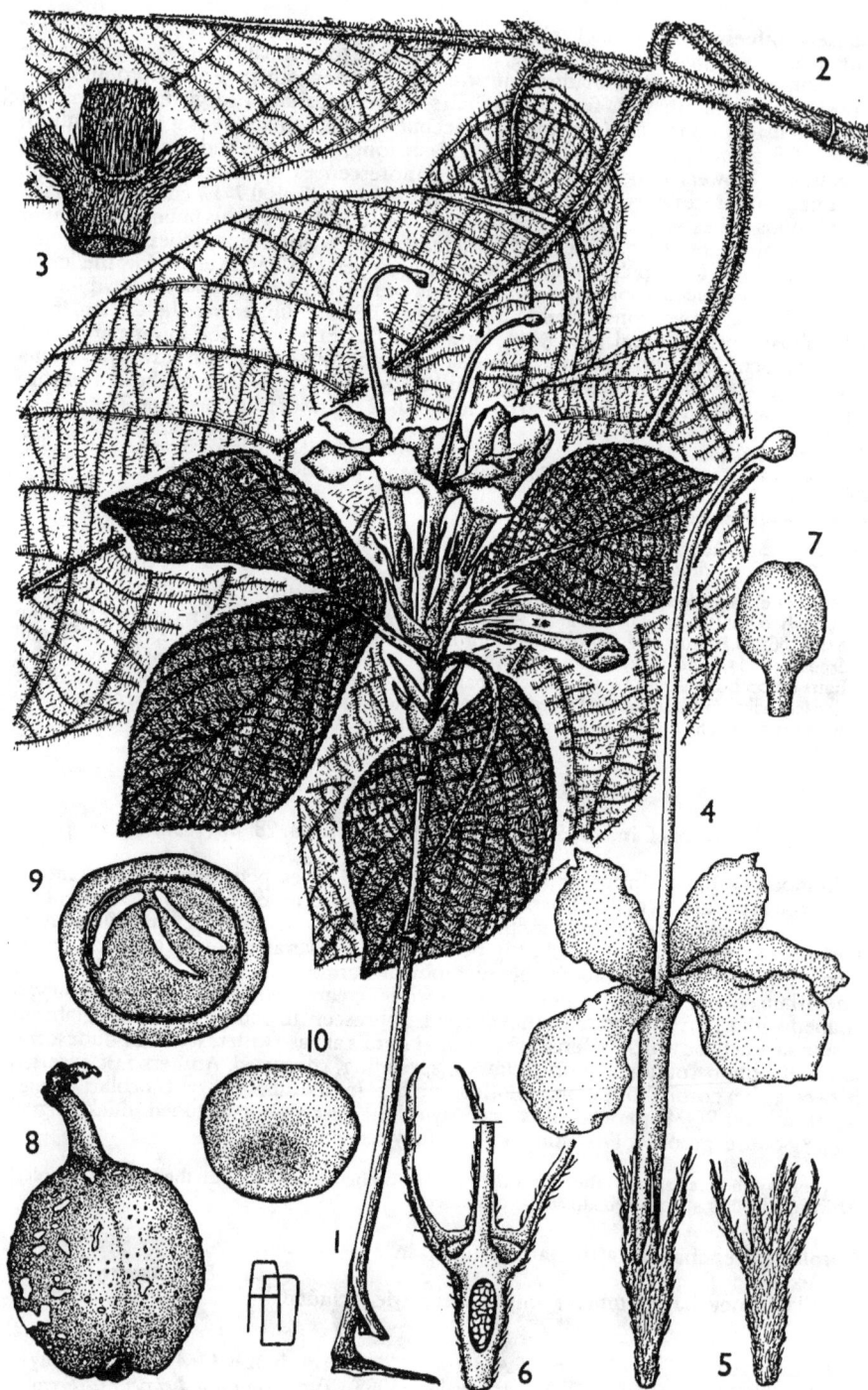

FIG. 78. *MACROSPHYRA LONGISTYLA* — 1, shoot with immature leaves and flowers, × ⅔; 2, shoot with mature leaves, × ⅔; 3, node with fallen stipule, × 4; 4, flower, × 2; 5, calyx, × 2; 6, calyx cut in half to show ovary, × 4; 7, stigma, × 4; 8, fruit, × 1; 9 transverse section of fruit, × 1; 10, seed, × 2. 1, 3–7, from *Dale* U860; 2, from *Onochie* in F.H.I. 22000; 8–10, from *Eggeling* 1658. Drawn by Ann Farrer.

sides but particularly beneath when young, later more sparsely hairy on both surfaces; petiole 0.8–7 cm. long; stipules oblong-lanceolate to oblong, 1–1.4 cm. long, 4–6 mm. wide, acute at the apex, ± adpressed setose outside above. Pedicels ± 5 mm. long, densely hairy. Calyx-tube hairy, narrowly obconic, 3–5 mm. long; lobes narrowly subulate-triangular, 3–8 mm. long, hairy. Corolla yellow or greenish yellow in bud and outside of tube, limb white in open flowers; tube 1.5–5.5 cm. long, 2–4 mm. wide, glabrescent to hairy outside; lobes 0.6–2.2 cm. long, 0.6–1.7 cm. wide. Anthers just exserted. Style and stigma purplish, the former 6–9.5(–12) cm. long, 1 mm. wide, the latter 3–5 mm. in diameter. Fruit subglobose, 3.5–4.5 cm. long and wide, with a slight median groove in dry state, slightly rugulose, lenticellate. Seeds rounded in outline, strongly compressed, ± 1 cm. in diameter, stuck together in a tight mass. Fig. 78.

UGANDA. Acholi District: Gulu area, Paicho, Awach, Feb. 1935, *Eggeling* 1658!; Teso District: Kyere, Feb. 1933, *Chandler* 1076!; Mubende District: 10 km. NW. of Katera, 16 Mar. 1969, *Lye* 2308!
DISTR. U 1–4; W. Africa from Senegal to Cameroun, Zaire, Sudan and Ethiopia
HAB. Rocky outcrops, forest edges, seasonally burnt grassland; 1050–1400 m.

SYN. *Randia longistyla* DC., Prodr. 4: 388 (1830)
    Gardenia paleacea A. Rich. in Mém. Soc. Hist. Nat. Paris 5: 241, 294 (1834); Hiern in F.T.A. 3: 105 (1877), in obs. Type: Senegal, *Leprieur* (P, holo.!)
    Oxyanthus villosus G. Don, Gen. Syst. 3: 494 (1834). Type: Sierra Leone, *Don* (BM, holo.)
    Gardenia longistyla (DC.) Hook. in Bot. Mag. 73, t. 4322 (1847)
    Macrosphyra paleacea (A. Rich.) K. Schum. in E. & P. Pf. IV. 4: 77 (1891), in obs.

2. M. brachystylis *Hiern* in Cat. Afr. Pl. Welw. 1: 463 (1898). Type: Angola, Golungo Alto, Ndelle, *Welwitsch* 3101 & by R. Delamboa, *Welwitsch* 3102 & Ambaca, between Izanga and Ngombe, *Welwitsch* 3104 (all LISU, syn., BM, K, isosyn.!) & Pungo Andongo, *Welwitsch* 3103 (LISU, syn., BM, isosyn.!)

Shrub or small tree 1.5–2.4 m. tall (*fide* Welwitsch), the ultimate branches ± climbing, or distinct liane up to 4.5 m. long; stems thick, pale, wrinkled, the bark somewhat peeling, at first pubescent but later glabrous; bud-scales (stipules) very sticky. Leaf-blades obovate or obovate-oblong to elliptic, 7.5–22.5 cm. long, 3.5–17 cm. wide, shortly acuminate at the apex, narrowly cuneate to obtuse or rounded or even sub-cordate at the base, often unequal, at first ± velvety pubescent on both surfaces, later shortly pubescent, particularly on the main venation beneath; petiole 1.5–7 cm. long, pubescent; stipules ovate-oblong to lanceolate, 1–1.7 cm. long, acute or shortly acuminate, pubescent. Inflorescences several-flowered; flowers 5–6-merous, fleshy; pedicels 4–5 mm. long, pubescent. Calyx-tube hairy, cylindrical-obconic, 0.7–1 cm. long; tubular part of limb 2 mm. long; lobes narrowly triangular-subulate, 4–6 mm. long. Corolla ± glabrous outside, purple outside, orange to yellow inside, spotted and streaked, or all yellow with red streaks; tube 2.5–3.5 cm. long, narrow at the base, 4.5 mm. wide, funnel-shaped above, 3.5 cm. wide; lobes rounded, 1.5–2 cm. long 1–2 cm. wide. Anthers included. Style included, 1.7–2.5 cm. long, swollen below, ribbed above; stigma ± 7–8 mm. diameter. Fruit ellipsoid, 9–14 cm. long, 2.5–4.3 cm. wide, glabrous, at first crowned with the remains of the calyx-limb. Seeds similar to last.

UGANDA. Toro District: Kibale Forest, 20 Sept. 1906, *Bagshawe* 1234!; Mengo District: near Entebbe, June 1934 & Mar. 1935, *Chandler* 1145! & Nagojje, Apr. 1916, *Dummer* 2808!
DISTR. U 2, 4; Zaire, Cabinda and Angola
HAB. Evergreen forest and forest edges; 1170–1350 m.

SYN. *Gardenia pomodora* S. Moore in J.B. 45: 264 (1907). Type: Uganda, Toro, *Bagshawe* 1234 (BM, syn.!) & Entebbe, *E. Brown* 363 (K, syn.!)

NOTE. The Uganda material seen mostly has much more narrowly cuneate leaf-bases than the Angolan material, but the amount seen is inadequate to comment on the possibility of there being two subspecies. Brown reports that it possesses a curious smell.

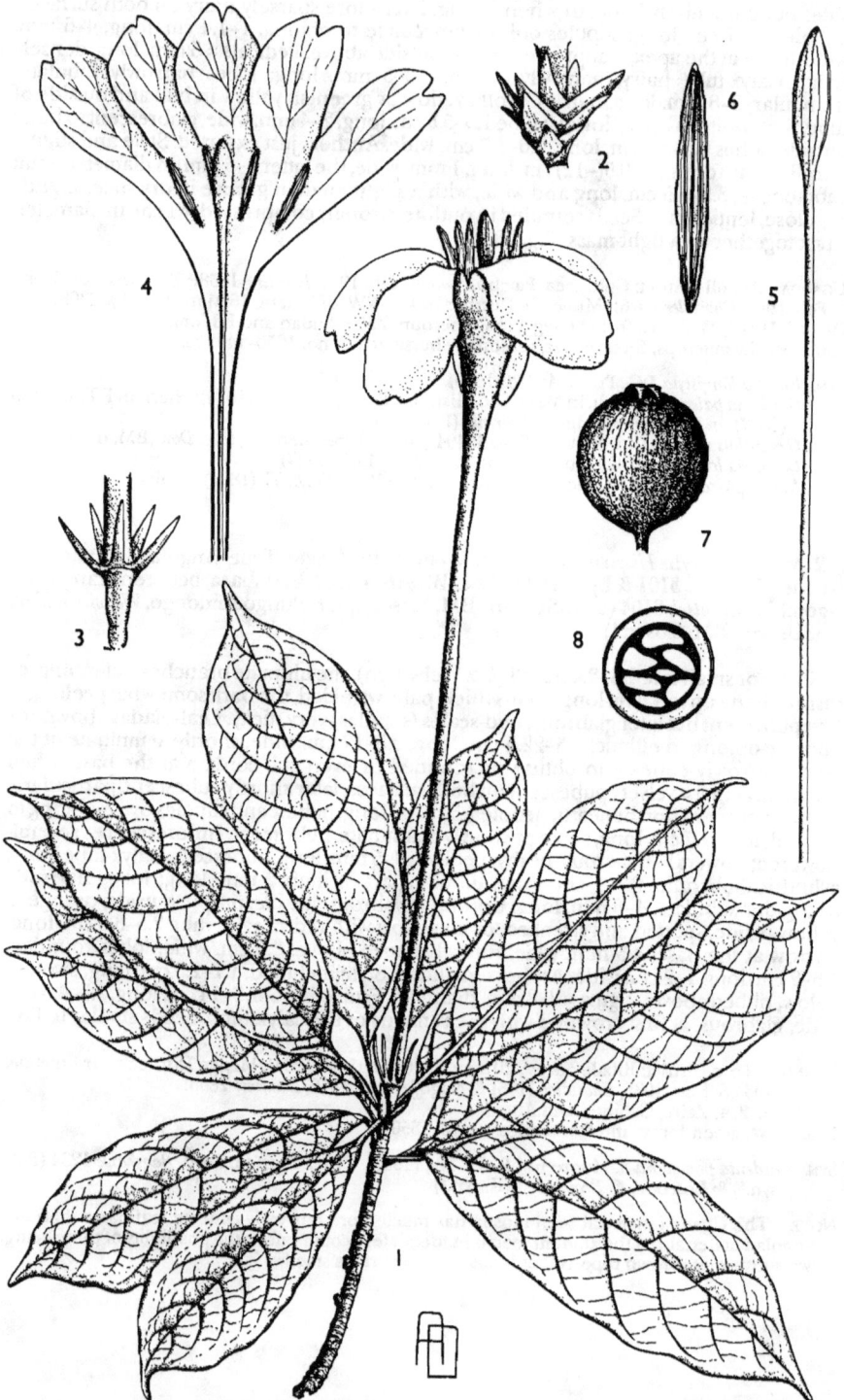

FIG. 79. *EUCLINIA LONGIFLORA* — 1, flowering branch, × ½; 2, stipules, × 1; 3, calyx, × 1; 4, upper part of corolla, opened out, × ½; 5, upper part of style with stigma, × 1; 6, stamen, × 1½; 7, fruit, × ½; 8, transverse section of fruit, × ½. 1, 4–6, from *Cult. Kew*; 2, from *Espirito Santo* 2159; 3, from *Vigne* 1446; 7, from *Deighton* 3099; 8, after illustration with *J.G. Adam* 21372. Drawn by Ann Farrer, with 2 & 3 by Sally Dawson.

61. EUCLINIA

Salisb., Parad. Lond., index sexualis et errata sub t. 93 (1808); Keay in B.J.B.B. 28: 39 (1958); Hallé in Fl. Gabon 17, Rubiacées, 2: 204 (1970)

Randia L. sect. *Euclinia* (Salisb.) DC., Prodr. 4: 388 (1830), pro parte; Hook.f., G.P. 2: 89 (1873), pro parte

Deciduous small shrubs or trees, the branches with very unequal internodes, some slender and elongated but the others at the apices very condensed; petiolar scars prominent. Leaves petiolate, tufted at the ends of the shoots, opposite, the blades thin, with pubescent domatia; stipules scarious, persistent. Flowers fragrant, solitary, terminal, 5-(6-8)-merous, surrounded at the base by scarious scales. Calyx-tube cylindrical-turbinate; free tubular part of limb very short; lobes elongate, persistent. Corolla turning black on drying; tube short to very long, cylindrical, the apical part narrowly to broadly funnel-shaped, tube pubescent towards the base; lobes contorted, overlapping to the left. Stamens sessile, the apices of the anthers just about reaching the throat mouth; pollen grains in tetrads, golden yellow. Ovary with 2 placentas forming 2 locules; ovules numerous; style swollen into an ellipsoid swelling at the level of the anthers with 2 closely adnate stigmatic lobes. Fruit globose or pyriform, crowned with the persistent calyx-lobes; peduncle accrescent. Seeds irregularly lenticular, compressed.

A genus of only 3 species confined to the forests of W. and Central Africa and Madagascar.

E. longiflora *Salisb.*, Parad. Lond., index sexualis et errata sub t. 93 (1808); F.W.T.A., ed. 2, 2: 121 (1963); Hallé in Fl. Gabon 17, Rubiacées, 2: 205, t. 47/8–14 (1970). Type: a specimen cultivated in Hort. Hibbert ex Sierra Leone material (BM-BANKS, holo.!)

Shrub or small tree 2.4–6 m. tall, the leafless lower parts of the branches with a peeling brown epidermis. Young leaves flushed reddish; leaf-blades oblong or oblong-obovate, 7–28 cm. long, 2.2–11 cm. wide, acuminate at the apex, cuneate at the base, glabrous above, pubescent beneath on main nerves, the midrib and in the nerve-axils; petiole 0.6–4 cm. long; stipules chaffy, narrowly ovate-triangular, 0.5–1.6 cm. long, 4 mm. wide, persistent. Pedicels up to 1.2 cm. long in fruit. Calyx-tube glabrous, 4–6 mm. long; tubular part of limb up to 1.5 mm. long; lobes linear- to triangular-lanceolate, 0.9–1.6 cm. long, 1.5–2.2 mm. wide, ± spreading. Corolla white turning cream; tube 16–24 cm. long, the main part narrowly cylindrical, the uppermost part widened out to form a funnel-shaped portion 2.5–6 cm. long, 2–6 cm. wide; lobes 5, elliptic-oblong, 1.5–5 cm. long, 1.2–3.5 cm. wide. Stigma 1.7–2 cm. long. Fruit subglobose or broadly ellipsoid, 3–3.5 cm. long, 2.8–3.2 cm. wide (Hallé gives 2.6–4 cm. diameter and a collector states 'size of a lime'). Fig. 79.

UGANDA. Toro District, without precise locality, 4 Oct. 1905, *Dawe* 542! & forest on hill N. of Mpokya, 15 July 1951, *Osmaston* 1051!
DISTR. U 2; Guinea Bissau to S. Nigeria, Cameroun and Angola, also in Sudan and Zaire; also at one time cultivated in England and Mauritius
HAB. Forest; ± 1200 m.

SYN. *Gardenia macrantha* Schultes, Syst. 5: 237 (1819). Type: as for *E. longiflora*
    Randia macrantha (Schultes) DC., Prodr. 4: 388 (1830); Hiern in F.T.A. 3: 97 (1877); F.W.T.A. 2: 78 (1931)
    Rothmannia macrantha (Schultes) Robyns, F.P.N.A. 2: 341 (1947); F.P.S. 2: 462 (1952)

NOTE. It is extraordinary that a plant with such conspicuous flowers should not have been collected more often in an area which has been worked over by so many excellent collectors.

62. CATUNAREGAM

Wolf, Gen. Pl.: 75 (1776)*

Ceriscus Gaertn., Fruct. & Sem. 1: [140], t. 28/4 (1788), *nom. inval.*
Ceriscus Nees in Flora 8: 116 (1825)
Xeromphis Raf., Sylva Tel.: 21 (1838)
Lachnosiphonium Hochst. in Flora 25: 236 (1842)
Lepipogon Bertol.f. in Mem. Accad. Sci. Ist. Bologna 4: 539, t. 21 (1853)
Randia L. sect. *Ceriscus* Hook.f. in G.P. 2: 88 (1873)

Mostly spiny shrubs or small trees, occasionally small pyrophytic subshrubs, the spines opposite or solitary and alternate. Leaves mostly clustered on short opposite axillary branchlets or cushion-like shoots; stipules interpetiolar, ovate-acuminate to triangular, apiculate, deciduous. Flowers 5-merous, rather small, 1–6 or more in terminal (sometimes appearing lateral) simple or branched cymes, each component ± 3-flowered, or 1–2(–3)-fasciculate or at least peduncle almost obsolete in species with very reduced cushion-shoots; bracts on the rachis absent or filiform; pedicels well developed. Calyx-tube ovoid or campanulate; limb-tube shortly cylindrical, ellipsoid or urceolate, the lobes short, mostly oblong, ovate or long and spathulate. Corolla subrotate, the tube shorter than the lobes, densely pubescent outside save for the base of the tube; tube with a distinct band of long hairs inside. Anthers exserted, oblong, apiculate; pollen grains single. Ovary 2(–3)-locular, the placentae attached to the septum; ovules numerous; style as long as tube or exserted; stigma ellipsoid or cylindrical, grooved or with 2 diverging transversely elliptic flattened lobes. Fruit with seed masses divided by the septum. Seeds discoid or compressed-ellipsoid with slightly raised reticulation on testa, embedded in a pulpy matrix.

A genus of 5–6 often very variable species occurring in Africa and Asia. *Xeromphis keniensis* Tennant has been removed to a separate genus *Tennantia* Verdc. (see p. 582).

1. Pyrophytic subshrub to 15 cm. tall . . . .   1. *C. pygmaea*
   Distinct shrub or small tree 2
2. Leaves glabrous, mostly small, 2–3(–7) cm. long, obtuse;
   flowers solitary or fasciculate; calyx glabrous; stems
   white or grey with alternate spines; fruits mostly under
   2 cm. long . . . .   2. *C. nilotica*
   Leaves pubescent, at least beneath on veins; calyx
   pubescent (in African material) 3
3. Leaves very densely velvety, spines almost always
   opposite . . . .   3. *C. spinosa*
                                                               subsp. *taylorii*
   Leaves not so densely velvety; spines alternate or opposite 4
4. Leaves mostly small, 2–3 cm. long, obtuse; flowers
   fasciculate, without common peduncles; fruit mostly
   under 2 cm. long . . . .   2. *C. nilotica*
                                                          (variants)
   Leaves mostly large, up to 13 × 5.8 cm., often acuminate;
   flowers often cymose, with common peduncles (in
   African material); fruit mostly larger, 2–3 cm. long   3. *C. spinosa*
                                                              subsp. *spinosa*

1. **C. pygmaea** *Vollesen* in Nord. Journ. Bot. 1: 735, fig. 1 (1981). Type: Tanzania, Kilwa District, Selous Game Reserve, Malemba, *Vollesen* in *M.R.C.* 4373 (C, holo.!, DSM, EA, iso.)

Pyrophytic subshrub with annual or biennial erect or ascending unarmed stems 5–10(–15) cm. tall from creeping woody subterranean stems; branches pubescent to tomentose, later ± glabrous. Leaves spaced on main stems but clustered on short lateral shoots, ovate, elliptic or obovate, 3–8.5 cm. long, (1–)1.5–3.5 cm. wide, acute, rounded or

* See Ross in Acta Bot. Neerl. 15: 156 (1966); Tirvengadum in Taxon 27: 513–517 (1979)

emarginate at the apex, narrowly attenuated at the base into the 4–10 mm. long petiole, pubescent-sericeous above, pubescent beneath, densely so on the nerves; stipules triangular, 4–7 mm. long, acuminate or cuspidate at the apex, persistent. Flowers solitary, terminal on the primary branches but appearing lateral due to sympodial growth of a branch from the axil of one of the uppermost leaves; pedicels ± 1 cm. long, densely pubescent. Calyx pubescent; tube narrowly campanulate, 3–4 mm. long; limb-tube 3–3.5 mm. long; lobes 10, unequal, the 5 large ones very variable, elliptic-spathulate, 2–4 mm. long, 1–1.5 mm. wide, acute, the 5 small ones triangular, 1 mm. long, 0.5 mm. wide, acute. Corolla white, turning yellow, densely silky pubescent outside; tube cylindrical, 5–6 mm. long; lobes rounded-elliptic, with wavy recurved margins and rounded incurved apex, 6–8 mm. long, 4–8 mm. wide. Style 6.5 mm. long; stigmatic lobes transversely elliptic, flattened, 1 mm. long, 1.5 mm. wide. Fruit ellipsoid, 1.3–1.8 cm. long, 1.1–1.5 cm. in diameter, with 10 distinct longitudinal nerves, reticulately veined between the nerves, pubescent, crowned by the persistent calyx.

TANZANIA. Kilwa District: Selous Game Reserve, Tundu Hills, Malemba, Nov. 1970, *Rodgers* in *M.R.C.* 1151! & 24 Jan. 1977, *Vollesen* in *M.R.C.* 4373 & Nahomba Valley, 9 Feb. 1978, *Vollesen* in *M.R.C.* 4913
DISTR. **T** 8; not known elsewhere
HAB. Woodland dominated by *Brachystegia spiciformis* and *Julbernardia globiflora* on red-grey sandy arid soils; 350–500 m.

2. C. nilotica *(Stapf)* Tirvengadum in Taxon 27: 515 (1979) & in Bull. Mus. Hist. Nat. Paris, Bot. 35: 21, t. 1/4 (1978).* Type: Ethiopia, Sennar, *Kotschy* 400 (K, holo.!)

Single or multi-stemmed shrub or small tree (1–)2–6 m. tall, little or much-branched, the branches often arching, nearly touching the ground; solitary alternate spines present, 1–3.5 cm. long, often dark at the tips and also short or cushion-like opposite spur-shoots; sometimes these spur-shoots are branched representing a much reduced lateral branched system; branchlets white or grey, glabrescent to sparsely or densely hairy at the tips; bark rough and wrinkled. Leaves clustered on the short shoots, obovate-spathulate to oblanceolate, 0.7–7 cm. long, 0.4–3 cm. wide, rounded at the apex, narrowed to the base into a petiole 1–10 mm. long, glabrous, glabrescent or densely pubescent; stipules rounded-ovate or triangular, 1–1.5 mm. long, soon falling. Flowers 1–2 at the ends of the short shoots, the peduncle obsolete; sometimes on spur-shoots which themselves bear very short branches, the flowers can appear terminal and axillary; pedicels 0.3–1.6 cm. long, glabrous. Calyx-tube cup-shaped, 1.5–2 mm. long, glabrous or puberulous; limb-tube cylindric-cup-shaped, longer and broader than the calyx-tube, 2.2–4 mm. long with well spaced mostly oblong or spathulate lobes 1–3.5 mm. long, glabrous or pubescent. Corolla sweet-scented, white or cream, turning yellow, adpressed silky or pubescent outside; tube short, 4–5 mm. long, glabrous at the base; lobes obovate-oblong to oblong-spathulate or round, 5–10 mm. long, 3.5–5 mm. wide. Stigma clavate, 2 mm. long, exserted. Fruit yellow-brown, ellipsoid to subglobose, 1.3–2.5 cm. long, 1–1.8 cm. wide, crowned with the calyx-limb, wrinkled and rugose, glabrous or slightly pubescent near the calyx. Seeds brown, compressed-ellipsoid, 4.5–5 mm. long, 3.5 mm. wide, 2 mm. thick, faintly rugulose.

UGANDA. W. Nile District: 4 Jan. 1906, *Dawe* 882! & West Madi, Amua, *Eggeling* 905!; Acholi District: Chua, Agoro, *Eggeling* 1708!
KENYA. Kwale District: between Kinango and Ngombani, 15 Dec. 1959, *Greenway* 9660!; Kilifi District: 32 km. W. of Malindi, 8 Dec. 1973, *Spjut* 3953!; Lamu District: Kiunga, 15 Dec. 1946, *J. Adamson* 287!
TANZANIA. Tanga District: Ngomeni, 2 Apr. 1968, *Faulkner* 4093!; Pangani District: Mfumoni, Madanga, 5 Mar. 1956, *Tanner* 2534!; Uzaramo District: Dar es Salaam University, 27 Dec. 1970, *Batty* 1175!
DISTR. **U** 1; **K** 4, 7; **T** 3, 6; Nigeria, Cameroun, Central African Republic, Somalia, Sudan
HAB. Riverine bushland, thicket edges, woodland, coastal bushland and scattered tree-grassland; 0–1200 m

SYN. *Randia nilotica* Stapf in J.L.S. 37: 519 (1906); T.T.C.L.: 526 (1949); Aubrév., Fl. For. Soud.-Guin.: 462, t. 100/5–6 (1950); I.T.U., ed. 2: 356 (1952)
Lachnosiphonium niloticum (Stapf) Dandy in F.P.S. 2: 441, fig. 159 (1952)
Xeromphis nilotica (Stapf) Keay in B.J.B.B. 28: 39 (1958); K.T.S.: 479 (1961); Keay in F.W.T.A., ed. 2: 121 (1963)

* Combination not legitimately made.

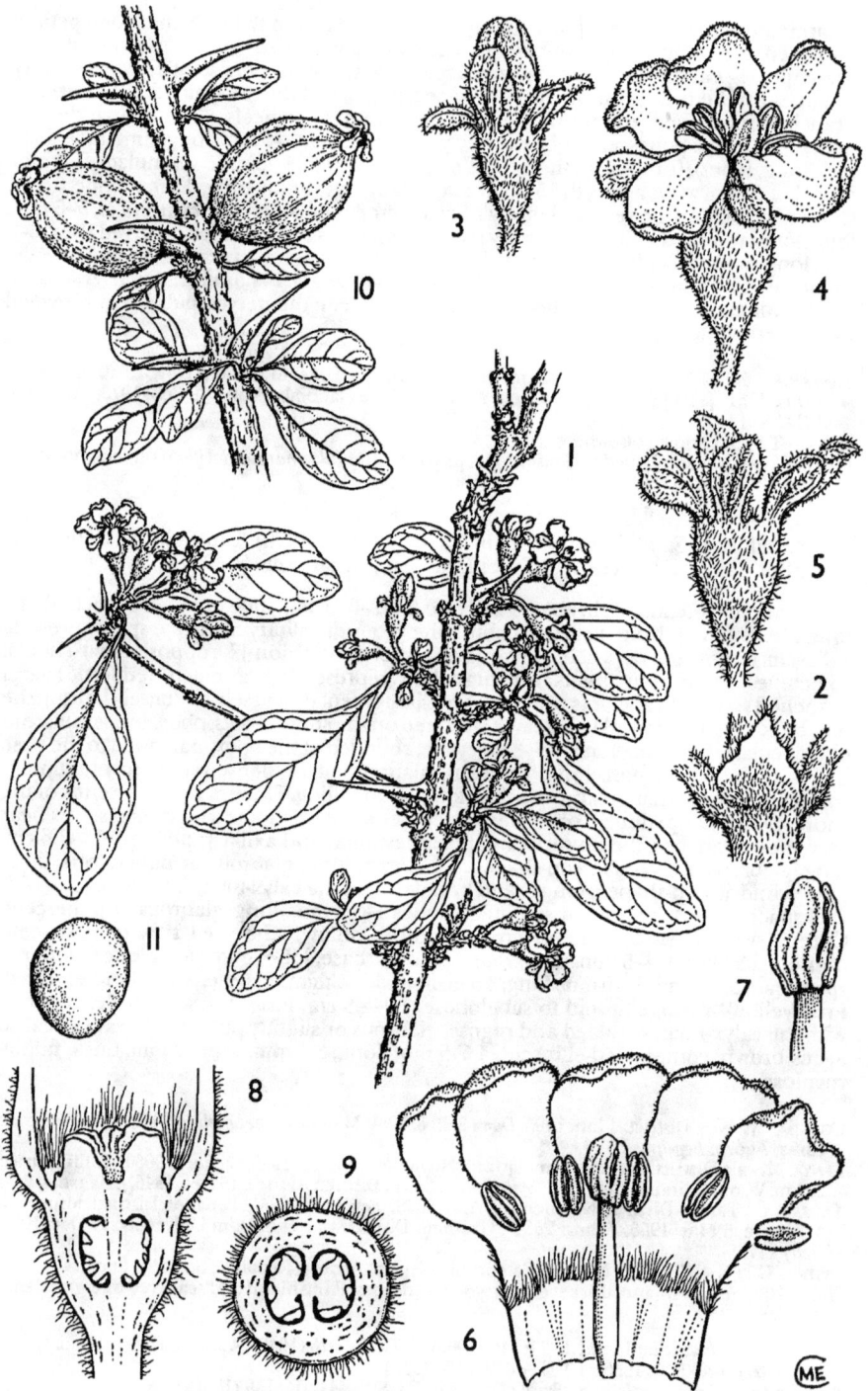

FIG. 80. *CATUNAREGAM SPINOSA* subsp. *TAYLORII* — 1, flowering branch, × ⅖; 2, stipules, × 6; 3, flower-bud, × 2; 4, flower, × 3; 5, calyx, × 3; 6, corolla opened out, × 4; 7, stigma, × 6; 8, longitudinal section of ovary, × 8; 9, transverse section of ovary, × 6; 10, fruiting branch, × ⅖; 11, seed, × 4. 1–8, from *Harris* 444; 9, from *Harris* 3818; 10, 11, from *Leippert* 5639. Drawn by Mrs M.E. Church.

NOTE. The variation in indumentum is considerable, the leaves and shoots varying from glabrous (typical form) to densely pubescent. Despite the differences between extremes this variation is too erratic and continuous for formal recognition of the hairy form to be worthwhile. The two certainly occur together in some areas, e.g. Kenya, Kitui. *Harris & Tadros* 5354 (Tanzania, Uzaramo District, 17 km. NNW. of Dar es Salaam, Kunduchi Saltworks, 7 Nov. 1970) is a hairy variant with the pedicels and calyx also pubescent, but it is not referable to *C. spinosa* subsp. *taylorii*. See general note at end of genus.

3. C. spinosa *(Thunb.) Tirvengadum* in Bull. Mus. Hist. Nat. Paris, Bot. 35: 13, pl. 1/3 (1978) & in Taxon 27: 515 (1979). Type: China, Macao, *Bladh* (BM, iso.!)

Several-stemmed, much-branched shrub or small tree 1.8–7.5 m. tall with spreading crown or drooping arcuate branches, very spiny, the spines either opposite and decussate or alternate, or sometimes practically to entirely spineless; stems slightly to densely velvety pubescent, eventually glabrescent, lenticellate; old branches with ± flaky grey-brown or brown bark; slash yellowish salmon. Leaves mostly clustered on spines or very short side branches which are often reduced to cushion-shoots, obovate to obovate-spathulate, 1.2–13 cm. long, 0.7–5.8 cm. wide, usually ± rounded at the apex, rarely acute, mostly very attenuate to the base, glabrous to densely velvety-woolly on both surfaces, sometimes discolorous and occasionally bullate; petiole 1–1.2 cm. long; stipules triangular, 5 mm. long, hairy, soon falling. Flowers terminal on short shoots, fasciculate or in short but distinct 1–2(–7)-flowered cymes or solitary; sometimes appearing falsely axillary and sometimes on leafless stems; peduncle up to 2 cm. long; pedicels 0.2–1.3 cm. long. Calyx-tube and limb-tube together 6.5–7.5 mm. long, glabrescent to densely hairy and often bristly; lobes very variable; mostly elliptic but tending to be more spathulate and larger in Asian material, 0.4–0.7(–1.3) cm. long, 0.3–0.5(–1.1) cm. wide, the margin sometimes recurved, but sometimes reduced to barely 1 mm. long. Corolla white or yellowish white, usually turning yellow, densely silky outside; tube 4–6 mm. long; lobes obovate, oblong-obovate or ± round, 0.6–1(–1.5) cm. long, 0.4–0.8(–1) cm. wide, tomentose at least near the margins inside. Fruits yellow-green to brown, oblong, ellipsoid or ± globose, 1.8–3.3 cm. long, 1.7–2.4(–3) cm. wide, wrinkled or slightly ribbed, glabrescent to yellow-brown tomentose. Seeds yellow-brown, compressed-ovoid, 5 mm. long, 4 mm. wide, 2.5–3 mm. thick, very finely reticulate.

subsp. **spinosa**

Leaf-blades very sparsely pubescent particularly on venation above and beneath or glabrescent or glabrous save for domatia. Spines present or absent, alternate or opposite. Lateral shoots up to 4 cm. long. Fruit up to 3 cm. long, 2.2 cm. wide (African specimens).

KENYA. Kwale District: Mrima Hill, 4 Sept. 1957, *Verdcourt* 1859! & 25 June 1970, *Faden* 70/245!; Tana River District: 3 km. N. of Wema, on E. side of R. Tana, 10 km. NE. of Garsen, 15 July 1972, *Gillett & Kibuwa* 19930!
TANZANIA. Morogoro District: Lusunguru Forest, 19 Sept. 1959, *Mgaza* 320!; Rungwe District: Masoko, 13 Jan. 1913, *Stolz* 1833!
DISTR. DISTR. **K** 7; **T** 6, 7; China to India and Sri Lanka, Mozambique, South Africa (Natal)
HAB. Lowland evergreen forest; 70–290 m.

SYN. *Gardenia spinosa* Thunb., Diss. Gard. No. 7 (1780); Willd., Sp. Pl. 1: 1229 (1798)
    G. spinosa L.f., Suppl. Pl.: 164 (1781). Type: India, Madras, *König* (LINN 297/10 & 11, syn., BM & LD ? iso.)
    G. dumetorum Retz., Obs. 2: 14 (1781); Willd., Sp. Pl. 1: 1229 (1798). Type: India, Madras, *König* (LD, holo., BM & LINN, ? iso.)
    Canthium coronatum Lam., Encycl. Méth. Bot. 1: 602 (1785). Types as for *Gardenia spinosa* L.f.
    C. chinense Pers., Syn. Pl. 1: 200 (1805). Type as for *Gardenia spinosa* Thunb.
    Randia dumetorum (Retz.) Poir. in Lam., Encycl. Méth. Bot., Suppl. 2: 829 (1811) & Tab. Encycl. Méth. Bot. 2: 227, t. 156/4 (1819); Hiern in F.T.A. 3: 94 (1877), pro parte
    R. spinosa (Thunb.) Blume, Bijdr.: 981 (1826)
    Xeromphis retzii Raf., Sylva Tel.: 21 (1838). Type as for *Gardenia dumetorum*
    Lachnosiphonium obovatum Hochst. in Flora 25: 238 (1842); Garcia in Mem. Junta Invest. Ultram., sér. Bot. 4: 28 (1958). Type: South Africa, Natal, Durban [Port Natal], *Krauss* 129 (K, iso.!)
    Randia lachnosiphonium Hochst. in Flora 25: 237 (1842), nom. provis.
    Lepipogon obovatum Bertol.f. in Mem. Accad. Sci. Ist. Bologna 4: 539, t. 21 (1853) & Illustrazione de piante Mozambicesi, Dissertazione 3: 7, t. 1 (1854); Hiern in F.T.A. 3: 247 (1877). Type: Mozambique, Inhambane, *Fornasini* (BOL, holo.†)
    Randia kraussii Harv., Thes. Cap.: 22, t. 33 (1859); Sond. in Fl. Cap. 3: 7 (1865). Type: South Africa, Natal, Durban [Port Natal], *Krauss* 129 (TCD, holo., K, iso.!)

R. monteiroae K. Schum. in E.J. 28: 63 (1899). Type: Mozambique, Maputo, Delagoa Bay, *Monteiro* 49 (B, holo.†, K, iso.!)

Xeromphis obovata (Hochst.) Keay in B.J.B.B. 28: 39 (1958); Palmer, Trees S. Afr. 3: 2044–5 (1973); Ross, Fl. Natal: 333 (1973); Palgrave, Trees S. Afr. 851 (1977)

NOTE. *Anderson* 781 (Lindi District, Nachingwea, 13 Oct. 1951) from *Brachystegia* woodland is said to be a prostrate woody perennial but Vollesen excludes it from his *C. pygmaea.*

subsp. **taylorii** *(S. Moore) Verdc.* in K.B. 36: 505 (1981). Type: Tanzania, between Zanzibar and Uyui, *W.E. Taylor* (BM, holo.!)

Leaf-blades velvety-woolly on both surfaces, discolorous. Usually very spiny with spines nearly always opposite and decussate and sometimes very closely placed. Cushion-shoots usually very short, 0.3–1(–2) cm. long. Venation very impressed above, sometimes bullate. Fruits 1.8–3.3 cm. long, 1.7–2.1 cm. wide, pubescent. Fig. 80.

TANZANIA. Mpanda District: Rukwa, Sonta, 5 Nov. 1963, *Richards* 18361!; Mpwapwa, 18 Oct. 1930, *Hornby* 323!; Morogoro District: Morogoro Fuel Reserve, Nov. 1954, *Semsei* 1875!; Lindi District: Rondo Plateau, Nahoro, 11 Dec. 1955, *Milne-Redhead & Taylor* 7615!
DISTR. T 1–8; Zaire, Mozambique, Malawi, Zambia, Zimbabwe
HAB. *Brachystegia* woodland, open bushland and scrub, grassland with scattered trees, sometimes on rocky ground; 100–1915 m.

SYN. *Randia vestita* S. Moore in J.B. 49: 150 (1911); T.T.C.L.: 526 (1949). Type: Zimbabwe, Harare, *Rand* 1395 (BM, holo.!)
   R. taylorii S. Moore in J.B. 49: 151 (1911); T.T.C.L.: 526 (1949)
   Lachnosiphonium vestitum (S. Moore) Garcia in Mem. Junta Invest. Ultram., sér. Bot., 4: 28 (1958)
   [*Xeromphis obovata* sensu F. White, F.F.N.R.: 425, fig. 70/D–G (1962); Verdc. & Trump, Common Poisonous Pl. E. Afr.: 147, fig. 12 (1969), *non* (Hochst.) Keay sensu stricto]

NOTE. Much used as a medicinal plant for snake bite and as a cure for gonorrhoea; also as a fish poison.

NOTE. (on species as a whole). Hiern included both the shrubby species dealt with above in *Randia dumetorum* (Retz.) Poir. and I must admit that he may ultimately prove to be correct; nevertheless there are, for practical purposes, two quite distinct entities in East Africa. Keay kept *obovata* apart from *spinosa* but many Asiatic specimens can be found which are virtually indistinguishable from African material and I do not think a case can be made for keeping them separate. Nevertheless the Asian material is far from uniform and may cover more than one taxon. Some Chinese specimens with glabrescent leaves and globose fruits are distinctly different from African material. Codd (Kirkia 1: 110 (1961)) combined *obovata* and *vestita* and certainly in South Africa the transition is gradual but in East Africa the division into two subspecies is much clearer. Some Asian *spinosa* comes close to *nilotica* in having glabrous leaves and alternate spines, but *nilotica* has a ± uniform facies, small obtuse leaves and small fruits. I have, however, seen two specimens which are somewhat intermediate, *Eggeling* 1708 (Uganda, Acholi district, Chua, Agoro) has brown stems, opposite spines, leaves pubescent with straight hairs, the midrib, margins and petiole spreading ciliate. It does not fit well with either of the taxa I am recognising.

63. GARDENIA

Ellis in Phil. Trans. Roy. Soc. 51(2): 935, t. 23 (1761); Keay in B.J.B.B. 28: 40 (1958); Hallé in Fl. Gabon 17, Rubiacées, 2: 218 (1970), *nom. conserv.*

Shrubs, small trees or occasionally small pyrophytic subshrubs, unarmed or lateral branches becoming stoutly spinescent; young parts often glutinous; branches sometimes in whorls of 3. Leaves opposite or verticillate, petiolate, thin to distinctly coriaceous, with or without domatia; stipules sheathing, often truncate; buds containing wax-secreting glands. Flowers sessile or shortly pedicellate, ⚥, solitary or in few-flowered fascicles, terminal or pseudo-axillary, usually white, mostly large and elegant, strongly perfumed. Calyx-tube variously shaped with the tubular part of the limb usually well-developed, truncate or with marginal or submarginal often decurrent lobes; limb sometimes unilaterally split. Corolla-tube funnel-shaped or cylindrical and then usually long and narrow; lobes 5–12, contorted, overlapping to the left in bud, obtuse or rounded at the apex. Anthers included or apices just exserted; pollen grains in tetrads. Ovary 1-locular, with 2–9 parietal placentas; style elongate, clavate at the apex, glabrous or pubescent; stigma exserted, bilobed or not, usually many-lamellate. fruit globose or ellipsoid, with a usually thick fibrous or woody wall. Seeds mostly very numerous, compressed, elliptic in outline, ± 3–10 mm. long, with thin smooth or slightly reticulate testa, stuck together into a solid mass with a pulp.

A large genus of the tropics of the Old World extending to New Caledonia; 10 species occur in the Flora area two of which belong to a group which has been a source of persistent difficulty during the past 150 years.

G. augusta (L.) Merr. (*G. jasminoides* Ellis, *G. florida* L.) the gardenia of horticulturists has been grown in Nairobi, Amani, Zanzibar (U.O.P.Z.: 272 (1949)) and probably elsewhere (Nairobi Arboretum, *Battiscombe* 942!; Tanzania, Lushoto District, Amani Nursery, 28 Oct. 1969, *Ngoundai* 407!; Zanzibar, Mbweni, Sir John Kirk's old garden, 8 Dec. 1930, *Greenway* 2677!). True *G. thunbergia* L. f. (*G. capensis* (Montin) Druce) is frequently cultivated in Nairobi (Nairobi Arboretum, 4 Feb. 1952, *G.R. Williams* 325! & Nairobi, Ngong Road, 12 Oct 1972, *Hansen* 723!). It is a native of South Africa and in the past *G. ternifolia* and *G. volkensii* have been much confused with it, but all three are unquestionably distinct species although very closely related.

1. Corolla very large, the tube narrowly funnel-shaped, (7–)8–24 cm. long, 3–5(–7.5) cm. wide at the throat, ± glandular-papillate outside; young foliage and buds very sticky . . . .   *11. G. imperialis*
   Corolla not so large or if as long then much narrower at the throat and glabrous to hairy outside; young foliage, etc. sometimes appearing shiny and often resin-coated but not so distinctly sticky 2
2. Desert shrub with small bullate densely pubescent leaves 1.2–4 cm. long, 0.5–1.7 cm. wide; corolla-tube 1.3–2 cm. long; fruit subglobose 8–10 mm. in diameter, hairy (**K** 1, 4, 7) . . . .   *3. G. fiorii*
   Not a desert shrub with above characteristics 3
3. Tertiary venation of leaf-blades very fine, ± parallel and at right-angles to the lateral nerves making a very characteristic pattern; young foliage covered with shiny resin; corolla small, campanulate, the tube 1.5 cm. long; fruit globose, 1–1.2 cm. in diameter . .   *1. G. transvenulosa*
   Tertiary venation not regular and characteristic as above; corolla larger, salver-shaped 4
4. Stemless subshrubs or under 30 cm. tall, usually flowering at ground level from a rosette of leaves . . . . .   *10. G. subacaulis*
   Shrubs or trees 5
5. Placentas 2–3 6
   Placentas 6–9 (rarely 3 in cultivated plant) 9
6. Cultivated plant; glabrous or young stem slightly pubescent; leaf-blades mostly under 10 cm. long; anthers almost completely exserted; corolla often double; placentas 2   *G. augusta*
   Wild plants; glabrous or densely hairy; only tips of the anthers exserted; placentas 2–3 7
7. Plant pubescent to densely hairy; leaves 1.5–10(–14) cm. long; corolla-tube 1.2–5 cm. long; placentas 2; fruits small, subglobose, 0.9–1.4 × 0.9–1.2 cm., densely pubescent   *2. G. resiniflua*
   Plant glabrous; leaves (2–)5–22(–30) cm. long; corolla tube 6–15 cm. long; placentas 3; fruits large, elongate, glabrous 8
8. Calyx-lobes 2.5–6.5 cm. long; fruit fusiform, 4–6 cm. long, 1.6–2.2 cm. wide, with 6 distinct corky ribs (E. Kenya and E. Tanzania) . . . .   *4. G. posoquerioides*
   Calyx-lobes 0.05–2 cm. long; fruit narrowly fusiform or cylindric, 9–12 cm. long, 1–1.5 cm. wide, irregularly striate but not ribbed or scarcely so (W. Uganda)   *5. G. vogelii*
9. Cultivated plant; calyx-tube together with the tubular part of the limb 3–4.5 cm. long, the tubular part usually over twice the length of the ovary and usually laterally split to form a spathe; calyx-lobes usually distinctly spathulate with narrow bases but sometimes only linear . . . .   *G. thunbergia*
   Wild plants; calyx-tube together with tubular part of the limb 1–4.5 cm. long but usually under 3 cm., the tubular part nearly always less than twice the length of the ovary 10

10. Leaf-blades distinctly scabrid-pubescent to velvety
      pubescent, particularly beneath 11
    Leaf-blades glabrous or only very slightly pubescent
      beneath at base 12
11. Leaf-blades crinkly at the margins and velvety or densely
      pubescent, 1.5–6(–18) cm. long, 1–5 cm. wide, but
      usually quite small; corolla-tube (1.2–)2.5–4 cm. long,
      densely adpressed hairy outside; fruit 2–4 cm. long,
      1.5–2.5 cm. wide, obscurely to distinctly and rather
      acutely ± 14-ribbed (U1) . . . .   6. G. aqualla
    Leaf-blades not crinkly at the margins, sparsely to fairly
      densely scabrid pubescent, 2–10 cm. long, 1.5–2.5 cm.
      wide; corolla-tube 3–7 cm. long, glabrous to hairy
      outside; fruit ± 5.5–7.5 cm. long, 3–5 cm. wide, not
      distinctly ribbed . . . .   9. G. ternifolia
                                                                            var. goetzei
12. Leaves drying a distinctive dull purple-grey above and
      paler and glaucous beneath, with the venation mostly
      dark reddish; calyx-lobes usually (but not always)
      subulate and drawn out to very fine points (U1)   7. G. erubescens
    Leaves drying green; calyx-lobes coarser and not finely
      subulate 13
13. Leaves slightly thinner in texture*, typically, when mature,
      rounded obovate-cuneate with rounded apex, typically
      ± 3.5 × 2.5 cm., with venation finer and less prominent
      but on some young shoots elongate lanceolate leaves
      8–18 cm. long, 1–5.4 cm. wide are also present; fruit in
      East Africa 7–10 × 6–10 cm., usually large, grey or
      whitish, usually with well-defined coarse ribs; seeds
      larger (4–)5.5–10 mm. long; bark usually pale grey and
      not breaking down into a powder; lenticels usually
      large and wart-like on fruit . . . .   8. G. volkensii
    Leaves thicker in texture, usually more elliptic or oblong or
      oblanceolate-obovate, 0.7–18 × 0.7–11 cm., with
      venation coarser and more prominent; young shoots
      not heterophyllous; fruit 4.5–7.5(–8) × 1.6–3.7(–5.6)
      cm., usually rather small, yellowish or reddish brown,
      not coarsely ribbed; seeds smaller, 3–4(–4.5) mm. long;
      bark often reddish brown and breaking down to a
      powdery surface; lenticels mostly smaller and obscure
      on fruit . . . .   9. G. ternifolia

1. G. transvenulosa Verdc. in K.B. 34: 347 (1979). Type: Kenya, Kilifi District, Sokoke
Forest, Musyoki & Hansen 997 (K, holo.!, C, EA, iso.)

Shrub or small tree 0.5–4(–9) m. tall, the young shoots shiny with secretion but
apparently not sticky, glabrous or with fine papilla-like indumentum; older stems grey,
ridged; bark smooth or rough. Leaves opposite; blades oblong-elliptic to elliptic, 4–12.5(–14)
cm. long, 2.4–8 cm. wide, shortly to distinctly acuminate at the apex, the actual tip obtuse,
gradually and then very abruptly cuneate at the base, thinly coriaceous, often pale
beneath, the younger ones covered with a shiny but apparently not sticky secretion,
glabrous save for fine papilla-like indumentum on and near the midrib and petiole above
and small hairy domatia in the axils beneath; lateral nerves rather close and prominent
with very numerous very close anastomosing veins at right-angles between them; petiole
0.5–1.8 cm. long, minutely papillate; stipules joined in a tube ± 6–8 mm. long, soon
splitting and becoming cup-like, minutely papillate. Flowers 5-merous, appearing solitary
or paired in the axils but actually terminating very abbreviated shoots; pedicels 2–3 mm.
long. Calyx covered with secretion, finely puberulous; tube subglobose or ovoid, 2–3 mm.
long; limb-tube 2–3 mm. long, angled by the decurrent lobes which are narrowly boat-shaped,

---

* It is possible to name these at a glance from foliage but the characters are subtle and difficult to
convey on paper; once appreciated no difficulty will be experienced. Field characters badly need
assessing since it is certain there are others not apparent from herbarium material.

7–10 mm. long, 1.8 mm. wide. Corolla white or greenish white, narrowly campanulate; tube 1.5 cm. long; lobes rounded, 5–6 mm. long and white. Fruit globose, 1–1.2 cm. in diameter, with slightly raised branching ribs, finely spreading puberulous, crowned by the persistent calyx-limb with the lobes accrescent to 1.5 cm.; fruiting pedicels 3–7 mm. long; placentas 2, nearly meeting. Seed-mass ± 8 mm. in diameter; seeds orange-brown, irregularly wedge-shaped with 2 large semicircular surfaces, a straight narrowly oblong surface and a curved margin, 2.5 mm. long, 1.8 mm. wide, 1 mm. thick, finely reticulate.

KENYA. Kilifi District: Sokoke Forest Station, 30 Jan. 1961, *Greenway* 9809! & 0.5 km. NE. of Sokoke Forest Station, along Kilifi-Vitengeni road, 24 Aug. 1971, *R.B. & A.J. Faden* 71/748! & Sokoke Forest, 12 Aug. 1936, *Moggridge* 135!

TANZANIA. Pangani District: Msubugwe Forest, 30 July 1936, *Tanner* 3034!; Uzaramo District: Pande Forest Reserve, 23 Apr. 1970, *Poćs & Harris in Harris* 4461!; Lindi District: Noto Plateau, Mtondoli, 9 Mar. 1935, *Schlieben* 6104!

DISTR. **K** 7; **T** 3, 6, 8; not known elsewhere

HAB. Dry lowland evergreen forest, woodland and bushland; 10–450(–700) m.

SYN. *G. sp.? resiniflua* sensu K.T.S.: 442 (1961)

NOTE. Closely allied to *G. succosa* Bak. from Madagascar. In Kenya apparently confined to the Kilifi–Sokoke forest area.

2. G. resiniflua *Hiern* in F.T.A. 3: 102 (1877); F.F.N.R.: 408 (1962); Drummond in Kirkia 10: 275 (1975). Types: Malawi, W. shore of L. Nyasa, Cape Maclear & Mozambique, near Tete and between Lupata and Tete, all collected by *Kirk* (K, syn.!)

Much-branched shrub or small tree 1.8–4.5 m. tall; young branchlets pubescent; bark grey, smooth, soon peeling to reveal a slightly rough glabrous surface yielding resinous secretion; wood hard; hairs on young leaves, calyces, etc. also with resinous secretion. Leaves usually ternate, mostly crowded on lateral shoots, deciduous, obovate, 1.5–10(–14) cm. long, 1–6.5(–9) cm. wide, abruptly acuminate at the apex, strongly narrowed to the base, the actual base cuneate, rounded or minutely cordate, sparsely to densely often rather scabrid-pubescent above, sparsely to densely pubescent or velvety beneath but even then with venation visible as a reticulation with reduced indumentum, often rugose above; petioles 0–1.5 mm. long; stipules forming an ovoid hairy tip 2–8 mm. long to shoots but soon splitting into ovoid parts. Flowers 5–6-merous, solitary or 1–3 in the upper axils of leafy shoots; pedicels 1–4 mm. long, pubescent. Calyx-tube ovoid, 2 mm. long, very densely spreading pubescent with pale hairs; limb-tube 1.5–2.5 mm. long, 5-winged by the decurrent lobes, pubescent; lobes leafy, oblong, 0.3–2.2 cm. long, 1–3.5 mm. wide, held radially to the tube, pubescent. Corolla white, sometimes becoming yellowish, strongly scented; tube narrow, 1.2–5 cm. long, 3.5–10 mm. wide, sparsely to densely pubescent outside; lobes oblong-obovate or spathulate, 0.7–4.5 cm. long, 0.4–1.2 cm. wide, sparsely pubescent outside. Fruit globose or ellipsoid, 0.9–1.4 cm. long, 0.9–1.4 cm. in diameter, obscurely ribbed, densely spreading pubescent, crowned with the persistent calyx. Seed-ball with seeds separated by pale matrix; seeds reddish brown, 3.5–4 mm. long, 3 mm. wide, reticulate.

subsp. **septentrionalis** *Verdc.* in K.B. 34: 348 (1979). Type. Tanzania, Iringa District, 15.5 km. on Trekimboga Track from Msembe [Msembi], *Greenway & Kanuri* 14784 (K, holo.!, EA, iso.!)

Calyx-lobes often longer, 0.8–2.2 cm. long, 3–3.5 mm. wide. Corolla-tube usually longer, (1.5–)2.1–5 cm. long, with larger lobes (1.5–)2.5–4.5 cm. long, (1–)1.2–1.6(–2) cm. wide; leaf-blades often more velvety beneath; fruit usually larger.

TANZANIA. Dodoma District: Manyoni area, near Ruwiri village, below Saranda scarp, 15 Dec. 1931, *B.D. Burtt* 3532!; Mpwapwa District: 32 km. along the Dodoma road from Mpwapwa, 12 Dec. 1935, *B.D. Burtt* 5430! & Mpwapwa, Dec. 1935, *Mr & Mrs Hornby* 571!; Iringa District: 96 km. on Iringa-Morogoro road, Nov. 1964, *Procter* 2685!

DISTR. **T** 5–8; not known elsewhere

HAB. *Commiphora-Acacia* thicket; mixed *Combretum* woodland; *Delonix-Adansonia* woodland, *Brachystegia* woodland, often on termite mounds or rocky hills; characteristic of grey hard pan soils; (175–)420–1260 m.

SYN. *G. sp. aff. resiniflua* Hiern; T.T.C.L.: 496 (1949)

NOTE. This has long been looked on as an undescribed species but it is at the most a rather weak subspecies of the southern plant. Despite the great differences between the very small-flowered populations and the very large-flowered ones, sometimes involving factors of 400%, there is a considerable overlap between the two and fairly large-flowered specimens do occasionally occur in Zimbabwe and small-flowered ones in Tanzania. The combination of characters mentioned usually serves to separate the two taxa; subsp. *resiniflua* occurs in Zaire (Shaba), Malawi, Mozambique, Zambia, Zimbabwe and Botswana.

3. G. fiorii *Chiov.*, Result. Sci. Miss. Stef.-Paoli, Coll. Bot.: 90 (1916); Cufod., E.P.A: 1001 (1965); Gillett in K.B. 21: 249 (1967). Types: Somalia, Golonle, *Paoli* 783 & between Audinle and Berdale, *Paoli* 976 & Iscia Baidoa, *Paoli* 1220 (all FT, syn.)

Shrub 2–4 m. tall with pale grey smooth bark; branches greyish white with short contorted branchlets, pubescent at apices but glabrous beneath. Leaves ± sessile often in whorls of 3; blades elliptic, obovate or oblanceolate, 1.2–4 cm. long, 0.5–1.7 cm. wide, ± acute at the apex, narrowly rounded, minutely subcordate or cuneate at the base, distinctly bullate, densely grey pubescent on both surfaces; stipules 3–4.5 mm. long, joined beneath to form a tube. Flowers solitary, sessile or subsessile at the apices of short branchlets which are occasionally reduced to mere nodules, appearing before the leaves or when the leaves are not fully expanded. Calyx-tube 1.5 mm. long, densely hairy; limb-tube ± 1 mm. long becoming 2–3 mm. long in fruit, pubescent; lobes linear to linear-spathulate, ± 4 mm. long, becoming 7–8 mm. long, 1.5–2 mm. wide in fruit, pubescent. Corolla strongly scented, cream; tube slender, 1.3–2.5 cm. long, widened at the throat, densely pubescent; lobes 5–6, elliptic-oblong, 0.7–1.6 cm. long, 4–8 mm. wide, glabrous save for the part exposed in bud which is pubescent, very obtuse. Placentas 2. Fruit subglobose, 9–10 mm. long, 8–9.5 mm. wide, puberulous and with long pubescence as well, thin-walled, glossy inside. Seeds brown, discoid or oblong, strongly compressed, 5 × 3.5–5 × 1 mm., strongly reticulate.

KENYA. Northern Frontier Province: Tanaland, Legumbisso, 28 Aug. 1945, *J. Adamson* 108! & Ramu to Mandera, 23 May 1952, *Gillett* 13304!; Tana River District: 1 km. S. of SKT 15 on Galole to Garissa road, 21 Dec. 1964, *Gillett* 16501! & Galole, Nov. 1964, *Makin* in *E.A.H.* 13049!
DISTR. **K** 1, 4, 7; Somalia, Ethiopia (Ogaden)
HAB. *Commiphora-Acacia* open scrub; 60–750 m.

SYN. *Randia fiorii* (Chiov.) Chiov., Fl. Somala 2: 234 (1932)

NOTE. The hard wood is used for spoons. Chiovenda's statement that the corolla is 5 mm. long is based on either material that does not belong or possibly buds (see K.B. 21: 249 (1967)).

4. G. posquerioides *S. Moore* in J.L.S. 40: 81 (1911); Verdc. & Greenway in K.B. 10: 602 (1956); K.T.S.: 442 (1961). Types: Zimbabwe, Chirinda Forest, *Swynnerton* 71 & 6504 (both BM, syn.!)

Glabrous shrub or small tree 1.8–6 m. tall with somewhat angular young branchlets; older stems covered with longitudinally fissured bark. Leaves rather glossy, opposite or ternate; blades oblong or ovate-oblong to narrowly oblong-obovate, (5–)10–22 cm. long, (2–)4.5–9.5 cm. wide, acuminate-cuspidate at the apex, cuneate at the base, rather thin, sometimes slightly undulate at the margin; petiole 0.5–1.3 cm. long; stipules ovate or ovate-triangular, 1–1.5 cm. long, obtuse, membranous, joined together on one side. Flowers terminal, solitary on short thick pedicels to 8 mm. long, 5–6-merous. Calyx-tube cylindric-fusiform, 1.5 cm. long, 4–5 mm. wide, longitudinally ribbed; tubular part of limb 1.2–2 cm. long; lobes linear-oblong, 2.5–6.5 cm. long, 1.5–5 mm. wide, bluntly acute. Corolla white; tube 9–12.5 cm. long, 2.5–5 mm. wide, enlarging to 1 cm. wide at the throat which is hairy within; lobes broadly oblong, 4–5.5(–7) cm. long, 1.3–2 cm. wide, obtuse, pubescent at the base inside. Anthers 2 cm. long, partly exserted. Ovary with 3 placentas; style pilose; stigma broadly club-shaped, 5–6 mm. long, sulcate, shortly lobed at the apex. Fruit fusiform, 4–6 cm. long, 1.6–2.2 cm. wide, with 6 distinct corky ribs, crowned with the persistent calyx. Seeds yellow-brown, compressed, 5–8 mm. long, 4–4.5 mm. wide, finely reticulate and shagreened.

KENYA. Kwale District: Shimba Hills, Mwele Mdogo, 25 Aug. 1953, *Drummond & Hemsley* 3875! & Lango ya [Longo] Mwagandi area, 23 Apr. 1968, *Magogo & Glover* 939!; Kilifi District: Rabai, Aug. 1937, *van Someren* 313!
TANZANIA. Morogoro District: Mkungwe Mt., 23 May 1933, *Schlieben* 3980! & same place, 5 July 1970, *Faden* in *Kabuye* 288! & Mtibwa Forest Reserve, Aug. 1952, *Semsei* 919!; Ulanga District: Magombera Forest Reserve, 7 Feb. 1977, *Vollesen* in *M.R.C.* 4435!
DISTR. **K** 7; **T** 6; Zimbabwe, also cultivated in Puerto Rico and Florida
HAB. Evergreen forest, *Brachystegia* woodland; 250–1000 m.

NOTE. The East African material sometimes has much wider calyx-lobes than material from Zimbabwe but is otherwise identical.

5. G. vogelii *Planch.* in Hook., Ic. Pl. 8, t. 782–3 (1848); Benth. in Hook., Niger Fl.: 381, t. 38 & 39 (1849); Hiern in F.T.A. 3: 103 (1877); F.P.S. 2: 437 (1952); Keay in F.W.T.A., ed. 2, 2: 123 (1963); Hallé in Fl. Gabon 17, Rubiacées, 2: 222, pl. 52/1–10 (1970). Type: S. Nigeria, Ibo country, *Vogel* 58 (K, holo.!, BM, iso.!)

Shrub or small bush 1–3(–5) m. tall with glabrous stems, the epidermis becoming scaly. Leaves almost always in whorls of 3, unequal in each whorl; blades elliptic to obovate-elliptic, oblong-elliptic or oblanceolate, (2–)5–20(–30) cm. long, 2–10 cm. wide, rounded to acute towards the apex which itself is acuminate, cuneate to rounded at the base, drying olive or brownish-olive, the margins sometimes undulate, or even slightly lobed near apex, glabrous; small domatia usually present, pilose; petioles 0.2–3 cm. long, glabrous; stipules connate into a tube 0.6–1.5 cm. long. Flowers terminal, sessile, held erect, sometimes solitary, sometimes 2–5 together supported by a kind of stipular involucre. Calyx-tube ± glabrous, ± 1 cm. long; tubular part of limb 1.2–3 cm. long, 5–7 mm. in diameter, sometimes splitting; lobes 6, linear to ovate, unequal, 0.05–2 cm. long, glabrous. Corolla white, the tube sometimes greenish towards the base; tube very narrowly cylindrical, 6–15 cm. long, 2–4 mm. wide, glabrous outside; lobes (5–)6(–7), oblong to narrowly oblong-lanceolate, 3.5–8.5 cm. long, 1–1.8 cm. wide, obtuse, spreading, glabrous; throat pilose. Anthers included or apical 2 mm. only exserted. Style about as long as tube; stigma green, very shortly expanded, ± exserted. Fruit cylindrical or narrowly fusiform, sometimes slightly curved, 9–12 cm. long, 1–1.5 cm. wide, irregularly striate, lenticellate, crowned by the persistent calyx-limb; pericarp fibrous. Seeds numerous, compressed, 4 mm. long, 3 mm. wide, 1 mm. thick.

UGANDA. Bunyoro District: Budongo, Feb. 1935, *Eggeling* 1636!; Masaka District: 6.4 km. SSW of Katera, Malabigambo Forest, 2 Oct. 1953, *Drummond & Hemsley* 4581!
DISTR. U 2, 4; Liberia, Ivory Coast, Ghana, Nigeria, Cameroun, Gabon, Zaire, Central African Republic, ? Sudan (see note) and Angola
HAB. Evergreen forest including swamp-forest; 1020–1150 m.

NOTE. A sheet from Zimbabwe (Melsetter) named as this species is almost certainly *G. posoquerioides*. I have not seen material from the Sudan.
    Pauwels (in B.S.B.B. 118: 110–111 (1985)) divided this species into two varieties, var *vogelii* and var. *seretii* (De Wild.) Pauwels, using calyx-limb characters (limb-tube unsplit, shortly to long-dentate for var. *vogelii* and limb-tube split, truncate to shortly dentate in var. *seretii*). Both varieties occur in Uganda, but the material available cannot be placed with certainty.

6. G. aqualla *Stapf & Hutch.* in J.L.S. 38: 427 (1909); Aubrév., Fl. For. Soud.-Guin.: 460, t. 99/1–2 (1950); F.P.S. 2: 437 (1952); Keay in F.W.T.A., ed. 2, 2: 123 (1963). Type: Sudan, Kuchuk Ali, *Schweinfurth* 1751 (K, lecto.!)

Branched shrub 1–3 m. tall, erect or branches sometimes subprostrate; youngest parts pubescent; older with scaly bark. Leaves elliptic, elliptic-oblong or sometimes oblanceolate, 1.5–6(–18) cm. long, 1–5 cm. wide, rounded to bluntly subacuminate at the apex, wrinkled at the margins, scabrid-pubescent above, mostly densely pubescent beneath; petioles 0–2 mm. long; stipules ovate-triangular, 4 mm. long, hairy. Flowers fragrant. Calyx-tube 4 mm. long, densely hairy; tubular part of limb (4–)6–8 mm. long, densely pubescent, divided into ± 6 linear to linear-oblong lobes, subulate and ± acute at the apex, 2–4 mm. long, pubescent. Corolla white or cream, turning yellow; tube cylindric, (1.2–)2.5–4 cm. long, slender, densely adpressed pubescent outside; lobes 6, elliptic, 2–2.5 cm. long, 1–1.5 cm. wide, sparsely pubescent outside. Fruit globose to ellipsoid, 2–4 cm. long, 1.5–2.5 cm. diameter, very obscurely to distinctly ± 14-ribbed, the epidermis often scaly.

UGANDA. Acholi District: Gulu, *S.*210! & Chua, near Pader, Mar. 1935, *Eggeling* 1757!
DISTR. U 1; Mali, Ghana, Upper Volta, Nigeria, Cameroun, Central African Republic, Sudan
HAB. 'Poor savannah'; ± 1000 m.

NOTE. Quite closely related to *G. ternifolia* but the small wrinkled hairy leaves and small usually strongly ribbed fruits will distinguish it.

7. G. erubescens *Stapf & Hutch.* in J.L.S. 38: 428 (1909); Aubrév., Fl. For. Soud. Guin.: 460–462, t. 99/3–4 (1950); F.P.S. 2: 437 (1952); Keay in F.W.T.A., ed. 2, 2: 123 (1963); E.P.A.: 1001 (1965). Type: N. Nigeria, Kontagora, *Dalziel* 224 (K, lecto.!)

Stout shrub or small tree 1.5–3(–6.5) m. tall with densely pubescent young branchlets; bark plane-like, smooth, pale grey; branching irregular. Leaves in whorls of 3; blades

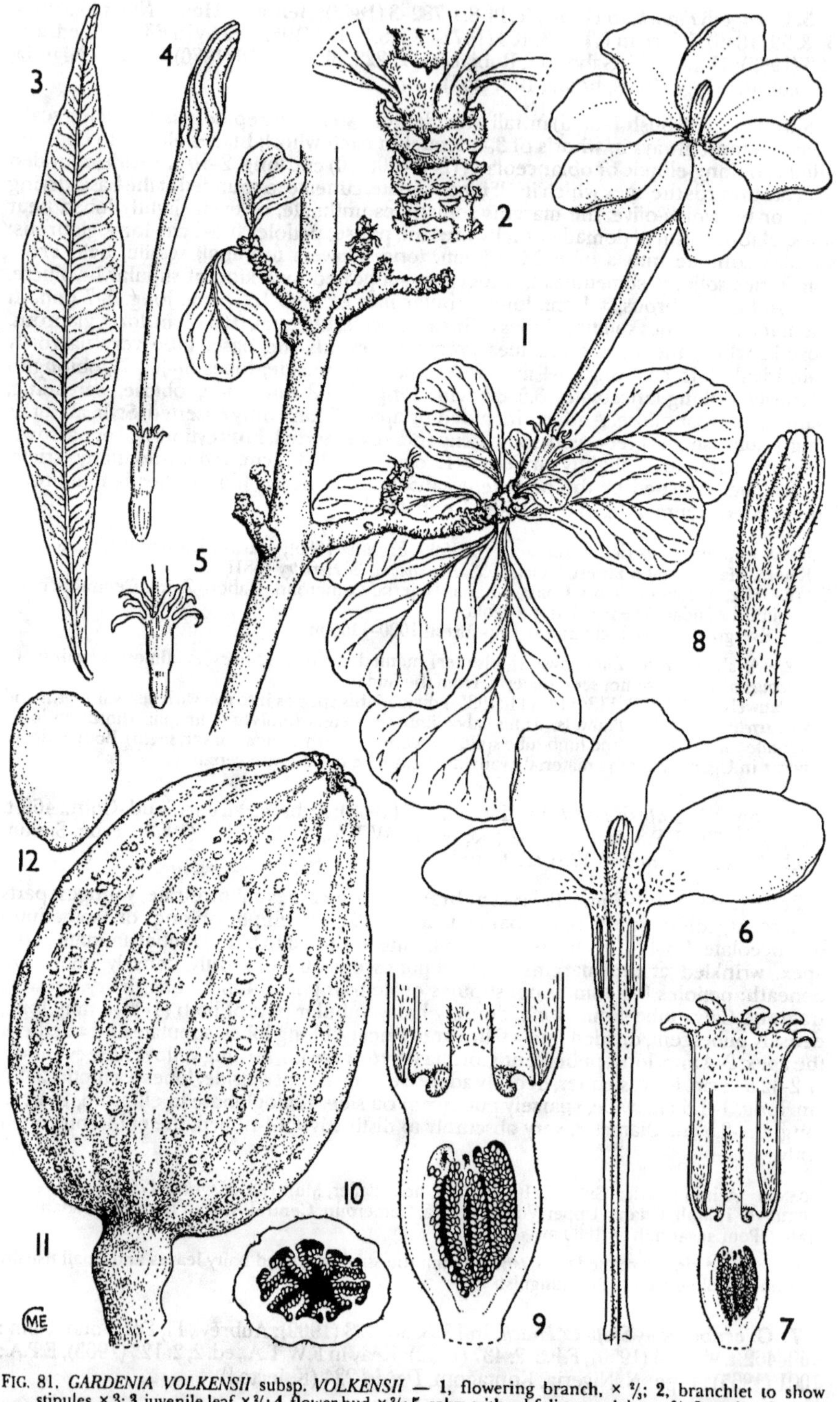

FIG. 81. *GARDENIA VOLKENSII* subsp. *VOLKENSII* — **1**, flowering branch, × ⅔; **2**, branchlet to show stipules, × 3; **3**, juvenile leaf, × ⅔; **4**, flower-bud, × ⅔; **5**, calyx, with subfoliaceous lobes, × ⅔; **6**, section through corolla, × ⅔; **7**, longitudinal section through calyx, showing ovary, × 2; **8**, stigma, × 2; **9**, longitudinal section through ovary, × 4; **10**, transverse section through ovary, × 4; **11**, fruit, × ⅔; **12**, seed, × 3. 1, 2, from *Glover et al.* 2011; 3, from *Bally* 12358; 4, 6–10, from *Glover et al.* 344; 11, from *Greenway & Kirrika* 10970; 12, from Zanzibar, collector unknown. Drawn by Mrs M.E. Church.

obovate to oblanceolate, (2.5–)4.5–20 cm. long, (1.2–)2–10 cm. wide, usually rounded at the apex, rarely acute, cuneate at the base, glabrous, nearly always drying purplish or purple-brown above, pallid and glaucous beneath, the venation mostly pale above in life, frequently drying dark reddish beneath; petiole up to 4 mm. long; stipules ± 3 mm. long, tomentose. Calyx-tube 4–8 mm. long, densely pubescent; tubular part of calyx-limb 0.6–1.2 cm. long, densely pubescent or finally nearly glabrous, pilose inside; lobes 6, usually filiform or at least ending in fine tips, 0.4–1.2 cm. long. Corolla cream turning yellowish, the tube mostly greenish white; tube 2.5–7.5 cm. long, sparsely to ± densely pubescent outside; lobes 6, elliptic or oblong-elliptic, (1.2–)2–4.7 cm. long, 0.7–1.7 cm. wide. Anthers well-included, up to 2 cm. long. Ovary with 6 placentas; style up to 7 cm. long; stigma-lobes yellow-green, 3–5 mm. long. Fruit grey to yellow, ellipsoid to narrowly oblong, 4.5–8 cm. long, 2.5–3 cm. wide, sometimes ridged when young.

UGANDA. W. Nile District: Aringa, Mt. Kei Forest Reserve, 26 Feb. 1955, *Dale* U861!
DISTR. U 1; Senegal to Nigeria, Central African Republic and Sudan
HAB. *Lophira, Combretum, Terminalia* wooded grassland; 1300 m.

SYN. *G. triacantha* DC. var. *parvilimbis* F. Williams in Bull. Herb. Boiss., sér. 2, 7: 378 (1907). Type: banks of the R. Gambia, *Whitfield* (BM, holo.!)

NOTE. This species is reported from Uganda (I.T.U., ed. 2: 343 (1952)) and three specimens cited. Two of these I have seen, *Thomas* 2238 and *Eggeling* 1807 are not correctly named although the latter shows some features of *G. erubescens*. They are better referred to *G. ternifolia*. Eggeling comments "found mixed with *G. jovis-tonantis* from which it is only distinguishable at close quarters; corolla-tube pubescent outside; corolla-lobes 5, 6 and 7 as opposed to 7, 8 and 9 in *G. jovis-tonantis*". Presumably he was comparing two populations of *G. ternifolia*.

8. **G. volkensii** *K. Schum.* in E.J. 34: 332 (1904); Stapf & Hutch. in J.L.S. 38: 422 (1909); T.T.C.L.: 497 (1949); K.T.S.: 442 (1961); E.P.A.: 1002 (1965). Type: Tanzania, "Sansibarküstengebiet", probably near Dar es Salaam, *Engler* (1902) 2199 (B, holo.†)

Tree or shrub 0.9–10 m. tall, with thick crown and occasionally arching branches which sometimes touch the ground; bark smooth, light grey, not breaking down to a powdery surface but sometimes flaking; branchlets pale, often greyish white, glabrous or pubescent or puberulous only at the apices. Leaves tufted at the ends of short spreading lateral shoots which are arranged ternately; blades obovate or obcuneate and usually small when mature, (0.8–)2.5–9.5 cm. long, (0.8–)1.7–5.5 cm. wide, mostly rounded at the apex, sometimes acute and occasionally crenate, distinctly cuneate at the base, thinner in texture than in next species and with the venation finer and less prominent, glabrous or finely puberulous to slightly scabridulous; some young branchlets (particularly in the Flora Area) are distinctly heterophyllous with some leaves of very different shape, lanceolate to rhomboid-lanceolate, 8–18 cm. long, 1–5.4 cm. wide; petiole 0–2(–3) mm. long; stipules semi-circular to ovate-triangular, 2–5 mm. long, pubescent and ciliate. Flowers solitary, sweet scented. Calyx-tube oblong, 0.6–1 cm. long, densely pubescent; tubular part of limb 0.5–1.4 cm. long, glabrous or pubescent outside; lobes 7–9(–10), extraordinarily variable, held at right-angles to tube and decurrent, varying from small linear teeth to spathulate lobes or distinctly leafy, 0.2–1.5(–2) cm. long, 0.5–8 mm. wide, occasionally 2 joined to form a bifid lobe, glabrous, or limb ± truncate. Corolla open at night, white or cream, turning yellow or orange, very variable in size; tube very narrowly cylindrical, usually more slender than in next species, (2.5–)5–12.5 cm. long, glabrous; lobes 6–9, elliptic to narrowly obovate, 2–5 cm. long, 1–3 cm. wide. Anthers 1.3–2 cm. long, inserted below the throat. Placentas 6–9. Style hairy; stigma yellow-green, 4–7 mm. long, 6–9-lobed, shortly exserted. Fruit very variable in size, usually whitish, grey or greyish orange, ellipsoid to subglobose, 4–11 cm. long, 2.7–10 cm. wide usually with a thick fibrous or woody wall 0.3–1 cm. thick (eventually becoming soft and pulpy, *fide* J.B. Gillett), usually with prominent wart-like lenticels, unribbed to very markedly coarsely 8–11-ribbed; apical and basal depressions in detached fruits wider than in next species, ± 1.5 cm.; stalk thick, up to 1.3 cm. wide. Seeds pale yellow-brown, more flattened than in next species, (4–)5–10 mm. long, finely reticulate-shagreened.

SYN. [*G. thunbergia* sensu Hiern in F.T.A. 3: 100 (1877), pro parte *non* L.f.]
[*G. jovis-tonantis* sensu Hiern, Cat. Afr. Pl. Welw. 1: 461 (1898), pro parte; Hallé in Fl. Gabon 17, Rubiacées, 2: 224 (1970), pro parte, *non* (Welw.) Hiern sensu stricto]

subsp. **volkensii**

Fruits usually large and white, (5–)7–10 cm. long, (5–)6–10 cm. wide, with 8–11 coarse ribs or occasionally only traces, very conspicuously lenticellate. Seeds larger and flatter than in *G. ternifolia*, (4–)5.5–10 mm. long, 5–7 mm. wide. Fig. 81.

UGANDA. Karamoja District: Morungaberu Pan, 20 July 1958, *Dyson-Hudson* 451! & Upe, Amudat, Nov. 1964, *Tweedie* 2930! & Nov. 1962, *Tweedie* 2501!

KENYA. Northern Frontier Province: NW. of Warges [Baraguess], Lugha Seya II, Mar. 1950, *Dale* K779!; Turkana District: Songot [Zingoute], May 1933, *Champion* T155!; Tana River District: Kurawa, 10 Oct. 1961, *Polhill & Paulo* 642!

TANZANIA. Musoma District: Ndabaka Plains, 19 Apr. 1962, *Greenway & Watson* 10615!; Mbulu District: between Ndala and Bagoyo Rivers, 5 Nov. 1963, *Greenway & Kirrika* 10970!; Nzega District: Wembere Plain, 1 Nov. 1960, *Richards* 13474!

DISTR. U 1; K 1–7; T 1–6; ? Z; Ethiopia (Omo valley), Somalia (S.), Mozambique, Zimbabwe, South Africa (Natal, Transvaal)

HAB. Grassland with scattered trees, thicket, dry thornbush and dry woodland especially in rocky places, along dry watercourses and also sandy places near the sea; 0–1950 m.

SYN. *G. somalensis* Chiov., Result. Sci. Miss. Stef.-Paoli, Coll. Bot.: 92, 213 (1916) & Fl. Somala 2: 238 (1932). Types: Somalia (S.), Urufle, *Paoli* 438 & Goriei to El Magu, *Paoli* 632 (both FT, syn.!)
   G. somalensis Chiov. var. *tubicalyx* Chiov., Result. Sci. Miss. Stef.-Paoli, Coll. Bot.: 93 (1916). Type: Somalia (S.), Hidlile, *Paoli* 666 (FT, holo.!)
   [*G. spatulifolia* sensu Codd, Trees & Shrubs Kruger Nat. Park: 173, t. 6 & figs. 159, 160 (1951); Marais in Fl. Pl. Afr. 32, t. 1241; Coates Palgrave, Trees Centr. Afr.: 389, figs. (1957); F.F.N.R.: 408 (1962); Letty, Wild Flowers Transvaal: 321, t. 161 (1962); Breitenb., Indig. Trees Southern Africa 5: 116, fig. (1965); De Winter et al., Sixty-six Transvaal Trees: 154 (1966); Palmer & Pitman, Trees S. Afr. 3: 2051–3, figs. (1972); van Wyk, Trees Kruger Nat. Park 2: 565, pl. 691 (1973); Ross, Fl. Natal: 333 (1973), pro parte; Drummond in Kirkia 10: 275 (1975), *non* Stapf & Hutch.]

G. volkensii K. Schum. var. *somalensis* (Chiov.) Cufod., E.P.A.: 1002 (1965)

NOTE. In southern Africa there has never been any confusion between *G. volkensii* and *G. ternifolia* subsp. *jovis-tonantis* and in general it has been recognised that *G. asperula* is only a minor variant of the latter; in E. Africa, however, the failure to realise that *G. volkensii* is widespread has led to great confusion. Although I have distinguished the large-fruited, large-seeded ribbed-fruited taxon (subsp. *volkensii*) from the small-fruited, small-seeded, unribbed fruited taxon (subsp. *spatulifolia*) the amount of fruiting material available from throughout the range of the species is scarcely adequate to justify the decision but the vast difference between the extremes is some justification for such a treatment. Var. *saundersiae* (N.E. Br.) Verdc. from S. Mozambique and South Africa (Natal) has exceptionally large calyx-lobes.

9. G. ternifolia *Schumach. & Thonn.*, Beskr. Guin. Pl.: 147 (1827); Stapf & Hutch. in J.L.S. 38: 425 (1909); Aubrév., Fl. For. Soud.-Guin.: 460, t. 99/5–9 (1950); Keay, in F.W.T.A., ed. 2, 2: 123 (1963). Type: Ghana, Accra [Gah] and Adampi, *Thonning* 140 (C, holo., FT, G-DC, iso.)

Shrub or small tree 1–6 m. tall, often stunted in appearance; bole up to 20 cm. wide and crown often broad, either entirely glabrous (save for style and domatia) or with stems, leaves and flowers variously pubescent or scabrid, sometimes only youngest parts of stem and upper stipules pubescent; bark smooth to rather rough, yellow-green or grey sometimes breaking down on the shoots to form a yellowish or reddish powder; branches often short and thick, arranged ternately. Leaves extremely variable, arranged ternately at the ends of short ternate shoots; blades oblanceolate to obovate; elliptic or oblong-obovate, (0.7–)4–18 cm. long, (0.7–)2–11 cm. wide, usually ± rounded at the apex, less often acute, very gradually elongate-cuneate at the base sometimes for over half the length or sometimes rounded, never of the distinct short rounded obovate-cuneate shape of the last species and distinctly thicker than in that species and with coarser more prominent venation, glabrous or sparsely to quite densely shortly scabrid-pubescent; hairy domatia usually present; petiole 0–5 mm. long; stipules broadly ovate, 2–4 mm. long, glabrous or pubescent, those at the nodes bearing flowers in the form of a cup at the base of the calyx-tube. Calyx-tube 0.4–1 cm. long, glabrous to rather scabrid-pubescent; tubular part of calyx 0.6–1.5 cm. long, ± truncate or with 6–9(–12) lobes which are very variable, linear, narrowly-oblong or elliptic to distinctly spathulate, 0.2–1.5 cm. long, 0.5–4 mm. wide, acute or obtuse; occasionally 2 joined together for half their length. Flowers strongly scented; corolla white at first, later bright then dull yellow; tube cylindric, (2.5–)4.5–11 cm. long, 1–2.5 cm. wide, glabrous to adpressed pilose but rarely densely hairy; lobes 6–9, elliptic to oblong-elliptic, 2–5.5 cm. long, 1–2.5 cm. wide, mostly glabrous. Anthers ± 1.5 cm. long, included in the tube. Stigma greenish yellow, fluted, 5–9-lobed, 5 mm.

long, shortly exserted. Fruit yellowish to reddish brown, narrowly to broadly oblong-ellipsoid or ellipsoid, less often subglobose or narrowly fusiform, 3.5–7.5(–8) cm. long, 1.6–3.3(–5.6) cm. wide, with wall fibrous or woody, 0.5–1 cm. thick, finely striate, and ± obscurely lenticellate but essentially smooth and without coarse ribs (although some ribbing has been indicated for young fruits in collectors' drawings). Seeds chestnut-brown, flattened ellipsoid, 3.5–4(–4.5) mm. long, 2–3 mm. wide, finely reticulate-shagreened.

subsp. ternifolia

Plant, even the youngest shoots, glabrous save for style and domatia. Flowers usually 6(–7)-merous.

UGANDA. W. Nile District: Madi Woods, 7 Feb. 1863, *Grant* 762!; Acholi District: Naam, Apr. 1943, *Purseglove* 1515!

DISTR. U 1; W. Africa from Senegal to Cameroun, Central African Republic, Sudan, Ethiopia

HAB. Grassland with scattered trees; 960 m.

SYN. [*G. thunbergia* sensu Hiern in F.T.A. 3: 100 (1877), pro parte *non* L.f.]
[*G. jovis-tonantis* sensu I.T.U., ed. 2: 344 (1952), pro parte *non* (Welw.) Hiern]

NOTE. It is with great doubt that I have referred these two rather fragmentary specimens to typical *G. ternifolia;* the calyx is certainly glabrous in both and the young shoots but the very young stipules have a few hairs in the *Purseglove* specimen.

subsp. jovis-tonantis (*Welw.*) *Verdc.* in K.B. 34: 354 (1979); Fl. Pl. Lign. Rwanda: 562, fig. 188/2 (1982); Fl. Rwanda 3: 164 (1985). Type: Angola, Golungo Alto, Serra de Alto Queta, *Welwitsch* 2573 (LISU, lecto., BM, K, isolecto.!)

Plant with at least youngest parts of stems sparsely to densely strigose-pubescent, also the stipules; calyx usually and corolla-tube less commonly pubescent. Flowers 6–9-merous. Fig. 1/3, 13.

var. jovis-tonantis (*Welw.*) *Aubrév.*, Fl. For. Soud.-Guin.: 460 (1950); Verdc. in K.B. 34: 354 (1979); Fl. Pl. Lign. Rwanda: 562 (1982); Fl. Rwanda 3: 164 (1985).

Leaf-blades entirely glabrous or with only a few sparse papilla-like hairs beneath near base and on main nerves.

UGANDA. Ankole District: Nyabushozi, 16 Sept. 1941, *A.S. Thomas* 3993!; Teso District: Serere, Mar. 1932, *Chandler* 529!; Mengo District: Maddu, Gumba, Mar. 1932, *Eggeling* 317!

KENYA. W. Suk District: Kapenguria, Mar. 1933, *Champion* T 189!; Trans-Nzoia District: Kitale, Milimani, Dec. 1969, *Tweedie* 3739!; Kwale District: Shimba Hills National Reserve, Buffalo Ridge, 22 Nov. 1971, *Bally & Smith* 14346!

TANZANIA. Tanga District: Kivindani, 28 Dec. 1958, *Faulkner* 2217!; Tabora District: Itigi–Chunya road, Rungwa, 6 Sept. 1962, *Boaler* 669!; Morogoro Fuel Reserve, Nov. 1954, *Semsei* 1864!; Songea District: Chandamara Hill, 16 Feb. 1956, *Milne-Redhead & Taylor* 8809!

DISTR. U 1–4; K 2–5, 7; T 1–4, 6–8; Nigeria, Cameroun, Cabinda, Zaire, Sudan, Ethiopia, Rwanda, Burundi, Mozambique, Malawi, Zambia and Angola

HAB. Grassland, moist grassland with scattered trees, bushland, *Brachystegia* and *Acacia* woodland; 0–2100 m.

SYN. *G. lutea* Fresen., Mus. Senckenb. 2: 167 (1837); Stapf & Hutch. in J.L.S. 38: 425 (1909); T.T.C.L.: 496 (1949); F.P.S. 2: 438, fig. 157 (1952); K.T.S.: 441 (1961); Keay in F.W.T.A., ed. 2, 2: 123 (1963); E.P.A.: 1002 (1965). Type: Ethiopia, '2 days N. of Gondar', *Rueppell* (FR, holo.)
Decameria jovis-tonantis Welw., Apont.: 579, nota 12 (1859)
Gardenia jovis-tonantis (Welw.) Hiern in F.T.A. 3: 101 (1877) & Cat. Afr. Pl. Welw. 1: 461 (1898), pro parte; Stapf & Hutch. in J.L.S. 38: 421 (1909); F.P.N.A. 2: 342 (1947); T.T.C.L.: 496 (1949); I.T.U., ed. 2: 344, fig. 71 (1952); F.P.S. 2: 436 (1952); K.T.S.: 441, fig. 84 (1961); F.F.N.R.: 408 (1962); Gomes e Sousa, Dendrol. Moçamb. 2: 675, t. 224 (1967); Hallé in Fl. Gabon 17, Rubiacées, 2: 224 (1970), pro parte; Nogueira in Bol. Soc. Brot. 49: 117, t. 1 (1975)
[*G. thunbergia* sensu Hiern in F.T.A. 3: 100 (1877), pro parte; P.O.A. C: 381 (1895); V.E. 1(1), fig. 233 (1910), *non* L. f.]

var. goetzei (*Stapf & Hutch.*) *Verdc.* in K.B. 34: 355 (1979); Fl. Pl. Lign. Rwanda: 562 fig. 188/2 (1982); Fl. Rwanda 3: 164, fig. 50/2 (1985). Type: Tanzania, Kissaki Steppe, *Goetze* 44 (K, holo.!)

Leaf-blades rather densely often ± scabridly pubescent, particularly beneath.

KENYA. Kitui, 18 Jan. 1942, *Bally* 1553! & *Gardner* in F.D. 3613!; Masai District: Chyulu Hills, Aug. 1931, *Gibbons* 2573!

TANZANIA. Mbulu District: Rift Wall Summit, 3 Jan. 1927, *B.D. Burtt* 1714!; Arusha District: near Usa R., 24 Oct. 1959, *Greenway* 9584!; Ufipa District: Muse to Sumbawanga road, 10 Nov. 1963, *Richards* 18395!

DISTR. **K** 4, 6; **T** 1, 2, 4–8; West Africa from Senegal to Nigeria, Mozambique, Malawi, Zambia, Zimbabwe, South Africa (Transvaal)

HAB. *Brachystegia* and mixed *Acacia* woodland; 250–2100 m.

SYN. *G. triacantha* DC., Prodr. 4: 382 (1830); Stapf & Hutch. in J.L.S. 38: 426 (1909); Aubrév., Fl. For. Soud.-Guin.: 460–1, t. 99/11–12 (1950); F.P.S. 2: 437 (1952); Keay in F.W.T.A., ed. 2, 2: 123 (1963). Type: Gambia, *Leprieur & Perrottet* (G, holo.)

[*G. thunbergia* sensu Hiern in F.T.A. 3: 100 (1877), pro parte *non* L.f.]

Randia torulosa K. Krause in E.J. 39: 529 (1907); T.T.C.L.: 527 (1949); Keay in B.J.B.B. 28: 70 (1958). Type: Tanzania, Morogoro District, near Liwale R., *Busse 561* (B, holo. †, EA, iso.!)

Gardenia asperula Stapf & Hutch. in J.L.S. 38: 423 (1909); Coates Palgrave, Trees Centr. Afr.: 385, figs. (1957). Type: Malawi, Lake Shirwa, *Meller* (K, lecto.!)

G. goetzei Stapf & Hutch. in J.L.S. 38: 427 (1909); T.T.C.L.: 496 (1949)

[*G. aqualla* sensu Chiov., Racc. Bot. Miss. Consol. Kenya: 53 (1935), *non* Stapf & Hutch.]

[*G. lutea* sensu T.T.C.L.: 496 (1949), *non* Fresen.]

[*G. jovis-tonantis* sensu Palmer & Pitman, Trees S. Afr. 3: 2050 (1972), *non* (Welw.) Hiern sensu stricto]

NOTE. I can find nothing of importance to separate *G. triacantha* from the other hairy taxa described by Stapf and Hutchinson but have seen very little material, particularly in fruit. De Candolle describes it as 'glabra' but the type was examined by Stapf and Hutchinson who describe the leaves as 'scabrida'. The field note of the *Burtt* sheet cited above mentions that the fruit is large and ribbed but no fruit accompanies the specimen. I strongly suspect he was thinking of the fruit of *G. volkensii*. Stapf and Hutchinson kept *triacantha*, *asperula* and *goetzei* apart on the characters of length of corolla and 8–9 corolla-lobes as against 6–7, now well-known to be so variable as to have no value. *R. torulosa* is a form with very short corolla. The isotype preserved at EA has the corolla-tubes about 1 cm. long but they are not fully developed; Krause gives 1.8 cm. as the upper limit and specimens with corolla almost this short have been seen recently. The leaves are also more densely pubescent but I think there is no doubt as to the identity.

10. G. subacaulis *Stapf & Hutch.* in J.L.S. 38: 420, t. 37 (1909); Brenan in Mem. N.Y. Bot. Gard. 8: 450 (1954); F.F.N.R.: 408 (1962); Medwecka-Kornás in Acta Bot. Acad. Sci. Hungar. 26: 131–137, figs. 1–4 (1980); Fl. Pl. Lign. Rwanda: 562 (1982); Fl. Rwanda 3: 164 (1985). Type: Malawi, Zomba, Namasi, *Cameron 77* (K, lecto.!)

Geophytic subshrub with stout erect woody stems buried in the ground or rhizomatous and prostrate but flowers usually appearing just above ground, not exceeding 30 cm. in height; young stems hairy but soon glabrous. Leaves in whorls of 3; blades elliptic to oblanceolate, 2–17.5 cm. long, 1–7 cm. wide, obtuse to ± acute at the apex, rounded to very narrowly attenuated at the base, rather coriaceous, scabrid to almost glabrous, the venation prominent and reticulate on both surfaces when dry; petioles 1–9 mm. long; stipules ovate, 4–7 mm. long, scabrid or pubescent. Calyx-tube 0.7–1 cm. long; tubular part of limb 0.8–1.4(–2.2) cm. long, sometimes slightly split on one side, glabrous to slightly pubescent, subtruncate or with 6–9 distinct linear, linear-oblong or linear-triangular lobes 1.5–8 mm. long, decurrent on the tube. Corolla at first creamy white but turning yellow after a day, sweetly scented; tube (3–)4–10 cm. long, glabrous to finely pubescent outside; lobes 6–8, obovate-elliptic, 2.5–4 cm. long, with the venation usually drying more darkly reticulate than is the case in the woodier allied species. Fruit brown, obovoid or ellipsoid to globose, 5–6.5 cm. long, ± 3.5–5 cm. in diameter, rather ridged when dry but not conspicuously lobed. Seeds discoid, ± 4–5 mm. wide.

TANZANIA. Ufipa District: Sumbawanga–Mkunde [Nkunde], 29 Nov. 1949, *Bullock 1952*! & road to Muse Gap, 22 Oct. 1960, *Richards 13375*!; Mbeya District: Mbosi Circle, Bomariva estate, 14 Jan. 1961, *Richards 13928*!; Tunduru District, *Allnutt 14*!

DISTR. **T** 1, 4, 7, 8; Malawi, Zambia, Rwanda, Mozambique

HAB. Open grassland and *Brachystegia* woodland on ground subject to burning; 780–1950 m.

SYN. [*G. thunbergia* sensu Hiern in F.T.A. 3: 100 (1877), pro parte, *non* L.f.]

NOTE. This differs in little but habit from *G. ternifolia*, the foliage and flowers being exactly similar to those of subsp. *jovis-tonantis*. The rhizomatous habit is distinctive but although seasonal burning has had much to do with the evolution of this species the specimens are not just variants of *ternifolia* due to burning. Experimental proof of this would be easy to obtain by protecting selected specimens from fire. The other stemless *Gardenia*, *G. tinneae* Kotschy & Heuglin, still only known from the type collected in the Sudan (Bahr-el-Ghazal, Bongoland) and a few specimens from Chad, is a quite different species as indicated by Stapf & Hutchinson. The small globose often ribbed fruits with peripheral sterile locules are very distinctive. The specimen reported by Lebrun from Chad (Bull. Soc. Bot. Fr. 118: 103, fig. 5 (1971)) as *G. subacaulis* is in fact *G. tinneae*.

11. G. imperialis *K. Schum.* in E.J. 23: 442 (1896); T.T.C.L.: 495 (1949); I.T.U., ed. 2: 343 (1952); Brenan in Mem. N.Y. Bot. Gard. 8: 450 (1954); Keay in B.J.B.B. 28: 41 (1958); F.F.N.R.: 407 (1962); Keay in F.W.T.A., ed. 2, 2: 122 (1963); F. Hallé, Et. Biol. et Morph. Gardén., Mém. O.R.S.T.O.M.: 101, etc. (1968); N. Hallé in Fl. Gabon 17, Rubiacées, 2: 220, t. 51 (1970). Type: Angola, Camboniederung, *Mechow* 495 (B, holo.†)

Shrub or small tree 3–12(?–15) m. tall with simple bole or several trunks up to 20(–45) cm. wide and often with horizontal annular rings; branchlets stout, spreading, grey; young parts of stem and young leaves viscid and puberulous to glabrescent; slash variable, often green, edged with cream or orange. Leaves opposite; blades large, elliptic to obovate-elliptic 5.5–56 cm. long, 3–26 cm. wide, gibbous at the base on either side of the midrib to form 2 cavities for ants 1–3 mm. deep, obtuse or less often shortly acuminate at the apex, cuneate to subcordate at the base, glabrous above, ± puberulous on the lower surface, olive-brown on drying, shiny; domatia absent; petiole 2–6 mm. long; stipules tubular, ± 1 cm. long, the margin incised-dentate; domatia absent. Buds dark purple-red. Flowers 1–3, erect or ± pendent, sessile, with strong sweet fragrance. Calyx sticky; tube oblong-ellipsoid, 1–1.5 cm. long; tubular part of limb 1–2 cm. long, ribbed, the ribs reaching the apices of 5(?–8) narrowly triangular lobes which are 5–10 mm. long and have thin margins. Corolla white, turning brown, or tube pinkish or brownish red outside, pinkish white inside and lobes white and pink, flecked crimson outside, the pink parts being those exposed in bud, subscabrous or glandular papillate outside; tube (7–)8–24 cm. long, 3–5(–7.5) cm. wide at the throat; lobes broadly ovate or oblong-falcate, 3.5–6 cm. long, 2.2–4 cm. wide, puberulous or papillate on the exposed parts. Anthers 2.5–4 cm. long. Style glabrous, gradually thickened into a yellow club-shaped apex split into 2 lobes for about 2–2.5 cm. Fruit red-brown, globose or broadly ellipsoid, or sometimes ± quadrangular in section, 5–7.2 cm. long, 3–5 cm. in diameter, the wall 2–3 mm. thick, the endocarp forming 2–3, pale, ± shiny, ± hemispherical pieces, crowned by the persistent calyx-limb. Seeds very numerous, yellow-brown, 3–5 mm. long, 3 mm. wide, compressed-ellipsoid, embedded in a pulp to form a solid mass, finely shagreened and wrinkled.

subsp. **imperialis**; Pauwels in B.S.B.B. 118: 114 (1985)

Leaf-blades smaller and relatively broader with 13–16(–20) lateral nerves. Corolla-tube wider and shorter, 11–14 cm. long. Fruit more cylindrical.

UGANDA. W. Nile District: Koboko [Kobboko], Mar. 1935, *Eggeling* 1834!; Masaka District: Katera, 1 Oct. 1953, *Drummond & Hemsley* 4503! & Lake Nabugabo, Aug. 1935, *Chandler* 1405!
TANZANIA. Bukoba District: Ruzinga Swamp, 19 Sept. 1934, *Gillman* 173!; Buha District: Bunganda, Kasulu, Buhoro, Oct. 1956, *Procter* 536!; Rungwe District: Kiwira R., 4 June 1907, *Stolz* 156!
DISTR. U 1, 4; T 1, 4, 7; Senegal to Cameroun, Central African Republic, Zaire, Burundi, Malawi, Zambia, Zimbabwe and Angola
HAB. Grassland, thickets of *Ficus*, etc. in seasonally burnt grassland, scrub and forest-patch edges by lakes, rivers and swamps; 1100–1300 m.

SYN. *G. viscidissima* S. Moore in J.L.S. 37: 158 (1905). Type: Uganda, Mengo District, Musozi, *Bagshawe* 144 (BM, holo.!)

NOTE. Subsp *physophylla* (K. Schum.) Pauwels (*Randia physophylla* K. Schum.) has a Guinea-Congolian distribution.

64. ROTHMANNIA

Thunb. in Vet. Acad. Handl. Stockholm 37: 65 (1776); Fagerl. in Arkiv Bot. 30A (7): 39 (1943), pro parte; Keay in B.J.B.B. 28: 47 (1958); Hallé in Fl. Gabon 17, Rubiacées, 2: 228 (1970)

Gardenia Ellis sect. *Rothmannia* (Thunb.) Endl., Gen. Pl. 1: 562 (1838)

Unarmed shrubs or small trees. Leaves opposite or occasionally ternate, petiolate, with or without domatia; stipules triangular, ± acuminate. Flowers ⚥, terminal, on very reduced branches, usually above a single leaf, sessile or shortly pedicellate or pedunculate, solitary or less often cymose, large or very large, often pendent. Calyx-tube usually turbinate to cylindrical; limb-tube velvety inside, often with colleters interspersed; lobes present or absent, erect. Corolla mostly white with coloured spots; narrowly to distinctly funnel-shaped or campanulate; lobes 5 (or in one species 7–8), contorted, overlapping either to the right or to the left according to the species. Stamens with anthers included or partly exserted; pollen grains simple. Ovary 1-locular, with 2 parietal placentas opposite to each other and sometimes partially fused; ovules very numerous; style glabrous subulate; stigma consisting of the usually club-shaped pollen receptacle with the stigmatic surface confined to the shortly bilobed apex. Fruit globular or ellipsoid, smooth or grooved, usually large, sometimes crowned with the persistent calyx-limb; pericarp thick or coriaceous. Seeds very numerous, sublenticular, 0.5–1.2 cm. long, embedded in the pulpy placental tissue and forming a solid mass.

A genus of about 30 species occurring in tropical Africa, Madagascar and Asia, doubtfully in America. 7 species occur in the Flora area, and one cultivated species *R. globosa* (Hochst.) Keay, a native of South Africa, has been recorded from Nairobi.

Calyx glabrous, whitish pubescent or with very short cream
   hairs; limb deciduous or persistent in fruit:
  Flowers subsessile, peduncle very reduced; pedicel short;
    corolla-lobes overlapping to right in bud:
   Leaves coriaceous, glossy above, glabrous; calyx and
     corolla entirely glabrous outside:
    Calyx-limb often splitting, truncate to repand or bearing
      filiform to subulate or occasionally linear lobes;
      bases of the lobes without well developed sinuses;
      corolla-tube gradually funnel-shaped or narrowly
      cylindrical at base then funnel-shaped or
      sometimes campanulate above; fruit globose;
      stipules 1–6 mm. long . . . .   *1. R. fischeri*
    Calyx-limb never splitting, always bearing filiform or
      sometimes narrowly linear lobes, 4–15 mm. long;
      bases of the lobes separated by well-developed
      sinuses; corolla-tube narrowly cylindrical at base
      then campanulate above; fruit broadly ellipsoid to
      ellipsoid; stipules 1–2.5 mm. long . . . . .   *2. R. ravae*
   Leaves chartaceous, glabrous to pubescent, with
    appressed hairs on nerves beneath; calyx and corolla
    pubescent outside . . . .   *3. R. urcelliformis*
  Flowers on very short branches, or in fascicles of 1–4
    flowers; peduncle 2–10 mm. long; pedicels absent;
    corolla-lobes overlapping to left in bud:
   Flowers solitary or very rarely in fascicles; corolla-tube
    2.5–14 cm. long, cylindrical part always exceeding the
    calyx-limb:
    Corolla-tube over 14 cm. long, funnel-shaped above;
     calyx-limb persistent in fruit . . . .   *4. R. longiflora*
    Corolla-tube 2.5–5.8 cm. long, campanulate above;
     calyx-limb eventually caducous in fruit   *5. R. manganjae*
   Flowers up to 4 in fascicles; corolla-tube 2–3.5 cm. long,
    cylindrical part scarcely extending beyond the
    calyx-limb, campanulate above (cultivated)   *R. globosa*
Calyx densely golden or reddish pubescent; limb persistent in
   fruit:
  Flowers in cymes of 3–17; corolla-tube 2.5–4.5 cm. long; fruit
    never ribbed, rusty coloured and flaking . . . .   *6. R. engleriana*
  Flowers solitary; corolla-tube 5.3–24 cm. long; fruit strongly
    10-ribbed or ± smooth, with rusty coloured pubescence
    only when young:
   Stigma small, included; upper portion of corolla-tube
    campanulate; leaves 4.5–14 cm. long, 2–6.8 cm. wide   *7. R. macrosiphon*
   Stigma large, exserted; upper portion of corolla-tube
    funnel-shaped; leaves 10–29 cm. long, 4–13 cm. wide   *8. R. whitfieldii*

1. R. fischeri (*K. Schum.*) *Bullock* in Oberm. in Ann. Transv. Mus. 17: 224 (1937); Keay in B.J.B.B. 28: 50 (1958), pro parte; F.F.N.R.: 419 (1962); Coates Palgrave, Trees of S. Afr.: 859 (1977); Bridson in K.B. 39: 68 fig. 1A (1984). Types: Tanzania, Mwanza District, Kayenzi [Kagehi], *Fischer* 318 (B, syn. † K, isosyn.!) & 296 (B, syn. †)

Shrub or small tree 3.4–8 m. tall; young branches glabrous, often with grey or fawn bark, which flakes off to reveal rusty coloured underlayer. Leaves glabrous, often drying blackish; blades narrowly to broadly elliptic or narrowly obovate to obovate, 2–11(–14) cm. long, 1–6 cm. wide, obtuse to acute or acuminate at apex, acute to cuneate or sometimes obtuse at base, subcoriaceous to coriaceous, shiny above; domatia present as glabrous to pubescent pits; petiole 0.1–1 cm. long; stipules triangular, 1–6 mm. long, acuminate, eventually caducous. Flowers solitary; peduncle very abbreviated; pedicel 1–4 mm. long; bracteoles small, scale-like. Calyx glabrous; tube 3–7 mm. long; limb-tube 0.4–1.1 cm. long, smooth or occasionally slightly 5-ribbed, often splitting for $\frac{1}{3}$ the length; lobes absent or filiform to subulate, rarely linear, up to 1.5 (–1.8) cm. long, triangular at base and not clearly separated by sinuses. Corolla cream with red or purple spots at the throat; tube funnel-shaped or narrowly cylindrical below then funnel-shaped or less often campanulate above, (3–)4.5–8 cm. long, 1–3 cm. wide at top, glabrous to glabrescent outside; lobes overlapping to right in bud, ovate or lanceolate, 1.3–3.5 cm. long, 0.7–2 cm. wide, acute to subacuminate or sometimes acuminate. Anthers ± $\frac{2}{3}$ exserted, 1.5–2.5 cm. long. Style 3–7.7 cm. long; stigma exserted, 1.3–1.7 cm. long, ± 3 mm. wide. Fruit green with pale green spots, blackish when dry, spherical, or somewhat pear-shaped (not in Flora area), 3–6.2 cm. in diameter, glabrous, smooth; calyx-limb deciduous leaving a pale pentagonal scar. Seeds brownish, 6–7 mm. long, ± 4 mm. wide and 2–3 mm. thick.

subsp. fischeri

Leaves elliptic to broadly elliptic or narrowly obovate, obtuse to subacuminate or less often rounded or acuminate at apex, cuneate or sometimes acute at base. Calyx-limb often splitting, with lobes 1–9(–18) mm. long. Corolla-tube gradually funnel-shaped or narrowly cylindrical below and funnel-shaped above, 4.3–8 cm. long, 1.3–3 cm. wide at top; lobes 1.8–2.5 cm. long.

TANZANIA. Mwanza District: Geita, Kasemeko, Uzinza, 5 July 1953, *Tanner* 1563!; Mpwapwa District: Kiboriani Mts., 3 Oct. 1938, *Greenway* 5796!; Iringa District: West Mufindi, Nov. 1960, *Procter* 1685!
DISTR. T 1, 4–7; Zaire (Shaba), Malawi, Zambia, Zimbabwe, South Africa (Transvaal)
HAB. Often on rocky hillsides in thickets or forest patches; (750–)1060–2100 m.
SYN. *Randia fischeri* K. Schum. in P.O.A. C: 380 (1895); T.T.C.L.: 527 (1949)

subsp. **verdcourtii** *Bridson* in K.B. 39: 68, fig. 1B (1984). Type: Kenya, Teita District, Bura, *Dale* in F.D. 3787 (K, holo.!, EA, iso.)

Leaves elliptic to broadly elliptic, acute to subacuminate at apex, obtuse to acute at base. Calyx-limb not splitting, truncate or repand; lobes not exceeding 1 mm. long. Corolla-tube gradually funnel-shaped, 5.5–7.2 cm. long, 1.2–1.5 cm. wide at top; lobes 1.3–1.8 cm. long.

KENYA. Machakos District: Kampi ya Ndege [Campi ya Ndege], 23 Jan. 1954, *Bally* 9458!; Masai District: Karibani [Garabani] Hill, 6 Mar. 1940, *van Someren* 3!; Teita District: Bura, July 1937, *Dale* in F.D. 3787!
TANZANIA. Pare District: Mankanya [Mankania] to Vudee [Wudee], 9 Mar. 1915, *Peter* 55758! & 14 Feb. 1926, *Peter* 41334! & top of Koko Hill, 2 Apr. 1972, *Wingfield* 1992!
DISTR. K 4, 6, 7; T 3; not known elsewhere
HAB. On rocky hillsides often in bushland; 910–2100 m.
SYN. *Randia fischeri* K. Schum. var. *major* K. Schum. in P.O.A. C: 381 (1895). Type: Kenya, Teita District, Ndi, *Hildebrandt* 2531 (B, holo.†)
[*R. fischeri* sensu Chiov., Racc. Bot. Miss. Consol. Kenya: 53 (1935); K.T.S.: 467 (1961), pro parte, quoad *Bursell* in E.A.H. 11651 & *Dale* in F.D. 3787, *non* K. Schum. sensu stricto]

NOTE. (on species as a whole). A third subspecies, subsp. *moramballa* (Hiern) Bridson has been recognised from Mozambique and Natal. This subspecies resembles *R. ravae* in its corolla which is narrowly cylindrical at base and campanulate above, but the cylindrical portion is not less than one-third the total length (as opposed to not more than one-quarter the total length in *R. ravae*). The calyx-limb often splits and the lobes are short (1–2 mm. long).

2. R. ravae *(Chiov.) Bridson* in K.B. 39: 71, fig. 1D (1984). Type: Somalia, Badadda [Baddada] *Senni* 272 (FT, holo.!)

Shrub or small tree, 2.5–10 m. tall; young branches slender, glabrous, with pale grey bark which eventually flakes off to reveal dark chocolate-brown underlayer. Leaves often drying blue-black, glabrous, narrowly elliptic to elliptic or sometimes broadly elliptic, 5–12 cm. long, 1.7–5 cm. wide, distinctly acuminate at apex, acute to cuneate at base, subcoriaceous to coriaceous, somewhat shiny above; domatia present, usually small, glabrous to pubescent pits, scarcely raised above; petioles (0.4–)0.5–1 cm. long; stipules triangular, usually rather small, 1–2.5 mm. long, caducous (or sometimes up to

5 mm. long, including acumen, on coppice shoots). Flowers solitary; peduncle very abbreviated; pedicels 1–4 mm. long; bracteoles resembling reduced stipules, sometimes with a subulate lobe up to 5 mm. long. Calyx glabrous; tube 5–8 mm. long; limb-tube 6–10 mm. long, usually with 5 distinct ribs (decurrent lobe bases), never splitting; lobes filiform or occasionally narrowly linear, 4–15 mm. long, somewhat widened at base but clearly separated by sinuses. Corolla white or cream with red to purple spots at the throat, glabrous outside; tube narrowly cylindrical at base then campanulate above, (3–)4.5–7 cm. long, 2.3–4.5 cm. wide at top, the cylindrical portion up to ¼ the total length; lobes overlapping to right in bud, ovate 2–3 cm. long, 1–2.2 cm. wide, rounded to acute or subacuminate. Anthers ⅓-exserted. Style up to ± 7.5 cm. long; stigmatic knob exserted, 1.7–2 cm. long, 0.2–0.3 cm. wide. Fruit green with pale green spots, ellipsoid, broadly ellipsoid or sometimes almost globose, 4–8 cm. long, 3.5–4.8 cm. wide, with a pentagonal scar at apex. Seeds brownish, 6–8 mm. long, ± 4 mm. wide, ± 2 mm. thick.

KENYA. Teita District: Maungu Hills, Nyangala [Nyangula] Hill, 15 Dec. 1969, *Archer* 619!; Kilifi District: Sokoke, *Dale* K 2026A!; Lamu District: 13 km. along Kiunga–Lamu road, 13 Aug. 1961, *Gillespie* 205!
TANZANIA. Handeni District: Kideleko [Kideliko], 22 Apr. 1954, *Faulkner* 1428!; Morogoro District: 13 km. NE. of Kingolwira Station, 4 Mar. 1954, *Welch* 211!; Uzaramo District: 96 km. W. of Dar es Salaam, 17 Aug. 1969, *Harris et al.* 3131!
DISTR. K 7; T 3, 6, 8; Somalia
HAB. In thicket or sometimes forest; 45–960 m.

SYN. *Randia ravae* Chiov., Fl. Somalia 2: 237, fig. 139 (1932)
[*Rothmannia fischeri* sensu K.T.S.: 467 (1961), pro parte quoad *R.M. Graham* in *F.D.* 1985 & *Gardner* in *F.D.* 3615; Vollesen in Op. bot. 59: 71 (1980), *non* (K. Schum.) Bullock]

NOTE. This species seems to be more closely related to *R. annae* (Wight) Keay, from the Seychelles, than to *R. fischeri*. It can easily be distinguished as *R. annae* is smaller in all its parts.

3. R. urcelliformis (*Hiern*) Robyns, F.P.N.A.2: 340 (1947); Keay in B.J.B.B. 28: 51 (1958); K.T.S.: 468, t. 28 (1961); Keay in F.W.T.A., ed. 2,2: 125 (1963). Types: Sudan, Nabambissoo, *Schweinfurth* 3034 (K, syn.!) & Zaire, Niamniam, *Schweinfurth* 3292 (BM, syn.!)

Shrub or small tree, (1–)4–8.5(–15) m. tall; young stems slender, rusty pubescent or less often glabrescent. Leaves paired or occasionally ternate; blades elliptic to obovate-elliptic, 5.7–18 cm. long, 1.2–10 cm. wide, apex distinctly acuminate, base cuneate, glabrous or less often glabrescent (? or very rarely pubescent) above, glabrescent or less often glabrous or pubescent beneath with denser appressed hairs on the nerves; domatia present; petiole 0.4–1.1 cm. long, glabrescent to pubescent; stipules triangular, 0.2–1.2 (–1.5) cm. long, long-acuminate, rusty pubescent to glabrescent outside, caducous leaving a line of hairs above the scar. Flowers solitary; peduncle very abbreviated; pedicel 3–7 mm. long; bracteoles filiform. Calyx sparsely to distinctly pubescent; tube 3–6(–11) mm. long, 5-angled; limb-tube 0.6–2.3 cm. long, slightly wider than the tube, usually splitting for ± half the length; lobes linear, 0.3–1.9(–2.7) cm. long, 1.5–4 mm. wide at base. Corolla white to cream with dark brownish-red blotches, pubescent outside; tube narrowly cylindrical below, funnel-shaped above, 3.1–7.7 cm. long, the funnel-shaped portion 1.8–5.1 cm. long, 0.8–1.8 cm. wide at top, 0.3–0.8 cm. wide at base, pubescent outside; lobes overlapping to right in bud, lanceolate, 1.2–4.5 cm. long, 0.8–1.8 cm. wide, acuminate, sparsely to densely covered with short stiff hairs. Anthers exserted for ⅓ their length, 0.6–2.3 cm. long. Style 5–7.5 cm. long, widening towards apex; stigma exserted, 1.2–2.6 cm. long, ± 5 mm. wide. Fruit green, spherical to ellipsoid, 2.5–7.6(?–10.2) cm. long, sometimes slightly ribbed, glabrous, the calyx-limb deciduous leaving a pale ± pentagonal scar. Seeds blackish, up to 8 mm. long, 6 mm. thick.

UGANDA. W. Nile District: E. Madi, Zoka Forest, Jan. 1952, *Leggat* 47!; Kigezi District: Malamagambo Forest, Feb. 1950, *Purseglove* 3293!; Mengo District: Kajansi Forest, 16 km. on Entebbe road, Feb. 1938, *Chandler* 2148!
KENYA. Kiambu District: Karura Forest, 18 Feb. 1966, *Perdue & Kibuwa* 8038!; Embu District: Njukiini Forest, 12 Feb. 1964, *Brunt* 1481!; N. Kavirondo District: NW. Kakamega Forest, 6 May 1971, *Mabberley & Tweedie* 1095!
TANZANIA. Arusha District: Ngurdoto National Park, Longil, 7 Oct. 1965, *Greenway & Kanuri* 11969!; Tanga District: c. 96 km. on Amani-Maruvera road, above International Business Combine saw-mill, 27 July 1969, *Magogo* 1264!; Iringa District: N. part of Gologolo Mts., 13 Sept. 1970, *Thulin & Mhoro* 946!
DISTR. U 1–4; K 3/5, 4–6; T 1–4, 6, 7; throughout W. tropical Africa, the Zaire Basin, Burundi, Sudan, Ethiopia, Angola, Zimbabwe and Mozambique
HAB. Forest; 850–1675(–2400) m.

SYN. *Gardenia urcelliformis* Hiern in F.T.A. 3: 104 (1877); T.S.K.: 139 (1936); T.T.C.L.: 496 (1949)
  G. riparia K. Schum. in P.O.A. C: 381 (1895). Types: Tanzania, Kilimanjaro, Marangu, Moonjo
  [Mondjo] stream, *Volkens* 1383 (B, syn. †, BM, isosyn.!) & near Mareales village, *Volkens* 1446
  (B, syn. †)
  G. tigrina Hiern, Cat. Afr. Pl. Welw. 1(2): 462 (1898). Types: Angola, Golungo Alto, *Welwitsch* 3098
  & 3099 & Cazengo, *Welwitsch* 3100 (all BM, K, isosyn.!)
  Randia stenophylla K. Krause, in E.J. 43: 140 (1909); F.W.T.A. 2: 78 (1931). Type: Togo, Agaua,
  Kersting 285 (B, holo. †)
  R. spathicalyx De Wild. in Ann. Mus. Congo, Bot., sér. 5, 3: 287 (1910). Type: Zaire, between
  Gumbari and Duru, *Seret* 735 (BR, holo., K, photo.!)
  R. urcelliformis (Hiern) Eggeling, I.T.U.: 202 (1940) & ed. 2: 356, t. 18 (1952)
  Rothmannia riparia (K. Schum.) Fagerl. in Arkiv Bot. 30A(7): 39 (1943)
  R. arcuata Bremek. in Exell, Suppl. Cat. Vasc. Pl. S. Tomé: 23 (1956). Type: S. Tomé, S. Vicente,
  Espirito Santo 120 (BM, holo.!, COI, iso.)

4. R. longiflora Salisb., Parad. Lond., t. 65 (1807); Keay in B.J.B.B. 28:53 (1958) & in
F.W.T.A., ed. 2, 2: 125 (1963); Hallé in Fl. Gabon 17, Rubiacées, 2: 238, t. 56/1–10, (1970).
Type: Ghana, *Brass* (BM, holo.!)

Bush or small tree 3–8.4(–9) m. tall, often climbing; young branches glabrous. Leaves
glabrous; blades elliptic or sometimes broadly elliptic, 6–14(–18) cm. long, (2.2–)2.5–5.8(–
7.8) cm. wide, apex acuminate, base acute; domatia present; petioles 0.3–0.8(–1) cm. long;
stipules triangular, ± 3 mm. long, acuminate, early caducous. Flowers sweetly scented,
solitary, terminal on short branches above a single leaf; peduncle (0.3–) 0.4–1 cm. long,
bearing ± 5–9 scale-like, ciliate bracteoles. Calyx glabrous outside; tube 0.5–0.8(–1) cm.
long; limb-tube (0.5–)0.7–1 cm. long, wider than the tube; lobes very shortly triangular or
linear, 1–2(–4) mm. long. Corolla purple to green outside, whitish, often with purple spots
inside, finely pubescent outside; tube narrowly cylindrical below, funnel-shaped above,
(?4.4–)14–18(–24) cm. long, the funnel-shaped portion 3.8–4(–5.5) cm. long, 1.2–4.2 cm.
wide at top, 0.2–0.4 cm. wide at base, with lines of pubescence in cylindrical portion
inside; lobes overlapping to the left in bud, ovate, (1–)1.3–2.5(–4) cm. long, (0.9–)1.2–1.7
cm. wide, obtuse, greyish tomentose above. Anthers included or with tips exserted,
2.7–3(–3.5) cm. long. Style 12–21.2 cm. long, gradually widened towards apex; stigma
partly exserted, (1.5–)2–2.9 cm. long, (0.2–)0.4–0.5 cm. wide. Fruit? green, drying blackish,
globose to ellipsoid, 3.5–5.2(–7) cm. long, with 10 indistinct ribs, glabrous, the calyx-limb
persistent. Seeds brownish, 6–8 mm. long, 5–6 mm. wide, 1–1.5 mm. thick.

UGANDA. Bunyoro District: Budongo Forest, Apr. 1932, *Harris* 735!; Masaka District: Malabigambo
  Forest, 6.4 km. SSW. of Katera, 2 Oct. 1953, *Drummond & Hemsley* 4558!; Mengo District: Kiagwe,
  Namanve Forest, Apr. 1932, *Eggeling* 401!
KENYA. N. Kavirondo District: Kakamega Forest, Apr. 1934, *Dale* 3281!
TANZANIA. Kigoma District: Masanza, Jan. 1955, *Forcus* 1! & Gombe Stream National Park, Mkenke
  valley, 14 May 1969, *Clutton-Brock* 229! & 23 Feb. 1970, *Clutton-Brock* 437!
DISTR. U 2, 4; K 5; T 4: also throughout W. tropical Africa, Zaire, Sudan and Angola
HAB. Forest or woodland; 900–1200(–1675) m.

SYN. *Randia maculata* DC., Prodr. 4: 388 (1830); Hiern in F.T.A. 3: 96 (1877); F.W.T.A. 2: 78 (1931).
  Type: as for *Rothmannia longiflora* Salisb.
  Gardenia speciosa A. Rich., Mém. Fam. Rub.: 160 (1830) & in Mém. Soc. Hist. Nat. Paris 5: 240
  (1834). Type: as for *Rothmannia longiflora* Salisb.
  G. stanleyana Hook. in Bot. Reg. 31, t. 47 (1845); Bot. Mag. 71, t. 4185 (1845). Type: cultivated at
  Turnham Green (London) in the nursery of Mr. Glendinning from a specimen from Sierra
  Leone collected by *T. Whitfield* (K, holo.!)
  Rothmannia stanleyana (Hook.) Benth. in Hook., Niger Fl.: 383 (1849); Fagerl. in Arkiv Bot. 30A
  (7): 39 (1943)
  Randia stanleyana (Hook.) Walp., Ann. 2: 794 (1852)
  R. longiflora (Salisb.) Th. Dur. & Schinz, Ét. Fl. Congo 1: 159 (1896), non Lam. (1789)
  R. sapinii De Wild., Comp. du Kasai: 419 (1910). Type: Zaire, Kasai-Kwilu, *Sapin* (BR, holo., K,
  photo.!)
  R. spathacea De Wild. in Ann. Mus. Congo. Bot., sér. 5, 3: 287 (1910). Type: Zaire, near Nala, *Seret*
  794 (BR, holo., K, photo.!)
  R. thomasii Hutch. & Dalz., F.W.T.A. 2: 78 (1931). Type: Sierra Leone, Yetaya, *N.W. Thomas* 2311
  (K, lecto.!)
  Rothmannia maculata (DC.) Fagerl. in Arkiv Bot. 30A (7): 39 (1943)

NOTE. Specimens in young bud are very difficult to distinguish from *R. manganjae* (Hiern) Keay, but
  that species tends to have slightly longer lobes on the calyx-limb than *R. longiflora*. Sterile
  specimens are almost impossible to distinguish, but the distributions of these two species are not
  known to overlap, although they are very close.

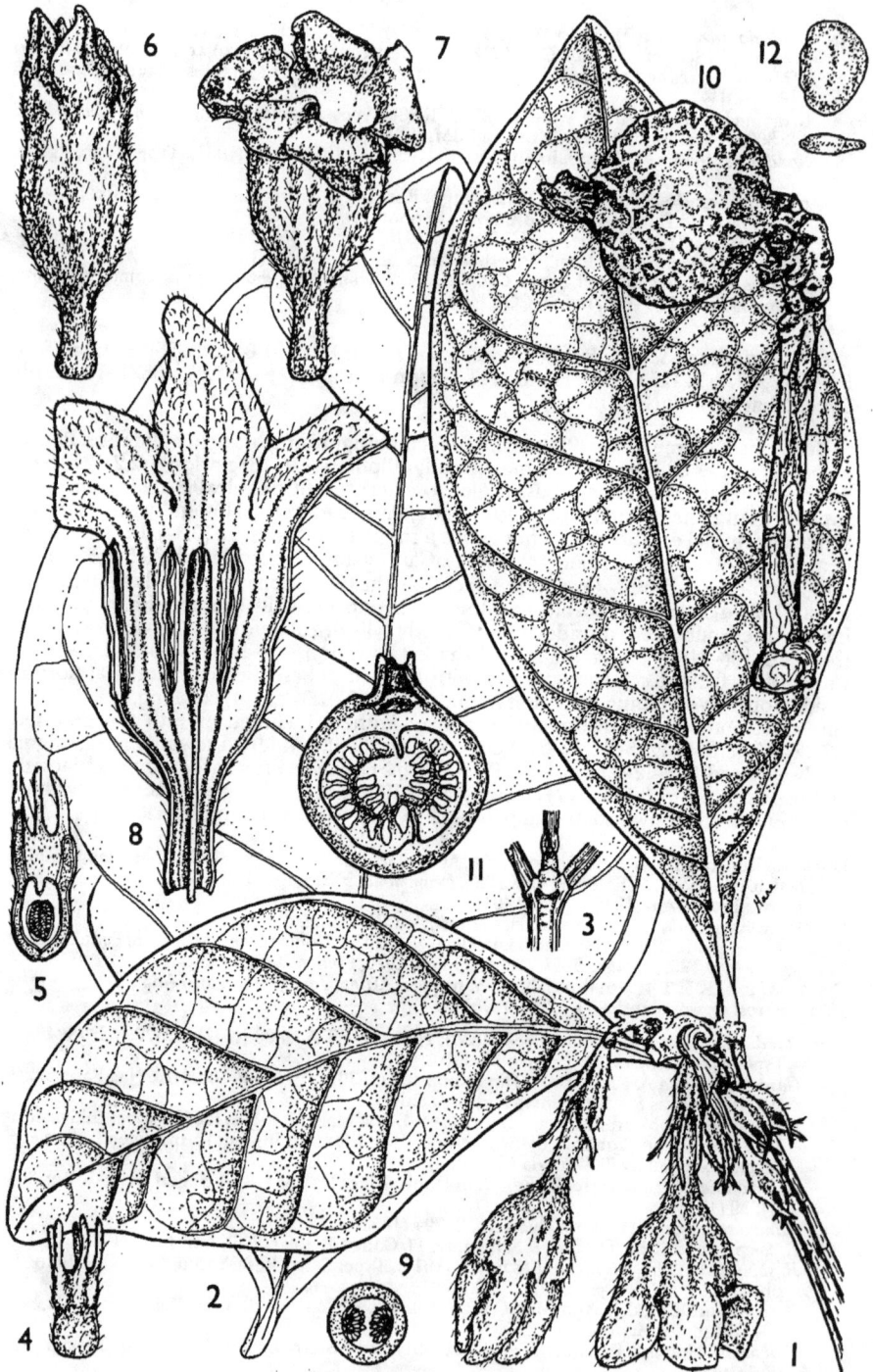

FIG. 82. *ROTHMANNIA ENGLERIANA* — **1**, flowering branch, × ⅔; **2**, leaf, × ⅔; **3**, nodes with stipule, × 1; **4**, calyx, × ⅔; **5**, section through calyx, × 1; **6**, corolla, immature, × ⅔; **7** mature corolla, × ⅔; **8**, section through corolla, × 1; **9**, transverse section of ovary, × 2; **10**, stalk with fruit, × ⅔; **11**, longitudinal section of fruit, × 1; **12**, seed, 2 views, × 2. 1, from *B.D. Burtt* 6567; 2, 10, from *Verdcourt* 2678; 3, from *Procter* 706; 4–8, from *Milne-Redhead* 2723; 9, from *Bullock & Shabane* 42; 11, 12, from *White* 2153. Drawn by Marie Bywater.

5. R. manganjae *(Hiern) Keay* in B.J.B.B. 28: 56 (March 1958); K.T.S.: 468 (1961). Type: Malawi, Mozambique, Manganja Hills, *Meller* (K, holo.!)

Tall shrub or small tree 2–12(–15) m. tall; young branches glabrous; older branches smooth. Leaves glabrous; blades elliptic or less often narrowly elliptic, (4.3–)8–15.2 cm. long, (2.1–)2.3–5.9 cm. wide, apex acuminate or acute, base cuneate, domatia sometimes present; petioles 0.3–1.6 cm. long; stipules triangular, acuminate, 1–3 mm. long, caducous. Flowers sweetly scented, solitary, terminal on short branches above a single leaf, or occasionally in fascicles of 2–3 flowers; peduncle 2–8 mm. long, bearing usually 3–7 scale-like or triangular glabrescent ciliate bracteoles; pedicels absent. Calyx glabrous to glabrescent outside; tube 0.4–1 cm. long; limb-tube 0.4–1.2 cm. long, a little wider than the tube; lobes shortly triangular or linear, up to 3 mm. long. Corolla white, flecked with reddish purple or rarely without markings, yellowish pubescent outside when dry; tube very short, narrowly cylindrical below, campanulate above, 2.5–5.8(–7.5) cm. long, the campanulate portion (1.7–)2.4–4.4(–6.5) cm. long, (1.2–)1.6–3.3(–3.8) cm. wide at top, 0.25–0.4 cm. wide at base, pubescent inside; lobes overlapping to the left in bud, ovate, 1–3.1(–4.3) cm. long, 0.6–2.1(–2.7) cm. wide, obtuse to acuminate, finely pubescent to tomentose above. Anthers entirely included or sometimes with tip exserted, 1.8–2(–3.3) cm. long. Style 1.5–3 cm. long, gradually widened; stigma included or up to ± half-exserted, 1.3–3(–3.7) cm. long. Fruit green, turning blackish when ripe, ± globose, 1.8–4 cm. long, the calyx-limb eventually caducous. Seeds brownish black, ± 5 mm. across.

KENYA. Trans-Nzoia/N. Kavirondo District: Mt. Elgon, June 1937, *Jex-Blake* in *C.M.* 6864!; S. Nyeri District: Kiganjo [Nyeri] station, Sept. 1930, *Gardner* in *F.D.* 2477!; Embu District: manyatta near Embu, 9 June 1932, *M.D. Graham* 1720!
TANZANIA. Lushoto District: Panusi [Fanusi], 29 Sept. 1936, *Greenway* 4626!; Ulanga District: Taveta, Sept. 1960, *Haerdi* 609/0!; Lindi District: Rondo Plateau, 1 Oct. 1951, *Bryce* 15!
DISTR. K 3–5; T 3, 6, 8; Mozambique, Malawi and Zimbabwe
HAB. Forest; (230–)360–1800 m.

SYN. *Gardenia manganjae* Hiern in F.T.A. 3: 103 (1877); T.S.K.: 139 (1936)
    Randia buchananii Oliv. in Hook., Ic. Pl. 14, t. 1356 (1881). Type: Malawi, Shire Highlands, Buchanan 41 (K, holo.!)
    R. fratrum K. Krause in N.B.G.B. 10: 604 (1929); Chiov., Racc. Bot. Miss. Consol. Kenya: 52 (1935). Type: Kenya, Meru, *R.E. & Th. C.E. Fries* 1631 (B, holo.†, K, S, iso.!)
    Rothmannia buchananii (Oliv.) Fagerl. in Arkiv Bot. 30A (7): 39 (1943)
    R. fratrum (K. Krause) Fagerl. in Arkiv Bot. 30A (7): 39 (1943)
    [*R. fischeri* sensu Brenan in Mem. N.Y. Bot. Gard. 8: 450 (1954), pro parte, quoad *Brass* 17832!, non (K. Schum.) Oberm.]
    R. manganjae (Hiern) Garcia in Mem. Junta. Invest. Ultram., sér. Bot., 4: 33 (1958)

NOTE. Most of the upper limits for the floral measurements are taken from *Jackson* 976 from Malawi. The specimens from Kenya tend to have consistently smaller hairier corollas than most of the material from other areas.

6. R. engleriana *(K. Schum.) Keay* in B.J.B.B., 28: 57 (1958); F.F.N.R.: 418 (1962). Type: Angola, *Mechow* 347 (B, holo.†)

Small tree (? 0.9–)1.8–8.4 m. tall; young branches puberulous or velutinous, older branches with cracked reddish bark. Leaves in pairs (or occasionally ? ternate); blades oblanceolate to elliptic or sometimes obovate to broadly obovate, 9.3–30 cm. long, 3.8–14.8 cm. wide, apex rounded to shortly acuminate, base cuneate, coriaceous, often drying yellowish-green, glabrous above, glabrescent or less often glabrous or pubescent beneath; domatia usually present; petiole 0.3–1.5 cm. long, glabrous to velutinous; stipules triangular, long acuminate, 4–5 mm. long, caducous early on. Flowers sweetly scented, pendulous in terminal cymes, 3–17-flowered; peduncle up to 1.5 cm. long, usually with primary and secondary branches present, yellowish pubescent when young, flaking and reddish when older; pedicels absent; bracteoles triangular, 1.5–5 mm. long. Calyx yellowish to amber, pubescent; tube ± 5-grooved and wrinkled when dry, 0.5–1 cm. long; limb-tube 0.4–1.4 cm. long, distinctly wider than the tube, with 5 ridges; lobes linear, 0.3–2.1 cm. long, 1.5–2.5(–4) mm. wide. Corolla white with red-purple spots at throat, golden pubescent outside when dry; tube short, narrowly cylindrical below, campanulate above, 4.7–7.2 cm. long, the campanulate portion 2.7–4.4 cm. long, 1.8–4.7 cm. wide at top, 0.3–0.4 cm. wide at base, glabrescent inside; lobes overlapping to left in bud, ovate, 1.1–3.4 cm. long, 1.2–2.3 cm. wide, acute or sometimes obtuse, whitish tomentose above. Anthers entirely included, 1.4–2.1 cm. long. Style 3.3–4.5 cm. long, gradually widening towards apex; stigma included, 1.3–2.1 cm. long, 2.5–4 mm. wide. Fruit rust-coloured, flaking, globose to ellipsoidal, 1.9–4.6 cm. long, the calyx-limb persistent. Seeds brown, 7 mm. long, 5 mm. wide, ± 1.25 mm. thick. Fig. 82.

TANZANIA. Biharamulo District: 110 km. from Geita on Biharamulo road, 16 July 1960, *Verdcourt* 2878!; Mpanda District: Mahali Mts., Mokoloka [Mkoloka], 19 Sept. 1958, *Jefford, Juniper & Newbould* 2502!; Iringa District: Iheme, 28 July 1933, *Greenway* 3405!
DISTR. T 1, 2, 4–8; Zaire, Burundi, Mozambique, Malawi, Zambia and Angola
HAB. *Brachystegia* woodland; (200–)1100–1850 m.

SYN. *Randia engleriana* K. Schum. in E. & P. Pf. IV.4: 76, fig. 27 (1891) & P.O.A. C: 380 (1895) & E.J. 23: 436 (1896); T.T.C.L.: 527 (1949)
    R. kuhniana F. Hoffm. & K. Schum. in P.O.A. C: 380 (1895); T.T.C.L.: 527 (1949). Type: Tanzania, Tabora District, Kakoma [Kagome], *Böhm* 46a (B, holo.†, K, iso.!)
    R. ternifolia Ficalho & Hiern in Hiern, Cat. Afr. Pl. Welw. 1: 459 (1898). Types: Angola, Pungo Andongo , Panda forest at R. Mangue, *Welwitsch* 2581b & Huila, between Lopollo and Quipungo, *Welwitsch* 2581 (both BM, K, isosyn.!)
    R. lemairei De Wild. in Ann. Mus. Congo, Bot., sér. 4, 1: 155, t. 39 (1903). Type: Zaire, Shaba, Lukafu, *Verdick* (BR, holo., K, photo.!)
    R. lacourtiana De Wild., Comp. du Kasai: 418 (1910). Type: Zaire, Kasai, Dilolo, *Sapin* D22 (BR, holo., K, photo.!)
    R. katentaniae De Wild. in F.R. 13: 140 (1914). Type: Zaire, Shaba, Katentania, *Homblé* 736 (BR, holo., K, photo.!)
    Rothmannia kuhniana (F. Hoffm. & K. Schum.) Fagerl. in Arkiv Bot. 30A (7): 39 (1943)

7. R. macrosiphon *(Engl.) Bridson* in K.B. 31: 180 (1976). Type: Tanzania, Tanga District, Duga, *Holst* 3179 (B, holo.†, K, P, iso.!)

Bush or small tree 1.8–7 m. tall; young stems glabrescent to finely velutinous. Leaf-blades oblanceolate to obovate, 4.5–14.4 cm. long, 2–6.8 cm. wide, apex acuminate, base cuneate, coriaceous, glabrous above, glabrous to glabrescent beneath with midrib sparsely pubescent; domatia absent; petioles 0.2–1.3 cm. long, sparsely pubescent; stipules triangular, acuminate, 3–5 mm. long, caducous. Flowers sweetly scented, solitary, terminal above a single leaf; peduncle 0.5–1.6 cm. long, pubescent, bearing 2–6 acute triangular bracteoles 2–6 mm. long. Calyx golden or reddish-brown pubescent; tube 10-grooved, 0.9–1.4 cm. long; limb-tube 0.4–1.1 cm. long, wider than the tube; lobes linear, 1.6–3.2 cm. long, 1.5–2 mm. wide, ± equal. Corolla white with reddish purple spots in throat, golden to buff pubescent outside when dry; tube long, narrowly cylindrical below, campanulate above, 11–24 cm. long, the campanulate part 3.8–6.2 cm. long, 3.1–4.5 cm. wide at top, 3–6 mm. wide at base, glabrous within, the cylindrical portion pubescent within; lobes overlapping to left in bud, ovate, 1.3–2.6 cm. long, (1.2–)1.8–3 cm. wide, rounded or obtuse, whitish tomentose above. Anthers well-included, 2–3 cm. long. Style ± 11–18 cm. long, gradually widening to apex; stigma well-included, 2–3 cm. long, 3–4 mm. wide. Fruit ± ovoid, 2.5–3.2 cm. long, 2–2.2 cm. wide, not or sometimes slightly ribbed, golden pubescent, the calyx-limb and sometimes corolla persistent (perhaps not fully mature). Seeds not known.

KENYA. Kwale District: Buda Mafisini Forest, 21 Aug. 1953, *Drummond & Hemsley* 3938! & Marenge Forest, Lungalunga–Msambweni road, 18 Aug. 1953, *Drummond & Hemsley* 3875!; Kilifi District: Arabuko, *R.M. Graham* in F.D. 1984!
TANZANIA. Lushoto District: Segoma Forest, 30 July 1966, *Faulkner* 3831!; Handeni District: Kwamarukanga Forest Reserve, 3 Feb. 1971, *Shabani* 640!; Uzaramo District: 22 km. on Pugu road, 22 June 1939, *Vaughan* 2823!
DISTR. K 7; T 3, 6, 8; not known elsewhere
HAB. Forest; 60–450 m.

SYN. *Randia macrosiphon* Engl. in Abh. Preuss. Akad. Wiss.: 28 (1894) & P.O.A. C: 381, t. 44 (1895); T.T.C.L.: 528 (1949)
    [*Rothmannia whitfieldii* sensu Keay in B.J.B.B. 28: 56 (1958), pro parte; K.T.S.: 468 (1961), pro parte *non* (Lindl.) Dandy]

NOTE. *Haerdi* 423/0 from Ulanga District, near Ifakara is closely allied to this species. The leaves are larger than is typical of *R. macrosiphon* and the flowers have a corolla-tube 12.5 cm. long, the campanulate portion being 7.3 cm. long, and 4.6 cm. wide and lobes 3.5 cm. long, 2.8 cm. wide with a subacuminate apex. The fruit (on EA sheet) is 4.5 cm. long, 3.4 cm. wide, not ribbed and covered with a yellow tomentum that wears off. The exact status of this specimen is best left undecided until further gatherings are obtained.

8. R. whitfieldii *(Lindl.) Dandy* in F.P.S. 2: 461, fig. 165 (1952); Keay in B.J.B.B. 28: 55 (1958); F.F.N.R.: 419 (1962); Keay in F.W.T.A., ed. 2, 2: 126, fig. 227 (1963); Hallé in Fl.

Gabon 17, Rubiacées, 2: 252, t. 59/7–10 (1970). Type: Sierra Leone, *Whitfield* 1844 (BM, holo.!, K, iso.!)

Shrub or small tree 1.8–8.4(–15?) m. tall; young stems pubescent. Leaf-blades elliptic to obovate, (9–)10–29 cm. long, 3.5–13 cm. wide, acuminate at apex, cuneate at base, coriaceous, glabrous above, pubescent to ± glabrous beneath; domatia absent; petioles 0.7–2.1 cm. long, puberulous or glabrescent; stipules triangular, acuminate, 2–6 mm. long, pubescent, deciduous. Flowers sweetly scented, solitary, pendent, terminal above a single leaf; peduncle 0.5–2.5 cm. long, reddish pubescent, bearing 2–6 acute triangular bracteoles 2.5–3 mm. long. Calyx golden or reddish brown pubescent; tube 10-grooved when dry, 0.6–1.3 cm. long; limb-tube 0.7–2.3 cm. long, wider than the tube; lobes linear, 0.9–5(–7.9) cm. long, 1.5–4 mm. wide, ± equal. Corolla white, or rarely said to be purple spotted, golden pubescent outside when dry; tube narrowly cylindrical below, funnel-shaped above, 5.3–23 cm. long, the funnel-shaped portion 2–7.8 cm. long, 2.6–8.2 cm. wide at top, 4–8 mm. wide at base, glabrous or pubescent within funnel-shaped portion, pubescent within cylindrical portion; lobes overlapping to left in bud, ovate, 1.6–6.8 cm. long, 1.6–5.8(–8) cm. wide, acute or rounded, creamish tomentose above. Anthers included or with tips exserted, 2.7–4.5 cm. long. Style 7.5–18.5 cm. long, abruptly widened at the stigma; stigma large, wholly or partly exserted, 3–7.5 cm. long, 0.6–11 mm. wide, distinctly bilobed at apex. Fruit globose, 2.8–7 cm. in diameter, smooth or strongly 10-ribbed, rusty pubescent when young, ± glabrous later on, crowned by a persistent calyx-limb. Seeds ± sublenticular, ± 0.7–1.1 cm. long, ± 3–4 mm. thick.

UGANDA. Acholi District: Zoka Forest, 19 Nov. 1941, *A.S. Thomas* 4038!; Bunyoro District: Bujenje, Feb. 1943, *Purseglove* 1266!; Mengo District: Budo, Mar. 1932, *Eggeling* 231!
TANZANIA. Rungwe District: Malawi border, Karolo Forest Reserve, 4 Nov. 1949, *Carmichael* in *F. H* 2954! & Kondeland, Nov. 1910, *Stolz* 413!; Lindi District: Rondo Plateau, Mchinjiri, Mar. 1952, *Semsei* 711!
DISTR. U 1, 2, 4; T 7, 8; throughout west tropical Africa, the Zaire basin, Sudan, Malawi, Zambia and Angola
HAB. Forest; (700–)1050–1675 m.

SYN. *Gardenia whitfieldii* Lindl. in Bot. Reg. 31, sub. t. 47 (1845).
  G. malleifera Hook. in Bot. Mag. 73, t. 4307 (1847). Types: Senegambia, *Heudelot* 809, Sierra Leone, *Turner & Whitfield* (all K, syn.!) cultivated specimen, ex *Whitfield* (not known)
  Rothmannia malleifera (Hook.) Benth. in Hook., Niger Fl.: 383 (1849)
  Randia malleifera (Hook.) Hook.f. in G.P. 2: 89 (1873); Hiern in F.T.A. 3: 98 (1877); P.O.A. C: 381 (1895); F.W.T.A. 2: 78 (1931); T.T.C.L.: 528 (1949); I.T.U., ed. 2: 355 (1952)
  R. eetveldiana De Wild. & Th. Dur. in B.J.B.B. 38: 194 (1899); & in Ann. Mus. Congo, Bot., sér. 4, 1: 155 (1903). Type: Zaire, Chinganga, *Dewèvre* 638a (BR, holo., K, photo.!)
  R. cuvelieriana De Wild. in Ann. Mus. Congo, Bot., sér. 5, 1: 79 (1903). Types: Zaire, Tshimbane, Djuma valley, *Gillet* 2096 & *Gentil* 82 (both BR, syn., K, photo.!)
  R. stolzii K. Schum. & K. Krause in E.J. 39: 526 (1907). Type: Tanzania, Rungwe District, Kondeland, near Isujuna, *Stolz* 87 (B, holo.†)
  R. homblei De Wild. in F.R. 13: 139 (1914). Type: Zaire, Shaba, Lubumbashi [Elisabethville], *Homblé* 298 (BR, holo., K, photo.!)
  Rothmannia eetveldiana (De Wild. & Th. Dur.) Fagerl. in Arkiv Bot. 30A (7): 39 (1943)

NOTE. The two specimens seen from T 8, *Rodgers* 683 and *Semsei* 711, are without flowers and as the leaf-size falls within the overlap between *R. whitfieldii* (Lindley) Dandy and *R. macrosiphon* (Engl.) Bridson, it is possible that they are better placed here until the collection of flowering specimens may prove the case.

65. PHELLOCALYX

Bridson, Gasson & Robbrecht in K.B. 35: 315 (1980)

Shrub or small tree; branches pubescent when young. Leaves opposite, petiolate, membranous, restricted to new growth at apex of stems; domatia present as white hairy tufts in the nerve-axils; stipules interpetiolar, ovate-triangular, with colleters and silky hairs inside. Flowers precocious, hermaphrodite, solitary, terminal on short lateral branches (but may be exceeded by the development of an axillary branch), sessile, large. Calyx 6–8-ribbed, covered with very short fleshy hairs when young, which soon erode to reveal a rather flaky corky underlayer; tube campanulate; limb entire in bud, acuminate, rupturing into 2–3 lobes at anthesis, densely covered with silky hairs with large interspersed colleters inside. Corolla-tube short in proportion to the lobes, cylindrical to

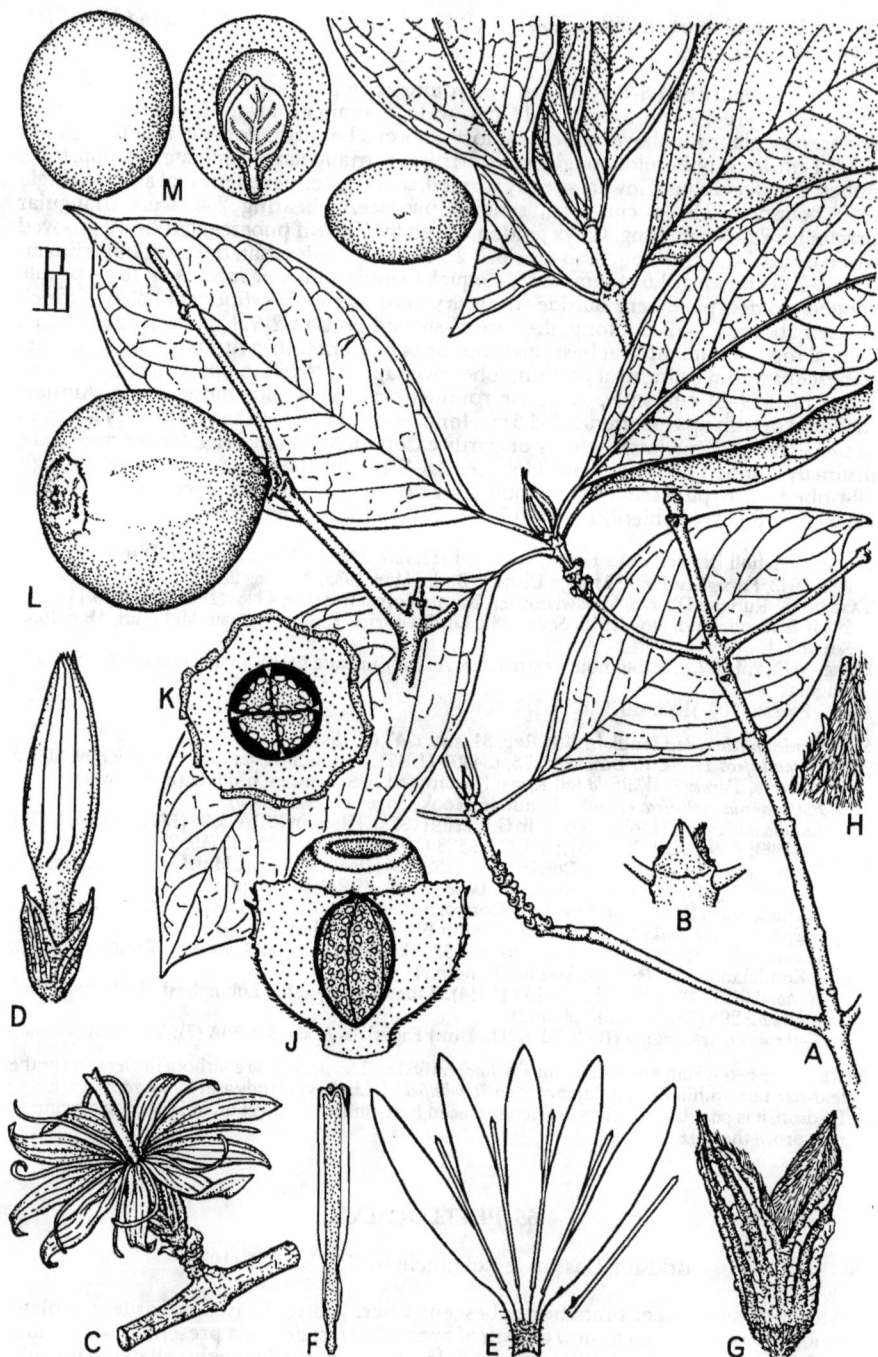

FIG. 83. *PHELLOCALYX VOLLESENII* — **A**, leafy branch with young flower-bud, × ⅔; **B**, stipule, × 2; **C**, flower on twig, × ⅔; **D**, mature flower-bud, × 1; **E**, section through young corolla (with one anther detached), × 1; **F**, style with stigma, × 1; **G**, calyx, × 2; **H**, detail of inside of calyx-limb, × 8; **J**, calyx-tube, with part removed to show ovary, × 8; **K**, transverse section of ovary, × 8; **L**, fruit on twig, × ⅔; **M**, seed, 2 views and section, × 4. A, from *Vollesen* in *M.R.C.* 2363; B, L, from *Rodgers* in *M.R.C.* 741; C–K, from *Vollesen* in *M.R.C.* 4265; M, from *Rodgers* in *M.R.C.* 595. Drawn by Diane Bridson.

funnel-shaped; lobes 8–11, narrowly oblong, spreading, overlapping to the right in bud (as seen from outside). Anthers 8–11, sessile, attached near the base, inserted at throat, exserted, ± two-thirds the length of the corolla-lobes, linear, shortly acuminate; pollen-grains simple. Ovary 4-locular (or ? incompletely so); septa thinly membranous, adhering to the placentas; placentas 4, elongate, fleshy, attached at the apex to the style-base; ovules numerous, impressed on the placentas; disc annular; style slightly swollen, glabrous; stigmatic club exserted, cylindrical, 2-winged, bifid at apex with each lobe emarginate. Fruit large, globular, smooth, 1-locular; calyx-limb caducous; pericarp thick. Seeds numerous, embedded in pulpy placental tissue, sublenticular; testa very thin, finely rugulose; endosperm entire, horny; embryo straight.

A monospecific genus.

P. vollesenii *Bridson* in K.B. 35: 316, figs. 1 & 2 (1980) & in Distr. Pl. Afr. 21, t. 723 (1980). Type: Tanzania, Kilwa District, c. 23 km. SW. of Kingupira, *Vollesen* in M.R.C. 4265 (K, holo.!, C, EA, iso.!)

Shrub or small tree 3–7 m. tall; young branches ± square, pubescent, older branches covered with thin greyish bark somtimes flaking to reveal red-brown under-layer. Leaves blue-black when dry, confined to new growth at apex of stem, mature leaf-blades narrowly obovate to obovate, 9–16 cm. long, 4–7 cm. wide, apex acute to shortly acuminate, base cuneate, chartaceous, glabrous above, glabrous or glabrescent with tertiary nerves conspicuously reticulate beneath; petioles slender 0.7–1.7 cm. long; stipules ovate-triangular, 2–4 mm. long, shortly apiculate, densely covered with silky hairs inside and with a few large colleters. Flower-buds borne on leafing branches; flowers borne on short lateral naked branches 0.8–6 cm. long; bracts stipule-like. Calyx-tube 3–4 mm. long; limb 1–1.3 cm. long. Corolla green or yellowish green with small purple flecks outside, drying blackish; tube 1–1.4 cm. long, ± 8 mm. wide at top, sparsely pilose outside, glabrous at throat but pubescent near base inside; lobes 2–3.3 cm. long, 6–13 mm. wide, obtuse or acute, glabrous. Style creamish, 9 mm. long; stigmatic club 1.5–2.4 cm. long. Fruit black when mature, 4–7(–10 from field notes) cm. in diameter, 1-locular; scar fawnish, surrounded by a narrow inner and broad outer fawnish rings. Seeds brown, 5 mm. long, 4 mm. wide, 3 mm. thick. Fig. 83.

TANZANIA. Kilwa District: Selous Game Reserve, Kingupira, 27 Aug. 1969, *Rodgers* in M.R.C. 593! & 4 Mar. 1970, *Rodgers* in M.R.C. 741! & 23 May 1975, *Vollesen* in M.R.C. 2363!
DISTR. T 8; Malawi, Mozambique
HAB. Thicket on sand or stony hill top; 270–365 m.

66. SHERBOURNIA

G. Don in Loudon, Encycl. Pl., Suppl. 2: 1303 (1855); Hepper in K.B. 16: 458 (1963); Hallé in Fl. Gabon 17, Rubiacées, 2: 126 (1970)

Amaralia Hook.f. in G.P. 2(1): 90 (1873); Wernham in J.B. 55: 1 (1917); Keay in B.J.B.B. 28: 57 (1958)

Scandent shrubs or lianes. Leaves petiolate, opposite, mostly thinly coriaceous; stipules rather large, oblong or elliptic, obtuse, deciduous. Flowers ♂ in 1–several-flowered inflorescences mostly appearing lateral (actually terminal) on one side of the stem at successive nodes or rarely appearing terminal, often ± sessile. Calyx-tube turbinate, sometimes ribbed; tubular part of free limb short; lobes 5, rather large, obtuse, contorted and overlapping to the right. Corolla rather large, the tube funnel-shaped to campanulate, constricted at the base, thickly silky outside, glabrous inside save for a densely barbate portion at base in constriction; lobes 5, short and broad, overlapping to the left. Anthers ± sessile, inserted in the throat of the corolla; pollen grains single. Ovary obconic, 2(–3)-locular; styles clavate, sulcate, the stigma scarcely separating into 2 lobes; ovules numerous, immersed in vertical rows on the 2 placentas. Fruit oblong-ellipsoid, many-seeded; seeds reticulate and wavy-rugulose.

A genus of about a dozen species, all tropical African.

S. bignoniiflora (*Welw.*) *Hua* in Bull. Soc. Hist. Nat. Autun 14: 396 (1901); Keay in

FIG. 84. *SHERBOURNIA BIGNONIIFLORA* var. *BIGNONIIFLORA* — 1, flowering branch, × ⅔; 2, stipules × 1; 3, flower-bud, × ⅔; 4, section through flower, × 1; 5, longitudinal section of ovary, × 3; 6, transverse section of ovary, × 3; 7, stigma, × 2; 8, fruit, × ⅔; 9, seed, × 8. 1, from *Scott Elliot* 5602; 2, from *Synott* 560; 3, from *Enti* in *F.H.* 6711; 4–7, from *Schweinfurth* 15; 8, 9, from *De Wilde* 3643. Drawn by Mrs M.E. Church.

F.W.T.A., ed. 2, 2: 127 (1963); Hallé in Fl. Gabon 17, Rubiacées, 2: 149, t. 34/1–2 (1970).
Type: Angola, Pungo Andongo, *Welwitsch* Carp. 642 (BM, syn.) & Golungo Alto, Serra de
Alto Queta, *Welwitsch* 2571 (LISU, syn., BM, K, isosyn.!)

Liane or scrambling shrub 3–4(–7) m. tall; young stems with adpressed yellowish brown
hairs, older ones glabrous and with the pale ridged bark peeling to some extent.
Leaf-blades elliptic or oblanceolate to oblong-elliptic, 4.5–19(–22.5) cm. long, 2.4–11.5 cm.
wide, bluntly acute to shortly acuminate at the apex, narrowly rounded to ± cordate or
rarely cuneate at the base, glabrous above, adpressed hairy on the main nerves beneath
and with shorter fine adpressed pubescence between them; petiole thickened, 0.8–2 cm.
long; stipules ovate, oblong or elliptic-oblong to ± lanceolate, 0.7–1.7 cm. long, 4–7 mm.
wide, venose, adpressed hairy, deciduous, the nodes hairy inside at the base. Flowers
solitary or inflorescences 1–3-flowered, axillary, subsessile or with peduncle and pedicels
4–6(rarely–25) mm. long, fragrant; bracts ovate, 2–3 at the base of the calyx, ± 4–9 mm.
long. Calyx-tube green, ± 0.5–1 cm. long, adpressed silky pubescent; limb 4 mm. long;
lobes mostly reddish, oblong or lanceolate, 1.4–2.3(–3.5) cm. long, 0.6–1.2 cm. wide, ± silky
or glabrescent, obtuse or ± pointed. Corolla creamy white, pink, pale purplish red or white
outside and purplish inside or orange variegated with purple; tube (2.5–)3–5.5 cm. long,
adpressed silky pubescent outside; lobes rounded-ovate, 1–2 cm. long, 1.2–1.8 cm. wide.
Fruit red or orange, subglobose, ovoid or ellipsoid, (6–)7.5–10 cm. long, 3–5.5 cm. wide,
bearing the persistent calyx-lobes, faintly ribbed. Seeds dark reddish brown, compressed-
ellipsoid, ± 3 mm. long, wavy-rugulose.

var. bignoniiflora

Leaf-blades mostly cordate or subcordate at the base. Calyx-lobes obtuse. Fig. 84.

UGANDA. Bunyoro District: Budongo Forest, 24 Apr. 1971, *Synnott* 560!
DISTR. U 2; Sierra Leone to S. Nigeria, Cameroun, Gabon, Central African Republic, Zaire, Sudan,
Zambia and Angola
HAB. Evergreen forest; 1050 m.

SYN. *Gardenia bignoniiflora* Welw., Apont.: 585, no. 13 (1859)
　　Amaralia bignoniiflora (Welw.) Hiern in F.T.A. 3: 112 (1877), pro parte; Keay in B.J.B.B. 28: 57
　　(1958); F.F.N.R. : 400 (1962)
　　A. heinsioides Wernham in J.B. 55: 5 (1917). Types: Sierra Leone, near Kafogo, *Scott Elliot* 5602
　　(K, syn.!) & Ghana, Finsenase, near Obuase, *Chipp* 149 (K, syn.!) & Zaire, Khor Kussumbo,
　　Schweinfurth 3142 (BM, syn.) & Khor Diamvonoo, *Schweinfurth* sér. 2, 15 (K, syn.!)

NOTE. Var. *brazzaei* (Hua) Hallé known from Gabon, Zaire and Angola has more cuneate leaf-
blades and acute calyx-lobes.

67. DIDYMOSALPINX

Keay in B.J.B.B. 28: 61 (1958) & in F.W.T.A., ed. 2, 2: 130 (1963); Hallé in Adansonia, sér. 2,
8: 367 (1968) & in Fl. Gabon 17, Rubiacées, 2: 258 (1970)

Erect or scrambling entirely glabrous shrubs (save for inside corolla-tube) with some
lateral branches modified into spines which are supra-axillary and held at right-angles.
Leaves opposite, petiolate; stipules slightly connate at the base, acute or acuminate-
subulate. Flowers rather large, ♂, 5-merous, axillary or supra-axillary, solitary on both
sides of the node, sometimes unilaterally subterminal; pedicels well developed. Calyx-
tube cylindric; limb very shortly tubular with 5 small erect or leafy lobes. Corolla-tube
funnel-shaped, strongly nerved in the midpetaline axes, pilose inside, glabrous outside;
lobes with contorted aestivation. Anthers sessile, included; pollen grains simple. Ovary
unilocular, with 2 opposing parietal placentas or sometimes ± joined to form a ± 2-locular
ovary; style expanded into a massive club of pithy tissue, with 2 apical stigmatic lips. Fruit
globose or ellipsoid, with thin walls, crowned with the persistent calyx-limb. Seeds 8–16
per fruit, lenticular, ± 1 cm. wide, embedded in a white tissue and forming a compact mass.

A small genus of 4 closely related species described from Africa, 2 of which occur in the Flora area.
A further undescribed plant also occurs but its status is uncertain. N. Hallé has excluded *D. parviflora*
Keay from the genus.

Calyx-lobes linear-lanceolate to lanceolate, 1–4.5 cm. long,
   1.8–8 mm. wide:
   Calyx-lobes lanceolate, 1.2–4.5 cm. long, 2–8 mm. wide, ± flat;
      corolla-tube 4–6 cm. long; style glabrous   . . . .   1. *D. lanciloba*
   Calyx-lobes narrowly linear-lanceolate, 1 cm. long, 1.8 mm.
      wide, strongly keeled; corolla-tube 7.6 cm. long   3. *D. sp. A*
   Calyx-lobes linear to linear-triangular, 3–6 mm. long, 1–1.5
      mm. wide; corolla-tube 4.5–7.5 cm. long; style pilose for
      about 1.5 cm. near the middle   . . . .   2. *D. norae*

1. D. lanciloba (*S. Moore*) *Keay* in B.J.B.B. 28: 62 (1958); Hallé in Fl. Gabon 17,
Rubiacées, 2: 262, t. 61/6–14 (1970); Robbrecht in B.J.B.B. 56: 150, fig. 1/A, B (1986). Type:
Uganda, Toro, Semliki Forest, *Bagshawe* 1281 (BM, holo.!)

Shrub 0.7–3.6 m. tall; spines paired or verticillate in threes. Leaf-blades elliptic, 5–18
cm. long, 1.8–7 cm. wide, sharply acuminate at the apex, cuneate at the base, olivaceous
on drying, shiny; domatia present, with very small hairs; petioles 2–9 mm. long; stipules
triangular, 2–8 mm. long, with subulate or apiculate apex. Flowers solitary in each axil of
each node; peduncle 0.8–2 cm. long. Calyx-tube obconic, 4–7 mm. long, 3 mm. wide;
tubular part of the limb 1.5–3 mm. long; lobes unequal, lanceolate, 1.2–4.5 cm. long, 2.8–8
mm. wide. Corolla cream, pale yellow-green or primrose-yellow; tube 4–6 cm. long,
1.8–3.5 cm. wide at the throat; lobes ovate, 1.7–7.4 cm. long, 1.1–2.8 cm. wide. Ovary with ±
16 ovules embedded in the 2 placentas; style glabrous, half as long to as long as the tube,
swollen into a club for ± 1 cm. at the apex. Fruit yellow, ellipsoid or subglobose, 3.3–3.5 cm.
long, 1.7–2.3 cm. wide, containing 1 to several seeds, smooth or feebly 5-ribbed. Seeds
1.1–1.3 cm. long, 1 cm. wide, 4 mm. thick, with a fibrous finely striate coat.

UGANDA. Toro District: Mwenge, Muhangi Forest, 10 June 1952, *Stuart-Smith* 33! & Semliki Forest,
   near Baranga, 28 Oct. 1906, *Bagshawe* 1281!
DISTR. U2; Cameroun, Central African Republic, Gabon, Zaire and Angola
HAB. Evergreen forest; 690 m.

SYN. *Gardenia lanciloba* S. Moore in J.B. 45: 264 (1907)
   G. tchibangensis Pellegr. in Bull. Soc. Bot. Fr. 85: 57 (1938) & Fl. Mayombe 3: 17, t.1/3, 4 (1938).
   Types: Gabon, Tchibanga, *Le Testu* 1478 & 1876 (P, syn.)

2. D. norae (*Swynnerton*) *Keay* in B.J.B.B. 28: 64 (1958); Verdc. in K.B. 14: 348 (1960);
K.T.S.: 438 (1961). Type: Zimbabwe, Chirinda Forest, *Swynnerton* 11 (BM, holo.!, K, iso.!)

Shrub 1–4.5(–8) m. tall or sometimes a small tree to 10 m.; branches often somewhat
scandent; spines paired. Leaf-blades elliptic, 4.5–14(–17) cm. long, 1.5–5.5(–7) cm. wide,
shortly acutely acuminate at the apex, cuneate at the base, olive-green on drying, often
shiny; domatia small, hairy; petioles 0.2–1 cm. long; stipules broad, 1–4 mm. long, mostly
with a short apiculum (Swynnerton's 14 mm. is presumably an error). Flowers solitary in
each axil of each node or frequently supra-axillary, said to be fragrant at night; peduncle
1–3.5 cm. long. Calyx-tube turbinate, 3–4 mm. long; tubular part of the limb 2 mm. long;
lobes linear or very narrowly linear-triangular, 3–6 mm. long, 1–1.5 mm. wide. Corolla
cream or white, tinged green; tube 4.5–7.5 cm. long; 2.8–3.3 cm. wide at the throat; lobes
broadly ovate-elliptic, (1–)2 cm. long, 2 cm. wide. Ovary with several ovules embedded in
the 2 placentas; style pilose for ± 2.5 cm. at its middle, 5–6 cm. long, swollen into a white
club for over 1.5 cm. at the apex. Fruit pale green or white with deep green longitudinal
lines, globose or ellipsoid, 2–3.3 cm. long, 1.7–2.3 cm. wide, ± 10-ribbed, crowned by the
persistent calyx-limb. Seeds pale ochraceous, irregularly orbicular, very compressed, 1.1
cm. in diameter, 3–4 mm. thick, covered with a fibrous coat. Fig. 85.

KENYA. Kwale District: Mrima Hill, 4 Sept. 1957, *Verdcourt* 1854! & Shimba Hills, Marere Hill area, 7
   Mar. 1968, *Magogo & Glover* 225! & Shimba Hills, Kirumoni Forest, 13 Apr. 1968, *Magogo & Glover*
   857!
TANZANIA. Lushoto District: Sigi Singali, 22 Apr. 1950, *Verdcourt* 165! & same locality, 25 June 1950,
   Verdcourt & Hesse 268! & Magunga Forest, 30 Apr. 1953, *Faulkner* 2114!; Lindi District: Rondo
   Plateau, Mchinjiri, Nov. 1951, *Eggeling* 6404!
DISTR. K 7; T 3, 6, 8; Mozambique, Zimbabwe
HAB. Evergreen forest, secondary forest, forest edges; 190–810m.

SYN. *Gardenia norae* Swynnerton in J.L.S. 40: 80 (1911)
   [*G. abbeokutae* sensu Brenan, T.T.C.L.: 495 (1949); Verdc. in K.B. 7: 360 (1952), *non* (Hiern)
   Keay]

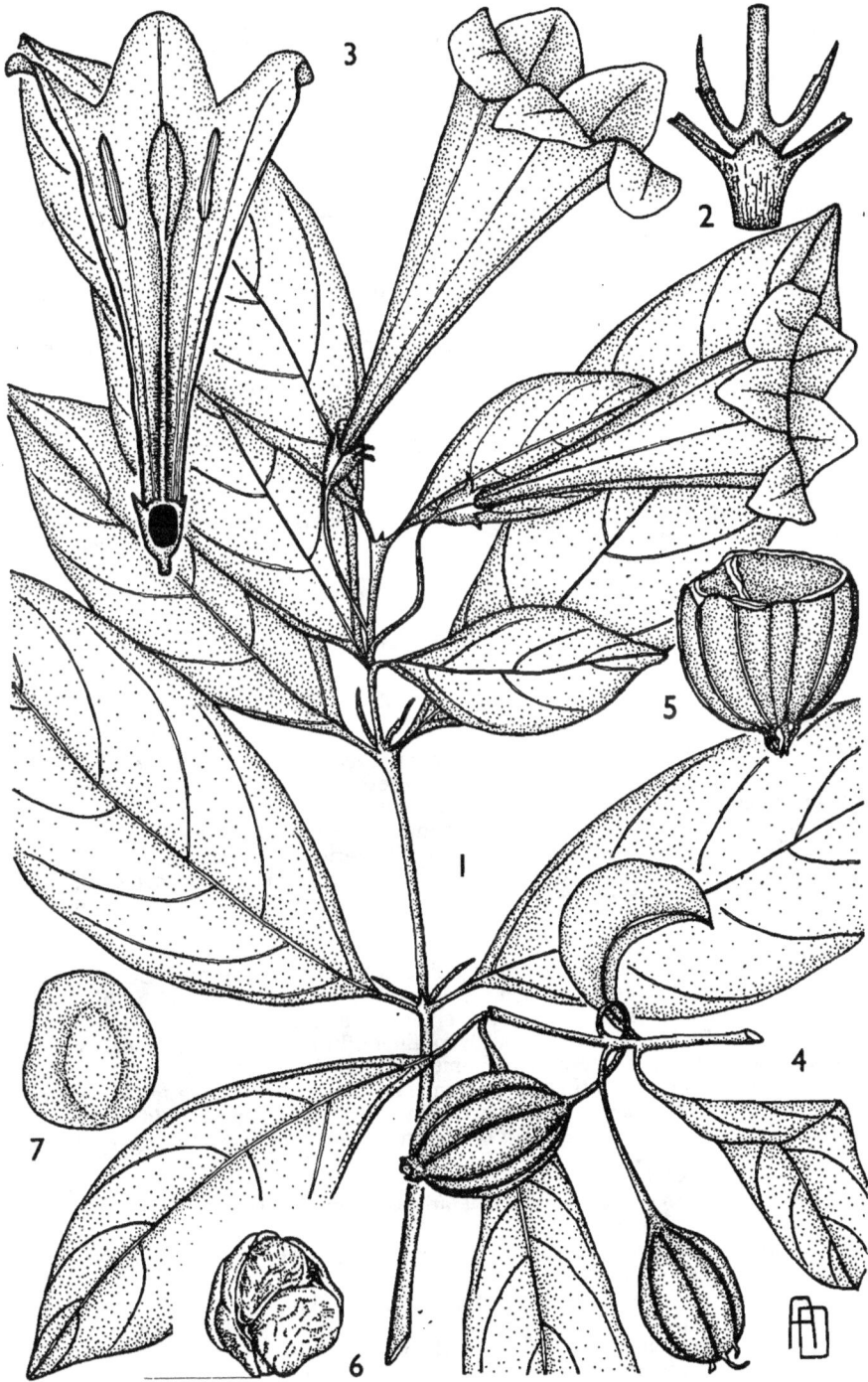

FIG. 85 *DIDYMOSALPINX NORAE* — **1**, flowering branch, × ⅖; **2**, stipule, × 2; **3**, half flower, × ⅖; **4**, branch
with fruit, × ⅖; **5**, part of fruit wall, calyx downwards, × 1⅓; **6**, seed mass, × 1; **7**, seed with testa removed, × 1⅓.
1, 2, from *Verdcourt* 165; 3, 4, from *Faulkner* 1214; 5–7, from *Magogo & Glover* 225. Drawn by Ann Farrer,
with 6 by Sally Dawson.

NOTE. Very closely allied to the W. African *D. abbeokutae* (Hiern) Keay but differing mainly in the hairy style. It was collected long ago by Kirk but the specimen has remained unnamed for nearly 100 years. The few flowering specimens seen from Zimbabwe, including the type, have the corollas much shorter than in East African material; this may be a constant feature but most of the Zimbabwe material available is in fruit. The fruits also are small and more globose. The possibility of there being two subspecies must await more material for confirmation.

3. D. sp. A

Shrub 1 m. tall. Leaf-blades broadly elliptic, 6–11 cm. long, 3–6.5 cm. wide, shortly acuminate at the apex, broadly cuneate at the base, olivaceous on drying, thinly coriaceous, very shiny; domatia apparently absent; petioles 5–8 mm. long; stipules broadly ovate to triangular, 4.5–6 mm. long, drawn out into an acumen from a broad base. Flowers solitary (only one seen); peduncle 2.3 cm. long. Calyx-tube narrowly obconic, 9 mm. long, 3.5 mm. wide; tubular part of limb 1.5 mm. long; lobes narrowly linear-lanceolate, 1 cm. long, 1.8 mm. wide. Corolla white; tube 7.6 cm. long, 1.5 cm. wide at the throat; lobes triangular, 1.3 cm. long, 6.5 mm. wide, narrowly pointed. Style hairy in the middle.

TANZANIA. Lindi District: Lake Lutamba, 10 Jan. 1935, *Schlieben* 5852!
DISTR. T 8; not known elsewhere
HAB. Open woodland; 240 m.

NOTE. I have refrained from dissecting the only flower seen; it is clearly very close to *D. norae*, but differs in the broader more coriaceous more shining leaves and much longer calyx-lobes. More material is needed to ascertain its status.

68. OXYANTHUS

DC. in Ann. Mus. Hist. Nat. Paris 9: 218 (1807); Keay in B.J.B.B. 28: 42 (1958); N. Hallé, Fl. Gabon, 17, Rubiacées, 2: 186 (1970)

Shrubs to small trees or rarely lianes, often sweetly scented in dry state. Leaves petiolate; domatia often present as hairy tufts in the nerve axils; stipules green in dry state, usually somewhat coriaceous, erect, persistent, sometimes ± connate at base. Flowers ♂, 5-merous, subsessile to pedicellate in lax or contracted, elongate (sometimes raceme-like) or corymbose panicles, lateral at successive nodes; bracteoles scarcely to well developed. Calyx-tube turbinate to ellipsoid or somewhat cylindrical; limb with distinct tubular part and short to long, usually narrow lobes. Corolla white; tube long, very narrowly tubular, the throat glabrous to sparsely pubescent; lobes narrow, contorted, overlapping to the left. Stamens fully or partly exserted, attached at the throat; anthers sessile or subsessile, apiculate or acuminate, sometimes prolonged at the base into sterile appendages. Pollen-grains in tetrads. Disc annular. Style very slender, glabrous, exserted, with a small club-shaped or fusiform swelling (stigmatic club) bearing 2 stigmatic lips at its apex. Ovary 1–2-locular, but the 2 locules are practically confluent, the two placentas joined only at the extremities; ovules very numerous, not immersed in the fleshy placentas. Fruit globose, ellipsoid or fusiform, ± fleshy in life, sometimes crowned with the persistent calyx-limb. Seeds sometimes held together with placental tissue, compressed-ellipsoid, quite large, 4–8 mm. long; testa ornamented with concentric horseshoe-shaped to fingerprint-like striations.

A moderate-sized genus confined to tropical Africa comprising about 40 species.

Shrubs with slender stems; stipules 4–5(–9) mm. long; leaf-blades not longer than 10 cm. long, with ± 6 main pairs of lateral nerves; flowers not known; fruit 1.6–2.2 cm. long (T6, Uzaramo District) . . . .   **4. O. sp. A**
More robust shrubs; stipules mostly more than 6 mm. long; leaf-blades mostly more than 10 cm. long, with at least (7–)8 main pairs of lateral nerves; fruit 1.7–6.5(–7.5) cm. long:
  Bracteoles conspicuous, narrowly lanceolate to lanceolate, 1.25–10(–12) mm. long; inflorescences narrowly pyramidal (± spike-like) to pyramidal:

Leaf-blades tomentose beneath, with bases strongly
    unequally cordate, calyx-tube and pedicels pubescent   *2. O. haerdii*
Leaf-blades glabrous to pubescent beneath or if tomentose
    then base sometimes slightly cordate, but never
    unequal:
    Inflorescences many-flowered, pyramidal or less often
        narrowly pyramidal; corolla-tube 1.8–6.5(–7) cm.
        long; calyx-lobes 1–3 mm. long; anthers with 2
        appendages at base . . . .   *1. O. speciosus*
    Inflorescence 1–11-flowered, narrowly pyramidal; corolla-
        tube 7–15 cm. long; calyx-lobes 2–8 mm. long;
        anthers not appendaged at base:
        Stigmatic club oblong-ellipsoid, 2–2.5 mm. long,
            obtuse; bracteoles 0.7–1 cm. long; calyx-lobes
            (4–)5–8 mm. long; fruit pear-shaped, mottled
            white . . . .   *3. O. zanguebaricus*
        Stigmatic club spindle-shaped, 4–5 mm. long,
            acuminate; bracteoles 4–9 mm. long; calyx-lobes
            2–4 mm. long; fruit ellipsoid to spindle-shaped
            or ovoid:
            Stipules triangular, acute at apex; apex of leaf-
                blades acute to shortly acuminate . . .   *5. O. goetzei*
            Stipules triangular at base, linear-subulate above;
                leaf-blades distinctly acuminate . . .   *6. O. lepidus*
Bracteoles inconspicuous or filiform to subulate, sometimes
    keeled at base, up to 5 mm. long; inflorescences a much
    branched or lax panicle, or occasionally ± congested
    and shortly pyramidal:
    Leaf-blades very large, 25–55 cm. long, 13–32 cm. wide,
        with bases equally or unequally cordate, often
        bullate; nerves pubescent beneath; inflorescence 20–
        80-flowered, much branched and crowded . . .   *11. O. unilocularis*
    Leaf-blades much smaller, cordate to cuneate or unequal
        at base, never bullate; nerves glabrous or sometimes
        pubescent beneath; inflorescence fewer-flowered or
        lax:
        Inflorescence 18–25-flowered, shortly pyramidal, ±
            congested to moderately lax; corolla-tube up to 2.2
            cm. long . . . .   *7. O. troupinii*
        Inflorescence lax or if moderately compact then only
            6–9-flowered; corolla-tube 3.4–22 cm. long:
            Leaf-blades broadly elliptic, ovate or obovate, if
                cordate at base then equal; calyx-lobes 1–4 mm.
                long:
                Inflorescence 6–9-flowered, moderately compact;
                    leaf-blades elliptic to obovate . . . .   *8. O. ugandensis*
                Inflorescence 10–30-flowered, lax; leaf-blades
                    elliptic to ovate . . . .   *9. O. pyriformis*
        Leaf-blades narrowly oblong to oblong, unequally
            cordate or rounded at base; calyx-lobes 0–1 mm.
            long . . . .   *10. O. formosus*

1. **O. speciosus** *DC.* in Ann. Mus. Hist. Nat. Paris 9: 218 (1807); Hiern in F.T.A. 3: 108
(1877); Keay, F.W.T.A., ed. 2: 129 (1963); Bridson in K.B. 34: 113 (1979). Type: Sierra
Leone, *Smeathman* (G, holo., BM, iso.!).

Shrub or small tree 2–6(–12) m. (or more *fide* I.T.U.) with horizontal glabrous branches;
bark smooth. Leaf-blades elliptic to oblong, 7.2–25.5 cm. long, 2.9–12.5 cm. wide,
acuminate at the apex, cuneate to rounded at the base, shining, entirely glabrous or
pubescent beneath; domatia present; petiole 0.7–1.5 cm. long; stipules triangular or
lanceolate, 0.6–2 cm. long, 0.3–1 cm. wide, acute. Inflorescences many-flowered, forming
oblong or conical panicles 2–5 cm. long (excluding the flowers); flowers fragrant;
peduncle 3–10 mm. long, glabrous or pubescent; pedicels (0–)1–5 mm. long; bracts and

bracteoles narrowly lanceolate, 1.25–9 mm. long, pubescent. Calyx glabrous or pubescent; tube 1.5–2 mm. long; limb-tube 0.75–1.75 mm. long, not or scarcely wider than tube; lobes triangular, acute, 1–2.5 mm. long, ciliate to pubescent. Corolla-tube glabrous, 1.8–7 cm. long, 1–1.5 mm. wide, pubescent inside; lobes oblong-lanceolate, 0.5–1.8 cm. long, 1.25–3 mm. wide, acute. Anthers exserted, long acuminate, with 2 well-marked sagittate basal appendages, usually different in colour to the thecae when dry. Stigmatic club pale green, fusiform, 1–1.5 mm. long. Fruit globose, ellipsoid, spindle-shaped or fusiform, 1.7–6.3 cm. long, 0.9–3 cm. wide, rounded, acute or constricted at apex to give a bottle-neck shape, glabrous, rather thick-walled; calyx-limb persistent. Seeds ochre or light brown, 4 mm. long, striated.

KEY TO INFRASPECIFIC VARIANTS*

Fruit ellipsoid to globose or rarely spindle-shaped, 1.8–4 cm.
   long:
   Fruit ellipsoid or rarely spindle-shaped, acute at apex;
      calyx-lobes 2–3 mm. long; bracteoles and calyx-lobes
      pubescent to ciliate; T 4; (subsp. *mollis*):
      Leaf-blades pubescent beneath . . . .   var. **mollis**
      Leaf-blades glabrous beneath . . . .   var. **glaber**
   Fruit subglobose to globose, rounded or occasionally obtuse
      at apex; calyx-lobes 1–2 mm. long; bracteoles and calyx-
      lobes ciliate; leaf-blades entirely glabrous (U 2, 4; T 1)   subsp. **globosus**
Fruit spindle-shaped to fusiform or less often ellipsoid, (2.5–)3–
   6.3 cm. long (U 1–3; K 3–5; T 2, 3, 5–8) . . . . . .   subsp. **stenocarpus**

subsp. **mollis** (*Hutch.*) *Bridson* in K.B. 34: 114 (1979). Type: Zambia, 8 km. E. of Chiwefwe, *Hutchinson & Gillett* 3689 (K, holo.!, BM, iso.!)

Leaf-blades glabrous or pubescent beneath; fruit ellipsoid or rarely spindle-shaped, (1.7–)2–3.8 cm. long; calyx-lobes linear-triangular to narrowly triangular, 2–3 mm. long; bracteoles 4–7 mm. long, bracteoles and calyx pubescent to ciliate.

var. **mollis**; Fl. Pl. Lign. Rwanda: 570 (1983); Fl. Rwanda 3: 182 (1985)

Leaf-blades finely pubescent beneath; bracteoles and calyx-lobes pubescent.

TANZANIA. Kigoma District: Malagarasi Pontoon Camp (Kibondo road), 27 Aug. 1950, *Bullock* 3222! & Gombe Stream Reserve, Kahama valley, 7 Apr. 1969, *Clutton-Brock* 23!; Mpanda District; between Pasagulu and Musenabantu, 8 Aug. 1959, *Harley* 9229!
DISTR. T 4; Zaire (Shaba & Kinshasa), Rwanda, Burundi, Zambia and Angola
HAB. Riverine forest or thicket; 800–1525 m.

SYN. [*O. rubifloris* sensu Hiern, Cat. Afr. Pl. Welw. 1: 465 (1898), pro parte, quoad *Welwitsch* 3078, *non* Hiern (1877)]
   O. mollis Hutch., Botanist in S. Afr.: 497 (1946); F.F.N.R.: 413 (1962)
   [*O. speciosus* sensu F.F.N.R.: 413 (1962), pro parte, quoad *White* 3360]
   O. huillensis Nogueira in Bol. Soc. Brot. 49, sér. 2: 119, t. 3 (1975). Type: Angola, Huila Province, *Henriques* 828 (LISC, holo., K!, LUAI, iso.)
   O. speciosus DC. var. *pubescens* Nogueira in Bol. Soc. Brot. 49, sér. 2: 120, t. 2 (1975). Type: Angola, Cuanza Norte, Camabatela, *Teixeira et al.* 12105 (COI, holo., K!, LISC, iso.)

NOTE. This taxon most likely occurs also in T 1, as it has been recorded from the banks of the Akagera R. on the Rwandan side.

var. **glaber** *Bridson* in K.B. 34: 114 (1979). Type: Zambia, Mpika District, 48 km. S. of Shiwa Ngandu on the Mpika road, *Angus* 873 (K, holo.!, FHO, iso.)

Leaf-blades glabrous beneath; bracteoles and calyx-lobes pubescent to ciliate.

TANZANIA. Buha District: Mkalinzi [Lussimbi] to Kwa Bikare [Bikare], 17 Mar. 1926, *Peter* 38848!; Kigoma District: Gombe Stream Reserve National Park, Mkenke valley, 3 Mar. 1970, *Clutton-Brock* 463!; Mpanda District: Mahali Mts., Dec. 1971, *Nishida* 84!
DISTR. T 4; Zaire (Shaba, Kivu), ?Burundi, Zambia, Angola

---

* Difficulty will be experienced in placing flowering specimens, but this may be done reliably by reference to their place of origin, except in the case of U 2 (see note after subsp. *stenocarpus*).

HAB. Forest; 800–1520 m.

SYN. [*O. speciosus* sensu Hiern, Cat. Afr. Pl. Welw. 1: 465 (1898); F.F.N.R.: 413 (1962), pro parte *non* DC.]

NOTE. This variety is very similar to subsp. *speciosus* (from Senegal to Nigeria), but the calyx-lobes and bracteoles tend to be larger and often have a somewhat denser indumentum.

subsp. **globosus** *Bridson* in K.B. 34: 115 (1979). Type: Uganda, Masaka District, Sese Is., Bugala, Sozi Point, *Eggeling* 88 (K, holo.!, BR, EA, iso.!)

Fruit subglobose to globose, 1.8–3.3 cm. long; calyx-lobes 1–2 mm. long; corolla-lobes 0.5–1(–1.2) mm. long; leaf-blades glabrous.

UGANDA. Ankole District: Kalinzu Forest, 26 Jan. 1970, *Synnott* 473!; Mengo District: Kiagwe, Magyo [Bukasa], Apr. 1932, *Eggeling* 387! & Entebbe, Sept. 1924, *Maitland* 196!
TANZANIA. Bukoba District: Kantale Forest, 16 Mar. 1935, *Ritchie* H11/35! & Kaigi, May 1935, *Gillman* 279! & Nyakato, 26 Mar. 1935, *Gillman* 219!
DISTR. U 2, 4; T 1; Central African Republic, Cameroun, Gabon, Zaire, Cabinda.
HAB. Forest; 1150–2200 m.

SYN. [*O. speciosus* sensu K. Krause in Z.A.E. 1907–8, 2: 321 (1911); Mildbraed, Z.A.E. 1910–11, 2: 14 (1922); De Wild., Pl. Bequaert. 2: 253 (1923); I.T.U. ed. 2: 353 (1952), pro parte, quoad *Dawkins* 468 & *Eggeling* 88 & 387; N. Hallé, Fl. Gabon, 17, Rubiacées, 2: 196, t. 45/1–11 (1970), *non* DC.]

subsp. **stenocarpus** *(K. Schum.) Bridson* in K.B. 34: 116 (1979); Fl. Pl. Lign. Rwanda: 570, fig. 190/3 (1983); Fl. Rwanda 3: 183, fig. 52/3 (1985). Type: Tanzania, E. Usambara Mts., Gonja, *Holst* 4265 (K, lecto.!)

Fruit spindle-shaped to fusiform or less often elliptic, (2.1–)3–6.5 cm. long, acute or constricted and bottle-necked at apex; corolla-lobes (0.6–)1–1.8 cm. long; leaf-blades glabrous or sometimes with spreading hairs along the nerves beneath or rarely pubescent.

UGANDA. Karamoja District: Napak, June 1950, *Eggeling* 5951!; Ankole District: Kalinzu Forest, July 1938, *Eggeling* 3768!; Mbale District: between Butandiga and Bulambuli, 1936, *Eggeling* 2436!
KENYA. Northern Frontier Province: Mt. Marsabit, 21 May 1962, *Kerfoot* 3708!; Embu District: Mt. Kenya, vicinity of Castle Forest Station, 18 Jan. 1973, *Spjut & Ensor* 3024!; N. Kavirondo District: Kakamega Forest, 3 Jan. 1968, *Perdue & Kibuwa* 9442!
TANZANIA. Arusha District: Mt. Meru, Kilinga Forest, 3 Feb. 1969, *Richards* 23904!; Morogoro District: Uluguru Mts., Bunduki, 10 Jan. 1934, *E.M. Bruce* 500!; Songea District: Matengo Hills, Lupembe Hill, 29 Feb. 1956, *Milne-Redhead & Taylor* 8907!
DISTR. U 1–3; K 1, 3–7; T 2, 3, 5–8; Rwanda, Ethiopia, Mozambique, Zimbabwe and South Africa (N. Transvaal).
HAB. Forest; 750–2300 m.

SYN. *O. gerrardii* Sond. var. (unnamed) sensu Oliv. in Trans. Linn. Soc., ser. 2, 2: 336 (1887); P.O.A. C: 381 (1895); T.T.C.L.: 507 (1949), based on Tanzania, Kilimanjaro, *Johnston* (K)
[*O. speciosus* sensu P.O.A. C: 381 (1895); T.T.C.L.: 508 (1949); I.T.U., ed. 2: 353 (1952), pro parte, quoad *Eggeling* 2436, *Sangster* 749 & *Snowden* 1082; K.T.S.: 453 (1961), pro parte majore]
O. stenocarpus K. Schum. in E.J. 33: 345 (1903); Chiov., Racc. Bot. Miss. Consol. Kenya: 53 (1935); T.T.C.L.: 508 (1949)
[*O. gerrardii* sensu S. Moore in J.L.S. 40: 82 (1911), quoad *Swynnerton* 116, *non* Sond.]

NOTE. There is a considerable amount of variation within this subspecies, as currently recognized. The specimens from T 3, tend to have much smaller corolla-lobes (6–7 mm. long) and bracteoles (2.5–4 mm. long) and also shorter pedicels than those from Uganda, Kenya and Ethiopia, to the north, and Tanzania (T 7, 8) and south tropical Africa to the south, while the specimens from T 2, 5 and 6 are somewhat intermediate in the above respects. Several specimens with fruit under 3 cm. long have been seen from T 6, Uluguru Mts. Subsp. *stenocarpus* and subsp. *globosus* are sympatric for part of their distribution. Both subspecies have been recorded from the Kalinzu Forest (Uganda, Ankole District) and one specimen *Synnott* 352 tends to be intermediate. Flowering material from this area cannot be determined to subspecies with certainty.

2. **O. haerdii** *Bridson* in K.B. 34: 119, fig. 1 (1979). Type: Tanzania, Ulanga District, Ifakara, *Haerdi* 252/0 (K, holo.!, EA, iso.!)

Shrub with finely pubescent stems. Leaf-blades ovate, 14–20.5 cm. long, 5.8–11.8 cm. wide, obtuse to shortly acuminate at apex, unequally auriculate at base, glabrous above, with soft short hairs beneath; domatia not seen; petiole 5–7 mm. long, pubescent; stipules triangular, 1–1.5 cm. long, 3–5.5 mm. wide, pubescent outside. Inflorescence ± 20-flowered, forming conical panicles; peduncle ± 1 cm. long, pubescent; pedicels 4–5 mm. long, pubescent; bracts 1.4 cm. long; bracteoles straw-coloured when dry, lanceolate, ± 5 mm. long. Calyx-tube 3 mm. long, pubescent; limb wider than tube, sparsely pubescent;

tubular part 2.5 mm. long; lobes very narrowly triangular, 2 mm. long. Corolla glabrous outside, tube (not fully mature) 6 cm. long, 2 mm. wide; lobes oblong, 2.4 cm. long, 8 mm. wide, obtuse. Anthers ± ½-exserted, acuminate, with two short appendages at base. Stigmatic club 6 mm. long. Fruit red-brown when dry, pear-shaped, gradually narrowing into the accrescent pedicel, 3.8 cm. long, 1.9 cm. wide, glabrescent; scar at apex concave. Seeds not known.

TANZANIA. Ulanga District: Ifakara, June 1959, *Haerdi* 252/0!
DISTR. T 6; known only from the type
HAB. Not known

SYN. [*O. natalensis* sensu Haerdi in Acta Tropica, Suppl. 8: 142 (1964) *non* Sond.]

NOTE. This species bears a strong resemblance to *O. latifolius* Sond. from Mozambique and South Africa (Natal). It is easily distinguished as *O. latifolius* has much smaller flowers and glabrous pedicels and calyces.

3. **O. zanguebaricus** *(Hiern) Bridson* in K.B. 34: 119 (1979). Type: Tanzania, Bagamoyo District, Ruvu [Kingani], *Hildebrandt* 1268 (K, holo.!, BM, iso.!)

Shrub or small tree 1.5–8.5 m. tall; young branchlets somewhat grooved, glabrous or less often glabrescent. Leaf-blades elliptic to narrowly ovate, 5.4–15.5 cm. long, 2.1–6 cm. wide, acute to shortly acuminate at apex, obtuse to rounded or occasionally acute at base, glabrous above and beneath or sometimes sparsely pubescent or pubescent beneath, thinly coriaceous, midrib straw-coloured when dry, raised above, tertiary nerves finely reticulate, raised above, obscure beneath; domatia absent or inconspicuous; petiole 3–6 mm. long, the margins and midrib raised above, glabrous or covered with short stiff hairs; stipules triangular at base, acuminate at apex, 0.7–1.1 cm. long, 4–6 mm. wide at base, glabrous or rarely glabrescent outside. Inflorescence (3–)5–11-flowered, forming compact narrowly pyramidal panicles occasionally at the top node, but usually at the next 1–5 nodes down; peduncles obscured by the stipules, up to 4 mm. long, the pedicels up to 5 mm. long; bracts and bracteoles lanceolate or linear-lanceolate, 0.7–1(–1.2) cm. long. Calyx glabrous; tube 2.5–3 mm. long; limb-tube ± 2 mm. long, ± as wide as the tube; lobes linear-subulate, (4–)5–8 mm. long. Corolla-tube (7.2–)9.7–11.4 cm. long, ± 2 mm. wide; lobes oblong-lanceolate, 1–2.2 cm. long, 3–5 mm. wide, acuminate. Anthers partly exserted, apiculate. Stigmatic club oblong-ellipsoid, 2–2.5 mm. long, obtuse at apex. Fruit pale brown when dry, flecked with white, pear-shaped, 3.5–5.3 cm. long, 2.4–3.2 cm. wide; calyx-limb eventually deciduous. Seeds blackish brown, 6 mm. long, strongly striated.

KENYA. Kilifi District: Sokoke, 6.4 km. from seashore, *Gisau* SOK 25!; Lamu District: Utwani Forest, Dec. 1956, *Rawlins* 225! & Witu district, Dec. 1956, *Rawlins* 259!
TANZANIA. Tanga District: Kange–Pongwe Forest area, 24 Dec. 1955, *Faulkner* 1779!; Uzaramo District: Pande Forest Reserve, 22 Nov. 1969, *Harris* 3605!; Lindi District: Lake Lutamba, 4 Dec. 1934, *Schlieben* 5684!
DISTR. K 7; T 3, 6, 8; Somalia, ? Mozambique
HAB. Forest; 60–240 m.

SYN. *Gardenia zanguebarica* Hiern in F.T.A. 3: 105 (1877); T.T.C.L.: 495 (1949)
   Randia scabra Chiov., Fl. Somala 2: 236 (1932). Types: Somalia, Baddada, *Senni* 225, 553 & 554 (FT, syn.)
   [*Oxyanthus goetzei* sensu K.T.S.: 452 (1961), pro parte, quoad *Dale* 1134 & *Donald* 453, *non* K. Schum.]

NOTE. The type, a poor fruiting specimen, has calyx-lobes shorter than normal (4–5 mm. long). The following specimens from Zanzibar, *W.E. Taylor* 546 & *Mturi* 190, possibly represent a different taxon. These specimens are generally smaller than *O. zanguebaricus*; the leaf-blades do not exceed 9.5 cm. long, the bracts and bracteoles are 3–5 mm. long and the calyx-lobes ± 2 mm. long, however, the corolla-tube is longer (12–13 cm.). In the absence of fruit it is difficult to decide if the true affinity of these specimens lies with *O. zanguebaricus* or with *O. sp. A*, only known from limited fruiting material.

4. **O. sp. A.**

Glabrous shrub. Leaf-blades narrowly oblong-elliptic, 7–10 cm. long, 1.8–4 cm. wide, acute or shortly acuminate at apex, acute at base, dark green and slightly shiny above, paler beneath, with ± 6 main pairs of lateral nerves, sometimes irregularly toothed at margin, domatia inconspicuous; petiole 3–6 mm. long; stipules triangular, 4–5(–9) mm. long, acute to acuminate. Inflorescence type and flowers unknown. Fruit ± ellipsoid, 1.6–2.2 cm. long, 0.8–1 cm. wide. Seeds not known.

TANZANIA. Uzaramo District: 32 km. S. of Dar es Salaam on road to Kilwa, 25 Jan. 1970, *Harris &* *Rodgers* 3989! & Kisiju, 10 Sept. 1982, *Hawthorne* 1781!
DISTR. **T** 6; not known elsewhere
HAB. Relict or undisturbed forest; 0–150 m.

NOTE. In the absence of flowering material there remains the possibility that this taxon could belong to *Mitriostigma*, or, on the other hand, it may possibly be related to certain specimens mentioned in the note after *O. zanguebaricus*.

5. O. goetzei *K. Schum.* in E.J. 28: 491 (1900); T.T.C.L.: 508 (1949), pro parte; Bridson in K.B. 34: 120 (1979). Type: Tanzania, Iringa District, Lofia R., *Goetze* 445 (B, holo.†, K, iso.!)

Shrub 1–4 m. tall; young branches sometimes slightly grooved, glabrous. Leaf-blades elliptic to broadly elliptic, 6.3–16.8 cm. long, 2.3–7.9 cm. wide, acute to acuminate at apex, acute to obtuse at base, glabrous above, often with a few stiff hairs on the midrib beneath or occasionally pubescent, thinly coriaceous to coriaceous, midrib straw-coloured, raised above, tertiary nerves finely reticulate, usually apparent on both faces; domatia often present; petioles 2–8 mm. long, channelled above, glabrous or with a few stiff hairs; stipules triangular, 0.6–3.5 cm. long, 3–6(–7) mm. wide at base, acute. Inflorescence 1–8-flowered, forming narrow compact panicles at the top node or next 1–3 nodes down, subsessile or with a peduncle not exceeding 3 mm. long; pedicels up to 5 mm. long; bracts and bracteoles lanceolate, 4–9 mm. long, often ciliate. Calyx glabrous; tube 2–5 mm. long; limb-tube 1.75–2.5 mm. long, slightly wider than the tube; lobes triangular at base, subulate above, 2–3.5(–5) mm. long, ciliate. Corolla yellowish or white; tube (5.8–)7.2–11.2 cm. long, 1.75–2.25 mm. wide; lobes oblong, 1.2–3 cm. long, 3–5 mm. wide, acuminate. Anthers mostly exserted, acuminate. Stigmatic club spindle-shaped, 4–5 mm. long, acuminate. Fruit brownish when dry, ellipsoid to spindle-shaped, 2.9–4.5 cm. long, 1.1–1.8 cm. wide; calyx-limb persistent. Seeds rusty brown, 7–8 mm. long, strongly striated.

subsp. **goetzei**

Shrub 1–1.8 m. tall. Stipules 0.9–3.5 cm. long. Inflorescence 3–8-flowered, frequently at uppermost leafy node or sometimes the next 1(–2) nodes down; bracteoles (4–)5–9 mm. long. Calyx-tube 3–5 mm. long.

TANZANIA. Iringa District: Uhehe, Lofia R., Jan. 1899, *Goetze* 445!
DISTR. **T** 7; Mozambique, Malawi, Zimbabwe
HAB. ? Forest; 400–600 m.

SYN. [*O. quirimbensis* sensu Hiern in F.T.A. 3: 109 (1877), quoad specim. *Kirk, non* Klotzsch]
    O. swynnertonii S. Moore in J.L.S. 40: 82 (1911). Types: Mozambique, Boka, Lower Buzi R., *Swynnerton* 580 & Zimbabwe, Chirinda Forest, *Swynnerton* 76, 135 & 6637 (all BM, syn!, K, isosyn.!)
    O. swynnertonii S. Moore var. *breviflora* S. Moore in J.L.S. 40: 83 (1911). Type: Zimbabwe, Chirinda Forest, *Swynnerton* 76a (BM, holo.!)

subsp. **keniensis** *Bridson* in K.B. 34: 120 (1979). Type: Kenya, Teita District, Kasigau, *Joanna* in C.M. 8883 (K, holo.!, EA, iso.!)

Shrub 2–4 m. tall. Stipules 0.6–1.6(–2.4) cm. long. Inflorescence 1–6-flowered, sometimes at the uppermost node but usually at next 1(–2) nodes down; bracteoles 4–6 mm. long. Calyx-tube 2–3 mm. long.

KENYA. Fort Hall District: Thika Falls, 15 Dec. 1932, *Napier* 2366!; Meru District: Meru National Park, 25 Dec. 1972, *Gillett & Ament* 20167!; Kitui District: Mutha Hill, 25 Jan. 1942, *Bally* 1646!
TANZANIA. Pare District: Gonja, near Maore village, 28 Nov. 1954, *Sangiwa* 58!; Morogoro District: Nguru Mts., Turiani, Nov. 1954, *Semsei* 1852!
DISTR. **K** 4, 7; **T** 6, 3: not known elsewhere
HAB. Forest; 610–1650 m.

SYN. [*O. oxycarpus* sensu K.T.S. 452 (1961), quoad *Joanna* in C.M. 8883 & *Napier* 2366, *non* S. Moore]

6. O. lepidus *S. Moore* in J.L.S. 37: 160 (1905); Bridson in K.B. 34: 121 (1979). Type: Uganda, Mengo District, Bwema [Wema] I., *Bagshawe* 594 (BM, holo.!)

Shrub or small tree 2–7 m. tall; young branches often strongly grooved, glabrous. Leaf-blades elliptic or less often broadly elliptic, 4–15.4 cm. long, 1.5–5.8 cm. wide, distinctly acuminate at apex, acute to cuneate at base, chartaceous, dull or moderately glossy above, midrib and lateral nerves raised on both faces, tertiary nerves indistinct, entirely glabrous or finely pubescent beneath; domatia present; petiole 2–7 mm. long, glabrous or sparsely pubescent; stipules triangular at base, linear to subulate above, 0.7–1.7 cm. long, 3–6 mm. wide at base. Inflorescence 2–4(–7)-flowered; peduncle up to 3(–4) mm. long; pedicels up to 5 mm. long; bracts and bracteoles lanceolate, 5–6 mm. long, ciliate or pubescent. Calyx glabrous; tube 2.5–3.5 mm. long; limb-tube 1.5–2.5 mm. long, wider than tube; lobes triangular at base, linear above, 3–4 mm. long. Corolla-tube 10.2–15 cm. long, 1.5–2 mm. wide; lobes oblong, 2–2.3(–3.2) cm. long, 2–3 mm. wide, acuminate. Fruit yellow when ripe, ellipsoid to ovoid, 2.2–7.5 cm. long, 1.1–1.5 cm. wide; calyx-limb persistent. Seeds straw-coloured or light to dark brown, thin, 5 mm. long, striated.

Key to Infraspecific Taxa

Bracts and bracteoles 5–6 mm. long; corolla-tube 10–15 cm.
    long; fruit 2.2–4 cm. long (subsp. *lepidus*):
    Leaf-blades finely pubescent beneath . . . .   **var. lepidus**
    Leaf-blades glabrous or occasionally with a few coarse hairs
       on midrib beneath . . . .   **var. unyorensis**
Bracts and bracteoles 2–4 mm. long; corolla-tube ± 8.5 cm.
    long; fruit up to 7.5 cm. long . . . .   **subsp. kigogoensis**

subsp. lepidus

Bracts and bracteoles 5–6 mm. long; corolla-tube 10–15 cm. long; fruit 2.2–4 cm. long.

var. lepidus

Leaf-blades finely pubescent beneath.

Uganda. Karamoja District: Malu, 1 Mar. 1930, *Paget-Wilkes KA.* 164!; Toro District: Bwamba Forest, 5 Mar. 1939, *A.S. Thomas* 2806!; Mengo District: Entebbe, Mar. 1923, *Maitland* 626!
Tanzania. Buha District: Gombe Stream Reserve, Kakombe valley, 24 Feb. 1964, *Pirozynski* 439!
Distr. U 1, 2, 4; T 4; Zaire (Haute Zaire, Kivu), Sudan
Hab. Forest; 820–1200 m.

Syn. *O. oxycarpus* S. Moore in J.B. 45: 265 (1907); F.P.S. 2: 455 (1952). Type: Uganda, Toro District, near Semliki R., *Bagshawe* 1300 (BM, holo.!)

var. **unyorensis** *(S. Moore) Bridson* in K.B. 34: 121 (1979); Fl. Pl. Lign. Rwanda: 570 (1983); Fl. Rwanda 3: 182 (1985). Type: Uganda, Acholi/Bunyoro Districts, Victoria Nile, Kabalega [Murchison] Falls, *Bagshawe* 1599 (BM, holo.!)

Leaf-blades glabrous or occasionally with a few coarse hairs on midrib beneath.

Uganda. Karamoja District: Mt. Debasien, Jan. 1936, *Eggeling* 2605!; Bunyoro District: Rabongo Forest, 15 June 1964, *G. Jackson* U163!; Mengo District: Bugabe, Sept. 1922, *Dummer* 5487!
Tanzania. Bukoba District: Rubare Forest Reserve, Nov. 1958, *Procter* 1061!; Kigoma District: near Tubila Railway Station, Nov. 1956, *Procter* 579!
Distr. U 1, 2, 4; T 1, 4; Rwanda, Burundi, Ethiopia
Hab. Forest; 650–1825 m.

Syn. *O. unyorensis* S. Moore in J.B. 46: 290 (1908)
    O. microphyllus K. Krause, Z.A.E. 1907–8, 2: 323 (1911). Type: Rwanda, Lake Kivu, Iwawu [Wau] I., *Mildbraed* 1150 [not 450 as on label] (B, holo.†, BR, fragment!)

subsp. **kigogoensis** *Bridson* in K.B. 42: 251 (1987). Type: Tanzania, Iringa District, Mufindi, Kigogo Forest, *Leedal* 5753 (EA, holo.!)

Bracts and bracteoles 2–4 mm. long; corolla-tube ± 8.5 cm. long; fruit up to 7.5 cm. long. Leaf-blades glabrous or sometimes glabrescent beneath with rather bristly hairs on midrib beneath.

Tanzania. Iringa District: Mufindi, Kigogo Forest, 2 Nov. 1979, *Leedal* 5753!, & 29 Apr. 1984, *Lovett & Congdon* 286! & 19 Mar. 1962, *Polhill & Paulo* 1820A!
Distr. T 7; not known elsewhere
Hab. Forest; 1700–1750 m.

Note. (on species as a whole). *O. lepidus* var. *unyorensis* is very close to the West African species *O. subpunctatus* (Hiern) Keay and to various other species which seem to belong to the same complex. It is possible that when revisionary work for Africa as a whole is undertaken the rank of *O. lepidus* may be altered.

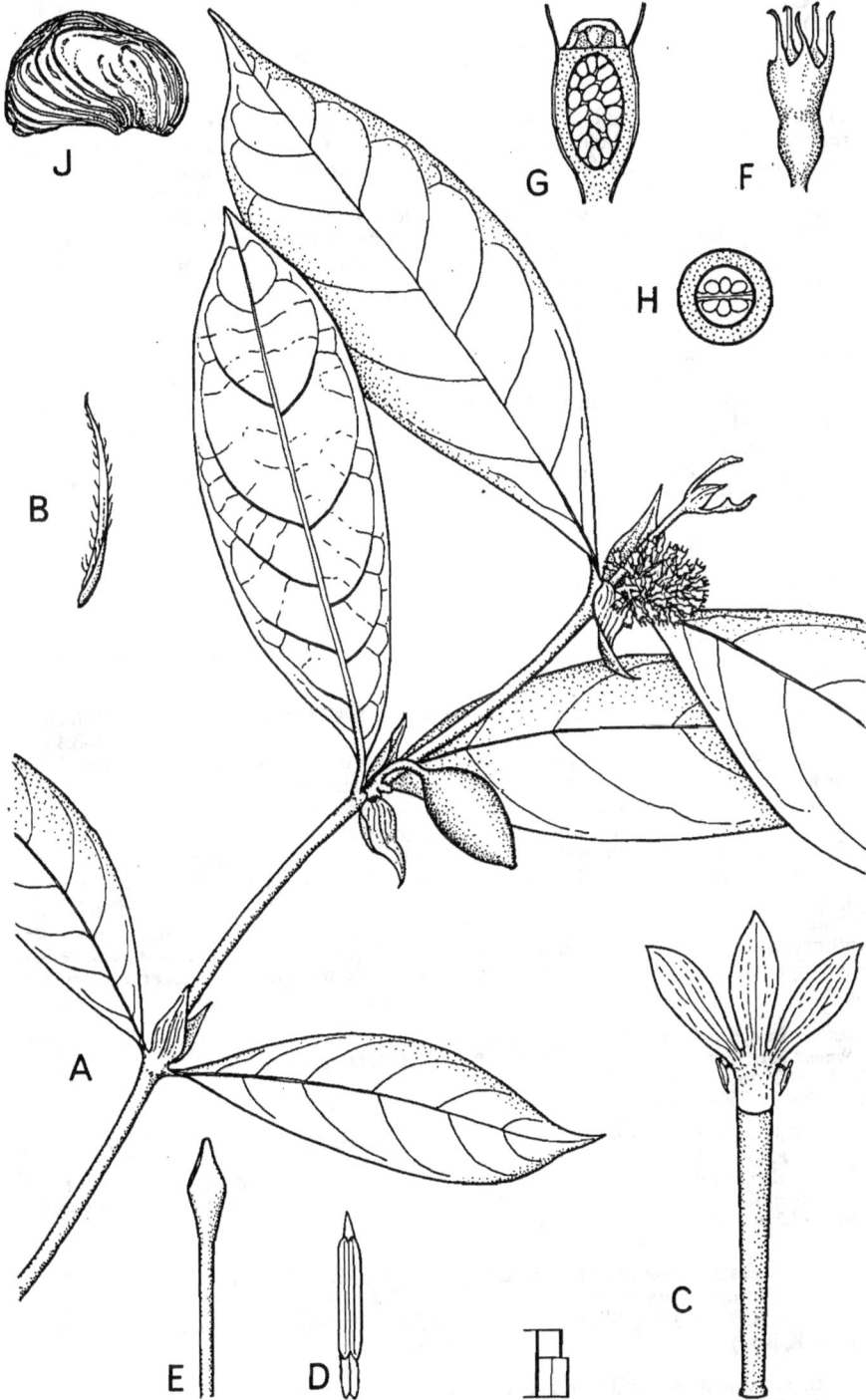

FIG. 86. *OXYANTHUS TROUPINII* — **A**, habit, × ⅔; **B**, bracteole, × 6; **C**, corolla with 2 lobes and portion of tube removed, × 1½; **D**, anther, × 8; **E**, stigma, × 6; **F**, calyx, × 4; **G**, longitudinal section through ovary, × 10; **H**, transverse section through ovary, × 10; **J**, seed, × 3. A, B, J, from *Katende* 1471; C–E, from *Troupin* 11328; F–H, from *Troupin* 11118. Drawn by Diane Bridson.

7. **O. troupinii** *Bridson* in K.B. 34: 122, fig. 2 (1979); Fl. Pl. Lign. Rwanda: 572, fig. 190/2 (1983); Fl. Rwanda 3: 183, fig. 52/2 (1985). Type: Rwanda, Cyangugu Prefecture, road from Butare [Astrida] to Cyangugu [Bukavu], Nyungwe region, *Troupin* 11282 (BR, holo.!, EA, iso.!).

A glabrous shrub or small tree or rarely a suffrutex, 1–12 m. tall; bark on young stems irregularly ridged. Leaf-blades whitish beneath (when fresh), narrowly elliptic to elliptic or less often oblanceolate, 6.2–14 cm. long, 2–6.1 cm. wide, acute to acuminate at apex, cuneate, obtuse or sometimes unequal at base, subcoriaceous, tertiary nerves finely reticulate, usually conspicuous on both faces; domatia present or absent; petiole 0.4–0.5 cm. long; stipules narrowly ovate to ovate or less often lanceolate, 1–2.2 cm. long, 0.4–0.7 cm. wide, acute. Inflorescence 18–25-flowered forming shortly conical panicle, scarcely longer than wide, compact to moderately lax; peduncle up to 3(–5) mm. long; pedicels up to 4 mm. long; bracts and bracteoles filiform above, somewhat keeled at base, 3–4 mm. long, ciliate. Calyx-tube 1.25–2 mm. long; limb-tube 1–1.5 mm. long, scarcely wider than the tube; lobes filiform to subulate or narrowly triangular, 1–1.5 mm. long. Corolla-tube 2.1–2.2 cm. long, 1–1.5 mm. wide, with a few hairs at throat; lobes lanceolate, 7–9 mm. long, 2–2.5 mm. wide, acuminate. Anthers exserted, long-acuminate, with 2 well marked basal appendages, usually different in colour to the thecae. Stigmatic club ellipsoid, distinctly beaked, ± 1 mm. long. Fruit ellipsoid or sometimes pyriform 2.5–3 cm. long, 1.2–1.5 cm. wide; calyx-limb persistent. Seeds orange-brown, 7 mm. long, rather fleshy, striated. Fig. 86.

UGANDA. Kigezi District: N. of Mafuga Forest Station, 14 Dec. 1971, *Katende* 1471!
DISTR. U 2; Rwanda, Burundi
HAB. Forest understorey; 2300 m.

8. **O. ugandensis** *Bridson* in K.B. 34: 124, fig. 3 (1979). Type: Uganda, Toro District, Semliki Forest, *Osmaston* 1376 (K, holo.!)

Shrub 1 m. tall; young stems glabrous tending to be grooved. Leaf-blades discolorous, sometimes very pale beneath, elliptic to narrowly obovate, 9.3–17.4 cm. long, 4–8.4 cm. wide, distinctly acuminate at apex, cuneate at base, chartaceous, glabrous above, finely pubescent beneath (at least when young); tertiary nerves finely reticulate, ± raised on both faces; domatia present; petioles 0.5–1 cm. long, sparsely pubescent beneath; stipules ovate, 0.6–1 cm. long, 5–6 mm. wide, acute, glabrous or sparsely pubescent outside. Inflorescence 5–9-flowered; peduncles very short, ± 1.5 mm. long; pedicels 2–4 mm. long; bracteoles inconspicuous. Calyx glabrous; tube 2–2.5 mm. long; limb-tube 2 mm. long, slightly wider than the tube; lobes narrowly triangular or linear-triangular, 2–3 mm. long. Corolla-tube 12.3–13.3 cm. long, 2 mm. wide; lobes oblong, ± 2.6 cm. long, 2 mm. wide. Anthers almost entirely exserted, long-acuminate. Stigmatic club 7 mm. long. Fruit yellow when ripe, ellipsoid, 3.2 cm. long, 1.7 cm. wide; calyx-limb persistent. Seed not known. Fig. 87.

UGANDA. Bunyoro District: Unyoro, 18 Oct. 1910, *Dawe* 1039! & Budongo Forest, 28 Sept. 1971, *Synnott* 681!; Toro District: Semliki Forest, 27 Oct. 1951, *Osmaston* 1376!
DISTR. U 2; not known elsewhere
HAB. Forest; 750–1050 m.

NOTE. This species is close to certain specimens from Zaire (Haute Zaire: *Germain* 8645, *Lebrun* 2721 & *Robyns* 1314 & Equateur: *Goosens* 4378) that have been determined *O. sp. cf. laurentii* De Wild. and to several undetermined *LeTestu* specimens from the Central African Republic. These specimens differ from *O. ugandensis* in the entirely glabrous leaf-blades, the shorter calyx-lobes (up to 1.5 mm. long) and the more acute stipules.

9. **O. pyriformis** *(Hochst.) Skeels* in U.S. Dep. Agric., Bur. Pl. Ind., Bull. 248: 56 (1912); Palmer & Pitman, Trees S. Afr. 3: 2067, photo. 2066 (1972); Ross, Fl. Natal: 333 (1973); Bridson in K.B. 34: 124 (1979). Type: South Africa, Durban [Port Natal], *Krauss* 110 (B, holo.†, K, iso.!)

Glabrous shrub or small tree 2–8.4 m. tall; young stems ± terete. Leaf-blades ovate, broadly elliptic or oblong-elliptic, 8.5–29 cm. long, 4.5–16.7 cm. wide, obtuse to shortly acuminate at apex, cuneate to truncate or less often cordate, and frequently unequal at the base, thinly coriaceous; tertiary nerves rather coarsely reticulate, often raised above; domatia obscure; petiole 0.5–1.3 cm. long; stipules ovate to triangular, 0.9–2.8 cm. long,

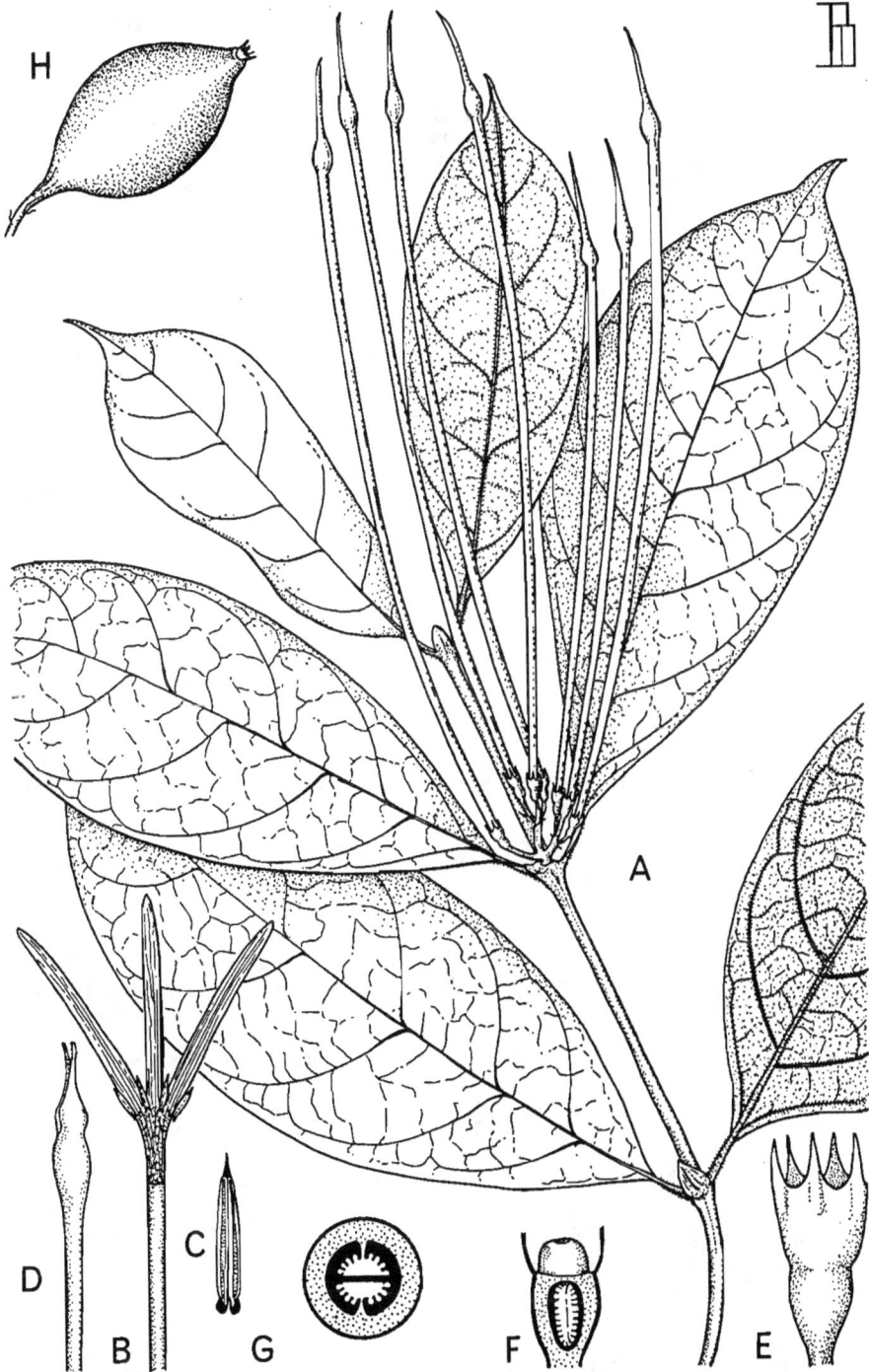

FIG. 87. *OXYANTHUS UGANDENSIS* — **A**, flowering branch, × ⅓; **B**, top of corolla with 2 lobes and portion of tube removed, × 1; **C**, anther, × 6; **D**, stigma, × 6; **E**, calyx, × 4; **F**, longitudinal section of ovary with one placenta removed, × 4; **G**, transverse section of ovary, × 6; **H**, fruit, × 1. A, from *Osmaston* 1376; B–G, from *Dawe* 1039; H, from *Synnott* 681. Drawn by Diane Bridson.

0.4–1.2 cm. wide, acute to acuminate. Inflorescence a moderately lax 10–30-flowered cyme (becoming lax in fruiting stage); peduncle short, exceeded by the stipule; pedicels 3–8 mm. long; bracteoles inconspicuous, ± subulate at tip. Calyx glabrous; tube ovate, 2–3 mm. long; limb-tube 1–2.5 mm. long, wider than the tube; lobes filiform to subulate, 1–4 mm. long. Corolla-tube 3.4–13.6 cm. long, 1.5–2.5 mm. wide at top; lobes oblong or narrowly obovate, 1–2.6 cm. long, 1.5–5 mm. wide, acute to acuminate. Anthers almost entirely exserted. Stigmatic club 1.75–4 mm. long. Fruit pale green, obovoid, 1.8–3.7 cm. long; calyx-limb caducous. Seeds pale brown, 5 mm. long, with obvious striations.

KEY TO INFRASPECIFIC VARIANTS

Corolla-tube 6.8–13.6 cm. long; calyx-lobes (1–)2–4 mm. long:
    Corolla-tube 6.8–11(–12) cm. long . . . . subsp. **tanganyikensis**
    Corolla-tube 11–13.6 cm. long . . . . subsp. **longitubus**
Corolla-tube 3.4–4 cm. long; calyx-lobes 1–2 mm. long   subsp. **brevitubus**

subsp. **tanganyikensis** *Bridson* in K.B. 34: 126 (1979). Type: Tanzania, Lushoto District, near Amani, Kwamkoro, *Greenway* 1016 (K, holo.!, EA, iso.!)

Bases of leaf-blades typically cuneate, often unequal or occasionally rounded and unequal; stipules up to 2.8 cm. long, triangular to triangular-ovate. Calyx with limb-tube 1–1.5(–1.75) mm. long and lobes 2–4 mm. long. Corolla-tube 6.8–11(–12) cm. long. Stigmatic club up to 3(–3.5) mm. long.

TANZANIA. Lushoto District: Shume-Magamba, July 1960, *Semkiwa* 108!; Ulanga District: near Kasawasawa, Aug. 1959, *Haerdi* 313/0!; Iringa District: Karenga [Gologolo] Mt., near Kidatu, Kilombero Scarp Forest Reserve, 28 Mar. 1970, *Harris & Pocs* 4235!
DISTR. T 3, 6–8; not known elsewhere
HAB. Forest and woodland; 350–1525 m.

SYN. [*O. goetzei* sensu T.T.C.L.: 508 (1949), pro parte, quoad *Greenway* 1016 & 4703, non K. Schum.]

subsp. **longitubus** *Bridson* in K.B. 34: 126 (1979). Type: Kenya, Kwale District, Shimba Hills, *Drummond & Hemsley* 3961 (K, holo.!, BR, EA, iso.!)

Bases of leaf-blades obtuse to cuneate; stipules up to 2.4 cm. long, triangular-ovate. Calyx with limb-tube (1–)1.5–1.75 mm. long and lobes 2–3 mm. long. Corolla-tube 11–13.6 cm. long. Stigmatic club 4 mm. long.

KENYA. Kwale District: Shimba Hills, Mwele Mdogo Forest, 18 km. SW. of Kwale, 23 Aug. 1953, *Drummond & Hemsley* 3961! & Lango ya Mwagandi [Longomwagandi] Forest, 11 Feb. 1968, *Magogo & Glover* 81! & 12 Apr. 1968, *Magogo & Glover* 812!
DISTR. K 7; not known elsewhere
HAB. Forest; 240–450 m.

SYN. [*O. goetzei* sensu K.T.S. 1452 (1961), pro parte, quoad *Battiscombe* 73 & 184, non K. Schum.]

NOTE. It is possible that this subspecies is not always distinct from subsp. *tanganyikensis;* additional gatherings are needed.

subsp. **brevitubus** *Bridson* in K.B. 34: 126 (1979). Type: Kenya, Masai District, Emali Forest, *V.G. van Someren* 92 (K, holo.!, EA, iso.!)

Bases of leaf-blades ± truncate to rounded or tending to be cordate; stipules up to 2 cm. long, tending to be ovate. Calyx with limb-tube 1–1.75 mm. long and lobes 1–2 mm. long. Corolla-tube 3.4–4 cm. long. Stigmatic club 2.25–3 mm. long.

KENYA. Kitui District: Mutito Hill, Jan. 1937, *Gardner* in F.D. 3616!; Masai District: Emali Forest, 3 Mar. 1940, *V.G. van Someren* 92!
TANZANIA. Arusha District: Ngurdoto Crater, SE. of Leopard Point, 27 Oct.1965, *Greenway & Kanuri* 12216! & Mt. Meru, Usa sawmills, Jan. 1952, *Watkins* 589!
DISTR. K 4, 6; T 2; not known elsewhere
HAB. Montane forest; 1200–1800 m.

SYN. [*O. speciosus* sensu K.T.S.: 452 (1961), pro parte, quoad *Gardner* in F.D. 3616, non DC.]

NOTE. The following fruiting specimen: Kenya, 28 km. SSE. of Embu, Kiangombe, *Abraham* in *Bally* 7382, probably belongs to this subspecies but cannot be placed with certainty. More collections from this area are needed.

NOTE. (on species as a whole). Subsp. *pyriformis* (formerly known as *O. natalensis* Sond.) can be separated from the subspecies occurring in the Flora area by the tubular part of the calyx-limb which is 2–2.5 mm. long in subsp. *pyrifomis* and only 1–1.75 mm. long in specimens from the Flora area.

10. O. formosus *Planch.* in Hook., Ic. Pl. 8, t. 785–6 (1848); Hiern in F.T.A. 3: 109 (1877); F.P.S. 2: 455 (1952); Keay, F.W.T.A., ed. 2, 2: 129 (1963); N. Hallé in Fl. Gabon 17, Rubiacées, 2: 188, t. 42/1–8 (1970); Bridson in K.B. 34: 127 (1979). Type: Liberia, Cape Palmas, *Vogel* 24 (K, holo.!)

Shrub or small tree 0.9–5 m. tall; young stems often grooved, glabrous. Leaf-blades usually turn brown or greyish brown when dry, discolorous, narrowly oblong to oblong or tending to be lanceolate or oblanceolate, 15–31 cm. long; 2.6–14 cm. wide, acuminate at apex, unequal and rounded to cordate with short side often acute at base, entirely glabrous or rarely pubescent beneath (not in Flora area); midrib raised above; tertiary nerves apparent or obscure but not raised; domatia absent; petiole 0.3–1.2(–1.5) cm. long; stipules lanceolate to ovate, (0.9–)1.3–3 cm. long, 0.55–1.3 cm. wide, acute. Inflorescence 6–30-flowered, lax, glabrous; peduncle 0.8–2 cm. long; pedicels 0.5–1.8 cm. long; bracteoles absent or vestigial. Calyx glabrous; tube 3 mm. long; limb-tube 2–4 mm. long, slightly wider than tube, truncate to repand or with short teeth up to 1(–2) mm. long. Corolla-tube 13–22 cm. long, ± 2.5 mm. wide; lobes narrowly oblong, 1.8–3.5 cm. long, 2–3(–4) mm. wide, acute. Anthers entirely exserted, acuminate. Stigmatic club fusiform, 5.5–8 mm. long. Fruit green, marked with white, turning to bright orange when ripe, ellipsoid to pear-shaped, 4–6 cm. long, 1.8–2.7 cm. wide; calyx-limb persistent. Seeds ? grey-brown, 7.5–8 mm. long, with distinct striations.

UGANDA. Bunyoro District: Bugoma Forest, 26 May 1910, *Dawe* 1021!; Toro District: Muhangi Central Forest Reserve, 11 June 1952, *Stuart-Smith* 38!; Kigezi District: S. Maramagambo Forest, 8 Nov. 1969, *Synnott* 421!
DISTR. U 2; Sierra Leone, Liberia, Ivory Coast, Ghana, Nigeria, Bioko [Fernando Po], Cameroun, Gabon, Zaire, Sudan, Angola
HAB. Forest; ± 1375 m.

SYN. *O. breviflorus* Benth. in Hook., Niger Fl.: 388 (1849); Hiern in F.T.A. 3: 109 (1877); F.W.T.A. 2: 80 (1931). Type: Bioko [Fernando Po], *Vogel* 209 (K, holo.!)
　O. bagshawei S. Moore in J.B. 45: 266 (1907). Type: Uganda, Toro District, bank of Semliki R., *Bagshawe* 1285 (BM, holo.!)
　O. longitubus Good, in J.B. 64, Suppl. 2: 16 (1926). Type: Angola, Cabinda, Buco Zau, *Gossweiler* 6870 (BM, holo.!, K, iso.!)

NOTE. The specimens from the Flora area characteristically all have narrowly oblong leaf-blades, few-flowered inflorescences and corolla-tubes towards the upper limits of the measurements given. An infraspecific taxon has not been recognised as several similar specimens have been noted from West Africa.

11. O. unilocularis *Hiern* in F.T.A. 3: 110 (1877); Aubrév., Fl. For. Côte d'Ivoire, ed. 2, 3: 249, t. 341 (1936); I.T.U., ed. 2: 353 (1952); F.P.S. 2: 455 (1952); Keay, F.W.T.A., ed. 2, 2: 129 (1963); N. Hallé, Fl. Gabon, 17, Rubiacées, 2: 193, t. 44 (1970). Types: N. Nigeria, Jeba on the Niger [Quorra], *Barter* 1075 & S. Nigeria, Bonny R., *Mann* 506 (both K, syn.!, P, isosyn.!)

Shrub or small tree, 1.5–6(–8) m. tall, with stout angular hollow branchlets, ± pubescent. Leaf-blades broadly ovate-elliptic or elliptic-obovate, 25–60 cm. long, 13–40 cm. wide, acuminate at the apex, equally or unequally broadly cordate at the base, hispid pilose when young, becoming glabrescent save for pubescence on the venation beneath, thin, often somewhat bullate; petiole 0.2–1.2 cm. long; stipules narrowly ovate, 1.5–3 cm. long, 0.9–1.5 cm. wide, acuminate, connate for ± 1 cm. in young stage, pubescent. Inflorescence a short 20–80-flowered corymb; peduncle 1–2 cm. long, pubescent; pedicels 2–5 mm.long, pubescent; bracteoles small, linear to filiform, 2–5 mm. long. Calyx-tube narrowly turbinate, 3–4 mm. long, glabrescent to sparsely pubescent; limb-tube 2 mm. long; lobes subulate, 3–6(–9) mm., ciliolate, at first erect then arched. Corolla-tube 13–20 cm. long, 1.5–3 mm. wide; lobes narrowly lanceolate, 1.5–3 cm. long, 3–5 mm. wide. Anthers exserted for most of their length. Stigmatic club green, 1 cm. long. Fruit broadly ellipsoid or subglobose, 2–3 cm. long, 1–2.5 cm. wide. Seeds light brown, 7–8 mm. long, striated.

UGANDA. Toro District: footslopes of Ruwenzori Mts., Bwamba Forest, 2 Feb. 1945, *Greenway & Eggeling* 7065!; Masaka District: NW. side of Lake Nabugabo, 10 Oct. 1953, *Drummond & Hemsley* 4696!; Mengo District: Kitubulu Forest, near Entebbe, Apr. 1935, *Chandler* 1201!
DISTR. U 2, 4; Sierra Leone to Nigeria, Cameroun, Central African Republic, Gabon, Zaire, Sudan, Angola

SYN. *O. macrophyllus* Hiern in F.T.A. 3: 110 (1877); K. Krause in Z.A.E. 1907–8, 2: 321 (1911). Type: Sudan, Djur, *Schweinfurth* 1734 (K, syn.!, BM, isosyn.!) & Niamniam, *Schweinfurth* 2832 (BM, syn.!)
　O. litoreus S. Moore in J.L.S. 37: 160 (1905). Type: Uganda, Masaka District, Shore of Lake Victoria at Misozi [Musozi], *Bagshawe* 95 (BM, holo.!)

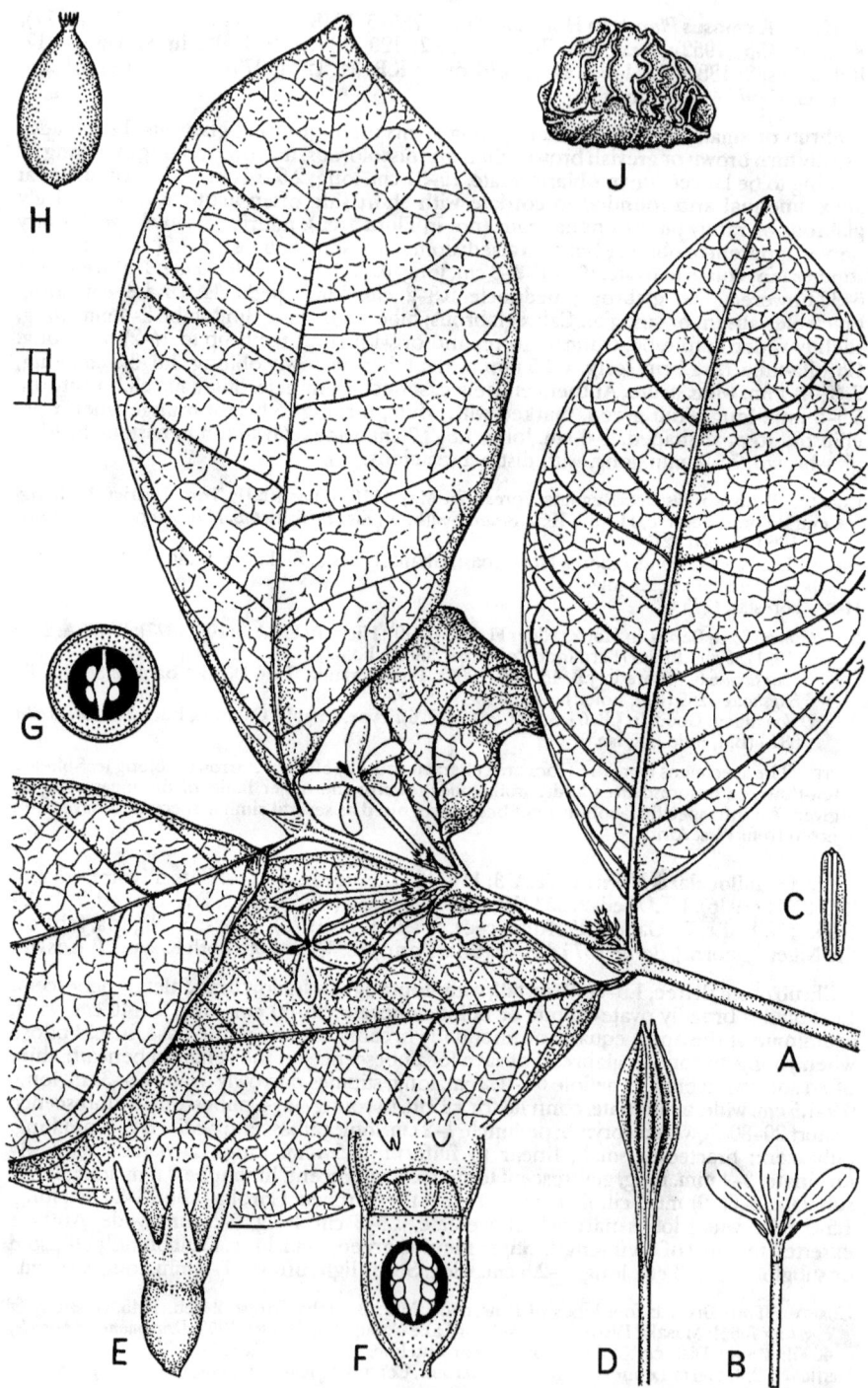

FIG. 88. *MITRIOSTIGMA GREENWAYI* — **A**, flowering branch, × ⅔; **B**, corolla with 2 lobes and portion of tube removed, × 1; **C**, anther, × 6; **D**, stigma, × 4; **E**, calyx, × 4; **F**, longitudinal section through ovary, × 10; **G**, transverse section through ovary, × 10; **H**, fruit, × 1; **J**, seed, × 6. All from *Greenway* 9639. Drawn by Diane Bridson.

69. MITRIOSTIGMA

Hochst. in Flora 25: 235 (1842)

Small shrubs. Leaves petiolate; domatia absent or present as small hairy tufts in the axils of the lateral nerves; stipules green in dry state, erect, persistent, scarcely connate at base. Flowers ☿ , 5-merous, subsessile to shortly pedicellate in small cymes, lateral at successive nodes or axillary, moderately lax or condensed; bracteoles present. Calyx-tube turbinate to ellipsoid; limb with distinct tubular part and narrowly triangular lobes. Corolla usually white, glabrous outside; tube not markedly long, cylindrical at base and campanulate or funnel-shaped above or cylindrical and widening only near apex, ± glabrous to sparsely pubescent inside; lobes short and erect or larger and erect or spreading. Stamens inserted in broad part of corolla-tube; anthers subsessile, attached near the base, entirely included or up to ⅔-exserted, shortly apiculate. Style slender, glabrous; stigmatic club included or exserted, winged, mitriform or scarcely so, shortly bifid at apex. Ovary 2-locular; placentas rather fleshy, subpendulous, bearing 6–12 ovules, not immersed in placental tissue. Fruit globose to ellipsoid, crowned by persistent calyx-limb. Seeds quite large, 4.5–6 mm. across, somewhat compressed or not, fleshy; testa ornamented with fine narrow reticulations.

A small genus of 5 rather dissimilar species, two in East Africa, two in W. Africa (one undescribed) and one in southern Africa.

M. axillare Hochst. extends from South Africa into Mozambique as far north as Cabo Delgado and quite probably occurs in Tanzania, although not yet recorded. It is readily distinguished from *M. usambarense* by its white flowers.

Inflorescences usually on both sides at each node; corolla
    white, cylindrical almost to the apex . . . . . .   *1. M. greenwayi*
Inflorescences on one side at each node and alternating;
    corolla purple, funnel-shaped . . . .   *2. M. usambarense*

1. M. greenwayi *Bridson* in K.B. 34: 127, fig. 4 (1979). Type: Kenya, Kwale District, Jadini, *Greenway* 9639 (K, holo.!, EA, PRE, iso.!)

A glabrous suffrutex, with several erect simple or branched stems, or a shrub, 0.6–1 m. tall; stems square and somewhat grooved when young; bark with very fine longitudinal ridges. Leaf-blades obovate, 7.5–22 cm. long, 3.8–10.5 cm. wide, apex obtuse, base obtuse, coriaceous; midrib shallowly crested above, tertiary nerves prominent above; petioles 0.3–1 cm. long, stout, channelled above; stipules triangular, 0.7–1.5 cm. long, acute or sometimes acuminate. Inflorescences one in each axil or occasionally only in one axil, few–15-flowered, subcapitulate, often subtended by leaf-like bracts up to 2 cm. long; peduncle 0.1–1 cm. long; bracteoles ovate, ± 4 mm. long, acuminate, ciliate. Calyx-tube ovate, 2–3 mm. long; limb-tube ± 1 mm. long; lobes narrowly triangular, 1.25–2 mm. long, acute, ciliate. Corolla white; tube 2 cm. long, 2 mm. wide at top; lobes narrowly obovate, 0.8–1.1 cm. long, 4 mm. wide, rounded. Anthers exserted for ± ⅔ their length. Stigmatic club 4 mm. long, winged. Fruit green, said to be globose or pyriform, 1.9–2.5 cm. long, 1–1.5 cm. wide, crowned by the persistent calyx-limb. Seeds red-black, 4 mm. across, wrinkled and with extremely fine reticulations. Fig. 88.

KENYA. Kilifi District: Mwarakaya [Nwarakaya], 21 Nov. 1978, *Brenan, Gillett et al.* 14669! & Kaya Forest, 20 Apr. 1981, *Hawthorne* 247 E! & Pangani, crossing of Lwandani stream on Chonyi–Ribe road, 17 Feb. 1977, *Faden et al.* 77/531!
DISTR. **K** 7; not known elsewhere
HAB. Coastal forest, on coral rock or limestone outcrops; near sea-level to 150 m.

2. M. usambarense *Verdc.* in K.B. 42: 245, fig. 1 (1987). Type: Tanzania, W. Usambara Mts., Mazumbai, *Lovett* 171 (K, holo.!, BR, MO, iso.!)

Shrub to 2 m. tall, with greenish brown longitudinally rugose branches; branching slightly supra-axillary. Leaf-blades elliptic to obovate-elliptic, 5.5–12 cm. long, 2.5–5 cm. wide, abruptly acuminate at the apex, cuneate at the base, glabrous save for sparsely pilose domatia beneath; petiole 5–8 mm. long; stipules triangular, 6 mm. long, acuminate, ± keeled, glabrous. Inflorescences slightly supra-axillary, placed unilaterally at successive nodes on alternate sides, 1 (or possibly –2 or –3)-flowered; peduncle short, 2 mm. long;

pedicels probably red, 1 cm. long; bracteoles lanceolate, 2–3 mm. long, ciliate. Calyx-tube ovoid, 2.5 mm. long; limb-tube 1.5 mm. long, with narrowly triangular acuminate ciliate lobes 2 mm. long, 1 mm. wide. Corolla purple, narrowly funnel-shaped with cylindrical base; tube 2.4 cm. long, 5–6 mm. wide at throat, finely pubescent inside save at base; lobes rounded, 5 mm. in diameter, glabrous. Anthers included. Stigmatic club divided into 2 lobes 2 mm. long reaching about halfway up the corolla-lobes. Fruit orange to red, ellipsoid, 2.2 cm. long, 1.4 cm. wide. Seeds pale brown, elliptic in outline, compressed, 5 mm. long, 4 mm. wide and 2 mm. thick.

TANZANIA. Lushoto District: W. Usambara Mts., Mazumbai, University Forest Reserve, 13 Sept. 1983, *Lovett* 171! & same area, Kwamshunde, 13 May 1981, *L. Tanner* 261! & Balangai West Forest Reserve, SE slope of Kilimandege, 28 Feb. 1984, *Borhidi et al.* 84298!
DISTR. T 3; not known elsewhere
HAB. Montane ridge forest; (?1400–)1500–1830 m.

70. TRICALYSIA

DC., Prodr. 4: 445 (Sept. 1830); A. Rich., Mém. Fam. Rub.: 144 (Dec. 1830); Robbrecht in B.J.B.B. 49: 239–360 (1979), 52: 311–339 (1982), 53: 299–320 (1983) & 57: 39–208 (1987)

Bunburya Hochst. in Flora 27: 553 (1844), *non Bunburia* Harv. (1830)

Natalanthe Sond. in Linnaea 23: 52 (1850)

Rosea Klotzsch in Ber. Bekanntm. geeign. Verhandl. Kön. Preuss. Akad. Wiss. Berlin 1853: 501 (1853); Peters in Reise Mossamb., Bot. 1: 293 (1862)

Empogona Hook.f. in Hook., Ic. Pl. 11: 72, t. 1091 (1871)

Diplocrater Hook.f. in G.P. 2: 96 (1873)

Probletostemon K. Schum. in E.J. 23: 450 (1897)

Neorosea N. Hallé in Fl. Gabon 17, Rubiacées, 2: 268 (1970) pro parte quoad speciem typicam

Mostly shrubs, less often small trees or (only a few species, not in the Flora area) large dominant trees; rarely (± 6 species) rhizomatous undershrubs. Leaves petiolate, often with domatia, never with bacterial nodules; stipules sheathing, crowned by two awns. Inflorescences axillary and opposite, sessile to subsessile, rarely 1-, mostly many-flowered and strongly contracted. Flowers (4–)5–9(–12)-merous, mostly sweetly scented, sessile or rarely pedicellate, ♂ (with the exception of a group of Madagascan species which is unisexual), occasionally dimorphic (sect. *Ephedranthera*); bracts (primary and secondary etc.) and bracteoles similar and arranged in a series or less often dissimilar, paired, fused, cupular and somewhat resembling the calyx-limb or less often dissimilar to calyx-limb, dentate, awned or occasionally with leafy appendages, subtending the usually condensed inflorescence branches and with 1 (–3 in simple inflorescences) cup subtending and frequently surrounding each calyx-tube; in some pedicellate species bracteoles free, either opposite or alternate. Calyx very variable, either with short limb-tube and distinct lobes (subgen. *Empogona)* or with well-developed, truncate or dentate limb-tube. Corolla mostly white, sometimes greenish, cream or rose, salver-shaped; lobes spreading or reflexed, contorted to the left. Stamens attached to the throat, mostly with distinct filaments; anthers medifixed, exserted, rarely (sect. *Ephedranthera*) sessile and half-included to almost completely included, with or without an inconspicuous or markedly conspicuous apical appendage formed by an extension of the connective. Disc annular. Ovary 2-locular, with 1–many ovules embedded in a placenta attached to the septum; style 2-lobed, exserted, lying between the anthers or just overtopping them, but deeply included in the short-styled form of sect. *Ephedranthera*. Fruits ± drupaceous, mostly less than 1 cm. in diameter, rarely (sect. *Probletostemon* and *T. allocalyx*) somewhat larger and with ± dry, sclerified indehiscent wall, mostly red, rarely orange, in subgen. *Empogona* first white then turning blackish; calyx mostly persistent; endocarp papery. Seeds 1–many per locule (rarely only 1 per fruit by abortion of the second chamber); seed-coat chestnut-brown, colliculate, striate or smooth; hilum long, linear, often curved; endosperm horny; embryo straight, attaining at most ½ the height of the seed; cotyledons ± equalling the basally oriented radicle.

A large genus of ± 95 African and ± 7 Madagascan species.

The Asiatic elements sometimes included in it, the Malaysian genus *Diplospora* DC. considered a section of *Tricalysia* by K. Schum. (in E. & P. Pf. 4(4): 82 (1891)), and the genus *Discospermum* Hook., from Sri Lanka, included in *Diplospora* by Hook.f. (G. P. 2: 96 (1873)), also made a section of *Tricalysia* by Schumann, are in need of in-depth investigation, but are probably both worthy of generic recognition (see Robbrecht & Puff in E.J. 108: 115 (1986)). Following a revision of the African species (see citations above) the author decided on the classification set out below. Subgen. *Empogona* ± matches Schumann's section *Kraussia*, but excludes those species now placed in the distinct genus *Kraussia* (genus 84). The Madagascan species, with the exception of *T. ovalifolia* (in East Africa as well as on Madagascar and several smaller Indian Ocean Islands), are probably best regarded as a separate section close to sect. *Tricalysia* but characterised by unisexual flowers.

Subgen. Tricalysia

Calyx (fig. 89/ 5-11) with well developed limb-tube, truncate, dentate or with subulate lobes, often showing one or several (fig. 89/10,11) longitudinal splits at anthesis, or scarcely developed in *T. allocalyx* (fig. 89/1). Corolla-throat glabrous or weakly hairy. Anthers often ± hairy, with or without an inconspicuous apical appendage. Style often ± hairy. Fruits red, rarely orange (colour unknown in sect. *Probletostemon*).

Sect. Probletostemon *(K. Schum.) Robbrecht* in B.J.B.B. 53: 307 (1983) & in Distr. Pl. Afr. 24, map 804 (1984).

Bracteoles free. Stamens with filaments present. Fruits larger than in other sections except for *T. allocalyx* in *Ephedranthera* (attaining 1.5-3 cm. long), ± dry. Species 1,2.

Sect. Tricalysia; Robbrecht in B.J.B.B. 57: 71 (1987) & in Distr. Pl. Afr. 31, map 1015 (1987)

Bracteoles fused. Calyx-limb remaining entire after anthesis (fig. 89/ 5-9), occasionally showing one longitudinal split. Stamens with filaments present. Species 3-14.

Sect. Rosea *(Klotzsch) Robbrecht* in B.J.B.B. 57: 180: (1987) & in Distr. Pl. Afr. 32, map 1063 (1987)

Bracteoles fused. Calyx-limb showing 3-4 very deep splits after anthesis, becoming spathaceous (fig. 89/10,11). Stamens with filaments present. Species 15-18.

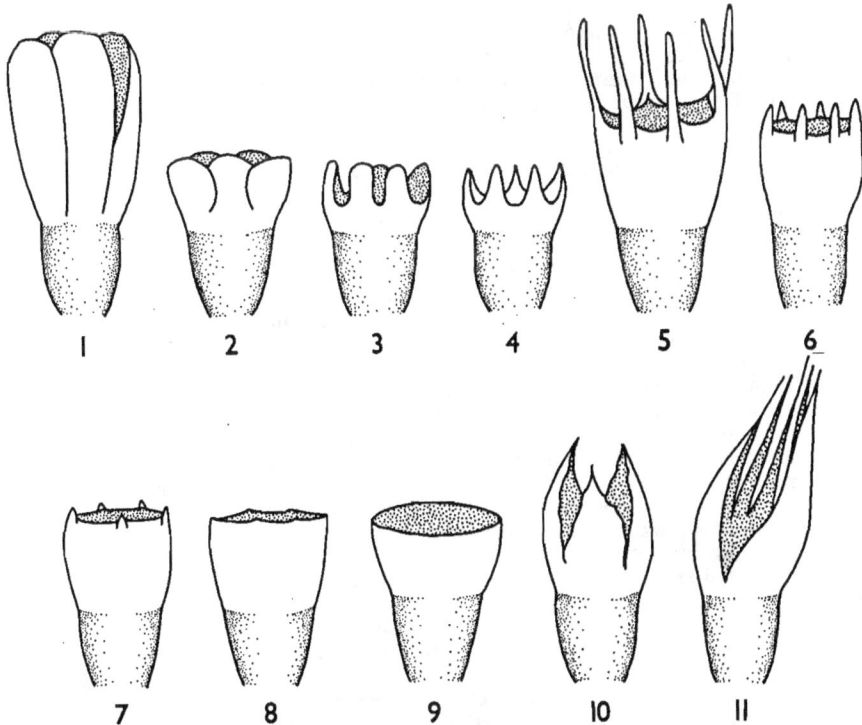

FIG. 89. *TRICALYSIA* — diagrammatic representation of calyx-types; 1-4, limb-tube short, lobes well developed (1, 2, lobes overlapping; 3, lobes ± rounded; 4, lobes ± triangular, separated by sinuses); 5-9, limb-tube well developed, lobes subulate, short or absent; 10, 11, limb ellipsoidal splitting longitudinally and ± irregularly at anthesis. Drawn by Sally Dawson after E. Robbrecht.

Sect. **Ephedranthera** *Robbrecht* in B.J.B.B. 52: 317 (1982) & in Distr. Pl. Afr. 23, map 759 (1982)
Bracteoles fused. Calyx-limb variable (fig. 89/1, 6, 7 & 10, 11). Anthers sessile, only partly exserted, or completely included. In this section several species have two floral forms, one with exserted, the other with included style-lobes, (see B.J.B.B. 52, fig. 2 (1982)). This type of heterostyly differs from that occurring in the *Rubioideae*, since in sect. *Ephedranthera* the anthers are similar in size and position in both forms. Some species have only one floral form (e.g. *T. vanroechoudtii* with only brevistylous flowers). Species 19–25.

Subgen. **Empogona** *(Hook.f.) Robbrecht* in B.J.B.B. 49: 259 (1979) & in Distr. Pl. Afr. 16: map 531 (1979)
Calyx with very short limb-tube and well-developed lobes (fig. 89/ 2–4). Corolla throat mostly bearded. Anthers with a conspicuous apical appendage. Style mostly entirely glabrous. Drupes first white, then turning purple and at complete maturity almost black.

Sect. **Empogona** *(Hook.f.) Brenan;* Robbrecht in B.J.B.B. 49: 267 (1979)
Flowers often (in Flora area always) distinctly stalked, mostly with free and alternate or subopposite bracteoles. Calyx-lobes not touching one another (fig. 89/3,4). Species 26–28.

Sect. **Kraussiopsis** *Robbrecht* in B.J.B.B. 49: 309 (1979)
Flowers sessile to subsessile; bracteoles either free and opposite, or fused. Calyx-lobes contorted (not in Flora area) or irregularly imbricate (fig. 89/2). Species 29–31.

NOTE.   The calyx (fig. 89) is the most variable and most important diagnostic character at species level in *Tricalysia*. In the majority of the species it is persistent and observable both in flowering and fruiting material.

1. Calyx with short limb-tube and 4 elongated lobes ± twice as
    long as the calyx-tube (fig. 89/1), silky outside; anthers
    sessile, hairy; corolla-tube exceeding 3 cm.; each
    placenta with more than 10 ovules; tertiary nerves
    conspicuous, perpendicular to midrib   . . . . .   *23. T. allocalyx*
    Characters not combined as above   2
2. Limb-tube of calyx well developed, truncate, dentate or
    with subulate lobes (fig. 89/5–11); fruits red, rarely
    orange (subgen. *Tricalysia*)   3
    Limb-tube of calyx short; lobes well developed, either ±
    rounded and overlapping, i.e. irregularly imbricate
    (fig. 89/2) or rounded to triangular and not touching
    one another (fig. 89/3,4); fruits first white, then purple
    to blackish (subgen. *Empogona*)   27
3. Stamens with distinct filaments, the anthers exserted; style-
    lobes always exserted *   4
    Stamens with sessile anthers, partly included; style-lobes
    either included or exserted (sect. *Ephedranthera*)   22
4. Bracteoles free, alternate or subopposite; fruits relatively
    large (2–3 cm. in diameter), with sclerified, dry,
    indehiscent wall; calyx-lobes not persisting as fruit
    matures (sect. *Probletostemon*)   5
    Bracteoles cupular; fruits with more fleshy walls, mostly
    1 cm. in diameter, with persistent calyx-lobes   6
5. Young twigs glabrous; tuft-domatia present; leaf-blades
    cuneate (rarely rounded) at base; calyx-lobes tri-
    angular, up to 4 mm. long; fruits ellipsoid, ± smooth,
    glabrescent   . . . .   *1. T. elliotii*
    Young twigs hairy; domatia absent; leaf-blades cordate
    (rarely rounded) at base; calyx-lobes subulate, 6–10
    mm. long; fruits spherical or transversally flattened, usually
    warty, pubescent   . . . .   *2. T. anomala*
6. Calyx-limb tubular, widened towards margin, not covering
    the corolla in bud, occasionally showing 1 longi-
    tudinal split when the flower opens (fig. 89/5–9; sect.
    Tricalysia)   7
    Calyx-limb ovoid or ellipsoid, ± entirely covering the
    corolla in bud, showing two or usually more deep
    longitudinal splits (becoming spathaceous when flower
    opens (fig. 89/10–11; sect. *Rosea*)   19

* Flowers not known for all taxa; if in doubt try both couplets.

7. Calyx truncate or at most with a vague indication of teeth (fig. 89/8, 9), glabrous outside . . . . *3. T. coriacea*
   Calyx dentate or with subulate lobes (fig. 89/5-7), mostly puberulous to pubescent outside 8
8. Calyx dentate, the lobes shorter than limb-tube (fig. 89/6, 7) 9
   Calyx with subulate lobes, the lobes longer than limb-tube (fig. 89/5) 18
9. Flower-stalks slender, over 10 mm. long . . . . . . *9. T. sp. D*
   Flowers sessile or with pedicels less than 5 mm. long 10
10. Vegetative parts densely yellowish pubescent (the leaf-blade above and beneath) . . . . *14. T. velutina*
    Plants less densely hairy (the leaf-blade above at most hairy on the midrib) 11
11. Leaf-blades in the dried state rather discolourous, brownish above and green beneath; woodland species, occasionally deciduous . . . . *13. T. niamniamensis*
    Leaf-blades in the dried state greenish or greyish green above and beneath, scarcely discolourous; forest species (also montane), evergreen 12
12. Leaf-blades coriaceous, glossy above . . . . *10. T. bagshawei*
    Leaf-blades papery or subcoriaceous, dull above 13
13. Total length of corolla (tube + lobes) 15-20 mm. . . . *11. T. microphylla*
    Total length of corolla (5-)8-12(-14) mm. 14
14. Leaf-blades stiffly chartaceous; tertiary nerves mostly obscure; domatia pitted, mostly ciliate or hairy; style (excluding arms) pubescent . . . . *4. T. pallens*
    Leaf-blades thinly chartaceous to chartaceous or occasionally subcoriaceous; tertiary nerves apparent; domatia tufted, pocket-like or at the most weakly pitted; style (excluding arms) glabrous or sparsely hairy (where known) 15
15. Leaves with (5-)6-8(-10) main pairs of lateral nerves 16
    Leaves with 3-5 main pairs of lateral nerves 17
16. Corolla-tube glabrous or with very few hairs inside *7. T. verdcourtiana*
    Corolla-tube distinctly pubescent inside . . . . . *8. T. sp. C*
17. Leaf-blades chartaceous; domatia tufted; only known in bud; corolla 6-merous; placenta 2-ovulate (W. Usambara Mts.) . . . . *5. T. sp. A*
    Leaf-blades thinly chartaceous; domatia pocket-like with a tuft of long hairs; flowers not known; fruit ovoid ± 8 mm. long, 6 mm. wide with 1-2 seeds (Ulanga District, Kwiro Forest Reserve) . . . . *6. T. sp. B*
18. Leaf-blades (2-)4-13 cm. long; flowers 6-merous . . . *11. T. microphylla*
    Leaf-blades 14-23 cm. long; flowers 5-merous . . . . *12. T. sp. E*
19. Evergreen 20
    Deciduous 21
20. Leaf-blades small, (2-)3.5-7 cm. long, (1.5-)2-3.5 cm. wide, mostly ovate; domatia absent; fruit subsessile or sessile with scar left by calyx-limb . . . . *15. T. gilchristii*
    Leaf-blades large, (5-)8-16 cm. long, (2-)3-6.5 cm. wide, mostly elliptic; domatia present; fruit borne on slender pedicels up to 30 mm. long and crowned by the persistent calyx-limb . . . . *16. T. pedicellata*
21. Inflorescences with several flowers; calyx-tube densely pubescent; limb scarious, splitting at anthesis into irregular withering lobes . . . . *17. T. schliebenii*
    Flowers solitary; calyx-tube glabrous; limb splitting at anthesis into 3 persistent lobes . . . . *18. T. sp. F*
22. Calyx-limb not becoming spathaceous (fig. 89/5-7) 23
    Calyx-limb soon showing one or more longitudinal splits then becoming spathaceous (fig. 89/10-11) 25
23. All vegetative parts entirely glabrous; corolla-tube 15-25 mm. long . . . . *22. T. elegans*

At least the young twigs and the stipules hairy; corolla-tube
    7.5–17 mm. long 24
24. Tip of the leaf-blade acuminate but not mucronate; leaf-
    blade glabrous above and beneath (except domatia);
    corolla-tube (7.5–)10–17 mm. long . . . . .   *19. T. vanroechoudtii**
    Leaf-midrib protruding in a small but distinct mucro; leaf-
    blade with midnerve hairy above and beneath; corolla-
    tube 7.5–11 mm. long . . . .   *20. T. acocantheroides*
25. Calyx-teeth and appendages of the bracteolar cup
    numerous, 3–5 mm. long, crowded at base of the
    flower; leaf-blades rounded at base . . . . .   *21. T. aciculiflora*
    Calyx-teeth and appendages of the bracteoles minute and
    rather indistinct; leaf-blades cuneate at base 26
26. Evergreen; young twigs, stipules and calyx-tubes glabrous;
    leaf-blades glabrous above and beneath . . .   *24. T. bridsoniana*
    Deciduous; young twigs, stipules and calyx-tubes hairy;
    leaf-blades hairy beneath and on the midrib above   *25. T. sp. G*
27. Flowers pedicellate; bracteoles 2, free and alternate or
    subopposite on most pedicels but sometimes 0–1 (or
    rarely 2 fused); margins of the calyx-lobes not touching
    (fig. 89/3, 4); sect. *Empogona* 28
    Flowers sessile; bracteoles of most flowers fused into cups,
    if free then opposite; calyx-lobes irregularly imbricate
    (fig. 89/2); sect. *Kraussiopsis* 30
28. Deciduous; most leaf-blades cordate or rounded at base   *26. T. junodii*
    Evergreen; leaf-blades cuneate at base (very seldom
    rounded) 29
29. Leaf-blades acute or shortly acuminate; domatia absent;
    corolla-throat densely shaggy; endosperm entire but
    sometimes with very few pieces of intruding seed-coat
    tissue near excavation) . . . .   *27. T. ovalifolia*
    Leaf-blades with acumen 7–20 mm. long; tuft-domatia
    present; corolla-throat sparsely hairy; endosperm
    deeply ruminate . . . .   *28. T. acidophylla*
30. Rhizomatous undershrubs . . . .   *31. T. cacondensis*
    Shrubs or small trees 31
31. Young twigs glabrous (rarely hairy on opposite sides);
    petioles glabrous or with very few short hairs; leaf-
    blades and stipules glabrous; leaf-blades (3–)5–8.5(–
    13) cm. long, (1.2–)2–3.5(–5) cm. wide, cuneate at base   *29. T. ruandensis*
    Young twigs, petioles, leaf-midnerves and stipules
    puberulous; leaf-blades 5–11 cm. long, 2.3–5.5 cm.
    wide, some rounded at base . . . .   *30. T. sp. H.*

1. T. elliotii *(K. Schum.) Hutch. & Dalz.*, F.W.T.A. 2: 83 (1931); Keay in F.W.T.A., ed. 2, 2: 151 (1963); N. Hallé in Fl. Gabon 17, Rubiacées 2: 286 (1970); Robbrecht in B.J.B.B. 53: 308 (1983) & in Distr. Pl. Afr. 24, map 805 (1983). Type: Guinée, near Sumbaraya, *Scott Elliot* 4937 (B, holo.†, BM, lecto.!, K, isolecto.!)

Tree, rarely a shrub, (4–)6–20 m. high, with the young twigs, like most of the vegetative parts, glabrous. Leaf-blades ovate, elliptic or obovate, 10–23(–28) cm. long, 3–9(–11) cm. wide, with an acumen up to 3 cm. long, rounded or cuneate at base, glabrous except the sparsely pubescent nerves and the tufted domatia beneath, subcoriaceous; 5–10 pairs of main lateral nerves; petiole ± 1 cm. long, with 2 lateral rows of hairs (which may be decurrent on the outside of the stipules); stipules sheathing, 1–2 mm. long, with awns 4–8 mm. long. Flowers 6-merous, sweetly scented, in rather lax, 5–10-flowered inflorescences; peduncle up to 5 mm.; bracts cupular but not resembling calyx-limb, often with conspicuous leafy appendages (these may attain ± the size of a normal leaf-blade); pedicel up to 5 mm. long, provided with 2 subopposite free bracteoles. Calyx glabrous,

* There is one record of a related species *T. longituba* De Wild. var. *longituba*, indigenous to Zaire and Zambia (differing most obviously by the presence of pit-domatia but keying out here): cultivated in Uganda, Kampala Plantation, Feb. 1931, *Snowden* 1969!

glabrescent or pubescent; limb a gradually widening tube with 6 teeth up to 4 mm. long, beset with conspicuous colleters (mostly ± in 2 horizontal rows) within. Corolla white, outside pubescent only on the lobes; tube 6–11 mm. long, 2–3 mm. in diameter, inside glabrous save for fascicles of long hairs between the insertion of the filaments; lobes 7–15 mm. long, 2–4 mm. wide. Filaments 4–11 mm. long; anthers versatile, with sparse long hairs. Placenta thin, with (5–)8–16 transversely flattened, densely placed ovules; style glabrous or with a few long hairs towards the tip, 14–25 mm. long including lobes of 2–5 mm. Fruits ? black, ellipsoid, 30 mm. long, 20 mm. wide, with a thick sclerified wall, ± smooth, glabrescent. Seeds few to ± 10 in each locule, with a smooth seed-coat; hilum linear.

SYN. *Probletostemon elliotii* K. Schum. in E.J. 23: 450 (1897)
   Tricalysia coffeoides R. Good in J.B. 64, Suppl. 2: 18 (1926). Type: Angola, Buco Zau, *Gossweiler* 6698 (BM, lecto.!, COI, isolecto.!)
   T. lecomteana Pellegrin in Bull. Soc. Bot. Fr. 85: 57 (1938); Hallé in Fl. Gabon 17, Rubiacées 2: 288 (1970). Type: Gabon, Tchibanga, *Le Testu* 1435 (P, lecto.!, BM, BR, isolecto.!)

   var. **centrafricana** *Robbrecht* in B.J.B.B. 53: 311, fig. 6 (1983) & in Distr. Pl. Afr. 24, map 807 (1983). Type: Zaire, Yangambi, *Louis* 6242 (BR, holo.!, B, K, MO, WAG, iso.!)

   Calyx pubescent, hairiness of the tube usually distinctly denser than that of the limb.

UGANDA. Mengo District: Nakiza Forest near Nansagazi, Dec. 1950, *Dawkins* 679! & near Mpigi, Mar. 1939, *Doughty* 16! & 15.5 km. on Masaka road, Aug. 1937, *Chandler* 1846!
DISTR. U 4; Cameroun, Gabon, Congo, Central African Republic, Zaire and Angola
HAB. Closed forest; 1065–1300 m.

SYN. *T. batesii* A. Chev., Caféiers du Globe 2, fig. 146 (1942) & 3: 243 (1947). Type: Cameroun, Bitye, near the River Ya, *Bates* 1752 (K, holo.!, FHO, iso.!)

NOTE. *T. elliotii*, although widely distributed, has remarkably constant characters. Only the indumentum of the calyx is distinctly variable, differing from completely glabrous (as in the types of *T. coffeoides* R. Good and *T. lecomteana* Pellegrin) to densely pubescent. Hallé (cit. supra) distinguished the fully glabrous specimens as *T. lecomteana*, but after a survey of all the material I included them in var. *elliotii* since there are too many intermediates (e.g. specimens with a glabrous calyx-tube and a sparsely pubescent limb). Var. *centrafricana* is virtually restricted to the Zaire Basin with some extensions — see distribution above; var. *elliotii* occurs from Guinée to Angola.

2. **T. anomala** *E.A. Bruce* in K.B. 1936: 485 (1936); Hallé in Fl. Gabon 17, Rubiacées 2: 289, fig. 67 (1970); Robbrecht in B.J.B.B. 53: 314 (1983) & in Distr. Pl. Afr. 24, map 809 (1983). Type: Tanzania, Morogoro District, Uluguru Mts., Tanana, *E.M. Bruce* 800 (K, holo.!, BM, BR, MO, iso.!)

Shrub or small tree to 10 m. high, with most parts (including the young twigs) densely hairy. Leaf-blades ovate, elliptic or obovate, 5–19 cm. long, 2–7 cm. wide, shortly acuminate at apex, rounded or cordate at base, nearly glabrous (except the nerves) to densely hairy, subcoriaceous; (6–)9–11 pairs of main lateral nerves; domatia absent; petiole 2–3 mm. long, hairy; stipules triangular, 2–3 mm. long, pubescent outside, fused at the base, with awns 4–8 mm. long. Flowers 6–8-merous, in rather lax, 1–5-flowered inflorescences; peduncle up to 5 mm.; pedicel 2–7 mm. long, hairy; bracts cupular but not resembling calyx-limb, sometimes with leafy appendages; bracteoles free, subulate, usually 1 per flower. Calyx-tube with dense long spreading hairs, yellowish in the dried state; limb a gradually widening tube 3–5 mm. long, pubescent, with subulate lobes 6–10 mm. long. Corolla white or greenish, pubescent outside only on the lobes; tube 10 mm. long, the upper half pubescent within; lobes 11–14 mm. long, 2 mm. wide, the bases pubescent inside. Filaments 2–3 mm. long; anthers ± 9 mm. long, with long hairs all over. Placenta with 4–7 ovules arranged in a U-shape; style hairy towards the tip, ± 18 mm. long including the 3–4 mm. long lobes. Fruits ? greyish, spherical or transversely flattened, sometimes ribbed towards the tip, 15 mm. long, 20 mm. wide, with a thick sclerified wall, usually rather warty and pubescent. Seeds 3–6 in each chamber, ± ellipsoid, with a colliculate testa and slightly curved hilum.

   var. **anomala**; Robbrecht in B.J.B.B. 53: 316 (1983) & in Distr. Pl. Afr. 24, map 810 (1983)

   Hairs on young twigs and midribs erect. Leaf-blades 13–19 cm. long, 5–7 cm. wide, sparsely pubescent between the nerves beneath.

TANZANIA. Lushoto District: Usambara Mts., Ambangulu Estate, Jan. 1947, *Wallace* 1255! & Kwamkoro, *Zimmermann!*; Morogoro District: Uluguru Mts., Tanana, Feb. 1935, *E.M. Bruce* 800!

DISTR T 3, 6; not known elsewhere
HAB. Montane forest; 900-1400 m.

var. **montana** *Robbrecht* in B.J.B.B. 53: 317 (1983) & in Distr. Pl. Afr. 24, map 812 (1983); Fl. Rwanda 3: 226 (1985). Type: Zaire, Kivu, Lusheni, *A. Léonard* 4711 (BR, holo.!, B, K, MO, WAG, iso.!)

Young twigs and midnerves beneath pubescent with appressed hairs. Leaf-blades 9-16 cm. long, 4-7 cm. wide, glabrous or very sparsely pubescent between the nerves beneath.

TANZANIA. Ulanga District: Kwiro Forest Reserve, Jan. 1979, *Cribb, Grey-Wilson & Mwasumbi* 11011!
DISTR T 6; Zaire (Kivu), Rwanda
HAB. Evergreen forest; 1300 m.

SYN. [*T. anomala* sensu Fl. Pl. Lign. Rwanda: 606 (1982), *non* E.A. Bruce sensu stricto]

NOTE. Var. *guineensis* Robbrecht is more widespread and occurs in West and Central Africa (from the Ivory Coast to Gabon and the Central African Republic). It has the leaf-blades densely hairy (soft to touch underneath) and much smaller (usually 5-11 cm. long). The hairs on the young twigs and midribs beneath are a very characteristic mixture of long and much shorter ones.

3. T. coriacea *(Benth.) Hiern* in F.T.A. 3: 120 (1877); Keay in F.W.T.A., ed. 2, 2: 151 (1963); N. Hallé in Fl. Gabon 17, Rubiacées 2: 300 (1970); Robbrecht in B.J.B.B. 57: 79, fig. 14 (1987) & in Distr. Pl. Afr. 31, map 1022, 1023 (1987). Type: Liberia, Grand Bassa, *Vogel* 172 (K, holo.!)

Shrub or small tree to 8 m. high, with mostly entirely glabrous vegetative parts (some sparse hairs may be observed on young twigs, petioles and stipules). Leaf-blades ovate to obovate (in the Flora area mostly the latter), (5-)6-20(-25) cm. long, (2-)3-8(-11) cm. wide, acute or shortly acuminate at apex, cuneate (rarely rounded, outside Flora area) at base, coriaceous; 3-6 pairs of main lateral nerves; domatia mostly hairy pits, sometimes absent; petiole (3-)5-15 mm. long; stipules sheathing, 1-2 mm. long, with awns 1-5 mm. long. Flowers (6-)7-8(-9)-merous, sessile, very variable in size (in the Flora area tending to the maximum dimensions given for all floral parts), in ± 5-flowered, sessile inflorescences; bracts and bracteoles cupular, similar to the calyx-limb. Calyx-tube glabrous; limb cup-shaped, markedly veined, truncate or obscurely dentate, glabrous or rarely sparsely pubescent outside, varying from glabrescent to densely covered with silky hairs within. Corolla white or rose, glabrous outside except the lobe-tips (rarely hairs becoming sparse and descending to the tube); tube 4-7(-9) mm. long and 1.5-3 mm. wide, throat and upper part of the tube hairy within; lobes (4-)5-9(-11) mm. long, 2-3 mm. wide. Filaments 1.5-3 mm. long; anthers 4-6(-8) mm. long, with a short apical appendage bearing a tuft of hairs. Placenta with 4-7 ovules arranged in a U-shape; style hairy except at base, ± 9-14(-17) mm. long including lobes of 3-4(-5) mm. Fruit red, spherical, 6-10 mm. in diameter. Seeds 2-5(-7) in each chamber, hemi-ellipsoid or angular with a minutely reticulate testa and linear often slightly curved hilum.

subsp. **coriacea**

Leaf-blades very variable in size and shape, but with their width always (¼-)⅓-½ of length. Fig. 90.

UGANDA. Masaka District: Namalala Forest, Aug. 1913, *Fyffe* 125!
TANZANIA. Mwanza District: Ibondo, July 1953, *Tanner* 1558!; Buha District: Mpemvi [Mpembe] R., 30 km. from Kibondo, Aug. 1950, *Bullock* 3114!; Rungwe District: Masebe near Chivanjee Tea Factory, Oct. 1977, *Mwasumbi* 11436!
DISTR U 4; T 1, 4, 6, 7; widespread in the Guineo-Congolian and Zambezian parts of tropical Africa
HAB. Mostly in association with water: swampy or seasonally inundated forests, river banks, fringing forest or bushland; also in deciduous woodland and termite-hill thicket; up to 1700 m.

SYN. *Randia coriacea* Benth. in Hook., Niger Fl.: 387 (1849)
  Tricalysia nyassae Hiern in F.T.A. 3: 121 (1877); K. Schum. in E.J. 30: 413 (1901); T.T.C.L.: 535 (1949); Brenan in Mem. N.Y. Bot. Gard. 8: 450 (1954); Garcia in Mem. Junta Invest. Ultram., sér. Bot., 4: 46 (1957); F.F.N.R.: 423 (1962). Type: Malawi, W. shore of Lake Malawi [Nyassa], near Roangiva, Sept. 1861, *Kirk* (K, holo.!)
  T. lastii K. Schum. in P.O.A. C: 382 (1895). Type: Mozambique, Namuli, *Last* (B, holo.†, K, lecto.!)
  T. katangensis De Wild. in Ann. Mus. Congo, Bot., sér. 4, 1: 156 (1902). Type: Zaire, Shaba, Lukafu R., *Verdick* 538 (BR, holo.!)
  T. petiolata De Wild. in Ann. Mus. Congo, Bot., sér. 5, 1: 202 (1905), 2: 75 (1907) & 3: 292 (1910), Miss. Laurent 1: 289 (1906), in B.J.B.B. 4: 208 (1914) & Pl. Bequaert. 3: 160 (1925). Type: Zaire, Kwilu, *Gillet* 2840 (BR, lecto.!)
  T. coriaceoides De Wild., Pl. Bequaert. 3: 160 (1925). Type: Zaire, Piali, *Bequaert* 1621 (BR, lecto.!)
  Polysphaeria brevifolia K. Krause in E.J. 48: 412 (1912); T.T.C.L.: 520 (1949); Verdc. in K.B. 35: 128 (1980). Type: Tanzania, Rungwe District, near Rutenganio, *Hasse* in Herb. Amani 619 (B, holo.†, EA, lecto.!)

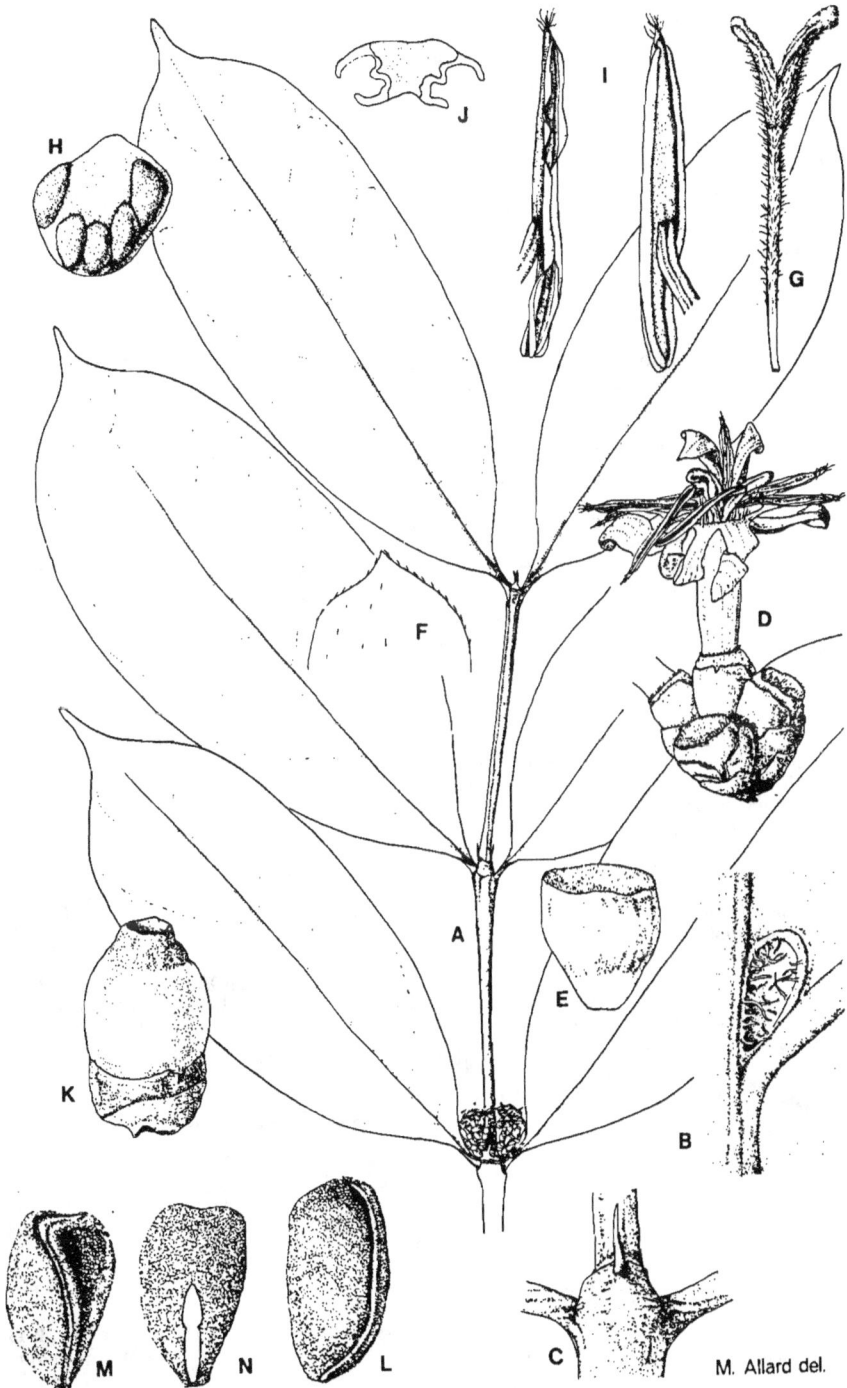

FIG. 90. *TRICALYSIA CORIACEA* subsp. *CORIACEA* — **A**, habit, × ½; **B**, domatium, × 18; **C**, stipule, × 3; **D**, portion of inflorescence, × 3; **E**, calyx, × 6; **F**, tip of corolla-lobe, from outside, × 12½; **G**, style and stigmatic arms, × 4½; **H**, placenta, × 25; **I**, stamen, 2 views, × 9; **J**, transverse section of anther, × 25; **K**, fruit, × 3; **L, M**, seed, ventral views, × 6; **N**, seed, abaxial view with position of embryo indicated, × 6. A, B, D–J, from *Louis* 2968; C, from *Louis* 13759; K–N, from *Germain* 7268. Drawn by M. Allard.

Tricalysia davyi S. Moore in J.B. 63: 146 (1925). Type: Zaire, Shaba, Lubumbashi, *Burtt-Davy* 17593 (K, holo.!)

T. vignei Aubrév. & Pellegr. in Bull. Soc. Bot. Fr. 83: 41 (1936); Keay in F.W.T.A., ed. 2, 2: 151 (1963). Type: Ghana, near Beyin, *Vigne* 1469 (P, lecto.!, K, isolecto.!)

NOTE. This widespread taxon is very variable as regards the size and the shape of the leaves, the size of the flowers, and some other minor characters (e.g. presence or absence of hairs on young vegetative parts, near the apex and on outside of corolla and calyx); I could not classify all this variation into formal taxa, although a sound biosystematic account could perhaps produce results: I have recognised only the narrow-leaved specimens occurring on the mountain ranges bordering Zimbabwe and Mozambique and on Mt. Mulanje (Malawi) as subsp. *angustifolia* (Garcia) Robbrecht. At any event the widely used name *T. nyassae* applied to the Zambezian material cannot be retained at specific level.

4. T. pallens *Hiern* in F.T.A. 3: 121 (1877); Keay in F.W.T.A., ed. 2, 2: 152 (1963); N. Hallé in Fl. Gabon 17, Rubiacées 2: 308, fig. 73 (1970); Robbrecht in B.J.B.B. 57: 114, fig. 19 (1987) & in Distr. Pl. Afr. 31, map 1037 (1987). Type: Bioko [Fernando Po], 1860, *Mann* (K, holo.!)

Shrub or sometimes a tree 1–10(–18) m. high, with puberulous or pubescent young twigs. Leaf-blades obovate or elliptic, rarely ovate, 6–11(–19) cm. long and 1.5–4(–8) cm. wide, cuneate at base and acute or acuminate at tip, with midrib puberulous or pubescent above and beneath (especially when leaves are young), papyraceous or subcoriaceous; 4–5 pairs of main lateral nerves; domatia with a pit- or cave-like excavation, ciliate or densely hairy; petiole 2–7(–10) mm. long, puberulous; stipules sheathing, 1–2 mm. long with awns 2–7 mm. long. Flowers (4–)5–6-merous, ± scented (sometimes described as unpleasant), subsessile, in 3–many-flowered inflorescences; bracts and bracteoles cupular. Calyx-tube glabrous, rarely pubescent; limb tubular, shortly toothed, pubescent, 1–1.5 mm. high. Corolla white, cream or greenish, pubescent outside (mostly only on the lobes) and ± hairy especially in throat and upper part of tube inside; tube 2–8 mm. long, 1–2 mm. wide; lobes acute at tip, 3–6 mm. long, 1–2 mm. wide. Filaments 0.6–3 mm. long; anthers 1.5–5 mm. long, mostly ± hairy, especially at the tip. Placenta with 2(–3) ovules; style pubescent, 4–12 mm. long including the 1–4 mm. long lobes. Fruits red, spherical, 3–6 mm. in diameter, with 1 hemispherical or 2, ¼-spherical seeds in each chamber. Seeds with curved hilum; seed-coat ± striate, thickened at the ventral side and weakly intruding into the endosperm.

KENYA. Meru District: Ngaia Forest, 12 June 1969, *Wachiori* 42!; Kitui District: Mutha Hill, *Bally* 1726!; Kwale District: Shimba Hills, Mwele Mdogo Forest, 10 Feb. 1953, *Drummond & Hemsley* 1197!
TANZANIA. Lushoto District: E. Usambara Mts., Mangubu, 21 Oct. 1936, *Greenway* 4685!; Buha District: Gombe Stream Reserve, Mkenke R., 28 Mar. 1964, *Pirozynski* 621!; Morogoro District: Manyangu Forest Reserve, Aug. 1959, *Paulo* 803!
DISTR. K 4, 7; T 3, 4, 6–8; Liberia, Ivory Coast, Ghana, Nigeria, Cameroun, Bioko, Rio Muni, Gabon, Congo, Central African Republic, Zaire, Angola, Zambia, Malawi, Zimbabwe and Mozambique
HAB. Evergreen forests, forest margins and thickets, sometimes riverine; 500–1200(–1500) m.

SYN. *T. gabonica* Hiern in F.T.A. 3: 122 (1877); Good in J.B. 64, Suppl. 2: 18 (1926). Type: Gabon R., *Mann* 930 (K, holo.!, P, iso.!)
   T. pallens Hiern var. *gabonica* (Hiern) Hallé in Fl. Gabon 17, Rubiacées 2: 310 (1970)
   [*T. griseiflora* sensu Hiern in Cat. Afr. Pl. Welw. 1(2): 469 (1898), pro parte, excl. *Welwitsch* 3128, *non* K. Schum.]
   T. griseiflora K. Schum. var. *longistipulata* De Wild. & Th. Dur. in Bull. Herb. Boiss., sér. 2, 1: 28 (1901). Type: Zaire, Kisantu, *Gillet* 887 (BR, lecto.!)
   T. longistipulata (De Wild. & Th. Dur.) De Wild. & Th. Dur. in Bull. Herb. Boiss., sér. 2, 1: 756 (1901); De Wild. in Ann. Mus. Congo, Bot., sér. 5, 1: 204 (1904), 2: 75 & 168 (1907) & 3: 292 (1910); Th. & H. Durand in Syll. Fl. Congo.: 265 (1909); De Wild., Pl. Bequaert. 3: 164 (1925); R. Good in J.B. 64, Suppl. 2: 17 (1926)
   T. sapinii De Wild. in Ann. Mus. Congo, Bot., sér. 5, 2: 168 (1907). Type: Zaire, Madibi, 2 July 1906, *Sapin* (BR, holo.!)
   T. myrtifolia S. Moore in J.L.S. 40: 83 (1911); Garcia, Mem. Junta Invest. Ultram., sér. Bot., 4: 43 (1957); K.T.S.: 476 (1961); F.F.N.R.: 424 (1962). Type: Mozambique, Maruma Mt., *Swynnerton* 684 (BM, lecto.!, K, isolecto.)
   T. ramosissima De Wild., Pl. Bequaert. 3: 169 (1925). Type: Zaire, Lubutu, *Bequaert* 6756 (BR, holo.!)
   [*T. microphylla* auct. sensu T.T.C.L.: 535 (1949), pro parte, quoad *Greenway* 5364, *non* Hiern]
   T. sp.? nov. aff. T. myrtifolia S. Moore sensu Vollesen in Op. Bot. 59: 72 (1980)

NOTE. Two sheets collected in Tanzania, Ulanga District, Uzungwa Mts., above Sanje (*Bridson* 600 & 611) have been annotated as a variant of *T. pallens* with large leaves. The following additional differences have also been noted: young twigs, petioles and midribs glabrous rather than puberulous or pubescent, calyx-limb 2 mm. long, corolla larger (towards upper limits given in description) with distinctly acuminate lobes. Unlike more typical specimens of *T. pallens* from the Sanje area (*Bridson* 595 and 630), *Bridson* 600 and 611 were growing directly on the river banks.

5. T. sp. A

Shrub 4 m. tall, with puberulous young twigs. Leaf-blades elliptic or narrowly elliptic, 4–10 cm. long, 1–4 cm. wide, acute or acuminate at tip, cuneate at base, glabrous save for the sparsely pubescent midrib beneath and the tuft-domatia, papyraceous; 3–5 pairs of main lateral nerves; petiole 3–5 mm. long, puberulous; stipules sheathing, with awns 2–3 mm. long. Flowers 6-merous, only known in bud, in 3–5-flowered inflorescences; pedicels up to 2 mm. long; bracts and bracteoles cupular, 2- or 4-appendiculate. Calyx-tube glabrous; limb tubular, shortly dentate, pubescent. Corolla pubescent outside, glabrous inside. Anthers glabrous. Placenta 2-ovulate; style glabrous. Fruit only known immature, shortly stalked, ± spherical, crowned by the persistent calyx-limb; seeds 1–2 in each chamber.

TANZANIA. Lushoto District: Usambara Mts., Zevigambo Peak, 18 June 1953, *Drummond & Hemsley* 2946! & Baga I Forest Reserve, Mar. 1984, *Borhidi et al.* 84475!
DISTR. T 3; not known elsewhere
HAB. Wooded grassland on steep rocky slope; microphyllous evergreen forest; 1600–1800 m.

SYN. *T. sp. 96* sensu Robbrecht in B.J.B.B. 57: 203 (1987)

6. T. sp. B

Shrub to ± 2 m. tall, with puberulous young twigs. Leaf-blades ovate, 4–7 cm. long, 1.8–2.7 cm. wide, acuminate at tip, rounded to somewhat cuneate at base, glabrous save for the midrib above and the domatia and midrib and lateral nerves beneath, thinly papyraceous; ± 4 pairs of main lateral nerves; domatia pocket-shaped, with a tuft of long hairs; petiole 2–4 mm. long, puberulous; stipules sheathing, 1.5 mm. long, puberulous, overtopped by an awn 2–4 mm. long. Flowers unknown; from immature fruit: calyx-limb tubular, 3 mm. long, puberulous, shortly dentate; placenta with 2 collateral ovules. Fruit single, ovoid, ± 8 mm. long, 6 mm. wide, crowned by the persistent calyx-limb. Seeds only known when immature (?1 or 2 per fruit, ellipsoid or hemi-ellipsoid).

TANZANIA. Ulanga District: Kwiro Forest Reserve, just below ridge on SW. side, 18 Jan. 1979, *Cribb, Grey-Wilson & Mwasumbi* 11010!
DISTR. T 6; not known elsewhere
HAB. Forest; 1400 m.

SYN. *T. sp. 94* sensu Robbrecht in B.J.B.B. 57: 203 (1987)

7. T. verdcourtiana *Robbrecht* in B.J.B.B. 55: 496 (1985) & 57: 102, fig. 17 (1987) & in Distr. Pl. Afr. 31, map 1031 (1987). Type: Tanzania, Mbeya District, Umalila, *Leedal* 3764 (EA, holo.!)

Tree or small shrub up to 11 m. high, with puberulous young twigs. Leaf-blades narrowly elliptic or narrowly obovate, 5–13 cm. long, 1.5–3.5(–5) cm. wide, rounded, acute or acuminate at tip (the midnerve protruding in a minute mucro less than 1 mm.), cuneate at base, glabrous except the puberulous midrib above and pubescent underneath, papyraceous or subcoriaceous; (5–)7–8(–10) pairs of main lateral nerves; domatia hairy, sometimes very weakly pitted (the hairs then arranged around the pit); petiole 3–5 mm. long, puberulous; stipules sheathing, ± 1 mm. long, with awns 3–6 mm. long. Flowers 4–6-merous, subsessile, in 1–5-flowered sessile inflorescences; bracts and bracteoles cupular, 2- or 4-apiculate, pubescent. Calyx-tube glabrous; limb tubular, 1.5 mm. long, shortly dentate, pubescent. Corolla of unknown colour; tube 4–6 mm. long, 1.5–2.5 mm. wide, glabrous or sometimes with very few hairs inside; lobes 5–7 mm. long, 2–2.5 mm. wide, acute (and sometimes laterally emarginate also) at tip, pubescent outside. Filaments flattened, 1.5–2.5 mm. long; anthers 4 mm. long, obtuse or shortly appendiculate at tip, glabrous or with few hairs near tip. Placenta with 2–4 ovules; style glabrous or with very few hairs, 4–6 mm. long, bearing 2 slender and hairy arms of ± the same length. Fruit red, spherical, ± 7 mm. in diameter. Seeds 1–3 in each chamber, angular at the inner side; hilum curved; seed-coat minutely striate.

TANZANIA. Mpanda District: Sitibi Mt., Kampisa R., 11 Dec. 1969, *Kielland* 31!; Rungwe District: Kyimbila, 3 Mar. 1914, *Stolz* 2564!; Njombe, 18 Dec. 1931, *Lynes* V.2!

TANZANIA. DISTR. T 4, 7; Malawi, Zambia
HAB. Submontane and montane rain forest, sometimes in ravines; ± 1675–2040 m.

8. T. sp. C

Tree or shrub up to 12 m. tall, with puberulous young twigs. Leaf-blades elliptic or narrowly obovate, 4–10 cm. long, 1.5–3 cm. wide, acuminate at tip (the midrib protruding in a minute mucro), cuneate at base, very sparsely pubescent above and beneath (more densely on the midrib), papyraceous; ± 5 pairs of main lateral nerves; domatia tufted; petiole 3–5 mm. long, puberulous; stipules sheathing, puberulous, with awns up to 6 mm. long. Flowers 4–5-merous, only known in bud, sessile in 3–5-flowered inflorescences; bracts and bracteoles cupular, 4-appendiculate, covered with short erect hairs. Calyx-tube pubescent; limb tubular, dentate, densely pubescent. Corolla greenish, densely pubescent outside and with long hairs inside the tube. Anthers rather sparsely pubescent. Placenta with 3 ovules; style pubescent. Fruits unknown.

TANZANIA. Njombe District: Lupembe, Ditima [Nditima], 15 & 22 Oct. 1931, *Schlieben* 1348! & 1371!
DISTR. T 7; not known elsewhere
HAB. Montane forest; 1800 m.

NOTE. This taxon is close to *T. kivuensis* Robbrecht and *T. verdcourtiana;* it especially approaches the former in that the calyx-lobes, anthers and style are pubescent to the same extent. More material is required. See Robbrecht in B.J.B.B. 57: 102 (1987).

9. T. sp. D

Slender shrub up to 2 m. tall. Leaf-blades elliptic to obovate, acuminate at apex, cuneate at base, 5–8 cm. long, 1.5–2.5 cm. wide, glabrous, subcoriaceous; ± 4 pairs of main lateral nerves; domatia absent; petioles 2–3 mm. long; stipules sheathing, with awns 4.5–6 mm. long, puberulous outside. Flowers known only in bud, 6-merous, in 1–3-flowered inflorescences, which are rather lax with long slender inflorescence branches; bracts subtending inflorescence branches fused, biapiculate; bracteoles subtending calyx-tube cupular, much smaller than calyx-limb. Calyx-limb-tube dentate, puberulous. Inside of corolla-tube glabrous. Style and anthers glabrous. Placenta probably 1-ovulate. Fruits red, elongate (± 14 × 5mm.), with 1 spindle-shaped seed; testa minutely colliculate; hilum straight.

TANZANIA. Mpanda District: Mahali Mts., 1967, *Nishida* 48! & Kungwe Mt., Kahoko, 22 July 1959, *Harley & Newbould* 4543! & Musenabantu, just below summit, 13 Aug. 1959, *Harley* 9330!
DISTR. T 4; not known elsewhere
HAB. Submontane evergreen forest; 1200–1800 m.

NOTE. Similar to *T. zambesiaca* Robbrecht, occurring in the eastern part of Zambia and near the borders of the Flora area, at the southern tip of Lake Tanganyika. *T. zambesiaca* differs from the above collections in the shorter stipule awns, presence of domatia and the more hairy petioles and pedicels. Presumably there are also floral differences, but mature flowering material from T 4 is lacking. See Robbrecht in B.J.B.B. 57: 143 (1987).

10. T. bagshawei S. *Moore* in J.B. 44: 84 (1906); T.T.C.L.: 535 (1949); Robbrecht in B.J.B.B. 57: 138, fig. 21 (1987) & in Distr. Pl. Afr., 31, map 1044 (1987). Type: Uganda, Mengo District, Entebbe, *Bagshawe* 792 (BM, holo!)

Shrub or sometimes a small tree, 1.5–4(–6) m. high, with puberulous or shortly hirsute young twigs. Leaf-blades mostly obovate, but sometimes elliptic or ovate, (3–)5–9(–11) cm. long, (1–)2–4(–5) cm. wide, acute or acuminate at apex, cuneate or rounded at base, glabrous except for the midrib bearing distinct hairs (at least at the base) above and beneath, coriaceous, often glossy above; 3–4(–5) pairs of main lateral nerves; domatia present or absent; petioles 1–5(–7) mm. long, puberulous or shortly hirsute; stipules sheathing, puberulous, with awns up to 5 mm. long. Flowers 6(–7)-merous, sweetly scented, in (1–)3–5-flowered subsessile inflorescences; pedicels up to 3 mm. long, shortly hirsute; bracts and bracteoles cupular, the proximal ones especially with 2 distinct appendages, upper cupules 2-dentate, hirsute. Calyx-tube hirsute; limb tubular, ± 2 mm. high, bearing short subulate lobes ± 0.5 mm. long. Corolla white; tube 3–9 mm. long, with rather sparse long hairs within and sometimes with sparse hairs outside; lobes 3–8 mm. long, 2–3 mm. wide, glabrous or pubescent outside (the margins always ciliate). Stamens with an attenuate filament 1.5–3 mm. long; anthers 3–4.5 mm. long, obtuse or acute at tip, when acute sometimes topped by a single hair. Placenta with 2 collateral ovules, rarely with a third ovule inserted between them; style glabrous or with a few hairs towards top and on the lobes, 8–12 mm. long, including lobes of ± 2 mm. Fruits red, ellipsoid, ± 7 mm. long, 5 mm. in diameter. Seeds 1 or 2 in each chamber, ± ellipsoidal; hilum linear and curved; testa colliculate.

subsp. bagshawei

Leaves with petioles 1–3 mm. long; domatia hairy, ± elongate, sometimes with a weak excavation. Corolla-tube 7–9 mm. long; lobes 4–8 mm. long, pubescent outside on the half of the lobes exposed in bud. Style 10–12 mm. long.

UGANDA. Karamoja District: Lokapeliethe Hill, 29 Oct. 1939, *A.S. Thomas* 3113!; Busoga District: Buwenda Plantation 7 km. N. of Jinja, 13 Nov. 1950, *G.H.S. Wood* 271!; Masaka District: Sese I., Bufumira I., 19 June 1937, *A.S. Thomas* 3010!
TANZANIA. Mwanza District: Ukerewe I., Rubya Forest Reserve, 17 June 1958, *Makwilo Semkiwa* 46! & Nyambiti, *Tanner* 1300! & Capri Point, *Tanner* 407!
DISTR. U 1, 3, 4; T 1; not known elsewhere
HAB. Forest and evergreen thickets; many records are from the shore forest of Lake Victoria, also in secondary formations, such as the understory of plantations; 1000–1200 m.

NOTE. *T. bagshawei* subsp. *malaissei* Robbrecht (in B.J.B.B. 57: 140 (1987) & in Distr. Pl. Afr. 31, map 1045 (1987)) occurs in Zaire (Upper Shaba) and Zambia; it is characterised by smaller corollas and almost constant absence of domatia. With one exception (the specimen from the Mafinga Mts., which belongs to *T. acocantheroides*) Zambian material cited as *T. bagshawei* (in F.F.N.R.: 424 (1962)) belongs to subsp. *malaissei*.

11. **T. microphylla** *Hiern* in F.T.A. 3: 123 (1877); T.T.C.L.: 535 (1949), pro parte, excl. *Greenway* 5364; Robbrecht in B.J.B.B. 57: 146, fig. 24 (1987) & in Distr. Pl. Afr. 32, map 1049 (1987). Type: Zanzibar, *Hildebrandt* 1163 (BM, holo.!)

Shrub 1–3 m. high, with puberulous or glabrous young twigs. Leaf-blades elliptic, ovate or obovate, (2–)4–13 cm. long, (1.2–)2–6.5 cm. wide, acuminate at apex, cuneate or rounded at base, subcoriaceous, (3–)4–7 pairs of main lateral nerves; hairy pocket-domatia; petiole 2–10 mm. long, puberulous when young; stipules sheathing, puberulous or glabrous, ± 1.5 mm. high, with awns 2–5 mm. long. Flowers 6-merous, sweetly scented, subsessile; inflorescences 3–7(–9)-flowered, subsessile; bracts and bracteoles cupular with 2 subulate lobes and 2 short teeth, the lower cups often with leafy appendages up to 2.5 cm. long and wide. Calyx pubescent or glabrous; limb with subulate lobes (0.5–)1–5 mm. long, often decurrent so that the limb-tube appears carinate, inside covered with silky hairs and colleters. Corolla white, pubescent outside, especially on the lobes; tube 8.5–10 mm. long, 2–4 mm. wide, inside with sparse long hairs in the lower part; lobes acute, 8–10 mm. long, 2.5–5 mm. wide. Filaments 1.5–2 mm. long; anthers 6–8 mm. long, glabrous. Placenta elongate, with 2 collateral ovules; style glabrous or with a few hairs on the outer side of its arms, 13–16 mm. long including the 2 slender ± 5 mm. long lobes. Fruit red, ellipsoid, ± 10 mm. long, 7 mm. wide. Seeds 1 or 2 in each chamber, their shape half or quarter of a sphere, with minutely colliculate seed-coat; hilum linear and curved.

KENYA. Kwale District: Muhaka Forest, 7 Mar. 1977, *Faden* 77/716! & Shimba Hills, Lango ya Mwagandi Forest, 23 Apr. 1968, *Magogo & Glover* 941!; Kilifi District: Chasimba, 16 Feb. 1977, *Faden, Gillett & Gachathi* 77/413!
TANZANIA. Tanga District: Merera, 28 Apr. 1926, *Peter* 39917!; Morogoro District: Uluguru Mts., Kitundu, *E.M. Bruce* 62! & Kanga Mt., 27 Feb. 1970, *Pocs* 6136R!; Zanzibar I., near Haitajwa Hill, 4 Dec. 1930, *Greenway* 2650!
DISTR. K 7; T 3, 6; Z; not known elsewhere
HAB. Forest; up to 1200 m.

SYN. *T. bussei* K. Schum. in E.J. 33: 346 (1904); T.T.C.L.: 535 (1949). Type: Tanzania, Pangani R. near Hale, *Busse* 327 (B, holo. †)

NOTE. *T. microphylla* is a very variable species in several respects. The young twigs, stipules, bracteolar cups and calyx-tubes vary from pubescent or puberulous (with intermediates, e.g. in Kenya) to glabrous. The cited specimen from the Kanga Mountain has these parts the most densely pubescent, combined with the occurrence of short erect hairs on the leaf-blade beneath. The leaf-blade is very variable in size and shape. Mostly the size of the blades approaches the maximum dimensions given in the descriptions, but small-leaved specimens (including the type) have been collected in Zanzibar. The calyx-lobes are conspicuously variable in length, (0.5–)1–5 mm. long, causing specimens at the extremes to appear quite dissimilar. Also in Zanzibar a variant with the leaf-blades rounded or truncate at the base has often been collected. *T. microphylla* possibly occurs in K 4 but this needs confirmation (one record in very young fruit only; Tana R.,

Seven Forks, Mar. 1974, *S.A. Robertson* 2010!). This species has been cultivated in the Agricultural Research Station at Amani (*Greenway* 3287!)

12. T. sp. E

Small shrub 1–2 m. tall, with glabrous young twigs. Leaf-blades obovate, 14–23 cm. long, 4–7 cm. wide, acuminate at tip, cuneate or rarely somewhat rounded and unequal at base, glabrous save for the pubescent nerves and hairy pocket-domatia beneath, papyraceous; ± 7 pairs of main lateral nerves; petiole 6–10 mm. long, pubescent on the lower side only; stipules triangular, ± connate at base, pubescent, 2 mm. long with awns 6–8 mm. long. Flowers 5-merous, only known in bud, sessile in 5-flowered inflorescences; bracts and bracteoles cupular, 4-appendiculate, pubescent. Calyx-tube densely pubescent; limb ellipsoid (? splitting at anthesis), pubescent, provided with 5 long teeth. Corolla pubescent outside and completely glabrous inside. Anthers glabrous. Style glabrous; placenta with 1 or 2 collateral ovules. Fruits of unknown colour, ellipsoid, ± 1 cm. long, with juicy wall and membranous endocarp, calyx not persistent, leaving an impressed scar around disc. Seeds 1 or 2 per fruit, ellipsoid or half-ellipsoid; hilum broad; seed-coat smooth.

TANZANIA Ulanga District: Uzungwa Mts., Sanje, 8 Sept. 1984, *Bridson* 647!
DISTR. T 6; not known elsewhere
HAB. Streamside forest; 750 m.

SYN. *T. sp. 95* sensu Robbrecht in B.J.B.B. 57 203 (1987)

13. T. niamniamensis *Hiern* in F.T.A. 3: 123 (1877); K.T.S.: 476 (1961); Robbrecht in B.J.B.B. 57: 171 (1987). Type: Sudan, SW. Gunango, *Schweinfurth* 2883 (P, lecto.!, BM, K, isolecto.!)

Shrub or sometimes a small tree, up to 6(–8) m. high, with puberulous or shortly hairy young twigs. Leaves occasionally deciduous; blades ovate, elliptic or obovate (from rather narrow to broader, even ± circular), 1–9(–12) cm. long, 0.7–2.5 cm. wide, rounded, acute or shortly acuminate at tip, cuneate, rounded or cordate at base, glabrescent (pubescent on midrib only) to more pubescent or even velutinous on both surfaces; 4–6 pairs of main lateral nerves; domatia hairy, often with a small pocket- or pit-shaped excavation; petiole 1–3(–6) mm. long, puberulous; stipules sheathing, 1.5–2.5 mm. long, with awns mostly short, 1.5–2.5(–5) mm., puberulous. Flowers 6-merous, strongly scented but not exactly pleasant, subsessile, in (3–)5-many-flowered inflorescences; bracts and bracteoles cupular, with 4 teeth (2 of which may sometimes be developed into leafy appendages). Calyx-tube pubescent, the extreme base sometimes glabrous; limb shortly tubular, pubescent, with short teeth, often showing 1 deep longitudinal split when the flowers open. Corolla white, cream or greenish, pubescent outside (becoming glabrous towards the base of the tube) and glabrous inside; tube 6–7.5 mm. long, 1–3 mm. wide; lobes 4–5 mm. long, 2–3 mm. wide, emarginate. Filaments 1.5–2.5 mm. long; anthers 3–4 mm. long, rounded and sometimes with a very few hairs at tip. Placenta with 2 collateral ovules; style glabrous or somewhat hairy at tip, 8–11 mm. long including ± erect lobes 2–3 mm. long. Fruits ? orange, spherical, ± 7 mm. in diameter. Seeds usually 1, hemispherical (rarely 2, ¼-spherical) in each chamber, with ± concave inner side and strongly curved hilum; seed-coat colliculate, striate around the hilum.

NOTE. Material from the Flora Zambesiaca area has previously been given the name *T. angolensis* A. Rich. (White, F.F.N.R.: 422 (1962) & Drummond in Kirkia 10: 275 (1975)), since the closely related species, *T. niamniamensis* and *T. griseiflora* were considered conspecific with *T. angolensis*. Following a revision of all available material (Robbrecht in B.J.B.B. 57: 159–176 (1987)) all three species are maintained; *T. angolensis* and *T. griseiflora* K. Schum. are geofrutescent species occurring in the western part of the Zambezian Region.

var. **niamniamensis**; Robbrecht in B.J.B.B. 57: 171 (1987) & in Distr. Pl. Afr. 32, map 1058 (1987)

Shrubs or small trees with straight lateral twigs and normal internodes. Leaf-blades (2.5–)4.5–9(–12) cm. long, (1–)1.5–2.5(–3.5) cm. wide.

UGANDA. W. Nile District: Metuli, Moyo, *Eggeling* 1241!; Busoga District: Seguru Hill, 19 Oct. 1949, *Osmaston* 259!; Mengo District: Bukomero, Aug. 1932, *Eggeling* 821!
KENYA. N. Kavirondo District: Mumias, 9 Mar. 1898, *Whyte!*; S. Kavirondo District: Bukuria, Sept. 1933, *Napier* 5290!; Kericho District: 60 km. E. of Kisumu, 7 Oct. 1971, *Gillett* 19343!
TANZANIA Mwanza District: Chole, 2 Aug. 1951, *Tanner* 374!; Shinyanga, May 1935, *B.D. Burtt* 5122!; Tabora, 30 Nov. 1976, *Shabani* 1129!

DISTR.   U 1–4; **K** 5; **T** 1, 4; Sudan, Zaire, Zimbabwe
HAB.   Woodland, thicket, bushland, riverine forest, often on rocky outcrops; 900–1530 m.

SYN. *Rosea sp.* sensu T. Thomson in Speke, Journ. Discov. Source Nile, App.: 636 (1863)

var. **nodosa** *Robbrecht* in B.J.B.B. 57: 174, fig.32 (1987) & in Distr. Pl. Afr. 32, map 1060 (1987). Type: Zimbabwe, S. of Mutare [Umtali], Dora R., *Chase* 982 (BR, holo.!, BM, COI, S, iso.!)

Stunted thickly-foliaged shrubs with knobbly, often upwardly-curved twigs with very short internodes. Leaf-blades 1–5(–6) cm. long, 0.7–2(–4) cm. wide.

UGANDA. Busoga District: Lake Victoria, Lolui I., 14 May 1964, *G. Jackson* U84!; Mengo District: Namiryango, May 1917, *Dummer* 3187!
DISTR.   U 3, 4; Zaire, Zambia, Malawi, Zimbabwe
HAB.   Woodlands and thickets, mostly in rocky situations; 1050–1200 m.

SYN. *T. seretii* De Wild., in Ann. Mus. Congo, Bot., sér. 5, 2: 168 (1907); Th. & H. Dur., Syll. Fl. Congol.: 265 (1909). Type: Zaire, Gombari, *Seret* 559 (BR, holo.!)
    [*T. angolensis* sensu *Drummond* in Kirkia 10: 275 (1975), *non* A. Rich.]

NOTE.   (on species as a whole). Var. *djurensis* (Hiern) Robbrecht occurs just outside the Flora area in S. Sudan and may be recognized by its leaf-blades mostly cordate or rounded at the base.

14. T. velutina *Robbrecht* in B.J.B.B. 57: 179 (1987) & in Distr. Pl. Afr. 32, map 1062 (1987). Type: Tanzania, Dodoma District, Manyoni, *B.D. Burtt* 3413 (K, holo.!, BR, FHO, iso.!)

Shrub ± 3 m. high, velutinous with yellowish hairs on all vegetative parts and outer parts of the flowers. Leaves deciduous (only known in ± young stage); blades elliptic, 2.5–5 cm. long, 1–2 cm. wide, acute at apex, ± rounded at base, papyraceous; domatia absent; nerves (except midrib) invisible because of hairiness; petiole ± 1 mm. long; stipules sheathing, 2 mm. long, with awns 1–1.5 mm. long. Flowers 5–6-merous, sweetly scented, appearing before or together with the new leaves, sessile, in 5–many-flowered sessile inflorescences on the leafless twigs of the previous season; bracts and bracteoles cupular, truncate . Calyx-tube pubescent with yellowish hairs; limb cupular, obscurely 5–6-toothed, often with 1(–2) splits after anthesis, pubescent outside and covered with silky hairs within. Corolla cream, velutinous outside (especially the lobes), glabrous inside; tube ± 5 mm. long, 1–2 mm. in diameter: lobes ± 4 mm. long, 1.5 mm. wide, rounded at apex. Filaments 2–2.5 mm. long; anthers 3–4 mm. long, obtuse at apex. Placenta with 2 collateral ovules; style glabrous, ± 10 mm. long including the 1.5–2 mm. long lobes. Fruits unknown.

TANZANIA. Shinyanga, Nov. 1928, *Koritschoner* 1765!; Dodoma District: Manyoni, 21 Nov. 1931, *B.D. Burtt* 3413!
DISTR.   T 1, 5; not known elsewhere
HAB.   Deciduous thicket; 1000–1200 m.

15. T. gilchristii *Brenan* in K.B. 3: 499 (1949); Robbrecht in B.J.B.B. 57: 195 (1987) & in Distr. Pl. Afr. 32, map 1069 (1987). Type: Tanzania, Iringa District, Mufindi, Kivere Forest, *Gilchrist* in F.H. 1728 (K, holo.!, EA, iso.,FHO, fragment!)

Small shrub; young twigs with short dense erect hairs. Leaf-blades ovate or ± elliptic, (2–)3.5–7 cm. long, (1.5–)2–3.5 cm. wide, obtuse at apex, cuneate, truncate, rounded or cordate, often somewhat unequal at base, puberulous on both sides when young, becoming almost glabrous above and sparsely puberulous (especially the midrib) beneath, papyraceous; 5–6 pairs of main lateral nerves; domatia absent; petiole (2–)4–7 mm. long, with short erect hairs; stipules triangular, with scarious margin, 1–2 mm. long and with awns up to 2 mm. long, puberulous. Flowers (4–)5-merous, subsessile, solitary or in many-flowered sessile inflorescences; bracts and bracteoles cupular, scarious, much resembling the calyx-limbs. Calyx-tube glabrous; limb first ovoid or ellipsoid and almost entirely covering the corolla-bud, with very short teeth at the tip, later when the flower opens, becoming deeply bilobed by longitudinal splitting, membranous and ± transparent, pubescent. Corolla white, sparsely pubescent outside and glabrous within; tube 3–3.5 mm. long, 1–1.5 mm. wide; lobes 4–5.5 mm. long, 2–2.25 mm. wide, emarginate and laterally apiculate at tip. Filaments flattened 1.5–2.5 mm. long; anthers 1.5–2 mm. long, with shortly apiculate tip. Placenta with 2 collateral ovules; style glabrous or with a few short hairs towards the tip, 7–8.5 mm. long including the 1–1.5 mm. long lobes. Fruit red, ± ovoid, 6 mm. long. Seeds 1–2 in each chamber, their shape respectively ½–¼-ellipsoid; testa minutely striate; hilum linear, straight.

TANZANIA. Iringa District: Mufindi, Kigogo Forest Reserve, 3 Oct. 1947, *Brenan, Greenway & Gilchrist* 8253! & 9 Oct. 1954, *Sangiwa* 41! & Livalonge Tea Estate, Nyalawa R. valley, 26 Aug. 1971, *Perdue & Kibuwa* 11257!

DISTR.   T 7; known only from Mufindi area.

HAB.   Upland forest; 1730–1900 m.

16. T. pedicellata *Robbrecht* in B.J.B.B. 57: 194 (1987) & in Distr. Pl. Afr. 32: map 1068 (1987). Type: Tanzania, Morogoro District, Kimboza Forest Reserve, *Polhill & Lovett* 4900 (K, holo.!, BR, iso.!)

Shrub or rarely a small tree 1.5–2.5 m. high; young twigs glabrous. Leaf-blades elliptic, more rarely ovate or obovate, (5–)8–16 cm. long, (2–)–3–6.5 cm. wide, acuminate at the apex, cuneate or rarely ± rounded at base, sparsely pubescent when young, later entirely glabrous, subcoriaceous; 3–5(–7) pairs of main lateral nerves; pocket-domatia tufted; petiole 5–10 mm. long, glabrous or with 2 lateral rows of hairs; stipules sheathing, ± 1.5 mm. long, with awns 1–2 mm. long. Flowers 6-merous, subsessile (but stalk elongating in fruit), in 3–several-flowered sessile inflorescences; bracts and bracteoles cupular, truncate or 2-apiculate, pubescent. Calyx-tube glabrous; limb cupular, 2 mm. long, showing deep longitudinal splits when the flower opens, pubescent outside. Corolla white, glabrous outside; tube 3.5 mm. long, 1.5 mm. wide, the upper half with sparse long hairs within; lobes 6.5 mm. long, 2.25 mm. wide, acute. Filaments 1.5 mm. long; anthers 3.5 mm. long, shortly apiculate, glabrous. Placenta with 2 collateral ovules; style glabrous, ± 10 mm. long including ± 2 mm. long lobes. Fruit red, ± spherical, ± 8 mm. in diameter, borne on long and slender pedicels (elongated in fruiting stage only, developed above the bracteolar cup) up to 30 mm. long. Seeds 1 or 2 in each chamber, ½- or ¼-spherical; hilum oblong, wide; seed-coat ± smooth.

TANZANIA. Kilosa District: Mikumi National Park, Vuma Hills, 29 June 1977, *Wingfield & Mhoro* 4091!; Morogoro District: Nguru Mts, Ruhamba [Koluhamba], Nov. 1953, *Semsei* 1450! & Kimboza Forest Reserve, 11 Apr. 1935, *E.M. Bruce* 1032!

DISTR.   T 6; not known elsewhere

HAB.   Forest; 300–700 m.

17. T. schliebenii *Robbrecht* in B.J.B.B. 57: 196 (1987) & in Distr. Pl. Afr. 32, map 1070 (1987). Type: Tanzania, Lindi District, Lake Lutamba, *Schlieben* 5683 (HBG, holo.!, B, BR, K, LISC, MO, WAG, iso.!)

Shrub or small tree, 2–6 m. high; young twigs with short erect hairs. Leaves unknown, deciduous; stipules sheathing, 1–2 mm. long, with awns 2–3 mm. long, especially the latter puberulous. Flowers 6-merous, sweetly scented, sessile, in 3–many-flowered sessile inflorescences; bracts and bracteoles cupular, 2- or 4-dentate. Calyx-tube pubescent, with golden yellow hairs; limb splitting when flower opens into 3–4 irregular and somewhat withering lobes, pubescent outside and within. Corolla white, pubescent outside (especially the lobes); tube 3–5 mm. long, 1–1.5 mm. wide, with long hairs in the upper half inside; lobes 4–4.5 mm. long, 2 mm. wide, acuminate. Filaments 1.5 mm. long; anthers 3–4.5 mm. long. Placenta with 2 collateral ovules; style ± 10 mm. long including the 1.5–2 mm. long lobes. Fruits unknown.

TANZANIA. Masasi, 12 Dec. 1942, *Gillman* 1204!; Lindi District: Lake Lutamba, Dec. 1934, *Schlieben* 5683!

DISTR.   T 8; not known elsewhere

HAB.   Deciduous woodland; 200–500 m.

18. T. sp. F

Multi-stemmed deciduous shrub, 1.8 m. high, with shortly hairy young twigs, flowering before the new leaves appear. Leaves unknown; stipules hairy, with a ± triangular base 1 mm. high and awns 1.5–2 mm. long. Flowers 6-merous, in 1-flowered inflorescences, sessile, with 2–3 puberulous 2-toothed bracteolar cups. Calyx-tube glabrous; limb ellipsoid, truncate at the tip and with only some minute teeth representing the lobes, puberulous, 3–4 mm. high, splitting deeply at anthesis. Corolla white, glabrous outside except the puberulous lobe-tips, with long hairs inside in the upper half of the tube; tube 5 mm. long, 2 mm. in diameter; lobes 6 mm. long, 2.5 mm. wide. Filaments 1.5 mm. long; anthers inframedifixed, 3.5 mm. long. Placenta with 3 ellipsoid ovules arranged in a U-shape; style glabrous, except for the shortly hairy upper part and lobes, 11 mm. long, deeply bilobed for 2.5 mm. Fruits unknown.

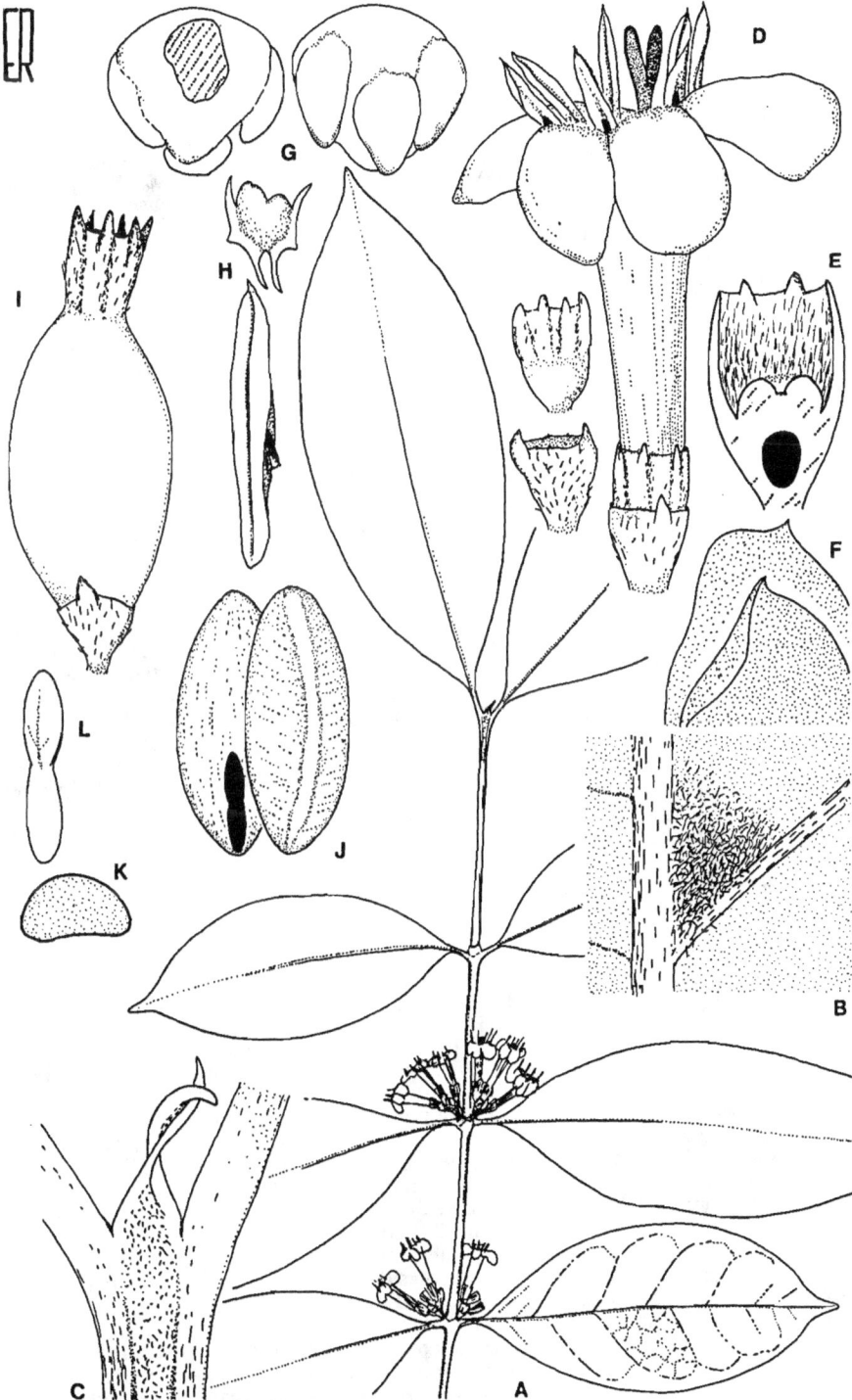

FIG. 91. *TRICALYSIA VANROECHOUDTII* — **A**, flowering branch, × ½; **B**, domatium, × 7; **C**, stipules, from apex of twig, × 7; **D**, flower, with calyx and bracteolar cup at side, × 3½; **E**, section through calyx, × 7; **F**, tip of corolla-lobes, × 7; **G**, placenta, 2 views, × 22½; **H**, stamen, lateral view, × 7 and transverse section of anther, × 15; **I**, fruit, × 3½; **J**, seed, 2 views with position of embryo indicated, × 3½; **L**, embryo, × 7. All from *Rammeloo* 4668. Drawn by E. Robbrecht.

TANZANIA. Newala District: Newala–Mikindani road, Nov. 1953, *Eggeling* 6737!
DISTR. T 8; not known elsewhere
HAB. Deciduous thicket; 500 m.

SYN. *T. sp. 91* sensu Robbrecht in B.J.B.B. 57: 197 (1987)

NOTE. This species belonging to section *Rosea* superficially resembles an equally imperfectly known species of section *Ephedranthera* (*T. sp. G.*), both being deciduous and having calyx-limbs that split. It differs from *T. sp. G.* in having 6-merous, much smaller flowers with glabrous calyx-tubes, 3 ovules on the placenta, and anthers with a distinct filament. *Schlieben* 5711A! (Lindi District, Lake Lutamba) possibly belongs to the same taxon but differs in some minor points (young twigs glabrous, 5-merous flower, style entirely glabrous). Another specimen from NE. Mozambique, Niassa, Cabo Delgado, *Mendonça* 915!, collected on the borders of the Flora area, certainly belongs near here, but in its much larger flowers resembles *T. jasminiflora* (Klotzsch) Hiern, (belonging to section *Rosea*) with a more southerly distribution (see Robbrecht in Distr. Pl. Afr. 32, map 1071 (1987)).

19. T. vanroechoudtii *(Van Roechoudt) Robbrecht* in B.J.B.B. 52: 324, fig. 5 (1982) & in Distr. Pl. Afr. 23, map 763 (1982). Type: Zaire, Kivu, Mushwere, *Lebrun* 5539 (BR, lecto.!, K, MO, WAG, isolecto.!)

Shrub 2–6 m. high; young twigs with short hairs on 2 sides. Leaf-blades mostly obovate, 6–14 cm. long, 2.5–6 cm. wide, shortly acuminate at apex, cuneate at base, glabrous above and beneath, coriaceous; 5–6 pairs of main lateral nerves; domatia hairy; petiole 5–10 mm. long, mostly puberulous on the adaxial side; stipules triangular, fused at the base, with awns 1–3 mm. long, puberulous, quickly becoming corky. Flowers 6-merous, sweet-scented, sessile; inflorescences (1–)2–3(–5)-flowered; bracts and bracteoles cupular, similar to the calyx-limb. Calyx-tube glabrous; limb ± dentate, at anthesis often with one longitudinal split, 2.5–5 mm. long, puberulous or rarely glabrous outside. Corolla white, glabrous outside and inside; tube (7.5–)10–17 mm. long, 2–5 mm. in diameter; lobes 6–7 mm. long, 3.5–4.5 mm. wide. Stamens subsessile; anthers 5–7 mm. long, shortly apiculate. Placenta with 3(–5) ovules arranged in a U-shape; style glabrous, exserted, 17–24 mm. long including 2 lobes 2–3 mm. long. Fruit of unknown colour, ellipsoid, 12 mm. long, 7 mm. wide. Seeds 1–2 in each chamber, hemi-ellipsoid; testa finely striate. Fig. 91.

UGANDA. Ankole District: Kalinzu Forest, June 1938, *Eggeling* 3677!; Masaka District: Byante Forest Reserve, June 1951, *Philip* 492! & Malabigambo Forest, 9 km. SSW. of Katera, Nov. 1953, *Drummond & Hemsley* 4601!
TANZANIA. Bukoba District: Kaigi, 1935, *Gillman* 353!
DISTR. U 2, 4; T 1; Zaire, Burundi
HAB. Upland rain-forest; (1200–)1500–2300 m.

SYN. *Coffea vanroechoudtii* Van Roechoudt, Revue Agric. Bot. Kivu 1: 30 (1932) & 2: 17–23 (1933); Chevalier, Caféiers du Globe 3: 217 (1947)

20. T. acocantheroides *K. Schum.* in P.O.A. C: 382 (1895); Robbrecht in B.J.B.B. 52: 327, fig. 6 (1982) & in Distr. Pl. Afr. 23, map 764 (1982). Types: Malawi, without locality, *Buchanan* 1469 (B, syn. †, K, lecto.!, BM, E, isolecto.!)

Shrub or sometimes a small tree, 1.5–4.5(–7) m. high, with puberulous young twigs. Leaf-blades mostly ovate, (2–)3.5–7.5(–9) cm. long, (1–)2–2.7(–4) cm. wide, pointed or shortly acuminate at apex (the actual tip mucronulate), cuneate or rarely rounded at base, glabrous above and beneath except for the puberulous midrib, rather coriaceous; 5–8 pairs of main lateral nerves; domatia evident, hairy; petioles 2–3(–5) mm. long, puberulous; stipules triangular, fused at the base, with awns 3–5 mm. long, puberulous. Flowers 5–6-merous, heterostylous, sweet-scented, sessile, inflorescences mostly 3- or 5-flowered; bracts and bracteoles cupular, those subtending inflorescence branches bearing 2 small foliaceous appendages and 2 short awns, those subtending flowers with 4 unequal awns. Calyx-tube glabrous; limb-tube 1.5–2.5 mm., puberulous; lobes subulate, mostly equalling the tube. Corolla white, the outside pubescent especially towards the top (except for that part of the lobes covered in bud) and inside glabrous; tube 7.5–11 mm. long, 1–3 mm. in diameter; lobes 4–5 mm. long, 2–3 mm. wide. Anthers 3 mm. long, shortly apiculate, sessile and almost completely included in the corolla-tube in short-styled flowers, or with a short filament (0.5–1 mm.) and ± exserted in long-styled flowers. Placenta mostly with 2 collateral ovules, sometimes 1- or 3-ovulate; style glabrous, included (the 2 lobes lying under the anther-bases) in short-styled flowers or shortly exserted in long-styled flowers. Fruit spheroidal, ?orange or red (only recorded from three collections), ± 8 mm. in diameter, with 2 chambers or sometimes 1 by abortion.

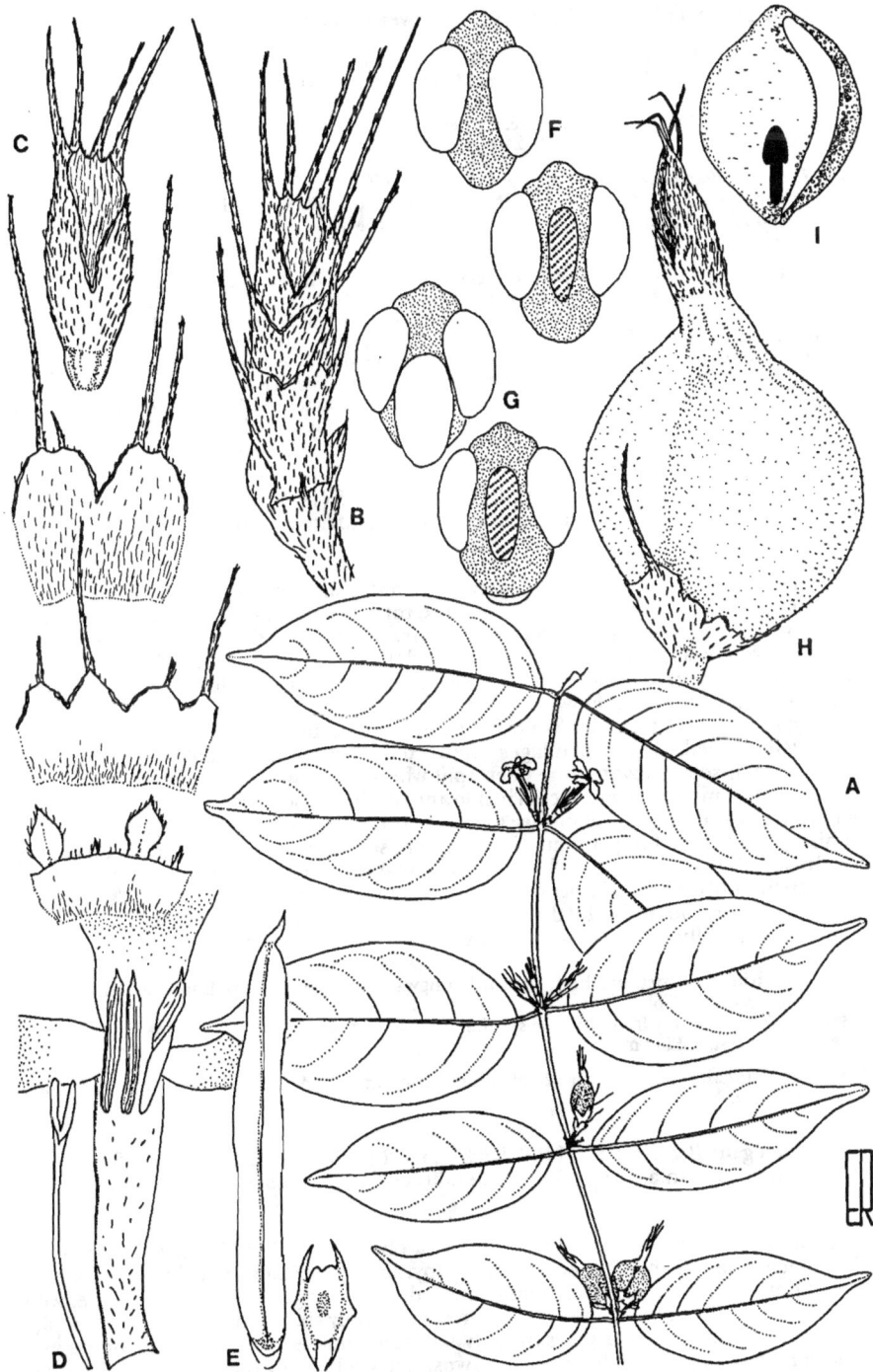

FIG. 92. *TRICALYSIA ACICULIFLORA* — **A**, habit, × ½; **B**, calyx in three bracteolar cups, × 4½; **C**, same, separated with cups opened out, × 4½; **D**, half corolla-tube with style and stigma, × 4½; **E**, anther, lateral view, × 15 and transverse section, × 22½; **F**, placenta, with 2 ovules, 2 views, × 22½; **G**, same with 3 ovules, × 22½; **H**, fruit, × 3⅓; **I**, seed, ventral view with position of embryo indicated, × 3⅓. All from *Mwasumbi* 11866. Drawn by E. Robbrecht.

Seeds 1–2 per chamber, hemi-ellipsoid; testa weakly striated.

Tanzania. Ufipa District: edge of Rukwa Rift escarpment, Nov. 1958, *Napper* 1107!; Mbeya District: Umalila, Usanga Mt., 23 Oct. 1899, *Goetze* 1365!; Songea District: Lupembe, 12 Nov. 1956, *Mgaza* 135!
Distr. T 4, 7, 8; Malawi, Zambia, Mozambique
Hab. Upland evergreen forest; 1500–2400 m.

Syn. *T. mucronulata* K. Schum. in E.J. 30: 413 (1901); T.T.C.L.: 535 (1949). Type: Tanzania, Mbeya District, Usanga Mt., *Goetze* 1365 (B, holo. †, Z, lecto.!, BR, E, P, isolecto.!)
       T. milanjiensis S. Moore in J.L.S. 37: 328 (1906). Type: Malawi, Mt. Mulanje, *Whyte* (BM, holo.!, K, iso.!)

Note. This is one of the few *Tricalysia* species occurring in the Flora area having heterostylous flowers; its floral organization does not correspond with the typical heterostyly of Rubiaceae-Rubioideae (compare many genera of the Hedyotideae, such as *Pentas* in Part 1 of the family). The short-styled flower of *T. acocantheroides* has the anthers almost completely included with the stigmatic lobes lying under their bases, while the long-styled flower exactly corresponds with the floral organization of non-heterostylous species of *Tricalysia*. The long-styled flower form is apparently very rare and has so far only been found a few times, viz. from the Nyika and Viphya Plateaus on the border between Zambia and Malawi, and one record from the Chongoni Forest Reserve, central Malawi. Field observations on the floral biology are greatly desired.

21. T. aciculiflora *Robbrecht* in B.J.B.B. 52: 332, fig. 8 (1982) & in Distr. Pl. Afr. 23, map 766 (1982). Type: Tanzania, Iringa District, Boma la Mzinga, *Mwasumbi* 11866 (K, holo.!, BR, iso.!)

Small shrub with puberulous young twigs. Leaf-blades ovate, sometimes elliptic, 6–11 cm. long, 2.5–4.5 cm. wide, acuminate at apex, rounded at base, glabrous above but with puberulous midnerve, very sparsely hairy beneath but more densely so on the nerves, papery; 5–8 pairs of main lateral nerves; domatia absent; petiole 3–5 mm. long, puberulous; stipules triangular, fused at the base, 1–2 mm. long, with awns up to 10 mm. long, puberulous outside. Flowers 5-merous, possibly heterostylous, subsessile, mostly solitary; bracts and bracteoles 3–4 pairs, cupular, puberulous, the upper cups provided with evident needle-like appendages often exceeding the calyx. Calyx-tube glabrous; limb-tube ellipsoid, showing one deep split when the flower-bud enlarges, appearing spathaceous, 5 mm. long, pubescent inside and outside; lobes 3–5 mm. long, needle-like, puberulous. Corolla white, glabrous; tube 9 mm. long, 2 mm. in diameter, sparsely hairy within; lobes 9 mm. long, 4 mm. wide. Anthers sessile, 4 mm. long, shortly apiculate, half-included. Placenta with 2–3 ovules; style glabrous, included; stigmatic lobes lying between the anther-bases. Fruit of unknown colour, spheroidal, somewhat didymous, ± 1 cm. in diameter, subtended by the persistent cups of bracteoles. Seeds mostly 1 in each chamber, hemi-ellipsoid; seed-coat with small elevated cells. Fig. 92.

Tanzania. Iringa District: northern part of Uzungwa [Udzungwa] Scarp, Boma la Mzinga, June 1979, *Mwasumbi* 11866!
Distr. T 7; known only from the type specimen
Hab. Rain-forest; ± 1200 m.

Note. Additional collections and field observations are needed to decide if the flowers are heterostylous.

22. T. elegans *Robbrecht* in B.J.B.B. 52: 330, fig. 7 (1982) & in Distr. Pl. Afr. 23, map 765 (1982). Type: Tanzania, Tanga District, Steinbrüch Gorge, *Faulkner* 2014 (K, holo.!, B, BR, iso.!)

Small glabrous shrub. Leaf-blades elliptic or obovate, 7–12 cm. long, 3–4.5 cm. wide, shortly acuminate at apex, cuneate at base, papyraceous; 5–7 pairs of main lateral nerves; pocket-domatia hairy; petiole 3–5 mm. long; stipules triangular, fused at the base, with awns 3–5 mm. long. Flowers possibly heterostylous (only short-styled form seen), (5–)6(–7)-merous, subsessile, (1–)2 pairs in each axil, common peduncle obsolete; bracts and bracteoles 2–4, cupular with 4 unequal awns. Calyx-tube glabrous; limb-tube 2–3 mm., ciliate at the margin; lobes subulate, 3 mm. long. Corolla pure white, delicate, short-lived, glabrous inside and outside except for cilia on the margins of the lobes towards the tips; tube 15–25 mm. long, 1–3 mm. in diameter; lobes 12–18 mm. long, 7–8 mm. wide, with pointed tip. Anthers sessile, 4–5 mm. long, with short hairy sterile appendage, more than half-included. Placenta with 3–5 ovules. Fruits unknown.

TANZANIA. Tanga District: Steinbrüch Gorge, July 1957, *Faulkner* 2014!
DISTR. T 3; known only from type specimen
HAB. Evergreen forest; 30 m.
NOTE. Field observations and additional collections are much needed.

23. T. allocalyx *Robbrecht* in B.J.B.B. 52: 334, fig. 9 (1982) & in Distr. Pl. Afr. 23, map 767 (1982). Type: Tanzania, Uzaramo District, Pande Forest Reserve, *Wingfield* 2812 (K, holo.!, EA, iso.!)

Shrub, sometimes scandent, 1.5–5 m. high, rarely with leaves and twigs in whorls of 3; young twigs puberulous. Leaf-blades ovate or elliptic, 6–12 cm. long, 3–5 cm. wide, acuminate at apex, rounded or cuneate at base, glabrous except for the puberulous midrib beneath, coriaceous, glossy above; 3–5(-6) pairs of lateral nerves; tertiary nerves perpendicular to midrib, ± conspicuous (reminiscent of *Mussaenda arcuata*); domatia hairy; petiole 5–10 mm. long, puberulous; stipules triangular, fused at the base, with awns 4 mm. long, puberulous outside, inside with silky hairs and colleters in a row. Flowers 6-merous, sessile; inflorescences 3- or 5-flowered; bracts and bracteoles cupular but not resembling calyx-limb, provided with 2(-4) short awns and 2 ± leafy appendages, pubescent outside and with silky hairs and a row of colleters at the base inside. Calyx-tube pubescent; limb 3 mm. long, deeply divided into 4 lobes, shallowly notched at the tip, similar to bracteoles in hairiness and colleters. Corolla white (only one withered one known), pubescent outside; tube ± 4 cm. long, sparsely hairy inside; lobes ± ⅕ the length of the tube. Anthers sessile, pubescent, 2 mm. long, almost included, apiculate. Placenta with ± 12 ovules; style pubescent, ? exserted. Fruit imperfectly known, almost spherical, ± 1.6 cm. in diameter, crowned by the persistent calyx-limb. Seeds numerous in both chambers; seed-coat with elevated isodiametric cells. Fig. 89/1.

TANZANIA. Uzaramo District: Pugu Hills, Oct. 1925, *Peter* 31457! & Pande Forest Reserve, Dec. 1978, *Mwasumbi* 11633!; Rufiji District: Kichi Hills, Dec. 1976, *Ludanga & Vollesen* in M.R.C. 4263!
DISTR. T 6; not known elsewhere
HAB. Lowland evergreen forest, coastal semi-deciduous thicket on sand; up to 450 m.

NOTE. Field observations and additional collections of flowers and fruit are much needed. This species possesses some characters unusual for the genus *Tricalysia;* these are indicated in couplet 1 of the key to species.

24. T. bridsoniana *Robbrecht* in B.J.B.B. 56: 146 (1986) & in Distr. Pl. Afr. 28, map 934 (1986). Type: Kenya, Kilifi District, Arabuko Forest, *Brenan, Gillett, Kanuri & Chomba* 14685 (K, holo.!, BR!, EA, K!, WAG!, iso.)

Shrub 3–5 m., with entirely glabrous twigs and leaves (except for the domatia). Leaf-blades obovate, elliptic or ovate, 5–10 cm. long, 1.5–3.5 cm. wide, acute at apex and with the midrib projecting as a small mucro, cuneate at base, coriaceous, shining above; 3–4 pairs of main lateral nerves; domatia hairy, mostly without excavation, sometimes with pocket-cavities; petiole 1–3 mm. long; stipules sheathing, ± 1 mm. long, with awns up to 4 mm. long. Flowers (?5–)6-merous, sweetly scented, solitary and sessile; bracts and bracteoles 3–4 pairs, cupular, the upper ones with 4 unequal awns, puberulous. Calyx-tube glabrous; limb ellipsoid, 4–5 mm. long, variable in hairiness outside, hairy inside, with 4–5 minute lobes at tip, when the flowers open developing 1 deep split and then becoming oblique with regard to corolla-tube, spathe-like. Corolla white; tube 13–27 mm. long, 1.5–3 mm. wide, with long slender hairs in the upper ⅔ inside; lobes 6–10 mm. long, 3–5 mm. wide, sparsely ciliolate, acute. Anthers sessile, 2.5 mm. long, shortly apiculate. Placenta with 2–4 ovules; style glabrous, short, included, the lobes positioned in the middle of the tube. Fruit red, ovoid, 3.5 mm. long, 2 mm. wide. Seeds few in each chamber, with curved linear hilum; testa colliculate, thickened near the hilum.

var. bridsoniana

Leaf-blades elliptic or sometimes ovate, 5–8(-10) cm. long, 1.5–3 cm. wide; petiole glabrous. Calyx-limb sparsely puberulous outside only at tip.

KENYA. Kilifi District: 3 km. E. of Jilore Forest Station, 15 Dec. 1972, *Spjut & Ensor* 2789! & Arabuko Forest, halfway between Mida and Jilore Forest Station, 24 Nov. 1978, *Brenan, Gillett, Kanuri & Chomba* 14685!; Coast Province, without locality, 11 Feb. 1969, *Padwa* 26!
DISTR. K 7; not known elsewhere
HAB. *Cynometra* forest; below 300 m.

SYN. *T. sp. 38* sensu Robbrecht in B.J.B.B. 52: 336 (1982)

NOTE. According to the collectors of the type, the flowers are 5-merous, but I could not confirm this. As for most other East African species of section *Ephedranthera*, further observations on the floral biology are needed.

var. **pandensis** *Robbrecht* in B.J.B.B. 56: 147 (1986) & in Distr. Pl. Afr. 28, map 935 (1986). Type: Tanzania, Uzaramo District, Pande Forest Reserve, *Harris, Tadros & Mwasumbi* 3597 (DSM holo.!, BR!, EA, K!, iso.)

Leaf-blades obovate, 5–9 cm. long, 1.5–3.5 cm. wide; petiole with ciliate margin. Calyx-limb puberulous outside.

TANZANIA. Uzaramo District: Pande Forest Reserve, 22 Nov. 1969, *Harris, Tadros & Mwasumbi* 3597!
DISTR. **T** 6; only known from the type
HAB. Evergreen lowland forest; 120 m.

SYN. *T. sp. 39* sensu Robbrecht in B.J.B.B. 52: 337 (1982)

25. T. sp. G

Shrub 3–4 m. tall, with puberulous young twigs. Leaves deciduous, only known in very young state; blades hairy beneath and on the midnerve above; domatia probably absent; stipules triangular, fused at the base, 2 mm. long, with awns of 3 mm. long. Flowers 7–8-merous, solitary, sessile; bracts and bracteoles 3 pairs, cupular, 4-awned. Calyx-tube puberulous; limb oblique with regard to corolla-tube, ovoid, puberulous, with minute teeth at the tip, developing one or more deep longitudinal splits. Corolla white; tube 12–16 mm. long; lobes 8 mm. long, 3 mm. wide. Anthers apiculate. Placenta with 5 ovules arranged in a U-shape. Fruits unknown.

TANZANIA. Lindi District: Lake Lutamba, 10 Dec. 1934, *Schlieben* 5711!
DISTR. **T** 8; not known elsewhere
HAB. Deciduous woodland; 240 m.

SYN. *T. sp. 40* sensu Robbrecht in B.J.B.B. 52: 337 (1982)

26. T. junodii (*Schinz*) *Brenan* in K.B. 2: 60 (1948); Garcia in Mem. Junta Invest. Ultram., sér. Bot., 4: 41 (1957); Ross in Bot. Surv. Mem. (S. Afr.) 39: 334 (1972); Robbrecht in B.J.B.B. 49: 267 (1979). Type: Mozambique, Delagoa Bay, Makororo Forest, *Junod* 311 (Z, holo.!)

Shrub (seldom a small tree) 0.5–4 m. tall, with hairy young twigs. Leaves deciduous; blades 2–14.5 cm. long, 1–5.2 cm. wide, mostly dimorphic, on lower parts of twigs, broader, ovate, acute at apex, cordate at base, near tips of twigs narrower, elliptic, acute to acuminate at apex, cuneate at base; 4–8 pairs of main lateral nerves; domatia absent; petiole 1–3 mm. long, puberulous; stipules sheathing, crowned by awns up to 4 mm. long, puberulous. Flowers (4-)5(-6)-merous, sweet-scented; inflorescence sessile, with 1–7 flowers; pedicels up to 7 mm. long, glabrous to velutinous; bracts fused into a spathe-like cup; bracteoles mostly 2, free, alternate, narrowly triangular. Calyx glabrous or pubescent; limb-tube short; lobes round to triangular, 0.5–1 mm. long, ciliate. Corolla white, cream or pink, glabrous outside except the ciliate lobe-tips, sometimes with the base of the tube puberulous; tube 3–6 mm. long, 1.5–2.5 mm. in diameter, shaggy at throat; lobes 3–5 mm. long, 1.5–3 mm. wide. Filaments 1–2 mm. long; anthers 3–4 mm. long, with a conspicuous apical appendage 1–2 mm. long. Style glabrous or sparsely hairy, 7–10 mm. long including lobes 1–2 mm. long. Placenta with 2 collateral ovules. Fruit first white, turning purplish and finally black, hairy, 8 mm. in diameter. Seeds 2 in each chamber, adaxially deeply excavated; seed-coat brown, with large ± polygonal elevated cells; endosperm entire.

SYN. *Empogona junodii* Schinz in Mém. Herb. Boiss. 10: 67 (1900)
    E. kirkii Hook.f. var. *australis* Schweickerdt in Bothalia 3: 255 (1937); Hutchinson, Botanist in S. Afr.: 671 (1946). Type: South Africa, Transvaal, Zoutpansberg, *Obermeyer, Schweickerdt & Verdoorn* 72 (PRE, holo.!, K, Z, iso.!)
    Tricalysia allenii (Stapf) Brenan var. *australis*(Schweickerdt) Brenan in K.B. 2: 62 (1947); van der Schijff, Check List Vasc. Pl. Kruger Nat. Park: 90 (1969); Ross, Bot. Surv. Mem. (S. Afr.) 39: 334 (1972)

var. **kirkii** (*Hook.f.*) *Robbrecht* in B.J.B.B. 49: 271, fig. 7 (1979) & in Distr. Pl. Afr. 16, map 533 (1979); Vollesen in Op. Bot. 59: 72 (1980). Type: Malawi, Cape Maclear, Oct. 1861, *Kirk* (K, holo.!)

Leaf-blades densely hairy (in Flora area velutinous beneath); pedicels and calyx-tubes velutinous.

TANZANIA. Kilwa District: Selous, Malemba Thicket, 1 Feb. 1971, *Ludanga* in *M.R.C.* 1183! & Nakalila Thicket, 14 Dec. 1975, *Vollesen* in *M.R.C.* 3079!; Masasi District: inselberg on Tunduru–Masasi road, Nov. 1951, *Eggeling* 6383!
DISTR. T 8; Zambia, Malawi, Zimbabwe, Mozambique, Caprivi Strip, Botswana and South Africa (Transvaal)
HAB. Woodland and deciduous coastal thicket; ± 400 m.

SYN. *Empogona kirkii* Hook.f. in Hook., Ic. Pl. 11, t. 1091 (1871); Hiern in F.T.A. 3: 115 (1877); Schinz, Pl. Menyharthianae: 75 (1905), *non Tricalysia kirkii* Hiern (1877)
Hypobathrum kirkii (Hook.f.) Baillon in Adansonia 12: 205 (1878)
Empogona allenii Stapf in K.B. 1906: 79 (1906). Type: Zambia or Zimbabwe, near the Victoria Falls, *Allen* 55 (K, holo.!)
Tricalysia allenii (Stapf) Brenan in K.B. 2: 60 (1947); F.F.N.R.: 423 (1962)
Tricalysia allenii (Stapf) Brenan var. *kirkii* (Hook.f.) Brenan in K.B. 2: 61 (1947); Garcia in Mem. Junta Invest. Ultram., sér. Bot., 4: 42 (1958); van der Schijff, Check-List Vasc. Pl. Kruger Nat. Park: 90 (1969); Drummond in Kirkia 10: 275 (1975)
[*T. junodii* sensu Drummond in Kirkia 10: 275 (1975), *non* (Schinz) Brenan sensu stricto]

NOTE. This widespread and very variable species (especially so in leaf- and flower-size) has been known as *T. allenii* since Brenan's revision of the *Empogona*-group (in K.B. 2: 53-63, (1947)). However, recent collections from South Africa show *T. allenii* is not specifically separable from the earlier described *T. junodii*. *T. junodii* is now regarded as a variety with more glabrous leaves and flowers limited to the south of Mozambique and South Africa (Transvaal and Natal). *T. ngalaensis* Robbrecht is a near ally occurring on the edge of Lake Malawi in the northern tip of Malawi. It differs in the almost glabrous leaves with tufted domatia, and by the number of seeds (1–2 per fruit). It may occur in the Flora area.

27. T. ovalifolia *Hiern* in F.T.A. 3: 119 (1877); Brenan in K.B. 2: 57 (1947); T.T.C.L.: 534 (1949); Robbrecht in B.J.B.B. 49: 278 (1979) & in Distr. Pl. Afr. 16, maps 535–537 (1979). Type: Zanzibar, *Kirk* (fruiting collection) (K, lecto.!)

Shrub or sometimes a small tree, 1–6 m. high; young twigs puberulous, rarely glabrous. Leaf-blades ovate, elliptic or obovate, ± pointed or acuminate at apex, cuneate at base, (3–)5–11 cm. long, (1–)2.5–4 (–6) cm. wide, glabrous or hairy, coriaceous; (6–)8–11 pairs of main lateral nerves; domatia rarely present (var. *A*); petiole 1–5 mm. long, almost glabrous to hairy; stipules sheathing, 1–2 mm. long, with awns 1–6 mm. long. Flowers 5-merous, sweetly scented, in 3–6-flowered inflorescences; pedicels 4–16 mm. long; bracts cupular, provided with 2 short teeth and 2 mostly ± leafy appendages; bracteoles free (1–)2, alternate, rarely with a bracteolar cup. Calyx-tube glabrous or hairy; limb-tube very short; lobes ± triangular, with rounded tips not touching one another. Corolla white, rarely pinkish, glabrous outside except for the often ciliate lobes and tube inside bearing a very conspicuous tuft of shaggy hairs in the upper part and the throat; tube 3–6.5 mm. long, 1–3 mm. wide; lobes 3.5–8 mm. long, 1–3.5 mm. wide, emarginate, rounded or rarely acute. Filaments 0.5–1.5 mm. long; anthers 3–4 mm. long, with an apical appendage 1–2 mm. long. Placenta with 2–5 ovules (outside Flora area: 1–7); style glabrous, 6.5–11 mm. long including lobes 2–3 mm. long. Fruits probably first white (not noted), at complete maturation purple to blackish, spherical, ± 7 mm. in diameter. Seeds 2–4 in each chamber, with convex outer side and excavated inner side; hilum oblong; seed-coat colliculate.

KEY TO INFRASPECIFIC VARIANTS

Leaf-blades glabrous, or hairy only on the nerves:	
Vegetative parts entirely glabrous, except the young twigs which may be puberulous and the upper side of the petiole which bears short hairs (often in two rows); flower-pedicels, bracteoles and ovary glabrous	a. var. **ovalifolia**
Petiole hairy all around; midnerve of leaf-blade puberulous above and beneath especially its base; flower-pedicels, bracteoles and sometimes also the ovary hairy	b. var. **glabrata**
Leaf-blades, at least underneath, densely hairy:	
Leaf-blades velutinous or softly hairy underneath; domatia absent; leaf-tip acute or somewhat rounded . . .	c. var. **taylorii**
Leaf-blades with stiff hairs beneath; tuft-domatia present; leaf-tip shortly acuminate	d. var. **A**

FIG. 93. *TRICALYSIA OVALIFOLIA* var. *OVALIFOLIA* — 1, flowering branch, × ⅔; 2, stipule, × 4; 3, flower with axes of triplet, × 3; 4, calyx, × 8; 5, half corolla, × 4; 6, stigmatic lobes, × 6; 7, stamen, × 6; 8, longitudinal section of ovary, × 10; 9, transverse section of ovary, × 12; 10, placenta ventral view, × 20; 11, placenta, dorsal view, × 20; 12, fruit, × 1; 13, seed, ventral view, × 6. 1–11, from *Vaughan* 1595; 12, 13, from *Drummond & Hemsley* 3255. Drawn by Mrs M.E. Church.

a. var. **ovalifolia**

Vegetative parts entirely glabrous, except the young twigs which may be puberulous and the upper side of the petiole which bears short hairs (often in 2 rows); domatia absent; flower-pedicels, bracteoles and calyx-tube glabrous. Fig. 93.

KENYA. Kwale District: coast 2 km. S. of Tiwi Mosque, 13 Apr. 1977, *Gillett* 21047!; Kilifi District: Watamu, 26 Apr. 1975, *Ng'weno* 10! & Kilifi, Jan. 1937, *Moggridge* 333!
TANZANIA. Tanga District: Kwamkembe, 13 Jan. 1937, *Greenway* 4827!; Pangani District: Madanga, 7 May 1957, *Tanner* 3508!; Uzaramo District: Dar es Salaam, Oyster Bay, 26 Nov. 1968, *Batty* 309!; Zanzibar I., Marahubi, 9 July 1963, *Faulkner* 3216!
DISTR. **K** 7; **T** 3, 6; **Z**; Somalia, Aldabra, Assumption, Comores (Moheli, Anjouan & Mayotte), Madagascar
HAB. Coastal evergreen or mixed formations (open to ± closed forest, bushland or thicket); also in secondary vegetation (disused plantations, roadsides); 0–80(-150) m.

SYN. *Hypobathrum albicaule* Baillon in Adansonia 12: 209 (1878); Drake in Grandidier, Hist. Pl. Madag. 31, Hist. Nat. Pl. 6, t.445 (1897). Type: Comores, Moheli, *Richard* 264 (P, lecto.!)
H. comorense Baillon in Adansonia 12: 210 (1878). Type: Comores, Mayotte, Pamanzi I., *Boivin* 3176 (P, holo.!, BR, iso.!)
Tricalysia cuneifolia Baker in K.B. 1894: 148 (1894); Schinz in Abh. Senckenb. Naturforsch. Gesellsch. 21: 91, pro parte, quoad typum (1897); Hemsley in J.B. 54, Suppl. 2: 18 (1916) & in K.B. 1919: 123 (1919). Type: Aldabra, *Abbott* (K, lecto.!)
Plectronia subopaca K. Schum. & Krause in E.J. 39: 538 (1907). Type: Tanzania, Dar es Salaam, *Stuhlmann* 233 (B, holo. †, K, fragment!)
Canthium subopacum (K. Schum. & Krause) Bullock in K.B. 1932: 380 (1932); T.T.C.L.: 485 (1949)

NOTE. *Ossent* 687! (Kenya, "Donyo Sabuk of Kijabe" — obscure locality — prob. Kijabe Hill), possibly represents a new taxon near or within *T. ovalifolia;* the vegetative parts match var. *ovalifolia,* but the corolla-lobes are very pointed. *Greenway* 9649!, Kenya, Kwale District, N. of Jadini, is very similar.

b. var. **glabrata** (*Oliv.*) Brenan in K.B. 2: 59 (1949) & T.T.C.L.: 534 (1949); K.T.S.: 476 (1961); Robbrecht in B.J.B.B. 49: 283 (1979) & in Distr. Pl. Afr. 16, map 536 (1979); Vollesen in Op. Bot. 59: 72 (1980). Type: Kenya/Tanzania, 60–100 km. from coast, *Johnston* (K, holo.!) - see discussion of type locality by Brenan, loc. cit.

Petiole hairy all around; midnerve of leaf-blade puberulous above and beneath, especially its base; domatia absent; pedicels, bracteoles and sometimes also the calyx-tube hairy.

KENYA. Teita District: 3 km. E. of Bura station, 17 Jan. 1972, *Gillett* 19571!; Kwale District: Mrima Hill, 20 Nov. 1978, *Brenan, Gillett, Kanuri & Chomba* 14620!; Kilifi District: Arabuko Forest, Jilori, Oct. 1965, *Tweedie* 3192!
TANZANIA. Handeni District: Kolokongi, 26 Nov. 1973, *Faulkner* 4832!; Morogoro District: Lusunguru Forest Reserve, 18 Sept. 1959, *Mgaza* 319!; Kilwa District: Selous Game Reserve, *Rodgers* 852!
DISTR. **K** 7; **T** 3, 6, 8; Madagascar
HAB. Dry thickets, wooded grassland and evergreen forest; up to 960 m.

SYN. *Empogona kirkii* Hook.f. var. ? *glabrata* Oliv. in Trans. Linn. Soc., Bot., ser. 2, 2: 331 (1887)
Tricalysia ovalifolia Hiern var. *acutifolia* Brenan in K.B. 2: 58 (1947) & T.T.C.L.: 534 (1949). Type: Lushoto District, Tanzania, E. Usambara Mts., Bombo–Daluni, *Greenway* 4103 (K, holo.!, EA, iso., FHO, iso.!)

c. var. **taylorii** (*S. Moore*) Brenan in K.B. 2: 59 (1947); K.T.S.: 476 (1961); Robbrecht in B.J.B.B. 49: 285 (1979) & in Distr. Pl. Afr. 16, map 537 (1979). Type: Kenya, Kilifi District, Giriama, *W.E. Taylor* (BM, holo.!)

Leaf-blade velutinous or softly hairy underneath (density of hairiness rather variable); domatia absent.

KENYA. Kilifi District: Arabuko, Oct. 1929, *R.M. Graham* in F.D. 2172! & Giriama, *W.E. Taylor*!; Teita District, Mzwanenyi, 17 Feb. 1982, *Kabuye* 82/94!
TANZANIA. Lushoto District: Mashewa, upper Bosha valley, 22 Aug. 1915, *Peter* 55791! & Mazumbai–Mashewa road, 17 Apr. 1916, *Peter* 55885!; Handeni District: Tamota, July 1950, *Semsei* 609!
DISTR. **K** 7; **T** 3; not known elsewhere
HAB. Thicket, dry forest and bushland with scattered trees; 0–1000 m.

SYN. *Empogona taylorii* S. Moore in J.B. 63: 145 (1925)
Tricalysia grahamii Dale, T.S.K.: 143 (1936), *nomen nudem*

NOTE. The leaf-blade above may be either glabrous except for the midribs (specimens from Tanzania) or puberulous (specimens from Kenya). Teratological dense tufts of hairs occur on the leaf-blades of the type specimen; such abnormalities (? galls) are rather commonly observed in Indian Ocean material of var. *ovalifolia.*

d. var. **A**

Leaf-blade with stiff hairs beneath; domatia tufted.

TANZANIA.   Uzaramo District: Pande Hill Forest Reserve, 5 Feb. 1973, *Harris et al.* 6761! & 8 Feb. 1976, *Wingfield* 3297! & Kibaha, 300 m. N. of Roman Catholic Mission, 27 Oct. 1970, *O. Flock* 490!
DISTR.   T 6; not recorded outside Uzaramo District
HAB.   Thicket and forest; 110–190 m.

28.   T. acidophylla *Robbrecht* in B.J.B.B. 49: 292, fig. 16 (1979) & in Distr. Pl. Afr. 16, map 539 (1979). Type: Tanzania, Lushoto District, Amani–Muheza [Muhesa] road, 3 km. E. of Sigi railway station, *Drummond & Hemsley* 3490 (K, holo.!, B!, BR!, EA, LISU, S!, iso.)

Shrub or small tree 2–6 m. high, with puberulous young twigs. Leaf-blades elliptic or obovate, sometimes narrowly so, 6–10(–15) cm. long, 2–4.5 cm. wide, with an acumen up to 2 cm. long, cuneate at base, glabrous above and beneath, papery; 4–7 pairs of main lateral nerves; domatia hairy, sometimes weakly pitted; petiole 3–5 mm. long, puberulous (but hairs often difficult to observe in adult leaves). Stipules fused into a short sheath overtopped by 2 interpetiolar awns 3–5 mm. long, glabrous. Flowers (4–)5-merous (except sometimes 6-lobed calyx); inflorescences with ± 5 flowers; pedicels 7–23 mm. long; bracts cupular with 4 appendages, 2 of which are often leafy; bracteoles at base of pedicel (1–)2, mostly free and alternate, narrowly triangular, sometimes cupular, 2-toothed. Calyx glabrous; limb-tube short; lobes round to triangular, 1 mm. long, sometimes with some hairs at tip. Corolla white, glabrous except the ciliate lobe-tips; tube 6.5 mm. long, 2–3 mm. in diameter, shaggy at top; lobes 4.5 mm. long, 3.5 mm. wide. Filaments 2 mm. long; anthers 2 mm. long, with distinct apical appendix 1 mm. long. Style glabrous, 11 mm. long including 2 mm. long lobes; placenta thin, with 3 deeply immersed ovules. Fruit of unknown colour, 8 mm. in diameter. Seeds 2 in each chamber, shape ± the quarter of a sphere; seed-coat brown, with large almost polygonal ± elevated cells; endosperm deeply ruminate.

TANZANIA.   Handeni District: Misufini, Oct. 1950, *Semsei* 586!; Morogoro District: Kimboza Forest Reserve, July 1952, *Semsei* 763! & Mtibwa Forest Reserve, Nov. 1954, *Semsei* 1952!
DISTR.   T 3, 6; not known elsewhere
HAB.   Forest, swamp-forest, riverine woodland; 300–400 m.
NOTE.   *Cribb, Grey-Wilson & Mwasumbi* 11070! from T 6, Ulanga District, Mzelezi is a somewhat more hairy variant probably not worthy of formal distinction (midnerves of leaves and lateral nerves beneath, stipules and fruit-pedicels puberulous).

29.   T. ruandensis *Bremek.* in B.J.B.B. 26: 253 (1956); Robbrecht in B.J.B.B. 49: 316 (1979) & in Distr. Pl. Afr. 16, map 550 (1979); Vollesen in Op. Bot. 59: 72 (1980); Fl. Pl. Lign. Rwanda: 608, fig. 206/1 (1982); Fl. Rwanda 3: 228, fig. 72/2 (1985). Type: Rwanda, Mutema, *Liben* 1416 (U, holo.!, BR, WAG, iso.!)

Small tree or shrub 1.25–5(–8) m. high, with the young twigs glabrous or with 2 decussate rows of short hairs; older twigs at first straw-yellow and smooth, later greyer and longitudinally fissured. Leaf-blades obovate or elliptic, (3–)5–8.5(–10) cm. long, (1.1–)2–3.5(–4) cm. wide, acute or with an acumen up to 1 cm. long at apex, cuneate at base, glabrous on both sides, coriaceous, shining above; (5–)7–10(–12) pairs of main lateral nerves; domatia absent; petiole 2–7 mm. long, glabrous; stipules sheathing, 1–2 mm. long overtopped by awns 1–3 mm. long, glabrous inside and outside. Flowers (4–)5-merous, sweetly perfumed; inflorescences 3–10-flowered; peduncles short, glabrous or covered with short hairs; pedicels at first very short but attaining 7 mm. in fruiting stage; bracts and bracteoles cupular, 2-toothed (the middle cup with 2 additional, opposite, leafy appendages), the cup subtending the flower 2-toothed, ciliate. Calyx-tube glabrous; limb-tube very short; lobes overlapping, rounded or elliptic, 0.5–1 mm. long, ciliate. Corolla white, outside glabrous except sometimes the ciliate lobe-tips; tube 2–5 mm. long, 1–1.25 mm. in diameter, hairy in the throat and the upper half; lobes 2.75–4.5 mm. long, 1.25–2 mm. wide. Anthers with apical appendix 0.25–0.50 mm. long. Placenta very thin, with 3–4 pear-shaped ovules; style glabrous, 4–6 mm. long, with 2 lobes 1–2 mm. Fruit white, then turning to purple and finally becoming blackish, 6–10 mm. in diameter. Seeds 1–4 per chamber, 3–4 mm. long, with adaxial excavation; seed-coat brown, with elevated ± rounded cells.

UGANDA.   Ankole District: close to border of Lyantonde, 20 Nov. 1951, *Trapnell* 2191!; Masaka District: Bigo [Biggo], 14 Apr. 1945, *A.S. Thomas* 4116!

Kenya. Kwale District: Jombo Mt., 11 Apr. 1978, *Verdcourt* 5289!

Tanzania. Bukoba District: 24 km. S. of Nyaishozi, Mar. 1958, *Procter* 852!; Dodoma District: Itigi–Singida road, Itigi Thicket, *Richards* 19879!; Lindi District: Rondo Plateau, Mchinjiri, *Semsei* 656!

Distr. U 2, 4; K 7; T 1, 3–5, 8; Zaire, Rwanda, Burundi, Mozambique, Malawi, Zambia and Zimbabwe

Hab. Sclerophyllous thicket, fire-protected woodland, riverine forest, often in rocky situations or on anthills, also on sandy soils; (400–)1000–1500 m.

Syn. [*T. cacondensis* sensu T.T.C.L.: 534 (1949), *non* Hiern]
     [*T. congesta* sensu F.F.N.R.: 423 (1962), ? *non* (Oliv.) Hiern]

Note. The name *T. congesta* was commonly in use for this species until the revision of *Tricalysia* subgen. *Empogona* in B.J.B.B. 49 (1979). The type material, however, cannot be identified with certainty, since it consists only of a short twig tip; there are no collector's notes about the habit, so it cannot be ascertained whether it came from a shrub or a geoxylic undershrub. The name *T. congesta* may thus apply either to *T. cacondensis* (geofrutex) or to the shrub described here. It is an older name than either but its true identity will probably remain obscure. Even the type locality is obscure. Oliver in J.L.S. 15: 95 (1876) described *Kraussia congesta* from a specimen collected by Cameron around the shores of Lake Tanganyika. According to Gillett in K.B. 14: 319 (1960) this collection may have come from Tanzania (T 4), Zambia or Zaire (Shaba). In view of this *Tricalysia congesta* (Oliv.) Hiern in F.T.A. 3: 120 (1877) has to remain a name of doubtful application.

Semsei 2824! (common tree growing in montane forest with *Newtonia-Markhamia*, collected at Mtowa Nguruwe, Lushoto District, 1400 m.) is an extreme variant with petioles up to 1 cm. long and leaf-blades up to 13 cm. long and 5 cm. wide.

30. T. sp. H

Small shrub with puberulous young twigs. Leaf-blades ovate, elliptic or obovate, 5–11 cm. long, 2.3–5.5 cm. wide, acute at apex, rounded or cuneate at base, glabrous above and beneath save for the puberulous midnerve, leathery, shining above; domatia absent; petiole 2–3 mm., puberulous; stipules sheathing 2 mm. long, overtopped by awns 5 mm. long, puberulous outside. Inflorescences 3–7-flowered; mature flowers unknown. Calyx with limb-tube very short and lobes rounded to elliptic, 0.5 mm. high, ciliate. Corolla unknown. Fruits unknown.

Tanzania. Lushoto District: Maramba, road to Tanga, Oct. 1918, *Peter* 56182!
Distr. T 3; not recorded elsewhere
Hab. Forest; 280 m.

Syn. *T. sp. aff. ruandensis* sensu Robbrecht in B.J.B.B. 49: 319 (1979)

31. T. cacondensis Hiern, Cat. Afr. Pl. Welw. 1(2): 467 (1898); K. Schum. in Baum, Kunene-Sambesi-Exped.: 385 (1903); Good in J.B. 64, Suppl. 2: 18 (1926); F.F.N.R.: 423 (1962); Launert in Prodr. Fl. SW. Afr. 115: 25 (1966); Robbrecht in B.J.B.B. 49: 320, fig. 23 (1979) & in Distr. Pl. Afr. 16, map 551 (1979). Type: Angola, road from Quipaca R. to fortress near Ferao, *Welwitsch* 3112 (LISU, lecto.!, BM, COI, K, isolecto.!)

Rhizomatous undershrub growing in dense patches, 10–30(–200) cm. high, with the young twigs rarely (but in Flora area always) glabrous. Leaf-blades obovate or sometimes elliptic, (3.2–)5–8(–11) cm. long, (0.7–)2–3(–5) cm. wide, rounded or acute at the apex, cuneate at base, glabrous except for short hairs on base of midnerve above, leathery, glossy above; (6–)7–8(–10) pairs of main lateral nerves; domatia absent; petiole almost absent to 2 mm. long, with the adaxial side shortly hairy or rarely (in Flora area) glabrous; stipules sheathing, 1–1.5(–3) mm. long, overtopped by 2 awns 1.5–3 mm. long, glabrous inside and outside except the often ciliate margin. Flowers 5-merous, sweetly perfumed; inflorescences (3–)4–5-flowered; peduncles short, shortly hairy; pedicels short and hairy; bracts cupular, often with 2 leafy appendages; bracteoles 2, triangular, almost free or cupular, ciliate. Calyx-tube glabrous; limb-tube very short; lobes overlapping, almost round, 1–1.5 mm. long, ciliate. Corolla white, glabrous outside except the ciliate lobe-tips; tube 2–5 mm. long, 1.5–2.5 mm. in diameter, hairy in the throat and the upper half within; lobes 5–7 mm. long, 2–2.5 mm. wide. Anther with a distinct apical appendage 1–2 mm. long. Placenta thin, with (2–)3–4 ovules often positioned in a horseshoe; style glabrous, 7–12 mm. long, including 2 lobes 2–3 mm. long. Fruit white, then turning purple and finally blackish, 6–8 mm. in diameter. Seeds 2–4 per chamber, 4–5 mm. long, with adaxial excavation; seed-coat brown, with elevated ± rounded cells.

TANZANIA. Tabora District: Simbo Forest Reserve, Oct. 1970, *Ruffo* 447!; Dodoma, Dec. 1921, *Swynnerton* 731!; District uncertain: Kabura, Nov. 1921, *Swynnerton* 732!
DISTR. T 4, 5; Zaire (Shaba), Angola, Zambia, Zimbabwe, Namibia, Botswana
HAB. Deciduous woodland, probably especially on seasonally waterlogged sandy soils; ± 1200 m.

SYN. [*T. buxifolia* sensu Good in J.B. 64, Suppl. 2: 18 (1926), pro parte, quoad *Gossweiler* 3741 & 3754, non Hiern]

NOTE. Specimens of *T. cacondensis* from the Flora area stand somewhat apart by their glabrous young twigs and petioles; they may be worthy of infraspecific distinction but more material is wanted. *T. congesta* may possibly provide an older name for this species (see note under *T. ruandensis*).

71. SERICANTHE

Robbrecht in B.J.B.B. 48: 27 (1978) & Distr. Pl. Afr. 13, maps 406–429 (1978)

[*Neorosea* sensu N. Hallé in Fl. Gabon 17, Rubiacées, 2: 268–277 (1970), excl. typum *N. jasminiflora* (Klotzsch) N. Hallé; Verdc. in F.T.E.A., Rub. 1: 11, 25 (1976)]

Shrubs or sometimes small trees (outside Flora area one suffrutescent species). Leaves mostly (and in Flora area always) provided with domatia as small tufts of hairs without apparent excavation; bacterial nodules often present on the underside of the leaf-blade, situated either along the midrib and down the petiole or (not in East Africa) scattered; stipules sheathing, overtopped by 2 awns, with silky hairs and colleters inside. Inflorescences axillary and opposite, with 1 or sometimes more flowers. Flowers (5–)7–8(–9)-merous, subtended by 2–4 bracteolar cups, each with 2 unequal pairs of awns. Calyx ellipsoid in bud, either open (not in East Africa), or closed, with the tip obtuse to apiculate, completely covering the young corolla, deeply splitting between some lobes when the flower opens. Corolla white, salver-shaped, usually (in East Africa always) with silky indumentum outside; lobes spreading, contorted to the left. Stamens inserted at the throat, with basifixed anthers and with the connective broadly flattened. Disc annular. Ovary 2-locular, each chamber having 1 pendulous placenta attached to the septum; placenta with (1–)2 ovules; style 2-lobed, hairy or glabrous, exserted. Fruit mostly orange, sometimes red, nearly spherical, at first crowned by the persistent calyx which often withers as the fruit ripens, bilocular; endocarp papery, shiny inside. Seeds either 1 per chamber and hemispherical or 2 per chamber and shaped like a quarter of a sphere; seed-coat brown and striate; hilar scar apical-adaxial, forming a ± elliptic plane, with a diameter of at least half the seed height; endosperm horny, entire; embryo straight, attaining half the diameter of the seed; cotyledons subcircular, placed in the centre of the seed, equalling the laterally oriented radicle.

A small genus of 15 species restricted to the tropics of the African mainland, previously referred to *Tricalysia*, but differing in flowers (especially stamens), pollen shape and exine and seed morphology.

Forest shrub; leaf-blade acumen 7–15 mm. long; corolla-tube
　　half as long as the lobes and shorter than the calyx-limb (at
　　least in var. *odoratissima*, uncertain in var. *ulugurensis*);
　　style glabrous . . . . 　*1. S. odoratissima*
Deciduous woodland shrub, with flowers usually appearing
　　before the leaves; leaf-blade acute or with an acumen
　　1–3(–6) mm. long; corolla-tube ± as long or longer than the
　　lobes and distinctly exceeding the tip of the calyx-limb;
　　style pubescent or glabrous . . . . 　*2. S. andongensis*

1. S. odoratissima *(K. Schum.) Robbrecht* in B.J.B.B. 48: 49, fig. 12 (1978) & in Distr. Pl. Afr. 13, maps 418, 419 (1978). Type: Tanzania, Lushoto District, Derema (? or Nguelo), *Scheffler* 56 (B, holo. †, BM, lecto.!, E, isolecto.!)

Shrub or sometimes a small tree to 3 m. high, with at least the young twigs sparsely puberulous. Leaf-blades elliptic or obovate, 8–15(–19) cm. long, 2.5–7(–8.8) cm. wide, with an acumen 7–15 mm. long, cuneate at base, puberulous or glabrous but with at least the midrib above and the midrib and lateral nerves beneath hairy (clearly so in young leaves); 4–6(–8) pairs of main lateral nerves; bacterial nodules (0–)2–3, linear, up to

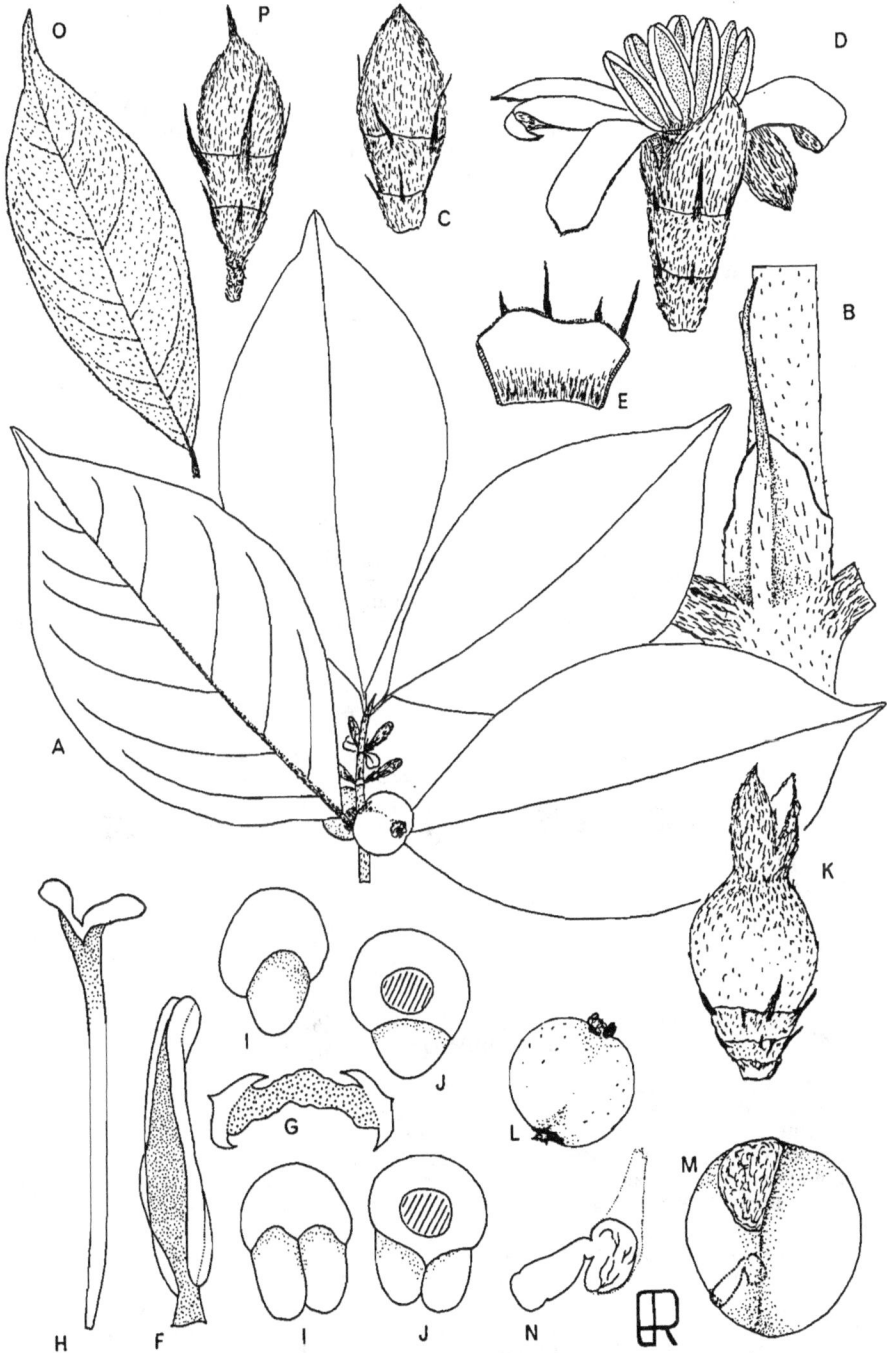

FIG. 94. *SERICANTHE ODORATISSIMA* var. *ODORATISSIMA* — **A**, habit, × ¼; **B**, stipule, × 7; **C**, flower-bud, × 4½; **D**, flower, × 3; **E**, upper bracteolar cup, opened out, × 3; **F**, stamen, × 10; **G**, transverse section of anther, × 22½; **H**, style with stigmatic arms, × 7; **I**, placenta, dorsal side, × 22½; **J**, placenta, ventral side, × 22½; **K**, young fruit, × 3; **L**, mature fruit, × 1; **M**, seed, ventral view with position of embryo indicated, × 3; **N**, embryo with suspensor, × 7. var. *ULUGURENSIS* — **O**, leaf, upperside, × ¼; **P**, flower-bud, × 3. A, L–N, from *Peter* 16819; B, from *Peter* 19047; C, from *Peter* 56678; D, E, from *Baagøe et al.* 165; F–J, from *Peter* 24624; K, from *Peter* 21866; O, from *Drummond & Hemsley* 1694, P, from *Harris* 5139. Drawn by E. Robbrecht.

3 cm. long, situated along the base of the midrib; petiole 4–8 mm. long, puberulous; stipular sheath 2–3 mm. long, puberulous outside, overtopped by 2 awns 2–3(–7) mm. long. Flowers solitary, 7–8-merous, sweetly perfumed, subsessile; bracteolar cups 2–3, 1–3 mm. long, each with 2 unequal pairs of awns, puberulous outside. Calyx-tube pubescent; limb ellipsoid, closed in bud, 5–7 mm. high, splitting into 2–3 lobes when the corolla opens, pubescent outside. Corolla inside with a ring of hairs covering the throat and the upper half of the tube; tube 3(–7) mm. tall and 1.5 mm. in diameter; lobes 7(–10) mm. long, 3 mm. wide. Stamens subsessile; anthers 4 mm. long. Placenta with (1–)2 free ovules hanging at the base; style glabrous, 7–8 mm. long, lobes 1 mm. long. Fruit red, 1–1.5 cm. in diameter. Seeds 8 mm. in diameter.

var. odoratissima

Leaf-blades (8–)12–15 cm. long, (4–)5–7 cm. wide, 2–3 times longer than wide, glabrous except for the midrib above and the midrib and lateral nerves beneath. Calyx-bud obtuse to acute. Fig.94/A–N.

TANZANIA. Lushoto District: E. Usambara Mts., 3 km. on Amani-Kwamkoro road, 18 Apr. 1968, *Renvoize & Abdallah* 1575! & Amani, 31 July 1975, *Baagøe, Danielsen & Vollesen* 165! & Ngua, 22 Dec. 1939, *Greenway* 5908!
DISTR. T 3; not known elsewhere
HAB. Rain-forest; 600–1000 m.

SYN. *Tricalysia odoratissima* K. Schum. in E.J. 33: 346–347 (1904); T.T.C.L.: 535 (1949), excl. ref. Ulugurus
Neorosea odoratissima (K. Schum.) N. Hallé in Fl. Gabon 17, Rubiacées. 2: 270 (1970)

NOTE. K. Schumann in his protologue cites the type as collected at Nguelo, however both isotypes are labelled Derema. In any event both places are very near to each other.
Two specimens from the W. Usambaras (near Ambangulu Tea Estate, *Polhill & Lovett* 5007A! & *Borhidi et al.* 841086!) fit var. *odoratissima* in their glabrous leaf-blades, which however, approach var. *ulugurensis* in dimension and shape.

var. ulugurensis *Robbrecht* in B.J.B.B. 48: 54, fig. 120/P (1978). Type: Tanzania, Uluguru Mts., Tegetero, *Drummond & Hemsley* 1694 (K, holo.!, B, BR, EA, iso.!)

Leaf-blades 8–11(–19) cm. long, 2.5–4.5 cm. wide, 3–4 times longer than wide, hairy (at least very sparsely so) on both surfaces. Calyx-bud with apiculate tip up to 3 mm. long. Fig.94/O, P.

TANZANIA. Morogoro District: Uluguru Mts., Bunduki Forest Reserve, Mar. 1953, *Paulo* 62! & Morningside–Bondwa, 12 Sept. 1972, *Harris* in *D.S.M.* 2728!; Rungwe District: Mwakaleli, Mar. 1954, *Paulo* 272!
DISTR. T 6, 7; Malawi
HAB. Rain-forest; 1100–1800 m.

NOTE. *Semsei* 937 (Morogoro District, Mtibwa Forest Reserve) has large leaf-blades (up to 19 cm. long, 8.8 cm. wide) which are glabrous above and sparsely puberulous beneath, and thus somewhat approaches var. *odoratissima*; the calyx is, however, distinctly apiculate in bud.

2. S. andongensis *(Hiern) Robbrecht* in B.J.B.B. 48: 31 (1978) & in Distr. Pl. Afr. 13, maps 408, 409 (1978). Type: Angola, Pungo Andongo, *Welwitsch* 3133 (? LISU, holo., K, iso.!)

Shrub or sometimes a small tree 1.5–6 m. high, with puberulous young twigs. Leaf-blades elliptic or obovate, (3–)5–6(–11.5) cm. long, (1.5–)2.5–4(–6) cm. wide, acute or with an acumen 1–3(–6) mm. long, cuneate or very seldom somewhat rounded at base, puberulous above and pubescent or (not in East Africa) velutinous beneath; 4–6(–8) pairs of main lateral nerves; bacterial nodules mostly 2 or sometimes more, linear, up to 3(–6) cm. long, situated along the basal half of the midrib but sometimes reaching almost the tip of the leaf; petiole (2–)3–8(–10) mm. long, puberulous; stipular sheath 1.5–3 mm. long, puberulous or glabrous, overtopped by 2 awns, 1–2(–3) mm. long, puberulous or glabrous. Flowers 1–2, sweetly perfumed, subsessile, usually appearing before the leaves; bracteolar cups 3, 1–3 mm. high, each with 2 unequal pairs of awns, puberulous outside. Calyx-tube glabrous or pubescent; limb ellipsoid and closed in bud, 6–10 mm. high, with the tip apiculate, 1–2 mm. long, splitting into 2–3 lobes when the corolla opens, pubescent outside. Corolla densely clothed with hairs inside the tube (save for its base); tube 8–14(–16) mm. tall, 1.5–3 mm. in diameter; lobes (7–)8–12 mm. long, 3–5 mm. wide. Filaments 0.5–1.5(–3) mm. long, glabrous or seldom pilose; anthers 4.5–6 mm. long. Placenta with 2 basal-lateral immersed ovules; style glabrous or pubescent, 13–20(–24) mm. long, with 2 lobes 1.5–3 mm. long. Fruit red, ± 1 cm. in diameter. Seeds ± 6 mm. in diameter.

var. andongensis

Leaf-blades beneath with only the nerves pubescent and with the surface between the nerves glabrous or nearly so.

TANZANIA. Mwanza District: Ibondo, 4 July 1953, *Tanner* 1559!; Singida District: Maw Hills, Sept. 1935, *B.D. Burtt* 5266!; Tunduru District: ± 11 km. E. of Songea District boundary, 7 June 1956, *Milne-Redhead & Taylor* 10677!
DISTR. T 1, 4, 5, 8; Zaire, Malawi, Zambia, Angola, South Africa (Transvaal)
HAB. Deciduous woodland, riverine forest; 800–1200(–1500) m.

SYN. *Tricalysia andongensis* Hiern, Cat. Afr. Pl. Welw. 1: 468 (1898); F.F.N.R.: 423 (1962), pro parte, excl. *White* 2475
 T. pachystigma K. Schum. in E.J. 33: 347 (1904); Good in J.B. 64, Suppl. 2: 18 (1926); T.T.C.L.: 535 (1949). Type: Malawi, Shire Highlands, *Buchanan* 96 (B, syn. †, K, lecto.!, E, isolecto.!)
 T. legatii Hutch., Botanist in S. Afr.: 306 (1946). Type: South Africa, Transvaal, Zoutpansberg near Louis Trichardt, *Legat* 23 (K, holo.!)
 Neorosea andongensis (Hiern) N. Hallé in Fl. Gabon 17, Rubiacées, 2: 270 (1970)

NOTE. *Harris & Rodgers* 3972! (EA) from a small relict forest-patch, T 6, 34 km. S. of Dar es Salaam on road to Kilwa, at only 150 m. altitude, most probably belongs here, though its leaves tend somewhat to those of *S. odoratissima* var. *ulugurensis*. The numerous characteristic insect-galls (± rounded, hairy excrescences) on the leaf-blades of this specimen have often been observed in *S. andongensis* but never in *S. odoratissima*.
 Var. *mollis* Robbrecht, with the leaves velutinous beneath, is found from Angola to Malawi and occurs just outside the Flora area in the Mbala District of Zambia.

Tribe 18. HYPOBATHREAE*

Flowers axillary, inflorescences congested, with cupular
 bracteoles; leaves petiolate, thinly coriaceous, not
 markedly shiny above . . . .   **72. Polysphaeria**
Flowers apparently in dense terminal inflorescences, but
 actually subterminal, the uppermost axillary components
 overtopping the inconspicuous terminal shoot; leaves
 subsessile, often ± cordate at base, coriaceous and very
 shiny above . . . .   **73. Lamprothamnus**

NOTE. *Feretia, Galiniera* and *Kraussia* are considered to belong to the *Hypobathreae* by Robbrecht, but in this work they are placed in *Coffeeae* sensu lato.

72. POLYSPHAERIA

Hook.f., G.P. 2: 108 (1873); Verdc. in K.B. 35: 97–130 (1980)

Glabrous or hairy shrubs or small trees with supra-axillary branches. Leaves usually thinly coriaceous with a characteristic divaricate venation, usually oblong-elliptic to lanceolate, less often more rounded; stipules triangular, acuminate or aristate with raised mid-line. Flowers sessile or rarely pedicellate, usually small in sessile or pedunculate bracteate clusters; bracts and bracteoles forming a complicated pattern of primary bracts, secondary bracts often supporting triads of flowers, and bracteoles often several per flower usually joined to form a cup at the base of the calyx. Calyx-tube campanulate or turbinate; limb-tube cup- or bowl-shaped, obconic or sometimes tubular, occasionally covering the corolla in bud, truncate or 4–5-toothed, persistent. Corolla-tube narrowly funnel-shaped or cylindrical, densely hairy at the throat; lobes 4–5, strictly contorted, overlapping to the left (as seen in bud), ovate to oblong, acuminate, spreading. Stamens 4–5, inserted in the corolla-tube, the filaments very short; anthers linear, fixed at the back near the base, included or partly exserted, sometimes minutely apiculate. Disc small but evident, fleshy. Ovary 2-locular; ovule solitary in each locule, pendulous; style filiform, exserted, shortly pubescent or hairy, or apparently glabrous in one imperfectly known species, with filiform or rarely ± ellipsoid undivided or slightly to distinctly bifid stigma. Fruits pea-like with rather tough epicarp, 1–2-locular, 1–2-seeded. Seeds subglobose to plano-convex with striate-sulcate testa and very ruminate endosperm.

* By B. Verdcourt

A genus of about 20 species confined to tropical Africa, Madagascar and the Comoro Is.

Brenan (K.B. 4: 85 (1949)) divided it into three sections — *Polysphaeria* (*Ephedranthae* S. Moore), *Stegnanthae* Brenan and *Dolichocalycinae* Brenan. His further subdivision of the first section into two series, *Polysphaeria* (*Ephedranthae* (S. Moore) Brenan) and *Cladanthae* (S. Moore) Brenan, based on the presence or absence of a peduncle has to be discarded as this character varies within one species (see Brenan in Mem. N.Y. Bot. Gard. 8: 451 (1954)). *Lamprothamnus* with an exactly similar testa sculpture and rumination is clearly closely related. Species 1–5 belong to sect. *Polysphaeria*, 6 and 7 to *Stegnanthae* and 9 to *Dolichocalycinae*. The genus is one of great complexity, some of the variation being difficult to understand. I am very dissatisfied with parts of the following treatment.
P. *brevifolia* K. Krause is a species of *Tricalysia*.

1. Bracteoles joined to form a beaked calyptra which entirely encloses the bud; calyx well developed and entirely enclosing the corolla in bud, usually becoming bilobed at anthesis . . . . 7. *P. dischistocalyx*
   Bracteoles ± joined or forming a cup but never entirely enclosing the bud 2
2. Calyx in young bud almost or quite enclosing the corolla 6. *P. cleistocalyx*
   Calyx never enclosing corolla even in young buds (save to some extent in intermediate forms of *P. braunii*) 3
3. Flowers and often but by no means always the stems, undersides of leaves and petioles rather coarsely ± spreading hairy, the hairs quite distinct; calyx short, usually distinctly toothed; leaves usually small, 0.4–9(–11) × 0.4–4.5 cm., those subtending branchlets sometimes ovate with subcordate bases; inflorescences never pedunculate . . . . 2. *P. parvifolia*
   Flowers glabrous outside or with adpressed hairs or if spreading then hairs minute; stems and leaves usually glabrous, less often puberulous, or hairy only in very immature plants 4
4. Calyx 0.7–1.3 cm. long; corolla-tube 1–1.1 cm long, with lobes 1.2 cm. long; fruit 1.2–1.4 cm. long (T 3, E. Usambaras) . . . . 9. *P. macrantha*
   Calyx and corolla (and often fruit) smaller 5
5. Style apparently glabrous; leaves large, 9–19 × 4.5–8.5 cm.; calyx, bracts and bracteoles pubescent outside; fruit 1–1.1 cm. diameter; calyx-limb 2.2 mm. long (T 3, E. Usambaras) . . . . 8. *P. sp. B*
   Style pubescent or quite hairy 6
6. Calyx, corolla, bracts and bracteoles glabrous or sometimes almost imperceptibly puberulous 7
   Calyx, bracteoles and bracts and often the corolla with a distinct indumentum outside 8
7. Inflorescences condensed, the bracteoles overlapping the calyx-tube (ovary), not pedunculate; calyx short, truncate when young but sometimes appearing erose or even toothed due to wear or rupturing, often dotted with white marks (at least in dry material) . . . . 1. *P. multiflora*
   Inflorescences laxer, the calyx-tube clearly visible, often very distinctly pedunculate; calyx shallowly undulate-toothed, not distinctly dotted with white marks 4. *P. lanceolata* subsp. *harleyi*
8. Bracts and bracteoles very reduced, not reaching beyond half-way up the calyx-tube (ovary); flowers tomentose with excessively fine very short but spreading hairs; stems with pale corky bark; inflorescences sessile; leaves ± obtuse or subacute; plant with a "*Tricalysia*-like" facies (Tanzania, Ulanga/Kilwa) . . . . 3. *P. sp. A*
   Bracts and bracteoles more developed, covering more of the calyx 9

9. Indumentum on calyx of ± spreading hairs shorter than in
   P. parvifolia; corolla with finer very short spreading
   hairs; inflorescences sessile or shortly pedunculate
   (**T 4**) . . . .   *4. P. lanceolata*
                                       subsp. *ellipticifolia*

   Indumentum on calyx at least and often on corolla
   adpressed and sometimes densely adpressed sericeous 10
10. Leaf-blades narrowly lanceolate, ± 4.5–5 times as long as
    wide, up to ± 2.3 cm. wide; corolla densely pubescent
    save base of tube (**K 4, T 3**) . . . .   *4. P. lanceolata* forma

    Leaf-blades not so narrowly lanceolate, mostly well under
    4 times as long as wide 11
11. Calyx usually ± 2 mm. long, sometimes covering a good
    deal of the corolla, mostly distinctly toothed;
    inflorescence sessile; flowers including corolla mostly
    densely adpressed sericeous; leaf-blades often large,
    3.3–18.5 cm. long, 1.2–6 cm. wide (Tanzania) . . .   *5. P. braunii*

    Calyx shorter, usually under 2 mm. long, never covering
    much of the corolla in bud, usually truncate; flowers
    often much less sericeous, the corolla only puberulous
    or glabrescent; inflorescences sessile or pedunculate   *1. P. multiflora* subsp.
                                                 pubescens

1. P. multiflora *Hiern* in F.T.A. 3: 127 (1877); P.O.A. C: 383 (1895); T.T.C.L.: 520 (1949); K.T.S.: 463 (1961); Verdc. in K.B. 35: 125, t.2A–C (1980). Types: Tanzania, Mafia I. ['Monfia'], *Frere* (K, syn.!) & Mozambique, ± 42 km. up the R. Rovuma, *Kirk* (K, syn.!)

Shrub or small tree 1–3(–4.5) m. tall, occasionally weak and almost trailing; stems glabrous, the bark peeling in long strips on older stems. Leaf-blades elliptic, oblong-elliptic or lanceolate, 2–11.5(–19) cm. long, 1–4.2 (–4.5) cm. wide, ± acute, the actual apex ± obtuse, less often acuminate (in some cultivated specimens), cuneate to subcordate at the base, thinly coriaceous, glabrous, often with pale minute white specks particularly near the midrib above; petiole 5–9 mm. long; stipules triangular, 2.5–4 mm. long, keeled, glabrous. Inflorescences sessile or practically so, the two opposite together forming a globose cluster up to 1.5 cm. diameter or peduncles up to 4 mm. long in one variant; components quite glabrous outside or almost imperceptibly puberulous; bracteoles cupular, tightly surrounding the calyx-base, ± 1 mm. tall. Calyx shallowly cup-shaped, 1.5(–2) mm. long, glabrous outside or imperceptibly puberulous or in one variant silky pubescent, glabrescent to densely pubescent inside, often with pale marks outside, usually truncate, sometimes undulate or faintly toothed or becoming lacerate. Flowers sweetly scented; corolla white; tube (2–)4 mm. long, glabrous outside or adpressed puberulous in one variant, densely white hairy at the throat; lobes ovate, 1.5–2 mm. long and wide. Style densely white hairy; disc prominent. Fruits globose, 7–9 mm. diameter, glabrous, ± white-dotted, somewhat didymous. Seeds ± semiglobose, ± 6 mm. long, 4 mm. wide.

subsp. **multiflora**

Inflorescences quite sessile; calyx glabrous outside or almost imperceptibly puberulous; corolla glabrous.

KENYA. Kwale District: Shimoni, *Drummond & Hemsley* 3913! & Cha Shimba [Pemba], 17 Mar. 1902, *Kassner* 362!; Kilifi District: Watamu, 23 Jan. 1968, *Bally* 13051!
TANZANIA. Bagamoyo, Sadani road, June 1964, *Procter* 2538!; Rufiji District: Mafia I., 2 Apr. 1933, *Wallace* 734!; Lindi District: Makonde Plateau, *Gillman* 1038!; Zanzibar, Prison I., 9 Feb. 1929, *Greenway* 1388!
DISTR. **K** 7; **T** 5, 6, 8; **Z**; ?**P**; Mozambique, Comoro and Aldabra Is.; cultivated in Mauritius
HAB. Margins of shrubby thickets and mangrove swamps, forest, woodland and on coral rock near high water mark; 0–900 m.
SYN. *Cremaspora congesta* Baill. in Adansonia, Sér. 1, 12: 283 (1879). Type: Comoro Is., Mayotte, *Boivin* 3175 (P, lecto.!, K, isolecto.!)
     Polysphaeria neriifolia K. Schum. in P.O.A. C: 383 (1895); T.T.C.L.: 520 (1949). Type: Tanzania, Tanga District, Amboni, *Holst* 2834a (B, holo.†, K, iso.!)
     P. squarrosa K. Krause in E.J. 39: 531 (1907); T.T.C.L.: 521 (1949). Type: Tanzania, Lindi District, Makonde Plateau, near Mtepera, *Busse* 1341 (B, holo. †)
     P. congesta (Baill.) Cavaco in Adansonia, Sér. 2, 8: 380 (1968)
NOTE. I agree with Brenan's pencilled note on the Kew cover of *P. neriifolia*, 'very doubtful if distinct from *P. multiflora*'. I have not seen authentic material of *P. squarrosa*, but the description does not

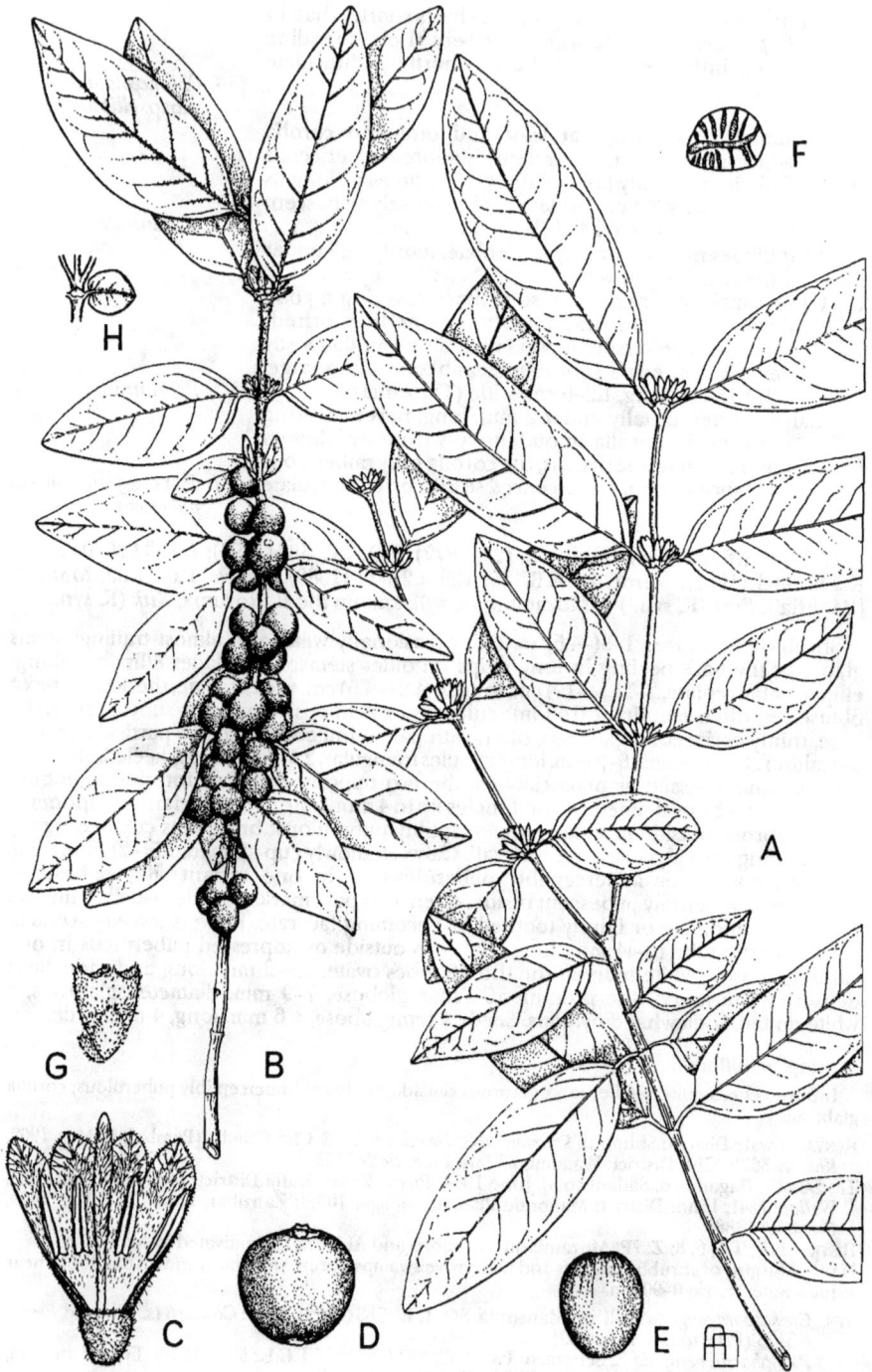

FIG. 95. *POLYSPHAERIA PARVIFOLIA* — **A**, flowering branch, × ⅔; **B**, fruiting branch × ⅔; **C**, flower, with corolla opened out, × 4; **D**, fruit, × 2; **E**, seed, × 2; **F**, transverse section of seed, × 2; **G**, calyx, × 6; **H**, modified leaf, × ⅔. **A**, from *Polhill & Paulo* 782; B, D, E, G, H, from *Graham* FD1522; C, from *Faulkner* 1099; F, from *Magogo* 110. Drawn by Ann Farrer, with F-H by Sally Dawson.

seem to fit any other species. One field note (*Ole Sayalel*) claims a height of 10 m. but I think this is an error. Certain specimens from Zanzibar (e.g. *Vaughan* 1830, *Faulkner* 2802 and 2312 and *Greenway* 2602) have the calyx almost imperceptibly puberulous or practically glabrous, thus forming a link with subsp. *pubescens*.

subsp. **pubescens** *Verdc.* in K.B. 35: 127 (1980). Type: Kenya, Tana River District, Bura, *Greenway* 9239 (K, holo.!, EA, iso.)

Inflorescences shortly pedunculate or subsessile; calyx distinctly puberulous to pubescent outside; corolla adpressed or spreading puberulous outside with minute hairs.

KENYA. Teita District: Mbololo Hill, 14 Feb. 1953, *Bally* 8578!; Tana River District: Mnazini, 26 Mar. 1973, *Homewood* 32! & Galole, near Oxbow Lake, 23 Mar. 1965, *Makin* 14!
TANZANIA. Pangani District: Mwera, Mseko, Mgandu, 15 Mar. 1956, *Tanner* 2655!; Zanzibar I., Mangapwani, 24 Jan. 1929, *Greenway* 1152!; Pemba I., Vitongoje, 14 Oct. 1929, *Vaughan* 787!
DISTR. K 7; T 3, ?6, ?8; Z; P (see note); Somalia
HAB. Riverine forest and bushland; 60–1140 m.

SYN. [*P. multiflora* sensu K.T.S.: 463 (1961), pro parte, *non* Hiern sensu stricto]
Plectronia guidottii Chiov., Fl. Somala 2: 246, fig. 145, (1932). Type: Somalia, Mansur on Juba R., *Guidotti* 212 (? BO, holo.)

NOTE. This subspecies links *P. multiflora* with *P. braunii* and *P. lanceolata* and is a temporary solution to the classification of the material concerned. I am not at all satisfied with it. *Procter* 25 (Tanzania, Rufiji District, Mohoro Forest Reserve, Feb. 1952), has the calyx distinctly pubescent but the corolla glabrous and is intermediate with typical subsp. *multiflora*. *Harris & Pocs* 4304 (Tanzania, Ulanga District, Kidatu to Sonjo, 29 Mar. 1970) is similar but has very pointed acuminate buds and is possibly a distinct variety. Other fruiting specimens which are probably pubescent variants of *P. multiflora* are *Peter* 45041 (Tanzania, Dar es Salaam to Bagamoyo), *Shabani* 39 and *Parry* 218 (Tanzania, Lindi District, Rondo Plateau), *van Rensburg* 626 (Tanzania, Kilosa District, Mkata R.), *Busse* 2827 (Tanzania, Lindi District, Nanya) and *Semsei* in F.H. 2170 (Tanzania, Masasi District, Mbangala). The leaves are more coriaceous and broader (up to 5.3 cm) in all these Tanzanian specimens and a third variant may be involved. All the material is poor and better collections are needed from all these localities. The recent finding of a specimen of undoubted *P. lanceolata* on the Kenya coast suggests subsp. *pubescens* may be a series of hybrids between that species and true *P. multiflora* the former having been almost eliminated.

    2. P. parvifolia *Hiern* in F.T.A. 3: 128 (1877); P.O.A. C: 383 (1895); T.T.C.L.: 520 (1949); F.P.S. 2: 459 (1952); K.T.S.: 464 (1961); Verdc. in K.B. 35: 117, fig.5 (1980). Types: Tanzania, Zanzibar I., *Hildebrandt* 1181 & *Kirk* 38 (K, syn.!)

Small tree or shrub 0.6–9(–?12) m. tall but probably rarely over 3.5 m.; stems slender, glabrous or more often shortly pubescent; bark reddish brown, peeling in long strips on the older stems. Leaf-blades oblong, elliptic, ovate-elliptic or sometimes ovate or almost round, 0.4–9(–11) cm. long, 0.4–4.5(–4.7) cm. wide, subacute to acuminate at the apex, cuneate to rounded or truncate at the base or in the case of leaves supporting lateral branchlets often subcordate to distinctly cordate, glabrous to distinctly pubescent or coarsely hairy particularly beneath; occasionally there are some pairs of very small leaves; petiole 3–5 mm. long, glabrous or more usually pubescent; stipules 1 mm. long with apiculum 1 mm. long, hairy. Inflorescences sessile or rarely peduncles up to 2.5 mm. long, the opposite parts usually joined to form a globose cluster at the node; paired basal bracts ovate-triangular, 3 mm. long, 2 mm. wide, acuminate; secondary bracts triangular, 1.5 mm. long, 2 mm. wide, each pair supporting 3 triads of flowers each with another set of bracts; main bracteoles at base of calyx joined to form a 2–3-lobed cup. Flowers practically scentless (?always). Calyx mostly with spreading white hairs; tube ± 0.5 mm. long, limb-tube about equal with distinct triangular teeth ± the same length. Corolla white or tube sometimes greenish, usually densely spreading pubescent, less often glabrous or glabrescent; tube funnel-shaped, 3–4.5 mm. long, sometimes constricted near the base; lobes ovate, 1.5–2 mm. long, 1.3 mm. wide. Style densely white hairy. Fruits globose, red or orange-red, 7–10 mm. diameter, glabrous or slightly pubescent. Fig.95.

KENYA. Kwale District: Shimba Hills, Mwele Mdogo Forest, 4 Feb. 1953, *Drummond & Hemsley* 1109!; Mombasa, Nov. 1884, *Wakefield*!; Kilifi District: Mida, *R.M. Graham* in F.D. 1522!
TANZANIA. Tanga District: Ngomeni, 30 July 1953, *Drummond & Hemsley* 3542!; Pangani District: Jasini, 15 Nov. 1955, *Milne-Redhead & Taylor* 7292!; Lindi District: Lake Lutamba, 10 Jan. 1935, *Schlieben* 5857!; Zanzibar I., Dole Ridge, 13 Apr. 1952, *R.O. Williams* 165!
DISTR. K 1, 7; T 1, 3, 6, 8; Z; P; Sudan, Ethiopia, Somalia
HAB. Rather dry evergreen forest, woodland, coastal bushland and scrub, old sisal plantations, cultivations, etc.; 0–500 m.

SYN. *P. parvifolia* Hiern var. ? *glabra* Hiern in F.T.A. 3: 128 (1877): Type: Tanzania, Bagamoyo, *Kirk* (K, holo.!)

NOTE. This is on the whole a well-characterised species, the indumentum being longer and coarser than in other species but it varies considerably in quantity. Var. *glabra* with glabrous leaves is not retainable. There is nearly always some trace of long hairs on the flowers but a few specimens are almost totally glabrous. The toothed longer calyx will distinguish them from *P. multiflora*. The Sudanese material is rather doubtfully named.

Harris 6368 is a puzzling plant said to be an erect shrublet 0.5 m. tall. The leaves are closely placed and very small, up to 2 × 1.6 cm., densely hairy beneath. The flowers appear to be solitary in the axils. Despite the totally different appearance I suspect it may be a very abnormal state of *P. parvifolia* rather than a distinct undescribed species.

3. P. sp. A

Shrub 2–4 m. tall; stems pale, covered with cracking corky bark, glabrous. Leaf-blades narrowly elliptic, 4–9 cm. long, 1.3–3.4 cm. wide, bluntly pointed at the apex, cuneate at the base, thinly coriaceous, glabrous, densely closely white-speckled above; petiole up to 8 mm. long, wrinkled; stipules ovate-triangular, 3 mm. long, obtuse, falling almost at once, puberulous. Inflorescences 5–10-flowered, sessile; lowermost bracteole 0.4 mm. long, upper 0.6 mm. long, both cupular and ± truncate, minutely spreading puberulous. Calyx similarly puberulous, cupular, up to 1.2 mm. long, truncate. Corolla white, ± 6 mm. long, with lobes ± 1.5 mm. long, finely puberulous. Fruit not seen.

TANZANIA. Ulanga/Kilwa Districts: confluence of Kilombero and Luwegu Rivers with the Rufiji R., 17 June 1932, *Schlieben* 2426! & Ulanga R., Tenende Ferry, 21 July 1971, *Rees* T 153!
DISTR. T 6/8; not known elsewhere
HAB. Riverine bushland; 250–300 m.

SYN. *Polysphaeria sp. C* sensu Verdc. in K.B. 35: 122 (1980)

NOTE. The leaves of this are rather distinctive in a way difficult to describe. It is very similar to *P. lanceolata* but differs in foliage and indumentum. It shows a deceptive resemblance to *Tricalysia* but the stamens are included.

4. P. lanceolata *Hiern* in F.T.A. 3: 128 (1877); P.O.A. C: 383 (1895); T.T.C.L.: 520 (1949); Verdc. in K.B. 35: 122 (1980). Types: Mozambique, Shupanga, *Kirk* (K, syn.!)

Slender spreading shrub 1.2–3.5 m. tall (see note after subsp. *harleyi*); stems glabrous, the older with grey fissured bark. Leaf-blades usually narrowly lanceolate, or narrowly elliptic to oblong-elliptic, 2.5–16(–19) cm. long, 1.3–5.7 cm. wide, mostly slenderly acutely acuminate at the apex but sometimes obtuse or the actual acumen rounded to subacute, cuneate to almost rounded at the base, rather thin, glabrous; petioles 0.4–1 cm. long; stipules triangular, 1.5–5 mm. long, 2–5 mm. wide, keeled, often apiculate, glabrous or pubescent, deciduous. Flowers not scented (?always), in sessile or shortly to distinctly pedunculate 3–8-flowered pubescent to glabrous inflorescences; when sessile the two at each side appearing as a single cluster; peduncles 0.2–1.8 cm. long; secondary peduncles 0–3 mm. long; bracts and bracteoles ovate-triangular, all rather similar, 1.5–3.5 mm. long, usually united into boat-shaped cups, acute, apiculate or caudate at the apex, mostly rather coarsely ± spreading pubescent or glabrous. Calyx with similar indumentum; tube 0.5–2.5 mm. long; limb-tube 1.3–2 mm. long, usually with distinct but distant acute or apiculate teeth, or shallowly undulate-toothed or even ± truncate. Corolla very pale lilac, white or pink-tinged, glabrous, slightly pubescent or usually puberulous all over with minute adpressed or slightly spreading hairs; tube 4.5–7 mm. long; lobes oblong-ovate to ovate, 1.7–2 mm. long and wide, obtuse or acute. Style up to 1.1 cm. long, densely hairy; stigmatic lobes 1.5–2 mm. long. Fruit often glaucous at first, becoming red-purple or black, subglobose, usually slightly compressed and obscurely didymous, 6–9 mm. long, 9–12 mm. wide, 7–8 mm. thick. Seeds ± shiny, chestnut, hemispherical, 6–7.5 mm. long, 3.5 mm. thick.

KEY TO INFRASPECIFIC VARIANTS

Calyx, bracts and bracteoles with some kind of indumentum outside; inflorescences sessile or peduncles short:
Leaves oblong-elliptic, lanceolate or narrowly elliptic, up to 14.5 × 4.3 cm. mostly acuminate . . . . subsp. **lanceolata** var. **lanceolata**

Leaves elliptic or oblong-elliptic, 2.2–6(–9) × 0.9–2.7(–3.7)
 cm., narrowed to an obtuse apex . . . .   subsp. **ellipticifolia**
Calyx, bracts and bracteoles mostly glabrous outside; peduncles
 (0–)0.2–3 cm. long . . . .   subsp. **harleyi**

subsp. lanceolata

Leaves oblong-elliptic, lanceolate or narrowly elliptic, mostly acuminate, 2.5–14.5 cm. long, 1.3–4.3 cm. wide. Inflorescences sessile or pedunculate. Flowers pubescent.

var. lanceolata

Inflorescences sessile or only shortly pedunculate, the peduncle under 4 mm. long.

KENYA. Embu/Kitui Districts: Tana R., Kindaruma Dam, Seven Forks, 30 Apr. 1967, *Gillett & Faden* 18240!; Kwale District: N. end of Mwachi Forest, Mwachi R., 17 Feb. 1977, *Faden et al.* 77/485!
TANZANIA. Pare District: Kisiwani [Kisuani], 6 Feb. 1930, *Greenway* 2168!; Kigoma District: Mkuti R., Nov. 1954, *Procter* 307!
DISTR. K 4, 7; T 3, 4; Mozambique, Malawi, Zambia
HAB. Riverine forest and *Cynometra, Gyrocarpus, Scorodophloeus* forest on steep rocky forested limestone slopes; 15–1100 m.
NOTE. Var. *pedata* Brenan, with much longer peduncles, occurs in Mozambique and Malawi. The identity of the K 4 specimen is dubious and is discussed on p. 577 (note to *P. cleistocalyx*). The Pare specimen *Greenway* 2168 (the basis for the T.T.C.L. record) is also dubious — the leaves are narrowly lanceolate but it seems intermediate with *P. braunii*; no flowers are available. *Peter* 10904A from nearby Buiko is similar. *P. schweinfurthii* Hiern described from the Sudan probably also belongs here.

subsp. ellipticifolia Verdc. in K.B. 35: 123 (1980). Type: Tanzania, Mpanda District, Kapapa, Richards & Arasululu 25856 (K, holo.!, BR, iso.!)

Leaves elliptic or oblong-elliptic, 2.2–6(–9) × 0.9–2.7(–3.7) cm., narrowed to an obtuse apex, rather more coriaceous than in typical subsp. *lanceolata.* Inflorescences shortly pedunculate. Calyx pubescent. Corolla tomentose with minute spreading hairs.

TANZANIA. Kahama District: 42 km. S. of Mbugwe, Moyowosi R., 29 July 1950, *Bullock* 3046!; Tabora District: Ugalla R., Isimbira [Isimbila], 25 Oct. 1960, *Richards* 13397!; Mpanda District: Kapapa, Msaginya R., 12 Sept. 1970, *Richards & Arasululu* 25856!
DISTR. T 4, 5, 7; not known elsewhere
HAB. Dry riverine bushland and seasonally wet valleys in *Brachystegia* woodland; 970–1350 m.

subsp. harleyi Verdc. in K.B. 35: 124 (1980). Type: Tanzania, Mpanda District, Kungwe-Mahali Peninsula, Belengi, Harley 9132 (K, holo.!, BR, EA, iso.!)

Leaves as in subsp. *lanceolata.* Inflorescences distinctly pedunculate; peduncles (0–)0.2–1.8 cm. long; bracts, bracteoles, calyx and corolla glabrous outside.

TANZANIA. Kigoma District: Makorongo, 20 Sept. 1963, *Azuma* 647!; Mpanda District: 19.2 km. N. of Kasogi, Belengi, 2 Aug. 1959, *Harley* 9132! & 104 km. S. of Kigoma, Mugombazi, 1 Sept. 1959, *Harley* 9490!
DISTR. T 4; not known elsewhere
HAB. Dry riverine forest, sometimes among rocks and bamboo woodland; evergreen forest with *Khaya;* 780–1350 m.
NOTE. I had at first thought this might be a distinct species, but apart from the lack of indumentum the plant dealt with above is little different from *P. lanceolata* var. *pedata* and, moreover, *Azuma* 672 from Kigoma District, 8 km. S. of Ilagara, described as a tall tree 15–20 m. tall, with smooth bark and greenish flowers, has undeveloped inflorescences which are obviously pedunculate but the bracteoles and calyx have very short spreading hairs; it was collected from coppice material and an error may have been made about the habit.

5. P. braunii *K. Krause* in E.J. 57: 33 (1920); T.T.C.L.: 519 (1949); Verdc. in K.B. 35: 112 (1980). Type: Tanzania, Pangani District, on R. Pangani, near Hale, *Braun in Herb. Amani* 1526 (B, holo.†)

Shrub 1.5–5(–6) m. tall with long slender glabrous branchlets (see note). Leaf-blades elliptic to oblong-lanceolate or lanceolate, 3.3–18.5 cm. long, 1.2–6 cm. wide, typically quite large, shortly acuminate at the apex, broadly cuneate to rounded at the base, glabrous; petioles 0.6–1.2(–1.8) cm. long; stipules triangular, 1.5–2 mm. long. Inflorescences sessile, the two axillary parts forming a globose cluster up to 2 cm. diameter appearing to be more bracteated than in other species with imbricated groups of bracts and bracteoles; bracts and secondary bracts semicircular, 2 mm. long and wide, adpressed pubescent; main bracteoles cupular, 1.5–2 mm. long, adpressed brown sericeous-pubescent, second and third bracteoles beneath the main cup also cupular and

FIG. 96. *POLYSPHAERIA DISCHISTOCALYX* — **A**, flowering branch, × 1; **B**, flower, × 3; **C**, calyx × 3; **D**, corolla opened out, × 3; **E**, gynoecium, × 3; **F**, same in more mature state, × 3; **G**, bracteolar calyptra bursting to reveal unopened calyx, × 3; **H**, the same splitting lengthways, × 3; **J**, unopened calyx with bracteolar calyptra removed, × 3; **K**, fruit, × 1; **L**, stipules × 2. Drawn from unstated source on 3 August 1950 by Miss D.R. Thompson.

± 2 mm. long. Calyx similarly pubescent; tube 1.2 mm. long; limb 2.5 mm. long, 4-toothed or ± truncate. Corolla creamy white, adpressed pubescent outside, acuminate in bud; tube 4.5–5 mm. long; lobes broadly ovate, 3 mm. long and wide, acute; throat densely white hairy. Style 7.5–8(–9.5) mm. long, densely white pubescent save near base. Fruit subglobose, 7–10 mm. wide, very slightly didymous with slight lateral groove, glabrous or puberulous around the base of the persistent calyx-limb remnant.

TANZANIA.   Pangani District: S. bank of R. Pangani between Hale and Makinyumbe, 1 July 1953, *Drummond & Hemsley* 3137! & Mruazi (? Mliwaza) R., near Mwera, Mar. 1925, *Procter* 2918! & island in R. Pangani near Hale, 31 Jan. 1915, *Peter* 55880! & 31 Jan. 1915, *Peter* 55629!; Morogoro District: plains, 480 m., *Wallace* 276!
DISTR.   T 3, 6, 8; not known elsewhere
HAB.   Riverine forest, rain-forest; 150–900 m.

NOTE.   I have seen no authentic material of this and there are certain marked discrepancies between Krause's original description and the plant described above, but since the type locality is accurately known and many sheets are available from there I am convinced my interpretation is correct. Krause's description of the corolla as 2–2.5 mm. long is doubtless a misprint; I have seen no *Polysphaeria* with such small flowers. The identification of *Schlieben* 3181 at Berlin as *P. cf. braunii* is also corroborative evidence for the present interpretation.
Sheet II of *Peter* 55950 is a very young shoot and both stems and leaves are spreading pubescent and the stipules have an apical appendage up to 4 mm. long; all the rest of the material seen has the stems and foliage entirely glabrous. The Uluguru material may represent a subspecies with rather shorter calyx-limb.

6. P. cleistocalyx *Verdc.* in K.B. 35: 112 (1980). Type: Tanzania, Kilosa District, Luhembe [Ruhembe], *Goetze* 389 (K, holo.!)

Shrub 1.5–6 m. tall; stems glabrous, the older ones covered with pale longitudinally fissured corky bark. Leaf-blades lanceolate or narrowly oblong-lanceolate, 5–14 cm. long, 2–4 cm. wide, narrowed to an acute apex, cuneate at the base, mostly thin, glabrous; petiole 8 mm. long; stipules triangular, 3 mm. long, 2.5 mm. wide, keeled, glabrous, ± persistent. Flowers in sessile or shortly pedunculate inflorescences; peduncles up to 4 mm. long; bracts and bracteoles 1.5–2.5 mm. long, variable, those next to the calyx joined to form a truncate or bilobed cup, sometimes apiculate, never completely enclosing the bud, finely densely adpressed silky; sessile inflorescences form a globose nodal cluster up to 1.5 cm. diameter. Calyx enclosing the bud in young flower, with similar indumentum, the subacute apex slightly toothed, splitting into 2 lobes as the flower develops; tube 1 mm. long, limb 4 mm. long. Corolla creamy white, glabrous or sparsely puberulous; tube funnel-shaped, 4 mm. long; lobes oblong-ovate, 2–2.5 mm. long and wide, rounded to an apiculate apex. Style densely hairy save at base, 6 mm. long; stigmatic lobes ± 1 mm. long. Fruits ripening from white to purplish black, subglobose or depressed globose, often somewhat didymous, 7–10 mm. long and wide, puberulous around the persistent calyx base.

var. cleistocalyx

Inflorescences sessile.

KENYA.   KENYA.   Kitui District: Tana R., Seven Forks, Kindaruma Dam, 30 Apr. 1967, *Gillett & Faden* 18241! & near W. end of the same dam, 26 Dec. 1970, *Gillett* 19268!
TANZANIA.   Kilosa District: Luhembe [Ruhembe], Dec. 1898, *Goetze* 389! & Hiwaga, 4 June 1921, *Swynnerton* 734! & Kilosa, 25 Nov. 1922, *Swynnerton* 735!
DISTR.   K 4; T 3, 5, 6; not known elsewhere
HAB.   Riverine forest, streamsides, rocky shores and lowland plains on black soil; 500–1190 m.

var. pedunculata *Verdc.* in K.B. 35: 113 (1980). Type: Tanzania, Mpwapwa, *Hornby* 335 (K, holo.!, EA, iso.!)

Inflorescences with short peduncles up to 4 mm. long.

TANZANIA.   Mpwapwa, 17 Dec. 1930, *Hornby* 335!
DISTR.   T 5; known only from type specimen
HAB.   Streamsides; 1050 m.

NOTE.   (on species as a whole). Very closely related to *P. dischistocalyx* but differing in bracteole structure. Uluguru material of *P. braunii* comes close. Gillett and Faden collected two *Polysphaeria* at Kindaruma Dam, Tana R., Seven Forks, one 18241 I have cited above as *P. cleistocalyx* and another, 18240, which is virtually identical in foliage, I have cited under *P. lanceolata* var. *lanceolata*; the collectors specifically state that 18241 is much like 18240 but distinct. Certainly the calyx is much shorter than 18241 and it resembles both *P. multiflora* subsp. *pubescens* and *P. lanceolata*. Much as I am loath to associate this single Kenya population with *P. cleistocalyx*, at this stage I see no alternative although the resemblance is probably due to extreme convergence. Much of the material from the Upper Tana is very inadequate and the exact relationships not at all clear. The presence of two identical-looking but distinct taxa in this area is puzzling and needs field investigation; it might throw light on the intricacies of this difficult genus.

7. P. dischistocalyx *Brenan* in K.B. 4: 81 (1949); T.T.C.L.: 520 (1949); Verdc. in K.B. 35: 113, fig. 4 (1980). Type: Malawi, Nyika Plateau, Mwanemba, *McClounie* 151 (K, holo.! & iso.!)

Shrub or less often a small tree, 1.5–7(–9) m. tall with graceful glabrous rather compressed branches, bark pale brown or grey, flaking. Leaf-blades lanceolate or oblong-lanceolate, 5.4–16.6(–21.5) cm. long, (1.5–)2–4.9(–5.8) cm. wide, acute, acuminate or obtuse at the apex, cuneate or somewhat rounded at the base, sometimes those below lateral branches reduced, ± round to oblong-lanceolate, glabrous, thinly coriaceous, shiny; lateral nerves 7–12; petiole 5–8 mm. long; stipules broadly triangular, 1.5–6 mm. long, 2.5–3 mm. wide, abruptly drawn out into an acumen 1 mm. long, deciduous. Inflorescences sessile many-flowered glomerules 1.2 cm. in diameter. Flowers scented, sessile, bracteolate; lower bracteoles connate into an irregularly bilobed cup, 1.5–2 mm. tall, puberulous outside, white-ciliate on the margin; upper forming an ovoid calyptra 4.5–5(–7) mm. long, 2.7–3 mm. wide, puberulous outside, strigose hairy inside with a curved beak at apex, at first closed and enclosing bud but later irregularly circumscissile or sometimes divided longitudinally into 2 lobes. Calyx pale pink turning brown, closed in bud, scarious, sparsely to rather densely puberulous outside either all over or only towards the apex, silky-strigose inside, 4–5.5 mm. long; tube 1.5–2.5 mm. long, 2-lobed, lobes broadly ovate, 2.25–3 mm. long, 3.25–3.5 mm. wide at the base, apiculate at the apex. Corolla white or whitish rose, obovoid in bud, 6 mm. long, acute, puberulous or glabrous outside; tube tinged greenish, 3.5–4.75 mm. long, 2.5–3.5 mm. wide at throat; lobes ovate-triangular, 2.5–2.8 mm. long, 1.7–2 mm. wide. Style white, ± 6.5 mm. long, densely pubescent save at the base, bifid at the apex. Fruit green, flushed red, becoming wine-red or black, globose, 6–8 mm. diameter, glabrous, crowned by the persistent calyx-limb; flesh green and sweet. Fig. 96.

TANZANIA. Ulanga District: Tabora, 23 Dec. 1931, *Schlieben* 1567!; Rungwe District: Kilambo, 6 Mar. 1913, *Stolz* 1911!; Songea District: ± 2.5 km. SW. of Kitai, by R. Nakawali, 8 Mar. 1956, *Milne-Redhead & Taylor* 9114!
DISTR. T 6–8: Malawi, ? Mozambique, Zambia
HAB. Riverine forest and *Brachystegia* woodland; 500–1400 m.
SYN. [*P. neriifolia* sensu K. Schum. in E.J. 30: 414 (1901), *non* K. Schum. (1895)]

8. P. sp. B; Verdc. in K.B. 35: 115 (1980)

Treelet to 3 m. with glabrous branches. Leaf-blades large, oblong, ovate-oblong or elliptic, 9–19 cm. long, 4.5–8.5 cm. wide, shortly acutely acuminate at the apex, broadly cuneate to more usually rounded or truncate at the base, glabrous, thinly coriaceous; petiole 5 mm. long. Inflorescences sessile but structure not ascertainable. Calyx, bracts and bracteoles adpressed pubescent outside. Calyx-tube 1 mm. long, narrower than the 2.2 mm. long limb-tube which is probably distinctly toothed. Styles apparently glabrous. Fruit subglobose, 1–1.1 cm. in diameter.

TANZANIA. Lushoto District: Kwamtili, 3 Oct. 1918, *Peter* 25227! & 25237!
DISTR. T 3; not known elsewhere
HAB. Primary rain-forest; 300 m.
NOTE. This seems to be a distinctive undescribed species, never recollected. The inflorescence remnants are totally inadequate for analysis but two persistent styles are glabrous.

9. P. macrantha *Brenan* in K.B. 4: 83 (1949); T.T.C.L.: 520 (1949); Verdc. in K.B. 35: 103 (1980). Type: Tanzania, Tanga Province, probably Lushoto District, Amani Arboretum, *Zimmermann* (K, holo.!)

Small tree up to 10 m. tall with glabrous compressed branches. Leaf-blades ovate-oblong to oblong-lanceolate or elliptic, (5.5–)9.5–17 cm. long, (1.7–)3–6.8 cm. wide, obtuse to acute or ± acuminate, rounded to broadly cuneate at the base, sometimes those below lateral branches smaller and broadly elliptic, glabrous, shining or dull, often drying brown beneath, midrib often slightly impressed above and prominent beneath, the venation mostly raised on both surfaces; lateral nerves ± 8–11; petiole (4–)6–10 mm. long;

stipules triangular or ovate-triangular, 3.5–6.5 mm. long, 3.5–5 mm. wide, shortly mucronate, puberulous outside, glabrous inside but with numerous small hair-like glands at the base, persistent or falling. Inflorescences sessile or shortly pedunculate, 2–4-flowered; peduncle 0–1.1 cm. long; bracts ovate or triangular, 2–4 mm. long, acute; pedicels 0–9 mm. long, with 1–2 pairs of bracteoles 3–7 mm. long, irregularly connate to form a cupule; bracts, bracteoles and peduncle densely shortly pubescent. Calyx-tube obconic, 2.5 mm. long, glabrous; limb at first oblong-ellipsoid to elongate-suburceolate, 0.72–1.3 cm. long, with ± 4 triangular teeth 1–2 mm. long, but during flowering becoming deeply split on one side, puberulous outside, densely adpressed pubescent inside. Corolla white, glabrous outside; tube 1–1.1 cm. long, 1.5–2 mm. wide and widened to the densely hairy throat; lobes spreading, ovate-oblong, 1.2 cm. long, 5 mm. wide, acute. Style 2.3 cm. long. Fruit subglobose, 1.2–1.4 cm. long, 1.2 cm. wide glabrous, obscurely to rather distinctly 8–9-ribbed in dry state, crowned by the persistent calyx-limb. Seeds 9 mm. long.

TANZANIA. Lushoto District: E. Usambara Mts., Monga, 26 Jan. 1939, *Greenway* 5830! & Sangarawe to Monga, *Zimmermann*! & Amani, 7 Mar. 1922, *Soleman* in *Herb. Amani* 5973!
DISTR. T 3; not known elsewhere
HAB. Rain-forest on steep slopes; 900–1020 m.

NOTE. Brenan described the fruit from T.T.F.H. 483 labelled Tabora District, Simbili Forest, but suspected there had been a confusion of labelling; no further material has turned up from T 4. I have seen material from Gonja Mt. above Mnyussi in the E. Usambaras which despite rather atypical foliage must represent fruiting *P. macrantha*; in this the fruits are rather distinctly ribbed. A very inadequate specimen *Greenway* 6574 from S. Pare Mts., Vidani at 1650 m., may represent this species.

73. LAMPROTHAMNUS

Hiern in Oliv. in Hook; Ic.Pl. 13,t. 1220 (June 1877) & in F.T.A. 3: 130 (1877)

Shrub or small tree. Leaves subsessile, glossy, rather coriaceous, ± cordate at the base; stipules interpetiolar, broadly triangular, apiculate. Flowers strongly sweetly scented, in apparently but not quite terminal and axillary panicles, the ultimate components dense and supported by small stipule-like triangular bracts; each flower supported by 2 ovate keeled bracteoles at the base of the calyx. Calyx-tube campanulate, the limb-tube exceeding the ovary, pubescent inside; lobes very transverse and rounded-truncate, usually 4 small ones in 2 contiguous pairs and 2 large ones or sometimes only 4–5, imbricate. Corolla white; tube narrowly funnel-shaped, with matted hairs inside save for the basal 1.5 mm.; lobes 6–7(-8), oblong-elliptic. Anthers linear, exserted for ⅔ or more of their length, apiculate. Disc annular, fleshy. Ovary thick-walled, with 1–2 small locules; ovules solitary, pendulous from a thickened funicle; style narrowly club-shaped, ± densely pubescent save at extreme base, the 2 arms ± fused but separable into flattened lobes for slightly over half its length. Fruit subglobose, presumably fleshy, crowned by the persistent calyx-limb. Seeds ovoid, very strongly ruminate, the exterior resembling hanks of rope, each longitudinal piece closely striated.

A monotypic genus restricted to East Africa, closely allied to *Polysphaeria* which has very similar testa sculpture.
  L. fosteri Hutch. described from Lagos, Nigeria, is *Morelia senegalensis* A. Rich.

L. zanguebaricus *Hiern* in Oliv. in Hook., Ic. Pl. 13, t. 1220 (June 1877) & in F.T.A. 3: 130 (Oct. 1877); T.T.C.L.: 503 (1949); K.T.S.: 449 (1961); Vollesen in Op. Bot. 59:69 (1980). Type: Tanzania, Dar es Salaam, *Kirk* (K, holo.!)

Evergreen usually much-branched shrub or small slender tree 1.8–9 m. tall; bark fissured or corrugated, reticulated, grey; slash pink, blaze red. Leaf-blades oblong, oblong-elliptic or oblong-ovate, 2–16 cm. long, 1.2–6.5 cm. wide, rounded to subacute at the apex, very slightly to distinctly cordate at the base, very often drying a leaden grey colour, glabrous; venation distinctly raised when dry, particularly above; petioles up to 2 mm. long; stipules 3–8 mm. long, with raised median line when dry, glabrous or upper ones slightly pubescent at base when young, subpersistent. Peduncles 1–2.5 cm. long, pubescent or nearly glabrous; bracts and bracteoles 2–4 mm. long, shortly pubescent. Calyx-tube and limb-tube together 3–5 mm. long; lobes up to 2 mm. long, ciliolate. Corolla-tube 0.5–1.2 cm. long; lobes 0.6–1.1 cm. long, 3.8–5 mm. wide. Anthers

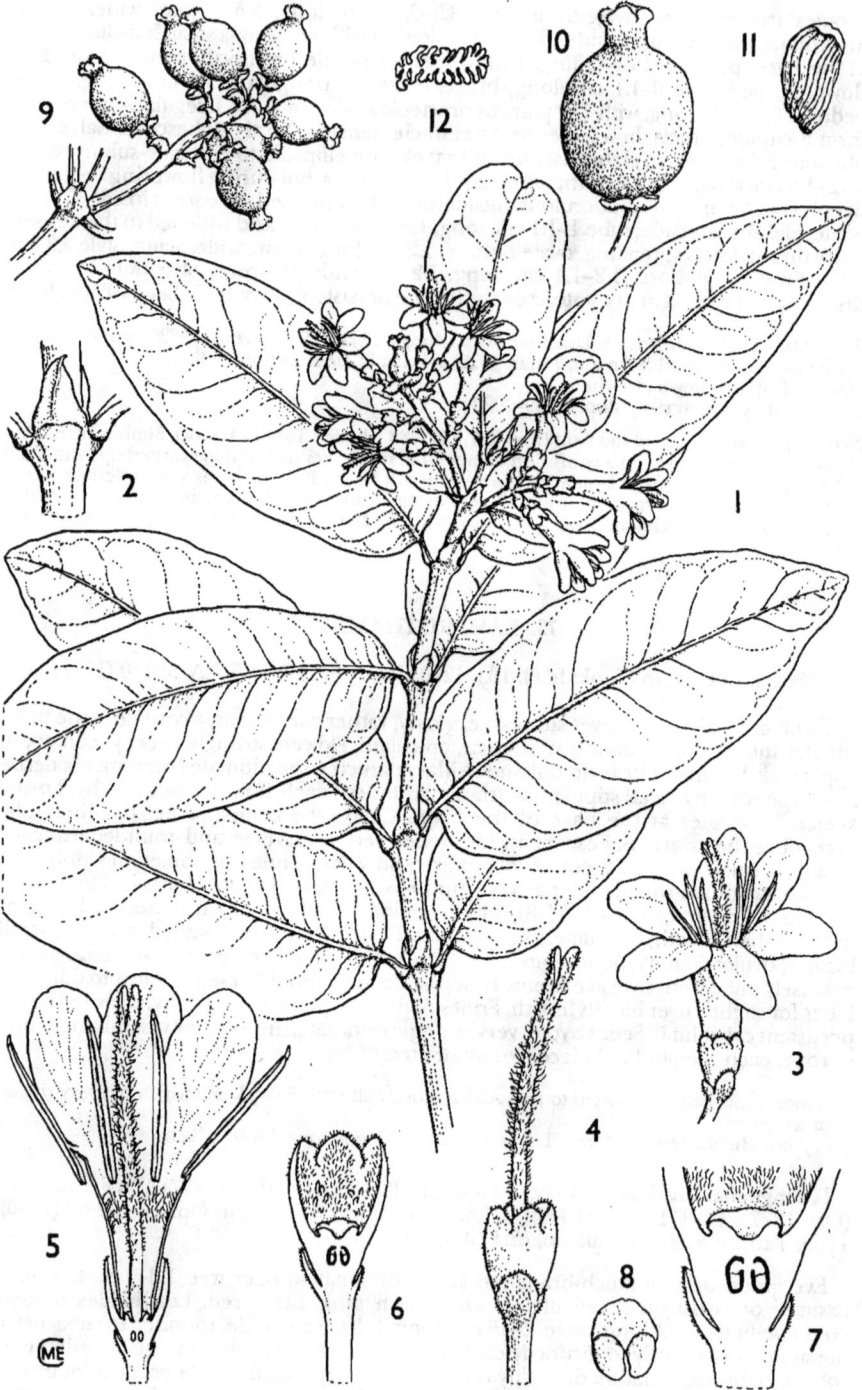

FIG. 97. *LAMPROTHAMNUS ZANGUEBARICUS* — 1, flowering branch, × ⅔; 2, stipules, × 2; 3, flower, × 2; 4, calyx with style, × 3; 5, section through flower, × 3; 6, section through calyx, × 4; 7, longitudinal section of ovary, × 8; 8, placenta with ovules, × 20; 9, part of infructescence, × 1; 10, fruit, × 2; 11, seed, × 2; 12, transverse section of seed, × 3. All from *Faulkner* 1524. Drawn by Mrs M.E. Church.

ochraceous or cream, 7–8 mm. long. Style cream, 1.2–2 cm. long, the arms mostly 7.5–10 mm. long when separated. Fruits green becoming red, 6–11 mm. long, 7–10 mm. wide. Seeds chestnut-coloured, somewhat shining, 5.5 mm. long, 4.5 mm. wide, 3 mm. thick; longest ruminations over 1 mm. deep. Fig. 97.

KENYA. Kwale District: 16 km. beyond Mariakani at 'mile 266' on Nairobi to Mombasa road, 8 Dec. 1961, *Polhill & Paulo* 908!; Mombasa District, Oct. 1947, *Jex-Blake* in *Bally* 5704!; Kilifi District: 6.4 km. N. of Malindi, Sabaki, 2 Nov. 1961, *Polhill & Paulo* 695!
TANZANIA. Tanga District: Tanga–Pangani road, 23 Feb. 1954, *Faulkner* 1524!; Morogoro/Rufiji Districts: Ukutu [Khutu] Steppe, Nov. 1898, *Goetze* 110!; Kilwa District: Selous Game Reserve, Kingupira Forest, 10 Sept. 1970, *Rodgers* 1126!
DISTR.   K 1, 7; T 3, 6, 8, Somalia
HAB. Wooded grassland, and particularly coastal bushland, thicket and woodland; 0–300 m.

NOTE. The fruit is said to be edible.

Tribe 19. PAVETTEAE*

Calyx-lobes subfoliaceous, 0.8–3 cm. long, persistent in fruit; corolla large (tube 0.8–11 cm. long), pubescent outside; fruit large, 0.8–2.2 cm. in diameter; seeds numerous; stipules conspicuous, either triangular and erect or ± rounded and reflexed:
  Corolla-tube shorter than or ± equal to lobes; anthers sessile but more than half-exserted; anther-thecae divided by numerous transverse septa (in Flora area); style exserted . . . .   **80. Dictyandra**
  Corolla-tube much longer than lobes; anthers included; anther-thecae undivided (in Flora area); style included or exserted . . . .   **81. Leptactina**
Calyx-lobes not nearly as conspicuous as above; corolla generally smaller, glabrous or occasionally pubescent outside; fruit smaller; seeds 1–several; stipules less conspicuous, never rounded and reflexed:
  Climbers with slightly to strongly recurved spines; stems square and slightly winged at corners; flowers borne in dense, often subcapitate corymbs; corolla-tube 2.5–3.5 cm. long; seeds 3–4 in each locule . . . .   **76. Cladoceras**
  If climbers then spines lacking and other characters not combined as above:
    Climbers; ovule 1, ascending from base of each locule; fruit with one spherical ruminate seed; stigma fusiform to clavate or broadly elliptic to subglobose; stipules fimbriate or entire . . . .   **77. Rutidea**
    Small trees, shrubs or sometimes climbers; ovules 1–several per locule, attached to septum; fruit with 1–several seeds, ruminate or entire (occasionally as above); stigma fusiform, occasionally clavate or with 2 divergent arms; stipules never fimbriate:
      Flowers 5–6-merous; fruit 1–several-seeded; seeds variously shaped, entire or ruminate; stipules often with central area darkened (dark green when fresh or black when dry) or stipules difficult to observe in taxa with reduced branches:
        Reduced dryland shrub, with leaves borne on cushion shoots; stigma clavate, winged; corolla-tube about half as long as the lobes; seeds ± 6 per fruit, orange-segment-shaped, angular, notched on inner straight margin . . . .   **74. Tennantia**

* Genera 75, 77–79 by D.M. Bridson, 74, 76, 80 and 81 by B. Verdcourt

Small trees or shrubs, occasionally reduced as above; stigma not winged; corolla-tube subequal to or sometimes longer than lobes; seeds 1–many per fruit, variously shaped but never as above   **75. Tarenna**

Flowers 4-merous; fruit 2-seeded; seeds ± hemispherical with excavation on ventral face; stipules never with central area darkened:

Bacterial nodules absent; inflorescence sessile or pedunculate, rachis sometimes white or tinted, often articulated; bracteoles linear, short; style with 2 divaricate arms; seeds dull, often rusty coloured, not rugulose . . . .   **78. Ixora**

Bacterial nodules usually present in the leaves, either scattered or arranged along the midrib; inflorescence sessile, laxly corymbose to subcapitate, never with rachis articulated; bracteoles stipule-like; style entire; seeds rugulose, moderately shiny . . . .   **79. Pavetta**

74. TENNANTIA

Verdc. in K.B. 36: 511 (1981)

Unarmed shrubs. Leaves and flowers borne on lateral abbreviated cushion-shoots. Stipules narrowly triangular, keeled. Inflorescences sessile at the ends of short shoots, ± 5-flowered, bracteate, the bracts presumably of stipular origin derived from suppressed internodes; pedicels obsolete; often flowering when leaves are very young. Calyx-tube cup-shaped, with limb-tube of equal length; lobes 5–6, ovate, separated by quite narrow sinuses. Corolla almost rotate, the tube very short; lobes oblong, much exceeding the tube. Anthers linear-oblong, apiculate, inserted at the throat in sinuses of corolla; filaments very short. Ovary subglobose or obconic, 2-locular; ovules 3(–4) per locule, arranged laterally on each placenta; style clavate, gradually narrowed from base to apex, winged, not divisible into two parts until very mature. Fruit small, globose, crowned with the persistent funnel-shaped calyx-limb. Seeds segment-shaped, angular, notched at middle on inner straight margin; testa glossy with very faint plane reticulation.

A monotypic genus formerly included in *Xeromphis*, i.e. *Catunaregam*. The very similar southern African *Coddia rudis* (Harv.) Verdc. (*Xeromphis rudis* (Harv.) L.E. Codd) has very different seeds and testa ornamentation and placentas. With the present dismemberment of the *Randia* complex it does not seem possible to accommodate them within one genus. Both are nearer to *Tarenna* than to *Catunaregam*.

T. sennii *(Chiov.) Verdc. & Bridson* in K.B. 36: 511, figs. 5 G–M & 6, t. 19 D–F & 20 D (1981); Bridson & Robbrecht in B.J.B.B. 55: 88, fig. 1 (1985). Type: Somalia (S.), Saar-Tumai, *Senni 437* (FT, holo.!, K, photo.!)

Shrub 1–3 m. tall, usually virgately branched, mostly unarmed but sometimes the abbreviated lateral shoots are spine-like; stems pale grey or whitish, rather smooth, only the younger shoots puberulous. Leaves borne on very short or cushion-like side-shoots, elliptic to narrowly obovate, 1–6.5 cm. long, 0.5–2.2 (–2.9) cm. wide, rounded at the apex and sometimes minutely apiculate, narrowed to the base into a petiole 0–7 mm. long, glabrous or puberulous; stipules 1–2 mm. long, becoming larger and ± corky, rendering the cushion-shoots scaly, usually puberulous and ciliate. Calyx-tube 1.2 mm. long; limb-tube 1.2 mm. long; lobes 0.8 mm. long. Corolla white, sometimes tinged pink; tube 1.5–2 mm. long; lobes green-tipped, 5 mm. long, ± 2 mm. wide. Anthers grey-buff, 4 mm. long. Style yellow, 8 mm. long. Fruit black, 5–6 mm. diameter, glabrous. Seeds 3.8–4.5 mm. long, 2 mm. wide. Fig. 98.

KENYA. Northern Frontier Province: Mathews Range, Mandasion [Mantachien], 7 Dec. 1960, *Kerfoot 2597*!; Machakos District: Mtito Andei, 30 Nov. 1959, *Greenway 9613*!; Kitui District: 16 km. N. of Mwingi on Tharaka road, 3 May 1960, *Napper 1550*!; Teita District: between Buchuma and Voi, 118 km. from Mombasa, 22 Sept. 1961, *Verdcourt 3226*!

DISTR. K 1, 4, 7; Somalia

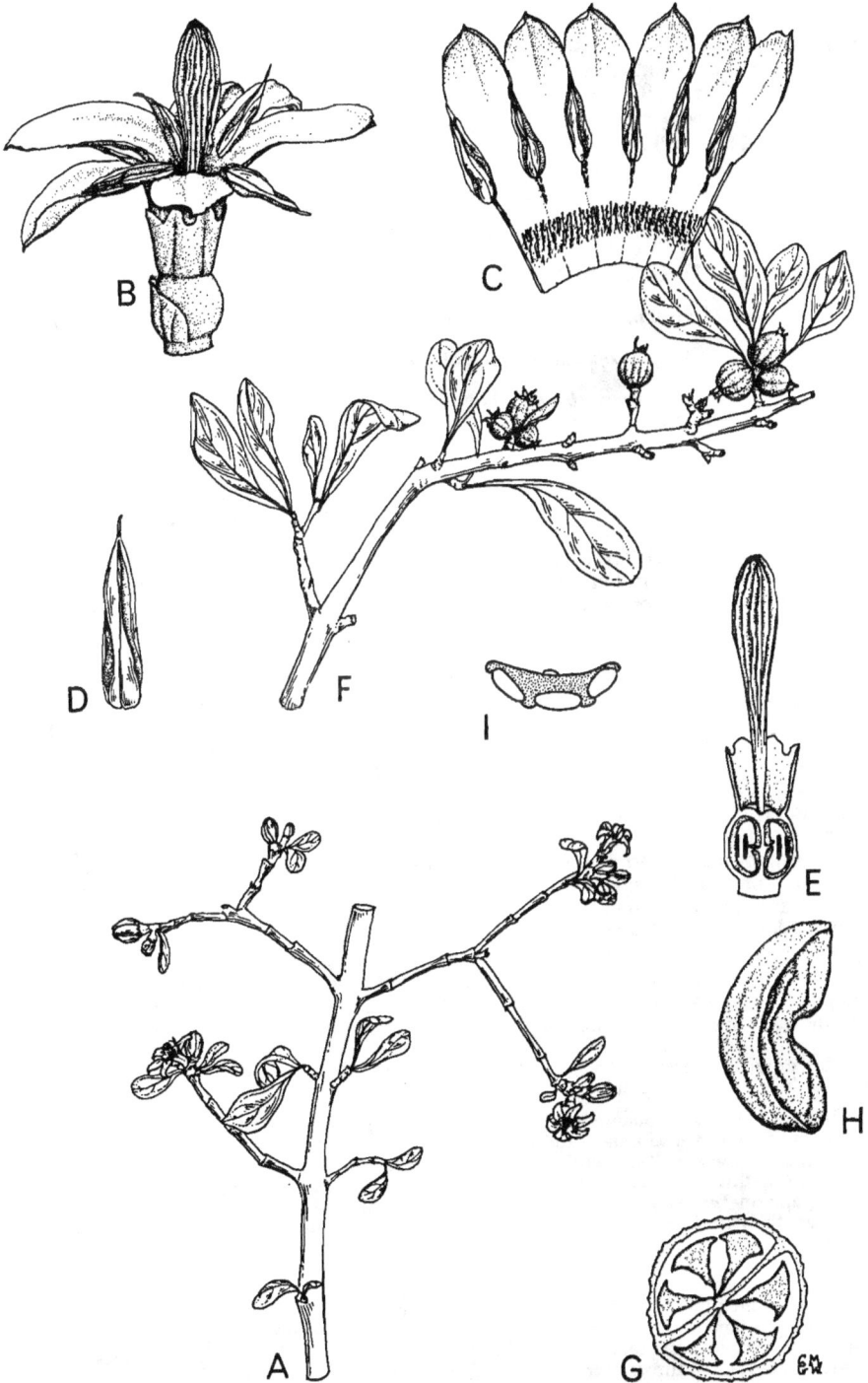

FIG. 98. *TENNANTIA SENNII* — **A**, flowering shoot, × ⅔; **B**, flower, × 4; **C**, corolla opened out, × 4; **D**, anther, × 6; **E**, gynoecium, × 4; **F**, fruiting shoot, × ⅔; **G**, transverse section of fruit, × 4; **H**, seed, × 6; **I**, transverse section of placenta to show disposition of ovules, × 17. A, from *Spjut* 4656; B–E, from *Verdcourt* 3226; F–I, from *Gillett* 19468. Drawn by Christine Grey-Wilson, with I by Sally Dawson.

HAB.   Dry bushland and woodland often on red sandy soil, also on bare rock outcrops; 420–1710 m.

SYN.   *Tricalysia sennii* Chiov., Fl. Somala 2: 240, fig. 141 (1932); E.P.A.: 1004 (1965)
   Xeromphis keniensis Tennant in K.B. 22: 435 (1968). Type: Kenya, Machakos District, Mtito
   Andei, *Greenway* 9740 (K, holo.!, EA, iso.!)

NOTE.   *Newbould* 2892 (Kenya, Meru District, Isiolo, 24 Nov. 1958) is described as having petals pale
   lilac above, brownish purple below, but no other labels mention such characteristics.

75. TARENNA

Gaertn., Fruct. & Sem. 1: 139, t. 28 (1788); Bridson in K.B. 34: 377 (1979)

Chomelia L., Gen. Pl., ed. 4: 58 (1752), *non* Jacq. (1760), conserved, *nec* Vell. (1827)

Coptosperma Hook.f. in G.P. 2: 86 (1873)

Enterospermum Hiern in F.T.A. 3: 92 (1877)

Zygoon Hiern in F.T.A. 3: 114 (1877)

Small trees, shrubs or sometimes lianes. Leaves opposite, petiolate or occasionally
subsessile, chartaceous to coriaceous, usually ± acuminate; domatia sometimes present;
stipules interpetiolar, ± truncate, triangular-lanceolate or ovate (not in Flora area), always
erect, often blackened at central area when dry, sometimes aristate. Flowers fragrant, ♀,
(4–)5-merous, usually rather small, pedicellate or less often sessile in mostly corymbose
terminal or infrequently axillary inflorescences; bracts sometimes present but
inconspicuous; bracteoles present, often on the pedicel. Calyx pubescent, puberulous or
glabrous; tube turbinate to ovoid; limb-tube usually short sometimes absent; lobes
overlapping or not, mostly under 5 mm. long. Corolla white to yellowish; tube cylindrical
or sometimes tending to be funnel-shaped, slightly shorter or longer than the lobes or
occasionally much longer, glabrous to bearded at throat; lobes ± oblong, contorted.
Stamens attached at the mouth of the tube, exserted and spreading or less often
incompletely exserted; filaments short; anthers dorsifixed near the base, linear or rarely
narrowly oblong, often shortly apiculate. Disc annular. Ovary 2-locular; placentas
attached near the top of the septum, either larger ± ovate with 1–15 ovules impressed on
them or small to moderately large with 2–5(–8) ovules pendulous from them; style
slender, pubescent or sometimes glabrous, with stigmatic club always exserted, bifid at
apex only or rarely completely bifid. Fruit usually pale green to green, white or black
when mature, small, scarcely fleshy, globose, mostly crowned by the persistent calyx-limb,
either 1-locular and 1-seeded or 2-locular with 1–several seeds in each locule. Seeds
either blackish, ± spherical with an irregular to ± regular hilar cavity, reticulate surface
and a ruminate or less often entire endosperm, or brown, hemispherical to variously
shaped with a ± round or occasionally elongate hilar cavity, surface smooth to the naked
eye (actually minutely reticulate) and an entire endosperm.

A large and diverse genus of about 180 species widespread in the tropics of the Old World, mostly
in Asia, Indonesia and Oceania but 50 or so in Africa.

The genus *Tarenna*, as outlined here, comprises six groups which are listed with brief descriptions
below, but not given any formal taxonomic status. As (in Africa at least) the generic limits between
Tarenna, *Enterospermum*, *Zygoon* and *Coptosperum* are far from clear-cut I have chosen to include them
in *Tarenna* as this would seem to be the most practical way of dealing with the matter until a world
monograph on *Tarenna* and related genera is available. Most of the existing generic descriptions
dealing with African species of *Tarenna* (e.g. F.W.T.A. and Fl. Gabon) have been based on group I
below.

Group 1. Leaves chartaceous to subcoriaceous or less often coriaceous, mature at time of
flowering or rarely immature. Inflorescence terminal on main and leafy lateral branches, sometimes
axillary. Calyx-limb with both tube and lobes apparent, lobes often overlapping. Corolla glabrous or
sometimes sparsely pubescent outside; tube shorter than or much longer than lobes, sparsely to
densely pubescent at throat. Placenta ± ovate, with 1–15 impressed ovules. Fruit 2-locular. Seeds
2-several per fruit, brown, smooth, variously shaped, least often hemispherical, with a distinct,
usually circular hilar cavity; endosperm entire. Species 1–8.

A cultivated 4-merous specimen (*Fernie* in *E.A.H.* 13755) from T 2, Moshi District, Lyamungu, has
been identified as *Tarenna richardii* Verdc. It belongs to a group of species (related to Group I) from
Madagascar, Aldabra and the Seychelles.

Group 2. Leaves chartaceous, glabrous or pubescent on both faces, usually immature at time of flowering. Inflorescence terminal on very short leafless lateral branches. Calyx-limb drying brown or fawn, both tubular part and lobes present. Corolla glabrous outside; tube a little longer than lobes, ± glabrous to sparsely pubescent at throat. Placenta ± ovate with 1 rather thick ovule impressed near the centre. Seeds 2 per fruit, light to dark brown, hemispherical, with pale circular hilar scar, smooth or wrinkled. Species 9–10.

The recently described species, *T. roseicosta*, links groups 1 and 2. It has the habit typical of group 2, but the calyx and several-ovulate placenta of group 1.

Group 3 (*Zygoon* Hiern). Leaves chartaceous, pubescent above, pubescent to tomentose beneath, sometimes immature at time of flowering. Inflorescence terminal on very short leafless spurs off the main and lateral branches. Calyx-limb drying fawnish, tubular part absent, very reduced or occasionally present; lobes often well spaced but occasionally overlapping. Corolla glabrous outside; tube shorter than lobes, sparsely to densely covered with longish hairs at throat. Placenta moderately small, bearing 3–7 pendulous ovules. Seeds 1 per fruit, dark brown to blackish, with an irregularly reticulate surface and an irregular hilar cavity; endosperm tending to be ruminate towards base. Species 11–12.

Tarenna zygoon Bridson (*Zygoon graveolens* Hiern) belongs to this group, and its distribution as known is Mozambique, Malawi, Zimbabwe and South Africa (Transvaal). It would seem likely that the East African record cited by K. Schumann in E.J. 28: 492 (1900) & Brenan, T.T.C.L.: 538 (1949), based on *Stuhlmann* 8886 from Tanzania, E. Uluguru Mts., Mgambo, Fisigo stream, (B †) was a misidentification, perhaps of a *Pavetta*.

Group 4 (*Enterospermum* Hiern). Leaves coriaceous or less often subcoriaceous. Inflorescence terminal on main and leafy lateral branches, or sometimes axillary, usually congested with rather short pedicels. Calyx-limb with both tube and lobes apparent. Corolla glabrous outside; tube slightly shorter to slightly longer than lobes, pubescent at throat. Placenta small to moderately large, bearing 2–8 pendulous ovules, or ovate with 2–3(–6) impressed ovules. Fruit 1(–2)-locular. Seeds 1(–2) per fruit, blackish, reticulate, ± spherical (or sometimes hemispherical) with an irregular or less often circular hilar activity; endosperm fully or partly ruminate. Species 13–18.

Group 5 (*Coptospermum* Hook.f. — see Bridson & Robbrecht in B.J.B.B. 55: 86 (1985)). Leaves coriaceous, very shiny above. Inflorescence terminal on main and leafy lateral branches, lax, with long pedicels. Calyx-limb tubular, repand or with short teeth. Corolla glabrous outside; tube slightly shorter than lobes, pubescent at throat. Placenta ± ovate, with 3 ovules impressed on the upper part. Seeds 1 per fruit, blackish brown, spherical, finely reticulate; hilar cavity absent; endosperm fully ruminate. Species 19.

Group 6. Leaves chartaceous, pubescent or glabrous. Inflorescences terminal on short leafy lateral branches and occasionally on the main branches. Calyx-limb with tubular part and triangular to sublinear lobes. Corolla salver-shaped; tube densely covered with adpressed hairs outside, pubescent within. Placenta small, bearing 2–3 pendulous ovules. Seeds 2–4 per fruit, reddish brown, with a shallow elongated hilar groove, smooth. Species 20. This group could be considered sufficiently distinct for removal from *Tarenna;* Leroy (pers comm., 1977) has suggested transference to the Madagascan genus *Homolliella* Arènes.

Flowers 4-merous; inflorescence terminal, trichotomously branched (cultivated species)	*T. richardii* (see note to Group 1)

Flowers (4–)5-merous; if inflorescence trichotomously branched, then axillary:
　Lateral branches typically short and patent; inflorescences frequently borne on short naked spurs; leaves often immature at time of flowering (except *T. kibuwae*, included in both parts of key):
　　Leaf-blades subcoriaceous, mature at time of flowering, small, up to 2.5 cm. long, 1.2 cm. wide; calyx-limb drying the same colour as the tube . . . . . *13. T. kibuwae*
　　Leaf-blades chartaceous, usually immature at time of flowering; calyx-limb usually drying brownish or fawn in contrast to the blackish tube:
　　　Leaves glabrous to glabrescent, sometimes hairy on nerves; placenta large, with 1–4 impressed ovules; seeds with circular excavation, smooth (where known); corolla-tube slightly longer than lobes (where known):

Inflorescence with slender rachis bearing usually (1–)2
   pairs of well-spaced patent branches; placenta
   with 2–4 ovules; fruit 4–5-seeded . . . . .   *8. T. roseicosta*
Inflorescence not as above; placentas 1-ovulate; fruit
   2-seeded:
   Inflorescences rather lax, not crowded into heads;
      calyx-lobes narrowly triangular . . . .   *9. T. burttii*
   Inflorescences crowded into spherical heads;
      calyx-lobes broadly ovate . . . .   *10. T. sp. A*
Leaves pubescent on both surfaces; placenta
   moderately small, bearing 3–5 ovules which exceed
   it; seeds (where known) with excavation small and
   irregular, reticulate; corolla-tube shorter than
   lobes:
   Inflorescence with branches and pedicels very
      reduced . . . .   *T. zygoon*
                                  (see note to Group 3)
   Inflorescence distinctly branched:
      Corolla-tube 2 mm. long; lobes 3 mm. long; flowers
         subsessile . . . .   *11. T. peteri*
      Corolla-tube 4 mm. long; lobes 7 mm. long;
         flowers pedicellate . . . .   *12. T. sp. B*
Lateral branches not typically short and patent;
   inflorescences terminal on leaf-bearing branches or
   occasionally axillary; leaves always mature at time of
   flowering:
   Corolla salver-shaped, densely covered with adpressed
      hairs outside; seeds 2–4 per fruit, with an elongate
      hilar groove, smooth . . . .   *20. T. trichantha*
   Corolla usually with lobes reflexed, glabrous outside;
      seeds not as above:
   Leaf-blades chartaceous or occasionally subcoriaceous;
      fruit containing 2–several seeds; seeds usually
      brown, smooth, with a ± circular hilar cavity;
      placenta ovate, with 1–several impressed ovules:
      Liane; leaf-blades with very pronounced acumen
         which is rounded or sometimes ± spathulate at
         tip; inflorescence very lax . . . . . .   *1. T. fusco-flava*
      Shrub or small tree; leaf-blades not having
        characteristic acumen as above; inflorescence
        only sometimes lax:
      Inflorescences mostly axillary, slender and lax,
        little branched; leaf-blades 6.5–10.5 cm. long   *2. T. uzungwaensis*
      Inflorescences terminal on main and lateral
        branches, sometimes with additional axillary
        inflorescences, not markedly slender or lax,
        usually clearly branched; leaf-blades 5–20 cm.
        long:
        Young stems not winged; leaf-blades not having
          ± linear acumen:
        Calyx-lobes lanceolate to triangular 1.75–2.3
          mm. long; secondary inflorescence-
          branches very reduced . . . . . .   *7. T. gossweileri*
        Calyx-lobes oblong to deltoid, shorter;
          secondary and tertiary inflorescence-
          branches clearly present:
          Stipules gradually narrowed or with subulate
            acumen 1–6 mm. long; calyx sparsely to
            densely pubescent; limb 1–2.25 mm.
            long, wider than tube; tertiary nerves
            scarcely raised above . . . . .   *3. T. pavettoides*

Stipules obtuse to acute with or without an apiculum up to 1 mm. long; calyx puberulous or sometimes glabrous; limb 1 mm. long, wider than tube; tertiary nerves raised and finely reticulate above . . . .   *4. T. drummondii*

Young stems somewhat winged; leaf-blades with acumen ± linear:

Shrub or small tree, 1.8–7 m. tall; leaf-blades oblanceolate to narrowly obovate, 7–15.5 cm. long, 1.7–5.2 cm. wide; stipules acute to shortly apiculate . . . .   *5. T. quadrangularis*

Tree to 15 m. tall; leaf-blades narrowly oblanceolate to oblanceolate, 11.5–18 cm. long, 2.5–4 cm. wide; stipules truncate to obtuse . . . .   *6. T. luhomeroensis*

Leaf-blades coriaceous or occasionally subcoriaceous; fruit containing 1(–2) seeds; seeds blackish or rarely brown with reticulate surface; endosperm partly to fully ruminate or seldom entire; placenta small to moderately large with 2–8 ovules, pendulous from it or less often large with 2–several impressed ovules:

Inflorescence markedly lax; bark very readily flaking; calyx-limb truncate to shortly toothed; stipules small, 1.5–2.5 mm. long, bearded within   *19. T. nigrescens*

Inflorescence congested to moderately lax; bark sometimes flaking; calyx-limb distinctly lobed; stipules not bearded within:

Inflorescences axillary, trichotomously branched; bracteoles ± cupular, surrounding the base of the calyx-tube . . . .   *17. T. supra-axillaris*

Inflorescence terminal, frequently with additional pairs of axillary inflorescences from the next 1–2(–3) nodes down; bracteoles not cupular:

Leaves subsessile, petioles not exceeding 2 mm. long:

Leaf-blades elliptic to oblong-elliptic, 0.8–2.4 cm. long, glabrous; corolla-tube shorter than lobes . . . .   *13. T. kibuwae*

Leaf-blades very narrowly elliptic, 1.5–3.5 cm. long, puberulous on both faces; corolla-tube equalling the lobes . . . . .   *14. T. wajirensis*

Leaves distinctly petiolate:

Domatia conspicuous; seed fully ruminate; placenta small . . . .   *18. T. littoralis*

Domatia absent; seed with endosperm only ruminate in vicinity of hilar cavity or entire; placenta $\frac{1}{8}$ the size of the ovules or much larger:

Stipules ovate, 3–7 mm. long, not apiculate; calyx puberulous to finely pubescent; placenta ovate, bearing up to 6 impressed ovules; seed with hilar cavity circular, usually surrounded by paler area; endosperm entire . . . .   *15. T. graveolens*

Stipules triangular, 2–3 mm. long, with a short arista; calyx glabrous; placenta ± $\frac{1}{3}$ the size of the 2(–3) pendulous ovules; seeds with hilar cavity somewhat irregular; endosperm ruminate towards base . . . . .   *16. T. neurophylla*

1. T. fusco-flava *(K. Schum.)* N. *Hallé* in Adansonia, sér. 2, 7: 506 (1967); Evrard in B.J.B.B. 37: 460 (1967); Tennant in K.B. 22: 437 (1968); N. Hallé, Fl. Gabon 17, Rubiacées 2: 98, t. 22/9–15 (1970); Bridson in K.B. 34: 379 (1979). Type: Cameroun, Yaoundé, *Zenker* 1519 (B, holo. †, K, P, iso.!)

Liane up to 5–20 m. long; branches slender, glabrous. Leaf-blades elliptic or oblong-elliptic, 5–9 cm. long, 2–4 cm. wide, with a very pronounced acumen 1–1.6 cm. long, 2–3(–4) mm. wide which is rounded or often ± spathulate, cuneate or rounded at the base, glabrous save for a few ± pilose domatia, the nerves lax and arcuate far from the margins. Corymbs glabrous, few–20-flowered; primary inflorescence-branches 1–3 cm. long, slender; pedicels (0.4–)0.8–2.2(–3) cm. long; bracts triangular, ± 1 mm. long, ciliate. Calyx-tube turbinate, 3 mm. long, glabrous; lobes imbricate, triangular to square, ± 1 mm. long and wide, rounded, ciliolate. Corolla white, glabrous outside; tube 0.9–1.3 cm. long; lobes oblong, 7–8(–10) mm. long, 2–3 mm. wide. Fruit grey-green becoming whitish, 6–8 mm. in diameter, with calyx-limb persistent 1 mm. long, 2-locular. Seeds light brown, 8–12 per fruit, smooth, ± 4 mm. long.

UGANDA. Bunyoro District: Budongo Forest, 6–8 Feb. 1935, *G. Taylor* 3331!; Mengo District: Mawokota, Feb. 1905, *E. Brown* 143! & Mpanga Forest, 5 Jan. 1955, *Byabainazi* 100!
DISTR. U 2, 4; Liberia, Ivory Coast, Ghana, Nigeria, Cameroun, Gabon, Zaire and Angola
HAB. Forest; 1150–1220 m.

SYN. *Chomelia fusco-flava* K. Schum. in E.J. 33: 339 (1903)
     C. laxissima K. Schum. in E.J. 33: 340 (1903). Type: Cameroun, Barombi-Kumba, *Preuss* 312 (B, holo.†)
     Tarenna flavo-fusca S. Moore in J.L.S. 37: 302 (1906); Keay in F.W.T.A., ed. 2, 2: 134 (1963)
     Ixora laxissima (K. Schum.) Hutch. & Dalz., F.W.T.A. 2: 87 (1931)

2. T. uzungwaensis *Bridson* in K.B. 42: 260 fig. 6 (1987). Type: Tanzania, Ulanga District, Uzungwe Mts., Sanje, *D. Thomas* 3785 (K, holo.!, MO, iso.!)

Glabrous shrub 2 m. tall; young stems slender, terete. Leaves paired, sometimes anisophyllous or with one not developed; blades narrowly elliptic or sometimes lanceolate, 6.5–10.5 cm. long, 2–3.2 cm. wide, acuminate at apex, acute to attenuate at base; domatia inconspicuous; petioles 3–5 mm. long; stipules triangular, 2 mm. long, obtuse to acute, turning dark, perhaps rather swollen when fresh. Inflorescences slender, apparently terminal (but axillary to apical bud), terminal on reduced lateral branches or axillary; inflorescence-branches present or suppressed; apparent pedicels 0.8–1.5 cm. long, but true pedicels much shorter; bracteoles small. Calyx-tube ± 1 mm. long; limb ± 1 mm. long, top third divided into rounded lobes, not overlapping. Mature corolla not known. Placenta bearing 3–4 impressed ovules. Fruit drying blackish, 5 mm. in diameter, crowned by persistent calyx-limb. Seeds light brown, ± 4 per fruit, ± 3 mm. long, smooth with oval excavation.

TANZANIA. Ulanga District: Uzungwa Mts., Sanje, 10 Oct. 1984, *D. Thomas* 3785!
DISTR. T 6; known only from type collection
HAB. Montane forest; 1400–1700 m.

3. T. pavettoides *(Harv.)* Sim, For. Fl. Cape Col.: 239 (1907); Cufod. in Phyton 1: 142 (1949); Ross, Fl. Natal: 333 (1973); Bridson in K.B. 34: 385 (1979). Type: South Africa, Natal, Durban [Port Natal], Field Hill, *Sanderson* 656 (TCD, holo., K!, MELB, iso.)

Small tree or shrub, 2.75–10 m. tall; young branches glabrous or with adpressed hairs, older stems square, covered with buff to red-brown sometimes flaking bark. Leaf-blades drying black or greenish to brown, oblanceolate to obovate or sometimes broadly elliptic, 5–20 cm. long, 2–8.8 cm. wide, acuminate or sometimes acute at apex, cuneate at base, sometimes with margins undulate, chartaceous to subcoriaceous, shiny above, glabrous above or sometimes becoming pubescent towards base of midrib, glabrous to glabrescent with the nerves glabrous or glabrescent to pubescent beneath; domatia present; petiole 0.5–3 cm. long, glabrous or with adpressed hairs; stipule-limbs with a blackened area when dry, deltoid, 2–6 mm. long, gradually narrowed or with a subulate acumen 1–6 mm. long, glabrous or less often with adpressed hairs outside, eventually deciduous. Inflorescences terminal on main and moderately short lateral branches, moderately lax to congested, sessile, usually with 1–3 pairs of pedunculate inflorescences arising from the next 1–2 nodes down, frequently with the supporting leaves reduced or absent, rarely with only the axillary branches present; primary inflorescence-branches 0.6–2 cm. long, covered with greyish to ± golden adpressed hairs; pedicels 0–8 mm. long, densely covered with adpressed hairs; bracteoles inconspicuous, subulate to filiform. Calyx sparsely to densely pubescent; tube 1–1.5 mm. long; limb wider than tube, 1–2.25 mm. long, divided for ± half its length into ovate to deltoid or oblong lobes, apiculate. Corolla white; tube 3–6.5 mm. long, glabrous outside, pubescent at throat; lobes oblong, 3.5–6 mm. long, 1.5–2.5 mm. wide, rounded, sometimes apiculate or tending to be emarginate. Placenta narrowly obovate to rotund, moderately fleshy to fleshy bearing 2–11, impressed ovules. Fruit glaucous green (or white or ? pale blue) turning black (when dry), 4–8 mm. in diameter, 2-locular, containing 2–15 seeds; calyx-limb persistent. Seeds brown, shiny, 2.5–4 mm. across, smooth.

KEY TO SUBSPECIES

Leaf-blades chartaceous, drying blackish; young stems glabrous or rarely pubescent (in specimens from Pemba); 2–7 ovules per placenta . . . . **a. subsp. affinis**

Leaf-blades subcoriaceous, drying greenish to dark brown; young stems glabrous to pubescent:

Young stems glabrous to sparsely pubescent; leaf-blades with nerves glabrous to glabrescent beneath; 7–11 ovules per placenta; fruit ± 15 seeded . . . . . **b. subsp. friesiorum**

Young stems densely covered with adpressed hairs or occasionally ± glabrous; leaf-blades with nerves glabrescent to pubescent beneath; 4–7 ovules per placenta; fruit 4–13 seeded . . . . **c. subsp. gillmanii**

a. subsp. **affinis** *(K. Schum.) Bridson* in K.B. 34: 385 (1979). Type: Tanzania, Usambara Mts., Mashewa [Mascheua], *Holst* 8731 (B, holo.†, K, iso.!)

Young stems glabrous or rarely pubescent. Leaf-blades drying blackish, chartaceous, acute or more often distinctly acuminate at apex, glabrous to glabrescent on nerves beneath; stipule-limb 2–4 mm. long with acumen 2–6 mm. long. Peduncles of axillary inflorescences slender, 1.3–8 cm. long, often rather acutely angled to main stem, glabrescent to pubescent; pedicels 0–3(–4) mm. long. Calyx-tube covered with short or somewhat longer greyish or sometimes ± golden hairs; limb 1–1.75 mm. long; lobes ovate to deltoid. Placentas bearing 2–7 ovules; fruit ± 5-seeded.

TANZANIA. Lushoto District: E. Usambara Mts., Korogwe, 21 Nov. 1947, *Brenan & Greenway* 8358!; Ulanga District: Tabora, 22 Jan. 1932, *Schlieben* 1654!; Songea District: ± 32 km. from Songea, by R. Mkurira, 8 Feb. 1956, *Milne-Redhead & Taylor* 8645!
DISTR. T 3, 6–8; P; Mozambique, Malawi and Zimbabwe
HAB. Forest; 0–1920 m.

SYN. *Chomelia affinis* K. Schum. in P.O.A. C: 380 (1895)
C. ulugurensis K. Schum. in E.J. 28: 61 (1899) & op. cit: 489 (1900); T.T.C.L.: 491 (1949). Type: Tanzania, Morogoro District, Uluguru Mts., Lussegwa, *Stuhlmann* 2724 (B, holo. †)
Tarenna affinis (K. Schum.) S. Moore in J.L.S. 37: 158 (1905); Garcia in Mem. Junta Invest. Ultram., sér Bot., 4: 25 (1958), pro parte
Pavetta swynnertonii S. Moore in J.L.S. 40: 99 (1911). Type: Zimbabwe, Chirinda Forest, *Swynnerton* 117 (BM, holo.!, K, iso.!)
[*Tarenna pavettoides* sensu T.T.C.L.: 533 (1949); Garcia in Mem. Junta Invest. Ultram., sér. Bot., 4: 25 (1958), quoad *Torre* 3590, *non* (Harv.) Sim pro majore parte, *non* quoad typum]

NOTE. Some minor differences can be found between specimens from T 3 and Zimbabwe; the T 3 specimens tend to have slightly longer hairs on the calyx and the peduncles of the axillary inflorescences are often longer and more acutely angled, while the Zimbabwe specimens generally have narrower leaves with a very distinct acumen. The specimens seen from T 6–8 and Malawi tend to be intermediate in the above respects, but show rather more variation as a group than do the two above groups. The specimens seen from Pemba have pubescent stems, but otherwise seem to belong to this subspecies.
It is possible that an isotype of *Chomelia ulugurensis* is still in existence, but I have been unable to trace it. Presumably Cufodontis saw one around 1949 when he made the combination *Tarenna ulugurensis* (Phyton 1: 142 (1949), also Bremekamp determined *Bruce* 933 as "*Tarenna ulugurensis* (K. Schum.) Bremek." in 1947, noting a few minor differences from the type.
The above subspecies is closest to subsp. *pavettoides* (from South Africa) which has a calyx with a shorter limb-tube (0.75–1 mm. long) and usually oblong lobes which tend to overlap, also the pedicels tend to be longer.

b. subsp. **friesiorum** *(K. Krause) Bridson* in K.B. 34: 386, figs 2A & 3A, B (1979). Type: Kenya, between rivers Meru and Nithi, *R. & Th.C.E. Fries* 1967 (B, holo. †, K, iso.!)

Young stems glabrous to sparsely pubescent. Leaf-blades drying greenish or brownish, subcoriaceous, acute to acuminate at apex, glabrous to glabrescent on nerves beneath; stipules with limb 3–5 mm. long and arista 2–3 mm. long. Peduncles of axillary inflorescences 1.8–4.2 cm. long, glabrescent to pubescent; pedicels 1–4 mm. long. Calyx-tube sparsely covered with grey hairs; limb 1.5–2 mm. long; lobes ovate to deltoid. Placentas thick and fleshy bearing 7–11 ovules; seeds ± 15 per fruit, 3 mm. across.

KENYA. N./S. Nyeri District: Mt. Kenya, *Rammell* 1105!; Fort Hall/Kiambu Districts: Thika, 12 Nov. 1966, *Faden* 66/51!; Meru District: 57.6 km. from Embu, Kithunguri [Kitanguri], 9 Jan. 1940, *Copley* in *Bally* 560!
DISTR. K 4, ?7; not known elsewhere
HAB. Forest; 1200–1525 m.

SYN. *Pavetta friesiorum* K. Krause in N.B.G.B. 10: 606 (1929)
     Tarenna friesiorum (K. Krause) Bremek. in F.R. 37: 192 (1934); Chiov. in Miss. Biol. Borana, Racc. Bot.: 227 (1939); K.T.S.: 474 (1961)
     [*T. graveolens* sensu Chiov., Racc. Bot. Miss. Consol. Kenya: 56 (1935), quoad *Balbo* 65, *non* (S. Moore) Bremek.]
     [*T. boranensis* sensu K.T.S.: 474 (1961), pro parte, quoad *Copley* in *Bally* 560 & *Patterson* in *Bally* 2205, *non* Cufod.]
     [*T. pavettoides* sensu K.T.S.: 475 (1961), *non* (Harv.) Sim sensu stricto]

NOTE. One rather poor fruiting specimen from K 7 (Kwale District, Shimba Hills, *van Someren* 245) appears to be typical of this taxon save that the branches are slightly more hairy.

c. subsp. **gillmanii** *Bridson* in K.B. 34: 386 (1979); Fl. Pl. Lign. Rwanda, fig. 205/3 (1982); Fl. Rwanda 3: 224 fig. 71/3 (1985). Type: Tanzania, Bukoba District, Nshamba, *Gillman* 565 (K, holo.!, EA, iso.!)

Young stems densely covered with adpressed hairs or occasionally ± glabrous. Leaf-blades drying brownish or sometimes greenish grey, subcoriaceous, acute to acuminate at apex, glabrescent to pubescent on nerves beneath; stipule-limbs 2–6 mm. long, with arista up to 4 mm. long. Peduncles of axillary inflorescences 2–6 cm. long, sometimes ± robust and obtusely angled to main stem, pubescent, often with supporting leaves suppressed giving the appearance of a large terminal inflorescence; pedicels 0–3 mm. long. Calyx-tube densely or less often sparsely covered with greyish or ± golden hairs; limb 1–2.25 mm. long, divided into ovate to deltoid lobes. Placentas bearing 4–7 ovules; seeds 4–13 per fruit, 3(–5) mm. across.

UGANDA. Kigezi District: Ishasha Gorge, May 1950, *Purseglove* 3417!; Masaka District: SW. of Lake Nabugabo, Bugabo, 1 Feb. 1969, *Lye et al.* 1899!; Mengo District: Kome I., near Bugombe, 27 Oct. 1968, *Lye* 112!
KENYA. Nandi District: Kaimosi Police Post, 13 Oct. 1981, *Gilbert & Tadessa* 6699!; N. Kavirondo District: Kakamega Forest, road from Kisieni, 30 Apr. 1979, *Bridson* 16! & 5 Jan. 1967, *Perdue & Kibuwa* 9457!
TANZANIA. Mwanza District: Ukerewe Is., Rubya Forest Reserve, 13 June 1958, *Willan* 329!; Kigoma District: Kungwe Mt., Kahoko, 23 July 1959, *Newbould & Harley* 4582!; Rungwe District: Masukulu [Mwasukulu], 15 Feb. 1913 *Stolz* 1887!
DISTR. U 2–4; K 3, 5; T 1, 4, 7; Zaire (Kivu, Shaba), Rwanda, Burundi, Sudan, Malawi, Zambia
HAB. At forest edges and in thickets or occasionally in *Protea* woodland on dry hillsides; (800–) 1125–1600 m.

SYN. [*Tarenna pavettoides* sensu I.T.U., ed. 2: 357 (1952); F.F.N.R.: 421 (1962), *non* (Harv.) Sim sensu stricto]

NOTE. While the majority of specimens of this subspecies have the young branches distinctly hairy, a few specimens from Uganda, Kenya and Tanzania (Kigoma District) have been seen which have ± glabrous stems. Both glabrous and hairy forms have been found growing together at Bugisu, Mbale District, and in the Kakamega Forest. Two specimens from Tanzania, Kigoma District (*Nishida* 164 in young fruit & 211, sterile) have petioles much shorter than usual for *T. pavettoides* but may perhaps belong to this subspecies.

4. **T. drummondii** *Bridson* in K.B. 34: 379, fig. 4 (1979). Type: Kenya, Kwale District, Mwasangombe Forest, *Drummond & Hemsley* 4024 (K, holo., EA, iso.!)

Shrub or small tree 2–7 m. tall, glabrous; young branches greenish brown when dry, ± terete; older stems covered with brownish grey bark. Leaf-blades greenish brown when dry, elliptic or narrowly obovate, 8–19 cm. long, 2.5–8.8 cm. wide, acute to acuminate at apex, acute to obtuse or rounded and sometimes unequal at base, moderately shiny above, duller beneath; domatia occasionally present; petiole 0.6–1.9 cm. long; stipules triangular or ± oblong, 3–4 mm. long, obtuse to acute, with or without an apiculum up to 1 mm. long. Inflorescences terminal, sessile, frequently with an additional 1–2 pairs of pedunculate inflorescences arising from the next 1–2 nodes down, lax; primary branches 1.2–3.5 cm. long, glabrous or sparsely pubescent; pedicels 0.5–1.2 cm. long, puberulous or less often glabrous; bracteoles absent or present inconspicuous, ciliate. Calyx

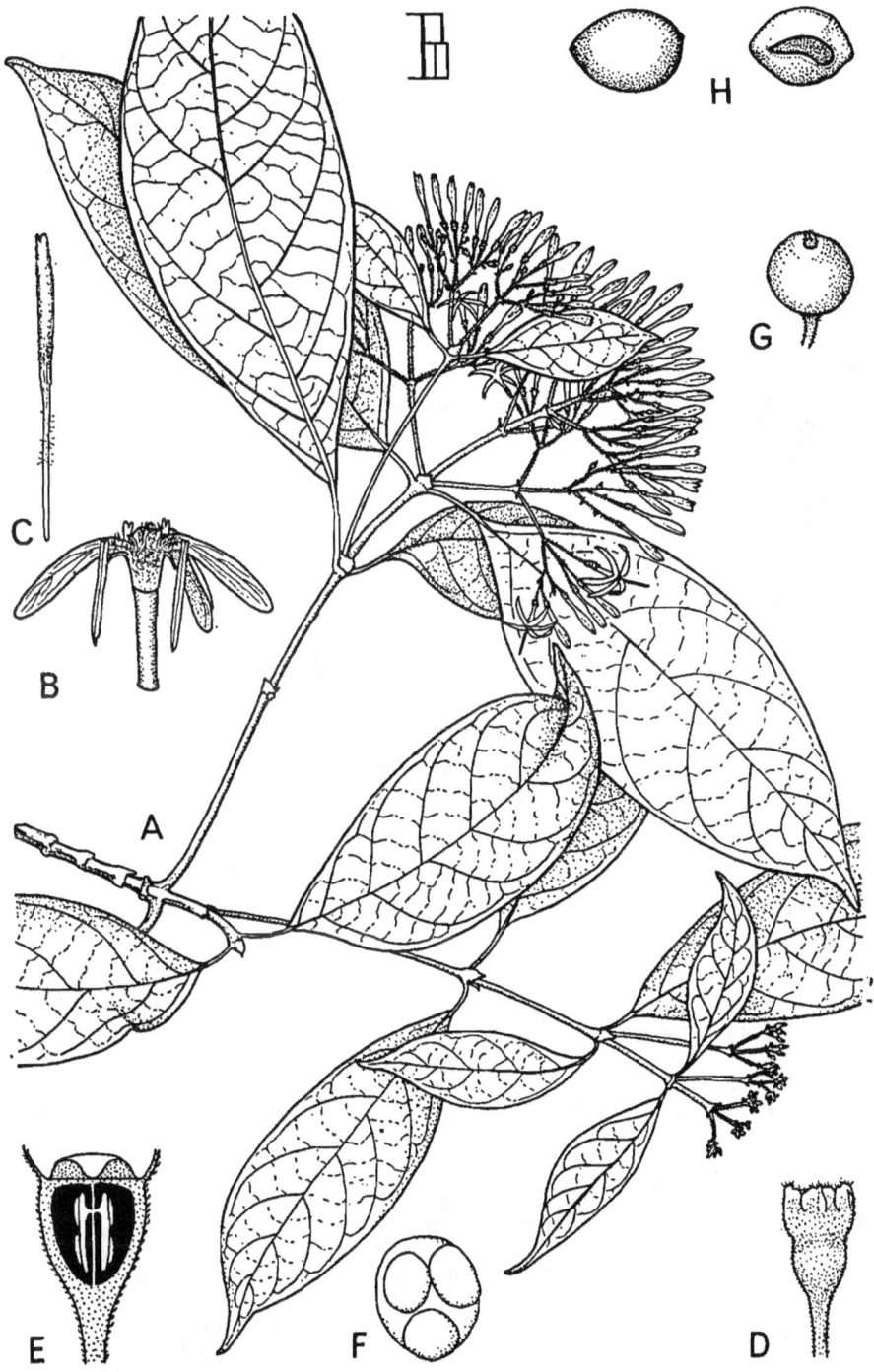

FIG. 99. *TARENNA DRUMMONDII* — **A**, flowering shoot, × ²⁄₃; **B**, corolla with 2 lobes removed, × 3; **C**, style with stigma, × 3; **D**, calyx, × 6; **E**, section through calyx-tube and ovary, × 14; **F**, placenta with ovules, × 20; **G**, fruit, × 1¼; **H**, seed (2 views), × 4. A, from *Drummond & Hemsley* 3876; B–F, from *Drummond & Hemsley* 4024; G–H, from *Shabani* 653. Drawn by Diane Bridson.

puberulous or less often glabrous; tube 1–1.25 mm. long; limb ± 1 mm. long, divided for ± ⅓ of its length into oblong to triangular lobes, apiculate, overlapping. Corolla white; tube ± cylindrical, 4–6 mm. long, ± 1.5 mm. wide, glabrous to sparsely pubescent outside, pubescent within; lobes narrowly oblong, 6–7 mm. long, 1–1.25 mm. wide, rounded or irregularly emarginate. Placenta ± ovate, rather thin, with 2–4 impressed ovules. Fruit black when ripe, ± 6 mm. in diameter, glabrous or sparsely puberulous, 2-locular, containing 5–7 seeds. Seeds brownish, 3–5 mm. long, smooth; hilar cavity circular; endosperm entire. Fig. 99.

KENYA. Kwale District: Marenge Forest, Lungalunga–Msambweni road, 18 Aug. 1953, *Drummond & Hemsley* 3876! & Shimba Hills, Pengo Hill, 27 Mar. 1969, *Magogo & Glover* 492!; Kilifi District: Chasimba, Massive Kambe, 16 Feb. 1977, *Faden et al.* 77/411!
TANZANIA. Handeni District: Mgambo Forest Reserve, 5 Feb. 1971, *Shabani* 653!; Uzaramo District: Pugu Forest Reserve, 10 Mar. 1964, *Semsei* 3701!; Rufiji District: Kichi Hills, 22 Dec. 1976, *Vollesen* in M.R.C. 4254!
DISTR. K 7; T 3, 6; not known elsewhere
HAB. Forest or open woodland; 100–460 m.

NOTE. The specimens from Uzaramo District have glabrous pedicels and calyces while those from Kenya and Handeni District have puberulous pedicels and calyces. *Semsei* 843 (an immature specimen) from Morogoro District, Kimboza Forest Reserve, may possibly also belong to this species.

5. T. quadrangularis *Bremek* in K.B. 1936: 484 (1936); T.T.C.L.: 533 (1949); Bridson in K.B. 34: 379 (1979). Type: Tanzania, Morogoro District, Uluguru Mts., Kisaki road, *E.M. Bruce* 317 (K, holo.!)

Shrub or small tree 1.8–7 m. tall, glabrous; young branches greenish brown and shiny when dry, square with small wings at angles; older stems with flaking brown bark. Leaf-blades greyish green to brown when dry, oblanceolate to narrowly obovate, 7–15.5 cm. long, 1.7–5.2 cm. wide; apex with ± linear cauda up to 2 cm. long, rounded at tip; base cuneate, moderately shiny on both faces; domatia present; petioles 1–2 cm. long, slender; stipules broadly ovate or somewhat triangular, 2–3.5 mm. long, acute or shortly apiculate, pilose inside. Inflorescences terminal, sometimes with an additional pair of pedunculate inflorescences arising from the node below, sessile, lax; primary branches 0.9–1.8 cm. long; pedicels 0.5–1 cm. long; bracteoles small, scale-like. Calyx glabrous; tube ± 1.5 mm. long; limb 1.25–1.5 mm. long, divided for ± ⅓ of its length into oblong lobes, apiculate and overlapping. Corolla white; tube cylindrical 6–7 mm. long, glabrous outside, pubescent within; lobes oblong, 6–7 mm. long, 2–3 mm. wide, apiculate or acuminate. Placenta obovate, moderately fleshy, with 4–5 impressed ovules. Fruit green, drying black, 5–6 mm. in diameter, crowned by the persistent calyx-limb, 2-locular, containing 4–8 seeds. Seeds brown, shiny, not fully developed but ± 4 mm. wide, smooth.

TANZANIA. Morogoro District: Uluguru Mts., 18 Dec. 1932, *Schlieben* 3116! & above Morningside, June 1953, *Semsei* 1255! & valley of Mwere R., 27 July 1972, *Mabberley* 1233!
DISTR. T 6; known only from Uluguru Mts.
HAB. Forest; 1350–1650 m.

6. T. luhomeroensis *Bridson* in K.B. 42: 257, fig. 4 (1987). Type: Tanzania, Iringa District, Uzungwa Mts., Luhomero Massif, Upper Lofia valley, *Rodgers & Hall* 4547 (K, holo.!)

Tree 15 m. tall, glabrous; young branches square, slightly winged. Leaves paired, infrequently anisophyllous, sometimes only one leaf developing at apex; blades greenish black when dry, narrowly oblanceolate to oblanceolate, 11.5–18 cm. long, 2.5–4 cm. wide, long acuminate at apex, attenuate at base, chartaceous, shiny on both faces, domatia present, glabrous or ciliate; petioles 1–2 cm. long slender; stipules sheathing, 2–3 mm. long, truncate to obtuse, turning black when dry. Inflorescences terminal, with an additional pair of inflorescences arising from node below; lax; peduncle absent or ± 1 cm. long in terminal inflorescence or up to 3 cm. long in axillary inflorescences; primary inflorescence-branches 0.8–1.2 cm. long; pedicels 6–8 mm. long in fruiting stage, glabrous; bracteoles small. Calyx only known of from immature bud and fruit; limb 1.5 mm. long, divided to the middle into rounded overlapping lobes. Corolla not known. Placenta with 3–4 impressed ovules. Fruit drying black, ± 1 cm. in diameter, crowned by persistent calyx-limb, 2-locular, containing (1–)2–6 seeds. Seeds pale brown, 4 mm. long, smooth with ± circular excavation.

TANZANIA. Iringa District: Uzungwa Mts., Luhomero Massif, upper Lofia valley, 18 Aug. 1985, *Rodgers & Hall* 4547!
DISTR. T 7; known only from type
HAB. Forest on rugged mountainous terrain of crystalline basement rocks; 1750 m.

7. T. gossweileri S. *Moore* in J.L.S. 37: 303 (1906); Bridson in K.B. 34: 381 (1979). Type: Angola, Malange, left bank of Quije R. near Quizol, *Gossweiler* 1258 (BM, holo.!, K, iso.!)

Small shrub 1.8–4.5 m. tall; young branches pubescent, older stems squarish, covered with reddish-brown flaking bark. Leaf-blades mostly drying dull greenish in colour, elliptic or less often narrowly or broadly elliptic, 3–13 cm. long, 1.6–6.4 cm. wide, shortly acuminate at apex, acute to ± rounded at base, somewhat coriaceous, glossy and glabrous above, save for the midrib, pubescent to glabrescent beneath; domatia usually present; petiole 0.4–1.3 cm. long, sparsely to densely pubescent; stipules with a brown corky area surrounding a central blackened area (in dry state), ovate to triangular, 2–4 mm. long, acute or with a linear acumen up to 3 mm. long. Inflorescences terminal on short leafy lateral branches, 4–14.5 cm. long, sessile; primary inflorescence-branches up to 9 mm. long, densely pubescent; secondary inflorescence-branches very reduced; pedicels up to 1.5 mm. long; bracteoles narrowly triangular. Calyx covered with adpressed hairs; tube ± 1.5 mm. long; limb-tube scarcely wider than tube, 1.5–2.5 mm. long; lobes triangular to lanceolate, 1.75–2.3 mm. long, obtuse, tending to arch back when mature. Corolla white, glabrous or rarely minutely pubescent outside; tube (0.6–)1.6–4 cm. long, pubescent at throat; lobes oblong, 6.5–10 mm. long, 2–3 mm. wide, rounded. Placenta fleshy, with 5–12 impressed ovules. Fruit green (blackish when dry), 6–7 mm. in diameter, glabrescent, 2-locular, containing ± 10 seeds; calyx-limb persistent. Seeds light brown, shiny, 3–4 mm. across, smooth.

var. **brevituba** *Bridson* in K.B. 34: 382 (1979). Type: Tanzania, Ufipa District, Kalambo R., Katundo village, *Richards* 10354 (K, holo.!)

Corolla-tube 6–7 mm. long.

TANZANIA. Ufipa District: Kalambo R., Katundo village, 15 Dec. 1958, *Richards* 10354!; Mbeya/Chunya District: Unyiha, Mlowo, 3 Oct. 1976, *Leedal* 3860!
DISTR. T 4, 7; Zambia
HAB. Riverine forest; 1520–1650 m.

NOTE. The only other specimen known of var. *brevituba* (*Fanshawe* 3856) was collected in Zambia at Kawambwa; it is possible that further collections could indicate either that this variety just represents the lower end of the range of variation of corolla-tube length, rather than a true variety or possibly that a subspecies is involved as the distribution of var. *gossweileri* (*T. rhodesiaca* Bremek.) as now known (Western Zambia and Angola), seems slightly disjunct from that of var. *brevituba*.

8. T. roseicosta *Bridson* in K.B. 42: 259 fig. 5 (1987). Type: Tanzania, Iringa District, near Kidabaga village, *Bridson* 583 (K, holo.!, BR, DSM, EA, MO, iso.!)

Scandent shrub or liane to 4 m. tall; young stems slender, terete, hollow, covered with adpressed hairs; lateral branches short, patent or sometimes a little reflexed, occasionally spine-like. Leaves not fully mature at time of flowering; blades elliptic, 4–10 cm. long, 1.5–3.4 cm. wide, acute to shortly acuminate at apex, cuneate at base, glabrous to glabrescent, with puberulous midrib above and beneath; midrib pink; domatia absent or inconspicuous; petioles 2–5 mm. long, sparsely pubescent; apical stipule triangular, acuminate, up to 4 mm. long, scarious; stipules on mature stems truncate with small triangular lobe, ± 2 mm. long, turning dark when dry. Inflorescences terminal on often very short lateral branches; inflorescence-branches (1–)2(–3) pairs, well spaced along the rachis and perpendicular to it, covered with short adpressed hairs; pedicels up to 2 mm. long (perhaps longer when mature), covered with short adpressed hairs; bracteoles small, present on pedicels. Calyx-tube ± 1 mm. long, glabrous becoming pubescent towards base; limb ± 1 mm. long, divided to midpoint into rounded, slightly overlapping lobes. Corolla only known from bud, glabrous outside, 4.5 mm. altogether. Placenta bearing 2–4 impressed ovules. Fruit drying black, 4–5 mm. diameter, crowned by persistent calyx-limb; pedicel lengthening to 11 mm. Seeds 4–5 per fruit, light brown, 3 mm. long, smooth, with ± circular excavation.

TANZANIA. Iringa District: Uzungwa Mts., Udekwa village, forest block to E. of West Kilombero Forest Reserve, Dec. 1981, *Rodgers & Hall* 1764!; Mufindi District, Lulando Forest, 10 May 1987, *Lovett & Congdon* 2149! & forest at Kidabaga village, 27 Aug. 1984, *Bridson* s.n.!

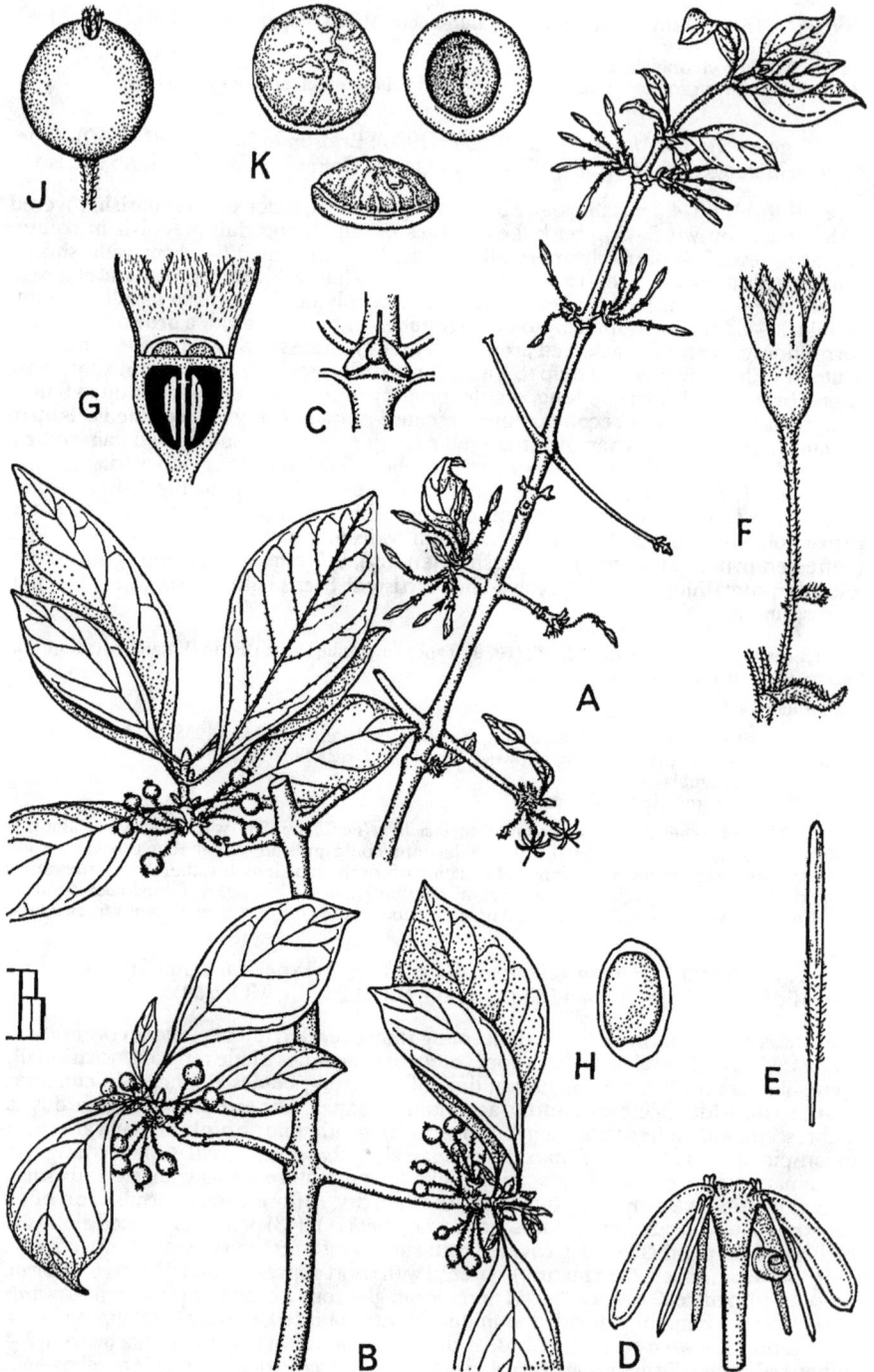

FIG. 100. *TARENNA BURTTII* — **A**, flowering shoot, × ⅔; **B**, fruiting shoot, × ⅔; **C**, stipule, × 3; **D**, corolla with 2 lobes removed, × 6; **E**, style with stigma, × 4; **F**, calyx with pedicel and bracteoles, × 8; **G**, section through calyx-tube and ovary, × 14; **H**, placenta with ovule, × 20, **J**, fruit, × 3; **K**, seed (3 views), × 4. A, D–H, from *Burtt* 5388; B, J, K, from *Leippert* 6298; C, from *Peter* 34300. Drawn by Diane Bridson.

DISTR.   T 7; not known elsewhere
HAB.   Moist montane forest or open areas of disturbed forest; 1450–2125 m.

9. T. burttii *Bridson* in K.B. 34: 388, fig. 5 (1979). Type: Tanzania, Dodoma District, Manyoni, *B.D. Burtt* 5388 (K, holo.!, BR, EA!, P!, iso.)

Shrub 2–4 m. tall; young stems terete, pubescent at first, later covered with pinkish-grey or sometimes light brown bark, not flaking; lateral branches short, patent. Leaves maturing after flowering; blades usually green when dry, broadly elliptic to obovate, 3.6–8.5 cm. long, 1.7–4 cm. wide, acute at apex, cuneate at base, chartaceous; nerves sparsely pubescent and often drying whitish beneath; domatia small; petiole 0.2–5 mm. long; stipules brownish, usually with the central area blackened and swollen when dry, 1–2.5 mm. long. Inflorescences terminal on short to very short spurs, 0–8 mm. long, usually towards the ends of short lateral branches or occasionally off the main branch, sessile, lax; inflorescence-branches reduced or ± suppressed; pedicels 3–6 mm. long, pubescent; bracts fawn, ovate, 2–3 mm. long, densely covered with silky white hairs inside; bracteoles fawn, pubescent, present on the pedicel. Calyx glabrescent; tube drying blackish, 1 mm. long; limb drying fawn, 0.75–1.75 mm. long, divided almost to base into narrowly triangular lobes, sparsely pubescent outside, pubescent within. Corolla glabrous outside; tube ± cylindirical, 3–4 mm. long, 1–1.25 mm. wide at top, sparsely pubescent at throat; lobes ± oblong, 4–5 mm. long, 1.5 mm. wide, rounded. Placenta ovate, rather fleshy, bearing a solitary ovule. Fruit green, 5–5.5 mm. in diameter, 2-locular, each locule containing 1 seed; calyx-limb persistent. Seeds pale brown, hemispherical, 4 mm. in diameter, with hilar cavity large and deep, ± irregularly sculptured on the convex face; endosperm entire. Fig. 100.

TANZANIA.   Mbulu District: Kisima cha Mungu, 10 Jan. 1928, *B.D. Burtt* 883!; Tabora District: near Kisengi, 11 Dec. 1979, *Lawton* 2226!; Kondoa District: Kondoa–Irangi at Iwanga, 15 Dec. 1925, *B.D. Burtt* 397!
DISTR.   T 2, 4, 5; not known elsewhere
HAB.   Deciduous bushland and thicket; 1150–1300 m.

10. T. sp. A; *Bridson* in K.B. 34: 388 (1979)

?Shrub, glabrous; branches ± terete, somewhat robust, covered with grey bark, tending to flake when older; lateral branches patent, moderately short. Leaves immature at time of flowering, borne on apex of main stem and upper lateral branches, subsessile; blades drying greenish brown, obovate, up to 4.2 cm. long, 2.8 cm. wide, rounded or emarginate at apex, cuneate at base papery; stipules brownish when dry, triangular, ± 4 mm. long. Inflorescence a spherical head, terminal at apex of short leafless lateral branches, 3.5–6.5 cm. long, and on very short spurs (up to 3 mm. long) near the apex of the lateral branches, sessile, crowded, ± 4.5 cm. in diameter; inflorescence-branches suppressed; pedicels 2–3 mm. long, pubescent; bracteoles ± 1.5 mm. long, densely crowded at base of pedicels, densely covered with brown hairs within. Calyx-tube 1.25 mm. long, pubescent; limb-tube dark-brown when dry, 1 mm. long, glabrous; lobes mid-brown when dry, broadly ovate, 1 mm. long, rounded, glabrous. Corolla glabrous outside; tube narrowly cylindrical, 1 cm. long, 0.75 mm. wide at top, ± glabrous at throat; lobes ± oblong, 5 mm. long, 2.5 mm. wide, rounded. Placenta ovate, bearing one rather thick ovule. Fruit not known.

TANZANIA.   Lindi District: Mlinguru, 26 Dec. 1937, *Schlieben* 5794!
DISTR.   T 8; known only from this gathering
HAB.   Bushland; ± 150 m.

NOTE.   A fruiting specimen, *Forbes* s.n. from Mozambique, seems to be allied to this species, but differs in its glabrous pedicels.

11. T. peteri *Bridson* in K.B. 34: 390 fig. 6 (1979). Type: Tanzania, Kilosa District, R. Rovuma, W. of Kidete, *Peter* 32674 (K, holo.!, B, iso.)

Shrub 1.5–3 m. tall; young stems pubescent, older stems terete, covered with dull greyish-buff flaking bark. Leaves immature at time of flowering, usually borne at apex of short lateral branches; mature blades drying bluish grey or olive-green, elliptic to broadly elliptic, up to 6.5 cm. long, 4 cm. wide, obtuse at apex, acute or obtuse at base, chartaceous to subcoriaceous, somewhat shiny and sparsely pubescent above, pubescent beneath;

domatia absent; petiole up to 1 cm. long, sparsely pubescent; stipules brown, triangular-lanceolate, 4–5 mm. long, apiculate, caducous. Inflorescence sessile, terminal on leafless (or occasionally with immature leaves from the axils) lateral branches, 2.5–4.8 cm. long, many-flowered; primary inflorescence-branches 2–8 mm. long (or up to 1.2 cm. long in fruiting stage), pubescent; pedicels reduced (or up to 4 mm. long in fruiting stage), pubescent; pedicels reduced (or up to 4 mm. long in fruiting stage); bracteoles brown when dry, ovate, ± 0.75 mm. long, sparsely pubescent outside, pubescent inside. Calyx glabrous; tube drying blue-black, 1 mm. long; limb 0.75–1 mm. long, divided almost to the base into overlapping, ± oblong lobes, dark brown with a pale margin or into well-spaced ovate lobes which are entirely light brown, ciliate. Corolla yellow-green in bud, glabrous outside; tube shortly cylindrical, 2 mm. long, 1.5 mm. wide at top, sparsely hairy at throat; lobes oblong, 3 mm. long, 1.25 mm. wide, rounded. Placenta moderately small, bearing 3–7 ovules which exceed it. Fruit black when dry, 7.5 mm. in diameter, 1-locular, crowned by persistent calyx-limb, 1-seeded. Seeds blackish, suborbicular, 5 mm. in diameter, reticulate, with hilar cavity small and somewhat irregular; endosperm somewhat ruminate.

TANZANIA. Mpwapwa District: Kongwa Pasture Research Station, 10 June 1967, *Wigg* 66!; Kilosa District: W. of Kidete, 2 Dec. 1925, *Peter* 32674!; Iringa District: NE. Kirimatonge Hill, *Greenway & Kanuri* 14799!
DISTR. T 5–7; known only from above cited specimens
HAB. Deciduous bushland; 650–1035 m.

NOTE. The two flowering specimens differ with regard to the calyx-lobes; *Peter* 32674 has lobes which overlap, while *Greenway & Kanuri* 14799 (in bud) has calyx-lobes which are well spaced, in all other respects the two specimens are a close match. However, further gatherings could indicate that *Greenway & Kanuri* 14799 represents a distinct taxon.

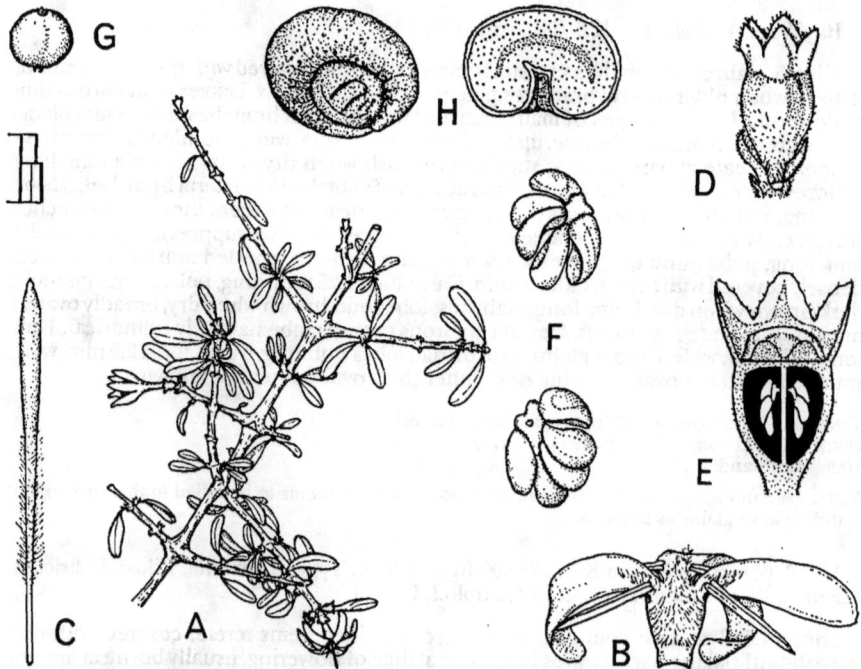

FIG. 101. *TARENNA KIBUWAE* — A, flowering shoot, × ⅔; B, section through corolla, × 3; C, style with stigma, × 4; D, calyx with bracteoles, × 8; E, section through calyx-tube and ovary, × 14; F, ovules on placenta (2 views), × 20; G, fruit, × 1; H, seed and section through seed, × 5. A–F, from *Gillett & Gachathi* 20598; G–H, from *Paulo* 473. Drawn by Diane Bridson.

12. T. sp. B; Bridson in K.B. 34: 390 (1979)

Shrub 6 m. tall; young stems green, ± square, sparsely pubescent, soon becoming covered with buff bark, rather rough on older branches. Leaves ? perhaps not fully mature at time of flowering, borne on new growth at apex of lateral branches and from axils of the older branches; blades dark bluish green when dry, ovate, 2.8–5.4 cm. long, 1.2–3.5 cm. wide, acute at apex, obtuse to rounded at base, chartaceous, pubescent on both faces; domatia absent; petioles up to 6 mm. long; stipule-limb triangular, 4 mm. long, corky when mature, bearing a caducous arista 6 mm. long. Inflorescences borne on short (± 2 cm. long) leafless lateral branches, forming ± spherical to ovoid heads, up to 4 cm. long, 3 cm. wide, not congested; inflorescence-branches up to 6 mm. long, pubescent; pedicels 2–3 mm. long, pubescent; bracteoles pale brown when dry, narrowly triangular, pubescent outside, present on the pedicels. Calyx 1.75 mm. long, glabrous; limb 0.75 mm. long, divided almost to base into well-spaced, brown ovate lobes, glabrescent. Corolla orange, black when dry, glabrous outside; tube ± cylindrical, 4 mm. long, 1.25 mm. wide at top, pubescent with long hairs at throat; lobes oblong, 7 mm. long, 2.5 mm. wide, rounded. Placenta moderately small, bearing 5 ovules which exceed it. Fruit not known.

TANZANIA. Mbulu District: Yaida Valley Escarpment, Endashali, 6.4 km. from the plain, 11 Jan. 1970, *Richards* 25073!
DISTR. T 2; known only from cited specimen
HAB. Not known; ± 1400 m.

13. T. kibuwae *Bridson* in K.B. 34: 395, fig. 7 (1979). Type: Kenya, Northern Frontier Province, Garissa–Dadaab road, 19 km. from Garissa, *Gillett & Gachathi* 20598 (K, holo.!, EA, iso.!)

Shrub 2–2.75 m. tall, glabrous; branches covered with pale, greyish bark. Leaves mostly borne on short or very short spurs off the main branches; blades green or blue-black when dry, elliptic to oblong-elliptic, 0.8–2.4 cm. long, 0.25–1.2 cm. wide, obtuse and occasionally mucronate at apex, cuneate at base, subcoriaceous, dull on both faces; lateral nerves obscure; domatia absent; petiole ± 0.75 mm. long; stipules pale brown when dry, triangular, 1(–1.5) mm. long. Inflorescences terminal on very short naked lateral spurs or less often on short leaf-bearing spurs, sessile, few-flowered; inflorescence-branches usually very reduced; pedicels up to 1 mm. long, pubescent; bracteoles brown when dry, small, ± triangular. Calyx sparsely puberulous; tube 1 mm. long; limb 0.75 mm. long, divided to ± halfway into broadly triangular lobes, paler at edges when dry. Corolla greenish cream, glabrous outside; tube funnel-shaped, 3.5 mm. long, 2.5 mm. wide at top, densely covered with long white hairs at throat; lobes (4–)5, ± oblong, 6 mm. long, 2.5 mm. wide, rounded. Placenta moderately small, bearing 5–8 ovules which exceed it in length. Fruit green (black when dry), 8 mm. in diameter, 1–2-locular, with 1 seed in each locule, crowned by persistent calyx-limb. Seeds blackish, suborbicular, 4 mm. in diameter, finely reticulate; hilar cavity circular or tending to be irregular; endosperm somewhat ruminate in vicinity of hilar cavity. Fig. 101.

KENYA. Northern Frontier Province: Garissa–Dadaab road, 19 km. from Garissa, 11 May 1974, *Gillett & Gachathi* 20598! & Korokora, N. bank of Garissa, 26 June 1960, *Paulo* 473!
DISTR. K 1; not known elsewhere
HAB. Deciduous bushland on sandy soil; 230–305 m.

14. T. wajirensis *Bridson* in K.B. 34: 398, fig. 8 (1979). Type: Kenya, Northern Frontier Province, Dadaab–Wajir road, 9 km. S. of Sabale, *Gillett & Gachathi* 2064 (K, holo.!, EA, iso.!)

Shrub 1–2 m. tall; very young branches finely pubescent, older branches covered with greyish bark. Leaves crowded on short spurs off the main branches or spaced on short lateral branches; blades olive-green or blue-black when dry, very narrowly elliptic, 1.5–3.5 cm. long, 0.5–1 cm. wide, obtuse and sometimes mucronate at apex, cuneate to obtuse at base, coriaceous, slightly shiny above, puberulous on both faces; lateral nerves ± apparent on both faces; domatia absent; petioles reduced, not exceeding 2 mm. long; stipules dark brown, triangular-ovate, 1.25–2 mm. long, puberulous outside. Inflorescences terminal on short leaf-bearing lateral branches, 0.4–1 cm. long, sessile, 5–25-flowered; primary inflorescence-branches 1–3 mm. long, finely pubescent, up to 1 mm. long; bracteoles brownish, small. Calyx finely pubescent; tube ± 1 mm. long; limb

± 1mm. long, divided to ± halfway into rounded lobes. Corolla white, glabrous outside; tube ± cylindrical, 4 mm. long, ± 1 mm. wide at top, pubescent at throat; lobes oblong, 4 mm. long, ± 1 mm. wide, rounded. Placenta ± ovate, bearing 2 impressed ovules. Fruit black when dry, 4.5 mm. in diameter, glabrescent, 1-locular, 1-seeded; calyx-limb persistent. Seeds not fully developed, but with a small circular hilar cavity surrounded by pale brown area.

KENYA. Northern Frontier Province: 40 km. N. of Wajir on road to Tarbaj, 19 May 1972, *Gillett* 19723! & Dadaab–Wajir road, 9 km. S. of Sabale, 2 May 1974, *Gillett & Gachathi* 2064!
DISTR. **K** 1; not known elsewhere
HAB. Deciduous bushland on red sand; 180–330 m.

15. T. graveolens (*S. Moore*) *Bremek.* in F.R. 37: 193 (1934); Chiov., Racc. Bot. Miss. Consol. Kenya: 56 (1935); T.S.K.: 137 (1936); T.T.C.L.: 532 (1949), pro parte; Cufod. in Phyton 1: 141 & 142 (1949); I.T.U., ed. 2: 357 (1952); K.T.S.: 475 (1961), pro parte; Bridson in K.B. 34: 394, fig. 3/J, K (1979); Fl. Pl. Lign. Rwanda: 603, fig. 205/1 (1982); Fl. Rwanda 3: 224, fig. 71/1 (1985). Type: Uganda, Toro District, near mouth of Mpanga R., *Bagshawe* 1173 (BM, holo.!)

Shrub or small tree, 1.8–7(–8.5) m. tall; young branches usually covered with grey to reddish bark, sometimes flaking when older. Leaf-blades yellowish green or blue-black when dry, narrowly elliptic to round, 3–13.5 cm. long, 1.2–6.3 cm. wide, obtuse or acute to acuminate at apex, cuneate at base, coriaceous, usually shiny above, glabrous or sometimes very finely pubescent to puberulous above and beneath; tertiary nerves raised on both surfaces; domatia absent; petiole 0.8–3 cm. long, glabrous or less often puberulous; stipules usually drying brownish, ovate, 3–7 mm. long, ± papery, caducous. Inflorescences terminal on main and less often short lateral branches, sessile, congested to moderately lax, occasionally with an additional pair of inflorescences arising from the node below; primary inflorescence-branches 0.5–1.5 cm. long, finely pubescent; pedicels 0–3 mm. long; bracteoles black or light brown, subulate or linear, up to 3 mm. long, usually present at base of calyx-tube. Calyx puberulous to finely pubescent; tube 1–1.5 mm. long; limb 0.25–1.5 mm. long, divided from half to almost its entire length into rounded lobes. Corolla white, cream or yellow; tube ± cylindrical, (1.75–)3–6 mm. long, glabrous outside, pubescent at throat; lobes oblong, (3–)4–5.5 mm. long, 1–3 mm. wide. Placenta ovate, bearing (1–)2–3(–6) impressed ovules. Fruit black, 4–7 mm. in diameter, (1–)2-locular, containing 1(–2) seeds per fruit; calyx-limb persistent. Seeds blackish or brown, ± spherical or sometimes compressed, 3.5–4.5 mm. in diameter, finely reticulate; hilar cavity circular, often with the surrounding area paler in colour; endosperm entire.

var. **graveolens**

Leaf-blades entirely glabrous, usually shiny above.

UGANDA. Karamoja District: Kidepo National Park, 14 May 1972, *Synnott* 972!; Bunyoro District: Bukumi Escarpment, 9 Apr. 1950, *Dawkins* 565!; Busoga District: Siavona Hill, 26 Mar. 1953, *G.H. Wood* 668!
KENYA. Naivasha District: Lake Naivasha, 24 Oct. 1966, *E. Polhill* 206!; Nairobi District: Karura Forest, May 1939, *Bally* in *C.M.* 9216!; Teita District: between Manyani and Voi, 1 Dec. 1959, *Greenway* 9625!
TANZANIA. Mwanza District: Nyegezi [Nyegese], 8 Mar. 1965, *Leippert* 5620!; Tanga District: Sawa, 21 Apr. 1970, *Faulkner* 4356!; Dodoma District: Manyoni kopje, 28 Dec. 1931, *B.D. Burtt* 3457!
DISTR. **U** 1–4, **K** 1–7; **T** 1–6; Zaire (Kivu), Rwanda, Burundi, Ethiopia, Somalia
HAB. Thicket and bushland, frequently among rocks; 0–2130 m.

SYN. [*Chomelia nigrescens* sensu K. Schum. in P.O.A. C: 380 (1895), pro parte, *non* (Hook.f.) Kuntze]
    Pavetta graveolens S. Moore in J.B. 45: 267 (1907); T.S.K.: 149 (1936)
    P. ligustriodora Chiov. in Result. Sci. Miss. Stef.-Paoli, Coll. Bot.: 96 (1916). Types: Somalia, *Paoli* 306, 486, 725 & 1154 (FT, syn.!)
    Tarenna edgardii Chiov., Racc. Bot. Mis. Consol. Kenya: 57 (1935), as "*edgardi*", nomen eventuale, & in Miss. Biol. Borana, Racc. Bot.: 227 (1939); Cufod. in Phyton 1: 141 (1949). Types: Ethiopia, Neghelli, *Zavattari* 209 (FT, syn.!) & Arero, *Zavattari* 356 (FT, syn.!)
    T. boranensis Cufod. subsp. *boranensis* in Nuov. Giorn. Bot. Ital., n.s. 55: 86 (1948) & in Phyton 1: 140, fig. 2/C (1949); K.T.S.: 474 (1961), pro parte, quoad *Hildebrandt* 2755. Types: Ethiopia, *Corradi*, several collections (FT, syn.!); Kenya, Kitui, *Hildebrandt* 2755 (WU, syn., BM!, K!, P!, W, isosyn.); Tanzania, Lake Victoria, *Conrads* 287 (WU, syn.) & Usambara Mts., Mashewa [Maschaua], *Holst* 3562 (W, syn., K, isosyn.!)

NOTE. This variety is polymorphic especially with regard to the size and shape of the leaf-blades, the size of the flowers, the degree of laxness of the inflorescence and the length of the bracteoles.

var. **impolita** *Bridson* in K.B. 34: 395 (1979). Type: Kenya, Northern Frontier Province, South Horr, *J. Adamson* 25 in *Bally* 3573 (K, holo.!)

Leaf-blades minutely puberulous on both faces or very sparsely so, dull above.

UGANDA. Karamoja District: Moroto, Feb. 1936, *Eggeling* 2974! & Lodoketemit [Lodoketeminit], 6 Nov. 1962, *Kerfoot* 4423a!
KENYA. Northern Frontier Province: Mathews Range, Olkanto, 10 Dec. 1944, *J. Adamson* 39 in *Bally* 4340!; Turkana District: Kaya Ejon, 29 Jan. 1965, *Newbould* 6848!; Baringo District: W. side of Lake Bogoria [Hannington], 19 Jan. 1969, *Faden & Napper* 69/67!
DISTR. U 1; K 1–4, 7; not known elsewhere
HAB. Stony and rocky hillsides or dry bushland; 930–1370 m.

NOTE. Several specimens from K 1, K 2 & K 4 (*Mathenge* 138 from Meru Game Reserve, Tana River) have leaf-blades which are less shiny above than is usual for var. *graveolens* and tend to this variety. *Eggeling* 2974, has the following field note — "stem and trunk spinose, spines on trunk becoming woody in age; occasionally totally absent".
     Subsp. *arabica* (Cufod.) Bridson, is known from the Yemen and N. Somalia; it differs from subsp. *graveolens* in its leaf-blades which are always narrowly elliptic, and dull to slightly shiny above and glabrous to minutely puberulous, in its inflorescences which are usually smaller than those of subsp. *graveolens* and in its calyx-lobes which are acute to ± rounded.

16. T. neurophylla (*S. Moore*) *Bremek.* in F.R. 37: 200 (1934); T.T.C.L.: 533 (1949), pro parte; Garcia in Mem. Junta Invest. Ultram., sér. Bot., 4: 26 (1958); F.F.N.R.: 421 (1962); Bridson in K.B. 34: 397, figs. 2D & 3L–N (1979). Type: Zimbabwe, near Bulawayo, *Eyles* 1140 (BM, holo.!, K, iso.!)

Shrub or tree 1.8–7(–9) m. tall, glabrous; young branches ± square, with grey or sometimes red-brown bark. Leaf-blades yellowish green to blue-black when dry, narrowly elliptic to elliptic or occasionally very narrowly elliptic, 4.2–11 cm. long, 0.8–4.3(–5.5) cm. wide, acute to acuminate at apex, cuneate at base, coriaceous, shiny above; tertiary nerves raised on both faces; domatia absent; petiole 0.5–1.5 cm. long; stipules triangular, 2–3 mm. long with a short arista, caducous. Inflorescences terminal, sessile or with a peduncle up to 0.8 cm. long, ± congested, sometimes with additional pairs of inflorescences arising from the node below; primary inflorescence-branches 0.3–2 cm. long, glabrous to pubescent; pedicels 0.3(–6) mm. long; bracteoles inconspicuous, scale-like to triangular or subulate, usually ciliate, usually present at the base of the calyx-tube. Calyx-tube (0.5–)0.75–1 mm. long, glabrous; limb often drying brownish or greenish in contrast to tube, divided nearly to base into rounded lobes, ciliate. Corolla whitish or yellowish green; tube narrowly funnel-shaped to cylindrical, 4–5 mm. long, glabrous outside, sparsely pubescent at throat within; lobes ± oblong, 4–6 mm. long, 1–1.75 mm. wide, rounded. Placenta ± ⅓ the size of the ovules or sometimes equalling them; ovules 2(–3) pendulous. Fruit black when ripe, 5–7 mm. in diameter, 1-locular, 1-seeded; calyx-limb persistent. Seed blackish, spherical, 3.5–4.5 mm. in diameter, finely reticulate; hilar cavity basal, small, somewhat irregular; endosperm somewhat ruminate.

TANZANIA. Kigoma District: Kabogo Mts., 27 Nov. 1962, *Kyoto University Expedition* 301!; Mpwapwa District: Kiboriani Mt., 15 Oct. 1938, *Hornby* 919!; N. Kilosa District, Jan. 1931, *Haarer* 1962!
DISTR. T 4–8; Zaire (Shaba); Mozambique, Malawi, Zambia and Zimbabwe
HAB. *Brachystegia* woodland; 760–1500 m.

SYN. [*Chomelia nigrescens* sensu De Wild. in Ann. Mus. Congo, Bot. sér. 4, 1: 226 (1903), quoad *Verdick* 266, non *C. nigrescens* (Hook.f.) Kuntze]
     Pavetta neurophylla S. Moore in J.B. 43: 47 (1905)
     Chomelia odora K. Krause in E.J. 57: 29 (1920). Type: Tanzania, Rungwe District, Kyimbila, *Stolz* 168 (B, holo.†, K, iso.!)
     Stylocoryne neurophylla (S. Moore) Bremek. in Ann. Transv. Mus. 13: 215 (1929)
     Tarenna odora (K. Krause) Cufod. in Phyton 1: 142 (1949); T.T.C.L.: 533 (1949) (but authority given as (K. Krause) Bremek.)

17. T. supra-axillaris (*Hemsley*) *Bremek.* in F.R. 37: 206 (1934); Bridson in K.B. 34: 397, fig. 3P, Q (1979). Type: Aldabra, Ile Esprit, *Fryer* (K, holo.!)

Shrub or small tree, 1.5–6.7 m. tall, glabrous; very young branches ± terete, covered with reddish-brown or light grey to fawn bark, becoming grey when older, scarcely flaking. Leaf-blades blue-black or less often yellow-green when dry, very narrowly elliptic to elliptic, 3–12.3 cm. long, 0.7–4.2 cm. wide, acute to subacuminate at apex, cuneate at base, coriaceous, shiny above; tertiary nerves raised above, scarcely raised beneath; domatia absent; petiole 0.3–1.1 cm. long; stipules turning black when dry, narrowly ovate to ovate or sometimes triangular, 2–5 mm. long, eventually caducous, sometimes with resinous deposits. Inflorescences axillary or occasionally terminal on the lateral branches, trichotomous; peduncle 0.7–2 cm. long; primary inflorescence-branches 2–5 mm. long; bracts scale-like or with triangular lobes; bracteoles ± cupular, somewhat irregularly lobed, ciliate, always surrounding the base of the calyx-tube. Calyx glabrous to minutely puberulous; tube 0.75–1.25 mm. long; limb 0.75–1 mm. long, divided to ± halfway or to near base into ovate lobes, rounded or sometimes slightly emarginate, ciliate. Corolla whitish; tube ± cylindrical to narrowly funnel-shaped, 2.5–5 mm. long, glabrous outside, sparsely to densely pubescent at throat; lobes ± oblong, 2–3.5 mm. long, 1–1.5 mm. wide, rounded or sometimes emarginate, minutely puberulous above and beneath. Placenta usually equalling the ovule or sometimes much smaller, bearing one ovule impressed to one side. Fruit black, 4–5 mm. in diameter, 1-locular, 1-seeded; calyx-limb persistent; pedicels somewhat accrescent, reaching 7 mm. long. Seed dark brown, 2–3 mm. in diameter, finely reticulate; hilar cavity irregular; endosperm ruminate towards base.

subsp. supra-axillaris

Very young branches with reddish brown bark; leaf-blades narrowly elliptic to elliptic, 1.2–3.4 cm. wide.

KENYA. Kwale District: Shimba Hills, Giriama Point, 31 Mar. 1968, *Magogo & Glover* 590/B!; Kilifi District: Arabuko, May 1929, *R.M. Graham* in *F.D.* 2129! & Malindi, May 1960, *Rawlins* 855!
TANZANIA. Uzaramo District: 96 km. W. of Dar es Salaam, 17 Aug. 1969, *Harris et al.* 3134!; Rufiji District: Mafia I., Utmaini (?Utuinaini), 3 Oct. 1937, *Greenway* 5382!; Pemba I., Ras Mkumbuu, 12 Dec. 1930, *Greenway* 2770!
DISTR. K 7; T 6; Z; P; Mozambique, Zimbabwe, Aldabra
HAB. In forest and also inner borders of mangrove swamps; 0–910 m.

SYN. *Pavetta supra-axillaris* Hemsley in J.B. 54, Suppl. 2: 19 (1916)
[*Tarenna graveolens* sensu K.T.S.: 475 (1961), pro parte, quoad *R.M. Graham* in *F.D.* 2129, non (S. Moore) Bremek.]

NOTE. Subsp. *barbertonensis* (Bremek.) Bridson is restricted to South Africa (Transvaal and Natal); it differs from subsp. *supra-axillaris* in having pale grey to fawn bark on the young branches and somewhat narrower leaf-blades.

18. T. littoralis (*Hiern*) *Bridson* in K.B. 34: 397, fig. 3R, S (1979). Type: Mozambique, Luame R., *Kirk* (K, lecto.!)

Shrub or small tree 1–9.6 m. tall, glabrous; young branches square, covered with off-white to very pale grey bark. Leaf-blades ovate to sometimes obovate, occasionally oblanceolate to narrowly oblanceolate, 3–10 cm. long, 1.3–6 cm. wide, rounded to obtuse at apex, cuneate at base, coriaceous, dark green and shiny above, paler beneath; tertiary nerves not raised above; domatia as conspicuous depressions lined with hairs; petiole 0.4–1.4 cm. long; stipules brownish when dry, ovate to triangular, 0.5–1 cm. long, caducous. Inflorescences terminal, usually sessile, forming a congested corymbose panicle; primary inflorescence-branches 1.2–2.5 cm. long; pedicels absent or very short; bracteoles small, scale-like to shortly subulate, always present at the base of the calyx-tube. Calyx-tube not exceeding 0.75(–1) mm. long; limb up to 0.5 mm. long, divided almost to base into rounded teeth. Corolla whitish; tube narrowly funnel-shaped, 3–4 mm. long, glabrous outside, sparsely hairy inside; lobes ± oblong, 1.75–3 mm. long, ± 1 mm. wide, obtuse. Placenta small, bearing 2–3 pendulous ovules. Fruit black when ripe, spherical or slightly longer than wide, 5–7 mm. long, 4–6 mm. wide, 1-locular, 1-seeded; calyx-limb persistent. Seed blackish brown, spherical, 4 mm. in diameter, finely reticulate; hilar cavity small, irregular; endosperm deeply ruminate.

KENYA. Kwale District: Vanga area, Ngowa [Ngoa], Nov. 1929, *R.M. Graham* in *F.D.* 2209!
TANZANIA. Tanga District: Kigombe beach, 12 July 1953, *Drummond & Hemsley* 3261!; Uzaramo District: Wazo Hill, 26 Apr. 1968, *Harris* 1654!; Zanzibar I., Chukwani, 7 May 1960, *Faulkner* 2550!
DISTR. K 7; T 3, 6; Z; Mozambique, Zimbabwe, South Africa (Natal)
HAB. Coastal bushland or foreshore; 0–105 m.

SYN. *Enterospermum littorale* Hiern in F.T.A. 3: 92 (1877); Oliv. in Hook., Ic. Pl. 13, t. 1269 (1878); P.O.A. C: 380 (1895); T.T.C.L.: 494 (1949); Garcia in Mem. Junta Invest. Ultram. sér. Bot. 4: 28 (1958); K.T.S.: 439 (1961); Ross, Fl. Natal: 333 (1973)
E. sp. sensu Garcia in Mem. Junta Invest., Ultram. sér. Bot., 6: 21 (1959)

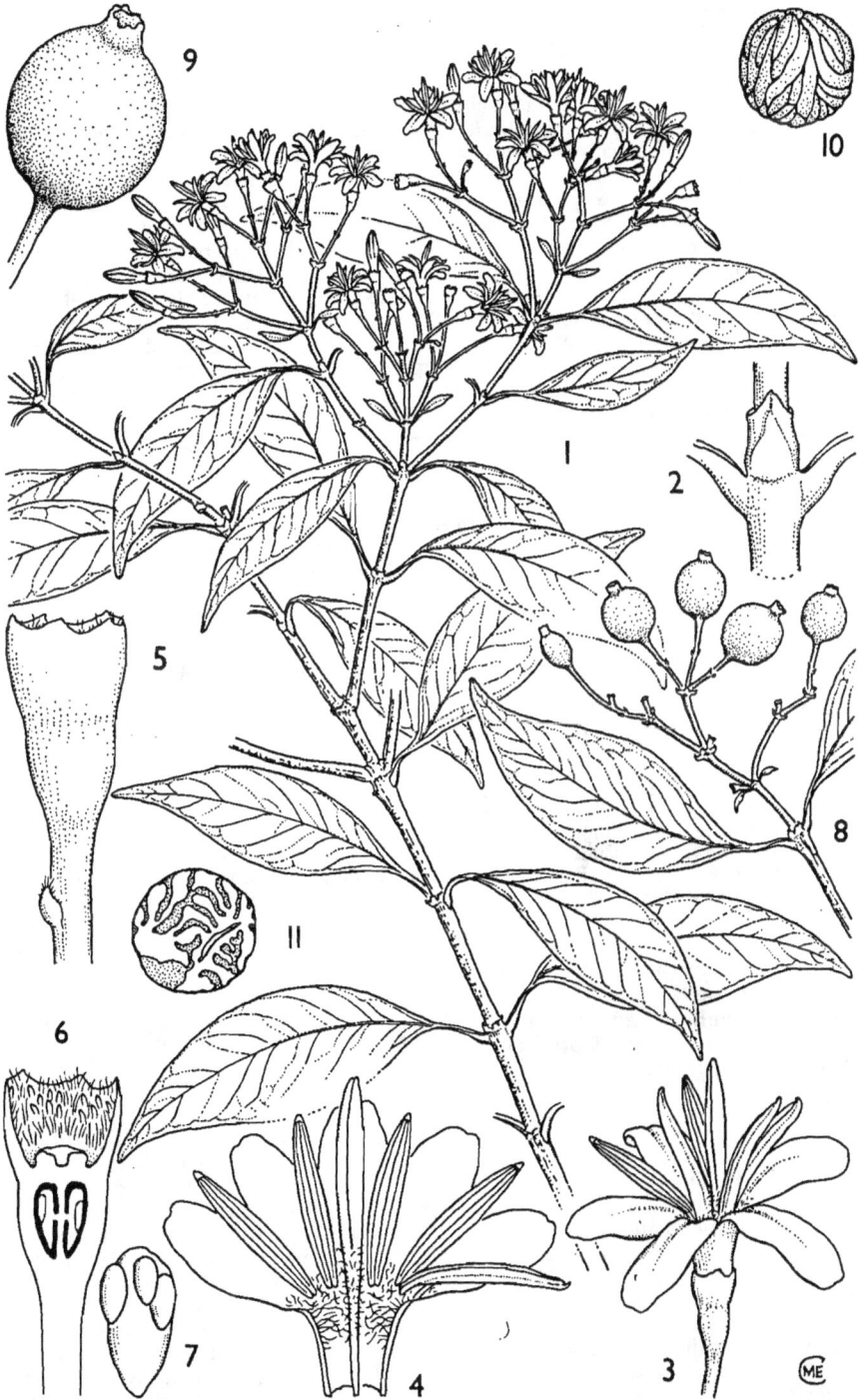

FIG. 102. *TARENNA NIGRESCENS* — **1**, flowering shoot, × ⅔; **2**, stipules, × 4; **3**, flower, × 3; **4**, corolla opened out, × 4; **5**, calyx, × 8; **6**, section through calyx and ovary, ×: 8; **7**, placenta with ovules, × 16; **8**, fruiting branch; × 1; **9**, fruit, × 3; **10**, seed, × 4; **11**, section through seed, × 4. 1–7, from *Dale* 3670; 8–11, from *Hawthorne* 948. Drawn by Mrs M.E. Church.

19. T. nigrescens (*Hook.f.*) *Hiern* in F.T.A. 3: 92 (1877); T.T.C.L.: 533 (1949), pro parte; K.T.S.: 475 (1961); Bridson in K.B. 34: 400, fig. 2E & 3T, U (1979). Type: Mozambique, Moramballa, *Kirk* (K, lecto.!)*

Shrub, small tree or climber, 3–6 m. tall, glabrous; young branches terete, covered with pale to dark reddish-brown shiny bark, sometimes becoming greyish with age, always flaking. Leaf-blades blackish when dry, narrowly elliptic to elliptic or lanceolate to narrowly ovate, 3.2–10.8 cm. long, 1–3.7 cm. wide, acute to acuminate at apex, cuneate or acute at base, coriaceous, very shiny above; tertiary nerves ± raised on both faces; domatia absent; petiole 0.5–1 cm. long; stipules inconspicuous, deltoid, 1.5–2.5 mm. long, obtuse, bearded within, readily caducous. Inflorescences terminal, sessile or with a peduncle up to 1.9 cm. long, often supported by rudimentary leaves; primary branches 0.6–1.4 cm. long; pedicels 0.3–1.9 cm. long; bracteoles scale-like, always 2 in middle of pedicel. Calyx glabrous or minutely puberulous; tube 1.5–2 mm. long; limb 1.25–2 mm. long, truncate to repand or sometimes with short teeth, ciliate. Corolla whitish or yellow; tube ± cylindrical, (1.5–)3–5 mm. long, 1.5–2 mm. wide at top, glabrous outside, pubescent at throat; lobes oblong, 5–6 mm. long, 1.75–2 mm. wide, obtuse, sometimes auriculated at base. Placenta ovate with usually 3 ovules impressed on the upper part. Fruit green, 5–8 mm. in diameter, 1-locular, 1-seeded; calyx-limb persistent. Seed blackish brown, spherical, 5 mm. in diameter, finely rugulose; hilar cavity inconspicuous; endosperm deeply ruminate. Fig. 102.

KENYA. Kwale District: Dzombo Mt., 8 Apr. 1968, *Magogo & Glover* 799! & Diani Forest, 11–13 July 1972, *Gillett & Kibuwa* 19857!; Kilifi District: Mida, Jan. 1937, *Dale* in *F.D.* 3670!
TANZANIA. Pangani District: Msubugwe Forest, 14 Mar. 1963, *Mgaza* 568!; Rufiji District: Kibiti, 19 Dec. 1968, *Shabani* 265!: Kilwa District: Selous Game Reserve, Kijawe, 2 Jan. 1970, *Ludanga* 1195!
DISTR. K 7; T 3, 6, 8; Somalia, Mozambique, Swaziland, Madagascar and Comoro Is.
HAB. Forest edges or coastal bushland; 0–350 m.

SYN. *Coptosperma nigrescens* Hook.f. in G.P. 2: 87 (1873); Garcia in Mem. Junta Invest. Ultram., sér. Bot., 4: 27 (1958)
Ixora pruinosa Baill., Adansonia 12: 217 (1878); Drake, Hist. Pl. Madag. 36, Atlas 4 tom. 6, t. 419 (1897). Type: Comoro Is., Mohilla, *Boivin* (P, syn.) & Mayotte, *Boivin* (P, syn., K, ?isosyn.! (3181 & 2 sheets s.n.))
Chomelia nigrescens (Hook.f.) Kuntze, Rev. Gen. Pl. 1: 278 (1891); K. Schum. in Engl., Abh. Preuss. Akad. Wiss.: 36 (1894), pro parte & in P.O.A. C: 38 (1895), pro parte; De Wild. in Ann. Mus. Congo, Bot. sér. 4, 1: 226 (1903), pro parte
Pavetta saligna S. Moore in J.L.S. 40: 98 (1911). Types: Mozambique, Boka, *Swynnerton* 1262 (BM, syn.!) & Lower Buzi, Chirinda, *Swynnerton* 1263 (BM, syn.!, K, isosyn.!)
Enterospermum pruinosum (Baill.) Dubard & Dop. in Journ. de Bot., sér. 2, 3; 13 (1925)
Coptosperma madagascariensis Garcia in Mem. Junta Invest. Ultram., sér. Bot. 4: 28 (1958), *nom. invalid*, based on Madagascan element of *C. nigrescens* Hook.f.

NOTE. The plate shown in Drake, Hist. Pl. Madag. 36, Atlas 4 tom. 6, pl. 420 (1897) under the name *Ixora nigrescens*, although a *Tarenna*, is nothing to do with this species.

20. T. trichantha (*Bak.*) *Bremek.* in F.R. 37: 207 (1934); Bridson in K.B. 34: 401, figs. 2F & 3V, W (1979); Bridson & Robbrecht in B.J.B.B. 55: 102, figs. 7–9 (1985). Type: Aldabra I., *Abbott* 23/26 (K, holo.!)

Shrub or small tree 1.2–4.5 m. tall; stems ± terete, frequently gnarled, sometimes pubescent when young, covered with grey or buff bark, readily peeling. Leaf-blades drying greyish or sometimes blue-black or brownish, usually discolorous, elliptic to round or ovate to broadly ovate, 2.8–9.9 cm. long, 1.6–6 cm. wide, obtuse to rounded or less often acute to shortly acuminate and usually mucronulate at apex, acute to obtuse or less often truncate or rounded and sometimes unequal at the base, glabrous to shortly pubescent above, glabrous with hairs restricted to nerves or pubescent beneath; domatia present as tufts of whitish hairs; petiole 0.7–1.6 cm. long, glabrescent to pubescent; stipules triangular, 1.5–3 mm. long, shortly apiculate, glabrescent to pubescent outside, caducous. Inflorescences terminal on short lateral branches and sometimes on the main branches, sessile, occasionally with an additional pair of branches arising from the next node down, often congested; primary branches 0.3–1.6 cm. long, sparsely to densely pubescent; pedicels absent or up to 3 mm. long; bracteoles filiform to subulate, usually surrounding

* Hiern based his combination on the flowering specimen *Kirk* s.n. from Mozambique and excluded the fruiting specimens from Madagascar that were cited by Hooker. Hooker also included the Seychelles in his distribution, but *T. nigrescens* has not been collected there.

the base of the calyx-tube, pubescent. Flowers (4–)5-merous. Calyx pubescent; tube 1–1.25 mm. long; limb 0.75–1.25 mm. long, equalling the tube in width, divided almost to base into triangular to sublinear lobes. Corolla white, salver-shaped, densely covered with adpressed hairs outside; tube 4–5 mm. long, 0.75–1 mm. wide at top, pubescent within; lobes ± oblong, 2 mm. long, 1–1.5 mm. wide, rounded, nerves conspicuous. Anthers oblong, only partly exserted. Placenta small, bearing 2–3 pendulous ovules. Fruit green, 4.5 mm. in diameter, sparsely covered with short hairs, 2-locular, usually 2–4-seeded; calyx-limb persistent; pedicels tending to lengthen. Seeds reddish brown, 2–3 mm. across, with a shallow hilar groove, smooth.

KENYA. Lamu District: Kitwa Pembe Hill, 15–16 July 1974, *Faden* 74/1104! & Pate I., July/Aug. 1980, *Marquis* P!
TANZANIA. Uzaramo District: Wazo Hill, 3 Aug. 1969, *Harris & Tadros* 3038! & 1 Aug. 1969, *Harris & Tadros* 3051!
DISTR. K 7; T 6; Mozambique, Comoro Is., Aldabra I.
HAB. Coastal thicket or sand-dunes; 0–76 m.

SYN. *Pavetta trichantha* Bak. in K.B. 1894: 148 (1894)

NOTE. The leaves vary from glabrous, save for the nerves beneath, to pubescent on both faces. Varieties have not been recognised as some specimens are intermediate.

Species of unknown position

T. mossambicensis *Hiern* in F.T.A. 3: 89 (1877). Type: Mozambique, Tete, *Peters* (B, holo. †)

Chomelia mossambicensis (Hiern) Kuntze, Rev. Gen. Pl. 1: 278 (1891); K. Schum. in P.O.A. C: 379 (1895)

It is possible that this is conspecific with *T. trichantha* (Bak.) Bremek., but in the absence of the type its position is best left uncertain.

76. CLADOCERAS

Bremek. in Hook., Ic. Pl. 35, t. 3411 (1940); Robbrecht & Bridson in B.S.B.B. 117: 247–251, figs. 1, 2, (1984)

Scandent shrub with short spreading axillary branchlets, some of which are reduced to short slightly to quite strongly recurved opposite spines. Leaves opposite, shortly petiolate, with elliptic blades having acarodomatia in the axils of the nerves; stipules simple, connate into a short sheath below, thickened above. Flowers ⚥, strongly scented, 5-merous, in dense corymbs or subcapitate at the ends of short axillary 1–few-noded branchlets, supported by a pair of normal leaves, reduced leaves or bracts. Calyx-tube ovoid, with very short free limb and short ovate-triangular lobes. Corolla white, glabrous outside; tube narrowly cylindrical with short sparse hairs inside; lobes elliptic to lanceolate, with contorted aestivation. Stamens with very short filaments, inserted at the throat; anthers exserted. Disc annular. Style filiform, glabrous; stigma-lobes lanceolate, flattened, adhering together, not exserted, with stigmatic papillae only on the outside, never in contact with the anthers in bud. Ovary 2-locular, each locule with 4–5 ovules immersed near the margins of fleshy placentas affixed to the middle of the septum. Fruit globose, fleshy with a thin endocarp, crowned by the persistent calyx-limb. Seeds few in each locule.

A monotypic genus confined to the coastal regions of East Africa and of very characteristic appearance.

Owing to the absence of the ixoroid pollination mechanism, Bremekamp excluded it from *Tarenna* which it much resembles in placentation; he also excluded it from the Gardenieae for the same reason, but I cannot place such paramount importance on this character; reduced forms of it could be evolved very easily. In his 1966 paper he was still uncertain of its position but still considered it closely related to *Posoqueria* which is certainly not correct.

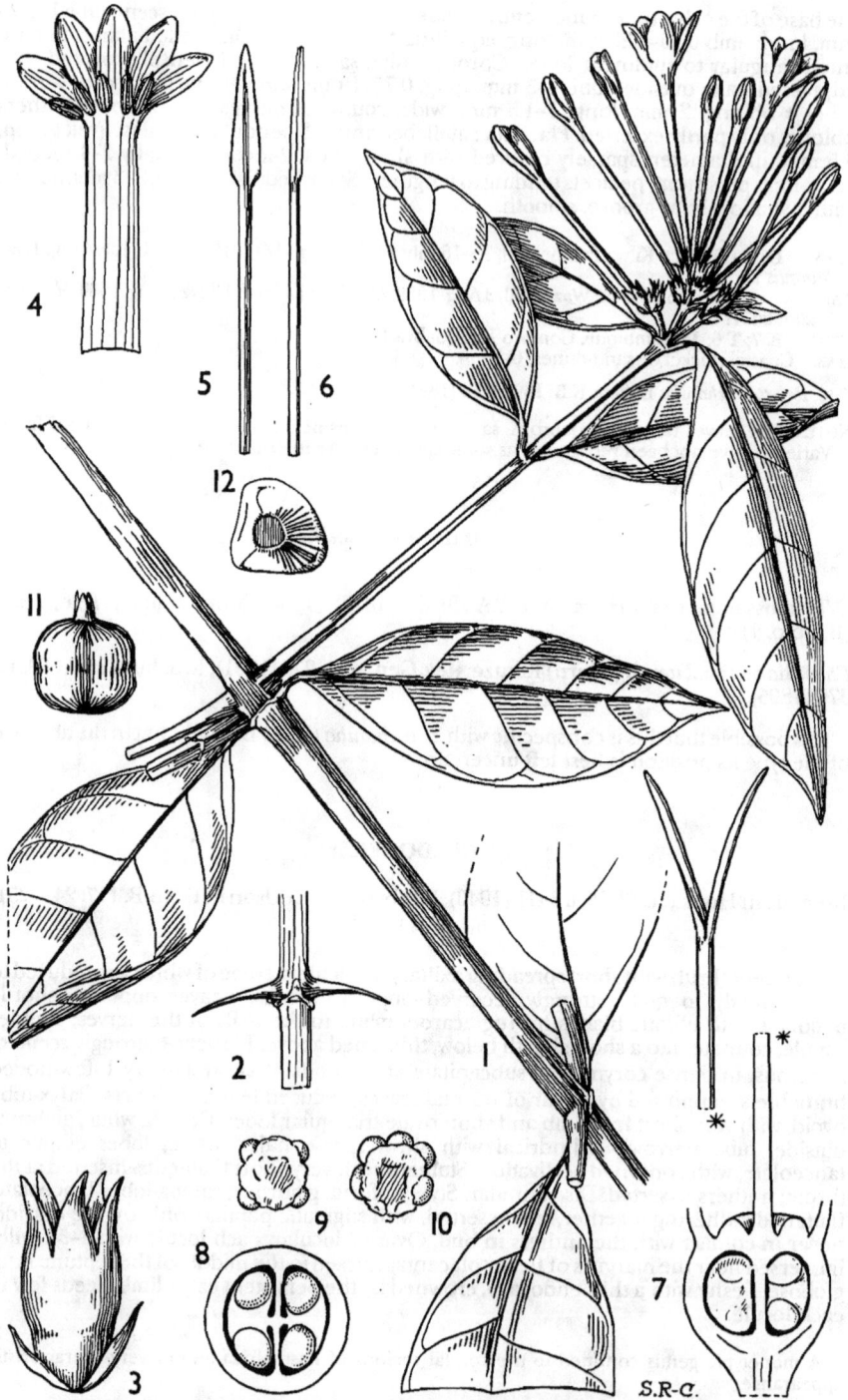

FIG. 103. *CLADOCERAS SUBCAPITATUM* — **1**, flowering branch, × 1; **2**, node showing spines and stipule, × 1; **3**, calyx with bracteoles, × 6; **4**, corolla opened out, × 2; **5**, style with stigma, dorsal view, × 6; **6**, same, lateral view, × 6; **7**, longitudinal section of gynoecium with stigmatic arms separated, × 8; **8**, transverse section of ovary, × 8; **9**, placenta with 4 ovules, × 12; **10**, placenta with 5 ovules × 12; **11**, fruit, × 2; **12**, seeds, × 4. 1–10, from *Greenway* 5392; 11, 12, from *Harris* 4464. Drawn by Stella Ross-Craig, with 11, 12, by Sally Dawson.

C. subcapitatum (*K. Schum. & K. Krause*) *Bremek.* in Hook., Ic. Pl. 35, t. 3411 (1940); T.T.C.L.: 491 (1949); Verdc. in K.B. 11: 449 (1957). Types: Tanzania, Rufiji District, Mafia I., *Busse* 426 (B, syn. †, EA, K, isosyn.!) & Uzaramo District, near Dar es Salaam, Mogo Forest [Sachsenwald], *Engler* 3241 & *Stuhlmann* 155 (B, syn. †)

Scandent or ? erect glabrous shrub to ± 2 m. tall, the stems quadrangular, with very shallow wings when young. Leaf-blades mostly elliptic or somewhat oblong-oblanceolate, (3.5–)4.5–12 cm. long, (0.9–)1.8–3.5 cm. wide, acute or shortly acuminate at the apex, cuneate at the base, slightly shining, scarcely coriaceous; petioles 2–4.5 mm. long; stipules 2.5–3.5 mm. long. Flowering branchlets 3–13 cm. long, the heads 9–15-flowered; peduncles 2–8(–?17) mm. long; pedicels very short or obsolete; bracteoles small. Calyx-tube 1.4 mm. long; limb 0.5 mm. long; lobes 2.5 mm. long. Corolla-tube 2.5–3.5 cm. long, ± 1 mm. wide; lobes 7 mm. long, 2–2.8 mm. wide, obtuse. Anthers 2–2.5 mm. long. Style 5.5–7.5 mm. long; stigma 2.5 mm. long. Fruits 8 mm. in diameter. Seeds yellow-brown, angular, 3–4 in each locule, 3.75 mm. long, with circular excavation, smooth and glossy. Fig. 103.

KENYA. Kilifi District: Rabai, Dec. 1933, *Joanna* in C.M. 5952!
TANZANIA. Lushoto District: near Amani, 13 May 1914, *Peter* 55515!; Uzaramo District: Pugu Hills, 10 Dec. 1939, *Vaughan* 2927!; Rufiji District: Mafia I., near Ndagoni, 4 Oct. 1937, *Greenway* 5392!
DISTR. K 7; T 3, 6; not known elsewhere
HAB. Evergreen forest and bushland; 0–900 m.
SYN. *Chomelia subcapitata* K. Schum. & K. Krause in E.J. 39: 525 (1907)

77. RUTIDEA

DC. in Ann. Mus. Hist. Nat. Paris 9: 219 (1807); G.P. 2: 116 (1873); Bridson in K.B. 33: 243–278 (1978)

Scandent shrubs, often with hairy stems. Leaves petiolate; domatia absent or present as hairy tufts in the axils of the lateral and sometimes tertiary nerves beneath (referred to in text as domatial hairy tufts); bacterial nodules absent; stipules, with 3–15 fimbriae or with a single lobe. Flowers ♂, 4–5(–6)-merous, sessile or pedicellate in pyramidal to racemiform or hemispherical to flat-topped panicles, terminal on the main or short lateral branches, sessile to pedunculate (the presence of rudimentary leaves frequently renders this character ambiguous); bracts and bracteoles present, either conspicuous and exceeding the calyx or inconspicuous. Calyx-tube ovoid to campanulate; limb-tube usually short, slightly wider than the tube, bearing 4–5(–6) ovate, subulate, filiform to linear or triangular lobes. Corolla white or cream, usually small, salver-shaped; tube cylindrical to funnel-shaped, glabrous to sparsely pubescent inside; lobes 4–5(–6), contorted in bud, spreading or reflexed. Stamens attached at the mouth of the tube, exserted, or at least partly so; filaments short; anthers attached near the base or occasionally near the middle, usually shortly apiculate. Disc fleshy, glabrous. Ovary 2(–3)-locular (or incompletely so); ovules solitary, inserted on a small fleshy placenta ascending from the base of the ovary; style slender, greatly or less often scarcely exceeding the corolla-tube, glabrous or nearly so or less often densely pubescent just above the base; stigma exserted or occasionally not fully exserted, fusiform, clavate or broadly elliptic to globose, entire or rarely 2–3-lobed. Fruit frequently orange to red, drupaceous, spherical, somewhat fleshy with a chartaceous endocarp, 1-locular; calyx-limb persistent or less often deciduous. Seed 1, globose, filling the cell, hilum basal, irregular; testa thin, reticulate; albumen horny, deeply ruminate; embryo curved.

A genus of about 22 species, confined to tropical Africa.

R. olenotricha Hiern a very distinctive and widespread species has been included in the key to species. It has been recorded from Zaire, Sudan and Zambia but not, so far, from the Flora area. The record for *R. membranacea* Hiern from Tanzania (T 4) given in K.B. 33: 273 (1978) is now known to have come from Kigoma in Zaire, Kivu Province and not Tanzania, this species has been left in the key as it could possibly occur in the Flora area.

Stipules divided into (3–)5–9 fimbriae . . . .   *1. R. orientalis*
Stipules with oblong to triangular base bearing 1 linear to subulate lobe:

Flowers 5-merous; stigma fusiform to clavate; panicles
  pyramidal:
  Both surfaces of leaf-blade with fine ridges (resembling
    fingerprints); bracteoles exceeding the calyx; stems
    setose . . . .   *2. R. dupuisii*
  Both surfaces of leaf-blade smooth; bracteoles shorter
    than calyx; stems not setose:
    Young branches pubescent to puberulous; leaf-blades
      chartaceous . . . .   *3. R. smithii*
    Young branches rusty-red tomentose; leaf-blades
      subcoriaceous . . . .   *R. olenotricha*
            (see note above)

Flowers 4-merous; stigma ellipsoid to broadly ellipsoid;
  panicles ± hemispherical to flat-topped:
  Short reflexed lateral branches frequently present, naked
    or bearing rudimentary leaves; corolla-tube short,
    3–5 mm. long; lateral veins acutely angled, not
    impressed above; base of leaf-blade cuneate to
    obtuse. . . . .   *R. membranacea*
          (see note above)

  Specialised lateral branches absent or occasionally
    present; corolla-tube longer, (4–)5–9 mm. long;
    lateral veins less acutely angled, often impressed
    above; base of leaf-blade obtuse or rounded . . .   *4. R. fuscescens*

1. R. orientalis *Bridson* in K.B. 33: 253, fig. 2 (1978); Fl. Pl. Lign. Rwanda: 594, fig. 202/1 (1982); Fl. Rwanda 3: 212, fig. 67/1 (1985). Type: Tanzania, Mpanda District, near head of Ntali R. below Kungwe Mt., *Harley* 9582 (K, holo.!, EA, iso.!)

Scandent shrub or climber or ? sometimes a suffrutex, 1.8–6 m. tall; young branches usually densely covered with goldish to rust-coloured hairs, becoming glabrous with age; older branches with greyish fawn bark. Leaf-blades drying green or brownish, somewhat discolourous, elliptic or less often broadly elliptic or oblanceolate, 4–15.3 cm. long, 2.1–7(–8.5) cm. wide, acuminate at apex, obtuse to rounded and often unequal at the base, frequently bullate or with the lateral nerves impressed above, very sparsely to sparsely strigose above, more densely or occasionally sparsely covered with hairs beneath; domatia frequently absent or domatial hairy tufts straw-coloured; petiole 0.2–1.8 cm. long, pubescent; stipules 7–15 mm. long, base broad, divided into (3–)5–9(–12) fimbriae above, with long hairs outside, glabrous inside, sometimes reflexing. Flowers 5-merous in hemispherical or less often shortly pyramidal, terminal panicles; lateral inflorescence-branches 1–2(–3) pairs, 0.3–2.4(–4 in fruit) cm. long, densely covered with gold to rust-coloured hairs; pedicels 0–1.5 mm. long; bracteoles numerous, linear-lanceolate to lanceolate, 2–11 mm. long, usually exceeding the calyx, pubescent. Calyx-tube 0.75–1 mm. long, glabrous; limb-tube 0.25–1 mm. long; lobes narrowly lanceolate or narrowly triangular, 1.25–4 mm. long, up to 1 mm. wide at base, sparsely pubescent or glabrescent. Corolla glabrous outside; tube 4–9 mm. long, 0.75–1 mm. wide at top; lobes ovate to rotund, 2.25–3.5 mm. long, (1–)1.25–2(–2.75) mm. wide, rounded or less often obtuse at apex. Stigma well-exserted, fusiform 2–3 mm. long. Fruit yellow, orange or red, 5–9 mm. in diameter, glabrous, with a corky scar left by the deciduous calyx-limb. Seed reddish black to blackish, 4.25–5 mm. in diameter, sometimes with irregular shallow longitudinal striations. Fig. 104.

UGANDA. Ankole District: Kalinzu Forest, May 1938, *Eggeling* 3627!; Mbale District: Elgon, Sipi, Feb. 1940, *St. Clair-Thompson* in *Eggeling* 3944!; Masaka District: Katera, 1 Oct. 1953, *Drummond & Hemsley* 4498!
KENYA. Nakuru District: Mau Forest, Endabarra, 28 Jan. 1946, *Bally* 4988!; N. Kavirondo District: Kakamega Forest, Feb. 1944, *Carroll* 4!; Kericho District: Itare R., Feb. 1940, *Copley* in *Bally* 784!
TANZANIA. Lushoto District: W. Usambara Mts., Mkuzi, 23 Feb. 1947, *Greenway* 7933!; Morogoro District: Uluguru Mts., 4.8 km. S. of Bunduki, Salaza Forest, 15 Mar. 1953, *Drummond & Hemsley* 1623!; Iringa District: Ihimbo, 7 Nov. 1952, *Carmichael* 155!
DISTR. U 2–4; K 3, 5; T 1, 3, 4, 6, 7; Zaire, Rwanda, Burundi, Mozambique, Malawi and Zimbabwe
HAB. Forest, frequently at edges or in thickets; 800–2250 m.

SYN. [*R. rufipilis* sensu K. Schum. in P.O.A. C: 390 (1895) & in E.J. 23: 465 (1897) & 28: 495 (1900); F.P.N.A. 2: 362 (1947), *non* Hiern]

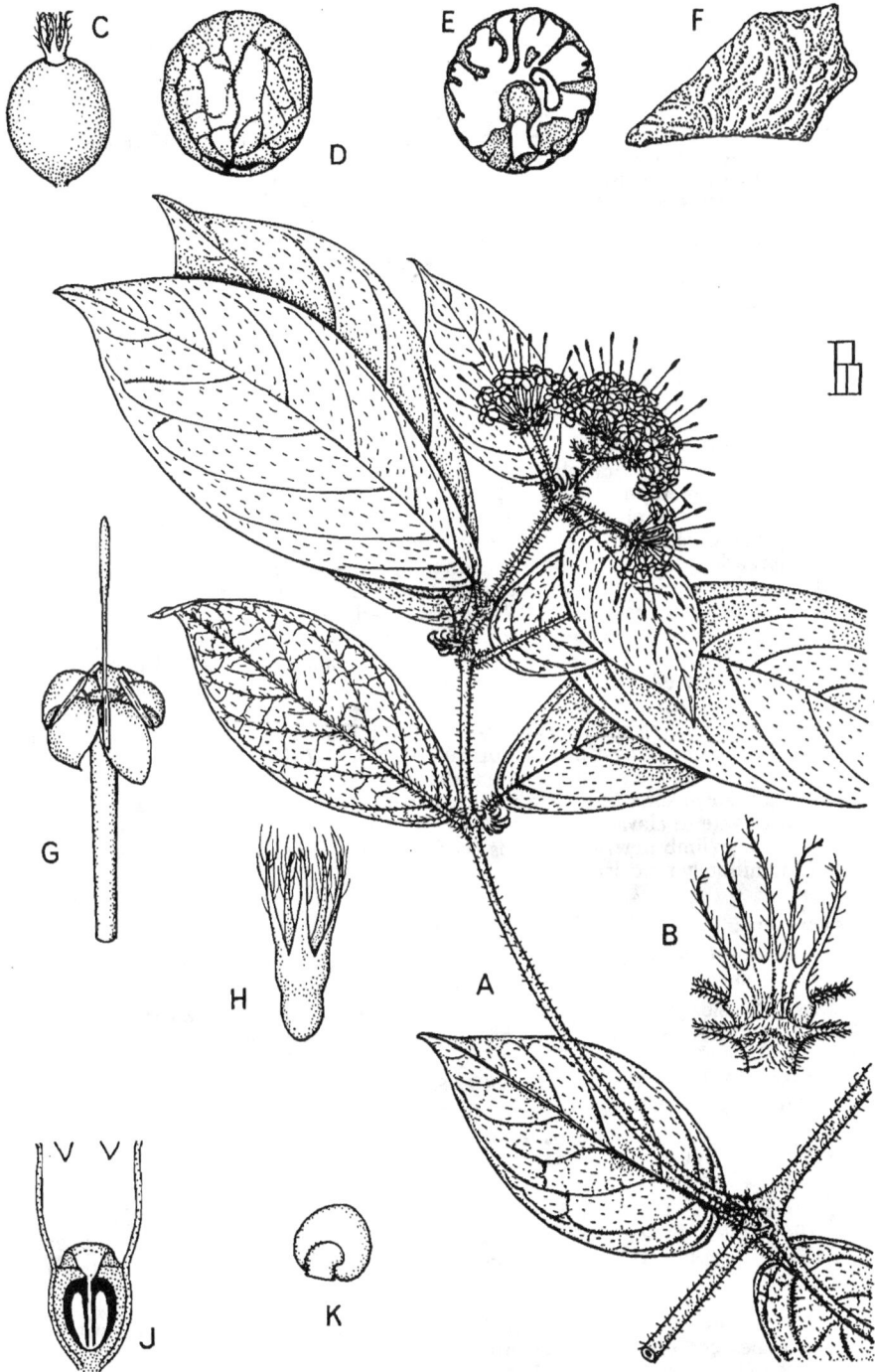

FIG. 104. *RUTIDEA ORIENTALIS* — **A**, flowering branch, × ⅔; **B**, stipule, × 2; **C**, fruit, × 2; **D**, seed, × 4; **E**, half seed, × 4; **F**, seed, part of surface, × 20; **G**, flower with calyx removed, × 4; **H**, calyx, × 6; **J**, section through calyx, × 14; **K**, ovule and placenta from outside, × 20. A, from *Drummond & Hemsley* 4498; B, G–K, from *Cribb & Grey-Wilson* 10352; C–F, from *Drummond & Hemsley* 1623. Drawn by Diane Bridson.

[*R. syringoides* sensu T.T.C.L.: 528 (1949); Garcia in Mem. Junta Invest. Ultram., sér. Bot. 6: 42 (1959); K.T.S.: 469 (1961); Hepper in F.W.T.A., ed. 2, 2: 146 (1963), pro parte, *non* (Webb) Bremek.]

NOTE. The following three species (also characterised by multifid stipules) — *R. parviflora* DC. (West Africa from Senegal to Nigeria), *R. rufipilis* Hiern (S. Nigeria & Cameroun) and *R. seretii* De Wild. (Cameroun & Zaire), together with *R. orientalis* Bridson have been much confused in the past. These species are closely related and a few ± intermediate forms have been seen; it is possible that additional collections may indicate that they would be better regarded as subspecies. *R. orientalis* can be separated from the other three species by its ± hemispherical inflorescence with usually only 1–2 pairs of lateral branches and its generally longer calyx-lobes and bracteoles. The nomenclature has been confused as the type of *R. rufipilis* Hiern was placed in synonymy with *R. syringoides* (Webb) Bremek. by Bremekamp in F.R. 37: 206 (1934) and this has been followed by subsequent authors. The type of *R. syringoides* (not seen) comes from Cape Verde Is. (or more likely from Cape Vert, Senegal) and is most probably a synonym of *R. parviflora* DC.

2. R. dupuisii *De Wild.* in Ann. Mus. Congo, Bot., sér. 5, 2: 174 (1907); Bridson in K.B. 33: 257, fig. 3 (1978). Types: Zaire, Kinshasa, Bingila, *Dupuis* (BR, syn.!) & Equateur, Injolo [Ingolo], *Laurent* 1439 (BR, syn.!)

Scandent shrub or climber 1–4 m. tall; young branches covered with rust-coloured spreading hairs; older branches with thin flaking bark. Leaf-blades elliptic, broadly elliptic or oblong-elliptic, 6–18 cm. long, 2.3–3.6 cm. wide, acuminate and sometimes mucronulate at apex, obtuse to rounded, often unequal or sometimes cordate at base, glabrous, glabrescent or sparsely pubescent above, glabrescent to sparsely pubescent with denser spreading hairs on the nerves beneath, shiny with surface forming fine ridges resembling fingerprints on both faces; domatia usually absent or if present then hairy tufts, inconspicuous; petiole 0.7–2.7 cm. long, densely covered with spreading hairs; stipules triangular at base, subulate above, 0.8–1.6 cm. long, covered with spreading hairs outside. Flowers 5-merous, sessile, in terminal pyramidal panicles; 2–4 pairs of main lateral inflorescence-branches covered with spreading hairs; bracteoles linear, up to 1.2 cm. long, exceeding and obscuring the calyx, covered with spreading hairs. Calyx ± covered with spreading hairs; tube 0.75–1 mm. long; limb-tube 0.25–1.25 mm. long; lobes ovate, 1–1.5(–2) mm. long. Corolla-tube funnel-shaped, 7–10 mm. long, ± 2 mm. wide at top, glabrous or sparsely pubescent outside; lobes ovate, 3–3.75 mm. long, 1.25–2 mm. wide, acuminate at apex. Style with a dense patch of pubescence just above base; stigma narrowly clavate to clavate, 2–3 mm. long. Fruit orange to red, 5–8 mm. in diameter, glabrous; calyx-limb persistent. Seeds reddish brown or reddish black, 4.5–5 mm. in diameter, minutely reticulate.

subsp. dupuisii

Leaf-blades glabrous or less often glabrescent above, usually with at least a few on each specimen reaching the upper limits for the measurements given above; stipules 1.2–1.6 cm. long.

TANZANIA. Mpanda District: Kabwe R., just south of Pasagulu, 8 Aug. 1959, *Harley* 9201!
DISTR. T 4; Cameroun, Gabon, Central African Republic, Zaire, Angola
HAB. Riverine forest; 1460 m.

SYN. *R. striatulata* Pellegrin in Bull. Soc. Bot. Fr. 83: 316 (1936). Types: Gabon, Tchibanga, *Le Testu* 1511 (P, syn., BM, K, isosyn.!) & Mayumba, *Le Testu* 1811 (P, syn., BM, isosyn.!)

NOTE. Subsp. *occidentalis* Bridson is restricted to W. Africa and differs in the leaves which are sparsely pubescent above and the shorter stipules.

3. R. smithii *Hiern* in F.T.A. 3: 189 (1877) & Cat. Afr. Pl. Welw. 1: 491 (1898); T.T.C.L.: 528 (1949); F.P.S. 2: 462 (1952); K.T.S.: 469 (1961); F.F.N.R.: 419 (1962); Hepper in F.W.T.A., ed. 2, 2: 146 (1963); Bridson in K.B. 33: 270, fig. 8A, B (1978). Type: Zaire, *Smith* (K, holo.!)

Climber or shrub 1–7 m. tall; young branches pubescent or less often puberulous or densely pubescent, older branches with brown or less often buff thin bark, flaking to reveal the red-brown underlayer. Leaf-blades usually drying blackish, elliptic to broadly elliptic, sometimes narrowly obovate or less often round, 3.2–15 cm. long, 1.5–9.2 cm. wide, usually acuminate at apex, cuneate, obtuse, rounded or cordate, sometimes tending to be unequal at base, scarcely shiny, glabrous to sparsely pubescent or rarely scabrid with midrib pubescent above, glabrous to sparsely pubescent or less often pubescent with midrib and lateral nerves pubescent or rarely sparsely pubescent beneath; domatial hairy tufts usually present in axils of tertiary nerves; petiole 0.6–3 cm. long, puberulous or pubescent; stipule-limb oblong to shortly triangular, 1.5–2(–4) mm. long, pubescent outside; lobe subulate, 0.6–1.2 cm. long. Flowers 5(–6)-merous, in pyramidal, often leafy panicles, terminal on main and on short lateral branches; inflorescence-branches pubescent; pedicels 0–1 mm. long or lengthening to 5(–7) mm. long in fruit; bracteoles subulate, not exceeding the calyx. Calyx sparsely or more often densely pubescent with greyish-white hairs; tube 0.75–1 mm. long; limb 0.5–1.25 mm. long, divided into ovate lobes. Corolla sparsely or densely covered with greyish-white hairs outside; tube (1.5–)3.5–4(–5) mm. long, 0.5–1 mm. wide at top; lobes ovate (1.25–)2–2.5(–3) mm. long, 1–1.5(–2) mm. wide, rounded, usually with hairs towards the throat. Stigma fusiform, 1–2.25 mm. long, well exserted. Fruit yellow to orange, 5–7 mm. in diameter, glabrous. Seeds reddish black, 4.25–5 mm. in diameter, minutely reticulate.

subsp. smithii

Plants not markedly robust; bark usually brown; leaf-blades never with lateral and tertiary nerves impressed above; stipule-limb usually oblong, seldom triangular; calyx and corolla sparsely to densely pubescent.

UGANDA. Masaka District: Minziro Forest, Oct. 1925, *Maitland* 1267!; Mengo District: Entebbe, Oct. 1931, *Eggeling* 33! & Sisa–Kisubi road, Sept. 1937, *Chandler* 1917!
KENYA. N. Kavirondo District: Kakamega Forest, Mar. 1944, *Carroll* 11! & near Forest Station, 25 Jan. 1982, *Gilbert* 6875!
TANZANIA. Bukoba District: Nyakato, 1935, *Gillman* 355!
DISTR. U 4; K 5; T 1; W. Africa from Sierra Leone to Angola, Zaire, Sudan and Zambia
HAB. Forest edges; 1150–1555 m.

SYN. *R. smithii* Hiern var. *welwitschii* Scott Elliot in J.L.S. 30: 82 (1894); Hiern, Cat. Afr. Pl. Welw. 1: 491 (1898); Good in J.B. 64, Suppl. 2: 28 (1926). Types: Angola, Golungo, *Welwitsch* 3168 (BM, syn.!, K, isosyn.!) & Berria, Tambakka Country, *Welwitsch* 5063 (BM, syn.!)
R. smithii Hiern var. *subcordata* Scott Elliot in J.L.S. 30: 82 (1894), *nom. invalid.* Types: Nigeria, Lagos, *Moloney* (K, syn.!) & Cabinda, Molémba-Landana, *Phillips* (K, syn.!)
R. albiflora K. Schum. in E.J. 28: 87 (1899). Type: Cameroun, near Bipinde, *Zenker* 1611 (B, holo. †, BM, K, iso.!)
R. brachyantha K. Schum. in E.J. 33: 359 (1903). Type: Angola, Lunda, between R. Luachima [Huachin] and R. Chiumba [Quihumbe], *Marques* 326 (not 325 as cited) (B, holo. †, LISU, iso.!)
R. breviflora De Wild., Pl. Bequaert. 5: 430 (1932); Robyns, F.P.N.A. 2: 363 (1947). Type: Zaire, Kivu, Lesse, *Bequaert* 3109 (BR, holo., BM, iso.!)
R. lomaniensis Bremek. in B.J.B.B. 22: 100 (1952). Type: Zaire, Bas Katanga, Kaniana, *Herman* 2292 (BR, holo.!)

subsp. submontana *(K. Krause) Bridson* in K. B. 33: 271 (1978); Fl. Pl. Lign. Rwanda: 594, fig. 202/3 (1982); Fl. Rwanda 3: 212, fig. 67/3 (1985). Type: Zaire/Uganda, Toro District, Ruwenzori Mts., Butahu [Butagu] valley, *Mildbraed* 2677 (B, holo. †, BM, iso.!)

Plants robust, tending to be in the upper limits for most of the characters given in the description; bark usually pale buff or greyish; leaves usually with lateral and often tertiary nerves impressed above; stipule-limbs tending to be triangular towards base; corolla and calyx never densely pubescent.

UGANDA. Toro District: Ruwenzori Mts., Mubuku [Mobuku] valley, Feb. 1932, *Humphreys* 1340! & Nyamwamba [Namwamba] valley, 3 Jan. 1935, *G. Taylor* 2794! & 17 Jan. 1935, *G. Taylor* 3150!
DISTR. U 2; Zaire (Haut Zaire & Kivu), Rwanda and Burundi
HAB. Forest; 1800–2590 m.

SYN. *Randia submontana* K. Krause in Mildbr., Z.A.E. 1907-8, 2: 319 (1911); F.P.N.A. 2: 340 (1947)

4. **R. fuscescens** *Hiern* in F.T.A. 3: 191 (1877); P.O.A. C: 390 (1895); T.T.C.L.: 528 (1949); F.F.N.R.: 419 (1962); Bridson in K.B. 33: 275, (1978). Type: Mozambique, Moramballa, *Kirk* (K, holo.!)

Shrub or liane, 1–9 m. high; branches occasionally short, patent to backwardly directed, leaf-bearing or sometimes naked, puberulous to finely pubescent when young; older branches with thin greyish bark, flaking to reveal red-brown underlayer. Leaf-blades frequently drying dark brown or blackish, elliptic to broadly elliptic, narrowly ovate to ovate or occasionally round, 1.6–10.8(–16) cm. long, 1.4–6.7 cm. wide, acute to acuminate or occasionally obtuse to rounded and mucronulate at apex, obtuse to rounded, sometimes unequal and less often cordate at base, shiny and glabrous to glabrescent above, glabrescent to sparsely pubescent with denser hairs on the nerves beneath; lateral nerves often impressed above; domatial hairy tufts present; petiole 0.8–1.5(–3.5) cm. long, puberulous to pubescent; stipule-limb oblong, 1–2 mm. long, puberulous to pubescent outside; lobe linear to subulate, 4–12 mm. long. Flowers 4-merous in terminal, ± hemispherical sessile panicles, 2–5.1 cm. across, occasionally with an additional pair of lateral panicles from the leaf-node below; usually with one pair of lateral inflorescence-branches, 0.6–3 cm. long, pubescent; pedicels up to 3 mm. long, puberulous to pubescent; bracteoles linear to subulate, present or absent on the pedicels, exceeding or shorter than calyx, up to 7 mm. long. Calyx glabrescent, puberulous or sparsely pubescent; tube 0.75–1 mm. long; limb up to 0.75 mm. long, ± truncate or with short triangular or subulate lobes. Corolla glabrous outside; tube (4–)5–9 mm. long, 0.5–1 mm. wide at top; lobes broadly ovate, 1.75–3 mm. long, 1–2 mm. wide, rounded at apex. Stigma ellipsoid to broadly ellipsoid, 1–2 mm. long, well exserted. Fruit yellow to red, 4.5–6 mm. in diameter, glabrous. Seed blackish, 3–5 mm. in diameter, minutely reticulate.

subsp. fuscescens; Bridson in K.B. 33: 275, fig. 9 D–E (1978)

Bracteoles absent from pedicels, or, if present, then never equalling or exceeding the calyces; pedicels 1–3 mm. long.

TANZANIA.   Lushoto District: Usambara Mts., Bumbuli, 9 May 1953, *Drummond & Hemsley* 2444!; Morogoro District: S. Uluguru Forest Reserve, edge of Lukwangule Plateau, 15 Mar. 1953, *Drummond & Hemsley* 1642!; Iringa District: Dabaga Highlands, Kibengu, 13 Feb. 1962, *Polhill & Paulo* 1465!

DISTR.   T 2, 3, 6–8; Zaire (Shaba), Mozambique, Malawi, Zambia and Zimbabwe
HAB.   Rain-forest or thicket; (800–)1000–2000 m.

SYN.   *Pavetta corynostylis* K. Schum. in P.O.A. C: 388 (1895). Type: Tanzania, Kilimanjaro, Kisuaberg, *Volkens* 1679 (B, holo. †, K, iso.!)
        P. zombana K. Schum. in E.J. 28: 85 (1899). Type: Malawi, Mt. Zomba, *Whyte* (B, holo. †, K, iso.!)
        Rutidea odorata K. Krause in E.J. 43: 150 (1909); T.T.C.L.: 528 (1949). Types: Tanzania, Lushoto District, Amani, *Warnecke* in Herb. Amani 393 (B syn. †, EA, K, isosyn.!) & *Zimmermann* in Herb. Amani 215 (B, syn. †)
        R. zombana (K. Schum.) Bremek. in F.R. 37: 206 (1934)
        [*R. obtusata* sensu Hutch., Botanist in S. Afr.: 533 (1946), quoad *Hutchinson & Gillett* 4099, *non* K. Krause]
        Psychotria nigrifolia Gilli in Ann. Naturhist. Mus. Wien 77: 23, t. 3 (1973). Type: Tanzania, Njombe District, Uwemba, *Gilli* 422 (W, holo.!)

subsp. bracteata *Bridson* in K.B. 33: 276, fig. 9F–K (1978); Fl. Pl. Lign. Rwanda: 594, fig. 202/1 (1982); Fl. Rwanda 3: 212, fig. 67/2 (1985). Type: Uganda, Mengo District, near Entebbe, *Chandler* 1224 (K, holo.!, EA, iso.!)

Bracteoles always present on the pedicels and exceeding the calyces; pedicels up to 1.5 mm. long.

UGANDA.   Bunyoro District: Budongo Forest, 26 May 1971, *Synnott* 612!; Mengo District; Kyewaga Forest near Entebbe, 17 Feb. 1950, *Dawkins* 522!; Mubende [Mubendi], 1913, *Snowden* 19!
KENYA.   C. Kavirondo District: Yala, 2 Jan. 1969, *Kokwaro* 1780!
TANZANIA.   Bukoba District: Nshamba, Sept.–Oct. 1935, *Gillman* 568!; Ufipa District: Mbisi Forest, 12 Aug. 1960, *Richards* 13088!; Mbeya District: Umalila, Sheyo, 4 Dec. 1971, *Leedal* 848!
DISTR.   U 2, 4; K 5; T 1, 4, 7; Zaire (Kivu), Rwanda, Burundi
HAB.   Thickets or forest edges; 1050–2100 m.

SYN.   [*R. zombana* sensu F.P.N.A. 2: 362 (1947), *non* (K. Schum.) Bremek.]

NOTE.   *Richards* 13088 (Ufipa District), *Leedal* 848 (Mbeya District) and *Hepper, Field & Mhoro* 5437 (Rungwe District) although clearly definable as this subsp. show tendencies to subsp. *fuscescens*. *Bjørnstad* 2419 from Iringa District, Ruaha Nat. Park seems to be intermediate between the two subspecies.

78. IXORA

L., Sp. Pl.: 110 (1753) & Gen. Pl., ed. 5: 48 (1754); Benth. & Hook.f., G.P. 2: 113 (1873); Bremek. in Bull. Inst. Bot. Buitenz. sér. 3, 14: 197–367 (1937)

Shrubs or small trees. Leaves opposite or rarely ternate, sessile or petiolate; blades usually drying greyish, greenish, brown or rarely black (not in Africa), chartaceous to coriaceous, entirely glabrous, less often pubescent beneath, very rarely puberulous above; domatia and bacterial nodules absent; stipules with a truncate to triangular limb, usually connate for most of the length, bearing a short or long cuspidate or aristate lobe,

rarely pubescent inside. Flowers usually fragrant, ☿, 4(–5)-merous, few–many, usually
borne in triads on terminal sessile to long-pedunculate ± lax cymes or rarely spherical
heads, occasionally terminal on very short branches so that they appear axillary (not in
Africa) or rarely cauliflorous (not in Africa); inflorescence-branches frequently coloured,
opposite and articulate or less often absent; pedicels present or absent;* inflorescence-
supporting leaves (or rudimentary leaves *fide* Bremekamp l.c.) often present at base of
peduncle, usually smaller and frequently differing in shape from the foliage leaves, but
with a fully developed stipule; bracts stipule-like; bracteoles present or rarely absent.
Calyx often reddish in colour, glabrous to shortly pilose or less often pubescent; tube
ovoid; limb short, sometimes almost absent, usually as wide as the tube, truncate or more
usually 4(–5)-toothed, shortly lobed or occasionally with well-developed lobes. Corolla
white, yellow, pink or red, glabrous outside or rarely pilose; tube cylindrical, usually
slender, only slightly widened at the throat, naked or somewhat bearded at throat; lobes
contorted in bud, spreading or reflexing, lanceolate, narrowly elliptic or ovate, much
shorter than or occasionally equalling the tube in length. Stamens attached at the mouth
of the tube, exserted and spreading or erect in mature flowers; filaments very short;
anthers attached near the base, linear, apiculate and sagittate, twisted when dehisced.
Disc annular, fleshy. Ovary small, 2-locular; placentas fleshy attached near the top of the
septum; ovules solitary, immersed in the placentas; style slender, equalling or slightly
exceeding the corolla-tube, glabrous or sometimes pilose; stigma exserted, 2-lobed; lobes
always completely separate when mature, usually recurved, equalling or a little shorter
than the anthers. Fruit a drupe, usually red (dull brown in dry specimens), spherical or
2-lobed, slightly fleshy or coriaceous, containing 1–2, 1-seeded thin-walled pyrenes;
calyx-limb persistent. Seeds frequently undeveloped, rusty brown in colour, 2,
hemispherical with a deep circular excavation in the centre of the plane ventral face,
convex dorsal face not sculptured but sometimes with minute protuberances which are
occasionally shiny; endosperm entire cartilaginous; embryo dorsal, curved; cotyledons
foliaceous; radicle pointing downwards.

A large genus of ± 300 species occurring throughout the tropics of the Old and New Worlds; ± 30
species are known from tropical Africa (none in South Africa).

Many species have been introduced into East Africa as ornamentals. The following are included
in the key below: — I. *chinensis* Lam. (*I. stricta* Roxb.) recorded from U 4, T 3; *I. coccinea* L. recorded
from U 4, T 3, 6, P & Z, *fide* Williams in U.O.P.Z.: 312, fig. 313 (1949) with var. *lutea* (Hutch.) Corner (*I.
lutea* Hutch.) recorded from T 6 (although the above species are easily distinguished the existence of
many horticultural forms may confuse the issue; furthermore, a third species *I.siamensis* G. Don (not
recorded from East Africa) may only be distinguished from *I. chinensis* with difficulty); *I. finlaysoniana*
G. Don recorded from K 4; T 3 & Z; *I. hookeri* (Oudem.) Bremek. (*I. odorata* Hook. *non* Spreng. *nec*
Boerl.) (the latter species is a native of Madagascar with one specimen (*Forbes* s.n.) recorded from
Mozambique and possibly occurs naturally in the Flora area. Of the two specimens seen from
Zanzibar, *Greenway* 2674 is definitely cultivated while *Toms* 28 M has no data. In the Flora area and
south tropical Africa the name *I. odorata* (e.g. K. Schumann in P.O.A. C: 387 (1895)) has been
frequently misapplied to *I. narcissodora* K. Schum. Most of the specimens can readily be distinguished
by the long pedicels and cuneate bases to the leaf-blades, however, a few specimens, especially those
from Pemba (see note after *I. narcissodora*) are harder to place. *I. hookeri* (Oudem.) Bremek. has also
been confused with *I. brachypoda* DC. (*I. radiata* Hiern) which occurs in West Africa, Zaire and Sudan,
but may be distinguished as *I. hookeri* has a glabrous calyx, subulate bracteoles and petiolate leaves,
whereas *I. brachypoda* has a puberulous calyx, ovate bracteoles and sessile to subsessile leaves;
I. pavetta Andr. (*I.parviflora* Vahl (1794) *non* Lam. (1792) & *I. arborea* Smith) is recorded from U 4 and
Z, *fide* Williams, U.O.P.Z.: 312 (1949); *I. javanica* (Blume) DC. (a complex aggregate, see Corner in
Gard. Bull. Str. Settl. 11: 206 (1941)), recorded from Africa, though not in the Flora area, is somewhat
similar to *I. chinensis* and *I. coccinea* but with generally larger flowers and longer petioles. '*I. alba*'
(presumably a var. of *I. stricta*) and '*I. congesta*' Roxb. = *I. griffithii* Hook. (see nomenclatural note by
Corner l.c.: 189) are cited in Jex Blake, Gardening in E. Africa, but no specimens have been seen.

KEY TO CULTIVATED SPECIES OF IXORA

Calyx with well-defined lobes, lanceolate, ± 4 times longer than
   the tube, membranous, drying paler than the tube with
   nerves apparent, persistent in fruit . . . .   *I. finlaysoniana*
Calyx-limb repand to dentate or with triangular lobes, never
   larger than the length of the tube:

* The presence of bracteoles around the base of the calyx-tube in many apparently pedicellate
species indicates that the flower is truly sessile. However, the entire length of the stalk immediately
below the calyx-tube, whether articulated or not, is included in the term pedicel for this account.

Corolla-tube short, not exceeding 0.8 cm. long, white;
inflorescences crowded, longer than wide . . . .   *I. pavetta*
Corolla-tube rarely as little as 0.8 cm. long, usually much
longer, white or coloured; inflorescences hemispherical or ± flat-topped:
Leaves sessile or shortly petiolate; corolla red, orange or
yellow; tube and lobes smaller than below:
Corolla scarlet or occasionally yellow; lobes acuminate;
leaf-blades with bases usually cordate . . . .   *I. coccinea*
Corolla orange to salmon-pink; lobes rounded; leaf-
blades with bases usually obtuse or subobtuse   *I. chinensis*
Leaves usually distinctly petiolate; corolla white, tinted
pink or reddish on the outside only; tube ± 6–8 cm.
long; lobes 1–1.6 cm. long . . . .   *I. hookeri*

KEY TO NATIVE SPECIES OF IXORA

Corolla-lobes 10–16 mm. long; tube ± 6–8 cm. long; leaf-blades
rounded to obtuse at base (if the majority of flowers are
distinctly pedicellate then see 6 (*I. narcissodora*) — possibly
native in coastal areas . . . .   *I. hookeri*
                                                                     (see note above)

Corolla-lobes 4–9.5 (–10) mm. long; tube 0.8–7.5(–8) cm. long;
leaf-blades cuneate to acute or obtuse at base:
Peduncle always present, (0.4–)1–10 cm. long; inflorescence-
supporting leaves usually present, sessile or subsessile;
blades usually rounded to obtuse or cordate at base;
pedicels frequently absent or not exceeding 4 mm. long:
Leaf-blades 1.2–11(–14.5) cm. long, drying brown or
sometimes greenish; corolla-tube up to 2.5 cm. long;
pedicels 0–1(–2) mm. long:
Lateral nerves not impressed above; calyx-lobes 0.25–
0.5 mm. long, corolla-lobes rounded to obtuse   *1. I. scheffleri*
Lateral nerves impressed above; calyx-lobes 1–1.25 mm.
long; corolla-lobes acuminate . . . . . .   *2. I. albersii*
Leaf-blades 6.5–17.5 cm. long, drying green or bronze-
green; corolla-tube 3.6–5.8 cm. long; pedicels
0–4 mm. long . . . .   *3. I. burundiensis*
Peduncle absent or very infrequently up to 1.5 cm. long;
inflorescence-supporting leaves absent or scarcely
differentiated from the foliage leaves; pedicels
frequently present, up to 12 mm. long:
Corolla-tube 1–1.7(–2.2) cm. long; calyx-limb 0.25–0.5 mm.
long, often distinctly lobed (Uganda):
Corolla-lobes (7–)8–10 mm. long; pedicels 2–11 mm.
long, glabrous . . . .   *4. I. mildbraedii*
Corolla-lobes (5–)7 mm. long; pedicels 0–4 mm. long,
pubescent . . . .   *5. I. seretii*
Corolla-tube 2.8–7.5(–8) cm. long; calyx-limb (0.5–)0.75–
1.25(–2) mm. long, repand to dentate or ± truncate
with very short lobes (Kenya, Tanzania):
Inflorescence-branches glabrous or sometimes
pubescent in generally pubescent plants; pedicels
mostly present, 0–10(–12) mm. long; corolla-tube
(3.2–)4–7.5(–8) cm. long . . . .   *6. I. narcissodora*
Inflorescence-branches pubescent in otherwise
glabrous specimens; pedicels absent or
occasionally up to 2 mm. long; corolla-tube 2.9–3.4
cm. long . . . .   *7. I. tanzaniensis*

1. **I. scheffleri** *K. Schum & K. Krause* in E.J. 39: 553 (1907); T.T.C.L.: 502 (1949); Bridson
in K.B. 32: 603 (1978). Type: Tanzania, E. Usambara Mts., Derema, *Scheffler* 218 (B, holo.†)

Small tree or shrub 6–17 m. tall; young branches glabrous; older branches covered with greyish-buff flaky bark. Leaves glabrous; blades drying brown or greenish, never markedly discolourous, chartaceous, elliptic to oblong-elliptic, 3–18 cm. long, 1.2–6 cm. wide, acute to acuminate or occasionally obtuse at apex, cuneate to occasionally rounded at base, slightly shiny or less often dull above, with 8–18 pairs of lateral nerves, not impressed above, tertiary nerves usually distinctly discolourous beneath; petiole 0.3–1.6 cm. long; stipule-limbs truncate, 2–5 mm. long; lobes keeled below, subulate to filiform above, up to 5 mm. long, caducous. Corymbs 2.5–13.5 cm. across; peduncle (0.4–)1–9.7 cm. long, glabrous or occasionally glabrescent; primary inflorescence-branches (0.5–)0.7–4.5 cm. long, glabrous or puberulous; pedicels 0–1(–2) mm. long; inflorescence-supporting leaf-blades oblong-elliptic to broadly oblong-elliptic, 1.2–9.3 cm. long, 1–5.8 cm. wide, rounded or obtuse at base; petioles 0.3(–5) mm. long; bracteoles triangular to subulate, present at base of calyx-tube, up to half as long as the tube. Calyx glabrous; tube 1–1.5 mm. long; limb-tube 0.25–0.75 mm. long; lobes shortly triangular, 0.25–0.5 mm. long, acute, obtuse or rounded. Corolla white flushed with pink; tube 0.8–2.5 cm. long, ± 1 mm. wide at top; lobes oblong-elliptic or narrowly obovate, 4–6(–7) mm. long, 2–3 mm. wide, rounded or sometimes obtuse. Fruit red, 0.8–1 cm. in diameter (often galled with corolla persistent). Seed pale brown, 6.5 mm. wide.

subsp. scheffleri

Calyx-lobes acute or less often obtuse-acute at apex; stipules with lobes 2–5 mm. long.

TANZANIA. Lushoto District: Kwamkoro Forest Reserve, 17 Jan. 1961, *Semsei* 3164!; Morogoro District: N. Uluguru Reserve above Morningside, June 1953, *Semsei* 1213!; Iringa District: Dabaga Highlands, Ihangana Forest Reserve, 15 Feb. 1962, *Polhill & Paulo* 1495!
DISTR. T 3, 6–8; Malawi
HAB. Forest or occasionally in abandoned cultivations; 910–1980 m.

SYN. *I. latituba* K. Krause in E.J. 57: 37 (1920); T.T.C.L.: 502 (1949). Type: Tanzania, SE. slope of Mt. Rungwe, *Stolz* 1100 (K, holo.!)
   I. ulugurensis Bremek. in K.B. 1936: 478 (1936); T.T.C.L.: 502 (1949). Type: Tanzania, Uluguru Mts., Tanana, Matombo Point, *E.M. Bruce* 804 (K, holo.!, BM, EA, iso.!)

NOTE. The type of *I. scheffleri* has not been seen, but *Schlieben* 3292 from Morogoro District (B, BM, P) has been annotated as *I. scheffleri*. This subspecies is quite variable , but not enough material is known at present (especially from T 3) to decide if the variation is constant enough to merit the recognition of varieties.

subsp. keniensis Bridson in K.B. 32: 603 (1978). Type: Kenya, Meru District, Marania, *J. Bally* 30 in *Bally* 3530 (K, holo.!, EA, iso.)

Calyx-lobes rounded, truncate (sometimes slightly retuse) or occasionally obtuse at apex; stipule-lobes not exceeding 1 mm. long.

KENYA. Mt. Kenya, *Battiscombe* 18! & S. Mt. Kenya, Jan. 1926, *Gardner* in F.D. 1335!; Embu, 11 Dec. 1948, *Abraham* in *Bally* 6514!
DISTR. K 4; not known elsewhere
HAB. Forest; 1950–2700 m.

SYN. [*I. ulugurensis* sensu Greenway, K.T.S.: 448 (1961), quoad *Gardner* in F.D. 1335, *non* Bremek.]
   [*I. latituba* sensu Greenway, K.T.S.: 448 (1961), quoad *Bally* 3530 & 6514 *non* K. Krause]

NOTE. (on species as a whole). This species is very close to *I. foliosa* Hiern which occurs in West Africa and the Cameroun. It may be distinguished by the less-branched inflorescence, with fewer and shorter bracteoles surrounding the calyx-tube (in *I. foliosa* the bracteoles equal the calyx-tube in length). *I. kavalliana* K. Schum. in P.O.A. C: 388 (1895), from the Kavalli Plateau, NE. Zaire, is said to be near *I. foliosa* Hiern, but no specimens so named have been found at K or BR.

2. I albersii K. Schum. in E.J. 33: 355 (1904); T.T.C.L.: 502 (1949). Type: Tanzania, W. Usambara Mts., Kwai, *Albers* 141 (B, holo. †, K, iso. fragment!)

Small tree 3.5–10 m. tall; young branches glabrous; older branches covered with greyish-buff corky gnarled bark. Leaves glabrous; blades drying brown and somewhat discolourous, chartaceous, elliptic, oblong-elliptic or obovate-elliptic, 3.5–11 cm. long, shortly acuminate or less often rounded to acute at apex, obtuse to acute at base, slightly shiny above, 8–11 pairs of lateral nerves, impressed above, tertiary nerves distinctly discolourous beneath; petiole 0.2–1 cm. long; stipule-limb truncate, 2–4 mm. long; lobe subulate, 1–2.5 mm. long. Corymbs 2.5–4.5 cm. across; peduncle 1.5–3.7 cm. long, glabrous; primary inflorescence-branches 0.4–2.2 cm. long, glabrous; pedicels 0–1 mm. long; inflorescence-supporting leaf-blades elliptic to rotund, 3–5.8 cm. long, 1.1–3.3 cm.

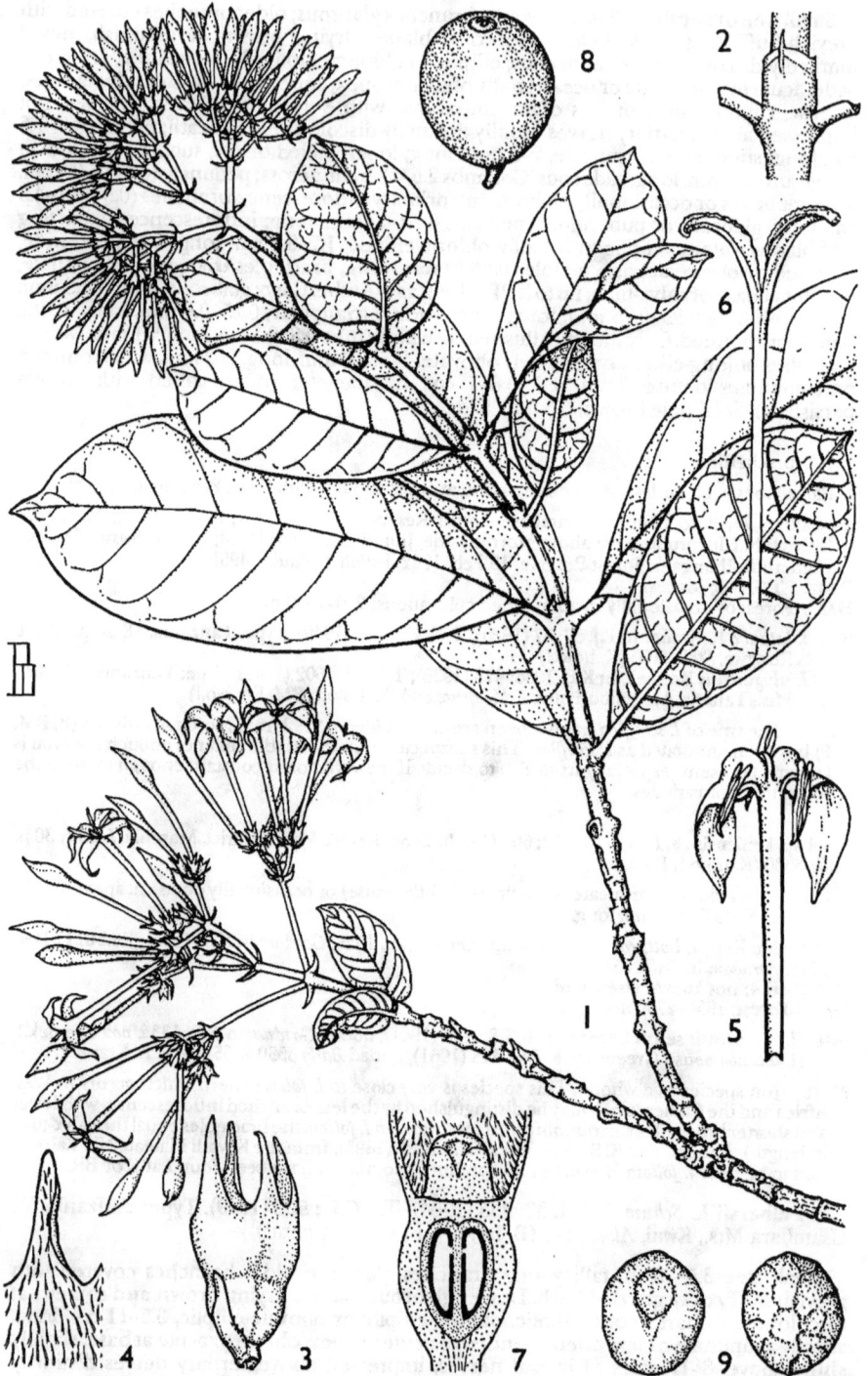

FIG. 105. *IXORA ALBERSII* — 1, flowering branch, × ⅗; 2, stipule, × 2; 3, calyx with bracteole, × 8; 4, portion of calyx-limb, from inside, × 14; 5, half corolla, × 2; 6, part of style with stigmatic arms, × 6; 7, longitudinal section through ovary, × 20; 8, fruit, × 2; 9, seed, 2 views, × 2. 1, 2, from *Drummond & Hemsley* 2611; 3–7, from *Parry* 92; 8, 9, from *Sangiwa* 70. Drawn by Diane Bridson.

wide, rounded or occasionally obtuse at base; petiole 1.5–2 mm. long; bracteoles triangular to subulate, present at the base of calyx-tube. Calyx glabrous; tube 2 mm. long; limb-tube 0.5 mm. long; lobes triangular to narrowly triangular or somewhat subulate above, 1–1.25 mm. long. Corolla red outside, white or pink inside; tube 1.3–2.2 cm. long, ± 1 mm. wide at top, lobes lanceolate to narrowly ovate, 6.5–8 mm. long, 3–4 mm. wide, acute to acuminate. Fruit ?black, 1.1 cm. wide. Seed (immature) rusty brown, 7 mm. in diameter. Fig. 105.

Tanzania. Lushoto District: W. Usambara Mts., Shagayu Forest near Sunga, 18 May 1953, *Drummond & Hemsley* 2611! & Shume-Magamba Forest Reserve, 29 Sept. 1955, *Benedicto* 66! & 14 July 1955, *Sangiwa* 70!
Distr. T 3; known only from W. Usambara Mts.
Hab. Evergreen forest, including *Ocotea-Podocarpus* forest; 1350–1850 m.

3. I. burundiensis *Bridson* in K.B. 32: 604, fig. 1 (1978); Fl. Pl. Lign. Rwanda: 563, fig. 190/1 (1982); Fl. Rwanda 3: 168, fig. 52/1 (1985). Type: Burundi, Bubanza, Mabayi, *Lewalle* 6238 (K, holo.!, EA, iso.!)

Shrub or small tree 4.5–16 m. tall; young branches glabrous, older branches covered with fawn slightly shiny gnarled bark. Leaves glabrous; blades drying green to greyish or bronze-green, coriaceous, oblanceolate to narrowly obovate, oblong-elliptic or occasionally elliptic, 6.5–17.5 cm. long, 2.3–5.6 cm. wide, obtuse or less often acute at apex, acute at base, midrib straw-coloured to fawn when dry, prominent beneath, with ± 12 pairs of lateral nerves, not impressed above, tertiary nerves apparent but not discolourous; petioles 5–8(–12) mm. long; stipule-limbs truncate, 2–4(–5.5) mm. long, often becoming corky; lobe keeled, 1–2 mm. long. Corymbs 2.7–8.7 cm. across; peduncle 2–6(–8.7) cm. long, glabrous; primary inflorescence-branches up to 2.4(–3.2 in fruit) cm. long; pedicels 0–4(–7) mm. long, glabrous; inflorescence-supporting leaf-blades broadly elliptic to ovate or occasionally elliptic, 1.6–9.8 cm. long, 0.9–4.4 cm. wide, obtuse or less often acute at apex, cordate or rounded to obtuse at base; petiole ± 1 mm. long; bracteoles lanceolate to narrowly ovate, usually with one pair at the base of each calyx-tube. Calyx glabrous; tube 2 mm. long; limb-tube 0.5–0.75 mm. long; lobes shortly triangular, 0.5–1 mm. long, acute. Corolla white or cream and pink at the base; tube 3.6–5.8 cm. long, 1–1.5 mm. wide at top; lobes oblong-elliptic, 7–8 mm. long, 3–3.5 mm. wide, obtuse. Fruit 0.9–1.1 cm. in diameter; calyx-limb persistent. Seeds (undeveloped) rusty brown, 7.5 mm. wide.

Tanzania. Mpanda District: below Kungwe Mt., Selimweguru Peak, 11 Sept. 1959, *Harley* 9584! & *Harley* 9598!
Distr. T 4; Zaire (Kivu), Rwanda and Burundi
Hab. Forest ridges and dry woodland; 2100–2250 m.

4. I. mildbraedii *K. Krause* in Z.A.E. 1907–8, 2: 332 (1911); Bridson in K.B. 32: 604 (1978). Type: Zaire, Ituri, between Irumu and Mawambi near El Musa, *Mildbraed* 2969 (B, holo. †, BR, fragment!)

Small tree or semi-scrambling bush 2–4.5 m. high; young branches glabrous; older branches covered with grey or brownish ± smooth bark. Leaves glabrous; blades drying greenish grey or brownish grey, chartaceous, elliptic or less often broadly or narrowly elliptic, 6.5–16 cm. long, 2.7–6.2 cm. wide, acute to acuminate at apex, cuneate at base, scarcely shiny above, lateral nerves not impressed above, tertiary nerves obscure or apparent but not discolorous; petioles 0.7–1.1 cm. long; stipule-limbs truncate, 2–3 mm. long; lobes subulate 1–2 mm. long. Corymbs lax, 5–10.5 cm. across, sessile or rarely with a peduncle up to 3 mm. long; primary inflorescence-branches 0.7–4.3 cm. long, glabrous; pedicels 2–11 mm. long or often with a few on each inflorescence undeveloped, glabrous; inflorescence-supporting leaves undifferentiated; bracts stipule-like; bracteoles small, triangular or triangular below and subulate above, 2 present at the base of each calyx-tube. Calyx-tube 1–1.25 mm. long, glabrous; limb 0.25–0.5 mm. long, ± truncate, dentate with the teeth shortly triangular, or somewhat irregular, glabrous or minutely puberulous. Corolla white or pale pink outside; tube 1–1.7(–2.2) cm. long, 1–1.25 mm. wide at top; lobes lorate, (7–)8–10 mm. long, 2–2.25 mm. wide, rounded or sometimes emarginate. Mature fruit not known, calyx-limb persistent in young fruit.

Uganda. W. Nile District: Amua, Dec. 1932, *Eggeling* 901!
Distr. U 1; Central African Republic, Zaire, Sudan
Hab. Stream sides; ± 900 m.

NOTE. This species has usually been labelled as *Ixora ureensis* Robyns, but this name was never published.

5. I. seretii *De Wild.* in Ann. Mus. Congo, Bot., sér. 5, 2: 178 (1907). Type: Zaire, Case Koko, left bank of Uele R., *Seret* 381 (BR, holo.!)

Small tree or shrub, 4.5 m. tall; young branches glabrous; older branches covered with greenish-buff ± smooth bark. Leaves glabrous; blades drying brown or greyish, slightly discolourous, elliptic or sometimes oblong, 8–19.5 cm. long, 3.3–6 cm. wide, acute to subacuminate at apex, obtuse or cuneate at base, nerves not impressed above, tertiary nerves apparent and sometimes discolourous beneath; petioles 0.4–1 cm. long; stipule-limbs triangular, 5 mm. long; lobes subulate, up to 2 mm. long. Corymbs lax, 6–8.3 cm. across, sessile; primary inflorescence-branches 1–3.9 cm. long, with short fine hairs; pedicels 0–4 mm. long, usually with the middle one of each triad undeveloped or shorter than the lateral ones, pubescent; inflorescence-supporting leaves undifferentiated; bracts stipule-like; bracteoles small, subulate to filiform, 1–2 present at base of calyx-tube. Calyx-tube 1–1.75 mm. long, with short fine hairs below, ± glabrous above; limb 0.25–0.5 mm. long, shallowly or somewhat irregularly lobed, glabrous or sparsely and minutely puberulous. Corolla white; tube 1.4–1.8 cm. long, ± 1 mm. wide at top; lobes ± oblong, 5–7 mm. long, 2 mm. wide, rounded to obtuse. Mature fruit not known.

UGANDA. Toro District: Bwamba, Nyaburogo [Nyaburongo] Forest, Feb. 1943, *Mukasa* in *Eggeling* 5249!
DISTR. U 1; Zaire
HAB. Forest; ± 900 m.

NOTE. This species is very close to *I. laurentii* De Wild. from Zaire, but I think should be regarded as a separate species, at least for the time being. *I. seretii* can be separated from *I. laurentii* chiefly by the larger stipule-limbs (5 mm. as opposed to 2–3 mm. long) and by the calyx-tube which tends to be slightly larger and ± straight-sided, whereas that of *I. laurentii* is turbinate; there is also a tendency for the majority of the pedicels to be suppressed in *I. laurentii*. These two species form a complex with the West African species *I. guineensis* Benth. (1849), *I. breviflora* Hiern (1877), *I. hiernii* Scott Elliot (1884) and *I. bauchiensis* Hutch. & Dalz. (1931), and it is possible that they are not truly distinct at specific level. I have decided not to make any changes at present for although the character distinctions are slight they do seem to coincide with geographical separation.

6. I. narcissodora *K. Schum.* in E.J. 33: 356 (1903); T.T.C.L.: 502 (1949); Bridson in K.B. 32: 606 (1978). Type: Tanzania, Pangani District, Pangani R., Makinyumbi, *Scheffler* 263 (B, holo. †)

Shrub or small tree (0.3–)1.5–10 m. tall; young branches glabrous or pubescent; older branches covered with greyish bark. Leaf-blades drying greyish green or greyish brown, chartaceous to coriaceous, narrowly oblanceolate to oblong-oblanceolate or elliptic to oblong-elliptic, 2.8–17.7(–20) cm. long, 0.8–8.1 cm. wide, acute to obtuse or occasionally rounded and usually with a long or short apiculum at apex, cuneate or occasionally obtuse at base, glabrous and dull to slightly shiny above, glabrous to finely pubescent beneath, lateral nerves not or scarcely impressed above, tertiary nerves usually ± apparent, seldom discolourous; petiole 0.1–1 cm. long, glabrous or pubescent; stipule-limbs truncate to triangular, (1.5–)2–3(–4) mm. long, glabrous or pubescent outside; lobe ± aristate, 0.5–3.5 mm. long, slightly keeled towards base. Corymbs (1.5–)3–11 cm. across, usually lax, sessile or rarely with a peduncle 0.5–1.5 cm. long; primary inflorescence-branches 0.2–2.8 cm. long, glabrous or pubescent; pedicels always present, with a few suppressed or rarely entirely suppressed, 0–10(–12) mm. long, glabrous or pubescent; inflorescence-supporting leaves absent or not markedly distinct from the foliage leaves; bracts stipule-like; bracteoles small, triangular to filiform, up to ¼ the length of the calyx-tube, usually 1–2 present at the base of the calyx-tube. Calyx glabrous, puberulous or pubescent; tube 1–2 mm. long; limb (0.5–)0.75–1.25(–2) mm. long, repand to dentate or ± truncate with very short subulate lobes. Corolla white to pink; tube (3.2–)4–7.5(–8) cm. long, 1–2 mm. wide at top; lobes narrowly oblong to oblong, 6–9(–10) mm. long, 2–3 mm. wide, obtuse to acuminate. Fruit 0.9 cm. in diameter, often galled with the corolla-tube persistent. Seed rusty brown, 5 mm. wide.

KENYA. Kwale District: Shimba Hills, Mwele Mdogo Forest, 6 Feb. 1953, *Drummond & Hemsley* 1153! & Lango ya Mwagandi [Longomwagandi] Forest, 10 Feb. 1968, *Magogo & Glover* 78!; Kilifi District: Pangani–Ribe road, Ribe Kaya Forest, 12 Jan. 1972, *Faden* 72/63!
TANZANIA. Lushoto/Tanga District: E. Usambara Mts., Sigi R., 22 Mar. 1917, *Peter* 19942a!;

Morogoro District: Turiani, Liwali R., Nov. 1954, *Semsei* 1855!; Songea District: Mkuanga Hill, 11 Apr. 1956, *Milne-Redhead & Taylor* 9576!; Pemba I., Jamvini, 15 Feb. 1929, *Greenway* 1451!

DISTR.   **K** 7; **T** 3, 6–8; **Z**; **P**; Mozambique, Malawi, Zimbabwe

HAB.   Usually in riverine forest or thickets, but also recorded from lowland wet evergreen forest and sea-shore; (2–50 m. Pemba and Zanzibar)240–1050 m.

SYN.   *I. stolzii* K. Krause in E.J. 57: 39 (1920); T.T.C.L.: 502 (1949). Type: Tanzania, Rungwe District, Songwe R., near Ulambya [Bulambya], *Stolz* 2248 (B, holo. †)
[*I. odorata* sensu Battiscombe, T.S.K.: 141 (1936); Brenan, T.T.C.L.: 502 (1949); R.O. Williams, U.O.P.Z.: 312 (1949), pro parte, *non* Hook.]
I. sp. near odorata Hook. sensu K.T.S.: 448 (1961), quoad *Battiscombe* 44, *Drummond & Hemsley* 1153, *Rawlins* 322 & *Wakefield* s.n.

NOTE.   No isotypes or named specimens of *I. narcissodora* K. Schum. are known, but *Schlieben* 4053 (B, BM, LISC, P) from Morogoro District has been annotated as *I. cf. narcissodora* K. Schum. The original description conforms quite well with that given above except that the leaves are said to be narrowly lanceolate to lanceolate-oblong and the upper measurements included in brackets for the length of the leaf and corolla-lobes are drawn from the description of *I. narcissodora* and not observed. Many specimens have been annotated as *I. chasei* Bullock, but this name was never published.
Rawlins 322 (the only specimen seen from Lamu District) has been tentatively included in this species although the inflorescence-branches and size of corolla are much less than is usual. *Harris & Pocs* 4269 (from Ulanga District, Kidatu to Sanje) has a very reduced inflorescence and an even shorter corolla-tube (2.4 cm. long); it is probably just an abnormal form. In Kilwa District a gradation from the lax inflorescence form (*Rees* T200) through forms with short to suppressed pedicels (*Vollesen* in *M.R.C.* 2689 and *M.R.C.* 4669) to a congested form in which the inflorescence branches are also very reduced (*Vollesen* in *M.R.C.* 2983) is known.
The material seen from Pemba is atypical in that the pedicels are more frequently absent and usually shorter (resembling *I. hookeri* (Oudem.) Bremek.); also of the three specimens seen, *Greenway* 1451 & *Vaughan* 528 have the inflorescence-branches more condensed than usual, while those of *Barraud* s.n. are lax. *Sacleux* 1040 from Zanzibar is similar to *Greenway* 1451 & *Vaughan* 528. The shorter length of the corolla-lobes and calyx-teeth would seem to indicate that such specimens are better placed here than in *I. hookeri* (Oudem.) Bremek.
The specimens from the Flora area are all glabrous with the exception of *Stolz* 1834 from the Rungwe District. Since glabrous, pubescent and intermediate specimens occur in south tropical Africa in ± equal numbers a variety has not been recognised.
This species is very close to *I. euosmia* K. Schum. (1903) from Cameroun, and the two taxa might better be treated as subspecies. The pedicels of *I. euosmia* tend to be typically shorter than those of *I. narcissodora* and the corolla-tubes do not exceed 5.3 cm. long. In my opinion these species should be kept separate until the genus is revised for Africa as a whole.

7. I. tanzaniensis *Bridson* in K.B. 35: 823 (1981). Type: Tanzania, Ulanga District, Kilombero, Kisawasawa, *Cribb, Grey-Wilson & Mwasumbi* 11213 (K, holo.!, DAR, EA!, iso.)

Shrub 0.3–1 m. tall; young branches glabrous, covered with greyish, somewhat shiny bark. Leaf-blades drying greyish green or bronzish, chartaceous, oblanceolate to narrowly obovate or elliptic, 9.8–23 cm. long, 4–8.5 cm. wide, acute to acuminate, and sometimes shortly apiculate at apex, cuneate at base, glabrous and dull above and beneath; lateral and tertiary nerves raised above, not discolourous; petiole 3–8 mm. long, glabrous; stipules 2–5 mm. long; limb triangular, glabrous outside; lobe subulate, ± 2 mm. long, keeled towards base. Corymbs sessile (but one pedunculate example seen on the same specimen as sessile ones); primary inflorescence-branches 0.4–2 cm. long, pubescent; pedicels absent or occasionally up to 2 mm. long; inflorescence-supporting leaves equalling or smaller than foliage leaves; bracts stipule-like, tending to be slightly larger than stipules; bracteoles inconspicuous. Calyx glabrous to puberulous; tube ± 1 mm. long; limb 0.75–1 mm. long, ± truncate, repand or with very short subulate lobes. Corolla white flushed pink; tube 2.5–3.4 cm. long, ± 1.5 mm. wide at top; lobes oblong, 7–8 mm. long, 3–3.25 mm. wide, obtuse to rounded. Fruit red, 8–9 mm. wide. Seed pale rusty brown, 5.5 mm. in diameter.

TANZANIA.   Morogoro District: Uluguru Mts., Kimboza Forest, 10 Sept. 1972, *Harris & Mwasumbi* in *D.S.M.* 2681! & 12 July 1972, *Pocs & Sharma* 6728/A!; Ulanga District: Kisawasawa, behind Catholic Mission, 27 Jan. 1979, *Cribb, Grey-Wilson & Mwasumbi* 11213! & Uzungwa Mts., Sanje Falls, 23 July 1983, *Polhill & Lovett* 5116!

DISTR.   **T** 6; not known elsewhere

HAB.   Forest; 275–750 m.

RUBIACEAE

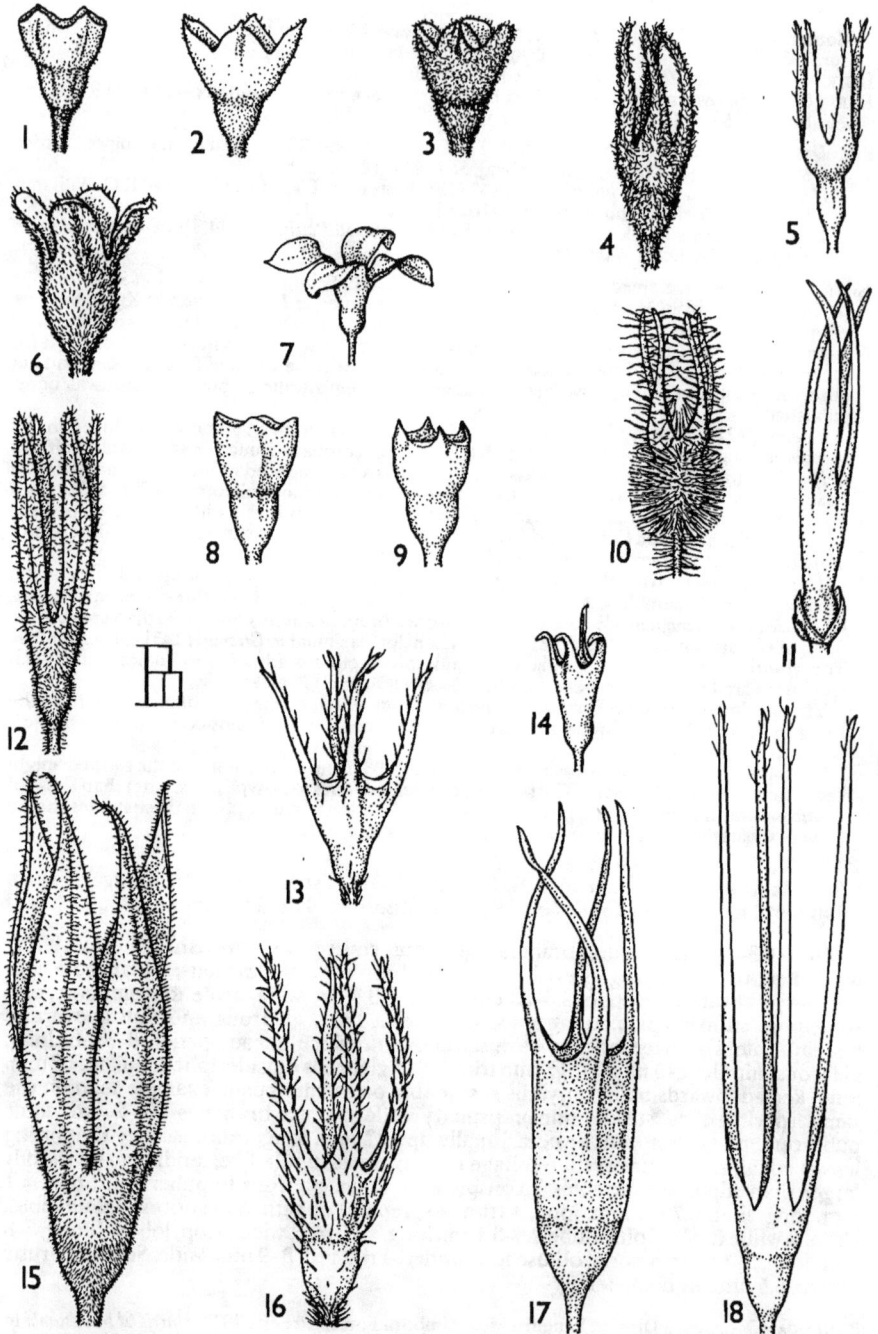

FIG. 106. *PAVETTA* — calyces, × 5. **1**, *P. ankolensis*; **2**, *P. manyanguensis*; **3**, *P. linearifolia*; **4**, *P. oliveriana* var. *oliveriana*; **5**, *P. crebrifolia* var. *crebrifolia*; **6**, *P. molundensis*; **7**, *P. sphaerobotrys* subsp. *tanaica*; **8**, *P. crassipes*; **9**, *P. lasiobracteata*; **10**, *P. sansibarica* subsp. *trichosphaera* var. *trichosphaera*; **11**, *P. uniflora*; **12**, *P. refractifolia*; **13**, *P. bruceana*; **14**, *P. sepium* var *merkeri*; **15**, *P. macrosepala* var. *puberula*; **16**, *P. bagshawei* var. *leucosphaera*; **17**, *P. ruwenzoriensis*; **18**, *P. comostyla* subsp. *nyassica* var. *nyassica*. 1, from *Osmaston* 2761; 2, from *Semsei* 1401; 3, from *Polhill & Paulo* 821; 4, from *Dawkins* 472; 5, from *Verdcourt* 1904; 6, from *Dawkins* 471; 7, from *Bally* 5851; 8, from *Hazel* 522; 9, from *Stolz* 2331; 10, from *Peter* 18723; 11, from *Graham* 2136; 12, from *Hornby* 157A; 13, from *Harris et al.* 3755; 14, from *J. Adamson* 1; 15, from *Shabani* 243; 16, from *Lock* 70/1; 17, from *Osmaston* 1212; 18, from *Richards* 17612. Drawn by Diane Bridson.

79. PAVETTA

L., Sp. Pl. 1: 110 (1753) & Gen. Pl., ed. 5: 48 (1754); Bremek. in F.R. 37: 1–208 (1934) & 47: 12–28, 81–98 (1939)

Shrubs or small trees, occasionally suffrutices. Leaves opposite, occasionally ternate, sessile or petiolate; blades with bacterial nodules usually present and domatia sometimes present in the axils of the nerves beneath; stipules, with a triangular to truncate limb, connate at base or up to the entire length, bearing a subulate arista or less often a short mucro, sometimes with silky hairs within. Flowers (1–)few–many, borne in sessile to subsessile corymbs, terminal on main and lateral leafy branches or terminal on short leafless spurs or branches (which in the absence of corky bark can resemble axillary pedunculate corymbs), occasionally entirely suppressed giving the inflorescence an axillary appearance, very lax to capitate; bracts connate, stipule-like; bracteoles free, usually small, sometimes absent. Flowers ♂, always 4-merous. Calyx-tube turbinate to campanulate; limb persistent or deciduous, rarely accrescent; tube very short or long, often wider than the tube; lobes short to long, ovate (often overlapping) to triangular, lanceolate to subulate, occasionally absent. Corolla white to creamy or greenish white, rarely red (*P. canescens* DC. from Angola and Zaire); tube cylindrical to funnel-shaped (but never greatly widened at the throat), throat bearded or pubescent to glabrous; lobes lanceolate, oblong or ovate, contorted in bud. Stamens attached at the mouth of the tube, exserted and spreading or reflexed in mature flower; filaments very short; anthers dorsifixed near the base, linear to oblong, ± equalling the corolla-lobes in length, apiculate and sagittate, usually twisted when dehisced. Disc annular, fleshy. Ovary small, 2-locular (? very rarely 4-locular); placentas fleshy, pendulous from the apex of the septum; ovules solitary or rarely paired, immersed or partially immersed in the cup-like placenta; style long-exserted, slender, glabrous to subglabrous, sometimes the exserted portion pubescent, the upper part (pollen receptacle) thickened, club-shaped, with 8 shallow ribs, always shortly hairy; stigmatic surface confined to the bidentate apex. Fruit a drupe, frequently black and shining, rarely white or coloured, spherical, slightly 2-lobed, scarcely fleshy, containing 1(–2) chartaceous pyrenes. Seeds 2 (or rarely 1 by abortion), attached to the centre of the septum, hemispherical with a wide circular excavation in the centre of the ventral face, convex dorsal face usually rugose; endosperm always entire and horny; embryo small, dorsal, curved.

About 400 species in the Old World tropics.

Bremekamp (in F.R. 37: 1–208 (1934) & 47: 13 (1939)) splits *Pavetta* into three subgenera, two of which occur in the Flora area (see first couplet of key for differential characters). Subgenus *Baconia* (DC.) Bremek. contains about 60 species and occurs only in tropical W. Africa, the Congo Basin, Uganda, Kenya and Tanzania and the northern borders of south tropical Africa. Subgenus *Pavetta* (*Eu-pavetta* Bremek.) contains about 330 species and is represented throughout Africa and the Old World tropics as far east as Melanesia and tropical Australia, but with the exclusion of Madagascar. The third subgenus *Dizygoön* Bremek. is restricted to the Zaire basin and Angola.

Corolla-tube broad, shorter than or equalling the lobes, always bearded at throat; calyx-lobes ovate-oblong, over-lapping in bud (figs. 106/6, 108/2) or shortly triangular to triangular (fig. 106/1), frequently puberulous; inflorescences always terminal on leafy branches; bacterial nodules linear, less often dot-like or sometimes absent; stipules frequently keeled, never membranous (subgen. *Baconia*) . . . . see Key 1

Corolla-tube slender, much longer than or rarely equalling the lobes, glabrous or pubescent within, but never bearded at throat; calyx-lobes sometimes absent or narrowly triangular, lanceolate to subulate, occasionally shortly triangular or often ovate-oblong (fig. 106/2–5, 7–18), never over-lapping, glabrous or pubescent but infrequently puberulous; inflorescences terminal on leafy or leafless branches or pseudo-axillary; bacterial nodules linear to dot-shaped, infrequently absent; stipules not or sometimes slightly keeled, papery to thinly membranous (subgenus *Pavetta*):

Flowers solitary, terminal on very reduced spurs, bracteate at base (fig. 106/11 and 116) . . . .   *75. P. uniflora*

Flowers never solitary:
- Plants leafless at time of flowering . . . .   *35. P. lutambensis*
- Plants not leafless at time of flowering:
  - Inflorescence terminal on branches bearing one or more pairs of leaves (the leaves can occur immediately below the inflorescence and are sometimes immature, or occasionally they may have fallen but the stipules and scars remain)   see Key 2
  - Inflorescence terminal on leafless branches which are often very short or occasionally entirely suppressed so that the inflorescence appears axillary . . . .   see Key 5 (p. 627)

KEY 1 (subgen. *Baconia*)

Flowers quite sessile, borne in tight heads ± 8 mm. across; calyx-limb 3 mm. long, divided to mid-point into triangular lobes, glabrous . . . .   *8. P. sp. A*

Flowers never sessile and borne in heads; if calyx-lobes triangular then pubescent:
- Calyx-lobes shortly triangular, not exceeding 0.75 mm. long, always persistent in fruit:
  - Leaf-blades elliptic to broadly elliptic, apex acute to subacuminate; bacterial nodules restricted to the midrib; stipules with subulate acumens (fig. 107)   *3. P. ankolensis*
  - Leaf-blades narrowly elliptic to oblanceolate, apex distinctly acuminate; bacterial nodules scattered; stipules with keeled acumens . . . .   *4. P. intermedia*
- Calyx-lobes oblong to ovate, overlapping in bud, 0.75–4 mm. long, sometimes caducous in fruit:
  - Bacterial nodules restricted to the nerves, rarely absent but then with leaf-blades puberulous on nerves beneath:
    - Leaf-blades coriaceous, shiny above, tertiary nerves prominent on both faces; stipules with acumens not or only slightly keeled; calyx-limb persistent in fruit   *1. P. nitidula*
    - Leaf-blades chartaceous, dull or slightly shiny above, tertiary nerves inconspicuous; stipules with keeled acumens; calyx-limb not persistent in fruit:
      - Leaf-blades with nerves finely puberulous, also with sparse whitish, longer, patent hairs beneath   *5. P. urundensis*
      - Leaf-blades with nerves glabrous or occasionally finely puberulous beneath . . . .   *6. P. hymenophylla*
  - Bacterial nodules absent, leaf-blades always glabrous on nerves beneath:
    - Leaves ternate or paired; blades 4.4–15.5 cm. long, 1.2–5 cm. wide, nerves impressed above; petioles 0.15–0.8(–1.3) cm. long (fig. 108) .   *7. P. ternifolia*
    - Leaves never ternate; blades (8–)12–27 cm. long, (2–)5–11 cm. wide, nerves not impressed above; petioles 1.3–6 cm. long (fig. 106/6) . . . .   *2. P. molundensis*

KEY 2

Leaves sessile; blades oblanceolate; calyx-lobes narrowly lanceolate, 1.1–1.3 cm. long; inflorescence a spherical head (T 4, Kigoma District) . . . .   *23. P. sp. B*

Leaves at least shortly petiolate (sometimes only 1.5 mm.); other characters not combined as above:
- Calyx-lobes lanceolate, large, 0.9–1.2 cm. long, 2–4 mm. wide (fig. 106/15):

Calyx-tube sparsely to densely pubescent; leaves not
markedly thin; corolla-tube 4.4 cm. long (fig. 106/15)   *12. P. macrosepala*

Calyx-tube covered with dense patent rust-coloured hairs;
leaves very thin, almost translucent; corolla-tube 1.3
cm. long . . . .   *13. P. schliebenii*

Calyx-lobes if lanceolate then never as long, nor exceeding
2 mm. wide:

Flowers quite sessile, borne in tight heads ± 8 mm. across;
calyx-limb 3 mm. long, divided to mid-point into
triangular lobes (**T 6**, Uzungwa Mts.) . . . . .   *8. P. sp. A*

Flowers if sessile and borne in heads then with calyx-limb
different:

Calyx-lobes ovate, ovate-triangular or transversely
oblong (fig. 106/2, 3):

Leaf-blades glabrous beneath; bacterial nodules
linear to dot-shaped, scattered (fig. 106/2)   *9. P. manyanguensis*

Leaf-blades finely tomentose beneath; bacterial
nodules punctate:

Leaf-blades elliptic, bullate; corymbs 9–15 cm.
across (fig. 109) . . . .   *10. P. teitana*

Leaf-blades linear, often revolute; corymbs up to
3.3 cm. across (fig. 106/3) . . . . . .   *11. P. linearifolia*

Calyx-lobes absent or filiform to lanceolate, or narrowly
or occasionally shortly triangular:

Leaf-blades oblong-oblanceolate, 22.5–31 cm. long,
glabrous; calyx-tube pubescent; limb sparsely
pubescent; lobes narrowly triangular; inflorescence-bearing branches 9.5–17.5 cm. long   *39. P. sp. D*

Leaf-blades smaller but if as above, then other
characters not combined:

Leaf-blades oblong to obovate, 16.5–30 cm. long,
sometimes with tertiary nerves turning blue-
black on drying; inflorescence-bearing
branches very short, 0.5–1.3 cm. long (**T 6**,
Nguru Mts.) . . . .   *58. P. axillipara*

Leaf-blades much smaller, or if large then
inflorescence-bearing branches longer:

Leaves small, 1–5.3 cm. long, mostly restricted to
spurs off the main branches or on young
stems only; inflorescence subumbellate   *74. P. gracilifolia*

Leaves usually larger and well dispersed along
the main branches; inflorescences various,
occasionally subumbellate:

Calyx-tube densely pubescent to tomentose (if
puberulous or adpressed pubescent, or if
obscured by bracts then see below); lobes
subulate, linear, narrowly triangular or
lanceolate (fig. 106/4, 10, 12); leaves
sparsely pubescent to pubescent above,
pubescent to tomentose or strigose beneath,
rarely glabrous . . . .   See Key 3

Calyx-tube glabrous, sparsely pubescent,
glabrescent or puberulous; lobes absent
or various but less often lanceolate (fig.
106/5, 7, 9, 13, 16–18); leaves glabrous
above, glabrous to sparsely pubescent
beneath . . . .   see Key 4 (p. 623)

Key 3

Calyx-tube covered with dense, patent, long, white or rust-
coloured hairs, usually in contrast to the limb and pedicels
(or very rarely with only a few hairs present on calyx-tube)
(fig. 106/10); hairs sparse but still discernible in fruit;
calyx-lobes mostly subulate to linear, sometimes triangular
at base; leaves somewhat strigose, hairs denser and
spreading on nerves beneath:
  Calyx-lobes 5–11 mm. long, 0.75–1 mm. wide at base; leaf-
    blades elliptic or rarely narrowly elliptic; petiole not
    exceeding 0.8 cm. long (fig. 110) . . . .   *14. P. olivaceo-nigra*
  Calyx-lobes up to 7 mm. long, if 0.75 mm. wide at base then
    only 2 mm. long; if leaf-blades elliptic then petiole
    exceeding 0.9 cm. long:
    Stipules with arista 9–14 mm. long; leaf-blades oblanceolate
      to narrowly obovate; petiole stout, 0.4–1.1 cm. long   *15. P. filistipula*
    Stipules with arista absent, up to 6 mm. long or if longer
      then with leaf-blades elliptic to oblanceolate and
      petiole slender, 0.9–4 cm. long:
      Corymbs not markedly lax, 3–4.5 cm. across; older stems
        gnarled, with nodes close together:
        Stipules with arista 5–6 mm. long; corolla-tube 2.1 cm.
          long; calyx-lobes 2 mm. long; patent hairs on
          calyx-tube dense . . . .   *16. P. coelophlebia*
        Stipules lacking an arista, apiculate at the most;
          corolla-tube 1–1.3 cm. long; calyx-lobes 2–4 mm.
          long; patent hairs on calyx-tube not dense   *17. P. lynesii*
      Corymbs lax to very lax, 5–10 cm. across; older stems not
        gnarled, nodes well spaced (fig. 106/10) . . .   *18. P. sansibarica*
Calyx-tube pubescent to tomentose, but indumentum seldom
markedly distinct from that of the limb and pedicel (fig.
106/4, 12); fruit usually glabrous or glabrescent; calyx-
lobes usually linear-lanceolate, lanceolate or narrowly
triangular; leaves seldom with indumentum strigose:
  Corymbs borne on very slender branches, ± 0.75 mm. wide,
    few-flowered; corolla drying pale with distinctive
    blackened apicula on lobes; leaf-blades 2–5.5 cm. long,
    obtuse at base, with stiff shiny hairs above; petioles
    0.3–1 cm. long . . . .   *24. P. lindiana*
  Corymbs borne on thicker branches, many-flowered;
    corolla drying blackish or pale; leaf-blades usually
    larger, if small then often cordate at base:
    Inflorescence-bearing branches short, 2–3 cm. long,
      pubescent to tomentose usually bearing one immature
      pair of leaves at apex; stipules papery, brown, 1.2–1.5
      cm. long . . . .   *57. P. bruceana*
    Inflorescence-bearing branches and stipules not as
    above:
      Leaf-blades small, 2.5–6.5 cm. long, cordate or less often
        rounded, obtuse or rarely cuneate at base; petioles
        0.2–0.6 cm. long; corymbs compact, not exceeding
        3 cm. across . . . .   *19. P. elliottii*
      Leaf-blades larger, never cordate at base; petioles 0.4–3
        cm. long; corymbs lax or compact:
        Leaves pubescent and sparsely strigose with denser
          hairs on midrib above, strigose with denser
          spreading hairs on nerves beneath; calyx-teeth
          narrowly linear, 0.5 mm. wide; stipules lacking
          arista, apiculate at the most . . . .   *17. P. lynesii*
        Leaves glabrous to finely pubescent above, sparsely
          pubescent to tomentose beneath; calyx-teeth
          narrowly linear, triangular or lanceolate, 0.75–
          1.5 mm. wide; stipules acuminate to aristate:

Calyx-lobes linear-lanceolate, 0.9–1.3 cm. long,
  almost equalling the corolla-tube in length;
  leaf-blades oblanceolate-oblong, 5.3–11.2 cm.
  long . . . .   *32. P. tendagurensis*
Calyx-lobes never over 0.9 cm. long, if almost equalling the corolla-tube then leaf-blades broadly
  elliptic, 11–18.5 cm. long:
  Calyx-lobes linear-lanceolate, rarely linear; leaf-
    blades elliptic or oblanceolate to obovate:
    Leaf-blades with nerves not noticeably drying
      whitish beneath; primary inflorescence-
      branches not exceeding 3 mm. long;
      corolla-tube pubescent or rarely glabrous
      outside; calyx-lobes 5–8 mm. long   *25. P. refractifolia*
    Leaf-blades with nerves often drying whitish
      beneath; primary inflorescence-branches
      1–10 mm. long; corolla-tube glabrous
      (rarely pubescent) outside; calyx-lobes 3–
      5 (–7) mm. long . . . .   *26. P. cataractarum*
  Calyx-lobes linear, narrowly triangular or
    triangular; leaf-blades narrowly elliptic to
    round or occasionally broadly obovate:
    Calyx-lobes 5–9 mm. long; leaf-blades broadly
      elliptic, round or broadly obovate:
      Corolla-tube 2 cm. long; leaf-blades round
        to broadly obovate, 6.7–15.3 cm. long   *21. P. burttii*
      Corolla-tube ± 1 cm. long; leaf-blades broadly
        elliptic, 11–18.5 cm. long   . . .   *22. P. greenwayi*
    Calyx-lobes 2–5 mm. long; leaf-blades narrowly
      elliptic to elliptic or rarely broadly elliptic:
      Inflorescence lax   . . . .   *20. P. oliveriana*
      Inflorescence compact, ± subspherical (T 4,
        Mpanda District)   . . . . . .   *23. P. sp. B*

Key 4

Inflorescences ± capitate or compact spherical heads; calyx-
  lobes filiform, linear, narrowly triangular or sometimes
  linear-lanceolate:
Calyx-lobes 2.3–3 mm. long; often flowering before leaves   *35. P. lutambensis*
Calyx-lobes 0.35–2 cm. long; flowers and leaves borne at the
  same time:
  Petioles very short, 1.5–3 mm. long; calyx-lobes linear-
    lanceolate, 0.9–1.3 cm. long; corolla-tubes only
    slightly exceeding calyx-lobes . . . . . .   *32. P. tendagurensis*
  Petioles not markedly short; calyx-lobes if linear-lanceolate
    then not more than 8 mm. long; corolla-tube
    distinctly longer than calyx-lobes:
    Inflorescences ± capitate, lacking branches or with
      rudimentary branches obscured by bracts; pedicels
      present or absent:
      Calyx-lobes linear-lanceolate or linear-triangular,
        0.5–1 mm. wide; bacterial nodules scattered:
        Calyx-lobes ± 4 mm. long; leaf-blades elliptic to
          obovate, 3.6–8 cm. long, subcoriaceous; pedicels
          1–2 mm. long . . . .   *31. P. richardsiae*
        Calyx-lobes 5–8 mm. long; leaf-blades oblanceolate
          to narrowly obovate; pedicels 2–4 mm. long   *33. P. mufindiensis*
      Calyx-lobes filiform; bacterial nodules mostly
        confined to nerves (fig. 111)   . . . . .   *34. P. stenosepala*
    Inflorescences ± globose heads, rudimentary inflor-
      escence-branches apparent, not obscured by bracts;
      pedicels always present:

Calyx limb-tube 0.5–1 mm. long; leaf-blades chartaceous, with short appressed hairs on nerves
beneath . . . .   *38. P. mshigeniana*
Calyx limb-tube 1–1.75 mm. long, somewhat wider
than the tube; leaf-blades chartaceous to subcoriaceous, glabrous or pubescent, but hairs not
adpressed:
Leaves glabrous to pubescent; stipules pubescent
outside (fig. 106/16) . . . .   *41. P. bagshawei*
Leaves glabrous or at the most glabrescent; stipules
always glabrous outside:
Bacterial nodules dot-like, scattered; stipules
triangular, 6–10 mm. long, filiform from a
triangular base, 0.35–2 cm. long calyx-lobes
filiform . . . .   *36. P. amaniensis*
Bacterial nodules elliptic to linear, mostly near
midrib with a few dot-like and scattered;
stipules with truncate-triangular base, 2–5
mm. long, and (? caducous) awn to 7 mm.
long; calyx-lobes linear-subulate (T 4,
Kigoma District) . . . .   *40. P. sp. E*
Inflorescences compact to lax corymbs or subumbellate; calyx-
lobes various:
Calyx-lobes always present, not obviously separated by
sinuses, filiform, subulate, linear, linear-triangular or
sometimes lanceolate (never shortly triangular or
dentate); limb-tube shorter than or equalling tube,
occasionally exceeding 1 mm. long, only slightly wider
than tube (fig. 106/5, 7, 18):
Lobes of calyx lanceolate, linear-lanceolate or narrowly
triangular, 0.5–2 mm. wide at base (if leaves with
puberulous nerves beneath see below):
Inflorescences lax; leaves subcoriaceous, oblong to
oblong-elliptic; stipules with lobe keeled . . .   *50. P. diversicalyx*
Inflorescences compact to moderately compact or
subumbellate; leaves different; stipules with lobe
not keeled:
Calyx-lobes lanceolate, 1–2 mm. wide, tending to
reflex, sometimes with nerves apparent (fig.
106/7); inflorescence compact to moderately
compact; secondary inflorescence-branches
always present; petioles 1–3.8 cm. long . . .   *28. P. sphaerobotrys*
Calyx-lobes linear-lanceolate to narrowly triangular,
0.5–1 mm. wide at base; inflorescence subumbellate; secondary inflorescence-branches
present or absent; petioles 0.3–2 cm. long:
Calyx-lobes (1.5–)2–2.5(–3) mm. long; corolla-
lobes 4–4.5 mm. long . . . .   *29. P. gerstneri*
Calyx-lobes 3.5–8.5(–12) mm. long (fig. 106/5);
corolla-lobes 5–8 mm. long . . . . . .   *30. P. crebrifolia*
Lobes of calyx filiform, subulate or linear (sometimes
from a triangular base), very narrowly triangular,
seldom as much as 0.5 mm. wide:
Inflorescence-bearing branches markedly slender,
erect or pendent; leaves absent or immature and
restricted to the apex:
Calyx-tube and pedicels puberulous; leaves
oblanceolate, glabrous above, glabrous to
puberulous beneath; inflorescence-bearing
branches pendent or erect, usually with
immature leaves . . . .   *55. P. constipulata*

Calyx-tube and pedicels pubescent; leaves ovate or
    infrequently oblanceolate, sparsely to very
    sparsely pubescent on both surfaces;
    inflorescence-bearing branches erect or rarely
    pendent, usually leafless . . . .   *56. P. sparsipila*
Inflorescence-bearing branches not markedly slender,
    always erect; leaves always present and often
    mature at flowering time, restricted to apex or not:
    Calyx limb-tube 0.5–1 mm. long; lobes 4–11 mm.
        long (fig. 106/18); leaves thinly chartaceous, 6.5–
        17.5 cm. long, glabrous to sparsely pubescent beneath   *37. P. comostyla* subsp.
                                    nyassica

    Calyx limb-tube 1–2 mm. long; lobes 3.5–7 mm.
        long; leaves chartaceous, 5–11 cm. long, glabrous
        or pubescent on nerves beneath:
        Leaf-blades with short adpressed hairs on nerves
            beneath; stipule-limbs with short adpressed
            hairs outside . . . .   *42. P. sp. F.*
        Leaf-blades glabrous or pubescent on nerves
            beneath; stipule-limbs pubescent or glabrous
            outside:
            Bacterial nodules dot-shaped, scattered; petioles
                and stipules pubescent . . . . . .   *43. P. aethiopica*
            Bacterial nodules linear, restricted to midrib and
                lateral nerves; petioles and stipules glabrous   *47. P. tarennoides*
Calyx-lobes sometimes absent, if present separated by
    sinuses, shortly triangular, dentate or subulate to linear,
    occasionally somewhat triangular at base; limb-tube
    equalling or often exceeding the tube, (0.75–)1–3(–5) mm.
    long and often distinctly wider (fig. 106/9, 13, 16, 17):
    Calyx with lobes 1–8 mm. long (fig. 106/13, 16, 17):
    Calyx limb-tube (1–)2–2.5(–5) mm. long, drying pale,
        circumscissile, not persistent in fruit (fig. 106/17);
        leaf-blades large, elliptic or oblanceolate, up to 21
        cm. long, glabrous or pubescent on both faces   *44. P. ruwenzoriensis*
    Calyx limb-tube not exceeding 2 mm. long, usually
        persistent in fruit; leaf-blades if as large then
        oblong-obovate and glabrous:
        Bacterial nodules elliptic to linear; confined to
            midrib and lateral nerves; inflorescence lax:
            Leaf-blades oblong-elliptic or oblong; stipules with
                arista not more than 4 mm. long . . . .   *50. P. diversicalyx*
            Leaf-blades narrowly elliptic to ovate; stipules with
                arista 8 mm. or more long:
                Calyx-lobes narrowly triangular or subulate to
                    linear above from a triangular base, 2–4
                    mm. long; leaf-blades acute or somewhat
                    acuminate at apex; bracteoles incon-
                    spicuous (**K** 7) . . . .   *47. P. tarennoides*
                Calyx-lobes ± subulate, scarcely widened at base;
                  leaf-blades gradually narrowing to a fine
                   point at apex; bracteoles apparent (**T** 6)   *52. P. mzeleziensis*
        Bacterial nodules dot-shaped, occasionally elliptic,
            mostly scattered, occasionally concentrated near
            midrib but at least a few scattered; inflorescence
            subumbellate to moderately compact corymbs:
            Inflorescence-bearing branches short, 2–3 cm.
                long, pubescent; stipules 1.2–1.5 cm. long;
                leaves pubescent or sparsely pubescent (fig.
                106/13) . . . .   *57. P. bruceana*

Inflorescence-bearing branches longer, covered
   with corky bark; stipules not more than 1 cm.
   long; leaves glabrous, sparsely pubescent or
   rarely pubescent:
Inflorescence-bearing branches markedly slender,
   often leafless or with only immature leaves;
   leaves glabrescent to sparsely pubescent,
   midrib adpressed pubescent beneath   *56. P. sparsipila*
Inflorescence-bearing branches not markedly
   slender, bearing one or usually more pairs
   of leaves; leaves glabrous, with a few
   spreading hairs on midrib or infrequently
   pubescent:
   Vegetative shoots with leaf-bearing nodes
     usually spaced; young developed stipules
     sheathing, often translucent, apiculate;
     inflorescences a compact to moderately
     compact corymb; calyx-lobes subulate,
     glabrous; bacterial nodules sometimes
     concentrated near midrib . . . . .   *45. P. abyssinica*
   Vegetative shoots with leaf-bearing nodes
     mostly bunched at apex; young developed
     stipules sheathing only at base, not
     translucent, acute to acuminate;
     inflorescence subumbellate, secondary
     branches usually reduced; calyx-lobes
     linear, often from a somewhat triangular
     base or infrequently subulate, sparsely
     pubescent, with just a few hairs at apex or
     occasionally glabrous; bacterial nodules,
     not near midrib, often conspicuous   *46. P. subumbellata*
Calyx with lobes 0–1 mm. long (fig. 106/9):
Bacterial nodules dot-like to elliptic, at least a few
   scattered . . . . *45. P. abyssinica*
Bacterial nodules linear or sometimes elliptic, confined
   to midrib and lateral nerves:
Leaves with midrib puberulous beneath; corolla-tube
   puberulous outside; calyx and pedicels
   puberulous (fig. 112) . . . . *49. P. mazumbaiensis*
Leaves with midrib glabrous or with a few longer
   hairs beneath; corolla-tube glabrous outside;
   calyx and pedicels glabrous or sometimes
   puberulous:
   Leaves chartaceous, elliptic to ovate, drying
     blackish; calyx small; limb-tube 0.75–1(–1.3)
     mm. long; lobes shortly triangular to dentate,
     subulate above, 0.25–0.75(–1) mm. long   *48. P. holstii*
   Leaves subcoriaceous, oblong to oblong-elliptic,
     usually drying green; calyx bigger; limb-tube
     at least 1 mm. long, ± truncate or with short
     subulate to dentate lobes:
     Leaf-blades large, 12–29 cm. long, very shiny;
       calyx-limb shortly dentate . . . . . *51. P. nitidissima*
     Leaf-blades not as large, 5.5–20 cm. long,
       moderately shiny; calyx-lobes truncate or
       shortly subulate:
       Stipules ovate-truncate, glabrous outside,
         becoming corky with age; lobe somewhat
         keeled; bracteoles inconspicuous   *50. P. diversicalyx*
       Stipules ± triangular, pubescent outside, not
         becoming corky with age, mucronulate;
         bracteoles conspicuous:

Inflorescences lax to moderately congested; primary inflorescence-branches 3–17 mm. mm. long; pedicels and calyx-tube always glabrous; corolla-lobes (10–)11–14 mm. long; calyx-limb 1.5–2 mm. long (fig. 106/9); leaf-blades with base acute or obtuse, frequently unequal-sided; stem bark somewhat shiny . . . . 53. *P. lasiobractea*

Inflorescences congested; primary inflorescence-branches (1–)2.5–8 mm. long; pedicels and calyx-tube puberulous, sparsely puberulous or infrequently glabrous or puberulous-pubescent; corolla-lobes 6–9 mm. long; calyx-limb 1–1.25 mm. long; leaf-blades with base usually cuneate, not or only slightly unequal-sided; stem bark dull . . . 54. *P. kyimbilensis*

KEY 5

Leaves crowded on short spurs off the main branches (or occasionally borne directly along the very young branches); blades 1–5.5 cm. long, often turning black when dry:

Inflorescence lax, 2.7–4.7 cm. across excluding the corollas; calyx-lobes triangular, 1–2 mm. long . . . . . . 73. *P. fascifolia*

Inflorescence subumbellate, 1.2–1.5 cm. across excluding the corollas; calyx-lobes ± filiform to linear, 2.5–5 mm. long 74. *P. gracilifolia*

Leaves borne along the main branches, often confined to the apices; blades large or small, drying green or blackish:

Calyx-lobes 5–7 mm. long; inflorescence-bearing branches 5–12 cm. long . . . . 56. *P. sparsipila*

Calyx-lobes not exceeding 3 mm. long; inflorescence-bearing branches frequently much shorter:

Leaf-blades linear to narrowly elongate-oblong or oblanceolate, up to 30 cm. long; calyx-limb truncate, undulate or slightly dentate (fig. 106/8); corymbs (3–)7.5–17 cm. across; stems thick . . . . . . 59. *P. crassipes*

Leaf-blades never linear; calyx-limb truncate or bearing lobes; corymbs not exceeding 7 cm. across; stems seldom thick:

Inflorescence-bearing branches 0.2–0.6(–3.3) cm. long, occurring directly beneath the leaf-bearing apices of the main branches; inflorescence-branches with stiffish hairs:

Leaf-blades with nerves closely spaced and acutely angled; calyx glabrescent (T 3, Pare District) 60. *P. sp. G.*

Leaf-blades with nerves neither closely spaced nor acutely angled; calyx pubescent . . . . . 61. *P. johnstonii*

Inflorescence-bearing branches absent or up to 10 cm. long, well below the apices of the branches; inflorescence-branches glabrous, softly pubescent or tomentose:

Calyx-limb truncate or undulate without well-defined lobes:

Calyx-limb 1.75 mm. long; inflorescence-bearing branches up to 0.5 cm. long . . . . . . 66. *P. haareri*

Calyx-limb (0.25–)0.5–1.25 mm. long; inflorescence-bearing branches up to 8.5 cm. long . . . 67. *P. gardeniifolia*

Calyx-limb with definite lobes (sometimes triangular and only 0.75 mm. long):

Inflorescences few-flowered, laterally compressed,
distinctly longer than wide; suffrutex or small
bush up to 0.45(–1.2) m. tall; branches slender;
leaves glabrous to pubescent (fig. 113)   *63. P. radicans*

Inflorescenses usually many-flowered, never
laterally compressed; if a slender- branched
suffrutex, then with leaves tomentose
beneath:

  Leaves tomentose or pubescent to tomentose
beneath; calyx densely pubescent or
tomentose, but if inflorescence-branches
lax with slender pedicels then see below:

    Leaf-blades 3–26 cm. long, tomentose
beneath, often bullate; stems
unbranched (save for the inflorescence-
bearing branches) or branched only
from young shoots:

      Leaf-blades oblanceolate to obovate;
primary inflorescence-branches always
clearly present, 0.2–2 cm. long; corolla-
tube pubescent or occasionally glabrous
outside   .   .   .   .   *62. P. schumanniana*

      Leaf-blades narrowly to broadly elliptic or
oblong; primary inflorescence-
branches usually suppressed (with at
least some pedicels coming directly
from the node) or up to 1 mm. long;
corolla-tube always glabrous outside:

        Stems slender, 2–3(–4) mm. wide; calyx
with limb-tube 0.5–1 mm. long;
inflorescence-bearing branches
suppressed or up to 2.1 cm. long   *64. P. decumbens*

        Stems thick, 0.7–1 cm. wide; calyx with
limb-tube 1–2 mm. long;
inflorescence-bearing branches
always suppressed (fig. 114)   *65. P. pseudo-albicaulis*

    Leaf-blades 2–6 cm. long, pubescent or
pubescent-tomentose beneath, never
bullate; stems with many short leafy
branches   .   .   .   .   *69. P. dolichantha*

  Leaves glabrous to pubescent beneath; calyx
glabrous to pubescent but never densely
pubescent:

    Calyx-lobes 2.5–3 mm. long; corymbs
compact   .   .   .   .   *70. P. grumosa*

    Calyx-lobes not exceeding 2.2 mm. long;
corymbs moderately compact to very lax
or subumbellate:

      Inflorescence moderately lax to very lax;
primary inflorescence branches 2–
18(–25) mm. long; pedicels 2–22 mm.
long:

        Corolla-tube not markedly slender, 1–2
mm. wide at top, always glabrous
outside, drying creamish; calyx-lobes
shortly triangular, up to 1 mm. long;
leaves drying greenish   .   .   .   .   *67. P. gardeniifolia*

        Corolla-tube slender, ± 1 mm. wide at top,
glabrous or pubescent outside, drying
blackish; calyx-lobes linear to
narrowly triangular, 1–2 mm. long
(fig. 106/14); leaves drying greyish to
black   .   .   .   .   *68. P. sepium*

Inflorescence subumbellate; primary
inflorescence-branches 0-2 mm. long;
pedicels 0.75-3.5(-4) mm. long:
Leaf-blades 1.8-7(-13.5) cm. long; petioles
0.1-0.3(-1) cm. long; calyx-lobes
shortly triangular to triangular, (0.75-)
-2.2 mm. long; corolla-tube (0.9-)
1.2-2.2 cm. long; primary
inflorescence-branches always
absent . . . .   *71. P. subcana*
Leaf-blades 6-9 cm. long; petioles 0.8-1.4
cm. long; calyx-lobes shortly
triangular, ± 0.75 mm. long; corolla-
tube 0.8-1.3 cm. long; primary
inflorescence-branches 0-2 mm.
long (fig. 115) . . . .   *72. P. ruahensis*

1. **P. nitidula** *Hiern*, Cat. Afr. Pl. Welw. 1(2): 486 (1898); Bremek. in F.R. 37: 69 (1934); F.F.N.R.: 414 (1962); Bridson in K.B. 32: 609 (1978). Type: Angola, *Welwitsch* 3189 (LISU, holo., BM, K, iso!)

Evergreen shrub or small tree 1-7 m. tall; young branches glabrous; older branches without corky bark. Leaves glabrous; blades oblanceolate or narrowly to broadly elliptic, 4-29 cm. long, 1.5-12.4 cm. wide, apex obtuse to acuminate or broadly to finely acuminate, base obtuse to cuneate, coriaceous, very shiny above, midrib and lateral nerves usually drying cream in colour, prominent beneath, tertiary nerves raised on both faces; bacterial nodules linear, restricted to the midrib and less often lateral nerves; petiole 0.2-5.5 cm. long; stipule-limbs truncate, 2-4 mm. long, sometimes corky around the margin; arista 1-6 mm. long, occasionally slightly keeled. Corymbs terminal on main and lateral leafy branches, moderately compact, 3-8.5 cm. across; primary inflorescence-branches 0.5-2.5 cm. long, glabrous or rarely puberulous; secondary to tertiary branches present; pedicels up to 5 mm. long, puberulous; bracts 3-4 mm. long, stipule-like; bracteoles inconspicuous. Calyx puberulous, rarely sparsely puberulous; tube 1 mm. long; limb-tube 0.75-1 mm. long ± 3 times wider than the tube; lobes transversely oblong-semi-circular, 0.75-2 mm. long, (1.2-)1.5-2.2 mm. wide, pubescent inside, overlapping in bud. Corolla-tube 5-7(-9) mm. long, (1.75-)2.5 mm. wide at top, glabrous save for the densely bearded throat; lobes ovate-oblong, 5-10 mm. long, 1.5-2.5(-3.5) mm. wide, obtuse to rounded. Fruit purplish red or black, (6-)7-8) mm. in diameter, shiny, glabrous or sparsely puberulous, calyx-lobes persistent, pedicels lengthening slightly. Seed blackish with light brown area around the excavation, 5 mm. wide, very finely rugulose.

TANZANIA.   Buha District: Kasulu [Kassulo] to Kivumba, 24 Feb. 1926, *Peter* 37518!; Kigoma District: Gombe Stream National Park, Mkenke valley, 17 Mar. 1970, *Clutton-Brock* 155!; Mpanda District: Mahali Mts., Katimba, 5 Sept. 1958, *Jefford & Newbould* 2347!
DISTR.   T 4; Cameroun, Zaire, Burundi, Zambia, Angola
HAB.   Deciduous woodland; 900-1500 m.
SYN.   [*P. baconia* sensu Hiern in F.T.A. 3: 176 (1877) pro parte, quoad, specim. *Smith* ex Zaire]
        P. baconia Hiern var. *congolana* De Wild. & Th. Dur. in Ann. Mus. Congo, Bot. Sér. 3, 1 Rel.
            Dewev.: 126 (1901). Type: Zaire, Kivu, to Kasongo Rivers, *Dewèvre* 1000 (BR, holo.)
        P. macrothyrsa K. Krause & Engl. in E.J. 57: 41 (1920), *non* Teysm. & Binn. (1867). Type:
            Cameroun, Molundu, *Mildbraed* 4921 (B, holo.†, HBG, iso.!)
        P. calothyrsa Bremek. in F.R. 37: 67 (1934). Type as for *P. macrothyrsa* Krause & Engl.
        P. congensis Bremek. in F.R. 37: 67 (1934). Type: Zaire, Equateur, Likimi, *Malchair* 64 (BR, holo.!,
            K, iso.!)
        P. coriacea Bremek. in F.R. 37: 69 (1934). Type: Zambia/Malawi, Stevenson Road, *Scott Elliott*
            8310 (K, holo.!, BM iso.)
        [*P. bidentata* sensu Bremek. in F.R. 37: 78 (1934), quoad Zaire, *Bequaert* 6781 & *Bruneel, non*
            Hiern]

NOTE.   This is a very variable species especially in regard to the shape and size of the leaf-blade and the petiole-length. Specimens from Angola and Zambia tend to have much narrower leaves and shorter petioles than those from Cameroun, Zaire and Tanzania, with both extremes occurring in Zaire, Kinshasa region. However, the coriaceous, shiny leaves with the tertiary nerves raised on both faces and the linear bacterial nodules along the midrib and sometimes, the lateral nerves make the species easy to recognise.

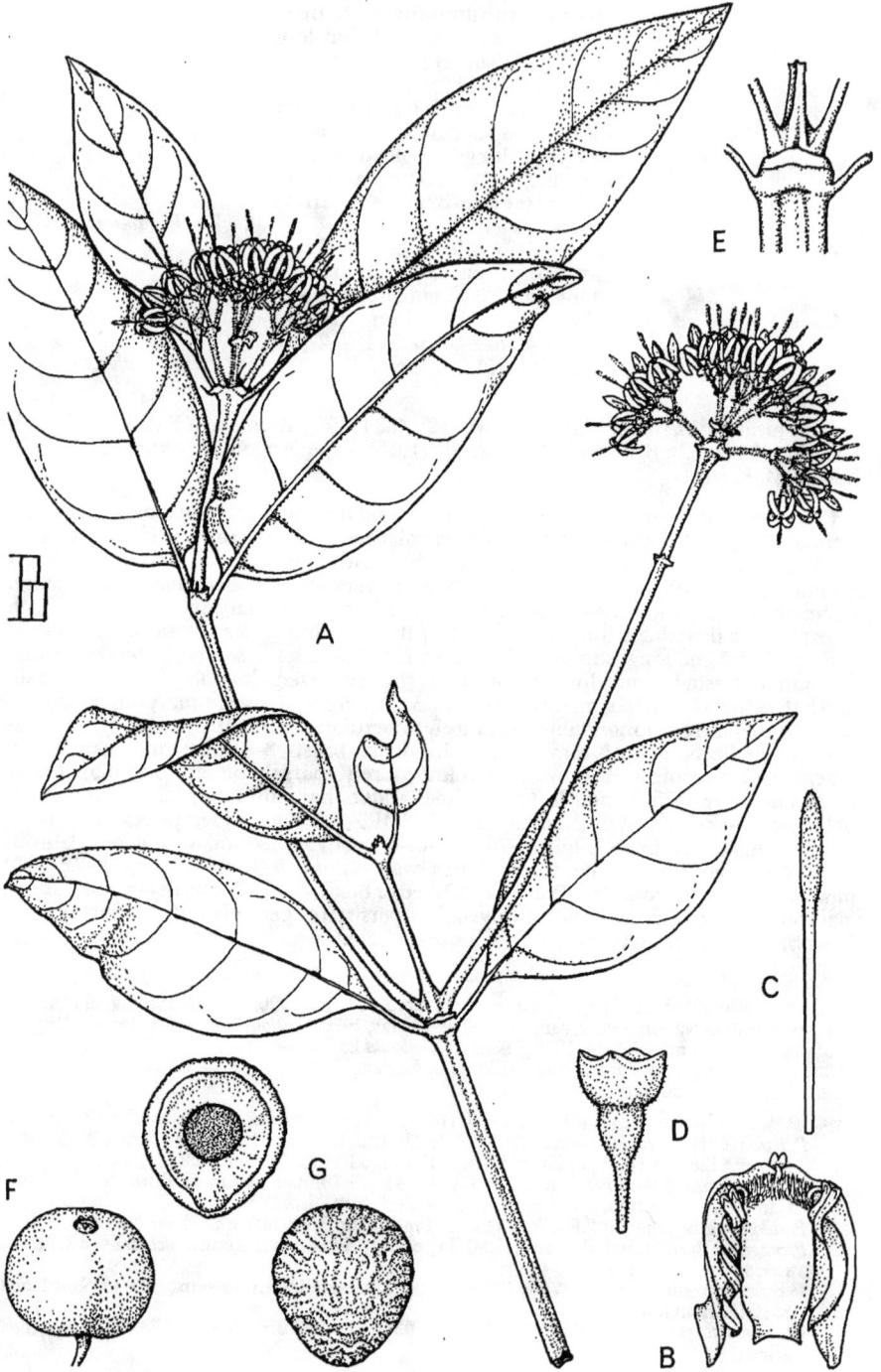

FIG. 107. *PAVETTA ANKOLENSIS* — **A**, flowering branch, × ⅓; **B**, half corolla, × 4; **C**, style and stigma, × 4; **D**, calyx, × 6; **E**, stipule, × 1¼; **F**, fruit, × 3; **G**, seed, 2 views, × 4. A–E, from *Eggeling* 3709; F, G, from *Pierlot* 3225. Drawn by Diane Bridson.

2. P. molundensis *K. Krause* in E.J. 57: 39 (1920); Bremek. in F.R. 37: 64 (1934); Bridson in K.B. 32: 610 (1978). Type: Cameroun, Molundu, *Mildbraed* 4673 (B, holo.†, HBG, iso.!)

Tree or small bush 2–7 m. tall; young branches glabrous; older branches without corky bark. Leaf-blades elliptic, oblong or obovate, (8–)12–28 cm. long, (2–)5–12 cm. wide, apex subacuminate to acuminate, base acute, glabrous, subcoriaceous, shiny above, midrib prominent beneath, tertiary nerves slightly raised on both faces, finely reticulate; bacterial nodules absent; domatia sometimes present; petiole 1–6 cm. long; stipule-limbs truncate, 2–4 mm. long, often turning corky at the base, with long hairs inside, eventually deciduous; arista 1–2 mm. long. Corymbs terminal on main and lateral leafy branches, moderately compact to lax, 4.5–10.5 cm. across; primary inflorescence-branches 0.6–2.3 cm. long, pubescent; 2nd–4th order branches present; pedicels 0.5–3 mm. long pubescent; bracts 2–3 mm. long, stipule-like, caducous; bracteoles inconspicuous. Calyx pubescent; tube ± 1 mm. long; limb-tube 1–2.5 mm. long, ± 3 times wider than the tube; lobes oblong or less often square, 1.5–3(–4) mm. long, 1.5–2 mm. wide, truncate or rounded, pubescent inside, overlapping in bud. Corolla-tube (4–)6–8 mm. long, ± 2.5 mm. wide at top, glabrous save for the densely bearded throat; lobes oblong, (4–)7–11 mm. long, 2–2.5 mm. wide, rounded. Fruit black, 8.5–9 mm. in diameter, glabrous to sparsely puberulous, calyx-lobes eventually falling. Seed blackish with a lighter area around the excavation, 5 mm. wide, slightly rugulose. Fig. 106/6.

UGANDA. Bunyoro District: Budongo Forest, 11 Apr. 1971, *Synnott* 550!; Kigezi District: Maramagambo Forest, 18 Sept. 1969, *Lye* 4078!; Mengo District: Semunya Forest, 9 Dec. 1949, *Dawkins* 471!
TANZANIA. Kigoma/Mpanda District: middle course of Luegele R., 2 July 1971, *L. (Japan)* 7!
DISTR. U 1, 2, 4; T 4; Cameroun, Zaire, Sudan
HAB. Forest undergrowth; 750–1200 m.

SYN. *P. insignis* Bremek. var. *glabra* Bremek. in F.R. 37: 65 (1934); F.P.N.A.: 357 (1947); F.P.S. 2: 456 (1952). Type: Uganda, Mengo District, Mawokota, *E. Brown* 217 (K, holo.!)

NOTE. *P. insignis* Bremek. var. *puberula* Bremek. l.c. is said to have puberulous young branches and puberulous leaves beneath; the type (Zaire Kivu, Beni, *Mildbraed* 2456 (B, holo.†)) may possibly belong to another species.

3. P. ankolensis *Bridson* in K.B. 32: 610, fig. 1 (1978). Type: Uganda, Ankole District, Kalinzu Forest, *Eggeling* 3709 (K, holo.!, EA, iso.)

Shrub 1.5–4 m. tall; young branches glabrous; older branches without corky bark. Leaves glabrous; blades elliptic to broadly elliptic, 6–13 cm. long, 2.2–5 cm. wide, apex acute to subacuminate, base acute to obtuse, subcoriaceous, moderately shiny above; bacterial nodules linear, restricted to the midrib, few; domatia present; petiole 0.8–1.2 cm. long; stipule-limbs truncate, 2–2.5 mm. long, becoming corky around the margin, arista 0.5–1 mm. long, caducous. Corymbs terminal on main and lateral leafy branches, occasionally secondary corymbs present, compact, 1.5–3 cm. across; peduncle absent or 1.2–2 cm. long; primary inflorescence-branches 0.6–1.1 cm. long, sparsely puberulous; secondary (–tertiary) branches present; pedicels 1–3 mm. long, sparsely puberulous to puberulous; bracts stipule-like; bracteoles inconspicuous. Calyx-tube puberulous, ± 1 mm. long; limb-tube glabrous, 0.75–1 mm. long, much wider than the tube; lobes shortly triangular, 0.5–0.75 mm. long, 1.25 mm wide, never overlapping. Corolla-tube 4–5 mm. long, 2 mm. wide at the top, glabrous outside, bearded at throat; lobes 6–6.5 mm. long, 2–2.5 mm. wide, obtuse to rounded. Fruit 7 mm. in diameter, shiny, glabrous; calyx-limb persistent, lobes becoming corky; disc conspicuously corky. Seeds blackish with a dark brown area around the excavation, 4 mm. wide, very finely reticulate on convex face. Figs. 106/1 & 107.

UGANDA. Ankole District: Kasunju Hill, 5 Jan. 1953, *Osmaston* 2761! & Kalinzu Forest, June 1938, *Eggeling* 3709!
DISTR. U2; Zaire (Kivu)
HAB. Forest; 1700 m.

4. P. intermedia *Bremek.* in F.R. 37: 71 (1934); Bridson in K.B. 32: 612 (1978). Type: Zaire, Kivu, Lesse, *Bequaert* 3136 (BR, holo.!)

Shrub 1.7–6 m. tall; young branches puberulous; older branches with fawnish bark. Leaf-blades narrowly elliptic to obovate, 7.7–22 cm. long, 2.7–9.8 cm. wide, apex acuminate to long acuminate, base cuneate, glabrous to sparsely puberulous or pubescent above, puberulous to pubescent especially on the nerves beneath, tertiary nerves slightly raised and finely reticulate above and beneath; bacterial nodules linear on the tertiary nerves, several to very few; domatia present or absent; petiole 0.7–5.5 cm. long, puberulous to pubescent; stipule-limbs truncate, 2–3 mm. long, keeled, keels extending as the lobes, 3–5 mm. long, caducous, leaving a corky rim to the limb, puberulous outside, with silky hairs in axils inside. Corymbs terminal on main and lateral leafy branches, moderately compact, 1.5–7 cm. across; primary inflorescence-branches 0.7–1.4 cm. long, puberulous; 2nd–4th order branches present; pedicels 0.5–3 mm. long, finely pubescent; bracts stipule-like. Calyx pubescent; tube ± 0.5 mm. long; limb-tube 0.75–1 mm. long, much wider than the tube; lobes shortly triangular to ovate-triangular, 0.5–0.75 mm. long, 1 mm. wide at base, slightly keeled, never overlapping. Corolla-tube 3.5–5 mm. long, ± 1.75 mm. wide at top, glabrous outside, densely bearded at throat; lobes 4–5(–8) mm. long, 1.5–1 mm. wide, acute. Fruit 7 mm. in diameter, shiny, sometimes striated when dry, glabrous; calyx-limb ± persistent. Seeds blackish with a brown area around the excavation, 5 mm. wide, scarcely rugulose on convex face.

UGANDA. Toro District: Kibale Forest, 23 Feb. 1955, *Dawkins* 838! & Aug. 1936, *Eggeling* 3134!
DISTR. U2; Zaire
HAB. Forest; 1500 m.

SYN. *P. baconia* Hiern var. *puberulosa* De Wild. in Miss. Em. Laurent 1: 346 (1906) & Ann. Mus. Congo, Bot. Sér. 5, 3: 296 (1910). Types: Zaire, Kalamu and Lukula, *Laurent* (BR, syn.)

5. **P. urundensis** *Bremek.* in F.R. 37: 66 (1934); Bridson in K.B. 32: 612 (1978); Fl. Pl. Lign. Rwanda: 581, fig. 195/3 (1983); Fl. Rwanda 3: 194, fig. 59/3 (1985). Type: W. Burundi, *Meyer* 1037 (B, holo.†)*

Shrub or tree, 2–5(–7) m. tall; young branches finely puberulous soon becoming glabrous; older branches with pale brown shiny bark. Leaf-blades papery, oblong-elliptic to broadly elliptic, 8.5–20 cm. long, 3–6.5 cm. wide, apex acute to acuminate, base cuneate, glabrous, slightly shiny and with the nerves usually impressed above, very sparsely hairy with nerves puberulous with white sparse long hairs beneath; bacterial nodules few, linear, restricted to the midrib or sometimes absent; domatia present; petiole 0.5–3 cm. long, pubescent-puberulous; stipule-limbs 2–3 mm. long, ± truncate, keeled, keel extending as narrowly triangular lobe, up to 8 mm long, slightly puberulous outside, sometimes with silky hairs in axils. Corymbs terminal on main and lateral leafy branches, moderately compact to lax, 2–6.5 cm. across; primary inflorescence-branches 0.5–1.6 cm. long, puberulous; tertiary branches present; pedicels 3–7 mm. long, puberulous with sparse white longer hairs; bracts stipule-like. Calyx puberulous-pubescent; tube 1–1.25 mm. long; limb-tube 1.5–2 mm. long, wider than the tube, circumscissile; lobes ovate to lanceolate, 2.25–3 mm. long, 1.5–2 mm. wide at base, rounded or sometimes acute, slightly keeled, margins usually glabrous outside, glabrous inside, overlapping in bud, often reflexing when mature. Corolla-tube 5–9 mm. long, ± 2 mm. wide at top, glabrous outside, densely bearded at throat; lobes oblong, 8–10 mm. long, 2 mm. wide, acute. Fruit black, ± 7 mm. in diameter, sparsely puberulous-pubescent; calyx-limb not persistent. Ripe seeds not known.

UGANDA. Kigezi District: Kasatora, *Cree* 204! & Impenetrable Forest, Shangi R., 21 June 1967, *Ball* 141!; Kigezi/Rwanda, between Sabinya & Ngahinga, June 1929, *Humbert* 8674!
DISTR. U 2; Zaire, Burundi, Rwanda.
HAB. Forest, valley bottoms; ± 2300 m.

SYN. *P. virungensis* Bremek. in B.J.B.B. 14: 307 (1937) & F.R. 47: 17 (1939); F.P.N.A.: 357 (1947). Type: Zaire, Kivu, Virunga Mts., Karisimbi, *Lebrun* 4952 (BR, holo.!)

NOTE. This species is very close to *P. hymenophylla* Bremek., but can be distinguished by the indumentum on the nerves of the leaves beneath.

6. **P. hymenophylla** *Bremek.* in F.R. 37: 68 (1934) & F.R. 47: 17 (1939); T.T.C.L.: 509 (1949); K.T.S.: 458 (1961); Bridson in K.B. 32: 612 (1978). Type: Tanzania, Lushoto District, Amani, *Grote* in Herb. *Amani* 3541 (B, holo.†, EA, K, iso!)

Shrub or bushy tree 1.5–12 m. tall; young branches glabrous or very sparsely puberulous, rarely puberulous; older branches with fawn bark. Leaf-blades papery, elliptic to obovate, 6.5–23.5 cm. long, 2.2–10.6 cm. wide, apex acute to acuminate, base

* The only other specimen cited by Bremekamp, *Humbert* 8674 (BR, P) has been seen.

cuneate, glabrous, rarely with nerves impressed above, glabrous or occasionally puberulous on the nerves beneath, dull or slightly shiny; bacterial nodules linear, usually restricted to midrib and occasionally lateral nerves, occasionally scattered, rarely absent; domatia present; petiole 0.8–3 cm. long, glabrous or puberulous; stipule-limbs truncate, 1.5–3 mm. long, keeled, keel extending as ± triangular lobe 2–7 mm. long, glabrous or slightly puberulous outside, with silky hairs in axils. Corymbs terminal on main and lateral leafy branches, usually lax, 3–10 cm. across; primary inflorescence-branches 1.3–4 cm. long, sparsely puberulous to puberulous; secondary and tertiary branches present; pedicels 5–9 mm. long, puberulous; bracts stipule-like. Calyx sparsely puberulous to finely pubescent; tube 1–2 mm. long; limb-tube 1–2 mm. long, wider than the tube, sometimes circumscissile; lobes ovate-oblong, sometimes ± square, (1–)1.5–3 mm. long, 1.5–3.5 mm. wide, truncate or rounded, slightly keeled, overlapping in bud, occasionally reflexing, glabrescent inside. Corolla-tube 3–7.7 mm. long, 1–2 mm. wide at top, glabrous outside, densely bearded at the throat; lobes oblong, 6–10 mm. long, 2–3 mm. wide, acute. Fruit black, 1 cm. in diameter, shiny, sometimes striate when dry, glabrous or occasionally puberulous; calyx-limb eventually falling. Seeds 5–5.5 mm. wide, brown-black, with a pale ring surrounding the excavation, shiny smooth.

KENYA. Meru District: Marania, 30 Apr. 1944, *J. Bally* in *Bally* 3534!; S.Nyeri/Embu District: S. slope Mt. Kenya, Oct. 1932, *Dale* in *F.D.* 3054! & Keria, 14 Feb. 1964, *Brunt* 1483!
TANZANIA. Lushoto District: E. Usambara Mts., Amani–Monga road ± 1.6 km. NNE. of Amani, 23 July 1953, *Drummond & Hemsley* 3419!; Morogoro District: Uluguru Mts., Salaza Forest, 15 Feb. 1953, *Drummond & Hemsley* 1622!; Songea District: E. Matagoro, 27 Mar. 1956, *Milne-Redhead & Taylor* 9409!
DISTR. K 4; T 3, 6–8; Malawi
HAB. Understorey of rain-forest; 600–2300 m.
SYN. *P. dalei* Bremek. in F.R. 37: 72 (1934); K.T.S.: 457 (1961). Type: Kenya, S. Mt. Kenya, *Dale* in *F.D.* 3054 (K, holo.!, EA, iso.)
 P. sp. (3) near *ternifolia* (Oliv.) Hiern sensu T.T.C.L.: 509 (1949), based on Tanzania, W. Usambara Mts., Magamba, *Wigg* 27 in *F.H.* 711 (FHO, K,!)
 P. sp. (4) sensu T.T.C.L.: 509 (1949), based on Tanzania, W. Usambara Mts., Shume-Magamba Forest, *Pitt-Schenkel* 430 (FHO!)
NOTE. There is a great deal of variation in the size and to a certain extent the shape of the calyx-lobes. Variations in the length of the corolla-tubes seem meaningless and have been noted within one inflorescence. I have found no justification for the maintenance of *P. dalei* in spite of the somewhat anomalous distribution.
 Wigg in *F.H.* 711 has been included in this species in spite of leaves with deeply impressed nerves above, as their papery texture and slender petioles exclude it from *P. ternifolia* (Oliv.) Hiern, which has not been recorded from the W. Usambaras.
 Fruiting specimens of this species are often confused with *P. holstii*, but may easily be distinguished by the puberulous inflorescence-branches and the keeled stipules.

7. **P. ternifolia** (Oliv.) Hiern in F.T.A. 3: 177 (1877); Bremek. in F.R. 37: 65 (1934); T.S.K.: 145 (1936); Bremek in F.R. 47: 17 (1939); T.T.C.L.: 509 (1949); K.T.S.: 462 (1961); Bridson in K.B. 32: 613 (1978); Fl. Pl. Lign. Rwanda: 580, fig. 194/2 (1983); Fl. Rwanda 3: 194, fig. 58/2 (1985). Type: Tanzania, Bukoba District, Karagwe [Karagué], *Grant* 422 (K, holo.!)

Shrub or small tree 1.2–4.5(–7) m. tall; young branches puberulous; older branches with smooth ± glossy bark. Leaves ternate or paired, glabrous; blades elliptic-obovate to oblanceolate, 4.4–15.5 cm. long, 1.2–5 cm. wide, apex subacuminate, base cuneate, shiny, usually with nerves impressed above, dull with nerves raised beneath, sometimes tending to be revolute; bacterial nodules absent; domatia often present; petiole 0.15–1(–1.3) cm. long; stipule-limbs subtruncate to truncate, 2–2.5 mm. long, shallowly keeled, keel extending as the arista, up to 4.5 mm. long, glabrous or puberulous outside, with long silky hairs inside. Corymbs terminal on main and lateral branches, moderately compact, 2–5 cm. across; primary inflorescence-branches 0.7–2.5 cm. long, puberulous; secondary to tertiary order branches present; pedicels 1–6 mm. long, puberulous to finely pubescent; bracts 1–1.5 mm. long, puberulous to finely pubescent; bracteoles very reduced or absent. Flowers fragrant, ?rarely 5-merous. Calyx puberulous; tube 1 mm. long; limb-tube 1.5–3 mm. long, ± 3 times wider than the tube; lobes ovate-oblong, 1.5–4 mm. long, 1.5–3 mm. wide, slightly keeled, overlapping when mature. Corolla-tube 4–6 mm. long, 2–2.5 mm. wide at top, glabrous save for the densely bearded throat; lobes ovate-oblong, 4.5–7 mm. long, 1.75–3 mm. wide, rounded. Fruit black, 6–8 mm. in diameter, shiny, very sparsely puberulous; calyx-lobes eventually falling. Seed dark red-brown, 5–5.5 mm. wide, smooth. Fig. 108.

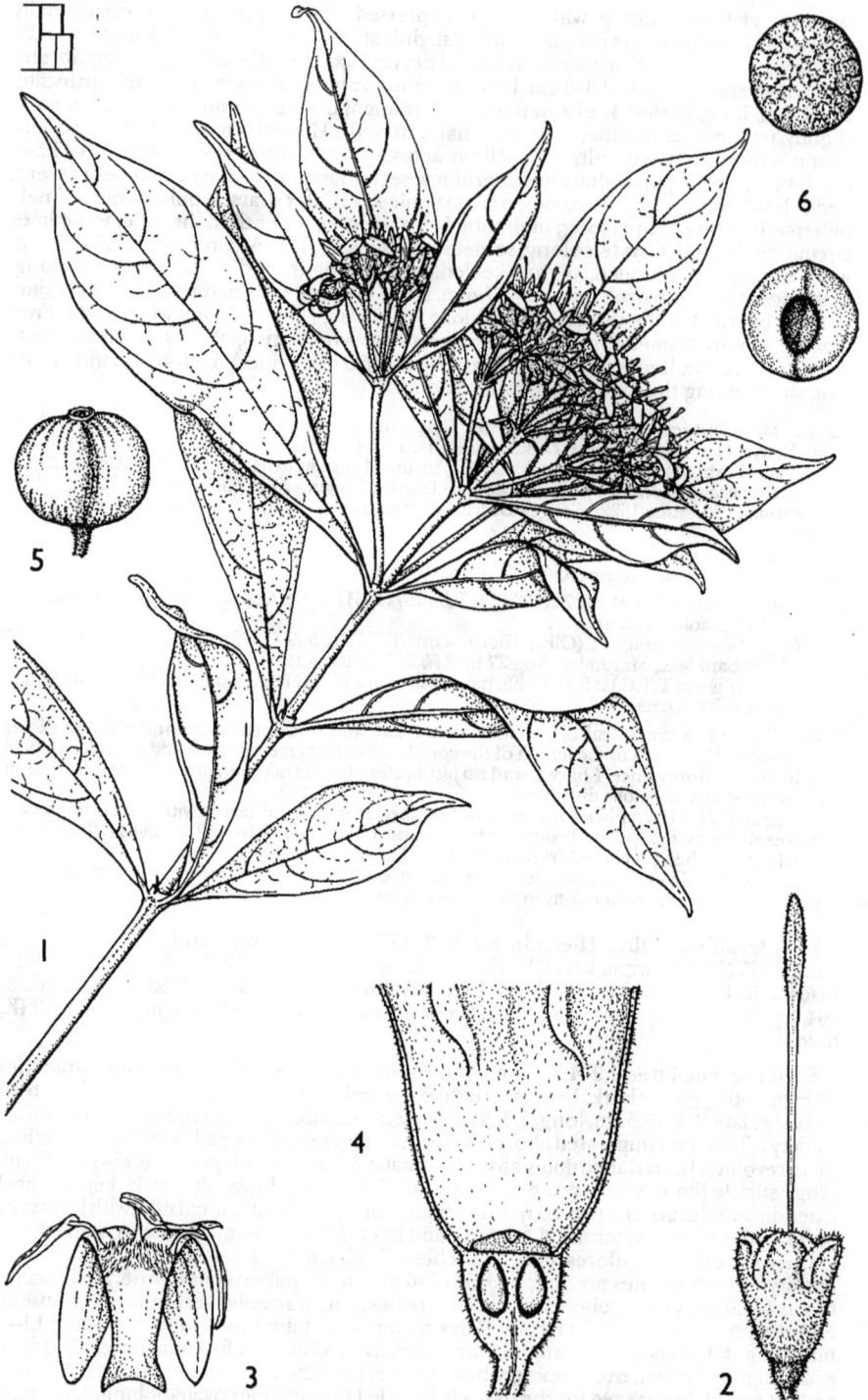

FIG. 108. *PAVETTA TERNIFOLIA*—1, flowering branch, × ⅔; 2, calyx with style and stigma, × 4; 3, half corolla, × 4; 4, section through calyx to show ovary, lobes partly shown, × 14; 5, fruit, × 3; 6, seed, 2 views, × 4. 1, from *Dawkins* 773; 2–4, from *Purseglove* 3418; 5, from *Eggeling* 4210; 6, from *Osmaston* 2818. Drawn by Diane Bridson.

UGANDA. Toro District: Kibale, near Ntuntu, Kicheche, 31 July 1970, *McNutt* 110!; Ankole District: Buhwezu, 16 Aug. 1929, *Snowden* 1414!; Masaka District: Sese Is., Bwendero, June 1925, *Maitland* 786!
KENYA. Trans-Nzoia District: Koitobos [Koittoboss], near Tyack's Bridge, Apr. ? 1972, *Tweedie* 4314!; N. Kavirondo District: Kakamega Forest, June 1961, *Lucas* 141!; Kericho District: Belgut Location, Cheptuiyet, Aug. 1960, *Kerfoot* 2156!
TANZANIA. Bukoba District: Ruiga River Forest Reserve, Oct. 1958, *Procter* 1035!; Ngara District: Bushubi, Keza, 25 May 1960, *Tanner* 4994!; Buha District: Kasulu Highlands, Kibanga, Nov. 1954, *Procter* 315!
DISTR. U 2, ?3, 4; K 3, 5; T1, 4; Zaire, Rwanda, Burundi
HAB. Forest (sometimes seasonally flooded), thickets, forest edges and grassland; 1150–1950 m.

SYN. *Ixora ternifolia* Oliv. in Trans. Linn. Soc. 29: 86, t.51 (1873)
      Pavetta niansae K. Krause in E.J. 43: 149 (1909). Type: Rwanda, Niansa Mt., *Kandt* 64 (B, holo.†, EA, iso.!)
      P. yalaensis Bremek. in K.B. 8: 501 (1954); K.T.S.: 463 (1961). Type: Kenya, N. of Yala R., *G.S. Rogers* 741 (K, holo.!. EA, iso.!)

NOTE. In spite of the slight discontinuity of distribution between the Kenya specimens and the rest there is little evidence to justify keeping *P. yalaensis* Bremek. as a subspecies. The Kenya specimens usually (with the exception of *Lucas* 141) have paired leaves and petioles 0.4–1 cm. long, while the rest generally have at least some ternate leaves and petioles 0.15–0.8(1–1.3) cm. long (with the exceptions of *McNutt* 110, Uganda, Toro District, & *Deville* 322 & 421, Zaire, Mahagi).

8. P. sp. A

Glabrous shrub to 2 m. tall. Leaf-blades drying dark green; elliptic, 6.5–11.2 cm. long, 2–3.8 cm. wide, acuminate at apex, acute and sometimes unequal at base, subcoriaceous, slightly shiny above, reticulate nervation raised on both faces; bacterial nodules dot-shaped to linear, situated along lateral nerves and with a few scattered; petioles 0.4–1 cm. long; stipule-limbs triangular, ± 3 mm. long, bearing an acuminate to linear somewhat keeled lobe, 2 mm. long. Inflorescences terminal on lateral branches; tight heads ± 8 mm. across; inflorescence-bearing branches 6–8.5 cm. long, with one pair of leaves 0.8–1.2 cm. below the head; flowers sessile; bracts stipule-like, up to 7 mm. long. Calyx-tube ± 0.75 mm. long; limb 3 mm. long, much wider than tube, divided to midpoint into acute triangular lobes. Corolla not known. Fruit black, ± 8 mm. in diamater; calyx-lobes persistent. Seed brown-black with slightly lighter area around excavation, ± hemispherical but tapering at base, 4.5 mm. long, 4 mm. wide, slightly rugulose.

TANZANIA. Ulanga District: Uzungwa Mts., Sanje, 14 June 1984, *Lovett* 291!
DISTR. T6; only known from above gathering
HAB. Forest with *Cephalosphaera*, *Myrianthus* and *Parinari*; 1150–1300 m.

NOTE. In the absence of flowers it is not possible to tell if this species belongs to subgenus *Pavetta*, *Baconia*, or even *Dizygoön*. It is a most distinctive species and the combination of calyx-type with inflorescence-type has not been noted elsewhere. Re-collection is urgently required.

9. P. manyanguensis *Bridson* in K.B. 32: 614; fig. 2 (1978). Type: Tanzania, Morogoro District, Turiani, Manyangu Forest Reserve, *Semsei* 1401 (K, holo.!, EA, iso.!)

Shrub or small tree; young branches glabrous, not becoming corky, angled and grooved in dry material. Leaf-blades lanceolate or oblong-elliptic, 15–24 cm. long, 4.3–9.4 cm., wide, apex acute to acuminate, base cuneate, tending to be coriaceous, moderately shiny above, glabrous save for the puberulous prominent midrib beneath; bacterial nodules dot-like, elliptic or linear, scattered, few to several; petiole 1.6–4 cm. long, ± puberulous below, glabrous above; stipule-limb truncate, 5–7 mm. long, glabrous outside; arista 3–7 mm. long. Corymbs terminal on leafy branches, ± crowded, up to 4.7 cm. across; primary inflorescence-branches 4–6 mm. long (or up to 1.8 cm. long in fruit), puberulous; secondary and tertiary branches present; pedicels 0–3 mm. long, puberulous; bracts stipule-like; bracteoles drying brown, scale-like, conspicuous. Calyx-tube 1 mm. long, puberulous; limb drying paler than tube; limb-tube 1 mm. long, distinctly wider than the tube, puberulous; lobes ovate, 1–1.25 mm. long, 1–1.5 mm. wide at base, apiculate, sparsely puberulous. Corolla-tube 1.2–1.4 cm. long, 1 mm. wide at top, glabrous outside, not bearded at throat; lobes oblong, 6 mm. long, 1.5 mm. wide, shortly apiculate. Fruit black, 7 mm. in diameter, not shiny, glabrous; calyx-limb persistent. Mature seeds not known. Fig. 106/2.

TANZANIA. Morogoro District: Nguru Mts., Manyangu Forest Reserve, Nov. 1953, *Semsei* 1401! & Mar. 1956, *Semsei* 2411!

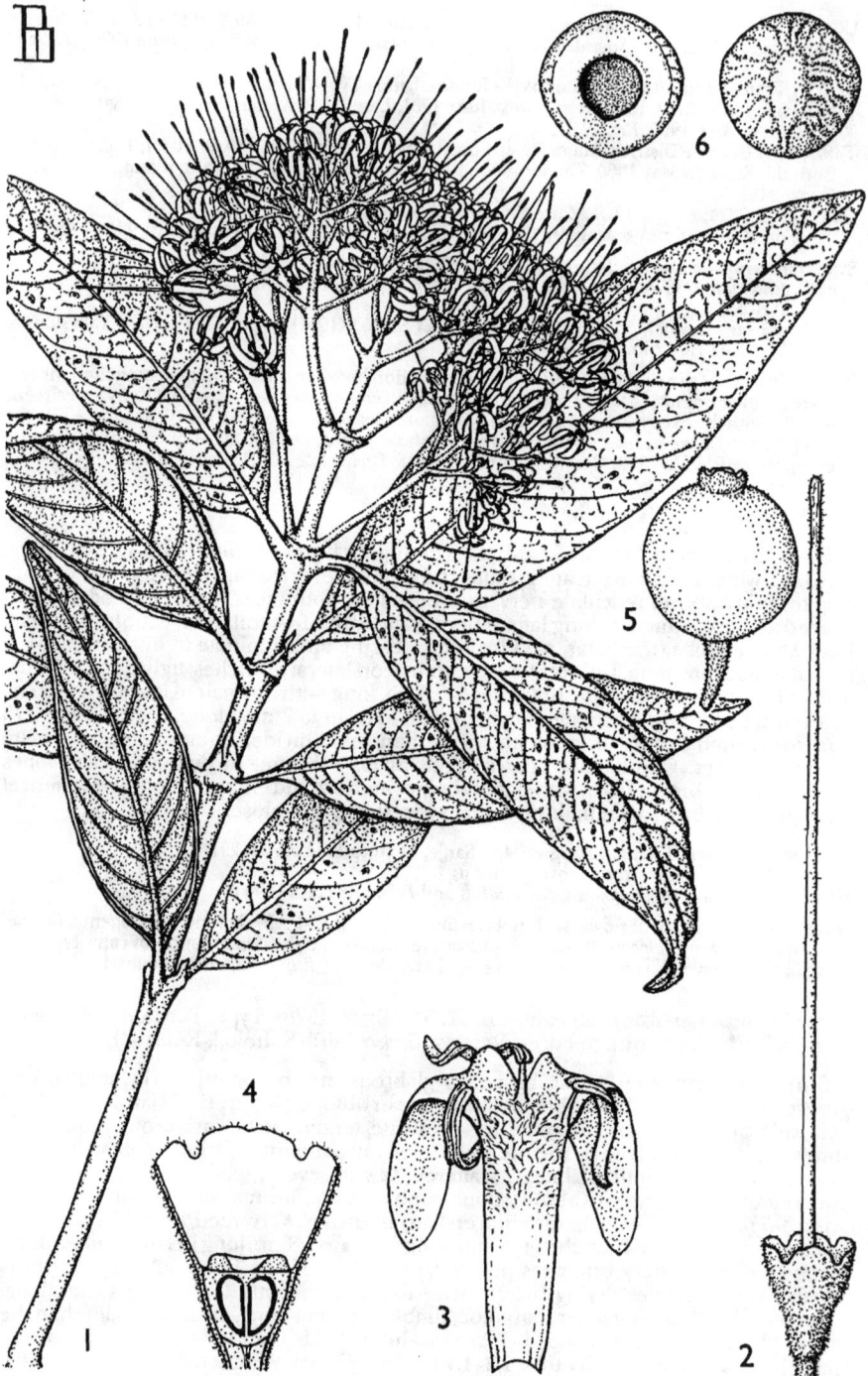

FIG. 109. *PAVETTA TEITANA*— 1, flowering branch, × ⅔; 2, calyx with style and stigma, × 4; 3, half corolla, × 4; 4, section through calyx to show ovary, × 10; 5, fruit, × 3; 6, seed, 2 views, × 4. 1–4, from *Kirrika* 211; 5, 6, from *Pole Evans & Erens* 1077. Drawn by Diane Bridson.

DISTR. T 6; known only from the Nguru Mts.
HAB. Forest; 1800 m.

10. P. teitana *K. Schum.* in P.O.A. C: 389 (1895); Bremek. in F.R. 37: 130 (1934); Chiov., Racc. Bot. Miss. Consol. Kenya: 57 (1935); K.T.S.: 461 (1961). Type: Kenya, Teita, *Hildebrandt* 2555 (B, holo.†)

Evergreen shrub or small tree 1.2–7 m. tall; young branches puberulous; older branches with thin grey or brown bark. Leaf-blades lanceolate-oblong, 9–18 cm. long, 2.5–5.5 cm. wide, apex acute to slightly acuminate, base acute to cuneate, bullate, shiny, glabrous above, finely tomentose with prominent nerves beneath, margins revolute; bacterial nodules small, slightly raised or punctate above, few to numerous, scattered; petiole 0.7–1.4 cm. long, puberulous, stipule-limbs pale buff-coloured, papery, truncate 3–5 mm. deep, puberulous outside, silky hairs absent inside; arista 4.6–8 mm. long. Corymbs terminal on main branches, often with additional paired inflorescence-branches from the next node down, many-flowered, crowded, 9–15 cm. across; primary inflorescence-branches 1–2.4 cm. long, sparsely puberulous to puberulous; 2nd–5th order branches present; pedicels 1–5 mm. long, puberulous; bracts 2–5 mm. long; bracteoles inconspicuous. Calyx puberulous; tube 1–1.25 mm. long; limb-tube 1–1.5 mm. long, wider than the tube; lobes transversely oblong, rounded to obtuse, ± 0.5 mm. long, 1–1.5 mm. wide. Corolla-tube (7–)8–9 mm. long, 2 mm. wide at top, glabrous or very sparsely puberulous outside, hairy at the throat, becoming glabrous towards the base inside; lobes oblong-lanceolate, 5–6(–7) mm. long, 1.5–2 mm. wide, obtuse to rounded, ciliate. Fruit black, 7–8 mm. in diameter, sparsely puberulous. Seed, 5 mm. wide; convex face slightly rugulose. Fig. 109.

KENYA. Kiambu District: Ruiru, 12 July 1952, *Kirrika* 211!; Kitui District: Yatta Plateau, Jan. 1937, *Gardner* in *F.D.* 3619!; Teita District: Mt. Kasigau, from Rukanga, 5 Apr. 1969, *Faden, Evans & Rathbun* 69/430!
DISTR. K4, 6, 7; not known elsewhere
HAB. Forest margins, especially on rocky ground; 740–1700 m.

SYN. *P. kaessneri* S. Moore in J.B. 43: 250 (1905). Type: Kenya, Machakos District, Mukaa [Muka], *Kassner* 920 (BM, holo.!, K, iso.!)

11. P. linearifolia *Bremek.* in F.R. 37: 128 (1934); K.T.S.: 459 (1961). Type: Kenya, Tana River District, Kinakomba [Kima-Kombo], *Leroy* 1039 (P, holo.!)

Shrub or small tree 1.5–4.5 m. tall; young branches glabrescent or more often pubescent; older branches with pale reddish-brown, thin, shiny bark. Leaf-blades linear, 6.8–12.5 cm. long, 0.8–1.6(–2) cm. wide, apex acute and often mucronulate, base obtuse or rounded, shiny, glabrous with midrib impressed above, finely greyish tomentose with midrib prominent and glabrescent to sparsely pubescent beneath, margins revolute; bacterial nodules punctate above, slightly raised beneath, few to numerous, scattered; petiole short, 2–3 mm. long, glabrescent to pubescent; stipule-limbs buff or brown, papery, truncate, 1.5–4 mm. long, sparsely hairy outside, silky hairs absent inside; arista 3.5–6 mm. long, hairy. Corymbs terminal on main and short leafy lateral branches, moderately compact, 2.3–3.3 cm. across; primary inflorescence-branches 0.4–0.9 cm. long, pubescent; secondary to tertiary branches present; pedicels 1–2.5 mm. long, pubescent; bracts ± 2 mm. deep, sparsely pubescent outside; bracteoles brown, papery. Calyx pubescent; tube 0.75–1 mm. long; limb-tube 0.75–1 mm. long; lobes ovate-triangular, 0.5–1(–1.5) mm. long, 1 mm. wide, obtuse to rounded or slightly acuminate. Corolla-tube 7–8.5 mm. long, 1.5 mm. wide at top, glabrous or with very few hairs outside, hairy at throat becoming glabrous towards the base within; lobes oblong-lanceolate, 4–5 mm. long, 1.25 mm. wide, obtuse to rounded, ciliate. Fruit black, 5–6 mm. in diameter, sparsely pubescent. Seed 4 mm. wide, immature. Fig. 106/3.

KENYA. Kilifi District: Marafa, 20 Nov. 1961, *Polhill & Paulo* 821!; Tana River District: Kinakomba [Kima Kombo], Dec. 1889, *Leroy* 1039!; Lamu District: Kitangani, 27 Dec. 1946, *Adamson* 308 in *Bally* 5999!
TANZANIA. Tanga District: Gombero Forest Reserve, 14 Nov. 1964, *Mgaza* 667!; Bagamoyo District: along Mapanda R., Kikoka Forest Reserve, Apr. 1964, *Semsei* 3802!; Uzaramo District: Msua, 1 Nov. 1925, *Peter* 31676!
DISTR. K 7; T 3, 6; not known elsewhere
HAB. In coastal bushland, beside rivers, in gullies; ± 50–100 m.

12. P. macrosepala *Hiern* in F.T.A. 3: 172 (1877); Bremek. in F.R. 37: 139 (1934); T.T.C.L.: 510 (1949); Vollesen in Op. Bot. 59: 70(1980). Type: Tanzania/Mozambique, Rovuma Bay, *Kirk* (K, holo.!)

Shrub or small tree up to 6 m. tall; young branches pubescent or glabrous, older branches covered with thin peeling corky bark. Leaf-blades elliptic or slightly obovate, 4.4–10 cm. long, 2.3–6.8 cm. wide, apex acute to rounded, sometimes shortly mucronulate, base acute to obtuse, entirely glabrous or pubescent on both faces; bacterial nodules linear, restricted to midrib, few or absent; petioles 1.2–2.8 cm. long, glabrous, or pubescent; stipules broadly ovate or occasionally subtruncate, mucronulate or shortly aristate, pubescent or glabrous outside, with few silky hairs inside. Corymbs terminal on leafy branches, or sometimes with additional subterminal corymbs, lax, 3–8 cm. across, few-flowered; primary inflorescence-branches 0.6–1.5 cm. long, glabrous or pubescent; secondary branches absent or present; pedicels 0.4–1.8 cm. long, glabrous or pubescent; bracts stipule-like; bracteoles very reduced or absent. Calyx-tube 2–3 mm. long, sparsely to densely pubescent; limb-tube (2–)3–5 mm. long; lobes lanceolate, 9–12 mm. long, 2–3 mm. wide, acuminate, costa prominent, sparsely pubescent to pubescent. Corolla-tube 4.4 cm. long, 1.25 mm. wide at top, glabrous or pubescent outside and within (see note); lobes ovate-oblong, 12–14 mm. long, 3.5 mm. wide, acuminate. Fruit straw-coloured, 1.3 cm. across, shiny, pubescent or ? glabrous, calyx-lobes persistent, accrescent; pyrene white, pithy. Seeds dull brown, 5 mm. wide, deeply wrinkled on convex face.

var.macrosepala

Leaves entirely glabrous; young branches and inflorescence-branches glabrous or pubescent; corolla glabrous outside.

TANZANIA. Kilwa District: Selous Game Reserve, Malemba, 15 Apr. 1970, *Rodgers* in *M.R.C.* 1030! & 25 Feb. 1976, *Vollesen* in *M.R.C.* 3279!; Mikindani District/Mozambique: Ruvuma [Rovuma] Bay, Mar. 1861, *Kirk*!
DISTR. T 8; ? Mozambique
HAB. Deciduous coastal thicket on sand; ± 400 m.

SYN. *P. macrosepala* Hiern var. *glabra* Bremek. in F.R. 37: 139 (1934), *nom. superfl.* Type as for species

var. puberula *K. Schum* in P.O.A. C: 388 (1895); Bremek. in F.R. 37: 139 (1934); T.T.C.L.: 510 (1949); Vollesen in Op. Bot. 59: 70 (1980). Type: Tanzania, Uzaramo [Usaramo], *Stuhlmann* 7611 (B, holo.†)

Leaf-blades sparsely pubescent above, pubescent beneath; petioles pubescent; young branches and inflorescence-branches always pubescent; corolla pubescent outside. Fig. 106/15.

TANZANIA. Uzaramo District: Bando Forest Reserve, Feb. 1965, *Procter* 2894!; Rufiji District: Utete, Kibiti, 18 Dec. 1968, *Shabani* 243!; Ulanga/Kilwa District: Selous Game Reserve, Nahomba, 27 Feb. 1971, *Ludanga* in *M.R.C.* 1283!
DISTR. T 6, 8; not known elsewhere
HAB. Thicket; ± 100 m.

13. P. schliebenii *Bremek.* in F.R. 49: 84 (1939); T.T.C.L.: 510 (1949). Type: Tanzania, Lindi District, Lake Lutamba, *Schlieben* 5845 (B, holo.†, BM, BR, P, iso.!)

Shrub 2–3 m. tall; branches with grey-buff bark, flaking, hairy when young. Leaf-blades ovate to obovate, 6.7–16 cm. long, 3.8–6.8 cm. wide, apex obtuse or rounded, with a short calloused mucro, base acute to cuneate, thinly herbaceous, sparsely hirsute on both faces, with denser spreading hairs on the nerves beneath; bacterial nodules linear, few, restricted to the midrib and bases of the lateral nerves; petiole 1–2 cm. long, hirsute; stipules ± ovate, up to 5 mm. long, shortly acuminate, sparsely pubescent outside, with dense silky hairs inside. Corymbs terminal on leafy branches, lax, 5–10 cm. across; primary inflorescence-branches 0.5–2 cm. long; third to fourth order branches present; pedicels 3–8 mm. long, covered with stiff rust-coloured hairs; bracts ovate, 4–5 mm. long, hirsute outside, with dense silky hairs inside; bracteoles inconspicuous. Calyx-tube 1.5–2 mm. long, densely covered with rust-coloured stiff spreading hairs; limb-tube 1.5–2 mm. long; lobes variable within each flower, lanceolate or less often linear-lanceolate or narrowly ovate, 10–12 mm. long, 2.5–4 mm. wide, obtuse, hirsute. Corolla-tube 1.3 cm. long, 4 mm. wide at top, with a few coarse hairs outside, sparsely pubescent within; lobes oblong-oblanceolate, 14 mm. long, 4.5 mm. wide, obtuse, sparsely hirsute beneath. Style up to 6 cm. long, ± 1 mm. wide, with 1–2 veins apparent. Fruit and seeds not known.

TANZANIA. Lindi District: Lake Lutamba, 8 Jan. 1935, *Schlieben* 5845! & Rutamba, 23 Mar. 1943, *Gillman* 1353!

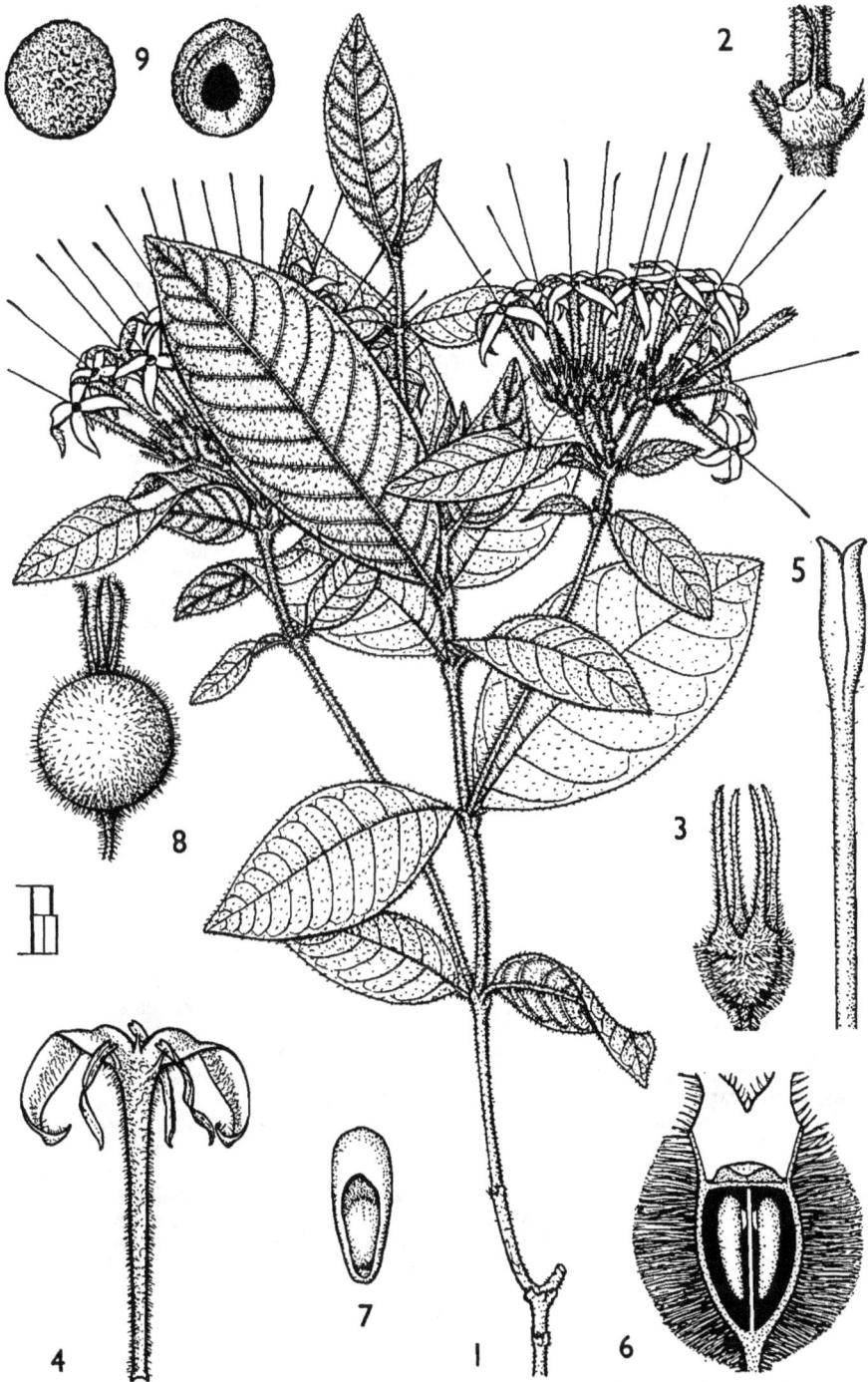

FIG. 110. *PAVETTA OLIVACEO-NIGRA* — 1, flowering branch, × ⅓; 2, stipule, × 3; 3, calyx, × 5; 4, half corolla, × 3; 5, part of style with stigma, × 20; 6, section through calyx-tube to show ovary, × 14; 7, placenta partly surrounding ovule, dorsal view, × 20; 8, fruit, × 2; 9, seed, 2 views, × 3. 1–7, from *Semsei* 3459; 8, 9, from *Muze* 35. Drawn by Diane Bridson.

DISTR.   **T** 8; not known elsewhere
HAB.   Deciduous woodland or thicket; 250 m.

NOTE.   Whitish to clear plastids (not to be confused with raphides) are visible in the leaves and calyx-lobes of the type specimen, further gatherings are needed to determine if this is a constant character.

14. P. olivaceo-nigra *K. Schum.* in P.O.A. C: 388 (1895); Bremek. in F.R. 37: 146 (1934); T.T.C.L.: 511 (1949); Bridson in K.B. 32: 614 (1978). Type: Tanzania, W. Usambara Mts., Heboma, *Holst* 2574 (B, holo.†, COI, K!, iso.)

Shrub 2–3.5 m. tall; young branches pubescent; older branches with thin fawn corky bark. Leaf-blades usually drying blackish, elliptic, rarely narrowly elliptic, 3.3–10.3(–14.5) cm. long, 1.2–4.7(–4.9) cm. wide, apex acute or more usually acuminate, base obtuse to acute, densely pubescent on both faces, lateral nerves usually impressed above; bacterial nodules absent; petiole 0.3–0.8 cm. long, densely covered with patent hairs; stipule-limbs brown, papery, truncate, 1.5–2.5 mm. long, pubescent outside, with fine silky hairs inside; arista usually drying blackish, 3–8 mm. long, pubescent. Corymbs terminal on lateral 3.5–11.5 cm. long branches bearing 1–3 pairs of leaves, moderately compact, (1.3–)2–4.5 cm. across; primary inflorescence-branches 0.3–1 cm. long, densely covered with patent hairs; secondary branches present; pedicels 1–3 mm. long, densely covered with patent hairs; bracts stipule-like; bracteoles inconspicuous. Calyx-tube 1.5–2 mm. long, very densely covered with white or less often rusty patent hairs; limb-tube 0.5–1 mm. long; lobes ± subulate above, wider at base, 5–11 mm. long, 0.75–1 mm. wide at base, covered with patent hairs. Corolla-tube 1.3–1.6 cm. long, 1.5 mm. wide at top, covered with spreading hairs outside, pubescent inside; lobes oblong, 7–9 mm. long, 1.75–2.5 mm. wide, obtuse, acute or slightly acuminate. Fruit black, 0.9–1.1 cm. in diameter, pubescent. Seed blackish, 5 mm. wide, convex face slightly rugulose. Fig. 110.

TANZANIA. Lushoto District: Lushoto, 5 May 1962, *Semsei* 3475! & Kitivo Forest Reserve, 19 Apr. 1962, *Semsei* 3459! & 23 June 1958, *Muze* 35!
DISTR.   **T** 3; not known elsewhere
HAB.   Forest, persisting on cultivated ground; 1350–1850 m.

SYN.   *P. corethrogyne* Bremek. in K.B. 3: 350 (1949); Verdc. in K.B. 11: 453 (1957). Type: Tanzania, Lushoto District, E. Usambara Mts., Mtai, *Zimmermann* in Herb. Amani 1771 (EA, holo.!)

15. P. filistipulata *Bremek.* in K.B. 1936: 483 (1936) & in F.R. 47: 83 (1939); T.T.C.L.: 510 (1949). Type: Tanzania, Morogoro District, Uluguru Mts., Morningside, *E.M. Bruce* 1120 (K, holo.!, BM, iso.!)

Shrub 1–2 m. tall; young branches covered with patent hairs; older branches with thin, fawn corky bark. Leaf-blades usually drying blackish, oblanceolate to narrowly obovate, 9–16.5 cm. long, 3–5.6 cm. wide, apex acuminate, base acute to cuneate, sparsely pubescent-strigose with midrib densely pubescent and lateral nerves often impressed above, pubescent with dense spreading hairs on nerves beneath; bacterial nodules absent; petioles 0.4–1.1 cm. long, densely pubescent; stipule-limbs fawnish, tending to be corky, triangular to subtruncate, 2.75–4 mm. long, pubescent outside; arista drying blackish, 9–14 mm. long, pubescent. Corymbs borne on lateral branches 12–16 cm. long bearing usually 1–2 pairs of leaves, moderately lax, 4–4.5 cm. across; primary inflorescence-branches 0.6–1.4 cm. long, covered with patent hairs; secondary branches present; pedicels 2–3 mm. long, covered with patent hairs; bracts ± stipule-like; bracteoles inconspicuous. Calyx-tube ± 2 mm. long, very densely covered with white to fawn patent hairs; limb-tube 0–2.5 mm. long; lobes subulate, 2.5–7 mm. long, less than 0.25 mm. wide at base, covered with patent hairs. Corolla-tube 1.4–2.1 cm. long, 2 mm. wide at top, pubescent outside, sparsely pubescent inside; lobes ± 0.7–1.2 cm. long, ± 2 mm. wide. Fruit blackish, 0.7 mm. in diameter, pubescent. Mature seeds not known.

TANZANIA. Morogoro District: Uluguru Mts., Tegetero, 20 Mar. 1953, *Drummond & Hemsley* 1691! & Morningside, June 1953, *Semsei* 1234! & without exact locality, 21 Nov. 1932, *Wallace* 472!
DISTR.   **T** 6; not known elsewhere
HAB.   Shrub layer of forest; 1100–1500 m.

16. P. coelophlebia *Bremek.* in K.B. 11: 171 (1956). Type: Tanzania, Morogoro District, S. Nguru Mts., Ruhamba Peak, *Drummond & Hemsley* 1967 (K, holo.!)

Shrub 2 m. tall; young branches covered with patent hairs; older branches covered with fawn, corky bark. Leaf-blades drying greenish, narrowly elliptic to oblanceolate-oblong, 9–15 cm. long, 3–5.5 cm. wide, apex acute, base acute, sparsely strigose above with denser hairs on the midrib, lateral and tertiary nerves impressed above, pubescent-strigose with denser spreading hairs on the nerves beneath; bacterial nodules absent; petiole 1–1.8 cm. long, densely pubescent; stipule-limbs fawn, corky, truncate or triangular, 3 mm. long, sparsely pubescent outside, arista drying blackish, 5–6 mm. long, pubescent. Corymbs borne on lateral branches 19 cm. long bearing 2 pairs of leaves, moderately compact, ± 3 cm. across; peduncle 0.5 cm. long; primary inflorescence-branches 5–6 mm. long, covered with spreading hairs; secondary branches present; pedicels 2–5 mm. long, covered with spreading hairs; bracts stipule-like; bracteoles inconspicuous. Calyx-tube ± 2 mm. long, very densely covered with white, patent hairs; limb-tube 1 mm. long, sparsely hairy; lobes very narrowly triangular to subulate, 2 mm. long, 0.75 mm. wide at the base, sparsely hairy. Corolla-tube 2.1 cm. long, 2 mm. wide at top, hairy outside, pubescent within; lobes oblong-lanceolate, 9–10 mm. long, 2.5 mm. wide, acuminate. Fruit not known.

TANZANIA. Morogoro District: S. Nguru Mts., Ruhamba Peak, 2 Apr. 1953, *Drummond & Hemsley* 1967!
DISTR. T 6; known only from the type gathering
HAB. Upland rain-forest; 1900 m.

17. **P. lynesii** *Bridson* in K.B. 42: 253, fig. 2 (1987). Type: Tanzania, Njombe, *Lynes* F.r. 63 (K, holo.!)

Small tree or shrub (or ? liana) 1–5 m. tall; young branches pubescent; older branches with fawn corky bark, gnarled. Leaf-blades drying greenish-brown, thinly chartaceous to chartaceous, oblanceolate to narrowly obovate or elliptic to oblong-elliptic, 6–18.5 cm. long, 2.5–7 cm. wide, apex acute or acuminate, base acute to cuneate, sparsely strigose with denser hairs on the midrib above, pubescent-strigose with denser spreading hairs on the nerves beneath; bacterial nodules elliptic near main and lateral nerves, often obscured by indumentum; petiole 0.4–1.7 cm. long, densely pubescent; stipule-limbs, triangular, 6–10 mm. long, acute to acuminate but not aristate, pubescent or densely strigose outside, becoming corky, eventually falling. Corymbs terminal on lateral branches 1–19 cm. long bearing 1(–2) pairs of leaves or with leaves fallen, ± compact to moderately lax, 4–7 cm. across; primary inflorescence-branches 0.4–1.2 cm. long, strigose; secondary and tertiary branches present; pedicels 3–6 mm. long, covered with spreading hairs; bracts ± stipule-like, densely or sometimes sparsely strigose outside; bracteoles inconspicuous or not. Calyx-tube 1–1.25 mm. long, densely covered with white to straw-coloured hairs, patent or not; limb-tube 0–0.5(–1) mm. long; lobes narrowly linear, 2–4(–5) mm. long, ± 0.5 mm. wide, pubescent. Corolla-tube 1–1.6 cm. long, 2 mm. wide at top, glabrous or pubescent outside; lobes narrowly oblong, 6–8 mm. long, 2–3 mm. wide, acute to acuminate or obtuse. Fruit 8–9 mm. in diameter, pubescent. Seeds blackish, area surrounding excavation dull brown, ± 6 mm. wide, convex face rugulose.

TANZANIA. Kilosa District: Ukaguru Mts., Mamiwa Forest Reserve, summit to E. of Ikwamba Peak, 15 Aug. 1972, *Mabberley & Alehe* 1471!; Iringa District: Uzungwa Range, Nyumbenito, Dec. 1981, *Rodgers & Hall* 1462! & Mufindi, Lupeme Tea Estate, 6 May 1968, *Renvoize & Abdallah* 1954!
DISTR. T 6, 7; not known elsewhere
HAB. Forest; 1200–2300 m.

SYN. [*P. trichosphaera* sensu Bremek. in F.R. 37: 138 (1934), pro parte; T.T.C.L.: 510 (1949), quoad *Lynes* F.r. 63, *non* Bremek.]

18. **P. sansibarica** *K. Schum* in E.J. 28: 79 (1899); Bremek. in F.R. 37: 137 (1934); Bridson in K.B. 32: 614 (1978). Types: Zanzibar, *Marseeler* 49 (B, syn.†) & 31 (B, syn.†, BR, isosyn.!)

Shrub 1.5–5 m. tall; young branches sparsely covered with long or short, spreading hairs; older branches covered in fawn or light brown bark. Leaf-blades oblanceolate or narrowly to broadly elliptic, 7.5–20 cm. long, 2.4–8 cm. wide, apex acuminate, base cuneate, sparsely strigose or pubescent with denser hairs on the midrib above and the nerves beneath; bacterial nodules present or absent; petiole 0.9–4 cm. long, pubescent; stipule-limbs subtruncate to triangular, 1–3 mm. long, sometimes corky around the margin, sparsely hairy outside, with stiff hairs inside; arista 0.4–1.2 cm. long, sparsely hairy. Corymbs terminal on lateral branches 8–19 cm. long bearing 2–3 pairs of leaves, or occasionally on the main branches, lax to very lax, 5–10 cm. across, few–many-flowered; primary inflorescence-branches 0.5–3 cm. long, pubescent; secondary to tertiary branches present; pedicels 2.5–11 mm. long, sparsely or densely covered in patent long or shortish hairs; bracts stipule-like; bracteoles inconspicuous. Calyx-tube 1.5 mm. long, very densely or densely covered in white to rusty patent hairs; limb-tube 0.5–1(–1.75) mm. long, sparsely pubescent to pubescent; lobes narrowly triangular to linear-subulate, 1–3 mm. long, ± 0.25–0.5 mm. wide at base, sparsely pubescent to pubescent. Corolla-tube (1.2–)1.6–2 cm. long, 2 mm. wide at top, very sparsely pubescent to pubescent outside, sparsely pubescent inside; lobes oblong, 8–11 mm. long, 2–2.5 mm. wide, acute to acuminate. Fruit blackish, ± 0.8 cm. in diameter, sparsely pubescent. Seed 6 mm. wide, slightly rugulose on convex face.

KEY TO INFRASPECIFIC VARIANTS

Leaves oblanceolate or oblanceolate-elliptic, 7.5–14 cm. long, 2.4–4.1 cm. wide, midrib usually drying blackish; pedicels 2–7 mm. long . . . . subsp.**sansibarica**
Leaves narrowly to broadly elliptic, 9–20 cm. long, 4.5–8 cm. wide, midrib not drying blackish; pedicels 5–11 mm. long (subsp. *trichosphaera*):
  Calyx-lobes 1.5–3 mm. long; calyx-tube covered with hairs 1–1.5 mm. long; hairs on leaves long . . . . var. **trichosphaera**
  Calyx-lobes not exceeding 1.2 mm. long; calyx-tube covered with hairs 0.5–0.75 mm. long; hairs on leaves short . . . var. **rufipila**

subsp. sansibarica

Branches slender. Leaf-blades oblanceolate or oblanceolate-elliptic, 7.5–14 cm. long, 2.4–4.1 cm. wide, apex narrowly acuminate, base narrowly cuneate, sparsely strigose above, strigose-pubescent with denser patent hairs on the nerves beneath; midrib usually drying blackish; bacterial nodules few, on the midrib and lateral nerves or scattered. Pedicels 2–7 mm. long. Calyx-tube densely or very sparsely (both extremes occur on one sheet) covered with whitish or rust-coloured patent hairs ± 1 mm. long; lobes 1–2 mm. long.

TANZANIA. Zanzibar I., Fumba, 16 Feb. 1930, *Vaughan* 1220! & Kombeni Caves, 7 Sept. 1930, *Vaughan* 1485! & without locality, Sept. 1895, *Sacleux* 858!
DISTR. Z; not known elsewhere
HAB. Coastal bushland; 2–10 m.

subsp. trichosphaera (*Bremek.*) *Bridson* in K.B. 32: 614 (1978). Type: Kenya, Kwale District, Shimba Hills, *Gardner* in F.D. 1445 bis (K, holo.!)

Branches not markedly slender. Leaf-blades narrowly to broadly elliptic, 9–20 cm. long, 4.5–8 cm. wide, apex acuminate, base cuneate, sparsely strigose or pubescent; midrib not drying blackish; bacterial nodules present or absent. Pedicels 5–11 cm. long.

var. trichosphaera

Leaf-blades elliptic to broadly elliptic, sparsely strigose above, pubescent-strigose with denser patent hairs on the nerves beneath. Pedicels pubescent. Calyx-tube covered with white to buff very dense patent hairs ± 1–1.5 mm. long; calyx-lobes 1.5–3 mm. long. Corolla-tube pubescent outside. Fig. 106/10.

KENYA. Kwale District: Shimba Forest, 1927, *Gardner* in F.D. 1445 bis! & Mwele Mdogo Forest, 5 Feb. 1953, *Drummond & Hemsley* 1131A!
TANZANIA. Lushoto District: Amani Institute, 10 Nov. 1971, *Furuya* 167!; Lushoto/Tanga District: Sigi–Longuza [Longusa], 6 Feb. 1918, *Peter* 22642!; Pangani District: Lagoni, Mwera, 26 June 1957, *Tanner* 3581!
DISTR. K 7; T 3; not known elsewhere
HAB. Forest; 45–950 m.

SYN. *P. trichosphaera* Bremek. in F.R. 37: 138 (1934) pro parte; K.T.S.: 462 (1961)
  P. shimbensis Bremek. in K.B. 11: 172 (1956); K.T.S.: 460 (1961). Type: Kenya, Kwale District, Shimba Hills, Mwelo Mdogo Forest, *Drummond & Hemsley* 1131A (K, holo.!)

NOTE. The type specimen of subsp. *trichosphaera*, *Gardner* in F.D. 1445 bis should not be confused with *Gardner* in F.D. 1445; the Kew sheet of the latter is *P. tarennoides*, while the EA sheet is a mixture of *P. tarennoides* and *Ixora*. The type of *P. shimbensis* Bremek. has broader leaves, a calyx with a longer limb-tube (1.75 mm.) and a shorter corolla-tube (12 mm.) than the rest of the specimens seen, but does not seem to be worth keeping up as an additional variety. One specimen, *Paulo* 827, while clearly belonging to this variety in regard to the length of the calyx-lobes, has an indumentum which approaches that of var. *rufipila* Bremek.

var. **rufipila** (*Bremek.*) *Bridson* in K.B. 32: 616 (1978). Type: Tanzania, Lushoto District, Amani, *Zimmermann* (K, holo.!)

Leaf-blades narrowly elliptic to elliptic, with sparse shorter stiff hairs above and sparsely pubescent with pubescent nerves beneath. Pedicels sparsely hairy. Calyx-tube covered with straw-coloured to rusty dense patent hairs 0.5–0.75 mm. long; calyx-lobes up to 1.2 mm. long. Corolla-tube sparsely pubescent outside.

TANZANIA. Lushoto District: Kwamkuyu–Amani, 18 Jan. 1917, *Peter* 19001! & 1 May 1914, *Peter* 55507! & 6 Jan. 1915, *Peter* 55603!
DISTR. **T** 3; not known elsewhere
HAB. Rain-forest; 600–650 m.

SYN. *P. rufipila* Bremek. in F.R. 37: 137 (1934); T.T.C.L.: 510 (1949)

19. **P. elliottii** *K. Schum. & K. Krause* in E.J. 39: 551 (1907); Bremek. in F.R. 37: 146 (1934); T.S.K.: 146 (1936); K.T.S.: 458 (1961); Bridson in K.B. 32: 616 (1978). Type: Kenya, ? Nairobi, *Elliott* 76 (B, holo.†, K, iso.!)

Shrubs, sometimes scrambling, 0.45–2 m. tall; young branches whitish tomentose; older branches with fawnish buff thin corky bark. Leaf-blades ovate, elliptic or rarely obovate, 2.5–6.5 cm. long, 1.5–4.7 cm. wide, apex obtuse to acuminate, rarely rounded, base cordate, less often rounded or obtuse, rarely cuneate, pubescent or scabrid-pubescent above, tomentose beneath; bacterial nodules small, dot-like, few-numerous, scattered; petiole short, 2–6 mm. long, tomentose; stipule-limbs brown, papery, truncate or occasionally broadly ovate, up to 5 mm. long, pubescent outside, with few silky hairs inside; arista 1.5–2 mm. long, hairy. Corymbs terminal on main and short lateral branches 0.8–15 cm. long bearing 1–3 pairs of leaves, compact, (0.8–)1.2–3 cm. across; primary inflorescence-branches up to 4 mm. long; secondary branches absent or very reduced; pedicels 0.5–2.5 mm. long, tomentose or less often sparsely pubescent; bracts 2–4 mm. long, pubescent outside; bracteoles brown, papery, small. Calyx-tube 1.5–2 mm. long, sparsely to densely pubescent; limb-tube 0.75–1 mm. long, slightly wider than tube; lobes narrowly triangular to narrowly lanceolate, rarely lanceolate or narrowly triangular below and subulate above, 2–5 mm. long, (0.25–)0.5–1(–1.5) mm. wide, densely pubescent. Corolla-tube 1.5–2.8 cm. long, 1.5–2.5 mm. wide at top, glabrous to pubescent, not bearded at throat; lobes lanceolate to oblong-lanceolate, 5–6(–8) mm. long, 1–3 mm. wide, acute to obtuse, often shortly acuminate. Fruit black, 8–9 mm. in diameter, shiny, slightly hairy. Seed 5 mm. wide, convex face slightly and finely rugulose.

var. elliottii

Leaf-blades ovate to elliptic; base cordate or less often rounded or obtuse. Calyx-lobes narrowly triangular, narrowly lanceolate or rarely lanceolate, 0.5–1(–1.5) mm. wide at base.

KENYA. Nairobi District: Ngong Rd, forest near Rowallan Scout Camp, 24 Feb. 1980, *Gilbert* 5830! & Muthaiga, Feb. 1938, *V.G. van Someren* in C.M. 7178!; Machakos District: E. of Kitandi, 23 Mar. 1965, *Gillett* 16671!
DISTR. **K** 4; not known elsewhere
HAB. Upland bushland and forest; 1650–1830 m.

NOTE. Bremekamp cites *Drake Brockmann* s.n. from Ethiopia, Harar, as belonging to this species; it would, however, seem better placed in *P. oliveriana* Hiern. *Verdcourt* 531 (Thika Road House) differs from the typical material in having the calyx-tube very sparsely hairy, while *Bally* 11471 (Nairobi District, Mbagathi) differs in the greater width (1.5 mm.) of the calyx-lobes. It is doubtful that either would merit varietal rank. *Prescott Decie* s.n. from North Nyeri, Nanyuki, has leaf-blades with obtuse bases and has been determined as *P. trichocalyx* Bremek. by Bremekamp, but is better placed in this variety.

var. **trichocalyx** (*Bremek.*) *Bridson* in K.B. 32: 616 (1978). Type: Kenya, Embu District, Chuka, *Fries* 1996 (S, holo.!)

Leaf-blades obovate; base cuneate. Calyx-lobes narrowly triangular at base, subulate above, 0.25–0.75 mm. wide at base.

KENYA. Embu District: Chuka, 26 Feb. 1922, *Fries* 1996!
DISTR. **K** 4; known only from the type

644 RUBIACEAE 79. PAVETTA

HAB. Moist montane forest remnants; ± 1400 m.

SYN. *P. trichocalyx* Bremek. in F.R. 47: 90 (1939); K.T.S.: 462 (1961)

20. P. oliveriana *Hiern* in F.T.A. 3: 174 (1877); K. Schum. in P.O.A. C: 388 (1895); Bremek. in F.R. 37: 147 (1934); T.S.K.: 146 (1936); F.P.N.A.: 359 (1947); T.T.C.L.: 511 (1949); F.P.S. 2: 457 (1952); K.T.S.: 460 (1961); Bridson in K.B. 32: 616 (1978); Fl. Pl. Lign. Rwanda: 578 (1983); Fl. Rwanda 3: 192 (1985). Type: Tanzania, Bukoba District, Karagwe, *Grant* 160 (K, holo.!)

Shrub or small tree, sometimes scrambling, 1.5–4.5(–7) m. tall; young branches pubescent, the older ones covered with thin buff to fawn bark. Leaf-blades elliptic or occasionally broadly or narrowly elliptic, (3–)4.7–18(–21.5) cm. long, 1.6–8.5(–9.4) cm. wide, apex usually shortly acuminate, base obtuse, truncate or sometimes unequal, finely pubescent above, tomentose beneath or rarely glabrous save for the midrib beneath and the ciliate margin; bacterial nodules dot-like, moderately small, scattered, usually numerous; petiole 0.4–2.8(–4) cm. long, tomentose; stipules triangular, 5–8 mm. long, acuminate, pubescent outside, without silky hairs inside. Corymbs terminal on main and lateral branches 2.5–14.5(–18.5) cm. long bearing 0–3 pairs of leaves at apex, sometimes with subsidiary corymbs from these nodes, lax or very lax, 2.2–9.5 cm. wide; primary inflorescence-branches 0.4–3 cm. long, tomentose; secondary and tertiary branches present; pedicels 1–8 mm. long, tomentose; bracts ovate, 4–8 mm. long, pubescent outside, with few silky hairs inside; bracteoles scale-like to filiform. Calyx tomentose; tube 1–2 mm. long; limb-tube 1 mm. long, scarcely wider than the tube; lobes triangular to narrowly triangular, 2–5 mm. long, 0.75–1 mm. wide at the base. Corolla-tube (1.4–)1.9–3.4 cm. long, 1.5 mm. wide at top, pubescent outside, not bearded at throat; lobes oblong, 4–6 mm. long, (1.25–)1.75–3 mm. wide, obtuse or acute. Fruit black, 0.8–1 cm. in diameter, sparsely pubescent; calyx-limb persistent; pedicels tending to be accrescent. Seeds brownish, 5 mm. wide, rugulose on convex face.

var. **oliveriana**

Leaf-blades finely pubescent above, tomentose beneath. Fig. 106/4.

UGANDA. Karamoja District: Mt. Moroto, June 1963, *Tweedie* 2640!; Ankole District: Kasyoha-Kitomi Forest, 28 Dec. 1969, *Synnott* 440!; Mengo District: Semunya Forest, 9 Dec. 1949, *Dawkins* 472!
KENYA. W. Suk District: 8.5 km. N. of Kapenguria, 28 July 1938, *Pole Evans & Erens* 1512!; Uasin Gishu District: Kapsaret Estate, Nov. 1948, *Bickford* in *Bally* 6537!; Masai District: Narok, Endama, 23 June 1961, *Glover et al.* 1973!
TANZANIA. Ngara District: Buseke, Keza, Bushubi, 20 May 1960, *Tanner* 4943!; Arusha District: Ngurdoto Crater National Park, 23 Feb. 1966, *Greenway & Kanuri* 12377!; Buha District: 8.5 km. N. of Kibondo, Feb. 1955, *Procter* 374!
DISTR. U 1–4; K 2, 3, 5, 6; T 1, 2, 4; Zaire, Burundi, Rwanda, Sudan, Ethiopia
HAB. Grasslands, thickets, evergreen forest and rock crevices; 900–2300 m.

SYN. *Pavetta sp.* sensu Thoms. in Speke's Nile Journ., App.: 636 (1863)
   [*Ixora (Pavetta) abyssinica* (Fresen.) Oliv. var. sensu Oliv. in T.L.S. 29: 87 (1873), *non* Fresen.]
   I. oliveriana (Hiern) Kuntze in Rev. Gen. Pl. 1: 287 (1891)
   Pavetta humbertii Bremek. in F.R. 37: 177 (1934); F.P.N.A.: 360 (1947). Type: Zaire, Kivu, Ninagongo, *Humbert* 7953 (BR, holo.!)
   P. lebrunii Bremek. in B.J.B.B. 14: 309 (1937). Type: Zaire, Kivu, Sake, *Lebrun* 5059 (BR, holo.!)
   P. sp. sensu Bremek. in Pich-Serm. in Ricerche Bot. 7, 1, Miss. Stud. Lago Tana: 143 (1951)

NOTE. *P. kiwuensis* K. Krause (in E.J. 43: 148 (1909); Bremek. in F.R. 37: 146 (1934); Type: Lake Kivu, *Keil* 219 (B, holo.†)) was probably just a narrow-leaved form of this species, as is *Gillman* 237 bis from Bukoba District, which had previously been determined as *P. kiwuensis* K. Krause by Bremekamp. *Wallace* 1322 is said to come from Southern Highlands Province, but the label is probably inaccurate.

var. **denudata** (*Bremek.*) *Bridson* in K.B. 32: 617 (1978). Type: Tanzania, Bukoba District, Nyakato, *Gillman* 237 (EA, holo.!)

Leaf-blades glabrous or with hairs on the midrib beneath or with a ciliate margin.

TANZANIA. Bukoba District: Nyakato, Apr. 1935, *Gillman* 237!
DISTR. T 1; Ethiopia
HAB. Fringing swamp; ± 1250 m.

SYN. *Pavetta denudata* Bremek. in K.B. 3: 353 (1949)

21. P. burttii *Bremek.* in F.R. 37: 147 (1934); T.T.C.L.: 511 (1949). Type: Tanzania, Kondoa District, Kinyassi Scarp, *B.D. Burtt* 948 (K, holo.!, EA, iso.!)

Shrub; young branches pubescent; older branches covered with gnarled buff bark. Leaf-blades rotund or broadly obovate, 6.7–15.3 cm. long, 4.6–10 cm. wide, apex rounded, apiculate or acute, base cuneate, pubescent above, tomentose beneath; bacterial nodules dot-like, small, scattered; petiole 0.5–2 cm. long, tomentose; stipules triangular, 0.6–1.2 cm. long, acuminate, pubescent outside, with few silky hairs at axils inside. Corymbs terminal on lateral branches 1–13.5 cm. long bearing 1–2 pairs of leaves at the apex, not markedly lax, 2.5–4.5 cm. across; primary inflorescence-branches 2–3 mm. long, densely pubescent; secondary branches present; pedicels 0.15–1 cm. long, pubescent; bracts ± ovate, 5–6 mm. long, pubescent outside; bracteoles scale-like to filiform. Calyx pubescent to tomentose; tube 1 mm. long; limb-tube 1–1.25 mm. long, slightly wider than the tube; lobes narrowly triangular to linear, 5–8 mm. long, 0.75–1 mm. wide at base. Corolla-tube 2 cm. long, 1.5 mm. wide at top, glabrous or pubescent outside, not bearded at throat; lobes ± oblong, 6 mm. long, 1.5 mm. wide, acute. Fruit not known.

TANZANIA. Mbulu District: Ufiome [Ufiume] Mt., on Irago Hill, 21 Jan. 1928, *B.D. Burtt* 1233!; Kondoa District; Kinyassi Scarp, 2 Jan. 1928, *B.D. Burtt* 948!
DISTR. T 2, 5; not known elsewhere
HAB. Forest or shaded upland wooded grassland; 1500–1700 m.

NOTE. The specimen from Mbulu District (apparently not seen by Bremekamp) has been included in this species rather than in *P. oliveriana* Hiern in spite of the pubescent corolla-tube. No specimens of *P. oliveriana* have been recorded from Mbulu District. See also note after next species.

22. P. greenwayi *Bremek.* in K.B. 3: 354 (1949). Type: Tanzania, Mbulu District, E. slopes of Mt. Hanang, Nangwa, *Greenway* 7582 (K, holo.!, EA, iso.)

A virgately branched shrub up to 2–4.5 m. tall; young branches tomentose; older branches with fawn-brown bark. Leaf-blades broadly elliptic, 11–18.5 cm. long, 6–11 cm. wide, apex obtuse to acute, base acute or unequal, pubescent above, tomentose to densely pubescent beneath; bacterial nodules small, scattered, few; petiole 0.5–2.5 cm. long, tomentose; stipules triangular, 0.8–1.5 cm. long, acuminate, pubescent outside, with silky hairs in axils. Corymbs terminal on lateral branches 9.3–15 cm. long bearing 1–2 pairs of leaves at the apex, moderately compact, ± 3 cm. across; primary inflorescence-branches up to 4 mm. long, tomentose-pubescent; secondary branches usually present; pedicels 5–8 mm. long, densely pubescent. Calyx-tube 1.25 mm. long, densely pubescent; limb-tube 1 mm. long, slightly wider than the tube, pubescent; lobes narrowly triangular to linear, 7–9 mm. long, pubescent. Corolla-tube 1.1 cm. long, ± 1.25 mm. wide at top, glabrescent outside, not bearded at throat; lobes ± oblong, 5(–7) mm. long, 1.75–2(–2.8) mm. wide, obtuse. Fruit 8 mm. in diameter, sparsely pubescent. Mature seed not known.

TANZANIA. Mbulu District: E. slope of Mt. Hanang, Nangwa, 4 Feb. 1946, *Greenway* 7582! & Tarangire National Park, Sangaiwe Hills, 8 June 1972, *Vesey-FitzGerald* 7410!
DISTR. T 2; not known elsewhere
HAB. Undergrowth of evergreen forest; 1600–2200 m.

SYN. *P. sp. 29* sensu T.T.C.L.: 513 (1949)

NOTE. This species and *P. burttii* Bremek. have been very tentatively maintained. It is possible that they may perhaps only merit infraspecific rank under *P. oliveriana* Hiern. More gatherings from Mbulu, Kondoa and neighbouring districts are required.

23. P. sp. B

Weak-stemmed shrub or climber, ± 1.8 m. tall; young stems densely pubescent; older stems with buff bark, sometimes gnarled. Leaf-blades narrowly elliptic to elliptic, 5.4–11 cm. long, 1.8–4.4 cm. wide, apex acuminate, base obtuse or sometimes acute, pubescent above and beneath; bacterial nodules dot-like to elliptic, scattered and along the nerves; petiole usually stout, 1.1–2.8 cm. long, densely pubescent; stipule-limbs truncate-triangular, 2–4 mm. long, pubescent outside, with few silky hairs inside; arista 2–3(–7) mm. long. Corymbs terminal on lateral branches 6.5–17 cm. long bearing 0–2 pairs of leaves at the apex, compact, 1.8–2 cm. across; primary and secondary inflorescence-branches ?1–2 mm. long; pedicels 1–2 mm. long, densely pubescent; bracts not seen. Calyx densely pubescent; tube 1 mm. long; limb-tube 0.75–1 mm. long, slightly wider than the tube; lobes narrowly triangular-linear, 2.5 mm. long, 0.75–1 mm. wide at base. Corolla and style not known. Fruit black, 9 mm. in diameter, sparsely pubescent; calyx-limb persistent; disc remaining pale; pedicels accrescent. Seeds brownish, 5.5 mm. wide, finely rugulose on convex face.

TANZANIA. Mpanda District: Kungwe-Mahali Peninsula, Musenabantu, 13 Aug. 1959, *Harley* 9327! & Kungwe Mt., Ntale R., 1959, *Newbould & Harley!*
DISTR. T 4; not known elsewhere
HAB. Forest; 1750–1825 m.

NOTE. The fruits of *Harley* 9327 have the disc accrescent and ± beaked but this is probably an abnormal condition. This species seems fairly closely related to *P. oliveriana* Hiern.

24. P. lindina *Bremek.* in K.B. 8: 504 (1953). Type: Tanzania, Lindi District, Rondo Plateau, scarp face below Mchinjiri, *Eggeling* 6038 (U, holo.!, EA, iso.!)

Scrambling undershrub; young branches densely pubescent; older branches covered with brown thin flaking bark. Leaf-blades drying greenish brown, elliptic to oblong, 2–5.5 cm. long, 1–2.7 cm. wide, apex acute and mucronulate, base obtuse, covered with stiff shiny hairs above, greyish tomentose beneath; bacterial nodules ± dot-shaped, along the midrib and lateral nerves; petioles 0.3–1 cm. long, tomentose; stipules ovate-triangular to triangular, gradually tapering to aristate, apex 4–7 mm. long, pubescent outside, with silky hairs inside. Corymbs terminal on very slender ± 0.75 mm. wide lateral branches 3.2–4.5 cm. long bearing 1–2 pairs of leaves near the apex, few-flowered, compact, ± 0.5–1 cm. across, sometimes with smaller secondary corymbs arising from the next node down; primary inflorescence-branches up to 2 mm. long, tomentose; secondary branches not present; pedicels 0.5–3 mm. long, tomentose; bracts stipule-like; bracteoles inconspicuous. Calyx-tube 1 mm. long, densely whitish tomentose; limb-tube 1 mm. long, not or scarcely wider than the tube, pubescent; lobes linear to subulate, 6–6.5 mm. long, 0.9–1 mm. wide at base, pubescent. Corolla-tube 1.4–1.5 mm. long, 1.25 mm. wide at top, sparsely pubescent outside not bearded at throat; lobes oblong, 5 mm. long, 1.75 mm. wide, with apiculum drying blackish, glabrous outside save for few hairs on the apiculum. Fruit not known.

TANZANIA. Lindi District: Rondo Plateau, below Mchinjiri, Feb. 1951, *Eggeling* 6038!
DISTR. T 8; known only from type gathering
HAB. Closed forest; 800–850 m.

25. P. refractifolia *K. Schum* in P.O.A. C: 388 (1895); Bremek. in F.R. 37: 158 (1934); T.T.C.L.: 513 (1949); Bridson in K.B. 32: 619 (1978); Vollesen in Op. Bot. 59: 70 (1980). Type*: Tanzania, Uzaramo District, Kikuli [Kikulu], *Stuhlmann* 6850 (B, lecto.†)

Small shrub 1–11.5 m. tall; young branches pubescent; older branches covered with buff, thin cracking bark. Leaf-blades elliptic to obovate, 4–11.6 cm. long, 2–5.8 cm. wide, apex usually shortly acuminate, occasionally rounded to acute, base cuneate, scabrid to pubescent above, pubescent to densely pubescent beneath; bacterial nodules, small dot-like, mostly along the lateral nerves plus a few scattered, numerous or few; petiole 0.4–1.5 cm. long, densely pubescent; stipules triangular, 0.4–1 cm. long, acuminate, pubescent outside. Corymbs terminal on lateral branches 2.3–15.3 cm. long bearing 1–2 pairs of leaves at the apex or rarely leafless, compact, 2–4.4 cm. across; primary inflorescence-branches 0–3 mm. long; secondary branches present or reduced; pedicels 0.5–3 mm. long, densely pubescent. Calyx-tube 1–1.25 mm. long, densely pubescent; limb-tube 0–0.5 mm. long, scarcely wider than the tube; lobes linear-lanceolate to lanceolate or occasionally linear, 5–8 mm. long, (0.75–)1–1.5 mm. wide, pubescent. Corolla-tube 0.9–1.7 cm. long, 1–2 mm. wide at top, pubescent outside or rarely glabrous, not bearded at throat; lobes oblong-lanceolate, 4–7 mm. long, (1–)1.5–2 mm. wide, acuminate, pubescent outside. Fruit black, 7–8 mm. in diameter, sparsely pubescent; calyx-limb persistent; pedicels thickening and slightly accrescent. Mature seeds not known. Fig. 106/12.

TANZANIA. Mpwapwa, 13 Jan. 1930, *Hornby* 157!; Rufiji District: Kibiti, 18 Dec. 1968, *Shabani* 245!; Kilwa District: Selous Game Reserve, W. of Lungonyo [Lungonya] R., 11 Nov. 1967, *Rodgers* 80!

* K. Schumann cited *Stuhlmann* 2615 from Zaïre, Muana, Duki River in addition to 6850, but 2615 falls outside the known geographical range of *P. refractifolia*; only the above syntype was cited by Bremekamp and this is accepted as lectotypification.

DISTR.   T 3, 5, 6, 8; not known elsewhere
HAB.   Wooded grassland, riverine thicket and forest; 150–1320 m.

SYN.   [*P. cooperi* sensu Bremek. in F.R. 37: 158 (1934), pro parte; Brenan, T.T.C.L.: 512 (1949), quoad
        Swynnerton 763 & *Brulz* 3123, *non* Harv. & Sond.]
        P. cephalotes Bremek. in K.B. 11: 176 (1956). Type: Tanzania, Mpwapwa, *Musomi* in van Rensburg
        596 (K, holo.!)
        P. handenina Bremek. in K.B. 11: 177 (1956). Type: Tanzania, Handeni District, 20 km. from
        Handeni on Mziha road, *Drummond & Hemsley* 1460 (K, holo.!)

NOTE.   Bremekamp cited two specimens from Tanzania as belonging to *P. cooperi* Harv. & Sond.,
        Swynnerton 763 from Morogoro [Mologolo] which belongs to this species and *Brulz* 3123 (B †) from
        Dar es Salaam which possibly also belonged to this species. *P. klotzschiana* sensu Brenan in
        T.T.C.L. and *P. gracilis* Klotzsch *non* A. Rich. recorded by Schumann in E.J. 28: 494 (1900) both
        referring to *Goetze* 107 (B †) also possibly relate to this species.
           P. incana Klotzsch (*P. klotzschiana* K. Schum.) from Mozambique, Malawi and Zimbabwe is very
        close to this species but may be distinguished as the corolla is glabrous outside. The type of *P.
        handenina* Bremek. has a corolla with the tube glabrous but the lobes hairy outside.

26. P. cataractarum S. *Moore* in J.B. 57: 89 (1917); Bremek. in F.R. 37: 158 (1934);
F.F.N.R.: 415 (1962); Bridson in K.B. 32: 622 (1978); Kok & Grobbelaar in S. Afr. J. Bot. 3:
136 (1984). Type: Zimbabwe, Victoria Falls, *F.A. Rogers* 5553 (BM, holo.! K, iso.!)

Shrub 0.5–4.5 m. tall; young branches sparsely pubescent to glabrescent or rarely
glabrous; older branches covered with thin grey or buff bark. Leaf-blades drying greyish
bronze, discolourous, obovate or occasionally broadly obovate or oblanceolate, 6–16.3
cm. long, 2.9–8.3 cm. wide, apex rounded and usually shortly acuminate, base cuneate,
sparsely pubescent to glabrous or scabrid above, pubescent to sparsely pubescent or
rarely glabrous beneath; nerves whitish, acutely angled, prominent beneath; bacterial
nodules dot-like to elongate, along the nerves and scattered, few–many; petiole 0.9–2.5
cm. long, pubescent to glabrescent; stipule-limbs usually greenish, triangular, 6–14 mm.
long, cuspidate, pubescent to sparsely pubescent outside, with silky hairs inside. Corymbs
terminal on branches 1–18 cm. long bearing up to 3 pairs of leaves near the apex,
compact to moderately lax, 2–4.7(–6) cm. across; primary inflorescence-branches 0.1–1
cm. long, pubescent; secondary and tertiary branches present or reduced; pedicels
(2–)4–7 mm. long, pubescent or rarely sparsely pubescent. Calyx-tube 1 mm. long, densely
pubescent; limb-tube 0.5–1 mm. long, slightly wider than the tube; lobes lanceolate or
linear-lanceolate, 3–5(–7) mm. long, 1–1.25 mm. wide, pubescent. Corolla-tube 0.8–1.5
cm. long, 1.5 mm. wide at top, glabrous or rarely hairy outside, not bearded at throat; lobes
oblanceolate-oblong, 3.5–6 mm. long, 1.25–2 mm. wide, rounded or sometimes
acuminate. Fruit 6–8 mm. in diameter, shiny, sparsely pubescent, calyx-limb persistent;
pedicels accrescent. Seeds greyish, 5–6 mm. wide, slightly rugulose on convex face.

TANZANIA.   Tabora/Dodoma/Chunya District: Rungwa Forest Reserve, 0.5 km. W. of Camp 2, 26
        Jan. 1969, *V.V. Gilbert* in *C.A.W.M.* 5281!; Dodoma District: Kisigo R., ± 10 km. W. of junction with
        Great Ruaha R., 21 Aug. 1970, *Thulin & Mhoro* 776!; Iringa District: Mpululu, by the Rangers' Post,
        17 Apr. 1970, *Greenway & Kanuri* 14373!
DISTR.   T ?4, 5, 7, ?8; Zaire, Malawi, Zambia, Zimbabwe and Caprivi Strip
HAB.   Riverine forest; 700–1000 m.

SYN.   *P. conflatiflora* S. Moore in J.B. 57: 90 (1917). Type: Zambia, Livingstone, *F.A. Rogers* 13535 (BM,
        holo.!)
        P. cataractarum S. Moore var. *hirtiflora* Bremek. in K.B. 8: 501 (1954). Type: Zimbabwe, Urungwe
        District, Chirindu, on Zambesi bank, *Whellan* 332 (SRGH, holo.!, K, iso.!)

NOTE.   One specimen *Busse* 1022, probably from Abdalah Haman's village, Rovuma, has been
        determined *P. cataractarum* by Bremekamp. This is probably correct but the specimen is atypical
        and shows some tendencies towards *P. refractifolia*.

27. P. sp. C

Shrub; young branches sparsely pubescent, older branches with reddish-brown bark.
Leaves sessile; blades drying bronze-grey, oblanceolate, 7.3–10 cm. long, 2.2–4.8 cm. wide,
apex rounded, sometimes shortly acuminate, base cuneate, glabrous to glabrescent
above, sparsely pubescent with denser pubescence on the nerves beneath; bacterial
nodules dot-like, scattered, numerous; stipule-limbs truncate to triangular, 4–7 mm. long,
sparsely pubescent outside; arista slender 2–3 mm. long. Corymbs terminal on lateral
branches 1.5–2.9 cm. long bearing 2 pairs of leaves at the apex, compact, 3–3.5 cm. across;
pedicels ± 2 mm. long, pubescent; bracts not known. Calyx-tube 1.25 mm. long, glabrous;

limb-tube 1 mm. long, slightly wider than the tube, glabrous; lobes narrowly lanceolate, 1.1–1.3 cm. long, 1–1.25 mm. wide, ciliate. Corolla-tube ± 1.6 cm. long, ± 1 mm. wide at top, glabrous outside, not bearded at throat; lobes ± 5 mm. long. Immature fruit black, 6 mm. in diameter.

TANZANIA. Kigoma District: Uvinsa, E. of Lugufu, 9 Feb. 1926, *Peter* 36610!
HAB. Unknown; 1060 m.
DISTR. T 4; known only from the one specimen

NOTE. This species is very close to *P. cataractarum* S. Moore, but can be easily distinguished by the sessile leaves, more compact inflorescence and longer calyx-lobes.

28. P. sphaerobotrys *K. Schum* in E.J. 28: 494 (1900); Bremek. in F.R. 37: 160 (1934); T.T.C.L.: 512 (1949); Bridson in K.B. 32: 623 (1978). Type: Tanzania, Kilosa District, Luhembe [Ruhembe], *Goetze* 403 (B, holo.†, K, fragment!)

Bush or shrubby tree 3–9 m. tall; young branches glabrous; older branches with pale brown or grey-brown peeling bark. Leaves glabrous; blades often drying grey or greyish-brown, very narrowly elliptic to elliptic or occasionally tending to be oblanceolate, 3.5–15 cm. long, 1.5–5.4 cm. wide, apex acuminate or less often acute, base narrowly cuneate to attenuate; bacterial nodules dot-like to linear, along the midrib and lateral nerves plus a few scattered; petiole 0.9–3.8 cm. long; stipule-limbs thinly or very thinly membranous and translucent, truncate, caducous, 2–7 mm. long, with caudate tip up to 3 mm. long. Corymbs terminal on main or lateral branches 4.3–13 cm. long with up to 5 nodes and ± 2 pairs of leaves near the apex; compact to moderately compact, 1.7–3.5 cm. across; primary branches up to 4 mm. long, glabrous; secondary branches usually present; pedicels up to 6 mm. long, glabrous; bracts 3–7 mm. long, membranous to thinly membranous, with very few silky hairs inside. Calyx glabrous; tube 1–1.25 mm. long; limb-tube 0.25–1 mm. long, slightly wider than the tube; lobes lanceolate, 2–7 mm. long, 1–2 mm. wide (in flowering specimens) tending to reflex, often with nerves apparent. Corolla-tube 0.5–1 cm. long, 1–1.5 mm. wide at top, glabrous outside, not hairy at throat; lobes oblong to oblong-lanceolate, 3–6.5 mm. long, 1–2 mm. wide, acuminate. Fruit, blackish grey, 6–7 mm. in diameter, glabrous; calyx-limb persistent, accrescent; inflorescence-branches and pedicels accrescent. Seeds ± 5 mm. wide, rugulose on convex face.

KEY TO INFRASPECIFIC VARIANTS

Corolla-tube 10 mm. long; lobes 3 mm. long (see note)   subsp. **sphaerobotrys**
Corolla-tube 5–8(–9) mm. long; lobes 4–6.5 mm. long:
   Calyx-lobes 5–7 mm. long (in flowering specimens); limb-
     tube 0.25–0.5 mm. long; filaments attached not more
     than 0.25 mm. from base of anthers . . . . . . subsp. **lanceisepala**
   Calyx-lobes 2–3 mm. long (in flowering specimens); limb-
     tube 0.75–1 mm. long; filaments attached 0.75–1 mm.
     from base of anthers . . . . subsp. **tanaica**

subsp. **sphaerobotrys**

Stipules membranous. Corolla-tube 1 cm. long; lobes 3 mm. long (see note). Calyx-lobes 4–5 mm. long, 1.25–1.5 mm. wide; limb-tube 0.75–1 mm. long.

TANZANIA. Kilosa District: Luhembe [Ruhembe], *Goetze* 403! & Vigude Kidode, Nov. 1952, *Semsei* 1039!
DISTR. T 6; not known elsewhere
HAB. Forest; ± 300 m.

NOTE. The length of the corolla-lobes is taken from the original description and unconfirmed from the limited material seen.

subsp. **lanceisepala** (*Bremek.*) *Bridson* in K.B. 32: 623 (1978). Type: Tanzania, Tanga District, Amboni, Mkulumuzi, *Geilinger* 605 (K, holo.!)

Stipules membranous. Corolla-tube 5–7 mm. long; lobes 4–6.5 mm. long. Calyx-lobes 5–7 mm. long, 1.25–2 mm. wide, usually with distinct venation. Filaments attached not more than 0.25 mm. from the base of the anthers.

TANZANIA. Lushoto District: Makuyuni area, *Koritschoner* 1428!; Tanga District: Amboni-Pande, 22 Sept. 1918, *Peter* 22887a! & 24 Feb. 1918, *Peter* 23885!

DISTR. **T** 3; not known elsewhere
HAB. Unknown; 25–?1000 m.

SYN. *P. lanceisepala* Bremek. in K.B. 8: 502 (1954)

subsp. **tanaica** (*Bremek.*) *Bridson* in K.B. 32: 623 (1978). Type: Kenya, Tana River District, Tana R., near rapids, *Sampson* 80 (K, holo.!)

Stipules very thinly membranous. Corolla-tube (6.5–)7–8(–9) mm. long; lobes 4–5.5(–6.5) mm. long. Calyx-lobes 2–3(–5 in fruit) mm. long, 0.75–1(–1.25 in fruit) mm. wide, ± obscurely veined. Filaments attached 0.75–1 mm. from the base of the anthers. Fig. 106/7.

KENYA. Tana River District: Munazini, 7 Oct. 1972, *Homewood* 8! & E. of Wenje, Kipendi Camp, 30 Aug. 1976, *Kibuwa* 2500! & Bura, 7 Mar. 1963, *Thairu* 45!
DISTR. **K** 7; Somalia
HAB. Riverine forest; 30–350 m.

SYN. *P. transjubensis* Chiov., Fl. Som. 2: 249, fig. 147 (1932). Type: Somalia, Mansur, on the Juba R., *Guidotti* 222 (? BO, holo.)
      P. tanaica Bremek. in F.R. 47: 27 (1939); K.T.S.: 461 (1961)
      P. manamoca Bremek. in K.B. 8: 502 (1954); K.T.S.: 459 (1961). Type: Kenya, Tana River District, Garissa, *Adamson* in *Bally* 5851 (K, holo.!, EA, iso.!)

NOTE. (For species as a whole) The treatment of this aggregate is very tentative as most of the material seen is either immature or in young fruit. More collections are needed, especially from Kilosa with mature corollas.
      P. microphylla Chiov., *P. sennii* Chiov. and *P. corradiana* Cuf. from Somalia are related to this species.

29. P. gerstneri *Bremek.* in K.B. 8: 501 (1954); Bridson in K.B. 32: 624 (1978). Type: S. Africa, Natal. Zululand, Native Reserve 1, at Chief uMavabela Mhlul's kraal, *Gerstner* 4992 (K, holo.!, PRE, iso.)

Shrub or small tree 1.8–3 m. tall; young branches glabrous or pubescent, older branches with greyish buff bark. Leaf-blades oblanceolate, narrowly obovate or elliptic, 2.3–10.5 cm. long, 0.8–3(–4.4) cm. wide, apex obtuse or acute to subacuminate, mucronulate, base cuneate, subcoriaceous, glabrous on both faces or sometimes sparsely pubescent especially on the nerves beneath, occasionally revolute; bacterial nodules dot-like to linear, scattered and along the midrib and lateral nerves; petiole 0.5–1.3(–2) cm. long, glabrous or sparsely pubescent; stipule-limbs brownish, subtruncate, 1.5–2 mm. long, sparsely pubescent or glabrous outside; arista up to 1.5 mm. long. Corymbs borne on short slender lateral branches 2–8(–10.5) cm. long bearing 1–4 pairs of leaves near apex or well spaced or occasionally on main branches, subumbellate, ± compact, 1.5–3 cm. across; primary inflorescence branches 1–6 mm. long, glabrous or pubescent; secondary branches present or absent; pedicels 3–6 mm. long, glabrous or pubescent; bracts 2–5 mm. long, papery with silky hairs inside; bracteoles minute. Calyx-tube 0.75–1 mm. long, glabrous to sparsely pubescent; limb-tube 0.25(–1) mm. long; lobes usually lanceolate-linear, lanceolate or sometimes narrowly triangular, (1.5–)2–2.25(–3) mm. long, 0.5–1 mm. wide, acuminate, glabrous, ciliate or sparsely pubescent, reflexing in young fruit. Corolla-tube 0.9–1.1 cm. long, 1–1.25 mm. wide at top, glabrous outside, not bearded at throat; lobes ± oblong, 3–4.5 mm. long, 1.25–1.75 mm. wide, acuminate. Fruit black, 6–7 mm. in diameter, shiny, the calyx-limb persistent. Seed blackish, 4.5 mm. wide, very finely rugulose on convex face.

TANZANIA. Zanzibar, Ufufuma, 23 Dec. 1930, *Vaughan* 1752! & Kizimkazi, 10 Jan. 1931, *Vaughan* 1805! & Kufile cave well, 23 Feb. 1936, *Vaughan* 2317!
DISTR. **Z**; Mozambique, South Africa (Natal Province)
HAB. Not recorded

NOTE. The specimens from Zanzibar are entirely glabrous, and tend to have the calyx-limb towards the upper limits of the measurements given in the description. The type has pubescent pedicels, calyx-tube and petioles, but as some ± intermediate specimens occur varieties have not been recognized.

30. P. crebrifolia *Hiern* in F.T.A. 3: 172 (1877); Bremek. in F.R. 37: 152 (1934); T.T.C.L.: 512 (1949); K.T.S.: 457 (1961); Bridson in K.B. 32: 624 (1978). Type: Kenya, Kilifi District, Sabaki R., *Kirk* (K, holo.!)

Evergreen shrub 1.2–3.4(–4.7) m. tall; young branches glabrous or rarely pubescent; older branches with buff to fawn bark. Leaf-blades elliptic to obovate, 3.5–12(–15) cm.

long, 1.5–5.3(–7) cm. wide, apex acute, sometimes with a short acumen or sometimes rounded, base cuneate, subcoriaceous, ± shiny above, glabrous or rarely pubescent beneath; bacterial nodules dot-like scattered, numerous or few; petioles 0.3–2 cm. long,glabrous or pubescent; stipule-limbs subtruncate to broadly ovate, 2–7 mm. long, usually with the marginal area buff and slightly corky; arista up to 3 mm. long. Corymbs terminal on leafy lateral branches 3–15 cm. long, subumbellate, ± compact, 2–4 cm. across; primary inflorescence branches 2–5.8(–10) mm. long, glabrous or rarely sparsely pubescent; secondary branches present or absent; pedicels 2–9 mm. long, glabrous or sparsely pubescent; bracts 3–7 mm. long, papery. Calyx-tube ? whitish when fresh (usually drying blackish in contrast to the lobes and pedicel), 1–1.25 mm. long, glabrous or rarely pubescent; limb-tube 0.75–1 mm. long; lobes linear to lanceolate or narrowly triangular, 3.5–12 mm. long, 0.5–1(–1.75) mm. wide, acute, glabrous, ciliate or rarely pubescent. Corolla-tube 1–2.2 cm. long, 1.5–2 mm. wide at top, glabrous outside, not bearded at throat; lobes oblong-elliptic, 5–8 mm. long, 1.75–2.5 mm. wide, acuminate. Fruit greenish black, 0.8–1.1 cm. in diameter, shiny; calyx-limb persistent. Seed greyish black, 5 mm. wide, slightly rugulose on convex face.

KEY TO INFRASPECIFIC VARIANTS

Plants entirely glabrous:
  Calyx-lobes not exceeding 8.5 mm. long . . . . **var. crebrifolia**
  Calyx-lobes 10–12 mm. long . . . . **var. kimbozensis**
Plants with young branches, leaves beneath, pedicels and
  calyces pubescent . . . . **var. pubescens**

var. crebrifolia

Plants entirely glabrous; calyx-lobes 3.5–8.5 mm. long. Fig. 106/5.

KENYA. Teita District: Mt. Kasigau, 1 June 1969, *Gillett* 18747!; Kwale District: Marenji Forest Reserve, 6 Sept. 1957, *Verdcourt* 1904!; Kilifi District: Marafa, 19 Nov. 1961, *Polhill & Paulo* 801!
TANZANIA. Lushoto District: W. Usambara Mts., Lushoto–Mombo road, 0.8 km. NE of Vuga turnoff, 14 June 1953, *Drummond & Hemsley* 2921!; Handeni District: Kwa Mkono, 18 June 1973, *Archbold* 1679!; Morogoro District: Mtibwa Forest Reserve, July 1952, *Semsei* 1950!
DISTR. K 7; T 3, 6; Somalia
HAB. Coastal scrub or forest; 0–1500 m.

SYN. *Ixora crebrifolia* (Hiern) Kuntze in Rev. Gen. Pl. 1: 286 (1891)
    Pavetta blepharosepala K. Schum. in P.O.A. C: 388 (1895). Type: Tanzania, Lushoto District, Usambara Mts., Mashewa [Maschewa], *Holst* 8851 (B, holo.†, COI, K, iso.!)

var. kimbozensis (*Bremek.*) *Bridson* in K.B. 32: 624 (1978). Type: Tanzania, Morogoro District, Kimboza Forest Reserve, *Semsei* 1398 (U, holo, EA, K, iso.!)

Plants entirely glabrous; calyx-lobes 1–1.2 cm. long.

TANZANIA. Morogoro District; Kimboza Forest Reserve, 24 Jan. 1976, *Cribb & Grey-Wilson* 10406! & 31 Dec. 1974, *Polhill & Wingfield* 4627! & Kibungo Forest, 31 Aug. 1981, *Leliyo* 104!
DISTR. T 6; not known elsewhere
HAB. Forest on limestone; 400–460(–600) m.

SYN. *P. kimbozensis* Bremek. in K.B. 11: 176 (1956)

NOTE. This taxon has only been recognised at varietal level as one other specimen has been seen from Kimboza Forest (*Semsei* 761) which has calyx-lobes of only 6 mm. long. Vollesen (in Op. Bot. 59: 70 (1980)) cites this variety as occurring in Selous Game Reserve (T6), but the specimen (*M.R.C.* 4388) has not been seen.

var. pubescens *Bridson* in K.B. 32: 624 (1978). Type: Kenya, Lamu District, Boni Forest, Marrarani, *Gillespie* 327 (K, holo.!)

Plants with young branches, leaves beneath, pedicels and calyces pubescent.

KENYA. Lamu District: Boni Forest, Marrarani, 9 Sept. 1961, *Gillespie* 327!
DISTR. K 7; known only from the type
HAB. Forest edge; 30 m.

31. P. richardsiae *Bridson* in K.B. 32: 624, fig. 4 A,B (1978). Type: Tanzania, Lushoto Escarpment, *Richards* 21973 (K, holo.!)

Bush 1.2–2.1 m. tall; young branches glabrous; older branches covered with dark brown thin bark. Leaves glabrous; blades elliptic to obovate, 3.6–8 cm. long, 1.5–4.3 cm. wide, apex shortly acuminate or acute, base cuneate, subcoriaceous, slightly shiny above; bacterial nodules dot-like, scattered; petiole 2–7 mm. long; stipule-limbs brown when dry, lanceolate, 12–13 mm. long, shortly apiculate, papery, glabrous outside. Corymbs terminal on short lateral branches 6.5–10 cm. long bearing usually 1–2 pairs of leaves, very compact, 1.5–2 cm. across; inflorescence-branches absent or obscured by the bracts; pedicels 1–2 mm. long, glabrous; bracts pale brown, ± ovate, 0.7–1 cm. long, sheathing the inflorescence base, membranous to papery, glabrous outside, with few silky hairs inside. Calyx-tube ± 1 mm. long, glabrous; limb-tube ± 1 mm. long; lobes linear-lanceolate, 4 mm. long, 1 mm. wide, ciliate. Corolla-tube 8 mm. long, 1.5–2 mm. wide at top, glabrous outside, not bearded at throat; lobes ± oblong, 5 mm. long, 1.5–1.75 mm. wide, acuminate. Fruit not known.

TANZANIA. Lushoto District: W. Usambara Mts., Soni–Mombo road, Jan. 1967, *Procter* 3521! & Lushoto Escarpment, 15 Jan. 1967, *Richards* 21973!
DISTR. **T** 3; not known elsewhere
HAB. In secondary growth near river or dry hillside; 900 m.

32. P. tendagurensis *Bremek.* in F.R. 37: 154 (1934); T.T.C.L.: 513 (1949); Bridson in K.B. 32: 626 (1978). Type: Tanzania, Lindi District, Tendaguru, *Migeod* 52 (BM, holo.!)

Bush or small tree, 1.2–2 m. tall; young branches glabrous to pubescent; older branches covered with thin reddish-brown or grey bark. Leaf-blades drying bronze or greyish-bronze, oblanceolate to oblong, 5.3–11.2 cm. long, 1.5–5 cm. wide, apex acuminate, base cuneate, glabrous to glabrescent above, glabrous save for the nerves or pubescent beneath; bacterial nodules dot-like, few, scattered; petioles short, 1.5–3 mm. long; stipules triangular with a stiff acumen 0.4–1.2 cm. long, glabrous or pubescent outside. Corymbs terminal on lateral branches 3–10.5 cm. long bearing 2–3 pairs of leaves at the apex, compact, 2.2–4 cm. across; primary inflorescence-branches 1.5–2 mm. long; secondary branches very reduced; pedicels 1–2 mm. long, glabrous or pubescent; bracts ± 2 mm. long, stipule-like. Calyx-tube 1–1.25 mm. long, glabrous or densely pubescent; limb-tube 0.75–1 mm. long, slightly wider than the tube, glabrous or pubescent; lobes linear-lanceolate, 0.9–1.3 cm. long, 0.75–1 mm. wide, pubescent or ciliate. Corolla white or ? lavender; tube 1–1.4 cm. long, 1–1.25 mm. wide at top, glabrous outside, not bearded at throat; lobes ± oblong, 4–6 mm. long, 1.5–2 mm. wide, acuminate. Fruit and seeds not known.

var. **tendagurensis**

Calyx with tube densely pubescent and limb pubescent. Leaves pubescent beneath.

TANZANIA. Lindi District: Tendaguru, 1 Jan. 1926, *Migeod* 52!
DISTR. **T** 8; known only from the type
HAB. Not recorded

var. **glabrescens** *Bridson* in K.B. 32: 626 (1978). Type: Tanzania, Lindi District, Mlinguru, *Schlieben* 5821 (BM, holo.!, B, LISC, P, iso.!)

Calyx with tube and limb-tube glabrous and lobes ciliate. Leaves glabrous save for the sparsely pubescent nerves beneath.

TANZANIA. Lindi District: Mlinguru, 3 Jan. 1935, *Schlieben* 5821! & Rondo Plateau, Mchinjiri, Jan. 1952, *Semsei* 632! & Tendaguru, 25 May 1929, *Migeod* 521!
DISTR. **T** 8; not known elsewhere
HAB. Wooded grassland, shrubland or open parts of forest belt; 210–810 m.

NOTE. This may perhaps be related to or conspecific with *P. mocambicensis* Bremek., but the type (Mozambique, Ilha de Ibo, *Rodrig. de Carvalho*) which is said to be at Kew has not been found either at K or COI.

33. P. mufindiensis *Bridson* in K.B. 32: 626, fig. 4C, D (1978). Type: Tanzania, Iringa District, Mufindi, Kigogo Forest Reserve, *Richards* 15746 (K, holo.!)

Shrub 1.5–3.5 m. tall; young branches glabrous with red-brown thin peeling bark; older branches greyish-buff. Leaf-blades oblanceolate or narrowly obovate, 5.5–12 cm. long, 1.8–4.4 cm. wide, apex acuminate, base cuneate, glabrous above, glabrescent with midrib pubescent to glabrescent beneath; bacterial nodules dot-like, scattered and along the

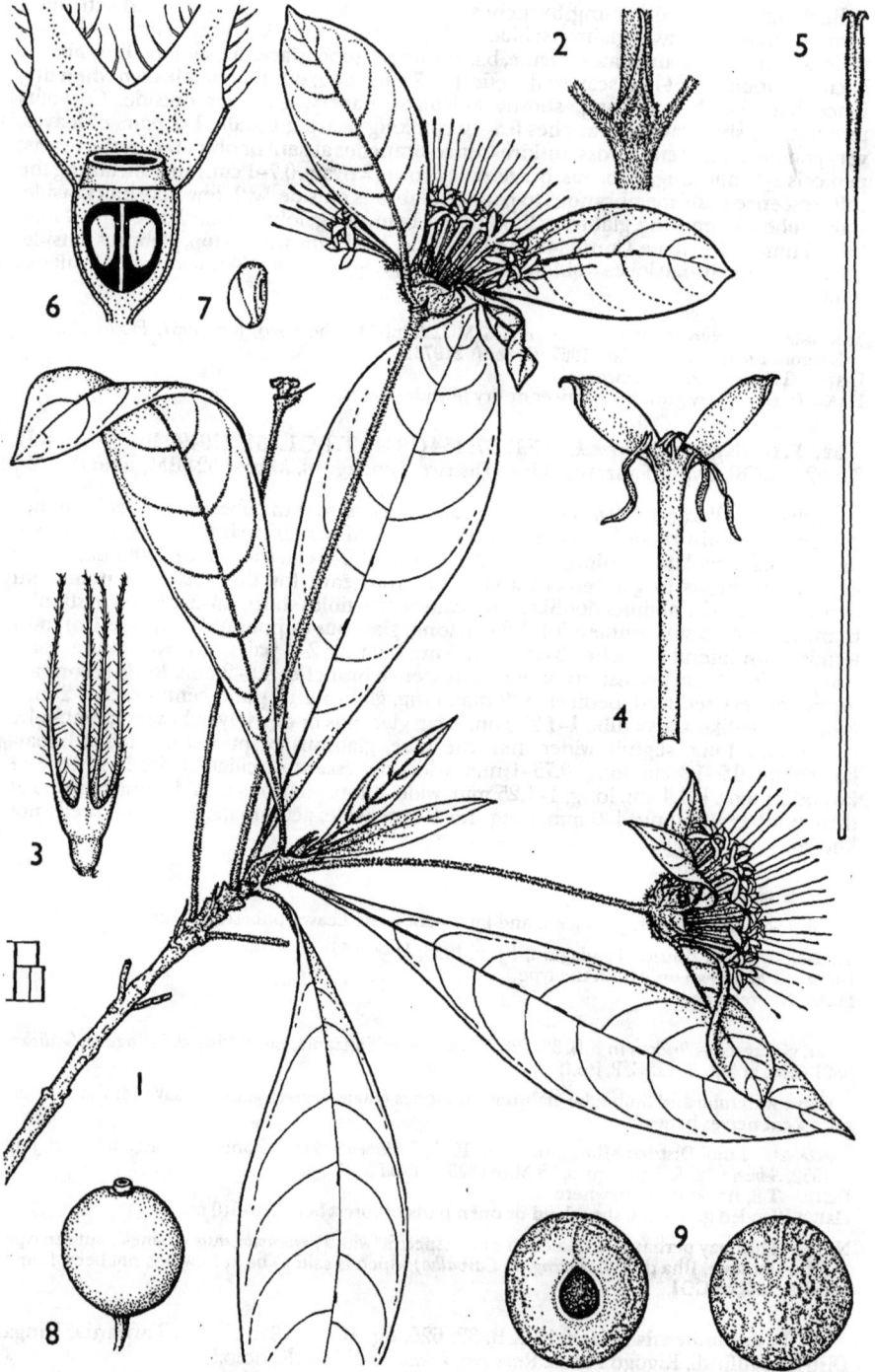

FIG. 111. *PAVETTA STENOSEPALA* subsp. *STENOSEPALA* — **1**, flowering branch, × ⅔; **2**, stipule, × 2; **3**, calyx, × 5; **4**, half corolla, × 3; **5**, style and stigma, × 3; **6**, section through calyx-tube to show ovary, × 20; **7**, ovule and placenta, lateral view, × 20; **8**, fruit, × 3; **9**, seed, × 4. 1, from *Renvoize & Abdallah* 1572; 2–7, from *Greenway* 3334; 8, 9, from *Dale* 3868. Drawn by Diane Bridson.

nerves; petiole 0.8–1 cm. long, glabrous to glabrescent; stipule-limbs triangular, 0.7–1.1 cm. long, shortly apiculate, glabrous outside, caducous. Corymbs terminal on short lateral branches 3.5–8.5 cm. long bearing 1–2 pairs of leaves, very compact, 1.5–2 cm. across; primary inflorescence-branches absent or rarely up to 3 mm. long; secondary branches absent; pedicels 2–4 mm. long, glabrous; bracts ovate, (2–)5–7 mm. long, papery, glabrous outside, with few silky hairs inside. Calyx glabrous; tube 1 mm. long; limb-tube 1 mm. long; lobes narrowly triangular to linear, 5–8 mm. long, 0.5–0.75 mm. wide. Corolla-tube 1.3–1.8 cm. long, 1.5–2 mm. wide at top, glabrous outside, not bearded at throat; lobes oblong or slightly oblanceolate, 4–7 mm. long, 1–2 mm. wide, acute. Style 3–3.7 cm. long. Fruit black, ± 0.8 cm. across, glabrous; calyx-limb persistent. Seeds not known.

TANZANIA. Iringa District: Mufindi District, Kigogo R. 19 Mar. 1962, *Polhill & Paulo* 1821A! & Kigogo Forest Reserve, 18 Dec. 1961, *Richards* 15746! & Nyalawa R. valley, Livalonge Tea Estate, 26 Aug. 1971, *Perdue & Kibuwa* 11249!
DISTR. T 7; not known elsewhere
HAB. Evergreen forest; 1650–1900 m.

34. P. stenosepala *K. Schum.* in P.O.A. C: 388 (1895); Bremek. in F.R. 37: 155 (1934); T.T.C.L.: 513 (1949); Bridson in K.B. 32: 627 (1978). Type: Tanzania, Tanga District, Amboni, *Holst* 2268 (B, holo.†)

An upright, straggling or creeping shrub 0.4–3(–4.5) m. tall; young branches pubescent or less often glabrous; older branches with red-brown, thin readily peeling bark. Leaves mostly confined to the apices of the branches; blades often drying bronze or bronze-grey above, narrowly oblanceolate to obovate, 3–13.5(–22) cm. long, 1.3–4.6(–8) cm. wide, apex acute to shortly acuminate, base cuneate to narrowly cuneate, scabrid, glabrescent or sometimes glabrous above, scabrid to pubescent or glabrous beneath; bacterial nodules dot-like or elliptic, mostly occurring on the midrib and lateral nerves but with a few scattered; petiole 0–2(–3) cm. long, pubescent or less often glabrous; stipules only shortly connate, triangular to ovate and gradually caudate, up to 1.4 cm. long, the caudate tip often equalling the limb, pubescent or less often glabrous outside, glabrous within. Corymbs terminal on lateral branches 1.8–15.5 cm. long bearing 2–3 pairs of leaves near the apex, capitate, 1–2 cm. across; inflorescence-branches absent; pedicels very reduced or up to 1 mm. long; bracts and bracteoles ± round, 0.3–1.5 cm. long with or without a caudate tip up to 0.5 cm. long, pubescent outside or less often glabrous, moderately to densely covered with white silky hairs inside, usually sheathing the base of the inflorescence. Calyx-tube ± 0.75 mm. long, glabrous or sparsely hairy; limb-tube up to 1 mm. long; lobes filiform or sometimes linear (in fruiting stage), 0.35–1.2 cm. long, sparsely ciliate to pubescent. Corolla-tube 1–2 cm. long, 0.75–1.5 mm. wide at top, glabrous or sparsely hairy towards top outside, not bearded at throat; lobes oblong-elliptic, 3–5.5 mm. long, 1–2 mm. wide, apex shortly acuminate, sometimes pubescent outside. Fruit greyish to black, 5.5–7 mm. in diameter, shiny; calyx-limb persistent; fruiting pedicels sometimes up to 3 mm. long. Seeds blackish, 5–6 mm. wide.

KEY TO SUBSPECIES

Young branches pubescent; leaves pubescent or scabrid beneath; calyx-lobes 0.5–1.2 cm. long, pubescent . . .	a. subsp. **stenosepala**
Young branches glabrous; leaves glabrous beneath; calyx-lobes 3–8 mm. long, sparsely ciliate, or glabrous:	
Shrub 0.75–2.4 m. tall; stipule-limbs triangular to ovate (2.5–)4–8(–14) mm. long; calyx-lobes 3.5–8 mm. long	b. subsp. **kisarawensis**
Shrub 0.3–0.5 m. tall; stipule-limbs truncate to triangular, not exceeding 2 mm. long; calyx-lobes ± 3 mm. long	c. subsp. **A**

a. subsp. **stenosepala**

Young branches always pubescent; leaves pubescent or scabrid beneath. Calyx-lobes 0.5–1.2 cm. long, filiform, pubescent. Fig. 111.

KENYA. Kwale District: Shimba Hills, Pengo Forest, 11 Feb. 1952, *Drummond & Hemsley* 1207!; Kilifi District: Arabuko, May 1929, *R.M. Graham* in F.D. 2149!; Lamu District: Witu, Utwani Ndogo Forest, Apr. 1957, *Rawlins* 425!
TANZANIA. Lushoto District: E. Usambara Mts., Nderema, by Kwamkuyu R., 24 Dec. 1956, *Verdcourt*

1717!; Pangani District: Bushiri, 31 Mar. 1950, *Faulkner* 550!; Uzaramo District: Banda Forest Reserve, Nov. 1964, *Procter* 2736!
DISTR. K 7; T 3, 6; Z; not known elsewhere
HAB. Evergreen forest or thickets and bushland; 0–1300(–1800) m.

SYN. *P. involucrata* Engl. in Abh. Preuss. Akad. Wiss.: 45 & 51 (1894), *nomen., non P. involucrata* Thw.
       P. crebrifolia Hiern var. *involucrata* K. Schum. in P.O.A. C: 388 (1895). Types: Tanzania, Lushoto District, Usambara Mts., Derema [Nderema], *Holst* 2268 (B, syn.†, K, isosyn.!) & Kwa Mshusa *Holst* 9199a (B, syn.†)
       P. mangallana K. Schum. & K. Krause in E.J. 39: 548 (1907); Bremek. in F.R. 37: 156 (1934); T.T.C.L.: 512 (1949); K.T.S.: 459 (1961). Type: Tanzania, Dar es Salaam, *Holtz* 370 (B, holo.†)

NOTE. Specimens from lower altitudes generally have the leaves pubescent beneath (corresponding to the description of *P. mangallana* K. Schum. & K. Krause), while those from higher altitudes are scabrid (corresponding to the description of *P. stenosepala* K. Schum.), somewhat intermediate or occasionally pubescent. Also the lower altitude specimens (with the exception of those seen from Uzaramo District) have bracts 0.3–0.7(–0.8) cm. long, while those from higher altitudes tend to have bracts (0.7–)0.8–1.5 cm. long. As there is a certain amount of character overlap *P. mangallana* K. Schum & K. Krause has not been recognised as a variety.

b. subsp. **kisarawensis** (*Bremek.*) *Bridson* in K.B. 32: 627 (1978); Vollesen in Op. Bot. 59: 70 (1980). Type: Tanzania, Uzaramo District, Kisarawe area, Vikindu Forest Reserve, *Paulo* 142 (U, holo., EA, iso.!, K, iso.!)

Bushes 0.75–2.4 m. tall; young branches glabrous; leaves glabrous beneath; stipule-limbs triangular to ovate, 4–8(–14) mm. long. Calyx-lobes 3.5–8 mm., filiform to linear (usually linear in fruit), sparsely ciliate or sometimes glabrous.

TANZANIA. Morogoro District: Dindili Hill, 30 June 1983, *Polhill, Lovett & Hall* 4952!; Ulanga District: Magombera Forest Reserve, 2 Nov. 1961, *Semsei* 3378!; Iringa District: Mwanihana Forest Reserve, 25 Nov. 1979, *Rodgers & Bulstrode* 17!
DISTR. T 6, 7; not known elsewhere
HAB. Forest; 100–300(–1700) m.

SYN. *P. kisarawensis* Bremek. in K.B. 11: 175 (1956)

NOTE. Both this subspecies and subsp. *stenosepala* occur in Uzaramo and Morogoro Districts. Field observations to determine whether the two taxa are sympatric or not would be of interest.

c. subsp. A

Shrubs 0.3–0.5 m. tall; young branches glabrous; leaves glabrous beneath; stipule-limbs truncate to triangular, not exceeding 2 mm. long; bracts ? smaller than in subsp. *kisarawensis*. Calyx-lobes ± 3 mm. long, ? glabrous.

TANZANIA. Rufiji District: Mafia I., Ras Mbisi, 27 July 1932, *Schlieben* 2599!
DISTR. T 6; known only from above specimen
HAB. Bushland; ± sea-level

NOTE. It is difficult to decide if the above specimen, which was apparently not seen by Bremekamp, truly represents a subspecies of *P. stenosepala* (as placed here) or a new species. Additional collections may elucidate this problem.

35. P. lutambensis *Bremek.* in F.R. 47: 87 (1939); T.T.C.L.: 512 (1949); Vollesen in Op. Bot. 59: 70 (1980). Type: Tanzania, Lindi District, Lake Lutamba, *Schlieben* 5593 (B, holo.†, B, BM, BR, P, iso.!)

Small shrub 1 m. high; branches leafless or with only young leaves at time of flowering, covered with fawn flaking bark, finely pubescent when young. Leaf-blades oblanceolate to obovate, 5.3–11.7 cm. long, 1.7–5 cm. wide, apex shortly acuminate, sometimes mucronate, base cuneate, sparsely scabrid to scabrid-pubescent above, sparsely pubescent to pubescent on nerves beneath; bacterial nodules absent or few, small, dot-like, mostly confined to the nerves; petiole 0.5–1 cm. long, sparsely pubescent; stipules truncate to triangular, 3–4 mm. long, with caudate tip caducous, densely scabrid-pubescent outside. Inflorescences terminal on lateral branches 1–9 cm. long eventually bearing 2 pairs of leaves near apex, capitate, 1–1.5 cm. across; pedicels up to 1 mm. long; bracts 0.3–1 cm. long, acuminate, pubescent-scabrid outside, with dense silky hairs inside; bracteoles with a membranous truncate limb bearing numerous fimbriae. Calyx pubescent; tube 0.75 mm. long; limb-tube 0.5 mm. long; lobes narrowly triangular-subulate, 2.3–3 mm. long, 0.3 mm. wide. Corolla-tube 0.8–1.2 cm. long, 0.5–1 mm. wide at top, glabrescent outside, sparsely pubescent at throat; lobes oblong, 4–5 mm. long, ± 1.5 mm. wide, obtuse, slightly acuminate. Fruit greenish grey or brownish when ripe, 5–8 mm.

in diameter, pubescent, shiny; calyx-limb persistent at first but eventually falling. Mature seeds not known.

TANZANIA. Kilwa District: Selous Game Reserve, Muvende road, 30 Dec. 1969, *Ludanga* in *M.R.C.* 861! & Kingipura Forest, 18 Dec. 1976, *Vollesen* in *M.R.C.* 4236!; Lindi District: Liho stream near Palihope, 13 Nov. 1903, *Busse* 2845!
HAB. Deciduous woodland and stream-banks; 120–210 m.
DISTR. DISTR. T 6, 8; not known elsewhere

36. P. amaniensis *Bremek.* in F.R. 47: 88 (1939); T.T.C.L.: 513 (1949); Bridson in K.B. 32: 627 (1978). Type: Tanzania, Lushoto District, Amani, *B.D. Burtt* 233 (K, holo.!)

Shrub, 1–1.5(–3.5) m. tall; young branches usually glabrous but sometimes very sparsely hairy or puberulous; older branches with pale mid-brown bark. Leaves mostly confined to the apices of the branches; blades often drying blackish blue, elliptic to broadly elliptic or sometimes oblanceolate, 3.5–15.5 cm. long, 2–6.4 cm. wide, apex acuminate, sometimes shortly mucronulate, base acute or less often cuneate, glabrous to glabrescent or occasionally sparsely puberulous above and beneath; bacterial nodules small, dot-like, few–many, scattered; petiole 0.3–2.5 cm. long, puberulous or glabrous; stipules triangular, 6–10 mm. long, only shortly connate, caudate or sometimes aristate, glabrous to glabrescent outside, glabrous or with few silky hairs inside. Corymbs terminal on lateral branches 2.5–15.3 cm. long bearing 0–2 pairs of leaves near the apex, very congested, 1.5–3 cm. across; primary inflorescence branches 0.5–3(–5) mm. long; secondary branches very reduced or absent; pedicels 1–1.5(–3) mm. long, glabrous to glabrescent; bracts pale brown, 4–6 mm. long, much smaller than width of inflorescence, papery, glabrous to glabrescent outside, with very few silky hairs inside. Calyx-tube 1–1.2 mm. long, glabrous or sparsely hairy; limb-tube 0.75–1.75 mm. long; lobes triangular at base, filiform above, 0.35–2 cm. long, up to 0.75 mm. wide at base, very sparsely pubescent to pubescent. Corolla-tube 1.4–2.2 cm. long, 1.5(–2) mm. wide at top, glabrous outside, not bearded at throat; lobes oblong-elliptic, 4.5–5.7 mm. long, 1.5–2(–2.5) mm. wide, acumen green, keeled. Fruit black, 8–9 mm. in diameter, glabrous. Seeds greyish, 5–6 mm. wide.

var. **amaniensis**
Calyx-lobes 0.35–7 mm. long. Stipules gradually tapered, or sometimes with a short arista. Leaves papery, elliptic to oblanceolate, seldom broadly elliptic.

TANZANIA. Lushoto District: Korogwe District, Potwe Forest, 30 Dec. 1960, *Semsei* 3143! & Amani, 5 Apr. 1922, *Soleman* in *Herb. Amani* 5970!; Tanga District: E. Usambara Mts., Mlinga Peak, 7 Mar. 1953, *Drummond & Hemsley* 1450!
DISTR. T 3; not known elsewhere
HAB. Shrub layer and edges of forests; 290–1250 m.
SYN. [*P. bagshawei* sensu Bremek. in F.R. 37: 151 (1934) pro parte, quoad specim. ex Tanzania; T.T.C.L.: 512 (1949), quoad *Soleman* in *Herb. Amani* 5970 & 5971 *non* S. Moore]

var. **trichocephala** (*Bremek.*) *Bridson* in K.B. 32: 628 (1978). Type: Tanzania: Lushoto District, Makuyuni area, *Koritschoner* 603 (K, holo.!, EA, iso.!)

Calyx-lobes 0.7–2 cm. long. Stipules usually with a distinct arista. Leaves thicker, elliptic to broadly elliptic.

TANZANIA. Lushoto District: W. Usambara Mts., Sakare–Bungu road, 4 July 1953, *Drummond & Hemsley* 3162! & valley near Mazumbai Forest Reserve, 27 Apr. 1975, *Hepper & Field* 5192! & Mazumbai, 1 Jan. 1981, *L. Tanner* 249!
DISTR. T 3; not known elsewhere
HAB. Shrub layer of forest; 400–1150 m.
SYN. [*Pavetta bagshawei* sensu Bremek. in F.R. 37: 151 (1934), pro parte, quoad specim. cit. ex Tanzania, *non* S. Moore]
P. trichocephala Bremek. in F.R. 47: 88 (1939); T.T.C.L.: 513 (1949)

NOTE. This species sometimes grows together with *P. stenosepala* K. Schum. subsp. *stenosepala* and mixed gatherings have occurred (e.g. *Verdcourt* 106). *P. amaniensis* is easily distinguished by the much smaller bracts, with very sparse silky hairs inside, the keeled green acumen on the corolla-lobes and the leaves which are glabrous to glabrescent beneath with scattered bacterial nodules.

37. P. comostyla *S. Moore* in J.L.S. 40: 98 (1911); Bremek. in F.R. 37: 150 (1934); Bridson in K.B. 32: 628 (1978). Type: Zimbabwe, Chirinda Forest, *Swynnerton* 75 (BM, holo.!, K, iso.!)

Shrub or small tree, sometimes scrambling, 1.8–3.6 m. tall; young branches glabrous to glabrescent; older branches covered with light brown to buff bark. Leaf-blades membranous to papery, usually drying blue-black, elliptic, broadly elliptic or obovate, 6.5–17.5 cm. long, 2.3–7.8 cm. wide, apex acuminate, base cuneate, attenuate or truncate then attenuate, glabrous to sparsely pubescent above and beneath; bacterial nodules dot-like to elliptic along the nerves and with a few scattered; petioles frequently drying blackish, (0.7–)1–2 cm. long, glabrous to glabrescent; stipules triangular, 3–9 mm. long, acuminate, glabrous to pubescent outside, with silky hairs inside. Corymbs terminal on lateral branches 6.2–17 cm. long bearing 2–3 pairs of leaves at the apex, compact to moderately lax, 2–6 cm. across, typically congested but rarely few-flowered; primary inflorescence-branches up to 1.4 cm. long, glabrous or pubescent; secondary and tertiary branches present or absent; pedicels 1.5–6 mm. long, glabrous or pubescent; bracts 4–5 mm. long, membranous. Calyx-tube usually drying blackish, 1–1.5 mm. long, glabrous or pubescent; limb-tube 0.5–1 mm. long, glabrous or pubescent; lobes linear to narrowly triangular, 0.4–1.1 cm. long, 0.25–1(–1.25) mm. wide, glabrous or pubescent. Corolla-tube 1.1–2 cm. long, 1.5–2 mm. wide at top, glabrous outside, not bearded at throat; lobes oblong, 5–8(–10) mm. long, 1.5–2.5 mm. wide, apex acuminate, often thickened and greenish. Fruit black, 7–9 mm. in diameter, glabrous to glabrescent; calyx-limb persistent. Ripe seeds not seen.

subsp. **nyassica** (*Bremek.*) *Bridson* in K.B. 32: 629 (1978). Type: Tanzania, Rungwe District, Kondeland, *Stolz* 85 (K, holo., not found)

Calyx-lobes linear, only slightly widened towards the base, 0.6–1.1 cm. long, 0.25–0.5(–0.75) mm. wide, always glabrous. Inflorescence-bearing branches covered in brown bark; inflorescence always moderately lax. Leaf-blades glabrous to glabrescent or sparsely pubescent above, sparsely pubescent or with sparse to very sparse hairs on nerves beneath.

var. **nyassica**

Leaf-blades broadly elliptic to obovate, 6.5–17.5 cm. long, 3.8–7.8 cm. wide, thinly papery; calyx-lobes 0.7–1.1 cm. long. Fig. 106/18.

TANZANIA.   Rungwe District: Kiwira R., Middle Fishing Camp, 26 Jan. 1963, *Richards* 17612! & Ikama, Mar. 1954, *Paulo* 289! & Mkenji R., Feb. 1954, *Paulo* 262!
DISTR.   T 7; Malawi
HAB.   Forest; 2100–2150 m.

SYN.   *P. nyassica* Bremek. in F.R. 37: 150 (1934); T.T.C.L.: 512 (1949)

NOTE.   This is very close to subsp. *comostyla* var. *comostyla* which occurs in the east province of Zimbabwe.

var. **matengoana** *Bridson* in K.B. 32: 629 (1978). Type: Tanzania, Songea District, Matengo Highlands, Lupembe Hill, *Zerny* 405 (W, holo.!)

Leaf-blades elliptic, 8–11.7 cm. long, 2.7–4 cm. wide, papery; calyx-lobes 0.6–0.65 mm. long

TANZANIA.   Songea District: Matengo Highlands, Lupembe Hill, 5 Feb. 1936, *Zerny* 405!
DISTR.   T 8; known only from the type
HAB.   Upland; ± 1900 m.

38. P. mshigeniana *Bridson* in K.B. 35: 823, fig. 2 (1981). Type: Tanzania: Ulanga District, Mahenge, Sali, Ngong Mt., *Cribb, Grey-Wilson & Mwasumbi* 11124 (K, holo.!, DAR, iso.)

Small, somewhat scandent shrub 1.8–4 m. tall; Young vegetative branches adpressed pubescent, inflorescence-bearing branches sparsely pubescent; older branches with dull buff coloured bark. Leaf-blades drying greenish, elliptic to oblanceolate or narrowly obovate, 6–14 cm. long, 2.2–6 cm. wide, acuminate at apex, cuneate at base, chartaceous, ciliate, glabrous above, nerves densely covered with short adpressed hairs and surface glabrous to sparsely covered with adpressed hairs beneath; bacterial nodules dot-like, scattered on either side of the midrib and lateral nerves; petiole 0.5–2.3 cm. long, densely covered with short adpressed hairs, stipules ± triangular, 0.5–1 cm. long, densely covered with adpressed hairs outside, caducous. Corymbs terminal on lateral branches (2–)5–16.5 cm. long bearing 2–3 pairs of leaves near the apex or occasionally terminal on main branches, compact, 2.6(–4 in fruit) cm. across; primary inflorescence-branches obscure in flowering stage, 1–2 mm. long in fruit; secondary and tertiary branches present, but rather obscure in flowering stage; pedicels up to 2 mm. long in flowering stage, densely covered with adpressed hairs, up to 4 mm. long and rather thickened in fruiting state; bracts papery, 4–8 mm. long. Calyx-tube 1–1.25 mm. long, densely covered with adpressed hairs; limb-tube 1–1.25 mm. long, slightly wider than tube, covered to sparsely covered with adpressed hairs; lobes filiform to linear or linear-triangular, 0.6–1 cm. long, 0.5–1 mm. wide at base, covered with adpressed hairs. Corolla-tube 1.8 cm. long, ± 2 mm. wide at top, glabrous outside, not bearded at throat; lobes ± oblong, 5–6 mm. long, 2 mm. wide, acuminate, and somewhat thickened beneath the apex. Fruit red-black, 7 mm. in diameter, glabrescent; calyx-limb persistent. Seeds blackish, 5–6 mm. wide, rugulose on convex face.

TANZANIA. Kilosa District: Ukaguru Mts., Mamiwa Forest Reserve, 15 Aug. 1972, *Mabberley & Salehe* 1477!; Ulanga District: Mahenge District, Kwiro Forest Reserve, 18 Jan. 1979, *Cribb, Grey-Wilson & Mwasumbi* 11007! & Uzungwa Mts., Sanje, 13 Sept. 1984, *D. Thomas* 3678!
DISTR. T 6; not known elsewhere
HAB. Upland forest; 1250–1800 m.

NOTE. *Eggeling* 6465 (at EA) from Morogoro District, Uluguru Mts., Morningside, possibly belongs to this species, however, the indumentum is more akin to *P. stenosepala* subsp. *stenosepala*. More collections are needed for confirmation.

39. P. sp. D

Shrub 1–3 m. tall; branches with fawn, shiny bark. Leaves large, glabrous; blades oblong-oblanceolate, 22.5–31 cm. long, 6.3–9 cm. wide, base cuneate, subcoriaceous, glossy on both faces; bacterial nodules dot-shaped to linear, small, scattered; petioles 1–5.3 cm. long; stipules truncate to triangular, ± 5 mm. long, with stiff arista, 4–6 mm. long, glabrous, outside becoming corky. Corymbs terminal on green lateral branches 9.5–17.5 cm. long bearing 1 pair of immature leaves at apex, moderately compact, ± 3 cm. across; primary inflorescence-branches ± 5 mm. long, sparsely pubescent; secondary and ? tertiary branches present, pubescent; pedicels very short, 0–1.5 mm. long; bracts ovate, ± 6 mm. long; bracteoles brown, papery when dry, ± linear. Calyx-tube ± 2 mm. long, pubescent; limb-tube 1.5 mm. long, slightly wider than the tube, sparsely pubescent; lobes narrowly triangular, 3–4 mm. long, 1–1.25 mm. wide at base, sparsely pubescent. Corolla, only known from young bud, glabrous outside, not bearded at throat; lobes acute. Immature fruit 7 mm. in diameter, glabrescent.

TANZANIA. Morogoro District: N. Uluguru Mts., Mzinga Falls, 18 Oct. 1970, *Pocs* 6265/B!; Iringa/Ulanga District: Mwanihana Forest Reserve, 23 Feb. 1980, *Rodgers, Homewood & Hall* 368!
DISTR. T 6; not known elsewhere
HAB. Riverine and upland forest; 800–1150 m.

NOTE. More gatherings of this little known species would be of great interest.

40. P. sp. E

Shrub or small tree 0.6–3 m. tall; young branches glabrous; older branches covered with brownish-buff bark. Leaves glabrous; blades sometimes drying blackish, tending to be coriaceous, elliptic to oblong-elliptic, 8–17 cm. long, 2.8–5.2 cm. wide, apex acuminate, base cuneate; bacterial nodules dot-like, elliptic or linear, near midrib and with a few scattered; petiole 0.8–2.8 cm. long; stipule-limbs truncate to triangular, 2–5 mm. long, becoming corky with age, glabrous outside, apex with ? caducous awn up to 7 mm. long. Corymbs terminal on lateral branches 7.7–15 cm. long bearing up to 3 pairs of leaves at the apex, ± 1.7–3 cm. across; primary and secondary inflorescence-branches ± reduced; pedicels 1.5–5 mm. long, glabrous. Calyx-tube 1 mm. long, glabrous to puberulous; limb-tube 1–1.75 mm. long, wider than tube, sometimes paler in colour; lobes linear to subulate 4–7 mm. long, 0.25–0.5 mm. wide, glabrous. Corolla not known. Fruit black, 9 mm. in diameter, shiny, glabrous; calyx-limb persistent. Mature seeds not known.

TANZANIA. Kigoma District: Kigoma, 24 Dec. 1963, *Azuma* in *Kyoto University Expedition* 1063! & Kasakati, Mar. 1965, *Suzuki* 85! & Kasangazi, 25 July 1958, *Mahinde* 143!
DISTR. T 4; not known elsewhere
HAB. Forest; 1200–1900 m.

NOTE. This species is extremely close to *P. angolensis* Hiern, from which it may be distinguished by the absence of adpressed hairs on the nerves of the leaves beneath and to *P. seretii* De Wild. which has glabrous but more papery leaves. *P. angolensis* and *P. seretii* are said to differ from each other in the length of the corolla-tube, a feature not represented among present gatherings of the Kigoma taxon. It is possible that both *P. seretii* and the Kigoma taxon could prove to be subspecies of *P. angolensis* but additional material is required to help elucidate this.

41. P. bagshawei S. *Moore* in J.L.S. 37: 163 (1905); Bremek. in F.R. 37: 151 (1934), pro parte; Bridson in K.B. 32: 629 (1978); Fl. Pl. Lign. Rwanda: 574, fig. 195/1 (1983); Fl. Rwanda 3: 190, fig. 59/1 (1985). Type: Uganda, Masaka District, Misozi [Musozi], *Bagshawe* 127 (BM, holo.!)

Shrub or small tree (or ? liana) 0.6–4.5 m. tall; young branches glabrous or pubescent, older branches covered with buff bark. Leaf-blades narrowly to broadly elliptic, 7.3–19.5 cm. long, (1.7–)2.3–9.3 cm. wide, apex acuminate or sometimes acute, base obtuse to cuneate, glabrous or occasionally glabrescent above, pubescent to glabrous beneath; bacterial nodules dot-like and scattered plus a few linear ones along the nerves; petioles 0.4–2 cm. long, pubescent to glabrous; stipule-limbs truncate to triangular, 2–6 mm. long, pubescent (or ? rarely glabrous) outside; arista up to 1 cm. long. Corymbs terminal on branches (? 3.5–)7–25 cm. long bearing 1–3 pairs of leaves usually near the apex, compact, (1.2–)2–4 cm. across; primary inflorescence branches up to 3(–5) mm. long, pubescent or glabrous; secondary and sometimes tertiary branches present or very reduced; pedicels 1–4 mm. long, glabrous or pubescent. Calyx-tube 1–1.5 mm. long, glabrous or pubescent; limb-tube 1–1.75 mm. long, wider than tube; lobes narrowly linear to subulate, (2.5–)4–8 mm. long, not exceeding 0.25 mm. wide, pubescent to sparsely pubescent. Corolla-tube 1.6–2.2 cm. long, ± 2–2.5 mm. wide at top, glabrous outside, not bearded at throat; lobes lanceolate-oblong, 4–8 mm. long, apiculate. Fruit black, shiny, 0.9–1 cm. in diameter; calyx-limb persistent. Seeds greyish black, 6 mm. wide, finely rugulose on convex face.

var. bagshawei

Young branches glabrous; leaf-blades glabrous above, glabrous or glabrescent or sometimes pubescent on nerves beneath, occasionally drying blackish. Calyx pubescent to sparsely pubescent; pedicels pubescent or glabrous.

UGANDA. Ankole District: Ruizi R., 12 Nov. 1950, *Jarrett* 43!; Kigezi District: Kayonza, Apr. 1948, *Purseglove* 2651!; Masaka District: Minziro Hill, Oct. ? 1927, *Maitland* 1082!
TANZANIA. Bukoba District: Kantare, 1935, *Gillman* 388! & Minziro Forest, July 1950, *Watkins* 462! & Feb. 1959, *Procter* 1169!
DISTR.   U 2, 4; T 1; Zaire, Rwanda, Burundi
HAB.   Forest; 1066–1700 m.

SYN.   *P. boonei* De Wild. in B.J.B.B. 5: 26 (1915); Bremek. in F.R. 37: 155 (1934). Type: Zaire, Orientale, Nala, *Boone* 12 (BR, holo.!)
       P. scaettae Bremek. in F.R. 37: 155 (1934). Type: Zaire, Kivu, between Tshifunzi and Botale, *Scaetta* 705 (BR, holo.!)
       P. stuhlmannii Bremek. in F.R. 37: 155 (1934); T.T.C.L.: 513 (1949). Type: Tanzania, Bukoba, *Stuhlmann* 1540 (B, holo.†)
       P. ugandensis Bremek. in F.R. 47: 86 (1939). Type: Uganda, Toro District, Kibale Forest, *Eggeling* 3111 (K, holo.!, EA, iso.)
       P. gillmanii Bremek. in K.B. 3: 351 (1949). Type: Tanzania, Bukoba District, Buyango Forest, *Gillman* 126 (EA, holo.!, K, iso.!)

NOTE.   There is a tendency for specimens from U 4/T 1, Burundi and Zaire (Kivu) to have more slender branches, narrower leaves and to be generally more glabrous than those from U 2, but there is too much overlap for the recognition of another variety. Specimens from T 3 cited as belonging to this species by Bremekamp are actually *P. amaniensis* Bremek.

var. **leucosphaera** (*Bremek.*) *Bridson* in K.B. 32: 630 (1978). Type: Uganda: Ankole District, Bunyaraguru, *Snowden* 1423 (K, holo.!, BM, P, iso.!)

Young branches pubescent; leaves glabrous or glabrescent above, pubescent beneath. Calyx pubescent; pedicels always pubescent. Fig. 106/16.

UGANDA. Toro District: Kibale Forest, June 1964, *Muwanga* 16!; Ankole District: Igara, Kalinzu Forest Reserve, Feb. 1953, *Osmaston* 2823(a)!; Kigezi District: Biroroho track, 5 km. from Ishasha–Katunguru road, 17 Jan. 1970, *Lock* 70/1!
DISTR.   U 2; Zaire
HAB.   Forest; 950–1500 m.

SYN.   *P. leucosphaera* Bremek. in F.R. 37: 148 (1934)
       P. acrochlora Bremek. in K.B. 3, 1948: 353 (1949). Type: Uganda, Ankole District, Igara, Kalinzu Forest, *Purseglove* 625 (K, holo.!)

NOTE.   The type sheet of *P. leucosphaera* Bremek. includes material referable to var. *bagshawei;* sheet II of *Purseglove* 625 not seen by Bremekamp when he described sheet I as *P. acrochlora* is also referable to var. *bagshawei.*

NOTE. (on species as a whole). It is possible that the types of *P. esculenta* De Wild. (from the same locality as the type of *P. boonei* De Wild.), with glabrous stipules and calyx, and *P. malchairii* De Wild. (from Zaire, Equatoria, Likimi), with glabrous stipules but hairy calyx-lobes, are but additional varieties of this species.

42. P. sp. F

Suffrutex to shrub 1–1.8 m. tall; young branches pubescent; older branches covered with thin greyish bark. Leaf-blades elliptic, 5.5–9 cm. long, 1.5–2.9 cm. wide, apex acute to acuminate, base acute to cuneate, glabrous above, save for hairs towards the base of the midrib, glabrescent save for the nerves covered with short adpressed hairs beneath; bacterial nodules dot-shaped to linear, mostly along the midrib; petiole 0.3–0.8 cm. long, covered with short adpressed hairs; stipule-limbs truncate, 1.5–2.5 mm. long, with short adpressed hairs outside, with a few silky hairs inside; arista 3–7 mm. long, somewhat keeled at base. Corymbs terminal on lateral branches 6–10 cm. long bearing 3–4 pairs of leaves towards the apex, ± lax, up to 4 cm. across; primary inflorescence-branches 2.5–5(–10) mm. long, covered with short adpressed hairs; secondary branches sometimes present; pedicels 1–2.5 mm. long, glabrous save for a few appressed hairs towards base; bracts stipule-like, 3.5 mm. long. Calyx glabrous; tube 1–1.5 mm. long; limb-tube 2 mm. long, wider than tube; lobes narrowly linear-triangular, 4–5 mm. long, 1 mm. wide at base. Corolla (known only from bud in Flora area); tube 1.8–2.3 cm. long, ± 1.5 mm. wide at top, glabrous outside, not bearded at throat; lobes 7–7.5 mm. long, acuminate. Fruit not known.

UGANDA. Toro District: Kasatoro Forest, Naiguru, Sept. 1947, *Dale* U507!
DISTR. U 2; Rwanda
HAB. Forest; 2400 m.

NOTE. *Troupin* 12201, from Rwanda, Cyangugu Territory, Butare-Cyangugu (Bukavú-Astrida) road, Uwinka, Banda R., seems to match the Uganda specimen quite closely, except that the calyx-lobes are only 5 mm. long, not 6 mm. as in the Ugandan material. This specimen also shows affinities to *P. pierlotii* Bridson, occurring in Zaire, (Kivu), Rwanda and Burundi, which may be distinguished by its subulate to linear-subulate calyx-lobes, and the usually narrowly obovate leaves with cuneate to attenuate bases. *P. bilineata* Bremek. from Sudan has a similar type of calyx to the Uganda species but is otherwise distinctive.

43. P. aethiopica *Bremek.* in K.B. 10: 174 (1956); Bridson in K.B. 32: 638 (1978). Type: Ethiopia, Sidamo Province, Mega Mt., *Bally* 9397 (K, holo.!, EA, iso.!)

Shrub 2–3 m. tall; young branches ± square, faces connecting the stipules densely pubescent and faces connecting the axils glabrous to sparsely pubescent; older branches covered with dull buff bark. Leaf-blades drying greenish grey or blackish, elliptic to narrowly obovate, 5–9 cm. long, 3–5 cm. wide, shortly acuminate at apex, acute to cuneate at base, slightly shiny and glabrous save for the midrib above, sparsely pubescent to pubescent beneath; bacterial nodules dot-shaped to elliptic, scattered but tend to concentrate near the midrib; petioles 0.3–1.4 cm. long, pubescent; stipules drying brown papery, lanceolate, 0.8–1 cm. long, aristate, pubescent outside, caducous. Corymbs terminal on lateral shoots 2–10.5 cm. long, the longer shoots with nodes and the shorter shoots usually just with mature or immature leaves beneath the inflorescence, rather compact, 3–4 cm. across; primary inflorescence-branches 2–4 mm. long, pubescent; secondary and sometimes tertiary branches present; pedicels 2–5 mm. long, glabrous at base becoming glabrescent at apex; bracts rather stipule-like; bracteoles pale brown, papery. Calyx-tube 1.5 mm. long, puberulous or with a few longer hairs as well; limb-tube 1.5 mm. long, a little wider than the tube; lobes linear from a triangular base, 3.5–5 mm. long, ± 0.5 mm. wide (or 1 mm. wide at base), glabrous or ciliate. Corolla glabrous outside; tube 1.5 cm. long, not bearded at throat; lobes lanceolate-oblong, 8 mm. long, 3 mm. wide, acuminate. Fruit black, 7 mm. in diameter, shiny, sparsely puberulous or with a few longer hairs; calyx-lobes persistent. Mature seeds not known.

KENYA. Northern Frontier District: Huri Hills, 11 Mar. 1979, *Synnott* 1879!
DISTR. K 1; Ethiopia
HAB. Thicket; ± 1000 m.

44. P. ruwenzoriensis *S. Moore* in J.L.S. 38: 255 (1908); Bremek. in F.R. 37: 142 (1934); F.P.N.A.: 359 (1947); F.P.S. 2: 456 (1952); Bridson in K.B. 32: 631 (1978); Fl. Pl. Lign.

Rwanda: 580, fig. 194/4 (1983); Fl. Rwanda 3: 194, fig. 58/4 (1985). Types: Uganda, Ruwenzori Mts., *Scott Elliot* 7399 & Toro district, Wimi, *Scott Elliot* 7833 (both K, syn.!, BM, isosyn.!)

Bush or scandent shrub, 1.2–3.4 m. tall; young branches pubescent or less often glabrous; older branches covered with fawn bark, occasionally pubescent. Leaf-blades elliptic to oblanceolate, (9.5–)12.5–21 cm. long, (3.4–)5.2–11 cm. wide, apex acute to acuminate, base obtuse to cuneate, sparsely scabrid-pubescent or occasionally glabrous above, scabrid-pubescent or less often glabrous or pubescent beneath; bacterial nodules dot-shaped, elliptic or linear, concentrated on the nerves or scattered, few, or sometimes absent; petiole 1–3(–6) cm. long, pubescent or glabrous; stipule-limbs truncate, 2–3(–4) mm. long, membranous, with arista 4–5 mm. long when young, thicker with a corky margin when older, glabrous to pubescent outside, caducous. Corymbs terminal on main and lateral leafy branches 9–25 cm. long, usually lax, 3.5–8.5 cm. across; primary inflorescence-branches 0.3–3.7 cm. long; secondary and often tertiary branches present; pedicels (0.5–)1–5 mm. long, glabrous; bracts up to 1 cm long, caducous. Calyx glabrous; tube 1–2.5 mm. long; limb-tube (1–)2–2.5(–5) mm. long, wider than the tube, drying pale, circumscissile; lobes linear 2–8 mm. long. Corolla-tube 1.4–2.2 cm. long, 1.5 mm. wide at top, glabrous outside, not bearded at throat; lobes oblong-lanceolate, 0.6–1.1 cm. long, 0.15–0.2 cm. wide, acuminate. Style 3.9–5.7 cm. long. Fruit black, 0.8–1 cm. in diameter, shiny; calyx-limb not persistent. Seeds blackish brown, 5 mm. wide, slightly rugulose on convex face. Fig. 106/17.

UGANDA. Toro District: NW. Ruwenzori, 1 Sept. 1951, *Osmaston* 1212!; Ankole District: Lutoto, Dec. 1938, *Chandler & Hancock* 2631!; Mengo District: Namanyonyi, June 1915, *Dummer* 2573!
DISTR. U 2, 4; Zaire, Rwanda, Sudan
HAB. Understorey of forest; 1500–2200 m.

SYN. *P. kiwuensis* K. Krause in Z.A.E.: 330 (1911). Type: Rwanda, Lake Karago, S. of Karisimbi, *Mildbraed* 1513 (B, holo.†), *non P. kiwuensis* K. Krause (1912)
     P. schubotziana K. Krause in Z.A.E.: 331 (1911); Bremek. in F.R. 37: 141 (1934); F.P.N.A.: 358 (1947). Type: Zaire, Kivu, Beni, *Mildbraed* 2182 (B, holo.†, BR, fragment!)
     P. kirschsteiniana K. Krause in E.J. 48: 422 (1912), *nomen*
     [*P. abyssinica* sensu Hutch. in Bot. Mag., t. 8838 (1920), *non* Fresen.]
     P. butaguensis De Wild., Pl. Bequaert. 2: 292 (1923). Type: Zaire, Kivu, Ruwenzori, Butagu Valley, *Bequaert* 3693 (BR, holo.!)
     P. kiloensis De Wild. in Pl. Bequaert 2: 295 (1923); Bremek. in F.R. 37: 141 (1934). Type: Zaire, Kilo, *Claessens* (BR, holo.!)
     P. ituriensis Bremek. in F.R. 37: 142 (1934); F.P.N.A.: 359 (1947). Type: Zaire, W. of Ruwenzori Mts., Ituri, *Humbert* 8823 (P, holo.!)

NOTE. Specimens from Uganda have leaves which are very sparsely to sparsely scabrid-pubescent above and sparsely scabrid-pubescent to pubescent beneath, however, specimens from Zaire show greater variation in the indumentum of the leaf.
     This species has also been cultivated in Kenya.

45. P. abyssinica *Fresen.* in Mus. Senck. 2: 166 (1837); Hiern in F.T.A. 3: 173 (1877); P.O.A. C: 388 (1895); Bremek. in F.R. 37: 144 (1934) pro parte maxima; T.S.K.: 145 (1936); T.T.C.L.: 511 (1949); F.P.S. 2: 457 (1952); K.T.S.: 455, fig. 86 (1961); Bridson in K.B. 32: 633 (1978). Type: Ethiopia, between Halei and Temben, *Rüppel* (FR, holo.!)

Bush or small tree, sometimes climbing, 1–7(–9) m. tall; young branches glabrous or pubescent, older branches covered with buff or greyish bark. Leaf-blades drying blackish or occasionally green, elliptic or less often broadly elliptic, oblong-elliptic or narrowly obovate, 3.5–16(–18) cm. long, 1.4–6.5(–7.7) cm. wide, apex acuminate, base cuneate, slightly shiny to shiny above, glabrous or rarely pubescent above, glabrous, with a few spreading hairs along the nerves, or rarely pubescent beneath; bacterial nodules dot-shaped to elliptic or sometimes linear, along the nerves or scattered; petiole 0.2–1.5 cm. long, entirely glabrous, puberulous above or less often pubescent; stipules drying translucent brown or rarely opaque, truncate, (0.2–)0.3–1 cm. long, usually aristate, membranous to papery, caducous. Corymbs terminal on lateral branches 0.9–18.4 cm. long bearing 2–5 leaf-bearing or naked nodes along their length, moderately compact to lax, 1.8–5.5 cm. across; primary inflorescence-branches 0.2–1.3 cm. long, glabrous, puberulous or less often pubescent; secondary and occasionally tertiary branches present; pedicels (1–)2–9 mm. long, glabrous to puberulous or rarely pubescent; bracts brown, 2–7 mm. long, very thinly membranous to membranous; bracteoles conspicuous or inconspicuous. Calyx glabrous, occasionally with very few hairs or less often pubescent; tube 1–2 mm. long; limb-tube (0.5–)1–3 mm. long, distinctly wider than tube; lobes ± subulate, sometimes widened at the base, 0.5–4(–5) mm. long (or 7–12 mm. long in one Ethiopian variety). Corolla-tube 0.8–2.2 cm. long (or 2.5–3.7 cm. long in 3 Ethiopian varieties), glabrous or rarely with short hairs outside, not bearded at throat; lobes ± oblong, 0.4–1 cm. long, up to 3 mm. wide, acuminate. Fruit black, up to 1 cm. in diameter, shiny, glabrous; calyx-limb persistent. Seeds blackish brown, 5 mm. wide, rugulose on convex face.

KEY TO INFRASPECIFIC VARIANTS

Leaf-blades papery or sometimes subcoriaceous; stipules often
     translucent, membranous to papery; calyx-lobes 0.5–5
     mm. long:
     Both calyx and corolla-tube glabrous outside (or occasionally
         calyx with very few hairs); young branches and petioles
         glabrous or occasionally pubescent or puberulous   **a. var. abyssinica**
     Either calyx or corolla-tube hairy outside; young branches
         and petioles densely pubescent:
         Calyx glabrous; corolla-tube shortly pubescent outside,
             leaf-blades elliptic   . . . .   **b. var. lamurensis**
         Calyx sparsely pubescent; corolla-tube glabrous outside;
             leaf-blades broadly elliptic to round   . . . .   **c. var. prescottii**
Leaf-blades coriaceous; stipules not translucent, thicker; calyx-
     lobes not exceeding 0.5 mm. long   . . . .   **d. var. usambarica**

a. var. abyssinica

Young branches glabrous or less often pubescent. Leaf-blades papery or sometimes subcoriaceous, glabrous, moderately shiny above, glabrous or with spreading hairs along the nerves or occasionally pubescent beneath; petioles glabrous, puberulous above or occasionally pubescent; stipules translucent. Inflorescence-branches and pedicels glabrous to puberulous or rarely pubescent. Calyx glabrous or occasionally with a very few hairs; lobes 0.5–4(–5) mm. long. Corolla-tube always glabrous outside.

UGANDA. Karamoja District: Napak, June 1950, *Eggeling* 5976! & Morongole, 11 Nov. 1939, *A.S. Thomas* 3327!; Mbale District; Bubungi, 5 July 1926, *Maitland* 1240!
KENYA. Northern Frontier Province: Marsabit, 4 Aug. 1957, *Verdcourt* 1822!; Nandi District: near Kapsabet, May 1933, *Dale* in F.D. 3122!; Kericho District: SW. Mau Forest, Sambret Catchment, 21 May 1962, *Kerfoot* 3729!
TANZANIA. Arusha District: Arusha National Park, Mt.Meru, Belo Forest, 4 Mar. 1970, *Richards* 25586!; Lushoto, 10 May 1962, *Semsei* 3476!; Kondoa District: Ghost Mt., 7 Feb. 1928, *B.D. Burtt* 1336!
DISTR. U 1, 3; K 1–6; T 2, 3, 5–7; ?Sudan (*fide* Andrews), Ethiopia
HAB. Forest or occasionally scrub; 1050–2500 m.

SYN. *P. abyssinica* A. Rich. (*non* Fresen.) var. *glabra* A. Rich. in Tent. Fl. Abyss. 1: 352 (1847); P.O.A. C: 388 (1895), *nom. illegit.* Type: Ethiopia, Tigre, Mt. Kubbi, *Schimper* 353 (K, holo.!)
   Ixora abyssinica (Fresen.) Oliv. in T.L.S. 29: 87 (1873)
   Pavetta ellenbeckii K. Schum. in E.J. 33: 354 (1903). Type: Ethiopia, Harar Province, Gara Muletta [Garu Mulata], *Ellenbeck* 519 (B, holo.†)
   P. kenyensis Bremek. in F.R. 37: 149 (1934); K.T.S.: 458 (1961). Type: Kenya, N. Nyeri District, Liki R., *Battiscombe* 1312 (K, holo.!, EA, iso.)
   P. maitlandii Bremek. in F.R. 37: 142 (1934); T.S.K.: 145 (1936); K.T.S.: 459 (1961). Type: Uganda, Mbale District, Bubungi, *Maitland* 1240 (K, holo.!)
   P. silvicola Bremek. in F.R. 37: 145 (1934); T.S.K.: 145 (1936); T.T.C.L.: 512 (1949); K.T.S.: 460 (1961). Type: Tanzania, Moshi District, Kilimanjaro, Usseri [Useri], *Volkens* 2007 (B, holo.†, K, iso.!)
   [*P. transjubensis* sensu Chiov. in Racc. Bot. Miss. Consol. Kenya: 57 (1935), quoad *Balbo* 548, *non* Chiov. 1932]
   P. trichotropis Bremek. in K.B. 3: 354 (1949); K.T.S.: 462 (1961). Type: Kenya, Elgeyo District, Kapsowar, *Gardner* 611 (EA, holo.!)
   [*P. spaniotricha* sensu K.T.S.: 461 (1961), quoad *Bally* 8814 & *Gillett* 15176, *non* Bremek.]

NOTE. All the specimens seen from the Flora area have glabrous or rarely pubescent petioles and inflorescence-branches with the exception of the type of *P. maitlandii* Bremek. (the only specimen seen from Mbale District), which has the petioles above and inflorescence-branches puberulous. The majority of the specimens examined from Ethiopia match the Mbale specimen in this respect or tend to be more pubescent with the remainder glabrous or intermediate.

All the specimens seen from the Ngong Hills (**K** 6) have smaller leaves (3.8–6(–7.4) cm. long, 1.6–2.9 cm. wide) and shorter internodes than typical material, but are probably not worth recognizing at varietal rank. Specimens from Nou Forest Reserve (Mbulu District) also tend to have smaller leaves.

The plant figured by Hutchinson in Bot. Mag. t. 8838 (1920) as *P. abyssinica* is I consider *P. ruwenzoriensis* S. Moore.

b. var. **lamurensis** (*Bremek.*) *Bridson* in K.B. 32: 636 (1978). Type: Kenya, Kiambu District, Limuru [Lamuru], *Scheffler* 314 (B, holo.†, BM, K, iso.!)

Young branches densely pubescent. Leaf-blades papery, narrowly elliptic to rhombic, glabrous above, sparsely pubescent beneath; petioles densely pubescent; stipules translucent. Inflorescence-branches and pedicels pubescent. Calyx glabrous. Corolla-tube puberulous outside.

KENYA. Kiambu District: Limuru [Lamuru], 6 July 1909, *Scheffler* 314!
DISTR. **K** 4; only known from type specimen
HAB. Upland forest; ± 2200 m.

SYN. *P. lamurensis* Bremek. in F.R. 37: 148 (1934); K.T.S.: 458 (1961)

NOTE. The specimen cited by Bremekamp in F.R. 47: 86 (1939) (*Mearns* 1315 from the west slopes of Mt. Kenya) is var. *abyssinica* not var. *lamurensis*.
Fayad 262 from **K** 6, Entasekera, perhaps approaches this variety, but as corollas are lacking a definite opinion cannot be formed.

c. var. **prescottii** *Bridson* in K.B. 32: 636 (1978). Type: Kenya, N. Nyeri District, Nanyuki, *Prescott Decie* (BM, holo.!)

Young branches densely pubescent. Leaf-blades chartaceous, broadly elliptic to round, sparsely pubescent above, pubescent beneath; petiole densely pubescent; stipules translucent. Calyx sparsely pubescent. Corolla-tube glabrous outside.

KENYA. N. Nyeri District: Nanyuki, 1926, *Prescott Decie* (two gatherings)!
DISTR. **K** 4; not known elsewhere
HAB. Forest; 1900 m.

d. var. **usambarica** (*Bremek.*) *Bridson* in K.B. 32: 636 (1978). Type: Tanzania, Lushoto District, W. Usambara Mts., Shagayu peak area, *Drummond & Hemsley* 2735 (K, holo.!, BR!, EA, iso.!)

Young branches glabrous. Leaf-blades coriaceous, shiny above, entirely glabrous; petioles glabrous; stipules not translucent. Inflorescence-branches and pedicels glabrous. Calyx glabrous; lobes not exceeding 0.5 mm. long. Corolla-tube always glabrous outside, not exceeding 1.3 cm. long.

TANZANIA. Lushoto District: W. Usambara Mts., Shume village, near World's View, 18 Mar. 1984, *Borhidi et al.* 841036/A! & Shagayu Forest, Apr. 1953, *Procter* 176! & between Oaklands and View Point, 17 Mar. 1970, *Batty* 957!
DISTR. **T** 3; known only from W. Usambara Mts.
HAB. Forest; 1675–2200 m.

SYN. *P. usambarica* Bremek. in K.B. 11: 173 (1956)

46. P. subumbellata *Bremek.* in F.R. 37: 176 (1934); T.T.C.L.: 515 (1949); Bridson in K.B. 32: 639 (1978). Type: Tanzania, Iringa District, Dabaga, *Lynes* 9 (K, holo.!)

Shrub or small tree, often scandent, 1.5–7 m. tall; young branches glabrous, usually drying black; older branches covered with buff to brown, scarcely flaking bark. Leaf-blades drying greenish or blue-black, narrowly obovate to obovate, less often elliptic, lanceolate-elliptic or rarely round, 3.7–9.5 cm. long, 1.5–4.7 cm. wide, apex rounded, obtuse, acute or slightly acuminate, base cuneate or rarely ± truncate. papery, dull, glabrous or with a few hairs on midrib above, sparsely pubescent to glabrous beneath; bacterial nodules dot-shaped to elliptic, large, scattered but usually absent from the marginal areas; petiole usually drying blackish, 2–4 mm. long, sparsely pubescent or glabrous; stipules truncate to triangular, 3–6 mm. long, only shortly connate when developed, acute to acuminate, caducous, often with a line of silky hairs along the scar. Corymbs terminal on lateral branches 1–9 cm. long bearing 0–2 pairs of leaves at the apex, subumbellate, moderately compact, 1–3 cm. across, flowers often pendulous; primary inflorescence-branches 1–6(–8) mm. long, glabrous; secondary branches absent or occasionally present; pedicels 1–6 mm. long, glabrous; bracts stipule-like, 2–5 mm. long, deciduous leaving a line of silky hairs; bracteoles inconspicuous. Calyx-tube 1–1.5 mm. long, glabrous or with very few hairs; limb-tube 1–2 mm. long, ± 1½ times wider than the tube, glabrous to sparsely hairy; lobes narrowly linear to linear, often from a triangular base or subulate, 1–4 mm. long, 0.25–0.75 mm. wide, covered with crisped or straight white hairs, with very few hairs, occasionally only one or two near apex or infrequently glabrous. Corolla-tube 0.9–1.6(–2) cm. long, 2 mm. wide at top, glabrous outside, not bearded at throat; lobes ± oblong, 4–6(–8) mm. long, 2–2.5 mm. wide, rounded or shortly acuminate. Fruit black, 1 cm. in diameter, shiny, glabrous; calyx-limb persistent. Seed reddish brown, 5 mm. wide, finely rugulose on convex face.

var. subumbellata

Leaf-blades chartaceous, narrowly obovate to obovate or sometimes elliptic, rarely rotund, sparsely pubescent to glabrous beneath. Calyx-lobes narrowly linear to linear, often from a triangular base or occasionally subulate, pubescent or sometimes with very few hairs at apex. Corolla-tube 0.9–1.6 cm. long; lobes 4–6 mm. long.

TANZANIA. Ufipa District: Sumbawanga, Mbisi Forest, 13 Mar. 1957, *Richards* 8699!; Iringa District: Dabaga Highlands, Kibengu, 13 Feb. 1962, *Polhill & Paulo* 1464!; Njombe District: Igeri, 19 Feb. 1968, *Robertson* 1014!
DISTR. T 4, 7; Malawi and Zambia
HAB. Forest or thicket edges; (1500–)2000–2600 m.
SYN. *P. spaniotricha* Bremek. in K.B. 3: 352 (1949). Type: Tanzania, Rungwe District, Bundali [Undalis], *R.M. Davies* 889 (EA, holo.!, K, iso.!)
    P. uwembae Gilli in Ann. Naturhistor. Mus. Wien. 77: 23, t. 2 (1973). Type: Tanzania, Njombe District, Uwembe, *Gilli* 414 (W, holo.!, K, iso.!)

var. subcoriacea *Bridson* in K.B. 42: 257 (1987). Type: Tanzania, Iringa District, Mufindi Tea Estate, Kibwele Forest, *Bridson & Lovett* 526 (K, holo.!)

Leaf-blades subcoriaceous, lanceolate-elliptic to narrowly obovate, always glabrous beneath. Calyx-lobes subulate, with very few hairs at apex or glabrous. Corolla-tube 2 cm. long; lobes 8 mm. long (corolla known only from one specimen from Malawi).

TANZANIA. Iringa District: Mufindi, Kibwele Forest, *Bridson & Lovett* 526! & Luiga Tea Estate, *Renvoize & Abdallah* 2064!
DISTR. T 7; Malawi
HAB. Forest; 1900 m.

47. **P. tarennoides** *S. Moore* in J.B. 43: 353 (1905); Bremek. in F.R. 37: 138 (1934); T.S.K.: 146 (1936); K.T.S.: 461 (1961). Type: Kenya, Kwale District, Mwelo Mdogo [Mele], *Kassner* 230 (BM, holo.!)

Shrub 2.4–5 m. tall; young branches glabrous; older branches covered with fawn shiny bark. Leaf-blades usually drying blackish, elliptic to ovate, 8.2–17 cm. long, 3.7–7.5 mm. wide, apex acute to slightly acuminate, base acute to cuneate, glabrous, slightly shiny above, dull beneath; bacterial nodules linear and confined to the midrib or sometimes the lateral nerves or rarely elliptic and between the nerves; petiole 1–3.5 cm. long; stipule-limbs drying black with fawn margin, truncate to slightly triangular, 2–7 mm. long, glabrous outside, with long silky hairs inside; arista up to 8 mm. long, usually caducous. Corymbs terminal on main and lateral leafy branches 9.5–19 cm. long, several-flowered, lax to very lax, 6–13 cm. across; primary inflorescence-branches 0.7–3 cm. long; secondary and tertiary branches present; pedicels 5–18 mm. long, glabrous; bracts, drying blackish, 3–4 mm. long; bracteoles very small. Calyx glabrous; tube 1–1.25 mm. long; limb-tube 1–2 mm. long, slightly wider than the tube; lobes narrowly triangular or subulate to linear above from a triangular base, 2–4 mm. long, ± 0.75 mm. wide at base. Corolla-tube 1.2–2.2 cm. long, 2–2.5 mm. wide at the top, glabrous outside and within; lobes oblong-lanceolate, 0.7–1.2 cm. long, 2.5–3 mm. wide, acute to shortly acuminate. Style 3.2–5.4 cm. long. Fruit black, ± 7 mm. in diameter, shiny, glabrous. Seeds ± 5 mm. wide, not fully mature.

KENYA. Kwale District: Kwale, 21 Dec. 1963 *Brown*, 736!, Shimba Hills, Longomwagandi Forest, 9 Feb. 1968, *Magogo & Glover* 40! & Shimba Hills, 5 Jan. 1966, *Christensen* 188!
DISTR. K 7; not known elsewhere
HAB. Undergrowth of coastal forest; 300–440 m.

48. **P. holstii** *K. Schum.* in P.O.A. C: 389 (1895); Bremek. in F.R. 37: 138 (1934); T.T.C.L.: 510 (1949). Type: Tanzania, Lushoto District, E. Usambara Mts., Derema [Nderema], *Holst* 2280 (B, holo.†, COI, K, iso.!)

Shrub, scrambling shrub or small tree 1.5–4.5 m. tall; young branches glabrous; older branches covered with buff, slightly corky bark. Leaf-blades usually drying blackish, elliptic to ovate, 7–17 cm. long, 3–8.4 cm. wide, apex obtuse, acute or subacuminate, base

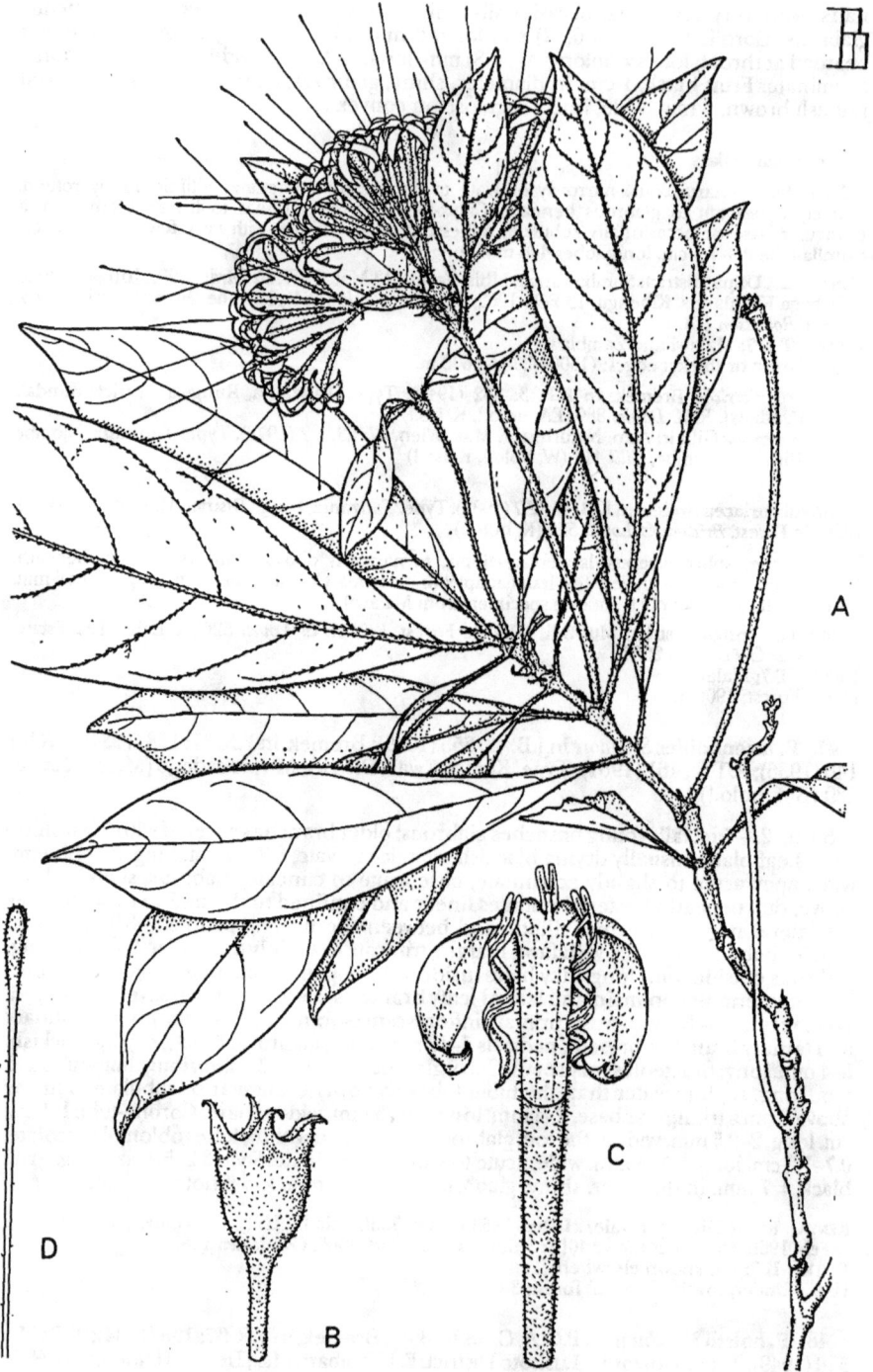

FIG. 112. *PAVETTA MAZUMBAIENSIS* — A, flowering branch, × ⅔; B, calyx, × 6; C, corolla with section removed, × 2; D, stigma with part of style, × 8. All from *Cribb & Grey Wilson* 10086. Drawn by Diane Bridson.

acute to cuneate, glabrous, slightly shiny above, dull beneath; bacterial nodules usually linear and restricted to the midrib or occasionally on lateral nerves or rarely elliptic and between the nerves; petiole 1.5–4.6 cm. long; stipule-limbs drying blackish at base and pale-brown to buff above, shallowly triangular to truncate, 2–3 mm. long, glabrous outside, with a few silky hairs inside; arista up to 3 mm. long, becoming keeled towards base. Corymbs terminal on main and lateral leafy branches, several-flowered, lax, 4.5–8 cm. across; primary inflorescence-branches 1–2.3 cm. long; secondary and tertiary branches present; pedicels 2–5 mm. long, glabrous; bracts drying with blackish base and brown margins, 5 mm. long; bracteoles brown, papery. Calyx glabrous; tube 1–1.25 mm. long; limb-tube 0.75–1(–1.3) mm. long, equalling or slightly exceeding the tube in width; lobes shortly triangular, subulate above, 0.25–0.75(–1) mm. long, sometimes with a few hairs. Corolla-tube (1–)1.2–1.7 cm. long, 1.25–2 mm. wide at top, glabrous outside, not bearded at throat; lobes oblong-lanceolate, 6–10 mm. long, 1.25–3 mm. wide, obtuse or sometimes shortly apiculate. Fruit black, 9 mm. in diameter, shiny. Mature seeds not known.

TANZANIA. Lushoto District: E. Usambara Mts., Monga, 11 Jan. 1947, *Greenway* 7917!; Tanga District: 9.6 km. from Amani, above I.B.C. Saw Mills, 20 Feb. 1971, *Magogo* 1477!; Morogoro District: Nguru Mts., Manyangu Forest, Liwale valley, 27 Mar. 1953, *Drummond & Hemsley* 1838!
DISTR. **T** 3, 6; not known elsewhere
HAB. Evergreen forest; 600–2000 m.
NOTE. *Pocs* 6139 from Morogoro District, Kanga Mt., has a larger calyx limb-tube (1.3 mm.) than usual, also the leaves have not blackened as much.

49. P. mazumbaiensis *Bridson* in K.B. 32: 631, fig. 5 (1978). Type: Tanzania, Lushoto District, W. Usambara Mts., Mazumbai Forest Reserve, *Cribb & Grey-Wilson* 10086 (K, holo.!)

Shrub; young branches pubescent; older branches covered with dull buff bark. Leaf-blades drying greenish brown, oblanceolate to narrowly obovate, 7.5–13 cm. long, 2.5–4.8 cm. wide, acuminate at apex, cuneate at base, glabrous to glabrescent with midrib pubescent towards base above, glabrescent with pubescent nerves beneath; bacterial nodules dot-shaped to linear, few, mostly confined to the midrib; petioles 0.4–1 cm. long, pubescent; stipule-limbs truncate to slightly triangular, 2 mm. long, pubescent outside; arista up to 4 mm. long, slightly keeled towards base. Corymbs terminal on lateral branches 5.4–8.5 cm. long bearing usually one pair of leaves near the apex, lax, 3.8–4.8 cm. across; primary inflorescence-branches 0.6–1.8 cm. long, finely pubescent; secondary and occasionally tertiary branches present; pedicels 2–5 mm. long, finely pubescent; bracts stipule-like; bracteoles inconspicuous. Calyx puberulous-pubescent; tube 1 mm. long; limb-tube up to 1 mm. long, slightly wider than tube; lobes narrowly triangular below, subulate at apex, ±0.75–1 mm. long. Corolla-tube 1.7–2 cm. long, 1.75–2 mm. wide at top, pubescent outside, not bearded at throat; lobes oblong, 7 mm. long, 2.5–3 mm. wide, acuminate and darkened at apex (when dry). Fruit greyish black, ± 1 cm. in diameter, shiny. Seeds not known. Fig. 112.

TANZANIA. Lushoto District: Mazumbai Forest Reserve, Feb. 1985, *Borhidi et al.* 85442! & 21 Aug. 1972, *L. Tanner* 12! & Matundsi-Mashindei ridge, SW. of Ambangulu Tea Estate, 5 Feb. 1985, *Borhidi et al.* 85471!
DISTR. **T** 3; known only from W. Usambara Mts.
HAB. Forest; 1200–1675 m.

50. P. diversicalyx *Bridson* in K.B. 42: 251, fig. 1 (1987). Type: Tanzania, W. Usambara Mts., Shagayu Forest Reserve, 2.5 km. ENE from Shagayu Saw Mills, *Borhidi et al.* 84849 (K, holo.!, VBI, iso.)

Shrub 2.5–5 m. tall; young branches glabrous; older branches covered in shiny buff-brown bark. Leaves glabrous; blades drying green, oblong or oblong-elliptic, 5.5–20 cm. long, 2.8–5.5 cm. wide, apex acute or acuminate, base obtuse, often unequal, coriaceous, slightly shiny above; bacterial nodules linear, restricted to the midrib and occasionally on the lateral nerves; petiole 0.3–2 cm. long; stipule-limbs ovate to truncate, 1.5–3 mm. long, becoming corky with age, slightly keeled, with keel extending as a thickened lobe 1–4 mm. long. Corymbs terminal on green (not corky) lateral branches 5.5–14.5 cm. long bearing 1–2 pairs of leaves at the apex, lax, 3.5–6 cm. across; primary inflorescence-branches 0.4–1.5 cm. long, glabrous; secondary branches present; pedicels 2–7 mm. long, glabrous;

bracts stipule-like, becoming corky; bracteoles inconspicuous. Calyx glabrous; tube 1.25–2 mm. long; limb-tube 1–2 mm. long, a little wider than the tube; lobes shortly subulate, or triangular, 0.5–2 mm. long. Corolla-tube 1.5–2.3 cm. long, 2–3 mm. wide at top, glabrous outside, not bearded at throat; lobes oblong, 0.8–1.1 cm. long, 3 mm. wide, acute. Immature fruit 6 mm. in diameter.

TANZANIA. Lushoto District: W. Usambara Mts., Lushoto, 20 May 1962, *Semsei* 3469! & Mukussu Forest Reserve, 13 Apr. 1970, *Ruffo* 351! & Shagayu Forest Reserve, 5 km. NW. of Mlalo Mission, on ridge of Mt. Sekukum on NE. side, 15 Feb. 1985, *Iversen et al.* 85743!
DISTR. T 3; known only from W. Usambara Mts.
HAB. Montane evergreen forest and forest margins; 1500–2000 m.

51. P. nitidissima *Bridson* in K.B. 42: 255, fig. 3 (1987). Type: Tanzania, Ulanga District, Uzungwa Mts., Sanje, *D. Thomas* 3851 (MO, holo.!, K, iso.!)

Small glabrous tree, 3 m. tall, with one or a few main branches and short lateral branches near apex. Leaves drying greenish; blades narrowly oblong to oblong-elliptic or elliptic, 12–29 cm. long, 5–10 cm. wide, acute to acuminate at apex, acute to attenuated, and often unequal at base, subcoriaceous, very shiny above; bacterial nodules linear, restricted to midrib; petioles 1.3–5.5 cm. long; stipule-limbs truncate, 3–4 mm. long, sometimes becoming corky, bearing a slightly keeled lobe, 2 mm. long. Corymbs terminal on green lateral branches 25.5–28 cm. long bearing 2–3 pairs of leaves near apex, lax, 5–10 cm. across; primary inflorescence-branches 0.7–2.2 cm. long; secondary and tertiary branches present; pedicels 2–7 mm. long; bracts ovate, 1 cm. long, apiculate; bracteoles inconspicuous. Calyx-tube ± 2 mm. long; limb-tube 1.25 mm. long, distinctly wider than tube; lobes shortly dentate, ± 0.5 mm. long. Corolla only known from bud; tube pubescent inside; lobes broadly lanceolate, ± 1 cm. long, ± 3 mm. wide. Fruit black when dry, 1 cm. in diameter. Seeds light brown with area surrounding cavity darker, ± 1 cm. in diameter, scarcely rugulose.

TANZANIA. Ulanga District: Uzungwa Mts., Sanje, 10 Oct. 1984, *D. Thomas* 3851!
DISTR. T 6; only known from the type collection
HAB. Montane forest; 1500–1600 m.

52. P. mzeleziensis Bridson in K.B. 35: 826, fig. 3 (1981). Type: Tanzania, Ulanga District, Mzelezi Forest Reserve, *Cribb, Grey-Wilson & Mwasumbi* 11060 (K, holo.!, DAR, EA!, iso.)

Glabrous shrub to 3 m. tall; older branches covered with fawn shiny bark. Leaf-blades drying olive-green to blackish, narrowly elliptic to elliptic, 7–18.5 cm. long, 1.5–7 cm. wide, gradually narrowing to a very fine point at apex, cuneate, sometimes tending to be unequal at base, shiny above; bacterial nodules linear along the midrib or sometimes lateral nerves or occasionally elliptic and positioned between the nerves; petiole 0.6–3.7 cm. long; stipule-limbs drying blackish, often with a corky margin, truncate to triangular, 1.5–2.5 mm. long, glabrous outside, with long silky hairs inside; arista 0.8–1.2 cm. long. Corymbs terminal on main and lateral leafy branches 10–12 cm. long, lax, 4.5–7 cm. across; primary inflorescence-branches 0.6–2.3 cm. long; secondary and tertiary branches present; pedicels 4–7 mm. long; bracts drying blackish with fawn margin, ± 3 mm. long, apiculate; bracteoles ± conspicuous, brown when dry. Calyx glabrous, tube 1 mm. long; limb-tube 1–1.5 mm. long, slightly to distinctly wider than tube; lobes subulate, scarcely widening at base, 1.5–2 mm. long, not exceeding 0.25 mm. wide. Corolla-tube 1.5 cm. long, ± 2 mm. wide at top, glabrous outside and pubescent within; lobes narrowly-oblong, ± 1 cm. long, 2–2.5 mm. wide, acuminate. Fruit black, 1 cm. in diameter, shiny, glabrous; calyx-limb not persistent. Seeds dark grey almost black, 6 mm. wide, rugulose on convex face.

TANZANIA. Ulanga District: Mzelezi Forest Reserve, 19 Jan. 1979, *Cribb, Grey-Wilson & Mwasumbi* 11060! & Uzungwa Mts., Sanje Waterfall, 5 Sept. 1984, *Bridson* 625!
DISTR. T 6; not known from elsewhere
HAB. Riverine forest; 725–850 m.

NOTE. This species is close to *P. tarennoides*, *P. holstii* and *P. lasiobractea*. *Bridson* 625, a fruiting sheet, has been included here, but since the calyx-limbs have all fallen certain identification is not possible.

53. P. lasiobractea *K. Schum.* in E.J. 30: 415 (1901); Bremek in F.R. 37: 140 (1934); T.T.C.L.: 511 (1949); Brenan in Mem. N.Y.B.G. 8: 453 (1954); Bridson in K.B. 32: 638 (1978). Type: Tanzania, Mbeya/Rungwe District, Umalila, *Goetze* 1466 (B, holo.†, BR, P, iso.!)

Shrub 1.5–2 m. tall; young branches glabrous, older branches covered with buff to pale brown, thin shiny bark. Leaves glabrous; blades drying greenish, narrowly oblong to oblanceolate, 7–14.5 cm. long, 1.8–4.9 cm. wide, apex acuminate, base often unequal, obtuse to acute, ± coriaceous, shiny above, midrib prominent beneath; bacterial nodules linear, situated along the midrib plus few dot-shaped and scattered; petiole 0.7–2 cm. long, glabrous; stipules triangular, 3–8 mm. long, mucronulate, puberulous outside, densely covered with silky hairs inside. Corymbs terminal on lateral branches 2.5–11.5 cm. long bearing up to 3 pairs of leaves, lax, 3.6–7 cm. across; primary branches 0.3–1.7 cm. long, secondary and often tertiary branches present; pedicels 1–9 mm. long, glabrous; bracts brown, papery, 5–9 mm. long, sparsely hairy outside, with dense silky hairs inside; bracteoles conspicuous. Calyx glabrous; tube 1.25–1.5 mm. long; limb 1.5–2 mm. long, almost twice as wide as the tube, truncate or with subulate lobes not exceeding 0.75 mm. long. Corolla-tube 1.4–1.9 cm. long, 2.25–3 mm. wide at top, glabrous outside, not bearded at throat; lobes oblong, 1.1–1.4 cm. long, 2.5–2.75 mm. wide, acuminate. Fruit black, 0.9–1 cm. in diameter, shiny, glabrous, calyx-limb persistent. Seeds blackish, 6 mm. wide, ± coarsely rugulose on convex face. Fig. 106/9.

TANZANIA. Mbeya District: Idunda, Umalila, Mar. 1972, *Leedal* 1308!; Rungwe District: Rungwe Mt., 23 Oct. 1956, *Richards* 6749! & Njombe District: Ukinga [Kinga], Madehani (Nadehani), 3 Dec. 1913, *Stolz* 2331!
DISTR. T 7; Malawi
HAB. Montane forest; 2100–2400 m.

54. P. kyimbilensis *Bremek.* in F.R. 37: 149 (1934); T.T.C.L.: 511 (1949); Bridson in K.B. 32: 638 (1978). Type: Tanzania, Njombe District, Madehani, *Stolz* 2332 (K, holo.!, BM!, BR, EA, P!, iso.)

Shrub or small tree, 1–8.4 m. tall; young branches puberulous or less often pubescent-puberulous, older branches covered with buff flaky bark, often gnarled. Leaf-blades black or green when dry, oblanceolate or sometimes narrowly elliptic, (4–)5–13 cm. long, 1.3–4.1 cm. wide, apex acute to slightly acuminate, base cuneate, subcoriaceous, glabrous or rarely very sparsely hairy above, often shiny, sometimes with sparse white hairs on nerves beneath, rarely ciliate; bacterial nodules elliptic to linear along the midrib; petiole 0.6–1.3 cm. long, glabrous or less often puberulous above, with few white hairs beneath; stipules ovate-triangular, 3.5–8 mm. long, obtuse, sometimes mucronulate, puberulous or pubescent outside, with white silky hairs inside. Corymbs terminal on lateral branches 1–8 cm. long leafless or bearing 1 pair of leaves at the apex, not markedly lax, 1.2–4 cm. across; primary inflorescence-branches (1–)2.5–8 mm. long, puberulous to ± glabrous or less often pubescent-puberulous; secondary branches present; pedicels 1.5–6 mm. long, puberulous to sparsely puberulous or pubescent-puberulous, occasionally glabrous; bracts brown, papery, 4–8 mm. long, pubescent to puberulous outside, with dense silky hairs inside. Calyx-tube 1–1.25 (–2) mm. long, glabrous or sparsely puberulous; limb 1–1.25 mm. long, ± 1½ times wider than the tube, glabrous or occasionally sparsely puberulous, truncate, slightly undulate or with lobes not exceeding 0.25 mm. long. Corolla-tube 1.2–1.5 cm. long, 2 mm. wide at top, glabrous outside, not bearded at throat; lobes ± oblong, 6–9 mm. long, 2–3 mm. wide, obtuse to slightly acuminate. Style 3–4.7 cm. long. Fruit spherical or compressed lengthwise, 0.8–1.2 cm. wide, glabrous, calyx-limb persistent. Seeds ± 5 mm. across.

var. kyimbilensis

Pedicels and young branches pubescent-puberulous. Leaf-blades drying blackish, sparsely hairy beneath and very sparsely hairy to glabrous above, ciliate.

TANZANIA. Njombe District: Madehani, 3 Dec. 1913, *Stolz* 2332!
DISTR. T 7; known only from the type gathering
HAB. Forest; 2100 m.

var. iringensis

var. **iringensis** *(Bremek.) Bridson* in K.B. 32: 639 (1978). Type: Tanzania, Iringa District, Idete, *Carmichael* 290 (U, holo.!; EA, K, iso.!)

Pedicels and young branches puberulous or less often sparsely puberulous to glabrous . Leaf-blades drying green or sometimes brownish, somewhat leathery, glabrous above, glabrous or sometimes with sparse white longish hairs along the midrib beneath.

TANZANIA. Iringa District: Dabaga, forest near Kidabaga village, 27 Aug. 1984, *Bridson* 585!; Rungwe District: W. slopes of Mt. Rungwe, 11 Mar. 1932, *St. Clair-Thompson* 1146!; Njombe District: Mdando Forest Reserve, 15 Nov. 1966, *Gillett* 17856!
DISTR. T 7; Malawi
HAB. Forest; 1800–2600 m.

SYN. *P. iringensis* Bremek. in K.B. 11: 175 (1956)

NOTE. (on species as a whole). This species is very close to *P. lasiobractea* but the several (although individually slight) differences included in the last couplet of key 4 to species indicate that two species are best maintained. Specimens of *P. kyimbilensis* from Malawi (all Nyika Plateau) have the pedicels sparsely puberulous to glabrous (tending to *P. lasiobractea*). No flowering material from this area has been seen and confirmation as to whether the corolla-lobes fall within the length range of *P. kyimbilensis* is still needed. The limited data available gives the flowering time of *P. kyimbilensis* as late October to mid-November, while extended periods of September to December and February to March are cited for *P. lasiobractea*. Additional collections and field information would be helpful, especially records indicating to what extent the two species can be sympatric.

55. P. constipulata *Bremek.* in K.B. 1936: 482 (1936) & F.R. 47: 28 (1939); T.T.C.L.: 510 (1949). Type: Tanzania, Morogoro District, Uluguru Mts., Morningside, *E.M. Bruce* 228 (K, holo.!, BM, iso.!)

Shrub or subshrub 0.3–2.5 m. tall; young branches glabrous or puberulous, older branches covered with buff shiny bark. Leaf-blades usually drying greyish or brownish black, oblanceolate, 12–20 cm. long, 3–7 cm. wide, apex acute to long acuminate, base cuneate, entirely glabrous or sparsely scabrid above and glabrous or puberulous beneath; domatia present; bacterial nodules dot-shaped, several, scattered; petiole 0.7–2 cm. long; stipules triangular, 0.6–1.3(–2.5) cm. long, aristate, papery when young, usually breaking off with the basal remains turning corky, glabrous or puberulous outside, with a few silky hairs in axils. Corymbs terminal on very slender to moderately thick green pendant or erect lateral branches 5.5–25 cm. long usually bearing 1–2 pairs of immature leaves at the apex, moderately lax, 1.8–4.6 cm. across; primary inflorescence-branches 2–9 mm. long, puberulous-pubescent; secondary and sometimes tertiary branches present; pedicels 2–5 mm. long, puberulous-pubescent; bracts 0.5–1 cm. long, stipule-like; bracteoles inconspicuous. Calyx-tube 0.8–1 mm. long, shortly pubescent; limb-tube 0.75–1 mm. long, slightly wider than the tube, glabrescent to shortly pubescent; lobes narrowly triangular, 2–3.7 mm. long, ± 0.5 mm. wide at base, glabrescent to shortly pubescent. Corolla-tube 1.3–1.6 cm. long, 1.5–2.5 mm. wide at top, glabrous outside, not bearded at throat; lobes lanceolate-oblong, 4–6 mm. long, 1.25–2 mm. wide, acute to shortly acuminate. Young fruit, drying black, ± 8 mm. in diameter, glabrescent; calyx-limb persistent. Seeds not known.

var. **constipulata**

Inflorescence-bearing branches 5.5–13.5 cm. long, very slender to slender, 1–2 mm. thick, pendent, glabrous; inflorescence 1.8–3.2 cm. across. Leaf-blades glabrous. Calyx-lobes glabrescent; corolla-tube 2.5 mm. wide at top.

TANZANIA. Morogoro District: Uluguru Mts., Bondwa [Bondua] Mt., 27 Nov. 1932, *Schlieben* 3009! & Lupanga, 14 Feb. 1970, *Harris & Pocs* 4103! & without precise locality, 23 Nov. 1932, *Wallace* 478!
DISTR. T 6; known only from the Uluguru Mts.
HAB. Forest; 1370–1740 m.

SYN. *P. constipulata* Bremek. var. *geoscopa* Bremek. in F.R. 47: 28 (1939), *nom superfl.* Type: as for species.

NOTE. *Pocs* 6804/A (DSM) from Uluguru Mts., Bondwa, tends to be more robust than usual and has stipules up to 2.5 cm. long.

var. **uranoscopa** *Bremek.* in F.R. 47: 28 (1939); T.T.C.L.: 510 (1949). Type: Tanzania, Morogoro District, Uluguru Mts., *Schlieben* 2920 (B, holo.†)

Inflorescence-bearing branches 12–25 cm. long, not markedly slender, at least 2 mm. thick, erect, puberulous; inflorescence 3.8–4.6 cm. across. Leaf-blades sparsely scabrid to glabrous above, puberulous on nerves beneath. Calyx-lobes shortly pubescent; corolla-tube 1.5 mm. wide at top.

TANZANIA. Morogoro District; Morogoro area, 8 Nov. 1932, *Schlieben* 2923! & without locality, *Rounce* 523!

DISTR   T 6; known only from Uluguru Mts.
HAB.   Forest; 900–1100 m.

56. P. sparsipila *Bremek.* in K.B. 1936: 484 (1936) & in F.R. 47: 88 (1939); T.T.C.L.: 513 (1949). Type: Tanzania, Morogoro District, Uluguru Mts., Bunduki, *E.M. Bruce* 437 (K, holo.!, BM, iso.!)

Shrub or small tree 1.8–3 m. tall; young branches glabrous, very soon covered with pale brown corky bark, flaking when older. Leaf-blades usually turning blue-black when dry, obovate or less frequently oblanceolate, 7–11(–12.4) cm. long, 2.4–4.5 cm. wide, apex acuminate, sometimes mucronulate, base cuneate to attenuate, sparsely to very sparsely hairy on both sides but with denser hairs on the nerves beneath; bacterial nodules dot-shaped, scattered; petiole 0.4–1.5 cm. long, sparsely hairy; stipule-limbs brown, transparent, ± triangular, 6 mm. long, membranous, caducous; arista up to 1.5 mm. long. Corymbs terminal on short leafless (but sometimes with nodes) lateral branches 5–12.3 cm. long, erect or less often pendent, 1–2 mm. thick, covered with a thin layer of cork, moderately compact, 2–4 cm. across; primary inflorescence-branches 0.3–1.2 cm. long, sparsely hairy; secondary and tertiary branches usually present; pedicels 1.5–5 mm. long, hirsute-pubescent; bracts brown or buff, 4–5 mm. long, papery; bracteoles fawnish brown, papery. Calyx-tube 1–1.25 mm. long, densely hirsute-pubescent to pubescent; limb-tube 1–1.75 mm. long; lobes very narrowly triangular to narrowly linear, 5–7 mm. long, 0.5–1 mm. wide at base, sparsely hairy. Corolla-tube 1.1–1.5 cm. long, 1.5–2 mm. wide at top, glabrous outside, not bearded at throat; lobes oblong, 5–6.5 mm. long, 1.5–2 mm. wide, apiculate. long. Fruit not known.

TANZANIA. Morogoro District: Uluguru Mts., above Morningside, Bondwa Peak, Jan. 1953, *Eggeling* 6462! & 30 Dec. 1974, *Polhill & Wingfield* 4617! & NW. Uluguru Mts., 18 Dec. 1932, *Schlieben* 3118a!
DISTR   T 6; known only from the Uluguru Mts.
HAB.   Forest; 1200–1500 m.

NOTE. In leaf-shape and size and colour when dry this species closely resembles *P. bruceana* Bremek. but may easily be distinguished by the longer corky inflorescence-bearing branches, and the shorter stipules and bracts. *Schlieben* 3118a resembles *P. constipulata* Bremek. var *constipulata* in its slender pendent inflorescence-supporting branches. However, the presence of cork on the latter and calyx and leaf type place this specimen in *P. sparsipila* Bremek.

57. P. bruceana *Bremek.* in K.B. 1936: 481 (1936) & in F.R. 47: 85 (1939); T.T.C.L.: 511 (1949). Type: Tanzania, Morogoro District, Uluguru Mts., Lukwangule, *E.M. Bruce* 704 (K, holo.!, BM, iso.!)

Shrub 1–1.8 m. tall; young branches densely pubescent, older branches covered with buff flaking bark. Leaf-blades turning blue-black when dry, obovate, 6.5–13 cm. long, 2–4.7 cm. wide, apex acuminate, sometimes mucronulate, base cuneate, sparsely pubescent to tomentose beneath; bacterial nodules dot-shaped, slightly raised and usually with a denser patch of hairs above and beneath, few, scattered; petiole 5–8 mm. long, pubescent to tomentose; stipules brown, triangular, 1.2–1.5 cm. long, acuminate to shortly aristate, papery, pubescent outside, caducous. Corymbs terminal on short lateral pubescent to tomentose branches 2–3 cm. long usually bearing 1 pair of immature leaves or occasionally leafless, moderately compact, 1.5–2 cm. across; primary inflorescence-branches 3–7 mm. long, pubescent; secondary branches absent; pedicels 2–4 mm. long, pubescent; bracts brown, 5–7 mm. long, pubescent outside with a few silky hairs inside; bracteoles brownish papery. Calyx-tube 1 mm. long, glabrescent to pubescent; limb-tube 1–1.25 mm. long, glabrescent; lobes narrowly linear, 3–5 mm. long, ± 0.25 mm. wide, sparsely pubescent. Corolla-tube 1–1.1 mm. long, 2 mm. wide at top, glabrous outside, densely pilose inside; lobes oblong, 6 mm. long, 1.5–2 mm. wide, acute. Fruit not known. Fig. 106/13.

TANZANIA. Morogoro District: Uluguru Mts., Lukwangule, Jan. 1930, *E.M. Bruce* 704! & 7 Dec. 1969, *Harris et al.* 3755!
DISTR   T 6; known only from the Uluguru Mts.
HAB.   Forest; 2040–2400 m.

58. P. axillipara *Bremek.* in K.B. 11: 171 (1956). Type: Tanzania, Morogoro District, Nguru Mts., Turiani, Koluhamba, *Semsei* 1451 (U, holo., EA, K, iso.!)

Small tree; young branches glabrous or puberulous; older branches with corky bark.

Leaf-blades oblong to obovate, 16.5–30 cm. long, 5–12 cm. wide, apex acuminate, base acute, papery, often with the tertiary nerves turning blue-black on drying, glabrous above, glabrous or with a puberulous midrib beneath; bacterial nodules dot-shaped or linear, along nerves and a few scattered; petioles 1.5–3 cm. long, glabrous or puberulous; stipule-limbs subtruncate to triangular, 0.2–2 cm. long, apex acute or subulate, often corky around the margin, puberulous or glabrous outside. Corymbs terminal on very short lateral branches 0.5–1.3 cm. long, bearing 1 pair of immature leaves, compact, 2 cm. across; primary inflorescence-branches up to 4 mm. long; secondary inflorescence branches present; pedicels up to 3 mm. long, glabrous; bracts brown, ± 3 mm. long, papery; bracteoles brown, papery, small. Calyx glabrous; tube 1–1.5 mm. long; limb-tube 1–1.5 mm. long; lobes narrowly triangular, 2–3 mm. long, ± 0.75 mm. wide at base, acute or attenuate. Corolla-tube 1.4–1.7 cm. long, 2 mm. wide at top, glabrous outside, not bearded at throat; lobes lanceolate-oblong, 7–8.5 mm. long, 2–2.5 mm. wide, apiculate. Fruit not known.

TANZANIA. Morogoro District: Nguru Mts., Turiani, Koluhamba, Nov. 1953, *Semsei* 1451! & Maskati to Mhonda, 12 Dec. 1966, *Robertson* 425!
DISTR. T 6; only known from Nguru Mts.
HAB. Below bamboo forest; 1200 m.

NOTE. *Semsei* 1451 has thick stems with crowded internodes, stout petioles and large triangular stipules, while *Robertson* 425, has slender stems with well spaced internodes, slender petioles and short, subtruncate stipules but all other characters match closely.

59. P. crassipes *K. Schum.* in P.O.A. C: 389 (1895); *Bremek.* in F.R. 37: 169 (1934); T.S.K.: 145 (1936); T.T.C.L.: 514 (1949); F.P.S. 2: 457 (1952); I.T.U., ed. 2: 355 (1952); K.T.S.: 457 (1961); F.F.N.R.: 415, fig. 67A, B (1962); Keay in F.W.T.A., ed. 2, 2: 140 (1963). Type: Kenya, South Kavirondo District, Karachonyo [Karataschongo], *Fischer* 313 (B, holo.†)

Shrub or small tree 1–8 m. tall; young branches glabrous, stout, angled; older branches covered with thin greyish, buff or rarely blackish cracking bark. Leaves usually clustered near the apices of the branches, paired, or occasionally ternate or quadrate, glabrous; blades linear to narrowly elongate-oblong or oblanceolate, 8–30 cm. long, 1.3–7.5 cm. wide, apex rounded or sometimes obtuse, base obtuse to attenuate; midrib straw-coloured, prominent beneath; bacterial nodules dot-shaped to linear, few to many or absent (in West African specimens); petiole short, up to 1.5 cm. long; stipule-limbs truncate, 2–6 mm. long, shortly acuminate, glabrous outside. Corymbs terminal on short leafless lateral branches 0.6–15.5 cm. long, crowded, (3–)7.5–17 cm. across; peduncles 0–7 mm. long; primary inflorescence-branches 0.6–3.3 cm. long, glabrous; secondary to quaternary branches present; pedicels 1.8–7 mm. long, glabrous; bracts 5–8 mm. long, usually slightly corky, caducous; bracteoles ± membranous. Calyx glabrous; tube 1–1.5 mm. long; limb-tube 1–1.75 (–2) mm. long, ± twice as broad as the tube, truncate or undulate to slightly dentate. Corolla-tube 0.8–1.5 cm. long, 1.5–2 mm. wide at top, glabrous outside, with few hairs within; lobes oblong, 4–6 mm. long, 1.5–2.5 mm. wide, obtuse to rounded. Fruit black, shiny, 6–8 mm. in diameter; pedicels slightly accrescent; calyx-limb persistent. Seeds greyish-brown, 5–5.5 mm. wide, slightly rugulose on convex face. Fig. 106/8.

UGANDA. Acholi District: SE. Imatong Mts., Agora slope, 8 Apr. 1945, *Greenway & Hummel* 7321!; Teso District: Serere, May–June 1932, *Chandler* 689!; Mengo District: Nakasongola, 17 Sept. 1954, *Langdale-Brown* 1319!
KENYA. Meru District: NE. of Mitinguu, 4 Jan. 1966, *Gillett* 17030!; Kitui District: Kitui, Boma, 18 Jan. 1942, *Bally* 1522!; N. Kavirondo District: Broderick Falls, Mar. 1959, *Tweedie* 1804!
TANZANIA. Musoma District: Rutenge, Buruma, Zanaki, 9 May 1959, *Tanner* 4233!; Morogoro District: N. Uluguru Reserve, above Morningside, June 1953, *Semsei* 1248!; Songea District: ± 5 km. E. of Lipumba, 6 Mar. 1956, *Milne-Redhead & Taylor* 8995!
DISTR. U 1–4; K 4, 4/6, 5; T 1, 4, 6–8; throughout W. Africa, Cameroun, Zaire, Burundi, Sudan, Ethiopia, Mozambique, Malawi and Zambia
HAB. Wooded grassland, *Combretum* or *Brachystegia* woodland or sometimes forest edge; 360–2000 m.

SYN. *P. barteri* Dawe in J.L.S. 37: 521 (1906). Type: Nigeria, Abo (Aboh), *Barter* 324 (K, holo.!)
      P. utilis Hua in Bull. Soc. Bot. France 54, Mém. 8: 22 (1907). Type: Mali, Bongouni, *Chevalier* 684 (P, holo., K, iso.!)
      P. crassipes K. Schum. var. *major* De Wild. in Pl. Bequaert. 2: 293 (1923). Type: Zaire, Shaba, Kapiri valley, *Homblé* 1181 (BR, holo.)

60. P. sp. G

Small tree; young branches pubescent; older branches covered with pale buff, dull bark. Leaves confined to the apices of the branches; blades drying bronze-brown, obovate, 6–10 cm. long, 2.4–6 cm. wide, apex shortly acuminate, base cuneate, sparsely pubescent above and beneath with denser pubescence on the nerves beneath; nerves closely spaced and steeply angled, ?impressed above; bacterial nodules small, dot-shaped, several, scattered; petiole very short, 2–3 mm. long, densely pubescent; stipules truncate-triangular, 7–9 mm. long, ? aristate, pubescent outside, with silky hairs inside. Corymbs borne on very short 0.2–0.5 (–3.3) cm. long leafless branches arising immediately below the leafing apices of the main stems, compact, 1.5–2.5 cm. across; primary inflorescence-branches ± 2 mm. long, pubescent; secondary branches absent or very reduced; pedicels 1.5–4 mm. long, pubescent, becoming glabrous towards apex; bracts ± stipule-like, sheathing the base of the inflorescence; bracteoles drying brown, ± ovate, papery. Calyx glabrescent; tube drying black, ± 1 mm. long; limb drying pale brown; limb-tube 1 mm. long, wider than the tube; lobes linear-subulate from a triangular base, 3 mm. long, ± 0.25 mm. wide. Withered corolla-tube 1 cm. long, 1 mm. wide, glabrous outside not bearded at throat; lobes ± 4 mm. long. Mature fruit not known.

TANZANIA. Pare District: Luji-Hemwera, 2 Mar. 1915, *Peter* 55664!
DISTR. T 3; known only from the one specimen
HAB. Not known; 1700 m.

61. P. johnstonii *Bremek.* in F.R. 37: 147 (1934). Type: Malawi, Mt. Zomba, *Whyte* (K, holo.!)

?Shrub or small tree 2–5 m. tall; young branches pubescent; older branches gnarled or smooth with buff, dull bark. Leaves confined to the apices of the branches; blades drying brownish, narrowly obovate to obovate, 8–17 cm. long, 3.2–8 cm. wide, apex rounded or sometimes acute, base cuneate, sparsely pubescent above, pubescent beneath with dense longer hairs on the nerves; bacterial nodules small, dot-shaped, few to several, scattered; petiole 1–10 mm. long, densely pubescent; stipules truncate to triangular, 4–8 mm. long, acuminate to long acuminate, pubescent or glabrous above and glabrescent at base outside, with silky hairs inside. Corymbs borne on leafless branches 0–11 cm. long arising shortly or immediately below the leafing apices of the main stems on both sides or sometimes only one, compact to very compact, 1.5–2.5 cm. across; primary inflorescence-branches up to 2 mm. long, pubescent; secondary branches very reduced; pedicels 2–5 mm. long, pubescent; bracts stipule-like; bracteoles inconspicuous. Calyx pubescent; tube ± 1 mm. long; limb-tube 1–1.5 mm. long, slightly wider than the tube; lobes linear-subulate from a triangular base, 1.5–4.5 mm. long, ± 0.25 mm. wide. Corolla-tube 1.2–1.4 cm. long, ± 1 mm. wide, glabrous outside, not bearded at throat; lobes 4–5 mm. long, 2–2.5 mm. wide. Fruit black (or red), 6–7 mm. in diameter, shiny, sparsely hairy; calyx-limb persistent. Mature seeds not known.

subsp. **breviloba** *Bridson* in K.B. 42: 253 (1987). Type: Tanzania, Iringa District, Mufindi, near Lake Ngwazi [Nkwazi], *Shabani* 1015 (K, holo.!, EA, iso.)

Calyx-lobes 1.5–2.5 mm. long. Inflorescence-bearing branches reduced, 0–3 mm. long or occasionally developed, up to 4.2 cm. long. Stipules always pubescent outside.

TANZANIA. Mbeya District: Poroto Mts., 8 Mar. 1932, *St. Clair-Thompson* 734a!*; Iringa District: Mufindi District, Ngwazi Air Strip, 20 Feb. 1986, *Bidgood & Lovett* 36! & Ruaha National Park, Isunkaviola Mt., Kangawalumba Forest, 26 Dec. 1972, *Bjørnstad* 2398!
DISTR. T 7; Zambia
HAB. Thicket associated with termite mounds, woodland or upland dry evergreen forest; 1300–1800 m.

NOTE. Subsp. *johnstonii* has been recorded from the Central and Southern provinces of Malawi. It has longer calyx-lobes, 3.25–4.5 mm. long, the inflorescence-bearing branches are usually present (1–11 cm. long) and the stipules can either be glabrous above and glabrescent at the base or pubescent. Unlike subsp. *breviloba* this taxon favours montane or submontane forest habitats.

* These data are only tentative and were taken from *St. Clair-Thompson* 734, *Vangueria sp. aff. linearisepala;* 734a was listed as being *Pavetta sp.* The actual sheet bears a number tag with 734, and no other data.

62. P. schumanniana *K. Schum* in P.O.A. C: 389 (1895); Hiern, Cat. Welw. Afr. Pl. 2: 488 (1898); Bremek. in F.R. 37: 171 (1934); T.T.C.L.: 514 (1949); F.F.N.R.: 415 (1962); Vollesen in Op. Bot. 59: 70 (1980); Fl. Pl. Lign. Rwanda: 580, fig. 194/3 (1983); Kok & Grobbelaar in S. Afr. J. Bot. 3: 186 (1984); Fl. Rwanda 3: 194, fig. 58/3 (1985). Type: Malawi, without locality, *Buchanan 734* (K, lecto.!)

Dwarf rhizomatous shrub or small tree 0.4–8.4 m. tall, branched or occasionally with unbranched stems; young branches pubescent; older branches covered with greyish or brown slightly corky cracking bark. Leaves paired or ? ternate, borne at the apex of the branches; blades oblanceolate to obovate, 3–20.2 cm. long, 1.3–8 cm. wide, apex rounded or less often obtuse or emarginate, base cuneate, often bullate, pubescent or scabrid above, greyish tomentose and with prominent nerves beneath; bacterial nodules, inconspicuous, dot-shaped, scattered; petiole 2–17 mm. long, tomentose; stipule-limbs triangular, gradually attenuated or caudate, up to 9 mm. long, pubescent outside. Corymbs terminal on suppressed or very abbreviated leafless lateral branches up to 1.7 cm. long, often crowded, moderately compact to lax, (1.1–)2–5 cm. across, occasionally shortly pedunculate; primary inflorescence-branches 0.2–2 cm. long, pubescent; secondary and tertiary branches present; pedicels 0.5–5 mm. long, pubescent; bracts stipule-like, 3–7 mm. long; bracteoles brown, membranous, inconspicuous. Calyx pubescent-tomentose; tube 1–2 mm. long; limb-tube 0.5–1.25 mm. long; lobes triangular to shortly triangular or occasionally ovate, 0.5–2 mm. long, 0.75–1.25 mm. wide. Corolla-tube 5–11 mm. long, ± 1.5 mm. wide at top, pubescent or sometimes glabrous outside, not bearded at throat; lobes oblong-oblanceolate, 3–5 mm. long, 1.5–2(–2.5) mm. wide, obtuse to rounded. Fruit black, 5–7 mm. in diameter, shiny, sparsely hairy; calyx-limb persistent; pedicels accrescent, reaching 15 mm. long. Seed dark grey, 5 mm. wide, convex face ± smooth.

TANZANIA. Mbulu District: Great North road, Pienaars Heights, 6 Jan. 1962, *Polhill & Paulo* 1071!; Dodoma District: 42.7 km. S. of Itigi Station on the Chunya road, 18 Apr. 1964, *Greenway & Polhill* 11633!; Songea District: ± 6.5 km. E. of Gumbiro, 25 Jan. 1956, *Milne-Redhead & Taylor* 8530!
DISTR. T 1, 2, 4–8; Cameroun, Zaire, Rwanda, Burundi, Mozambique, Malawi, Zambia, Zimbabwe, Angola, Namibia, South Africa (Transvaal and Natal)
HAB. In wooded grassland and *Brachystegia* woodland; 360–2400 m.

SYN. *P. arenicola* K. Schum. in Warb., Kunene Zambesi Exped.: 391 (1903). Type: Angola, *Baum* 522 (B, holo.†)
    P. stipulopallium K. Schum. in Warb., Kunene Zambesi Exped.: 392 (1903). Type: Angola, *Baum* 948 (B, holo.†)
    [*P. canescens* sensu Bremek. in Ann. Transv. Mus. 13: 209 (1929), pro parte, *non* DC., (1830)]

NOTE. Several sheets from T 6, Kilosa, Ulanga and Morogoro Districts, have very reduced calyx-lobes and a somewhat less robust habit than most of the material seen, but would not seem to merit varietal rank.
    Several other species (mostly restricted to Angola) are closely related to *P. schumanniana*, and one, *P. loandensis* (S. Moore) Bremek., may prove to be only a variety.
    Schlieben 5593a, Lindi District, Lake Lutamba, shows tendencies towards *P. decumbens* and *P. pseudo-albicaulis* with its unbranched stems ± 1 m. tall. However, the branched inflorescences, much shorter corolla-tubes with sparse hairs outside and the darker lenticellate bark would seem to indicate that it is better thought of as an aberrant form of this species.
    This species is reported to cause "Gousiekte" cattle poisoning in South Africa (see Codd & Voorendyk in Bothalia 8 (suppl. 1): 57 (1965)).

63. P. radicans *Hiern*, Cat. Afr. Pl. Welw. 1(2): 487 (1898); Bremek. in F.R. 37: 173 (1934); F.F.N.R.: 416 (1962); Bridson in K.B. 32: 640 (1978). Type: Angola, *Welwitsch* 3184 (BM, holo.!, EA, K, iso.!)

Erect or prostrate rhizomatous suffrutex or small shrub, (7.5–)15–45(–120) cm. tall; young branches pubescent; older branches with greyish-brown thin flaking bark. Leaves confined to the apical regions of main and short lateral branches, sometimes immature at time of flowering; blades oblanceolate, 1.7–6.7 cm. long, 0.5–1.8(–2.1) cm. wide, apex obtuse, base cuneate, very sparsely pubescent to pubescent above and beneath; bacterial nodules small, dot-shaped, very few to few, scattered; petiole up to 8 mm. long, pubescent; stipules with triangular membranous limbs and usually with a short arista, 1–3 mm. long, pubescent outside, with silky hairs absent within. Flowers sweet-scented; corymbs terminal on abbreviated leafless lateral branches 0–3.2 cm. long, few-flowered, laterally compressed, distinctly longer than wide, 0.5–1(–1.5) cm. wide; primary inflorescence-branches 0–8 mm. long; secondary branches present or absent; pedicels 1–8 mm. long,

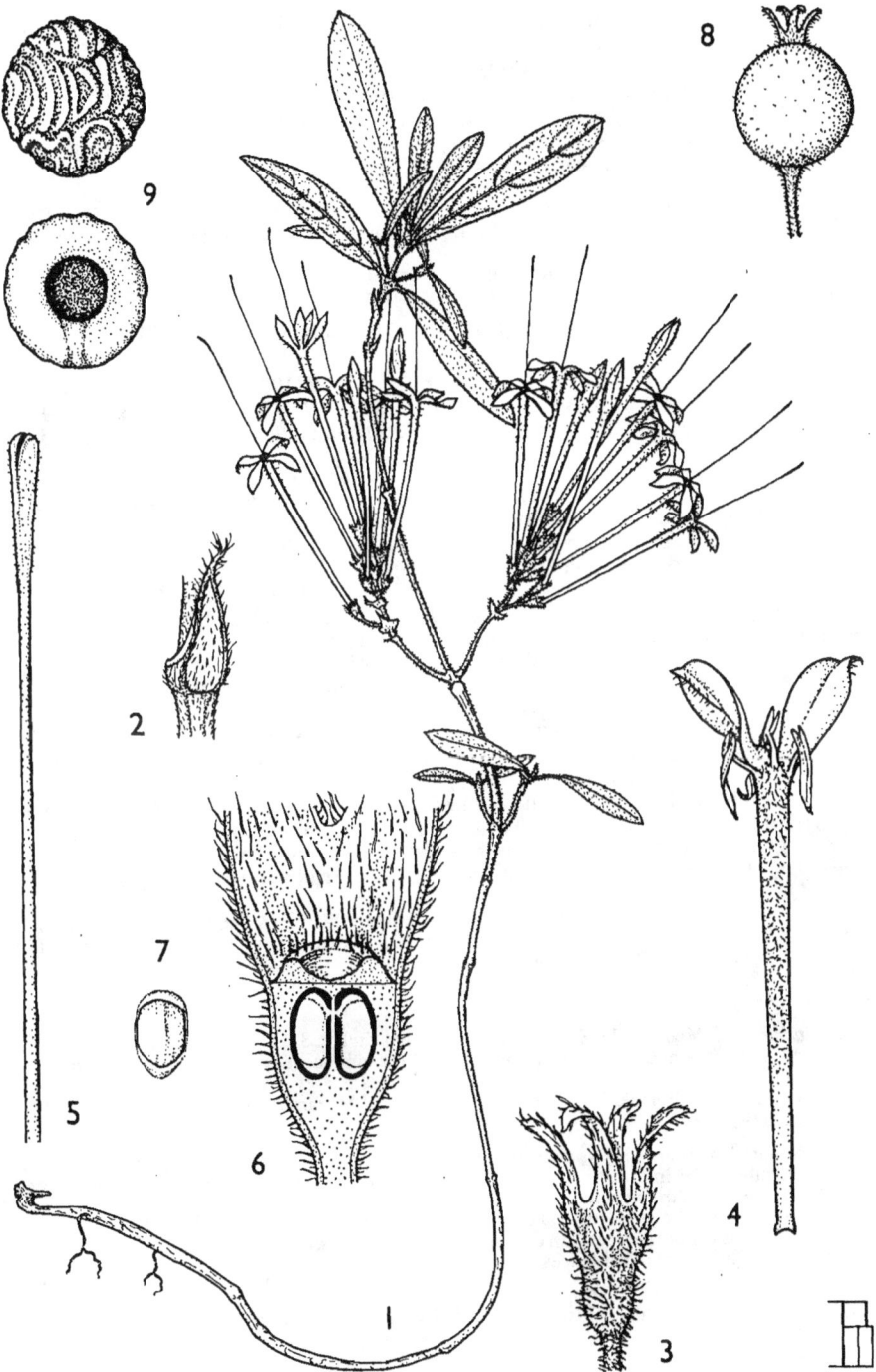

FIG. 113. *PAVETTA RADICANS* — 1, habit, × ⅔; 2, stipules, × 4; 3, calyx, × 8; 4, half corolla, × 2; 5, part of style with stigma, × 6; 6, section through calyx-tube to show ovary, × 20; 7, placenta partly surrounding ovule, dorsal view, × 20; 8, fruit, × 3; 9, seed, 2 views, × 4. 1, from *Pawek* 6258; 2–7, from *Bullock* 1975; 8, 9, from *Pearson* 2091. Drawn by Diane Bridson.

pubescent; bracts 2–3 mm. long, pubescent; bracteoles brown, papery, small. Calyx pubescent; tube 1 mm. long; limb-tube 0.6–1.3 mm. long; lobes triangular, 0.5–1(–2.5) mm. long, 0.5–1 mm. wide at base. Corolla-tube (1.5–)1.9–4.5 cm. long, 1.5 mm. wide at top, glabrous or sparsely pubescent outside, not bearded at throat; lobes lanceolate, 6–8 mm. long, 1.5–2 mm. wide, obtuse, usually with a short greenish mucro, glabrous or sparsely hairy outside. Fruit black, 5–7 mm. in diameter. glossy, glabrescent, occasionally with the pedicels lengthening to 1.5 cm. Seeds ± 5 mm. wide. Fig. 113.

TANZANIA. Buha District: 112 km. from Kasulu to Kibondo, 15 Nov. 1962, *Verdcourt* 3322!; Ufipa District: Chapota, 2 Dec. 1949, *Bullock* 1975!; Iringa District: Sao Hill-Madibira road, 8 Dec. 1962, *Richards* 17342!
DISTR. T 4, 7; Zaire, Malawi, Zambia, Zimbabwe, Angola
HAB. *Brachystegia* woodland, sometimes on rocky scarps; 1650–1830 m.

SYN. *P. cecilae* N.E. Br. in K.B. 1906: 106 (1906). Type: Zimbabwe, Selukwe, *Cecil* 124 (K, holo.!)

64. P. decumbens *K. Schum. & K. Krause* in E.J. 39: 547 (1907); Bremek. in F.R. 37: 171 (1934); T.T.C.L.: 514 (1949); Bridson in K.B. 32: 640 (1978); Vollesen in Op. Bot. 59: 70 (1980). Type: Tanzania, Kilwa District, Donde, Mangatana, *Busse* 608 (B, holo.†, EA, iso.!)

Erect little-branched suffrutex 20–40 cm. tall or shrub 1.2 m. tall; young branches pubescent; older branches 2–3(–4) mm. thick covered with buff flaking bark. Leaves at flowering time immature and restricted to the apex of the main stem; immature blades narrowly to broadly elliptic, 3.5–6.5 cm. long, 0.9–4.5 cm. wide, apex ± acute, base cuneate, pubescent above, densely tomentose beneath; petiole 3–6 mm. long; mature leaves drying greyish-black, very narrowly elliptic to elliptic, 9.8–16.4 cm. long, 2.5–4.8 cm. wide, acute to acuminate at apex, cuneate at base, densely scabrid-pubescent above, tomentose beneath; bacterial nodules dot-shaped, raised beneath; petiole 0.3–1 cm. long; stipule-limbs triangular, 2–6 mm. long, pubescent outside, with silky hairs inside; arista 2–8 mm. long. Fascicles terminal on abbreviated or virtually entirely suppressed leafless lateral branches up to 2.1 cm. long, compact to ± lax, 1.1–2.2 cm. across; primary inflorescence-branches usually suppressed, so that at least some pedicels arise directly from the node; secondary and tertiary branches 1–4 mm. long; pedicels 2–7 mm. long, pubescent; bracts ± ovate, 3–9 mm. long, pubescent outside, with dense silky hairs inside. Calyx densely pubescent; tube 0.75–1 mm. long; limb-tube 0.5–1 mm. long, slightly wider than the tube; lobes triangular, 0.25–1 mm. long, 0.5–0.75 mm. wide at base. Corolla-tube 0.85–1.7 cm. long, ± 1.25 mm. wide at top, glabrous outside, not bearded at throat; lobes oblong-lanceolate, 4–5 mm. long, 1.25–1.5 mm. wide, obtuse. Fruit and seed not known.

TANZANIA. Ulanga District: Selous Game Reserve, 7 km N of Mlahi, 13 Oct. 1975, *Vollesen* in M.R.C. 2782!; Kilwa District: Selous Game Reserve, Tunda Hills, Lung'onyo R., 14 Apr. 1968, *Rodgers* 218!; & Nakilala Thicket, 14 Dec. 1975, *Vollesen* in M.R.C. 3084!
DISTR. T 6, 8; Mozambique
HAB. *Brachystegia* woodland; 100–750 m.

SYN. *P. albicaulis* S. Moore in J.L.S. 40: 97 (1911); Bremek. in F.R. 37: 170 (1934), pro parte; T.T.C.L.: 514 (1949) (see note after next species). Type: Mozambique, Beira, *Swynnerton* (BM, holo.!, K, iso.!)
      P. stephanantha Bremek. in K.B. 8: 504 (1954). Type: Mozambique, Moebede road, *Faulkner* K.346 (K, holo.!)

NOTE. Mature leaves are only known from one specimen, *Rodgers* 218; the length of the corolla-tube is a little longer in this specimen and it is said to be a shrub 1.2 m. tall, but otherwise agrees well with the rest of the material seen.
      This species may be distinguished from suffruticose forms of *P. schumanniana* (also in T 6, 8) as that species has inflorescences with distinct primary branches, corolla-tubes usually hairy outside and the usually rounded leaf-apices.

65. P. pseudo-albicaulis *Bridson* in K.B. 32: 640, fig. 6 (1978); Vollesen in Op. Bot. 59: 70 (1980). Type: Tanzania, Lindi District, Lake Lutamba, *Schlieben* 5324 (K, holo.!, B, BM, BR, EA, P, iso.!)

?Small shrub 0.8–2.5 m. tall, unbranched; young branches pubescent; older branches 0.7–1 cm. thick, covered with white to pale fawn corky, cracking bark. Leaves restricted to the apical region of the main stem, sometimes ternate; blades bullate when young, narrowly elliptic or oblong to elliptic, 15.5–26 cm. long, 4.7–11.6 cm. wide, apex acute, base cuneate to obtuse, pubescent above, densely tomentose beneath; bacterial nodules

RUBIACEAE

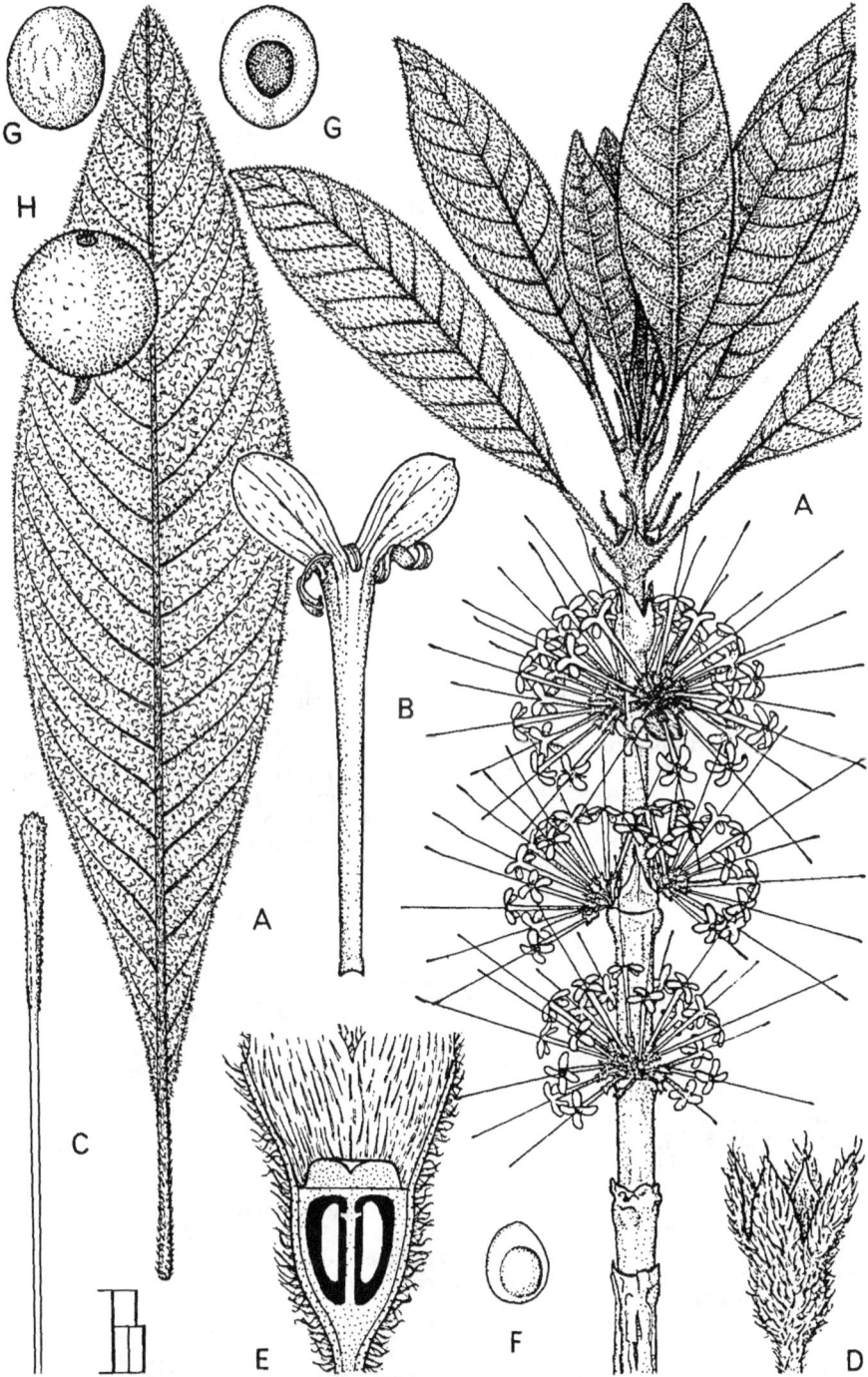

FIG. 114. *PAVETTA PSEUDO-ALBICAULIS* — **A**, shoot with flowers and immature leaves and mature leaf, × ½; **B**, half corolla, × 3; **C**, part of style and stigma, × 8; **D**, calyx, × 6; **E**, section through calyx, × 12; **F**, ovule, dorsal view, × 12; **G**, seed, 2 views, × 2; **H**, fruit, × 2. A–G, from *Schlieben* 5324; H, from *Rodgers* 908. Drawn by Diane Bridson.

obscure, small, dot-shaped, ?mostly near the midrib; petiole 3–7.5 cm. long, densely pubescent; stipule-limbs truncate to triangular, 3–7 mm. long, pubescent outside, with silky hairs inside; arista 0.2–1.3 cm. long. Inflorescence-bearing branches always entirely suppressed; fascicles compact, 2–2.3 cm. across, peduncle absent or rarely up to 1 mm. long; inflorescence-branches very reduced, ± 1 mm. long; pedicels 1–5 mm. long, tomentose; bracts ± ovate, 1.1–1.2 cm. long, pubescent outside, with dense white silky hairs inside. Calyx tomentose; tube 1–1.5 mm. long; limb-tube 1–2 mm. long, slightly wider than the tube; lobes triangular, 0.25–1.25 mm. long, 1–1.25 mm. wide at base. Corolla-tube 1.5–2 cm. long, 2 mm. wide at top, glabrous outside, not bearded at throat; lobes oblong to broadly obovate, 5–6 mm. long, 2–2.5(–3.5) mm. wide, obtuse. Fruit black, 8–9 mm. in diameter, shiny, sparsely pubescent; calyx-limb persistent. Ripe seeds not known. Fig. 114.

TANZANIA. Kilwa district: Selous Game Reserve, Lihangwa R. source, 12 Feb.. 1970, *Rodgers* in *M.R.C.*908!; Lindi District: Lake Lutamba, 12 Apr. 1934, *Schlieben* 5324! & without data, *Gillman!*
DISTR. T 8, Mozambique
HAB. Clearings or old cultivations; 140–600 m.

SYN. [*P. albicaulis* sensu Bremek. in F.R. 37: 170 (1934), pro parte, quoad *Stocks*, *non* S. Moore]

NOTE. *Busse* 2515a from Tanzania, Lutamba [Gegelam Letamba] (B †), cited by Bremekamp, loc. cit., and Brenan, T.T.C.L.: 514 (1949), could have belonged to either this species or *P. decumbens* K. Schum. & K. Krause.

66. P. haareri *Bremek.* in F.R. 37: 171 (1934); T.T.C.L.: 514 (1949). Type: Tanzania, N. Kilosa District, Kibedya, *Haarer* 1991 (K, holo.!, EA, iso.!)

?Shrub; young branches pubescent; older branches with light brownish-grey thin bark. Leaves confined to the apices of the main branches and short lateral branches off the new wood only; blades oblong to oblanceolate, 6–9 cm. long, 2.2–4.6 cm. wide, apex rounded, base cuneate, glabrescent above and beneath; bacterial nodules dot-shaped, sometimes raised, sparse, scattered; petiole 3–10 mm. long, glabrescent; stipules with a triangular limb and short arista, up to 5 mm. long, pubescent outside, with longer silky hairs inside. Corymbs terminal on very short lateral branches (up to 5 mm. long) off the older wood only, few-flowered, moderately lax, 1.5–3 cm. across; primary inflorescence-branches 0.5–1.6 cm. long; secondary and tertiary branches present; pedicels 1–7 mm. long, pubescent; bracts 4–5 mm. long, pubescent outside, with long silky hairs inside; bracteoles inconspicuous. Calyx-tube 1 mm. long, pubescent; limb-tube 1.75 mm. long, wider than the tube, undulate to shortly and broadly quadrate, glabrescent. Corolla-tube 8 mm. long, 1.5 mm. wide at top, glabrous outside, the upper ⅔ hairy within; lobes lanceolate, 5 mm. long, 2 mm. wide, rounded, sometimes shortly mucronulate. Fruit unknown.

TANZANIA. Kilosa District: Kibedya, Jan. 1931, *Haarer* 1991!
DISTR. T 6; known only from the type gathering
HAB. Not known; ± 760 m.

67. P. gardeniifolia *A. Rich.*, Tent. Fl. Abyss. 1: 351 (1847); Hiern in F.T.A. 3: 177 (1877); Becc. in Martelli, Fl. Bogos: 43 (1886); Armari in Pirotta, Fl. Eritrea 1(2): 153 (1904); Bremek. in F.R. 37: 180 (1934); F.P.S. 2: 458 (1952); Bridson in K.B. 32: 641 (1978); Fl. Pl. Lign. Rwanda: 567, fig. 194/1 (1983); Fl. Rwanda 3: 190, fig. 58/1 (1985). Type: Ethiopia, Maundet, *Schimper* 1141 (P, holo., K, LE, iso.!)

Shrub or small tree (0.6–)1.2–7 m. tall; young branches glabrous to densely pubescent; older branches covered with buff to greyish bark. Leaf-blades usually drying green or brown, narrowly to broadly elliptic or oblanceolate to obovate, 1.6–12.5 cm. long, 1–6.1 cm. wide, apex acute, obtuse or rounded, occasionally slightly emarginate, often with small horny callous at the tip, base cuneate to attenuate, entirely glabrous to pubescent; bacterial nodules dot-shaped to linear, scattered, usually several; petiole 0.1–2 cm. long, glabrous to pubescent; stipule-limbs greenish, opaque, subtruncate to triangular, 2–5 mm. long, shortly acuminate, glabrous or pubescent outside, with silky hairs inside. Corymbs terminal on short spurs (which are sometimes entirely suppressed) or on short leafless lateral branches up to 8.5 cm. long, moderately compact to very lax, 1.7–7 cm. across; peduncles 0–7 mm. long; primary inflorescence-branches 0.3–1.8(–2.5) cm. long, glabrous to pubescent; secondary branches very short or ± equalling primary branches; tertiary branches occasionally present; pedicels 0.2–2.2 cm. long, slender or ± robust, glabrous or pubescent; bracts buff-coloured, 2–5 mm. long, membranous, with sparse to dense silky hairs inside, caducous; bracteoles buff-coloured, ± translucent. Calyx-tube ± 1 mm. long, glabrous or pubescent; limb (0.25–)0.5–1.25 mm. long, truncate, repand or with short triangular lobes up to 1 mm. long, glabrous or occasionally sparsely hairy. Corolla cream to yellow, usually drying buff or brown; tube 0.4–1.5 cm. long, 1–2 mm. wide at top, glabrous outside, not bearded at throat; lobes lanceolate to ovate, (3–)5–7 mm. long, 2–3 mm. wide, obtuse to shortly acuminate. Fruit black, 6–8 mm. in diameter, shiny. Seed greyish black, 4–6 mm. wide, rugulose on convex face.

var. **gardeniifolia**; Bridson in K.B. 32: 641 (1978); Kok & Grobbelaar in S. Afr. J. Bot. 3: 186 (1984).

Inflorescence-branches and pedicels always glabrous; young branches usually glabrous or occasionally pubescent; stipules glabrous or sometimes pubescent outside; leaf-blades glabrous or rarely glabrescent with midrib pubescent above, glabrous or less often sparsely pubescent to pubescent beneath. Calyx always glabrous.

UGANDA. Karamoja District: near Moroto, May 1948, *Eggeling* 5823! & Mt. Moroto, 4 Sept. 1956, *Hardy & Bally* in *Bally* 10718!; Ankole District: Mulema, *Bagshawe* 285!
KENYA. Northern Frontier Province: Moyale, 2 Sept. 1953, *Bally* 9072!; Machakos District: Kiuu Hill [Kilima Kiu], 16 Feb. 1972, *Kokwaro* 3020!; Masai District: 35.2 km. on Kajiado–Namanga road, 27 Nov. 1960, *Archer* 209!.
TANZANIA. Musoma District: Serengeti, Seronera, 1 Apr. 1961, *Greenway* 9948!; Pare District: Same, Kiko Hill, 2 Apr. 1972, *Wingfield* 1989!; Iringa District: Ruaha National Park, 2 km. on Msembe-Kimiramatonge track, 4 Mar. 1970, *Greenway & Kanuri* 14014!
DISTR. U 1, 2; K 1–4, 6, 7; T 1–7; Togo, Nigeria, Cameroun, Central African Republic, Zaire, Rwanda, Burundi, Sudan, Ethiopia, Somalia, Malawi, Zambia, Zimbabwe, Botswana, Namibia, South Africa (Transvaal, Natal)
HAB. Bushland to forest, often on rocky ground; 600–2100 m.

SYN. *P. adelensis* Delile in Rochet d'Héricourt, Sec. Voy. Choa: 343 (1846); Bremek. F.R. 37: 180 (1934), nom. nud., based on Ethiopia, Choa, *Rochet d'Héricourt* 40 (P!)
P. assimilis Sond. in Harv. & Sond., Fl. Cap. 3: 20 (1865); Bremek. emend. in Ann. Tranv. Mus. 13: 209 (1929) & F.R. 37: 181 (1934); F.P.N.A.: 360 (1947); T.T.C.L.: 515 (1949); K.T.S.: 457 (1961); F.F.N.R.: 416 (1962), pro parte. Type: South Africa, Durban [Port Natal], *Gerrard & McKen* 1355 (K, holo.!)
P. gardeniifolia A. Rich. var. *breviflora* Vatke in Oesterr. Bot. Zeitschr. 25: 231 (1875); Warburg, Kunene-Sambesi-Exped., Baum: 391 (1903). Types: Ethiopia, Habab, *Hildebrandt* 495 (B, syn.†) & Somalia, *Hildebrandt* 889 (B, syn.†)
Ixora assimilis (Sond.) Kuntze in Rev. Gen. Pl. 1: 286 (1891)
I. gardeniifolia (A. Rich.) Kuntze in Rev. Gen. Pl. 1: 286 (1891)
Pavetta gardeniifolia A. Rich. var. *laxiflora* K. Schum. in P.O.A. C: 389 (1895). Types: Kenya, Teita District, *Hildebrandt* 2570 (B, syn.†) & Tanzania, Mwanza/Kwimbwa District, Magu to Kagehi, *Fischer* 311 (B, syn.†)
P. gardeniifolia A. Rich. var. *angustata* A. Rich. sensu Armari in Pirotta, Fl. Eritrea 1(2): 153 (1904), but no original reference found, in Tent. Fl. Abyss. 1: 351 (1847)
P. krauseana K. Krause in E.J. 48: 420 (1912). Type: Namibia, near Waterberg, Karstfeld, *Dinter* 1793 (B, holo.†)
P. saxicola K. Krause in E.J. 48: 421 (1912); Bremek. in F.R. 37: 180 (1934); F.W.T.A. ed. 2, 2: 140 (1963). Type: Togo, Kumondi, *Kersting* 740 (B, holo.†, K, fragment!)
P. assimilis Sond. var. *glabra* Bremek. in Ann. Transv. Mus. 13: 210 (1929) & in F.R. 37: 182 (1934); T.T.C.L.: 515 (1949) nom. superfl. Type as for *P. assimilis* Sond.
P. assimilis Sond. var. *brevituba-glabra* Bremek. in Ann. Transv. Mus. 13: 210 (1929); & in F.R. 37: 182 (1934). Type: South Africa, Transvaal, Magaliesberg, Silkaatsnek, *Bremekamp* 1214 (PRE, holo.!)
P. hochstetteri Bremek. var. *glaberrima* Bremek. in F.R. 37: 182 (1934); F.P.S. 2: 458 (1952). Type: Ethiopia, Meeguetjot, *Schimper* 210 (K, holo.!)
P. hochstetteri Bremek. var. *mollirama* Bremek. in F.R. 37: 183 (1934). Type: Ethiopia, Galla, Bidduma, *Ruspoli & Riva* 122 (FT, holo.!)
P. hochstetteri Bremek. var. *graciliflora* Bremek. in F.R. 47: 95 (1939); K.T.S.: 458 (1961). Type: Tanzania, Mwanza District, Capri Point, *Rounce* 251 (K, holo.!, EA, iso.!)
P. petraea Bremek. in F.R. 47: 93 (1939); T.T.C.L.: 516 (1949); K.T.S.: 460 (1961). Type: Tanzania, Masai District, Merkerstein, *Greenway* 4326 (K, holo.!, EA, iso.!)
P. termitaria Bremek. var. *glabra* Bremek. in F.R. 47: 94 (1939); K.T.S.: 462 (1961). Type: Zambia, Mwinilunga District, Matonchi Farm, *Milne-Redhead* 4479 (K, holo.!)
P. fossorum Bremek. in K.B. 3: 357 (1949). Type: Somalia, Ahl Hills, Sugli, *Collenette* 284 (K, holo.!)
P. somaliensis Bremek. in K.B. 3: 357 (1949). Type: Somalia, Golis Range, Gan Libah, *Glover & Gilliland* 1160 (EA, holo.!, K, iso.!)

NOTE. *P. appendiculata* De Wild. (Type: Mozambique, Marrumbula, *Luja* 376 (BR, holo.!)) has the corolla-lobes minutely auriculate on one side, but otherwise agrees well with var. *gardeniifolia;* its status has been left unaltered for the present as no other specimens from Mozambique have been seen.

Gillett 13423, from Kenya, **K** 1, Dandu, has a much more compact inflorescence (inflorescence branches not exceeding 3 mm. long) than the rest of the material from East Africa, but it is hardly worth recognising a separate variety on the evidence of one specimen.

var. **subtomentosa** *K. Schum.* in E.J.: 28: 494 (1900); T.T.C.L.: 516 (1949); Bridson in K.B. 644 (1978); Pl.Fl. Lign. Rwanda: 546 (1983); Kok & Grobbelaar in S. Afr. J. Bot. 3: 178 (1984); Fl. Rwanda 3: 192 (1985). Type: Tanzania, Iringa District, Kilima Plateau, *Goetze* 661 (B, holo.†, K, iso.!)

Inflorescence-branches and pedicels sparsely to densely pubescent; young branches pubescent; stipules pubescent outside; leaf-blades sparsely pubescent, pubescent, scabrid-pubescent or rarely glabrous or glabrescent above, pubescent beneath; calyx-tube glabrous to densely pubescent; limb glabrous.

UGANDA. Ankole District: Lutoto, Oct. 1940, *Eggeling* 4125!; Kigezi District: Kamwezi, Feb. 1948, *Purseglove* 2590! & May 1950, *Eggeling* 5879!
TANZANIA. Bukoba District: Rwakarindini, 28 Sept. 1948, *Ford* 720!; Kondoa District: scarp between Kolo and Chungai, 24 km. N. of Kolo, 13 Jan. 1962, *Polhill & Paulo* 1157!; Iringa District: 6.4 km. E. of Iringa on Morogoro road, 4 Feb. 1962 *Polhill & Paulo* 1347!
DISTR. U 2; ?K 1, T 1–5, 7; ?Nigeria, Zaire, Rwanda, Malawi, Zambia, Zimbabwe, Botswana, Angola, Namibia, S. Africa (Transvaal, Natal)
HAB. Dry rocky hillsides, in thickets or grassland; 840–1800 m.

SYN. *P. heidelbergensis* Bremek. in Ann. Transv. Mus. 13: 210 (1929) & in F.R. 37: 182 (1934). Type: South Africa, Transvaal, Johannesburg, Jeppestown Ridge, *Gilfillan* 138 (PRE, holo.!)
  P. assimilis Sond. var. *pubescens* Bremek. in Ann. Transv. Mus. 13: 209 (1929) & in F.R. 37: 181 (1934). Type: South Africa, Transvaal, Pretoria, Fountains, *Leendertz* (540) *T.M.* 8674 (PRE, holo.!)
  P. assimilis Sond. var. *brevituba-pubescens* Bremek. in Ann. Transv. Mus. 13: 210 (1929), & in F.R. 37: (1934). Type: South Africa, Pretoria, road to Fountains, *Leendertz* (541) *T.M.* 8675 (PRE, holo.!)
  P. ovaliloba Bremek. in F.R. 37: 181 (1934); T.T.C.L.: 515 (1949). Type: Tanzania, Pare District, Kirangi, *Phillips* (K, holo.!)
  P. rhodesiaca Bremek. in F.R. 37: 181 (1934). Type: Zimbabwe, Harare (Salisbury), *Eyles* 4645 (K, holo.!)
  P. assimilis Sond. var. *scabrida* Bremek. in F.R. 47: 95 (1939); T.T.C.L.: 515 (1949). Type: Tanzania, Iringa, *Lynes* 65 (K, holo.!, EA, iso.!)
  P. assimilis Sond. var. *puberula* Bremek. in F.R. 47: 95 (1939); T.T.C.L.: 515 (1949). Type: Zimbabwe, Mt. Selinda, Inyamadzi R., *Michelmore* 289 (K, holo.!)
  P. assimilis Sond. var. *tomentella* Bremek. in F.R. 47: 95 (1939); T.T.C.L.: 515 (1949). Type: Tanzania, Shinyanga, *B.D. Burtt* 5101 (K, holo.!, BM, EA, iso.!)
  P. termitaria Bremek. var. *pubescens* Bremek. in F.R. 47: 94 (1939). Type: Angola, Mexico, between R. Nkoki and R. Lukusa, *Milne-Redhead* 3988 (K, holo.!)
  P. pleiantha Bremek. var. *glabrifolia* Bremek. in K.B. 3: 356 (1949). Type: Tanzania, Bukoba District, Nshamba, *Gillman* 536 (K, holo.!, EA, iso.! (as 536a))
  P. pleiantha Bremek. var. *velutina* Bremek. in K.B. 3: 356 (1949). Type: Tanzania, Bukoba District, Nshamba, *Gillman* 536b (EA, holo.!)
  P. tomentella Bremek. in K.B. 3: 356 (1949). Type: Tanzania, Bukoba District, Nshamba, *Gillman* 528 (EA, holo.!, K, iso.!)

NOTE. The record from **K** 1 is based on *Ali Mohamed* in *E.A.H.* 16338 from near Wajir; the calyx-lobes are more developed than typical and it is possible that a separate taxon may be involved, the corolla is not known.

NOTE. (on species as a whole). This is a very variable species and the degree of the laxness of the inflorescence and length of corolla-tube seem especially variable. The calyx tends to become more lobed in specimens from Zambia, Zimbabwe and adjacent areas, but this is scarcely consistent enough for the recognition of varieties.
Peter 32787 from Kilosa District and *Sabani* s.n. from Iringa District are ± intermediate between the two varieties, having pubescent leaves but pedicels with very few hairs.

68 P. sepium *K. Schum.* in P.O.A. C: 389 (1895); Bremek. in F.R. 37: 178 (1934); T.T.C.L.: 515 (1949); Bridson in K.B. 32: 645 (1978). Type: Tanzania, Kilimanjaro, *Volkens* 2193 (B, holo. †, BM, iso.!)

Scandent or virgately branched shrub or small tree, 1–1.8 or 4.5–5.5 m. tall; young branches glabrous or pubescent; older branches with greyish-white bark. Leaf-blades drying grey-green to blue-black, narrowly to broadly elliptic, 3.5–11 cm. long, 1–4.3 cm. wide, apex acute to obtuse or sometimes tending to be acuminate, base obtuse to cuneate, glabrous or pubescent; bacterial nodules dot-shaped, scattered, several; petiole 0.5–1.5 cm. long, glabrous or pubescent; stipule-limbs translucent, greenish or fawn, truncate, 2–5 mm. long, membranous, pubescent or glabrous outside; arista blackish, up to 1 mm. long. Corymbs terminal on short spurs or short leafless (or rarely with few immature leaves) lateral branches 0.2–10 cm. long, lax, 1.7–3 cm. across; primary inflorescence-branches 2–10 mm. long, glabrous or pubescent; secondary branches present or absent; pedicels (3–)5–11 mm. long, slender, glabrous or pubescent; bracts stipule-like, 2–3 mm. long; bracteoles inconspicuous. Calyx pubescent or glabrous; tube ± 1 mm. long; limb-tube 0.5–1 mm. long; lobes narrowly triangular to linear, 1–2 mm. long, 0.25–0.75 mm. wide at base. Corolla drying blackish; tube slender, 0.45–1.3 cm. long, ± 1 mm. wide at top, glabrous or pubescent outside, not bearded at throat; lobes narrowly oblong to oblanceolate, 3–6 mm. long, 1–1.5 mm. wide, obtuse. Fruit greyish, 5 mm. in diameter, sparsely pubescent or ?glabrous; pedicels ?accrescent. Mature seeds not known.

KEY TO VARIETIES

Corolla-tube pubescent or sparsely pubescent outside; leaves, calyx-tubes and pedicels pubescent	a. var. **sepium**
Corolla-tube glabrous outside; never with leaves, calyx-tubes and pedicels all pubescent:	
Calyx-tube and pedicels densely pubescent; shrub or tree 4.5–5.5 m. tall	c. var. **massaica**
Calyx-tube and pedicels sparsely pubescent to glabrous; shrubs up to 2 m. tall:	
Corolla-tube 4.5–6 mm. long; leaf-blades sparsely hairy on both faces	b. var. **glabra**
Corolla-tube (6–)8–12 mm. long; leaf-blades glabrous	d. var. **merkeri**

a. var. **sepium**

Corolla-tube 6–13 mm. long, pubescent or sparsely pubescent outside; calyx, pedicels and inflorescence-branches pubescent; leaf-blades pubescent on both faces; stipules pubescent outside.

KENYA. Masai District: Laitokitok [Loitokitok], 31 Jan. 1970, *Rauh* Ke 235!; Teita District: Teita Hills, below Chawia, 15 May 1931, *Napier* 1139!
TANZANIA. Mbulu District: Lake Manyara National Park, Mtowa Mkindu, 29 May 1965, *Greenway & Kanuri* 12096! & 8 Jan. 1927, *B.D. Burtt* 625!; Arusha District: Arusha National Park, Momela Lakes, 17 Feb. 1972, *Mbane & Willy C.A.W.M.* 5941!
DISTR. K 6, 7; T 2; not known elsewhere
HAB. Ground-water forest, bushland on hillsides; 900–1370 m.
SYN. *P. scandens* Bremek. in F.R.: 177 (1934); T.T.C.L.: 515 (1949); K.T.S.: 460 (1961). Type: Kenya, Teita Hills, below Chowea [Chawia], *Napier* 1139 (K, holo.!, EA, iso.)
P. sepium K. Schum. var. *pubescens* Bremek. in F.R. 37: 178 (1934); T.T.C.L.: 515 (1949), *nom. superfl.* Type as for *P. sepium*
NOTE. The type specimen (with immature flowers) and *Mbane & Willy* 5941 have corolla-tubes 6–7 mm. long, but this is probably not very significant as both long and short corolla-tubes occur on the type of *P. scandens* Bremek.

b. var. **glabra** *Bremek.* in F.R. 37: 178 (1934); T.T.C.L.: 515 (1949); Bridson in K.B. 32: 645 (1978). Type: Kenya, Machakos District, Kibwesi, *Scheffler* 103 (B, holo.†, BM, K, P, iso.!)

Corolla-tube 5 mm. long, glabrous outside; calyx-limb sparsely hairy; calyx-tube, pedicels and inflorescence-branches glabrous; leaf-blades sparsely hairy on both faces; stipules hairy outside.

KENYA. Machakos District: Kibwesi, 28 Jan. 1906, *Scheffler* 103!
DISTR. K 4; known only from type
HAB. Bushland on rocky soil; 1000 m.
SYN. *P. squarrosa* K. Krause in E.J. 43: 145 (1909). Type as above

c. var. **massaica** *Bridson* in K.B. 32: 645 (1978). Type: Tanzania, Masai District, Embagai, *St. Clair-Thompson* 340 (K, holo.!)

Shrub or tree 4.5–5.5 m. tall. Corolla white (or ?lilac); tube 4.5–6 mm. long, glabrous outside; calyx, pedicels and inflorescence-branches densely pubescent; leaf-blades glabrescent save for the sparsely pubescent mid-rib above and sparsely hairy nerves beneath; stipules pubescent outside.

TANZANIA. Masai District: Embagai, 6 Feb. 1932, *St. Clair-Thompson* 340! & 5 Feb. 1932, *St. Clair-Thompson* 1242!
DISTR. T 2; not known elsewhere
HAB. Forest; 2000–2400 m.

d. var. **merkeri** (*K. Krause*) *Bridson* in K.B. 32: 645 (1978). Type: Tanzania, 'Massaisteppe', *Merker* 822 (B, holo. †)

Corolla-tube (6-)8-12 mm. long, glabrous outside; calyx, pedicels and inflorescence-branches glabrous to very sparsely pubescent or occasionally sparsely pubescent; leaves entirely glabrous or with the petioles sparsely pubescent above; stipules glabrous to slightly hairy outside. Inflorescence very lax to moderately lax. Fig. 106/14.

KENYA. Northern Frontier Province: Ndoto Mts., track from Ngurunit Mission, 11 June 1979, *Gilbert, Kanuri & Mungai* 5624!; Meru District: Meru Game Reserve, 10 Jan. 1966, *J. Adamson* 66/1!; Masai District: Tsavo National Park West, Mzima Springs, 31 Jan. 1977, *Gillett* 21016!
TANZANIA. Masai district: Engaruka road, 25 Feb. 1970, *Richards* 25524!; Pare District: Kisiwani, 2 Feb. 1936, *Greenway* 4568! & Same Rest House, 27 Jan. 1935, *R.M. Davies* 988!
DISTR. K 1, 4, 6; T 2, 3; not known elsewhere
HAB. Bushland; 550-1005 m.

SYN. *P. merkeri* K. Krause in E.J. 43: 146 (1909); Bremek. in F.R. 37: 179 (1934); T.T.C.L.: 515 (1949)
     P. capillipes Bremek. in K.B. 8: 505 (1954); K.T.S.: 457 (1961). Type: Kenya, Kitui District, Ikutha, *Bally* 1605 (K, holo.!, EA, iso.!)

NOTE. The type of *P. merkeri* has not been seen, but the description fits the specimens fairly well and *Greenway* 4568 & *R.M. Davies* 988 have both been annotated as *P. merkeri* K. Krause by Bremekamp in 1939.
     The two specimens from **K** 4 (the type of *P. capillipes* and *J. Adamson* 66/1) and one from **K** 1 (*Gilbert et al.* 5624) have laxer inflorescences than the specimens from **K** 6, **T** 2, 3, but *Gillett* 21016 from **K** 6 is intermediate in this respect. Additional collections would help determine if recognition of *P. capillipes* at varietal level would be possible.

69. P. dolichantha *Bremek.* in F.R. 37: 175 (1934); T.T.C.L.: 515 (1949); K.T.S.: 458 (1961); Bremek. in F.R. 47: 90 (1939). Type: Tanzania, Mpwapwa, *Hornby* 96 (K, holo.!, EA, iso.!)

Shrub, 1-2 m. tall; young branches tomentose, older branches with buff to grey flaking bark. Leaves confined to apices of main stems and short branches off the main stems; blades oblanceolate to obovate, 2.2-6 cm. long, 1.2-3.2 cm. wide, apex obtuse, rounded or emarginate often with a short callous at tip, base cuneate, scabrid-pubescent above, softly pubescent to tomentose beneath; bacterial nodules, small, dot-shaped, few to many, scattered; petiole 2-4 mm. long, tomentose; stipule-limbs truncate or triangular, usually very shortly mucronulate, 1.5-5 mm. long, tomentose outside, with silky hairs within. Corymbs terminal on short leafless branches 0.9-6.7 cm. long, few-flowered, compact, 1-1.8 cm. across; primary inflorescence-branches 0-5 mm. long; secondary branches present or suppressed; pedicels 2-5 mm. long, tomentose; bracts 3 mm. long, tomentose outside, with long silky hairs inside; bracteoles pale, inconspicuous. Calyx tomentose; tube 1-1.5 mm. long; limb-tube 0.5-1 mm. long; lobes triangular 0.75-2 mm. long, 0.75-1 mm. wide at base. Corolla-tube 1.2-3.1 cm. long, 1.25-1.5 mm. wide at top, glabrous or pubescent outside, with sparse silky hairs within; lobes narrowly ovate to ovate, 3-4.5 mm. long, 2-2.25 mm. wide, rounded or apiculate, often drying blackish with pale margins or with just the venation blackened. Fruit black, ± 5-7 mm. in diameter, glabrescent to pubescent. Mature seeds not known.

KENYA. Naivasha District: 32 km. from Kikuyu on road to Narok, 20 Jan. 1963, *Verdcourt* 3558!; Masai District: Ol Lorgosalie [Lorgasailie], 3 Aug. 1943, *Bally* 2657!; Teita District: Sagala Hill, 11 Dec. 1961, *Polhill & Paulo* 966!
TANZANIA. Mbulu District: Msasa River Gorge, Lake Manyara National Park, 8 June 1965, *Greenway & Kanuri* 11829!; Arusha District: Mt. Meru National Park, 1 km. W. of Lake Momella, 15 Jan. 1970, *Katende & Lye* in *Lye* 4885!; Kondoa District: Sambala, 27 Mar. 1929, *B.D. Burtt* 2573!
DISTR. K 1, 3, 6, 7; T 2, 5; not known elsewhere
HAB. Bushland and forest edges; 760-1675 m.

70. P. grumosa *S. Moore* in J.L.S. 37: 162 (1905); Bremek. in F.R. 37: 176 (1934), pro parte. Type: Uganda, Ankole District, Rufuha [Rufwua] R., *Bagshawe* 511 (BM, holo.!, K, iso.!)

Shrub, young branches pubescent; older branches covered with thin greyish bark. Leaves drying blackish; blades narrowly elliptic to elliptic, 2.5-7.5 cm. long, 1-3 cm. wide, apex acute or slightly acuminate, base acute, glabrous becoming sparsely pubescent towards the base above and beneath; bacterial nodules small, dot-shaped, few, scattered; petioles 1-2 mm. long, pubescent; stipule-limbs drying greenish grey, membranous, truncate 2-3 mm. long, glabrous. Corymbs terminal on short leafless (or occasionally with immature leaves) branches 1.8–5.5 cm. long, compact, 2–2.4 cm. across; primary inflorescence-branches 1–4 mm. long, pubescent; secondary branches very reduced; pedicels 1.5–3 mm. long, pubescent; bracts buff-coloured, 2 mm. long; bracteoles black with pale brown margin (on dry specimens). Calyx-tube 1 mm. long, pubescent; limb-tube 0.75–1 mm. long, wider than tube, sparsely pubescent; lobes narrowly triangular, 2.5–3 mm. long, sparsely pubescent, ± 0.75 mm. wide at base. Corolla-tube 2 cm. long, ± 1 mm. long at top glabrous outside, sparsely hairy inside; lobes oblong-elliptic, 4.5–6 mm. long, 1.75 mm. wide, acute to acuminate. Fruit not known.

UGANDA. Ankole District: Rufuha [Rufwua] R., 11 Dec. 1903, *Bagshawe* 511!
DISTR. U 2; known only from the type
HAB. On ant-hill in swamp; ± 1450 m.

NOTE. This species is very close to *P. subcana* Hiern var. *subcana* but may easily be distinguished by the denser indumentum, laxer inflorescence and the longer calyx-lobes. *E. Brown* 4731, cited as *P. grumosa* S. Moore by Bremekamp, is in fact *P. subcana* Hiern var. *subcana*

71. P. subcana *Hiern* in F.T.A. 3: 172 (1877); Bremek. in F.R. 37: 175 (1934); F.P.S. 2: 457 (1952); K.T.S.: 461 (1961); F.W.T.A. ed. 2, 2: 140 (1963); Bridson in K.B. 32: 646 (1978). Type: Sudan, Bahr el Ghasal, Jur (Ghattas Zeriba), *Schweinfurth* 3250 (K, holo.!, BM, P, iso.!)

Shrub 1–3.5(–?8) m. tall; young branches glabrous or pubescent; older branches with grey or buff bark, occasionally flaking. Leaves not or rarely restricted to the apices of the branches; blades narrowly to broadly elliptic, rarely round, 1.8–7(–13.5) cm. long, 0.5–2.5(–6) cm. wide, apex obtuse or sometimes acute, occasionally mucronulate, base cuneate, glabrous or pubescent on both faces; bacterial nodules small, dot-shaped, few to many, scattered and on the lateral nerves; petioles 0.1–0.8(–1) cm. long; stipule-limbs 2–5 mm. long, truncate, membranous, glabrous or pubescent outside, glabrous or with silky hairs within; arista often drying blackish, up to 4 mm. long. Flowers fragrant; inflorescences subumbellate, few–several-flowered, 0.8–1.5(–2.5) cm. across, terminal on leafless spurs or branches up to 2.5(–7) cm. long or occasionally on leafy branches; primary and secondary inflorescence-branches always absent; pedicels 1–2.5(–4) mm. long, pubescent or glabrous; bracts inconspicuous. Calyx pubescent or glabrous; tube 1 mm. long; limb-tube 1 mm. long, very slightly wider than the tube; lobes shortly triangular to triangular, (0.75–)1–2.2 mm. long, 0.75–1 mm. wide at base. Corolla-tube (0.9–)1.2–2.2 cm. long, 1.5 mm. wide at top, glabrous or occasionally pubescent outside, pubescent within; lobes oblong-elliptic, 4–6(–7.5) mm. long, 1.25–1.75(–2.2) mm. wide, acute, obtuse or sometimes slightly apiculate. Fruit black, 5–8 mm. in diameter, shiny, glabrous or sparsely hairy; pedicels somewhat accrescent, reaching 8 mm. Seeds blackish, 5 mm. wide, slightly rugulose on convex face.

var. subcana

Corolla-tube sparsely pubescent to pubescent outside; young branches and calyces always pubescent; leaves pubescent or occasionally glabrous.

UGANDA. Bunyoro District: Kitoba, May 1943, *Purseglove* 1566!; Ankole District: Masaka–Mbarara road, near Sanga, 23 Nov. 1952, *Ross* 1245!; Mubende District: 160 km. NW. of Kampala on Mubende road, Sept. 1915, *E. Brown* 2731!
KENYA. Northern Frontier Province/Turkana District: Lake Rudolf, Sept. 1899, *Welby!*
DISTR. U 2–4; K 1/2; N. Nigeria, Chad, Central African Republic, ?Zaire, Sudan
HAB. Wooded grassland or valley forest, sometimes on old ant-hills; 800–1220 m.

SYN. *Ixora subcana* (Hiern) Kuntze, Rev. Gen. Pl. 1: 287 (1891)
Pavetta pubiflora Bremek. in F.R. 47: 91 (1939). Type: Uganda, Teso District, Serere, *Chandler* 593 (K, holo.!, EA, iso.!)

NOTE. Many specimens determined as *P. subcana* by Bremekamp are in fact *P. dolichantha* Bremek. The glabrous leaves on the type of *P. pubiflora* and on *A.F. Broun* 71 from Sudan show a tendency towards var. *longiflora*.

var. longiflora (*Vatke*) *Bridson* in K.B. 32: 646 (1978). Types: Ethiopia, Habab, *Hildebrandt* 436 (not found) and Eritrea, Cheren [Keren], *Beccari* 148 (FT, syn.!)

Corolla-tube always glabrous outside; young branches always glabrous; calyces glabrous or pubescent; leaves always glabrous.

UGANDA. Karamoja District: Lodoketeminit, 21 May 1963, *Kerfoot* 5000!; Toro District: Katwe, 20 June 1945, *A.S. Thomas* 4143!; Mengo District: Gomba, Madu, Mar. 1932, *Eggeling* 548!

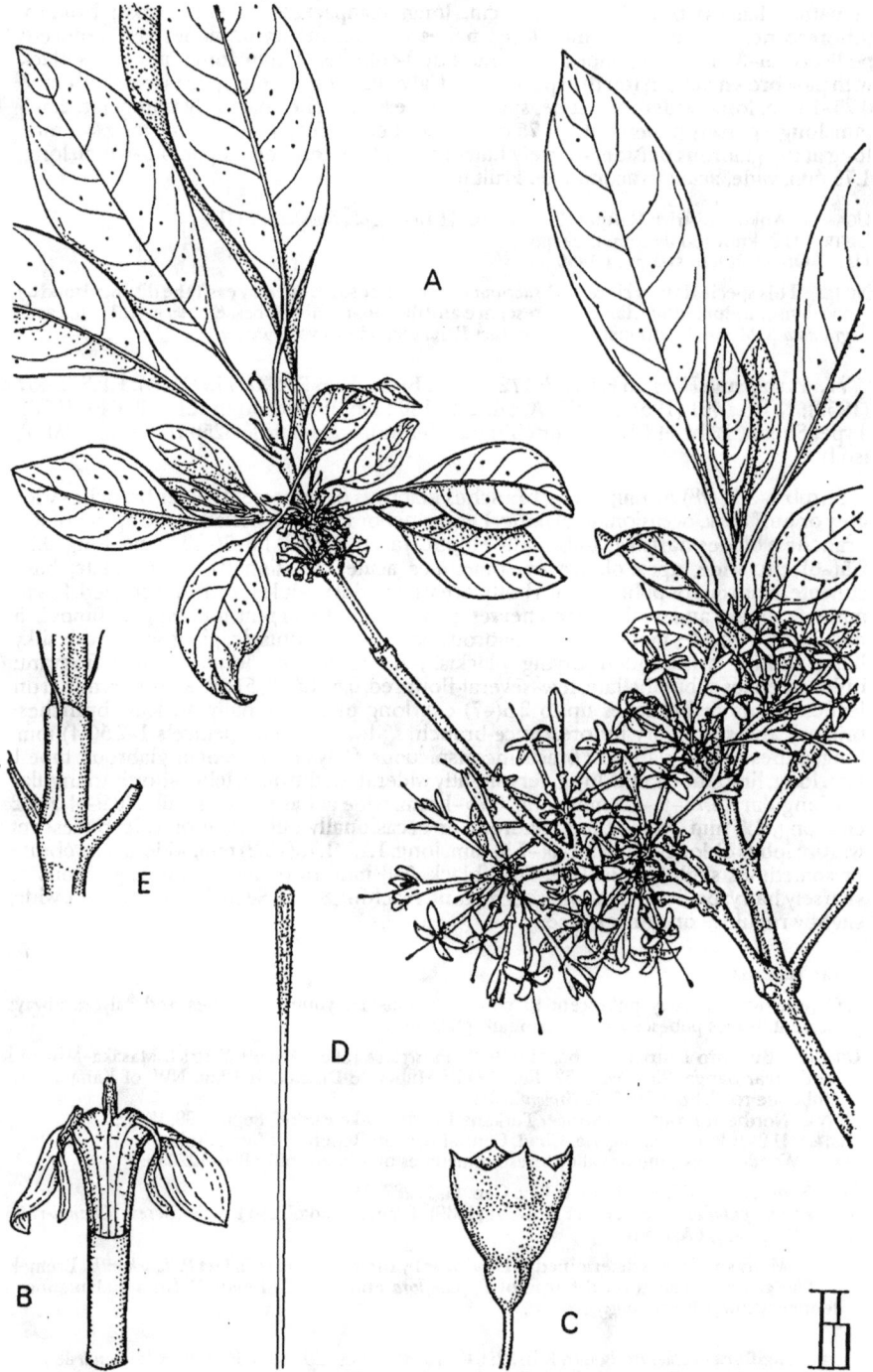

FIG. 115. *PAVETTA RUAHAENSIS* — **A**, habit, × ⅔; **B**, corolla with section removed, × 3; **C**, calyx, × 6; **D**, style and stigma, × 8; **E**, stipule, × 3. A, from *Richards* 21074; B–D, from *Bjørnstad* 1344, E, from *Richards* 21084. Drawn by Diane Bridson.

KENYA. W. Suk District: Suam R., below Kongelai, Mar. 1965, *Tweedie* 3029!; Kericho District: Sotik, Jan. 1960, *Dale* 1047!; Masai District: Mara Masai Reserve, Talek [Telek] R., 27 Sept. 1950, *Kirrika* in *Bally* 7749!
TANZANIA. Shinyanga District: Uduhe area, on road to Mango, 24 Jan. 1936, *B.D. Burtt* 5502!; Musoma District: near Moru Kopjes, 9 Feb. 1968, *Greenway, Kanuri & Braun* 13161!; Kahama District: Ngaya road, 8 Jan. 1933, *B.D. Burtt* 4522!
DISTR. U 1–4; K 2, 3, 5, 6; T 1, 2, 4, 5; Central African Republic, Zaire, Sudan, Ethiopia
HAB. Bushland and woodland, often associated with river and lake banks or ant-hills; 600–1830 m.

SYN. *P. gardeniifolia* A. Rich. var. *longiflora* Vatke in Oesterr. Bot. Zeitschr. 25: 231 (1875); Hiern in F.T.A. 3: 178 (1877); K. Schum. in P.O.A. C: 389 (1895); T.T.C.L.: 516 (1949)
    P. kerenensis Becc. in Martelli, Fl. Bogos: 43 (1886); Bremek. in F.R. 37: 176 (1934). Type: Ethiopia, Eritrea, Cheren [Keren], *Beccari* 148 (FT, holo.!)
    P. albertina S. Moore in J.B. 45: 267 (1907); Bremek. in F.R. 37: 177 (1934); T.T.C.L.: 514 (1949); F.P.S. 2: 457 (1952); K.T.S.: 455 (1961). Type: Uganda, Lake Albert, *Bagshawe* 1318 (BM, holo.!)
    P. kabarensis Bremek. in B.J.B.B. 14: 310 (1937) & in F.R. 47: 92 (1939); F.P.N.A. 360, t. 36 (1947). Type: Zaire, Lake Kivu, Kabare, *Bequaert* 5353 (BR, holo.!)
    P. unguiculata Bremek. in F.R. 47: 91 (1939); K.T.S.: 463 (1961). Type: Kenya, ? Sirta Plains [?Ukambani], *Curtis* 1073 (A, holo.!)
    P. kotschyana Cuf. in Nuovo Giorn. Bot. Ital. 55: 91 (1948). Type: Sudan, Fassoglu, *Kotschy* 474 (W, holo., K, iso.!)
    P. rudolphina Cuf. in Nuovo Giorn. Bot. Ital. 55: 89 (1948). Types: Ethiopia, between Lakes Chew Bahir and Turkana [Rudolf], Caschei R., *Corradi* 2789, 2795–2798 (FT, syn.!)
    P. rudolphina Cuf. var. *robusta* Cuf. in Nuovo Giorn. Bot. Ital. 55: 90 (1948). Type: Ethiopia, between Lakes Chew Bahir and Turkana, Caschei R., *Corradi* 2799 (FT, holo.!)

NOTE. Several sheets of sterile or fruiting material from Sudan and Ethiopia (not annotated by Bremekamp) at Kew and *Pappi* 7687 (FT) (on which Bremekamp's description of *P. kerenensis* was based) have much larger leaves than typical material from the Flora area. The leaves on the type are moderately large and a good match for several sheets within the Flora area.

72. **P. ruahensis** *Bridson* in K.B. 32: 647, fig. 7 (1978). Type: Tanzania, Iringa District, Ruaha National Park, Mbagi, *Richards* 21074 (K, holo.!, P, iso.!)

Shrub 1–2 m. tall; young branches glabrous; older branches with greyish bark, occasionally flaking. Leaves glabrous, mostly restricted to the apices of the branches; blades elliptic to broadly elliptic, 6–9 cm. long, 1.8–4 cm. wide, apex obtuse to acute or sometimes shortly acuminate, base narrowly cuneate or attenuate; bacterial nodules dot-shaped, few to several, scattered; petioles 0.8–1.4 cm. long; stipule-limbs truncate, 4–6 mm. long, membranous, glabrous outside, glabrous or with silky hairs within; arista up to 5 mm. long. Inflorescence sometimes branched or with several close together, terminal on spurs or short leafless branches which are almost suppressed or up to 1.2(–6) cm. long, subumbellate, ± 20-flowered, 1–1.3 cm. across; primary inflorescence-branches absent or up to 2 mm. long; pedicels 0.75–3.5 mm. long, glabrous; bracts inconspicuous, 3–4 mm. long, pubescent inside. Calyx glabrous; tube 1 mm. long; limb-tube 1–1.5 mm. long, wider than the tube; lobes shortly triangular, ± 0.75 mm. long, 1–1.25 mm. wide at base. Corolla-tube 0.8–1.3 cm. long, 1.5–2 mm. wide at top, glabrous outside, not bearded at throat; lobes oblong-elliptic, 4–6 mm. long, 1.75–2.5 mm. wide, obtuse, sometimes slightly apiculate. Fruit ± 7 mm. in diameter, shiny, glabrous; pedicels somewhat accrescent, up to 7 mm. long. Mature seeds not known. Fig. 115.

TANZANIA. Iringa District: Ruaha National Park, track between Mbagi and Ibuguziwa, 31 Jan. 1966, *Richards* 21084! & near mouth of Mwagusi sand-river, 7 Aug. 1970, *Thulin & Mhoro* 635! & Great Ruaha R., 5 km. NE. of Msembe, 5 Feb. 1972, *Bjørnstad* 1344!
DISTR. T 7; not known elsewhere
HAB. Riverine forest and woodland; 800–1050 m.

NOTE. This species is very close to *P. subcana* var. *longiflora* but may be distinguished by the larger leaf-blades, longer petioles, larger stipules and the generally shorter calyx-lobes and corolla-tube.

73. **P. fascifolia** *Bremek.* in F.R. 37: 177 (1934); Bridson in K.B. 32: 649 (1978); Vollesen in Op. Bot. 59: 70 (1980). Type: Mozambique, Nyasa, mouth of Messalo [M'salu], *Allen* 117 (K, holo.!)

Shrub 0.6–5 m. tall; young branches glabrescent to pubescent; older branches covered with whitish-grey bark. Leaves crowded on short spurs off the main stem or occasionally borne directly along the young branches only; blades usually turning black when dry, oblanceolate to obovate, 1.4–5.5 cm. long, 0.6–2.7 cm. wide, acute to rounded, sometimes mucronulate, base narrowly cuneate, glabrous to scabrid above, glabrous or finely pubescent beneath; bacterial nodules small, dot-shaped, several, scattered; petiole 0.3–1.5 cm. long, sparsely to densely pubescent; stipule-limbs truncate, 1–2 mm. long, pubescent outside; arista 0.5–1.5 mm. long. Corymbs terminal on main branches or on slender 1–1.25 mm. wide leafless lateral branches 1.1–5.5 cm. long, lax, 2.7–4.7 cm. across; primary inflorescence-branches 0.1–1 cm. long; secondary to tertiary branches present; pedicels 2–7 mm. long, pubescent; bracts 4–7 mm. long, glabrous or pubescent outside, with silky hairs inside; bracteoles drying brownish. Calyx pubescent; tube 0.75–1 mm. long; limb-tube 0.5–1 mm. long, slightly wider than the tube; lobes triangular, 1–2 mm. long, 0.5–0.75 mm. wide at base. Corolla-tube 1.3–1.6 cm. long, ± 1 mm. wide at top, glabrous outside, not bearded at throat; lobes oblong-lanceolate, 5–6 mm. long, 1.75 mm. wide, obtuse to acute. Fruit black, 7–8 mm. long, glabrescent, calyx-limb persistent. Seeds greyish brown, 5 mm. wide, slightly rugulose on convex face.

TANZANIA.   Kilosa District: Mikumi National Park, Kikoboga, 7 Feb. 1976, *Ole Sayalel* 1203!; Uzaramo District: 2 km. from Kibaha on Soga road (± 40 km. W. of Dar es Salaam), 28 Mar. 1972, *Flock* 253!; Kilwa District: Selous Game Reserve, Kingupira Camp, Jan. 1971, *Ludanga* in *M.R.C.* 1207!
DISTR.   T 6, 8; Mozambique
HAB.   Riverine thicket or on termite mounds; 100–520 m.

NOTE.   Two specimens, *Flock* 253 and *Ludanga* 1207 have leaves which are scabrid above and pubescent beneath. Varieties have not yet been described as little material has been seen and intermediate forms probably occur.

74. P. gracilifolia *Bremek.* in Ann. Transv. Mus. 13: 203 (1929) & in F.R. 37: 162 (1934); Kok & Grobbelaar in S. Afr. Journ. Bot. 3: 186 (1984). Type: South Africa, Natal, Umlaas Drift, *Wood* 340a (Nat. Herb. 1585) (NH, holo., K, iso.!)

Shrub 0.3–3 m. tall; young branches pubescent or glabrescent, slender; older branches covered with whitish-grey bark. Leaves crowded on short spurs off the main stem or occasionally borne directly along young branches; blades occasionally turning black when dry, narrowly elliptic to oblanceolate, 1–5.3 cm. long, 0.3–1.8 cm. wide, apex acute to obtuse or sometimes rounded, rarely subacuminate, base narrowly cuneate, glabrous, scabrid or sparsely pubescent above, pubescent or glabrous beneath; bacterial nodules dot-shaped, scattered; petiole 0–4 mm. long, pubescent; stipule-limbs truncate, 1–2 mm. long, membranous with a few hairs outside; arista 0.75–1 mm. long. Corymbs terminal on lateral branches, ± subumbellate, 1.2–2.3 cm. across (excluding the corollas); lateral branches slender. 0.5–1 mm. wide, 0.2–5(–10) cm. long, leafless or less often with 1(–2) pairs of immature leaves; primary inflorescence-branches 0–1(–2) mm. long; pedicels 1–6 mm. long, glabrous to pubescent; bracts 1.5–4 mm. long, with silky hairs inside; bracteoles sometimes present. Calyx-tube 0.5–1 mm. long, glabrous to pubescent; limb-tube 0.5–1 mm. long, glabrous to sparsely pubescent; lobes ± filiform to linear, 2.5–6 mm. long, glabrous or more often ciliate to sparsely pubescent. Corolla-tube 1.2–2 cm. long, ± 1 mm. wide at top, glabrous or sometimes sparsely pubescent outside, not bearded at throat; lobes 4–7 mm. long, 1.5–2 mm. wide, apiculate. Fruit drying black, 8 mm. in diameter, crowned by persistent calyx-limb. Mature seeds not known.

TANZANIA.   Ulanga District: Selous Game Reserve, Luhanyando Camp, 10 Feb. 1971, *Rees T* 115!; Kilwa District: Selous Game Reserve, Lihangwa R. source, 14 Feb. 1970, *Rodgers* in *M.R.C.* 928!
DISTR.   T 6, 8; Mozambique, South Africa
HAB.   Thicket; 400–700 m.

SYN.   *P. breyeri* Bremek. in Ann. Transv. Mus. 13: 203 (1929) & in F.R. 37: 162 (1934). Types: South Africa, Nelspruit, *Breyer* in *Transv. Mus.* 17716 & in *Transv. Mus.* 18805, White R., *F.A. Rogers* in *Transv. Mus.* 20712, Louws Creek, *Thorncroft* 2043 & Mozambique, Maputo area, *Maputoland Expedition* in *Transv. Mus.* 14417 (all PRE, syn.)
       P. delagoensis Bremek. in Ann. Transv. Mus. 13: 203 (1929) & in F.R. 37: 153 (1934); Bridson in K.B. 32: 649 (1978); Vollesen in Op. Bot. 59: 70 (1980). Type: Mozambique, Maputo [Lourenço Marques], *Borle* 62 (PRE, holo., K, iso.!)
       P. woodii Bremek. in Ann. Transv. Mus. 13: 203 & in F.R. 37: 163 (1934). Type: South Africa, Natal, Nonoti, *Wood* 885 in *Nat. Herb.* 9273 (PRE, holo., NH, iso., K, photo.!)
       P. breyeri Bremek. var. *pubescens* Bremek. in F.R. 37: 162 (1934), *nom superfl.* Type as for *P. breyeri*
       P. breyeri Bremek. var. *glabra* Bremek. in F.R. 37: 162 (1934). Type: South Africa, Natal, Alexandra, *Rudatis* 1204, pro parte (K, holo.!)
       P. gracilifolia Bremek. var. *glabra* Bremek. in F.R. 37: 162 (1934). Type: South Africa, Natal, Inanda, *Wood* 1048a (K, holo.!)
       P. gracilifolia Bremek. var. *pubescens* Bremek. in F.R. 37: 162 (1934), *nom. superfl.* Type as for *P. gracilifolia*
       P. divaricata Bremek. in K.B. 8: 503 (1954). Type: Mozambique, Maputo, *Hornby* 2576 (K, holo.!, PRE, iso.)

FIG. 116. *PAVETTA UNIFLORA* — 1, habit, × 1; 2, leaf, upper surface, showing bacterial nodules, × 2; 3, calyx, × 2; 4, upper part of corolla, opened out, × 4; 5, stamen, × 6. All from *Graham* 1856. Drawn by Stella Ross-Craig.

NOTE.   The synonomy adopted by Kok & Grobbelaar (in S. Afr. Journ. Bot. 3: 186 (1984)) has been
followed with the addition of *P. divaricata* from Mozambique. This species seems quite variable
and only subtly distinct from the South African taxa *P. barbertonensis* Bremek. and *P. capensis*
(Houtt.) Bremek. subsp. *komghensis* (Bremek.) Kok.

75. P. uniflora *Bremek.* in Hook., Icon. Pl. 32, t. 3194 (1933) & in F.R. 37: 168 (1934);
K.T.S.: 463 (1961); Bridson in K.B. 32: 649 (1978). Type: Kenya Kilifi District, Arabuko,
R.M. Graham in *F.D.* 1856 (K, holo.!, EA, iso.!)

Small tree or ? scandent shrub 1.8–7 m. tall; young branches glabrous, older branches
with buff thin corky bark. Leaves confined to very short spurs off the main and lateral
branches; blades obovate, 1.2–4.8 cm. long, 0.45–1.8 cm. wide, apex rounded or obtuse,
often with a thick mucro, base tapered, glabrous; bacterial nodules dot-shaped to linear,
very sparse to several, scattered; petiole 0–3 mm. long; stipule-limbs buff-coloured,
bearing a longer ± subulate cusp up to 2 mm. long, glabrous, hairy at base within. Flowers
solitary, sessile on the leafy spurs; bracts not exceeding 2 mm. long, hairy within. Calyx
glabrous; tube 1–2 mm. long; limb-tube 1.8–2 mm. long; lobes very narrowly triangular to
linear, 4–9 mm. long, up to 1 mm. wide. Corolla-tube 2.6–4.3 cm. long, 1–2.5 mm. wide at
the top, glabrous outside and within; lobes lanceolate, 4.5–6 mm. long, 2–3 mm. wide,
acute to acuminate. Fruit straw-coloured to dark grey, 5–6 mm. in diameter, shiny,
glabrous. Seeds 4–5 mm. wide, convex face with very shallow longitudinal striations. Figs.
106/111 & 116.

KENYA.   Kilifi District: Galana Ranch, SE. Dakabuko Hill, 4 May 1975, *Bally* 16739! & Arabuko, May
    1929, *R.M. Graham* in *F.D.* 2136! & Arabuko-Sokoke Forest, Jilori, 26 Nov. 1961, *Polhill & Paulo* 859!
TANZANIA.   Tanga District: Kigombe, Dhali Wood, 20 May 1958, *Faulkner* 2150!; Bagamoyo District:
    Kiko Forest Reserve, 27 Mar. 1964, *Semsei* 3728!; Uzaramo District: Kibaha, 4 Feb. 1971, *Flock* 303!
DISTR.   K 7; T 3, 6, 8; Somalia, Mozambique
HAB.   Undergrowth in *Brachylaena*, *Pandanus* or other coastal woodland or in coastal bushland;
    0–150 m.

SYN.   *Plectronia sennii* Chiov. in Fl. Somala 2: 243, fig. 144 (1932). Type: Somalia, Juba region, Jack
    Omisso, *Senni* 479 (FT, holo.!)
    [*Canthium sennii* (Chiov.) Cuf., E.P.A.: 1010 (1965), *nom non rite publ.*]
    [*Pavetta sennii* (Chiov.) Bridson in K.B. 38: 320 (1983), *non Pavetta sennii* Chiov. (1932)]

80. DICTYANDRA

Hook.f., G.P. 2: 85 (1873); Hallé in Fl. Gabon, 17, Rubiacées 2: 86 (1970); Robbrecht in Pl.
Syst. Evol. 145: 105–118, figs. 1–3 (1984)

Shrubs or small trees. Leaves petiolate, opposite, with domatia; stipules flattened,
triangular and erect or rounded and bent back. Flowers 5-merous, pedicellate, ♂, in
terminal corymbs. Calyx-tube ovoid or turbinate; tubular part of limb short; lobes ± erect,
contorted. Corolla with tube about equalling the lobes; lobes narrowly lanceolate,
contorted. Anthers exserted, subsagittate, the thecae simple, or locellate with 4 vertical
lines of small compartments. Ovary thick-walled, 2-locular with an ovate peltate placenta
covered with very numerous ovules; style rather thick, with an exserted bilobed stigma.
Fruits globose or ellipsoid, crowned with the persistent calyx-lobes. Seeds squarish,
angular, small, black, shining and smooth, with a hilar pit.

A genus of only 2* species confined to West and Central tropical Africa, one occurring in Uganda.

D. arborescens *Hook.f.*, G.P. 2: 85 (1873); Hiern in F.T.A. 3: 86 (1877) & Cat. Afr. Pl.
Welw. 1: 456 (1898); I.T.U., ed. 2: 342 (1952); Hepper in F.W.T.A., ed. 2, 2: 132 (1963); Hallé
in Fl. Gabon, 17, Rubiacées 2: 87, t. 20/1–12 (1970). Type: Angola, Golungo Alto, Serra de
Alto Queta, *Welwitsch* 2561 (LISU, lecto., BM, K, P, isolecto.!)**

---

* Robbrecht, loc. cit. (1984), transferred the well known species *D. involucrata* back to *Leptactina*
and described a new species *D. congolana* Robbrecht.
** Hooker cites the type localities as "Africa occidentalis a flumine Gabon ad Angolam incola";
Hiern gives Angola, Golungo Alto and Old Calabar. It is clear that more than one sheet was used by
Hooker but I am unable to trace the Gabon sheets and have chosen the Welwitsch specimen cited
above as a lectotype. Hiern mentions three other Welwitsch 'carp. colls.' which may all be syntypes.

FIG. 117. *DICTYANDRA ARBORESCENS* — 1, flowering branch, × ⅔; 2, stipules, × 2; 3, flower-bud, × 1; 4, flower, × 1; 5, half corolla, × 2; 6, dehisced anther, × 4; 7, stigmatic lobes, × 4; 8, longitudinal section of ovary, × 6; 9, transverse section of ovary, × 10; 10, fruits, × ⅔; 11, seed, × 10. 1–9, from *Maitland* 219; 10, from *Eggeling* 533; 11, from *Thomas* 4079. Drawn by Mrs M.E. Church.

Shrub or small tree 2–10.5 m. tall; trunk sometimes with stout blunt conical spines. Leaves glossy dark green, drying blackish, mostly clustered towards the ends of the branches; blades elliptic to obovate-elliptic, 7–25 cm. long, 3–10 cm. wide, shortly acuminate at the apex, cuneate at the base, glabrous save for the densely pubescent axillary domatia; petiole 0.5–4.5 cm. long; stipules triangular, 7–14 mm. long, 5–7 mm. wide, sharply narrowly acuminate, with a median rib, the stems grooved beneath them. Flowers fragrant in several–many-flowered inflorescences at the extremities of the lateral branches; peduncle ± 2.5 cm. long; secondary branches ± 0.5 cm. long; pedicels 0.5–1.5 cm. long, all axes glabrous or puberulous; bracts 4–8 mm. long, with 2 lateral lobes derived from stipules. Calyx-tube turbinate, 3 mm. long, hairy; lobes ovate, 0.8–1.2 cm. long, 6–9 mm. wide, held erect, ± glabrous or ciliate, the border enrolled on the side covered in the bud, recurved and acute at the apex. Corolla white, adpressed buff silky pilose outside, glabrous inside; tube 1–1.8 cm. long; lobes oblanceolate to narrowly elliptic, 1.6–2.5 cm. long, 4–7 mm. wide, narrowly acute at the apex. Anthers pale yellow, subsessile, ± basifixed, exserted, 1.1–1.5 cm. long, very narrowly acute at the apex which is recurved, locellate, the individual compartments under 1 mm. wide. Disc annular, yellow. Style yellow, up to 3 cm. long, densely hairy; lobes ± 1 cm. long, very attenuate, strongly divergent. Fruit subspherical, 1–2.2 cm. in diameter, the black surface finely sinuous in dry state, crowned with the erect contorted calyx-lobes. Seeds very numerous, ± 1.8–2 mm. long, glossy, angular. Fig. 117.

UGANDA.   Bunyoro District: Budongo Forest, Nov. 1932, *Harris* 175!; Kigezi District: Malamagambo Forest, Feb. 1950, *Purseglove* 3296!; Mengo District: Kipayo, Dec. 1913, *Dummer* 523!
DISTR   U 2, 4; widespread in western Africa from Sierra Leone to Angola, Central African Republic, Gabon and Zaire
HAB.   Evergreen forest; 1050–1500 m.

81. LEPTACTINA*

Hook.f., Ic. Pl. 11: 73, t. 1092 (1871); Hallé in Fl. Gabon 17, Rubiacées 2: 70 (1970); Robbrecht in Pl. Syst. Evol. 145: 105–118, figs. 2, 3 (1984)

Shrubs or small trees, nearly always erect, rarely slightly scrambling or one species a prostrate shrubby herb. Leaves petiolate, opposite, sometimes slightly unequal, usually with distinct pubescent domatia; stipules usually conspicuous. Flowers mostly rather large, ♂, 4–6-merous in few–many-flowered cymes at the extremities of terminal shoots and lateral branches. Calyx-tube ellipsoid or obconic; tubular part of limb ± obsolete or only evident in fruit; lobes contorted, foliaceous, subequal, mostly erect during flowering, venose, persistent. Corolla-tube long, narrowly cylindrical; lobes linear or linear-lanceolate to elliptic, usually long and pointed, spreading, contorted in bud. Stamens sessile, inserted below the level of the throat, included or apex of anthers exserted, occasionally locellate**; pollen grains single. Ovary 2-locular; ovules numerous on placentas attached to the septum; style usually with pollen-collecting hairs at the level of the anthers; stigma with 2 linear lobes exserted or mostly included. Fruit subglobose or oblong to oblong-conic, scarcely or slightly fleshy, often longitudinally ribbed. Seeds numerous, small, mostly angular, smooth and glossy, with a hilar pit.

A genus of about 25 species restricted to tropical Africa or just outside the tropics at the SE. limit of its range; most of the species occur in evergreen forest but several extend to bushland.

A revision of the species is much needed but many described by K. Schumann are problematical owing to the destruction of the types.

Stipules relatively large and leafy, narrowed to the base, 0.8–3.5
    cm. long, 1–2.6 cm. wide:
Stipules obtuse; leaves petiolate with finer indumentum;
    bark not so characteristic but perhaps peeling   .   .   *1. L. platyphylla*

---

* The spelling *Leptactinia* was used in G.P. 2: 85 (1873) and by many subsequent authors but there seems no etymological reason for altering the original spelling.
** Robbrecht has transferred *Dictyandra involucrata* (Hook.f.) Hiern back to *Leptactina* despite its locellate anthers; all other characters suggest this is correct.

Stipules cuspidate; leaves almost sessile, with bristly hairs on
   the venation beneath; bark separating from older
   branches in thick pieces consisting of very many
   compressed papery layers . . . .   2. *L. papyrophloea*
Stipules smaller, not leafy, not narrowed to the base, 0.5–1.6
   cm. long, 4–6 mm. wide, usually small, always acute:
Erect shrubs or small trees; leaf-blades mostly densely
   pubescent on both surfaces:*
Style described as glabrous; leaf-blades up to 10 × 6 cm.
   (T 6) . . . .   3. *L. oxyloba*
Style sparsely to densely pubescent; leaf-blades usually
   2–7 × 1.5–2.5 cm. but sometimes larger (T 8)   4. *L. delagoensis*
Prostrate shrublet or woody herb, occasionally erect but then
   scarcely exceeding 1 m.; leaf-blades either glabrous on
   both surfaces save for hairs on main nerves or sparsely
   to densely pubescent beneath . . . .   6. *L. benguelensis*

1. **L. platyphylla** (*Hiern*) *Wernham* in J.B. 51: 278 (1913); F.P.S. 2: 442 (1952); K.T.S.: 450 (1961); Vollesen in Op. Bot. 59: 69 (1980); Fl. Pl. Lign. Rwanda: 566, fig. 191 (1982); Fl. Rwanda 3: 174, fig. 53 (1985). Type: Zaire, "Monbuttuland", Bongwa's village, *Schweinfurth* 3626 (K, holo.!, BM, iso.)

Shrub (sometimes scrambling) or small tree (1.5–)2.5–7.5 m. tall; young stems densely to rather sparsely pubescent and with a fine understorey of much shorter indumentum. Leaf-blades elliptic to oblong-elliptic, 6.6–29.5(–38) cm. long, 4.3–15 cm. wide, shortly to distinctly acuminate at the apex, ± rounded to cuneate at the base, almost glabrous save for short hairs on the venation to rather sparsely pubescent above, sparsely to densely or almost velvety pubescent beneath, with small axillary tufts of hairs visible in the more glabrescent leaves; petiole 0.3–1.7 cm. long, hairy like the stems, with longer hairs on the lateral angles; stipules broad, obtuse at the apex, narrowed beneath, the narrowest part just above the base, somewhat mushroom-shaped, with a basal triangular part pubescent outside tapering into a median line of hairs at its apex or pubescent all over, 0.8–2.3(–3) cm. long, 1.1–2.6(–3) cm. wide, reflexing. Inflorescence subcapitate, terminating lateral shoots; peduncle 0–1.2(–2) cm. long, ± densely pubescent; pedicels 0.2–1.4 cm. long; bracts opposite, 3-fid, leafy, the divisions lanceolate; bracteoles 1.1–1.5 cm. long. Flowers 5–6-merous. Calyx densely pubescent; tube 2–5 mm. long, 1–3 mm. wide, the limb-tube very short or absent; lobes lanceolate, 0.9–2.1 cm. long, 2–6 mm. wide, acute. Corolla white, scented, densely hairy outside; tube 2.2–11 cm. long, 5–8 mm. wide at the apex, 1–4 mm. wide at the base; lobes lanceolate, 1.6–5.2 cm. long, 0.2–1 cm. wide, acute. Style included; anthers with tips exserted. Fruits oblong-ellipsoid to ovoid-subglobose, 1.1–2 cm. long, 0.7–1.4 cm. wide, ridged in dry state, at first densely pubescent. Seeds dark brown, angular, 1.5–2 mm. long and wide, slightly glossy, with a hilar pit. Fig. 118.

UGANDA. NW. Kigezi, Amahingo, Aug. 1949, *Purseglove* 3053!; Masaka District: Sese Is., Nov. 1904, *E. Brown* 115!; Mengo District: Kyagwe [Kiagwe], Namanve, June 1932, *Eggeling* 448!
KENYA. N. Kavirondo District: Kakamega Forest, June 1986, *Lucas* 114!; Kwale District: Kwale, *R.M. Graham* in *F.D.* 1771! & Shimba Hills, Lango ya Mwagandi [Longo Mwagandi], 12 Apr. 1968, *Magogo & Glover* 811!
TANZANIA. Tanga District: Kange Estate, 20 Nov. 1951, *Faulkner* 848!; Lushoto District: Mashewa, 25 May 1943, *Greenway* 6691!; Uzaramo District: Pugu Hills, 17 Feb. 1971, *Batty* 1230!; Ulanga District: Kisawasawa, 19 Sept. 1959, *Haerdi* 318/0!
DISTR. U 2, 4; K 5, 7; T 3, 4, 6, 7, ?8; Cameroun, Central African Republic, Zaire, Burundi, Rwanda, Sudan, Mozambique, Malawi
HAB. Evergreen forest, woodland, secondary bushland; 45–1650 m.
SYN. *Mussaenda? platyphylla* Hiern in F.T.A. 3: 70 (1877)
    Leptactina hexamera K. Schum. in E.J. 33: 341 (1903); T.T.C.L.: 505 (1949). Type: Tanzania, Uluguru Mts., Kikurungu Mt., *Stuhlmann* 9252 (B, holo.†)
    L. surongaensis De Wild. in Ann. Mus. Congo, Bot., Sér. 5, 2: 73 (1907). Type: Zaire, Suronga, *Seret* 422 (BR, holo.!)

NOTE. There is much variation in this species but subspeciation does not seem to be clear enough to justify names. Material from Kenya and north-east Tanzania, i.e. at the top eastern end of the forest migration loop from the West, has the foliage much more densely velvety, the leaf-blades more cuneate, the calyx-lobes broader and the fruits often more narrowly oblong compared with

* If almost glabrous see *5. L. sp. A.*

FIG. 118. *LEPTACTINA PLATYPHYLLA* — 1, flowering branch, × ⅖; 2, flower, pentamerous, × ⅖; 3, half corolla × ⅖; 4, stigmatic arms, × 2; 5 longitudinal section of ovary, × 5; 6, transverse section of ovary, × 5; 7, fruits, × 1; 8, seed, × 10. 1, 3–6, from *Faulkner* 2213; 2, from *Lucas* 114; 7, 8, from *Batty* 1230. Drawn by Mrs M.E. Church.

Uganda and Kakamega material which also often has longer broader corolla-lobes. Material from W. Africa annotated *L. surongaensis* by Hallé tends to have smaller calyx-lobes, flowers and leaves but the type locality of *L. surongaensis* at 27°18'N 03°48'E and that of *L. platyphylla* at 28°00'N, 04°00'E are sufficiently close to be certain they represent the same taxon.

Some material from Ulanga District is distinctive in having the leaves glabrous beneath save for the nervation.

What may be a distinct variant or perhaps even a distinct species occurs at Ndanda, Masasi District; *Scheven* 33 has fruiting calyx-lobes up to 1 cm. wide and *Gillman* 1437 has younger calyx-lobes 6 mm. wide, but neither is in flower and the material is inadequate for a decision.

2. L. papyrophloea *Verdc.* in K.B. 36: 506 (1982). Type: Tanzania, Lindi District: Rondo Plateau, *Schlieben* 5983 (BR, holo.!, B, K, PRE, iso.!)

Multi-stemmed bush or small tree to 10 m.; shoots rather stout, with purplish black bark, the leaves so closely placed that branchlets appear clothed with dry brown stipules; internodes 1–2 cm. long; bark separating from older branches in thick pieces consisting of very many compressed papery layers. Leaves thinly coriaceous, ± sessile or petiole very short; blades obovate-oblong or elliptic, 16–?25 cm. wide, acute at the apex, narrowed to a rounded base, with rather long ± bristly hairs on the venation beneath and very sparsely on midrib and main nerves above; venation prominent on both surfaces; stipules leafy and nervose, up to 3.5 cm. long, 2.5 cm. wide, cuspidate at the apex, narrowed to base, bristly hairy and ciliate. Flowers few at apices of shoots contained in an involucre of calyx-like bracts hidden in the leaf tufts. Corolla white; tube ± 4–5 cm. long, hairy outside; lobes oblong-lanceolate, ± 3 cm. long, 8–9 mm. wide. Fruit not known.

TANZANIA. Lindi District: Rondo [Muera] Plateau, 14 Feb. 1935, *Schlieben* 5983! & same area, Mchinjiri, Dec. 1951, *Eggeling* 6417!
DISTR. T 8; not known elsewhere
HAB. Bushland; 550–750 m.

3. L. oxyloba *K. Schum.* in E.J. 28: 488 (1900); T.T.C.L: 505 (1949). Type: Tanzania, Uzaramo District, Kisaki (?Kisangire) Steppe, *Goetze* 45 (B, holo. †)

Arborescent shrub to 5 m. tall; stems slender, rusty tomentose when young, later glabrescent and with red peeling bark. Leaf-blades oblong, elliptic or oblong-elliptic, 3.5–9(–10) cm. long, 1.5–4.5(–5) cm. wide, acuminate or somewhat obtuse at the apex, narrowed to the base, densely softly subtomentose on both surfaces; petiole 2–5 cm. long, pubescent; stipules triangular, 6(–8) mm. long, acute, slightly joined at the base with petioles, subtomentose outside, pilose and glandular within. Flowers several at the apices of abbreviated leafy branchlets, sessile; bracts similar to the stipules and sometimes bilobed (or with several lanceolate teeth); leaves supporting inflorescence 0.9–1.2 cm. long. Calyx yellow-grey pubescent; tube turbinate, 4 mm. long, subtomentose; lobes lanceolate, leafy, 2.2 cm. long, acuminate. Corolla ?white; tube slender, 5.5–6.5 cm. long, dilated above, subtomentose outside; lobes 2.2–2.5 cm. long, acuminate, glabrous inside. Style 7 cm. long, shortly bifid, glabrous, the stigmas 4 mm. long. Young fruit oblong-ellipsoid, 1.3 cm. long, 6 mm. wide, hairy, ribbed.

TANZANIA. Uzaramo District: Kisaki (?Kisangire), Oct. 1898, *Goetze* 45 & near Banda Forest Reserve, Jan. 1965, *Procter* 2848! & near Mfyoza, 12 Nov. 1969, *Ruffo* 297!
DISTR. T 6; not known elsewhere
HAB. Evergreen thicket, grassland with scattered trees; 250 m.
NOTE. I have not seen the type nor any duplicates of it so its identity remains uncertain. It is with considerable doubt that I have referred *Ruffo* 297 and *Procter* 2848 here. It may merely be a broad-leaved variant of the next species, but the glabrous style does not fit.

4. L. delagoensis *K. Schum.* in E.J. 28: 60 (1900); Verdc. in K.B. 33: 492 (1979); Vollesen in Op. Bot. 59: 69 (1980). Type: Mozambique, Maputo [Lourenço Marques], *Schlechter* 11654 (B, holo.†, Z, iso.!)

Small tree or erect shrub (? rarely scandent), (0.5–)1–4 m. tall, with sparse branching; branches slender, at first puberulous to densely pubescent, at length glabrescent and bark peeling from the shoots in strips. Leaves borne towards the apices of the branches, often not fully developed at flowering stage; blades elliptic to ovate-elliptic, 2–7(–16.5) cm. long, 1.5–2.5(–5.2) cm. wide, shortly acuminate at the apex, cuneate at the base, glabrescent to velvety pubescent on both surfaces but particularly beneath and hairy on the midrib and main nerves; petiole 0.3–1(–2.5) cm. long, pubescent. Stipules triangular-ovate, 0.5–1.2 cm. long, long acuminate at the apex, ± joined at the base, yellowish or rusty pubescent, ± persistent. Flowers scented, sessile, clustered at the apices of the branches; bracts stipule-like, toothed. Calyx silky pubescent; tube elongate-top-shaped, 2–3 mm.long; limb-tube short, becoming 2 mm. long in fruit; lobes lanceolate, 1–2 cm. long, 2–3.5 mm. wide, acute, glabrous inside. Corolla opening in the evening, sparsely to very densely silky outside, white; tube 2.5–6.5 cm. long; lobes lanceolate-oblong, 2–3.3(–4.5) cm. long, 0.7–1.1(–1.5) cm. wide, glabrous inside. Anther-tips just exserted. Style sparsely to densely pilose, the stigma-lobes 6 mm. long, just to well-exserted for 1 cm. Young fruit subglobose to oblong, 0.8–1.3 cm. long, 5–9 mm. wide, ridged, hairy.

SYN. [*L. benguelensis* sensu Ross, Fl. Natal: 332 (1973), *non* (Benth. & Hook.f.) R. Good]

subsp. **bussei** *(K. Schum. & K. Krause) Verdc.* in K.B. 33: 493 (1979). Type: Tanzania, Kilwa District, Mbarangandu, *Busse* 669 (B, holo.†, EA, iso.!)

Leaf-blades densely hairy on both surfaces, usually quite velvety.

TANZANIA. Lindi District: between Lake Lutamba and the base of the Rondo Scarp, Nov. 1953, *Eggeling* 6759! & Lake Lutamba, 13 Sept. 1934, *Schlieben* 5327! & 10 Dec. 1934, *Schlieben* 5709! & Mingoyo, 23 Mar. 1943, *Gillman* 1340!
DISTR. T 8; not known elsewhere
HAB. *Brachystegia* woodland on sand; 200–450 m.

SYN. *L. bussei* K. Schum. & K. Krause in E.J. 39: 523 (1907); T.T.C.L.: 505 (1949)

NOTE. Although easily distinguished from *L. benguelensis* in East Africa, further south some intermediates with its subsp. *pubescens* do occur, but on the whole it seems to be distinguished by its habit. It is certainly not just a state of subsp. *pubescens* protected from fire. The type of *L. delagoensis* is a specimen which differs from the bulk of the material in having the leaf-blades more or less glabrous save for the nerves beneath. Even the bulk of the material which I have referred to subsp. *delagoensis*, which occurs in Mozambique, Zimbabwe and South Africa (Natal and Transvaal), is usually densely pubescent at least beneath.

5. L. sp. A

Shrub to 3 m.; stems slender, glabrous, with grey-brown peeling bark which reveals a dark reddish brown undersurface. Leaves tufted at the ends of shoots; nodes densely congested, roughened with old stipule bases; blades elliptic to narrowly elliptic-oblong, 5.5–11 cm. long, 1.5–3.5 cm. wide, acuminate at the apex, cuneate at the base, glabrous above and shining, but with scattered white marginal hairs and sparse hairs on the venation beneath and tufts of hairs in the axils of the lateral nerves; petiole 5 mm. long, sparsely pilose; stipules narrowly triangular, 1 cm. long, 3 mm. wide, subulate-acuminate, sharply keeled. Flowers apparently axillary; pedicels ± 3–4 mm. long; bracteoles at base of calyx ± 6 mm. long, 1 mm. wide, or sometimes trifid, 1.2 cm. long, 4 mm. wide. Calyx-tube oblong, 5 mm. long, densely adpressed pilose; lobes narrowly lanceolate, 1.8–2.2 cm. long, 2–3.5 mm. wide, ciliate and with some sparse hairs. Corolla and fruits unknown.

TANZANIA. Uzaramo District: Pande Forest Reserve, 28 Dec. 1978, *Mwasumbi* 11636!
DISTR. T 6; known only from the one gathering
HAB. Lowland forest; 120 m.

6. L. benguelensis *(Benth. & Hook.f.) R. Good* in J.B. 64, Suppl. 2: 9 (1926); T.T.C.L.: 504 (1949); F.F.N.R.: 412 (1962); Verdc. in K.B. 33: 493 (1979). Type: Angola, Huila, Humpata to Mumpulla, *Welwitsch* 2563, pro parte (LISU, holo., BM, K, iso.!)

Perennial wiry shrubby herb or shrublet, ± erect and much branched or prostrate and forming cushions 0.3–2 m. in diameter and 0.3–1.2 m. tall from a thick woody rootstock; stems glabrous or with small scattered hairs on the youngest parts or densely pubescent; bark peeling on the older parts. Leaf-blades oblong-elliptic, narrowly elliptic or somewhat oblanceolate, 1.8–13(–15) cm. long, 0.6–5 cm. wide, acute to ± obtusely acuminate at the apex, cuneate at the base, slightly coriaceous, ± shiny above, glabrous save for a few hairs on the venation beneath, or puberulous to pubescent on both surfaces; free part of petiole ± 3–7(–12) mm. long; stipules adnate to petiole for up to 3 mm., triangular, 0.6–1.6 cm. long, 4–6 mm. wide, with midrib raised, narrowly acute. Flowers solitary or 3–several together, sessile, terminal and/or terminating slender axillary branches, fragrant; bracts closely adpressed to calyx and stipules, toothed. Calyx adpressed puberulous to pubescent; tube oblong, 4 mm. long; tubular part of limb wider than tube, 1.5(–5 in fruit) mm. long; lobes linear-oblong or oblong-lanceolate, 1.2–3 cm. long, 2.5–5(–7.5) mm. wide, adpressed puberulous on and near the prominent midrib outside, ciliolate, or pubescent all over. Corolla white; tube slender, 2.5–6.5 cm. long, densely covered outside with minute curved hairs which are inflated near the base or adpressed pubescent; throat pilose; lobes 4–5, elliptic, 1.2–3.8 cm. long, 1.1–1.5 cm. wide, with similar hairs outside. Style covered with rather spreading hairs; stigma bearing 2(–3) lobes 8 mm. long, just to slightly exserted. Fruits eventually orange-yellow or cream, subglobose to ellipsoid, fleshy, 1–2 cm. long, 0.8–1.2 cm. wide, ribbed, glabrous and shiny or with minute scattered hairs. Seeds dark brown, irregularly rhomboid, ± 3 mm. long, sharply angular, with unequal facets, and a marked hilar pit.

subsp. **benguelensis**

Leaf-blades ± glabrous save for sparse hairs on the venation beneath; corolla-tube with shorter hairs in Flora area.

TANZANIA. Mwanza District: Geita, Ruamagaza Forest Reserve, Bulolwa, 20 Oct. 1964, *Carmichael* 1139!; Kigoma District: Uvinza [Uvinsa], 29 Aug. 1950, *Bullock* 3247!; Dodoma District: Chaya, 20 Nov. 1961, *Semsei* 3424! & Manyoni, on escarpment on road to Kilimatindi, 16 Dec. 1935, *B.D. Burtt* 5395!
DISTR. **T** 1, 4–6; Burundi, Zaire, Zambia, Angola
HAB. *Brachystegia* woodland, rocky hillsides, riverine thickets; 910–1400 m.
SYN. *Heinsia benguelensis* Benth. & Hook.f., G.P. 2: 77 (1873); Hiern in F.T.A. 3: 82 (1877)
   Leptactina heinsioides Hiern in F.T.A. 3: 88 (1877); T.T.C.L.: 505 (1949). Type: Tanzania, S. part of Lake Tanganyika, *Cameron* (K, holo.!)
   L. tetraloba N.E. Br. in Gard. Chron. 24: 391 (1885); T.T.C.L.: 505 (1949). Type: specimen grown at Kew from material sent from Tanzania, Morogoro District, Usagara Mts., Kwa Chiropa, by *Hannington* (K, holo.!)
   [*L. lanceolata* sensu K. Schum. in Warb. Kunene-Sambesi-Exped.: 382 (1903), pro parte, *non* K. Schum. sensu stricto]
   L. benguelensis (Benth. & Hook.f.) R. Good var. *glabra* R. Good in J.B. 64, Suppl. 2: 9 (1926). Type: Angola, Benguella, Cassuango, *Gossweiler* 3251 (BM, holo., K, iso.!)
NOTE. Several collectors mention that the fruits are edible but they must be thinly fleshy and are full of stony seeds. When unburnt, shoots can reach about 1 m. tall.

subsp. **pubescens** *Verdc.* in K.B. 33: 493 (1979). Type: Zimbabwe, Umvukwes, E. of Imshi Mine, *Leach* 11274 (K, holo.!, SRGH, iso.)

Leaf-blades mostly densely pubescent beneath; calyx-lobes and stipules more pubescent than in typical subspecies; corolla-tube with longer hairs.

TANZANIA. Rungwe District: Masukula, 4 June 1907, *Stolz* 149!; Njombe District: Msima Stock Farm, *Emson* 376!; Songea District: Unangwa Hill, 13 Feb. 1956, *Milne-Redhead & Taylor* 8686! & 72 km. E. of Songea, Namtundo, 21 Sept. 1960, *Hay* 105!
DISTR. **T** 7, 8; Malawi, Mozambique, Zambia, Zimbabwe, Angola
HAB. *Brachystegia* woodland, dry slopes; 870–1300(–1650, ?1950) m.
SYN. *L. lanceolata* K. Schum. in E.J. 23: 433 (1896) & in Warb., Kunene-Sambesi-Exped.: 382 (1903), pro parte. Type: Angola, Malange [Malandsche], *Mechow* 229 (B, holo.†)
   [*L. heinsioides* sensu K. Schum. in E.J. 30: 412 (1901), *non* Hiern]
NOTE. The possibility that the species exhibits a very slight degree of heterostyly with styles of slightly different lengths needs investigation. The separation into two taxa on the basis of leaf indumentum is fairly clearly linked to geography in Tanzania and the area of Flora Zambesiaca but in Angola the two occur together, in fact one of the plants on the type gathering at the British Museum is subsp. *pubescens* and the rest subsp. *benguelensis*. For the purposes of the Flora of Tropical East Africa I consider recognition of the two valid. There is also variation in the indumentum on the outside of the corolla-tube. In Tanzania in subsp. *benguelensis* the hairs are short and very obviously thickened at the base, but in subsp. *pubescens* they are much longer and more slender at the base. Elsewhere the hairs are longer and less thickened at the base in both subspecies, particularly in Angola, but variation is such that I have not recognised a third subspecies. If a name is needed for the Tanzanian populations placed under subsp. *benguelensis* here then one based on *L. heinsioides* Hiern could be employed.

Tribe 29. **COFFEEAE** sensu lato*

Much research on tribal limits is still required and *Coffeeae* is here used in a very wide and artificial sense. Current opinion (Robbrecht & Puff in B.J.S. 108: 63–137 (1986)) restricts *Coffeeae* sensu stricto to two genera, *Coffea* and *Psilanthus*, while *Feretia, Galiniera*

---

* Genera 82–87 by D.M. Bridson, genera 88, 89 by B. Verdcourt

and *Kraussia* are considered to belong to the *Hypobathreae*, *Heinsenia* and tentatively *Belonophora* to the *Aulacocalyceae* and *Calycosiphonia* to the *Gardenieae* subtribe *Diplosporinae*.

Calyx-limb reduced to a rim, usually shorter than the disc, but
   occasionally truncate and ± equalling disc; fruit 2-
   seeeded; seeds with a well-defined groove on ventral face
   (*Coffeeae* sensu stricto):
   Corolla-tube ± equal to the lobes; anthers and style exserted;
     fruit longer than wide, evergreen or deciduous shrubs
     with monopodial or rarely sympodial growth   . . .   **85. Coffea**
   Corolla-tube longer than lobes; anthers and style included;
     fruit wider than long or as above; deciduous shrubs
     usually with sympodial growth . . . .   **86. Psilanthus**
Calyx-limb clearly apparent, truncate, dentate or lobed, mostly
   but not always persistent in fruit; fruit 1–many-seeded;
   seeds not as above (*Coffeeae* sensu lato):
   Flowers 1–many, subtended by a single leaf, clustered at
     ends of very reduced branches, often appearing lateral
     (fig. 75/3); corolla campanulate, tubular at extreme
     base (0.7–)0.9–1.2 cm. long, excluding lobes; fruit 1–
     several-seeded . . . .   **89. Heinsenia**
   Flowers axillary, sometimes appearing terminal in
     deciduous species; corolla-tube cylindrical or
     occasionally campanulate, but then not more than 7
     mm. long:
     Deciduous shrubs; calyx-limb with tubular part ± obsolete;
       lobes linear-lanceolate to lanceolate, deciduous in
       fruit; stipules and bracts chaffy . . . . . .   **83. Feretia**
     Evergreen shrubs; calyx-limb truncate or dentate, mostly
       persistent in fruit; stipules and bracts not chaffy:
       Corolla subrotate; tube ± 1 mm. long; seeds (1–)2-per
         locule, held together at apex by an arilloid
         structure (placental remains), longitudinally striate
         and ruminate . . . .   **82. Galiniera**
       Corolla distinctly tubular; seeds not as above:
         Inflorescence many-flowered, long-pedunculate,
         laxly branched; corolla funnel-shaped to
         campanulate; fruit 2–6-seeded . . . . .   **84. Kraussia**
         Inflorescence 1–10-flowered, stalked or subsessile;
         corolla-tube cylindrical; fruit 2–4-seeded:
           Calyx-limb well developed, 2.5–7 mm. long, truncate
             with rudimentary teeth, characteristically
             split down one side; fruit 2-seeded;
             leaf-acumen very distinctive, 0.8–4 cm. long   **87. Calycosiphonia**
           Calyx-limb 1–2 mm. long, lobed; fruit 2–4-seeded;
             leaf-acumen sharp but not as long as above   **88. Belonophora**

82. GALINIERA

Del. in Ann. Sci. Nat., sér. 2, 20: 92, (1843); G.P. 2: 91 (1873)

Shrub or small tree; branches glabrous or pubescent. Leaves opposite, petiolate; domatia present or absent; stipules triangular. Flowers sweetly scented, ♂, 5-merous, pedicellate in pedunculate axillary cymes; bracts and bracteoles present. Calyx glabrous or pubescent; tube extremely short; lobes ovate. Stamens attached at the top of the corolla-tube, exserted and spreading when mature; filaments short; anthers dorsifixed near base, narrowly oblong, with connective distinctly apiculate. Disc annular. Ovary 2-locular; placentas small, attached near the top of the locules, bearing 2 pendulous ovules or rarely one ovule on a larger placenta; style short, subulate, pubescent; stigmatic club deeply bifid, but two halves often held together; 5 membranous ciliate wings present

FIG. 119. *GALINIERA SAXIFRAGA* — 1, flowering branch, × ½; 2, stipule, × 2; 3, calyx, × 4; 4, corolla, with one lobe cut, × 3; 5, style with stigmatic arms, × 4; 6, longitudinal section through ovary, × 8; 7, fruit, × 2; 8, seed, 2 views, × 3; 9, seed cut in half, × 3; 10, detail of seed surface, × 18. 1–6, from *Newbould & Jefford* 1741; 7–10, from *Ash* 1474. Drawn by Miss Mary Millar Watt.

on each lobe. Fruit fleshy, globose, 2-locular; calyx-limb persistent. Seeds (2–)4 per fruit, ±
¼(–½)-spherical, finely reticulate; endosperm ruminate; placenta forming a small aril-
like structure at apex of the pair of seeds.

A small genus of 2 species, 1 species restricted to Madagascar (not seen) and 1 to tropical Africa.

G. saxifraga *(Hochst.) Bridson,* comb. nov. Type: Ethiopia, Aber, near Addeselam,
Schimper 863 (B, holo.†, K, iso.!)

Shrub or small tree 1.8–14 m. tall; young branches ± terete, glabrous to densely
pubescent. Leaf-blades elliptic to oblong-elliptic, 5.5–20 cm. long, 2–8 cm. wide,
acuminate at apex, cuneate at base, chartaceous, glabrous above, glabrous to sparsely
pubescent with densely pubescent nerves beneath; nerves red beneath; domatia
conspicuous or obscure; petioles 0.6–1.8 cm. long, glabrous or sparsely pubescent;
stipules triangular, 0.5–1 cm. long, acuminate or ± abruptly lobed. Cymes ± compact;
peduncle 0.5–2 cm. long, glabrous to pubescent; pedicels 2–4 mm. long, sparsely to
densely pubescent; bracts and bracteoles small. Calyx sparsely to densely pubescent; tube
± 1 mm. long; limb 0.25–0.5 mm. long, divided nearly to base into acute triangular lobes.
Corolla waxy white, often tinged pink or red; tube very short, scarcely exceeding 1 mm.
long, hairy within; lobes ovate, 0.5–1 cm. long, 2.5–4 mm. wide, obtuse. Fruit red or
purple-brown, 7.5–9 mm. in diameter, dull. Seeds dark reddish brown, 4–5 mm. long,
finely reticulate; endosperm longitudinally deeply ruminate. Fig. 119.

UGANDA. W. Nile District: Zeu [Zeio], Mar. 1935, *Eggeling* 1674!; Kigezi District: Muko–Kumba road,
    31 Oct. 1929, *Snowden* 1627!; Mbale District: between Butandiga and Bulambuli, 1936, *Eggeling*
    2437!
KENYA. Nakuru District: Solai Forest, *Brasnett* 1381!; Fort Hall District: Kimakia, 16 Jan. 1960,
    Greenway 9682!; Kericho District: SW. Mau Forest Reserve, 8 Aug. 1949, *Maas Geesteranus* 5628!
TANZANIA. Moshi District: southern slopes of Mt. Kilimanjaro, 23 Feb. 1953, *Drummond & Hemsley*
    1289!; Mpanda District: below Kungwe Mt., 11 Sept. 1959, *Harley* 9595!; NW. Mt. Rungwe, lower
    Kiwira [Kiwera] R., 24 Oct. 1947, *Brenan & Greenway* 8201!
DISTR. U 1–4; K 3–6; T 2–4, 6, 7; Zaire, Rwanda, Burundi, Sudan, Ethiopia, Malawi and Zambia
HAB. Forest; (760–)1700–3000 m.

SYN. *Pouchetia saxifraga* Hochst. in sched. sub. num. *Schimper* 863 (1842); A. Rich, Tent. Fl. Abyss. 1:
    355 (1848)
    Galiniera coffeoides Del. in Ann. Sci. Nat., sér. 2, 20: 92 t.1, fig. 6 (1843); Ferret & Galinier, Voy.
    Abys. 3: 138, t.6 (1848); Hiern in F.T.A. 3: 114 (1877); K. Schum. in P.O.A. C: 382 (1895); Chiov.,
    Racc. Bot. Miss. Consol. Kenya: 54 (1935); I.T.U.: 195 (1940); T.T.C.L.: 495 (1949); F.P.S. 2: 435
    (1952); K.T.S.: 439 (1961); Fl. Pl. Lign. Rwanda: 560, fig. 188/1 (1982); Verdc. in K.B. 36: 500,
    fig. 20A (1981); Fl. Rwanda 3: 159, fig. 50/1 (1985). Type: Ethiopia, between Addis and
    Maitalo, *Ferret & Galinier* 23 (MPU, holo.!)
    Ptychostigma saxifraga (Hochst.) Hochst. in Flora 27: 23 (1844) *nom. illegit.*

NOTE. Under the entry for *Ptychostigma* the Index Nominum Genericorum states, "Hochstetter
    contends that his name, printed on a label in 1842, has priority over *Galiniera* Delile 1843". I have
    not accepted *Ptychostigma* as valid since doubt concerning the distinctness of *Ptychostigma* from
    Pouchetia was indicated by Hochstetter's use of the word "forsan" and failure (at that date) to make
    the combination *Ptychostigma saxifraga*. However, the specific epithet does appear valid and
    predates *coffeoides*.

83. FERETIA

Del. in Ann. Sci. Nat., sér. 2, 20: 92 (1843); Bridson in K.B. 34: 367 (1979)

Shrubs, usually hairy at the extremities or less often glabrous. Leaves opposite; stipules
usually ovate, brown and papery when dry. Flowers fragrant, ♂, sessile and appearing
terminal on short leafless spurs (but actually axillary pairs with very reduced internodes)
or sometimes pedicellate (referred to as pedunculate by some authorities) in pairs from
the axils of young leaves, 5-merous, but the corolla can occasionally be 4- or 8-merous;
bracts and bracteoles brown and papery when dry. Calyx-tube narrowly ovoid; tubular
part of limb ± obsolete; lobes linear-lanceolate to lanceolate. Corolla white to pink; tube
funnel-shaped above, cylindrical below, glabrous to pubescent outside, pubescent at
throat; lobes ± oblong, usually slightly exceeding the tube, spreading to reflexed,
overlapping to the left in bud. Stamens inserted at throat; anthers sessile, attached ⅓ of
the length from the base, ± ⅔ exserted. Disc annular. Ovary 2-locular; ovules 2–10,

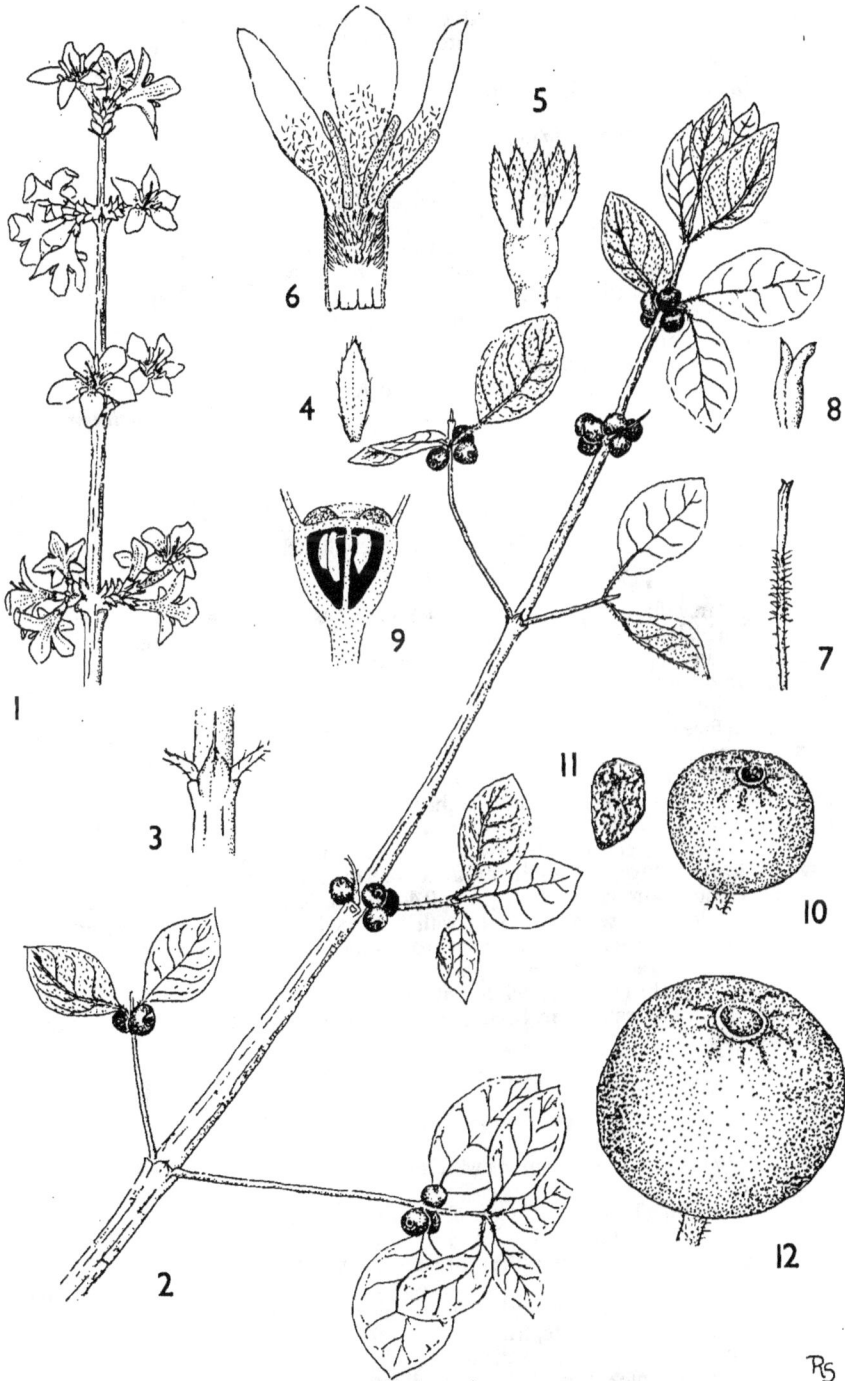

FIG. 120. *FERETIA APODANTHERA* subsp. *KENIENSIS* — 1, flowering branch, × ⅔; 2, branch with leaves and fruit, × ⅔; 3, stipule, × 2; 4, bract, × 6; 5, calyx, × 6; 6, half corolla, × 4; 7, style and stigma, × 4; 8, stigma, × 6; 9, longitudinal section of ovary, × 10; 10, fruit, × 4; 11, seed, × 6. Subsp. *TANZANIENSIS* — 12, fruit, × 4. 1, 4-9, from *Graham* 1752; 2, 3, 10, 11, from *Bally* 1914; 12, from *Renvoize & Abdallah* 2319. Drawn by R. Strachan.

pendulous in two rows from a small fleshy placenta; style slender, pubescent or with very few hairs; stigmatic club fusiform, entirely bifid, but usually cohering save at apex, sparsely pubescent to pubescent. Fruit globose, fleshy; calyx-lobes eventually deciduous. Seeds compressed, with finely reticulate testa.

A genus of 2 species, confined to tropical Africa.

Leaves pubescent or glabrous beneath, 1.5–7 cm. long, often subtending fruit or flowers; fruit 3–7 mm. in diameter or if 1.1–1.5 cm. in diameter then calyx-scar small, not exceeding 3 mm. in diameter; calyx-tube glabrous, bracts translucent except for central area . . . .   *1. F. apodanthera*

Leaves always pubescent, 3–13.5 cm. long, restricted to apical area of stem above the fruit; fruit 1.2–2 cm. in diameter with large (4–8 mm. in diameter) pale calyx-scar; calyx-tube pubescent; bracts tending to be translucent only at margins . . . .   *2. F. aeruginescens*

1. F. apodanthera *Del.* in Ann. Sci. Nat., sér. 2, 20: 92, t. 1, 4 (1843); Hiern in F.T.A. 3: 115 (1877); F.P.S. 2: 434 (1952); Keay in F.W.T.A., ed. 2, 2: 147, fig. 230 (1963); Bridson in K.B. 34: 368 (1979). Type: Ethiopia, Djeladjeranne [Tchellakchekenne] on the bank of R. Takazze, *Ferret & Galinier* (MPU, holo.!)

A shrub 2–9 m. tall; young branches slender, glabrous to pubescent, covered with thin reddish bark which readily flakes off to reveal the rough greyish under-layer. Leaf-blades elliptic to broadly elliptic or ovate, 1.5–7 cm. long, 0.9–3.4 cm. wide, obtuse or occasionally rounded, sometimes apiculate at apex, acute or obtuse to rounded at base, entirely glabrous or with hairs on the nerves beneath or sparsely to densely covered with fine hairs on both faces; domatia present; petiole 0.1–1.1 cm. long, pubescent above; stipules papery and brown when dry, ovate or triangular, 2–5 mm. long, acuminate or apiculate. Flowers either precocious and sessile to shortly pedicellate or, if borne in leaf-axils, then either pedicellate with pedicels attached a short distance above the node and decurrent to node or sessile; pedicels 0–3.6 cm. long; bracts brown and papery when dry, often translucent except for central area, ovate, 2–4 mm. long; bracteoles linear. Calyx-tube 1.25–2 mm. long, glabrous; lobes linear-lanceolate to lanceolate, 1.5–5 mm. long, acute, sometimes ciliate. Corolla white or pink, 0.4–1.4 cm. long, 3.5–7.5 mm. wide at top, 0.75–1.5 mm. wide at base, glabrous or with 5 sparsely pubescent to pubescent lines outside, sparsely to densely pubescent at throat; lobes 0.5–1.5 cm. long, 2–5 mm. wide, obtuse to rounded or acute. Placenta bearing 8(–10) ovules; stigma and style sparsely to densely pubescent. Fruit red or white with purple streaks, 0.3–1.7 cm. in diameter, glabrous, with a small calyx-scar. Seeds light to mid-brown, flattened, 3–7 mm. across, finely reticulate.

KEYS TO INFRASPECIFIC VARIANTS

(Specimens with leaves and fruit)

Petioles 1–4 mm. long; base of leaf-blade obtuse to rounded or sometimes acute; apex acute or less often obtuse; pedicellate fruit (or flowers) often present in the axils; fruit 3–7 mm. in diameter . . . .   a. subsp. **apodanthera**

Petioles 2–11 mm. long; base of leaf-blade acute; apex obtuse to rounded or sometimes acute; fruit (or flowers) very rarely pedicellate; fruit 0.5–1.7 cm. in diameter:

Fruit white with purple streaks, 5–6 mm. in diameter; young stems sparsely pubescent to pubescent; leaves glabrous to pubescent . . . .   b. subsp. **keniensis**

Fruit pink to red, 1.1–1.7 cm. in diameter; young stems glabrous or less often sparsely pubescent; leaves glabrous   c. subsp. **tanzaniensis**

(Specimens with precocious flowers)

Ovules 2–5 per placenta; calyx-lobes 1.5–4 mm. long; corolla
　sparsely pubescent to pubescent at throat:
　Calyx-lobes 1.5–3 mm. long; corolla-lobes obtuse to rounded
　　or less often acute (**U** 1)　. . . . .　a. subsp. **apodanthera**
　Calyx-lobes 2.5–4 mm. long; corolla-lobes acute or less often
　　obtuse (**K** 7)　. . . . .　b. subsp. **keniensis**
Ovules 4–8(–10) per placenta; calyx-lobes (2–)4–5 mm. long;
　corolla pubescent to densely pubescent at throat (**T** 1, 4, 7)　c. subsp. **tanzaniensis**

a. subsp. **apodanthera**

Young stems sparsely pubescent to pubescent. Leaf-blades acute or less often obtuse at apex, obtuse to rounded or sometimes acute at base, glabrous or occasionally glabrescent to sparsely pubescent on both faces; petiole 1–4 mm. long. Flowers either precocious and sessile or pedicellate and borne in the leaf-axils; pedicels 0.3–3.6 cm. long. Calyx-lobes 1.5–3 mm. long. Corolla creamy white to white suffused pink or with tube purple outside and lobes white, sparsely pubescent to pubescent at throat; lobes obtuse to rounded or less often acute. Ovules 2–4 per placenta; stigma and style sparsely pubescent to pubescent. Fruit white with purple streaks or red, 3–7(–8) mm. in diameter. Seeds 3–3.5 mm. across.

UGANDA. W. Nile District: near Koich R., Rumogi, Mar. 1935, *Eggeling* 1817! & 1819!; Karamoja District: Nabilate, 11 Jan. 1937, *A.S. Thomas* 2245!
DISTR. U 1; Mauritania, Senegal, Gambia, Mali, Ghana, N. Nigeria, Cameroun, Central African Republic, Sudan, Ethiopia, Somalia
HAB. Riverine forest; ± 1200 m.

SYN. *Feretia ? canthioides* Hiern, in F.T.A. 3: 116 (1877). Types: Senegal, *Heudelot* 436 & N. Nigeria, Nupe, *Barter* 1242 (both K, syn.!)

b. subsp. **keniensis** *Bridson* in K.B. 34: 368 (1979). Type: Kenya, Kwale District, N. of Jadini, *Greenway* 9624 (K, holo.!, EA, iso.!)

Young stems sparsely pubescent to pubescent. Leaf-blades acute at base, obtuse at apex, glabrous to sparsely pubescent on both faces; petiole 3–8 mm. long. Flowers usually precocious, sometimes appearing with immature leaves or less frequently with mature leaves, but then not from the axils, sessile or rarely shortly pedicellate. Calyx-lobes 2.5–4 mm. long. Corolla creamy white to white suffused pink, sparsely pubescent to pubescent at throat; lobes acute or sometimes obtuse. Ovules 2–5 per placenta; style sparsely pubescent to pubescent. Fruit white with purple streaks, 5–6 mm. in diameter. Seeds 3.75–4 mm. across. Fig. 120/1–11.

KENYA. Kwale District: behind Diani Beach, 9 July 1968, *Gillett* 18651!; Kilifi District: Kikambala, 2 · Apr. 1960, *Bally* 12195!; Lamu District: Kiunga, Apr. 1931 *MacNaughtan* 9!
DISTR. K 7; Somalia
HAB. Coastal forest and bushland, often on coral rags; 0–30 m.

SYN. [*F. apodanthera* sensu K.T.S.: 439 (1961), *non* Del. sensu stricto]

NOTE. This subspecies is very close to subsp. *apodanthera*, but I feel worthy of recognition as it occupies a different habitat and phytochorion.

c. subsp. **tanzaniensis** *Bridson* in K.B. 34: 368 (1979). Type: Tanzania, Dodoma District, Kazikazi, *B.D. Burtt* 4326 (K, holo.!)

Young stems glabrous or less often sparsely pubescent. Leaf-blades acute to obtuse or rounded at apex, acute at base, glabrous; petiole 2–11 mm. long. Flowers precocious or appearing with immature leaves. Calyx-lobes (2–)4–5 mm. long. Corolla white flushed pink or cherry pink, pubescent to densely pubescent at throat; lobes obtuse. Ovules 4–8(–10) per placenta; style pubescent to densely pubescent. Fruit pale pink to bright red, 1.1–1.7 cm. in diameter. Seeds 5–7 mm. across. Fig. 120/12.

TANZANIA. Shinyanga, *Koritschoner* 1764!; Mpwapwa, 2 Feb. 1930, *Hornby* 169!; Iringa District: Ruaha National Park, 21 May 1968, *Renvoize & Abdallah* 2319!
DISTR. T 1, 4, 5, 7; not known elsewhere
HAB. Bushland, often near rivers or on dry hillsides; 700–1375 m.

SYN. [*F. apodanthera* Del. var. *australis* sensu T.T.C.L.: 495 (1949), pro parte, quoad *B.D. Burtt* 3426 & 3815, *non* K. Schum.]

NOTE. In fruit this subspecies is very distinct from both subsp. *apodanthera* and subsp. *keniensis*, but as the flowering stage is less distinct I have not recognized it at specific level.

2. F. aeruginescens *Stapf* in K.B. 1906: 79 (1906); F.F.N.R.: 407 (1962); Bridson in K.B. 34: 370 (1979). Type: Zambia, Victoria Falls, *Allen* 57 (K, holo.!)

Straggling shrub, 1.8–9 m. tall; very young branches pubescent, soon glabrescent, covered with thin reddish bark which readily flakes off to reveal the characteristic rough grey bark of the older stems. Leaves borne on new stems; blades narrowly obovate to obovate or sometimes ovate, 3–13.5 cm. long, 1.7–6 cm. wide, acute or less often obtuse at apex, obtuse or sometimes rounded at base, glabrescent to pubescent above, pubescent to densely pubescent beneath, nerves usually pinkish beneath; domatia absent; petiole 2–3 mm. long, pubescent; stipules brown and papery when dry, triangular, 3–7 mm. long, acuminate. Flowers precocious, always borne on very reduced spurs off the older stems, 1(–2) pairs, or occasionally with 2–3 spurs occurring close together to give a multi-flowered appearance, sessile; bracts brown and papery when dry, ovate, 3–7 mm. long, ± translucent at margin; bracteoles linear. Calyx-tube ± 2 mm. long, covered with silky white hairs; lobes brown and papery when dry, lanceolate, 2–4 mm. long, acute or with an apiculum, ciliate. Corolla white tinged with pink, glabrous or with very few hairs outside; tube 0.8–1.5 cm. long, 0.9–1.1 cm. wide at top, 1–1.5 mm. wide at base, pubescent at throat; lobes 1.1–2 cm. long, 3–7 mm. wide, obtuse to rounded. Placenta bearing 6–7 ovules; stigma ± sparsely pubescent. Fruit bright red (or rose pink), 1.2–2 cm. in diameter, glabrescent, with large pale circular calyx-scar. Seeds reddish, flattened, 6–7.5 mm. across, irregularly rugulose.

TANZANIA. Mbeya District: Unyiha [Unika] Plateau, Oct. 1899, *Goetze* 3123!; Rungwe District: Ulambya [Bulambya] to Songwe R., 3 Mar. 1914, *Stolz* 2557!
DISTR. T 7; Zaire (Shaba), Mozambique, Malawi, Zambia, Zimbabwe, Botswana and Caprivi Strip
HAB. Not recorded, but elsewhere frequently in riverine woodland or near termite mounds; ± 900–1200 m.

SYN. *F. apodanthera* Del. var. *australis* K. Schum. in E.J. 30: 412 (1901); T.T.C.L.: 495 (1949). Type: Tanzania, Mbeya District, Unyiha [Unyika] Plateau, Sante [Saube], Niamba [Yambe] R., *Goetze* 1407 (B, holo.†, BR iso.!)

84. KRAUSSIA

Harv. in Hook., Lond. Journ. Bot. 1: 21 (1842); Sond. in Fl. Cap. 3: 22 (1864–65), pro parte; G.P. 2: 95 (1873), pro parte; Bullock in K.B. 1931: 255 (1931) & 1934: 231 (1934)

Rhabdostigma Hook.f. in G.P. 2: 109 (1873); Hiern in F.T.A. 3: 130 (1877); K. Schum. in P.O.A. C: 383 (1895) & in E. & P.Pf. 4 (4): 87 (1897)

Shrubs; branches glabrous or less often pubescent. Leaves opposite, petiolate; blades usually drying grey to mahogany-brown, obtuse to acuminate, chartaceous to subcoriaceous; domatia often present; stipules triangular to lanceolate, sometimes apiculate. Flowers ⚥, 5-merous; inflorescence axillary, pedunculate, usually lax with well-developed pedicels; bracts and bracteoles present. Calyx glabrous to puberulous; tube ovate to turbinate; limb divided nearly to base into triangular lobes, rounded to acute, sometimes pubescent within. Corolla-tube very shortly cylindrical at base, funnel-shaped or campanulate above, densely bearded at throat; lobes usually exceeding the tube in length, oblong-lanceolate or narrowly obovate. Stamens attached at the mouth of the tube, exserted and spreading; filaments short; anthers dorsifixed near the base, linear, with connective distinctly apiculate. Disc annular. Ovary 2-locular; placentas attached near the top of the septum, large or moderately small, with 1–3 impressed ovules; style short, fleshy, swollen and usually spindle-shaped, glabrous; stigmatic club long, exserted, fusiform, with 10 membranous ciliate wings. Fruit globose, somewhat fleshy, 2-locular, containing numerous succulent red granules; calyx-limb persistent. Seeds 2–6 per fruit, ± ovoid, usually with one flattened face, finely reticulate; hilar cavity or groove not present; endosperm entire.

A genus of 3 species, restricted to tropical and South Africa.

Inflorescence with 4–5 pairs of branches, some at least alternate; calyx-limb 0.5–1.5 mm. long; fruit 2-seeded; leaf-blades subcoriaceous; tertiary nerves not apparent   *1. K. kirkii*

Inflorescence with 1–2 pairs of opposite branches; calyx-limb
   3–4 mm. long; fruit 4–6 seeded; leaf-blades chartaceous;
   tertiary nerves apparent beneath . . . .   *2. K. speciosa*

1. K. kirkii *(Hook.f.) Bullock* in K.B. 1934: 231 (1934); T.T.C.L.: 509 (1949); K.T.S.: 448 (1961). Type: Tanzania, Kilwa [Quiloa], *Kirk* 105 (K, holo.!)

Shrub 1.5–4.5(–8.4) m. tall, glabrous, young branches square, covered with chocolate or dark reddish bark. Leaf-blades drying greyish to mahogany-brown, elliptic to oblong-elliptic, 5.3–14(–17) cm. long, 2–5(–6.3) cm. wide, obtuse to acute or slightly acuminate at apex, acute to obtuse at base, subcoriaceous, slightly shiny above; tertiary nerves obscure beneath; domatia present; petiole 0.5–1 mm. long; stipules triangular, 2.5–6 mm. long, not or very shortly apiculate, central rib absent or infrequently present. Inflorescence a lax axillary panicle; peduncle slender, 3–7 cm. long; lateral branches with the first pair opposite or nearly opposite and the next 2–4 pairs of branches ± alternate; pedicels 0.3–1.2 cm. long; bracteoles up to 2 mm. long. Calyx-tube 1.5–2 mm. long; limb 0.5–1.5 mm. long, divided nearly to the base into triangular lobes, acute to rounded. Corolla creamish to yellowish; tube (2.5–)3–4 mm. long, 4–5 mm. wide at top, ±1 mm. wide at base, bearded at throat;lobes oblong-lanceolate, 4–7 mm. long, 2–3 mm. wide, acute. Ovules solitary, impressed to one side of the placenta; style spindle-shaped; stigma with wings ciliate. Fruit ?green, spherical or slightly bilobed, 5–5.5 mm. in diameter, crowned by the persistent calyx-limb. Seeds 2 per fruit, dark reddish brown, 4 mm. long, finely reticulate.

KENYA. Kilifi District: Marafa, Jan. 1937, *Dale* in F.D. 3668!; Lamu District: Kitangani, 27 Dec. 1946, *J. Adamson* 309 in *Bally 5815!* & Iwezo, 23 Feb. 1955, *Power!*
TANZANIA. Tanga District: near Ngomeni railway station, 12 Dec. 1959, *Semsei* 2951!; Uzaramo District: Dar es Salaam, Ubungo, 3 Dec. 1968, *Mwasumbi* 10422!; Rufiji District: Kilwa Kivinje, 4 Dec. 1955, *Milne-Redhead & Taylor* 7550!
DISTR K 7; T 3, 6, 8; not known elsewhere
HAB. Forest, coastal bushland, wooded grassland or mangrove communities; 0–480 m.

SYN. *Rhabdostigma kirkii* Hook.f., G.P. 2.: 109 (1873); Hiern in F.T.A. 3: 131 (1877); Oliv. in Hook., Ic. Pl. 23, t. 2275 (1893); K. Schum. in P.O.A. A: 17 (1895) & P.O.A. C: 383 (1895) & in E & P. Pf. 4(4): 87 (1897).

2. K. speciosa *Bullock* in K.B. 1931: 256 (1931); T.T.C.L.: 503 (1949); K.T.S.: 449 (1961). Type: Tanzania, Morogoro District, Kimboza, Mikese-Kisaki road, *Greenway* 2516 (K, holo.!, EA, iso.)

Shrub or small tree 2–9 m. tall; young branches greenish (or black when dry), ± square, glabrous or finely pubescent; older branches covered with buff smooth bark. Leaf-blades drying greyish to mahogany-brown, oblong-elliptic, 9.5–16(–21) cm. long, 3.9–7.8 cm. wide, acuminate or sometimes acute at apex, rounded or rarely auriculate at base, chartaceous, tertiary nerves usually apparent beneath, glabrous above, glabrous or with nerves puberulous beneath; domatia occasionally present; petiole 0.7–1.2 cm. long, glabrous or puberulous; stipules triangular or ovate-triangular, 0.5–1 cm. long, acute sometimes exceeded by an apiculum up to 2 mm. long. Inflorescence a lax axillary panicle; peduncle slender, 2.5–5 cm., long, glabrous or puberulous; lateral branches 1–2 pairs, opposite; pedicels 4–9(–20 in fruit) mm. long, glabrous or puberulous; bracteoles 4–7 mm. long. Calyx sparsely puberulous to puberulous outside; tube 1.5–1.75 mm. long; limb 3–4 mm. long, divided almost to base into acute triangular lobes, densely pubescent within. Corolla white to yellow; tube 4.5–7 mm. long, 4–5.5 mm. wide at top, ± 1 mm. wide at base, densely bearded at throat; lobes lanceolate to narrowly ovate, 7–13 mm. long, 2.5–5 mm. wide, acute to acuminate. Ovules 2 on a larger placenta or 3 on a smaller placenta; style spindle-shaped; stigma with wings ciliate. Fruit ? green (brownish when dry), spherical or slightly bilobed, 8–9 mm. in diameter; calyx-limb persistent. Seeds 4–6 per fruit, maroon, 5 mm. long, finely reticulate. Fig. 121.

KENYA. Kwale District: Shimba Hills, Lango ya [Longo] Mwagandi area, 23 Mar. 1968, *Magogo & Glover* 435!; Lamu District: Utwani Forest Reserve, Mambosasa [Mombasasa], 16 Oct. 1957, *Greenway & Rawlins* 9348! & Jan. 1957, *Rawlins* 326!
TANZANIA. Lushoto District: Longuza Hill, above the Sigi Gorge, 19 Nov. 1947, *Brenan & Greenway* 8343!; Morogoro District: Mtibwa Forest Reserve, Nov. 1953, *Semsei* 1434!; Ulanga District: Lukoga Forest Reserve, 6 Nov. 1961, *Semsei* 3390!
DISTR. K 7; T 3, 6; not known elsewhere
HAB. Forest; 10–900 m.

FIG. 121. *KRAUSSIA SPECIOSA* — 1, flowering branch, × ⅔; 2, stipules, × 2; 3, flower, × 2; 4, calyx with style, × 3; 5, longitudinal section of flower, × 3; 6, longitudinal section of ovary, × 5; 7, longitudinal section of ovary, × 8; 8, placenta with ovules, × 20; 9, part of infructescence, × 1; 10, fruit, × 2; 11, seed, × 2. 1–8, from *Faulkner* 1074; 10, 11, from *Magogo & Glover* 435. Drawn by Mrs M.E. Church.

85. COFFEA

L., Sp. Pl.: 172 (1753) & Gen. Pl., ed. 5: 80 (1754); G.P. 2: 114 (1873)

Shrubs or small trees. Leaves opposite, petiolate, glabrous; domatia usually present, the cavity completely or partially defined, glabrous or pubescent; stipules very shortly united above the axils, obtuse to aristate, usually with colleters within. Flowers ⚥, (4–)5–8(–12)-merous, borne with leaves or precociously, 1–many per axil or rarely terminal on spurs; inflorescence-stalks (including pedicel) individual or grouped on common peduncle, glabrous or pubescent (but this character is often obscured by resinous exudate); bracteoles 1–4, usually cupular, stipule-like with rudimentary lobes or sometimes bearing 2 linear or subfoliaceous lobes; occasionally additional free, scale-like bracteoles are borne towards apex of inflorescence-stalks. Calyx-tube campanulate to turbinate; limb usually ± obsolete and shorter than the disc, occasionally equalling or rarely exceeding the disc, truncate to dentate, usually beset with colleters. Corolla white or occasionally pink, glabrous; tube cylindrical and widened at throat or somewhat funnel-shaped, shorter than, subequal to or, occasionally, a little longer than the lobes; lobes contorted in bud, spreading. Stamens attached at mouth of corolla-tube, exserted, erect; filaments ± $\frac{1}{3}$ the length of the anther; anthers linear, attached dorsally up to $\frac{1}{3}$ from the base. Disc annular. Ovary 2-locular; ovules solitary, subpeltately attached to the middle of the septum; style slender, glabrous, exceeding the corolla-tube; stigma exserted, with 2 usually divergent arms. Fruit a ± ellipsoidal drupe, usually fleshy, containing (1–)2, ± coriaceous, 1-seeded pyrenes. Seed oblong-ellipsoid, grooved on inner face; testa thin with compressed cellular sculpturing that superficially appears striated, shiny (referred to as silver-skin by coffee growers); endosperm pale in colour, horny, asymmetrically folded from the groove; embryo erect, somewhat curved.

A genus of about 90 species native to tropical Africa, Madagascar and the Mascarenes, with 2(–3) economic species widely cultivated throughout the tropics.

Leroy (in Ass. Sci. Intern. Café, 9ᵉ colloque, Londres: 475 (1980)) recognised three subgenera of *Coffea*:— subgen. *Coffea*, subgen. *Baracoffea* (Leroy) Leroy (*Paolia* Chiov.) and subgen. *Psilanthopsis* (A. Chev.) Leroy. The characters distinguishing the first two (as far as the African species are concerned) are indicated in couplet 1 of the key to species. Subgen. *Psilanthopsis*, distinguished by a dentate calyx-limb, is monotypic, consisting of the Angolan species *Psilanthopsis kapakata* A. Chev. (the combination in *Coffea* never having been validated); however, tendencies towards certain east African species of subgen. *Coffea* have been noted by Bridson (in K.B. 36: 818 (1982)) and Leroy (in *tom. cit.*, 10ᵉ colloque, Salvador: 419 (1982)).

The subsections of *Coffea* sect. *Coffea* proposed (but not validly published) by Chevalier (in Comptes Rendus Acad. Sci., Paris 210: 359 (1940) & Caféiers du Globe 3: 102 (1947)) have not been found of any practical taxonomic value.

Apart from the more important economic species, *C. arabica*, *C. canephora* and *C. liberica* (dealt with in the text), *C. stenophylla* G. Don has been grown at Lyamungu Research station near Moshi. This species is frequently cultivated under the name "Highland Coffee of Sierra Leone" and is reported to have an excellent flavour. It is sometimes confused with some narrow-leaved forms of *C. arabica* such as var. *angustifolia* auctt. (the usually quoted authority (Roxb.) Froehn. is incorrect since *C. angustifolia* Roxb. has been identified as *Pittosporum moluccanum* (Lam.) Miq.) or var. *monosperma* Ottolander & Cramer. True *C. stenophylla* has a violet-black fruit with a somewhat accrescent disc and the flowers are 6–8-merous (unlike *C. arabica* which are typically 5-merous). The figure of *C. stenophylla* in Bot. Mag. 122, t. 7475 (1896) and the copies based on it in K.B. 1896: 190 (1896); De Wild., Miss. Laurent: 62 (1906); Cheney, Coffee Monogr. Econ. Sp., t. 11 (1925) & A. Chev., Caféiers du Globe 1, fig. 17 (1929), show a terminal solitary inflorescence in addition to the axillary ones. This would appear to have been based on an anomalous specimen of a different species (in my opinion *C. togoensis* A. Chev.). The illustrations in De Wild. Miss. Laurent 1, t. 63 (1906), Cramer in Meded. Dept. Landb. Ned. Ind. 11: 608 (1913) & Cheney, *cit. supr.*, t. 12 (1915) seem to be more accurate. Hybrids between *C. stenophylla* and *C. liberica* and *C. congensis* Froehner have been recorded.

1. Flowers terminal on short lateral spurs; scarious bracteoles present; small much branched deciduous bush of dry areas (**K** 1) (subgen. *Baracoffea*) . . . . 22. *C. rhamnifolia*
   Flowers always axillary (rarely borne in apical axils so as to appear terminal); scarious bracteoles absent; small trees or bushes of forests or thickets, evergreen or sometimes deciduous (subgen. *Coffea*) 2
2. Very young stems glabrous 3
   Very young stems sparsely pubescent to pubescent or puberulous 22

3. Flowers precocious; leaves borne at apices of branches; blades broadly elliptic; flowers and fruit usually borne in fascicles or 3 or more per axil . . . . *20. C. zanguebariae*
  Flowers not precocious; leaves well spaced along the stems; blades seldom broadly elliptic 4
4. Stipules obtuse or occasionally acute, rarely apiculate 5
  Stipules apiculate to distinctly aristate or occasionally acute 8
5. Leaf-blades with (7–)8–13 main pairs of lateral nerves, mostly 14–37 cm. long; domatia usually situated across the lateral nerves; flowers 4–many per axil . . . *1. C. liberica*
  Leaf-blades with not more than 7 main pairs of lateral nerves, not exceeding 12.5 cm. long; domatia situated in the nerve-axils; flowers 1–3 per axil; fruit lacking accrescent disc 6
6. Leaf-blades subacuminate at apex; fruit with inflores-cence- stalks 1.3–1.5 cm. long . . . . *4. sp. A*
  Leaf-blades rounded to obtuse at apex; fruit with inflorescence-stalks 5–9 mm. long 7
7. Leaf-blades broadly elliptic to round; fruit 1.6–2.2 cm. long; lower bracteole slightly larger than upper bracteole, somewhat lobed (K 7) . . . . *2. C. fadenii*
  Leaf-blades obovate to elliptic; fruit 1.3–1.7 cm. long; lower bracteole slightly smaller than upper bracteole, unlobed (T 3) . . . . *3. C. mongensis*
8. Leaf-blades oblanceolate or narrowly elliptic; flowers 6–8-merous; fruit violet-black with somewhat enlarged disc (cultivated) . . . . *C. stenophylla (see note above)*

  Leaf-blades elliptic to broadly elliptic; fruit orange to red, brownish or rarely black 9
9. Never more than 3 flowers (or fruit) per axil, usually borne separately 10
  More than 3 flowers (or fruit) per axil, borne separately and in fascicles or both at same axil 18
10. Calyx with limb equalling or slightly exceeding disc; bracteoles with linear to subulate lobes 3–4 mm. long 11
  Calyx with limb usually shorter than disc; bracteoles if lobed then either much shorter or subfoliaceous 12
11. Calyx-tube not ribbed; limb equalling disc; stipules distinctly aristate; domatia large (T 6, Morogoro District) . . . *14. C. sp. E*
  Calyx-tube ribbed; limb exceeding disc; stipules apiculate; domatia small (Mafia I.) . . . . *15. C. sp. F*
12. Bark on young twigs ± maroon; leaves ± coriaceous, distinctly shiny above (T 8, Lindi District) . . . . *13. C. sp. D*
  Bark on young twigs buff to mid-brown; leaves papery to subcoriaceous or occasionally coriaceous, dull to moderately shiny above 13
13. Leaf-blades acute to subacuminate or occasionally obtuse at apex, always dull above; flowers (5–)6–8-merous or if 5(–6)-merous then subsessile; fruit beaked or not (coastal areas) 14
  Leaf-blades acuminate to long-acuminate, dull to moderately shiny above; flowers always 5-merous, and borne on an inflorescence-stalk; fruit not beaked 15
14. Inflorescence-stalks 2–6 mm. long, not obscured by bracteoles; flowers 6(–8)-merous; stipules 2–4 mm. long, shortly aristate; fruit distinctly beaked . . . *9. C. pseudo-zanguebariae*

  Inflorescence-stalks not exceeding 2 mm. long, obscured by bracteoles; flowers 5(–6)-merous; stipules 3.5–5.5 mm. long with well-developed arista; fruit unbeaked or sometimes beaked . . . . *10. C. sessiliflora*

15. Leaf-blades with 7–10 main pairs of lateral nerves; fruit 1–1.7 cm. long (**K** 1, Mt. Marsabit) . . . . . .   6. *C. arabica*

     Leaf-blades with 5–7(–8) main pairs of lateral nerves; fruit 0.7–1.2 cm. long 16

16. Leaf-blades elliptic to narrowly obovate, acute to cuneate at base, subcoriaceous, moderately shiny above; bark on young stems mid-brown (**T** 7) . . . .   12. *C. sp. C*

     Leaf-blades elliptic to broadly elliptic, acute at base, papery to subcoriaceous, dull or somewhat shiny above; bark (on all but youngest stems) buff to light brown 17

17. Domatia absent or glabrous; stipules with arista 0.5–3.5 mm. long; lower bracteoles frequently with subfoliaceous lobes; fruit red . . . .   7. *C. eugenioides*

     Domatia sparsely pubescent to pubescent; stipules acute or with arista not exceeding 1 mm. long; lower bracteoles with rudimentary lobes; fruit green with rose striations, becoming black when mature . . . .   8. *C. salvatrix*

18. Lateral nerves 7–14(–17) main pairs, frequently strongly impressed above; fruit 0.9–2 cm. long 19

     Lateral nerves 5–7(–8) main pairs, seldom impressed above; fruit not exceeding 1 cm. long (where known) 20

19. Bracteoles bearing subfoliaceous lobes up to 2.2 cm. long; pedicels usually very short (so that calyces do not exceed bracteoles at anthesis); leaves 12–35(–40) cm. long, oblong-elliptic, obovate or broadly elliptic; lateral nerves in (8–)11–15(–17) main pairs; domatia absent or pubescent . . . .   5. *C. canephora*

     Bracteoles bearing subfoliaceous lobes not exceeding 0.5 cm. long; pedicels 1–2(–3) mm. long (so that the calyces usually exceed the bracteoles at anthesis); leaves 7–18 cm. long, mostly elliptic to broadly elliptic; lateral nerves in 7–10 main pairs; domatia absent or small and glabrous . . . .   6. *C. arabica*

20. Bark on young twigs reddish brown; flowers 6–7-merous (**T** 3, E. Usambara Mts.) . . . .   11. *C. sp. B*

     Bark on young twigs buff to light brown; flowers 5-merous 21

21. Domatia absent or glabrous; stipules with arista 0.5–3.5 mm. long; lower bracteoles frequently with subfoliaceous lobes; fruit red . . . .   7. *C. eugenioides*

     Domatia sparsely pubescent to pubescent; stipules acute or with an arista not exceeding 1 mm. long; lower bracteoles with rudimentary lobes; fruit green with rose striations, ripening to black . . . .   8. *C. salvatrix*

22. Young stems with puberulous line on both sides; domatia absent or glabrous . . . .   7. *C. eugenioides*

     Young stems evenly puberulous or pubescent; domatia pubescent, less often glabrescent or occasionally absent 23

23. Inflorescence-stalks reduced; bracteoles with linear-subulate lobes 1–8 mm. long; calyx-limb equalling or exceeding disc or sometimes shorter than disc (only fruiting stages known) 24

     Inflorescence-stalks clearly apparent; bracteoles with lobes scarcely developed or tending to be spathulate; calyx-limb shorter than disc 26

24. Leaves coriaceous, possibly borne after the flowers; flowers often borne in apical axils of short lateral shoots   21. *C. sp. J*

     Leaves papery to subcoriaceous, borne with flowers; short lateral shoots not present 25

25. Leaves crisped at margins; bracteoles 4–8 mm. long; domatia with an incompletely defined cavity, pubescent   16. *C. sp. G*

     Leaves not crisped at margins; bracteoles 1–2.5 mm. long; domatia small ± glabrous pits . . . .   10. *C. sessiliflora* subsp. *mwasumbii*

26. Leaf-blades small, 2–7(–10) cm. long, 1–3.5(–4) cm. wide
    (mostly towards lower limits), crowded, often crisped;
    stems pubescent; flowers always 5-merous . . . .   *19. C. mufindiensis*
    Leaf-blades larger, 3–15 cm. long, 1.5–6.2 cm. wide (mostly
    towards upper limits), not crowded, not or slightly
    crisped; stems sparsely pubescent to puberulous;
    flowers 6(–8)-merous or unknown 27
27. Leaf-blades acute to subacuminate or occasionally obtuse
    at apex; domatia conspicuous or sometimes small;
    flowers 6(–8)-merous; fruit beaked . . . . . .   *9. C. pseudo-*
                                                             zanguebariae
    Leaf-blades shortly acuminate to acuminate; domatia
    absent or small; flowers not known; fruit not beaked,
    where known 28
28. Young branches covered with brownish bark; leaf-blades
    broadly elliptic (T 6, Ulanga District) . . . . .   *17. C. sp. H*
    Young branches covered with maroon-red bark; leaf-
    blades elliptic to narrowly obovate (T 4, Kigoma District) *18. C. sp. I*

1. **C. liberica** *Hiern* in Trans. Linn. Soc., Bot. sér. 2, 1: 171, t. 24 (1876) & in F.T.A. 3: 181 (1877), pro parte, excl. pl. ex Angola; Froehner in N.B.G.B. 1: 233 (1897) & in E.J. 25: 269 (1898); De Wild. in Actes Congr. Intern. Bot., Paris: 237 (1900), Les Caféiers: 39 (1901), Miss. Laurent: 338, t. 104 (1906) & in Ann. Jard. Buitenz., suppl. 3, 1: 373 (1910); Cheney, Coffee, Monogr. Econ. Sp.: 76, t. 28–35 (1925); A. Chev., Caféiers du Globe 1: 75, fig. 11, 12 (1929); Lebrun in Mém. Inst. Roy. Col. Belge, Sect. Sci. Méd. 8°11 (3): 153, t. I/12, III/7, XV, XIX (1941); A. Chev., Caféiers du Globe 2: t. 1 (1942) & 3: 170 (1947); F.P.N.A.: 354 (1947); U.O.P.Z.: 204 (1949); T.T.C.L.: 492 (1949); Wellman, Coffee: 76 (1961); Haarer, Modern Coffee Prod.: 22 (1962); F.W.T.A., ed. 2, 2: 154, fig. 231 (1963); Bridson in K.B. 40: 806 (1985). Type: Sierra Leone, cultivated on Mr Effenhausen's farm, *Danielle* (BM, lecto.!)

Tree up 20 m. or sometimes a bush 3–8 m. tall; young branches thick, glabrous, covered with light brown shiny bark. Leaf-blades narrowly obovate to obovate or elliptic to broadly elliptic, 14–38 cm. long, 5.5–20.5 cm. wide, rounded, obtuse, rounded and then shortly acuminate or rarely acute at apex, acute to cuneate at base, stiffly papery to moderately coriaceous; dull to shiny above, somewhat undulate; lateral nerves (7–)8–13 main pairs; domatia frequently situated across the base of the lateral nerves or occasionally in the axil, glabrous to sparsely pubescent; petiole 0.8–2 cm. long; stipules triangular-ovate or sometimes almost truncate, 2–4.5 mm. long, obtuse or sometimes acute. Flowers 5-merous or 6–9-merous, 4–30(–50) per axil, borne in 1–3 or more congested fascicles; peduncle ± obsolete at flowering stage; usually 1–3 compressed branches present; bracteoles up to 2 mm. long, unlobed or rarely with small subfoliaceous lobes. Calyx-tube 1.75–2.5(–3.6) mm. long; limb reduced to a rim. Corolla sometimes slightly fleshy; tube 4–13 mm. long, up to 4 mm. wide at throat; lobes oblong, 8–16 mm. 3–6 mm. wide, rounded or obtuse. Fruit red or streaked, oblong-ellipsoid or rather broadly oblong, sometimes narrowing towards base, 1.2–2.2 cm. long, 0.9–1.6 cm. wide; disc accrescent and prominent or scar-like in fruit. Seeds fawnish green, 0.7–1.5 cm. long, 0.6–1 cm. wide.

SYN. [*C. arabica* sensu Benth. in Hook., Niger Fl.: 413 (1849) pro parte, *non* L.]
    C. dewevrei De Wild. & Th. Dur., Mat. Fl. Congo, fasc. 6: 32 (1899), in B.S.B.B. 28: 202 (1899), in Actes Congr. Intern. Bot., Paris: 233 (1900), Reliq. Dewèvr.: 128 (1901), Les Caféiers: 38 (1901), Miss. Laurent: 325, t. 75 (1906) & in Ann. Jard. Bot. Buitenz., Suppl. 3(1): 370 (1910); A. Chev., Caféiers du Globe 1: 79 (1929), 2, t. 2 (1942) & 3: 179 (1947). Type: Zaire (province doubtful *fide* Lebrun 1941), *Dewèvre* 1149 (BR, holo.)
    C. arnoldiana De Wild. in Actes Congr. Inter. Bot., Paris: 236 (1900), Les Caféiers: 36 (1901) & Miss. Laurent: 325, t. 74 (1906); Z.A.E. 2, 1907–8: 329 (1910); De Wild. in Ann. Jard. Buitenz., suppl. 3(1): 366 (1910). Type: Zaire, Bas-Congo, cultivated at Eala, *Dewevre* 377 (BR, holo.)
    C. klainii De Wild., Les Caféiers: 13 (1901), Miss. Laurent: 300, t. 102 (1906) & in Ann. Jard. Buitenz., suppl. 3(1): 372 (1910); A. Chev., Caféiers du Globe 1: 78 (1929), 2, t. 11 (1942) & 3: 174 (1947). Type: Gabon, Libreville, *Klaine* 1838 (P, holo., K, iso.!)
    C. dybowskii De Wild., Les Caféiers: 14 (1901), Miss. Laurent: 325, t. 105 (1906) & in Ann. Jard. Buitenz., suppl. 3(1): 370 (1910); A. Chev., Caféiers du Globe 3: 183 (1947) as race of *C. dewevrei*. Type: Central African Republic, Kemo R., *Dybowski* 672 (P, holo.)
    C. excelsa A. Chev. in Rev. Cult. Colon. 12: 258 & 259 (1903) & in Comptes Rendus Acad. Sci.,

Paris 140: 517 (1903); De Wild. in Ann. Jard. Buitenz., suppl. 3(1): 371 (1910); Cheney, Coffee, Monogr. Econ. Sp.: 93 t. 37 (1925); A. Chev., Caféiers du Globe 1: 80 (1929) & 3: 181 (1947) as race of *C. dewevrei*; T.T.C.L.: 492 (1949); U.O.P.Z.: 204 (1949); I.T.U., ed. 2: 341 (1952); F.P.S. 2: 432 (1952); Wellman, Coffee: 77 (1961); Haarer, Modern Coffee Prod.: 23 (1962). Types: Central African Republic, Chari, Ndelle region, between 8°–8°30'N & Ubangi, along rivers Bata and Kotto (Boro de la Plata, de la Gounda, du Mamfa), *Chevalier* 7000 (P, syn., BM, isosyn.!), 7238, 7355 & 7686 (P, syn.)

C. sylvatica A. Chev., in Rev. Cult. Colon. 12: 258 & 259 (1903); De Wild. in Ann. Jard. Buitenz., suppl. 3(1): 379 (1910), as *C. silvatica*; A. Chev. Caféiers du Globe 1: 80, fig.60 (1929) & 3: 183 (1947), as race of *C. dewevrei*. Type: Central African Republic, Bangui, near the Elephant Rapids (Bonjos), *Chevalier* 5200 (P, syn., K, isosyn.!), 5251 & 5254 (P, syn.)

C. aruwimiensis De Wild., Miss. Laurent: 321, t. 64–66 (1906); Z.A.E. 2, 1907–8: 329 (1910); De Wild. in Ann. Jard. Buitenz., suppl. 3(1): 366 (1910); A. Chev., Caféiers du Globe 3: 182 (1947), as a race of *C. dewevrei*. Type: Zaire, Wanie Rukula, Monga (Ubangi) & cultivated at Basoko, Lirange (Ubangi), *E. & M.Laurent* (BR, syn.)

C. royauxii De Wild., Miss. Laurent: 326, t. 78 (1906) & in Ann. Jard. Buitenz., suppl. 3(1): 378 (1910). Type: Zaire, Banzyville (Ubangi), *Royaux* (BR, holo.)

C. abeocuta De Wild. in Ann. Jard. Buitenz., suppl. 3(1) 359 (1910), *nomen*

C. zenkeri De Wild. in Ann. Jard. Buitenz., suppl. 3(1): 382 (1910), *nomen*; A. Chev., Caféiers du Globe 3: 182 (1947), as race of *C. dewevrei*, invalid. Based on Cameroun, Bipinde, *Zenker* 3335 (BM, K) & Mamfia, *Zenker* 245

C. abeokutae Cramer in Meded. Dept. Landb. Ned. Ind. 11 : 425, fig. opp. p. 396 (1913); A. Chev., Expl. Bot. Afr. Occ. Fr.: 334 (1920), Caféiers du Globe 1: 82 (1929); Hutch. & Dalz., F.W.T.A. 2: 96 (1931); A. Chev., Caféiers du Globe 3: 175 (1947). Type: cultivated at Bogor [Buitenzorg], Indonesia, presumably from material sent from W. Nigeria, communicated by *Cramer* (BO, holo., K, iso.!)

[*C. macrochlamys* sensu A. Chev. in Rev. Bot. Appliq. 6: 669 (1926), *non* K. Schum.]

C. excelsoidea Portères, in Ann. Agric. Afr. Occ. 1: 220 (1937), *nom invalid;* A. Chev., Caféiers du Globe 3: 185 (1947), as a race of *C. dewevrei*. Based on material from Guinea and Ivory Coast.

C. neoarnoldiana A. Chev., in Comptes. Rendus Acad. Sci., Paris 207: 654 (1938), *nomen* & Caféiers du Globe 3: 185 (1947), as race of *C. dewevrei*, invalid. Based on Zaire, cultivated at Eala from seeds from Aruwimi, *Laurent* sec. *De Wildeman* 1904 (BR, holo.)

C. oyemensis A. Chev., in Rev. Bot. Appliq. 19: 403 (1939), Caféiers du Globe 3: 174 (1947). Type: Gabon, Oyem, Woleu-Ntam, *Le Testu* (P, holo., BM, iso.!)

C. liberica Hiern var. *dewevrei* (De Wild. & Th. Dur.) Lebrun, in Mém. Inst. Roy. Col. Belge, Sect. Sci. Nat. Méd. 8°, 11(3): 168 (1941)

C. ituriensis A. Chev., Caféiers du Globe 3: 184 (1947), as a race of *C. dewevrei*. Based on Zaire, between Ruwenzori Mts. & SW. of Pays des Monbouttous and N. of Nepoko-Ecyulu, *Leplae* 5 (P) & Dunju cult.,? s.n. (P!)

NOTE. In addition to the above list of synonyms the following varieties are recognised by Chevalier, Caféiers du Globe 3 (1947):— *C. liberica* var. *pyriformis* Fauchère (1908); var. *liborensis* Sibert (1932); var. *liberiensis* Sibert (1932); var. *aurantiaca* A. Chev. (1947); var. *gossweileri* A. Chev. (1947); var. *grandifolia* A. Chev. (1947); *C. abeokutae* var. *longicarpa* Portères (1937); var. *sphaerocarpa* Portères (1937); var. *camerunensis* A. Chev. (1947); var. *indeniocarpa* (Sibert) A. Chev. (1947); var. *macrocarpa* A. Chev. (1947) & var. *microcarpa* A. Chev. (1947). Also the following combinations are used on the plates by Chevalier in Caféiers du Globe 2 (1942), although not adopted in the text in volume 3: t. 3 *C. dewevrei* var. *dybowskii;* t. 4 *C. dewevrei* var. *sylvatica;* t. 5 & 6 *C. dewevrei* var. *excelsa;* t. 7 *C. dewevrei* var. *ituriensis;* t. 8 & 9 *C. dewevrei* var. *neoarnoldiana;* t. 10 *C. dewevrei* var. *aruwimiensis* & t. 13 *C. dewevrei* var. *zenkeri.*

C. liberica sensu lato represents a complex group of taxa (specific or intraspecific) and although certain groups can easily be recognised the disjunction between them is very difficult to ascertain solely from the morphological evidence present on herbarium specimens. Since Liberian and Excelsa coffees are widely cultivated, much of the original distribution patterns has been obscured and it has become difficult to relate possible taxonomic characters to geographical areas in many instances. Also the further complications of cultivated forms hybridising with indigenous forms cannot be dismissed. A detailed multidisciplinary study of this group is required.

With regard to the species cultivated in East Africa, I think it is helpful (although not entirely satisfactory) to adopt Lebrun's simplification of recognizing two varieties. These he separated as follows (excluding his characters concerning the disc which I have found confusing):—

Corolla 6–9-merous; tube distinctly widened at throat; lobes wide, usually 5–10 mm. wide; fruit generally 20–25 mm. long, 17–21 mm. wide . . . . var. **liberica**

Corolla 5–6(–8)-merous; tube somewhat widened at throat; lobes narrower, usually 2.5–7 mm. wide; fruit smaller 12–20 mm. long, 8–16 mm. wide . . . . var. **dewevrei**

In addition to the difference in the size of the fruit, there is also a difference in shape, texture and colour. In general the fruit of var. *liberica* has a thicker, more leathery pericarp and is often more tapered towards the base than var. *dewevrei;* the disc in var. *liberica* is prominent and ± obconic, while

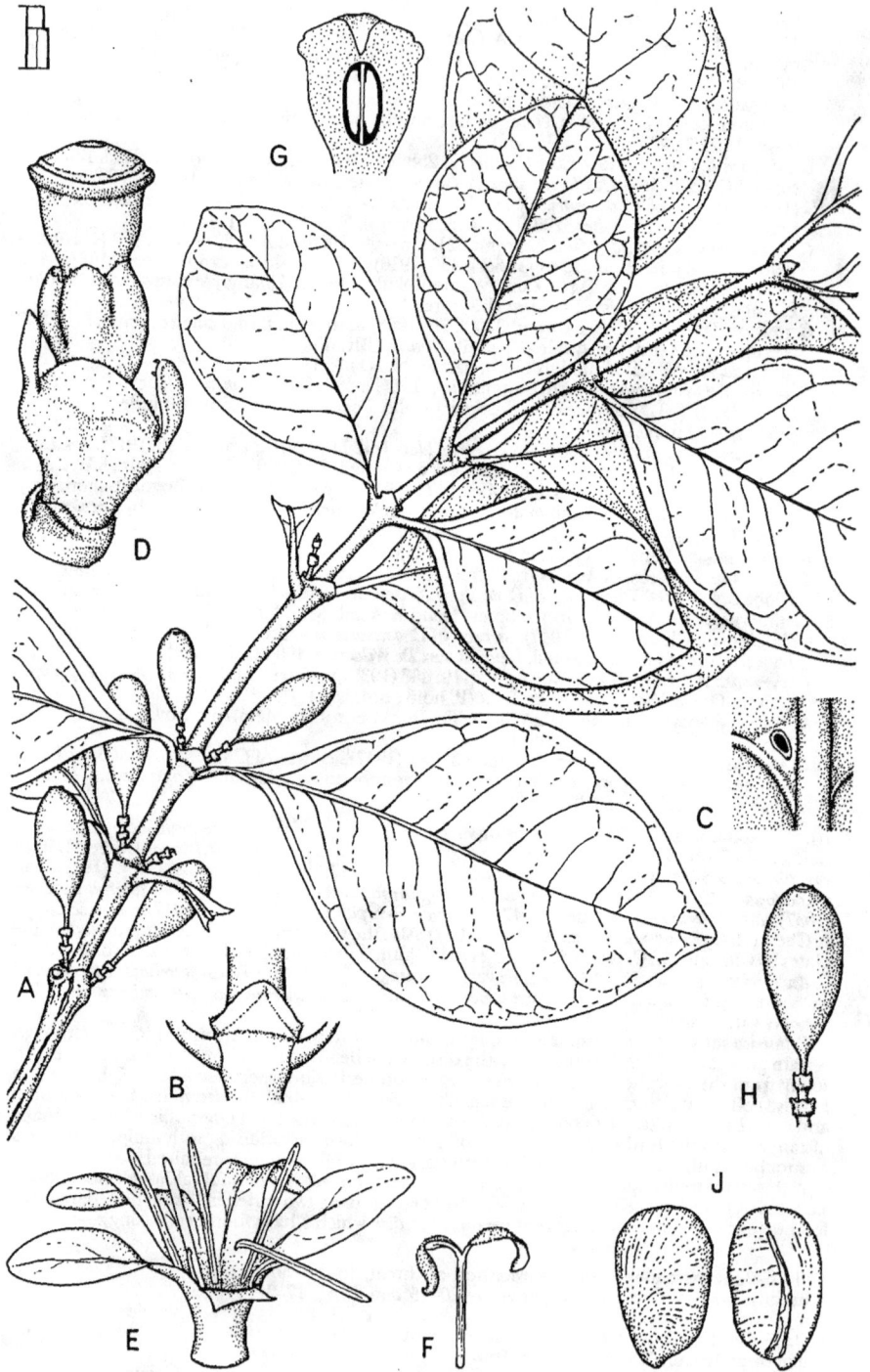

FIG. 122. *COFFEA FADENII* — **A**, fruiting branch, × ⅔; **B**, stipule, × 3; **C**, domatium, × 6; **D**, single-flowered inflorescence, × 8; **E**, corolla with one lobe cut, × 3; **F**, style and stigma, × 3; **G**, longitudinal section of ovary, × 8; **H**, fruit, × 1; **J**, seed, 2 views, × 2. A–C, H, from *Faden 72/269*; D–G, J, from *Faden 71/56*. Drawn by Diane Bridson.

in var. *dewevrei* it is either prominent and cylindrical or flat and 'scar-like'; it has been reported that the fruits in var. *dewevrei* are red and those of var. *liberica* vary from yellow to dark reddish brown often with spots or flecks of red but there may be intermediates. *C. excelsa* falls within the range of variation of var. *dewevrei*. It is not cited in synonymy by Lebrun as the type specimen comes from outside the Zaire area but Chevalier included it as a race of *C. dewevrei* in Caféiers du Globe 3: 181 (1947).

Chevalier (Rev. Bot. Appliq. 6: 671 (1926)) states that all *C. excelsa* introduced into agriculture comes from one locality in the valley of the Boro (Central African Republic). Assuming this to be true, it is perhaps possible to recognise *C. excelsa* among the other forms of var. *dewevrei*. I have found this difficult, partly because of the lack of authenticated herbarium specimens of *C. excelsa*.

Coffee growers are said to prefer Excelsa coffee as the fruit is easier to pulp and the flavour is less bitter than Liberian coffee. Useful discussion on the various forms and cultivars of *Coffea liberica* sensu lato is given in Portères in Ann. Agric. Afr. Occ. Fr. 1, 2: 219 (1937); Cramer, Rev. Lit. Coff. Res. Indonesia: 105 (1957); Wellman, Coffee: 76 (1961) & Haarer, Modern Coffee Prod.: 22 (1962).

Herbarium specimens of var. *liberica* have been recorded from Tanzania T 3, Zanzibar and Pemba, and var. *dewevrei* (De Wild. & Th. Dur.) Lebrun from Uganda (U 4) Kenya (K 5) and Tanzania (T 3), but both varieties are doubtless more widely grown.

var. **dewevrei** (*De Wild. & Th. Dur.*) *Lebrun* forma **bwambensis** *Bridson* in K.B. 37: 314 (1982). Type: Uganda, Toro District, Bwamba, Ntandi [Muntandi], *Eggeling* 3388 (K, holo.!, EA, iso.!)

Leaves elliptic to broadly elliptic or sometimes narrowly obovate, mostly towards upper limits given in general description, dull above; lateral nerves 10–12 main pairs. Flowers 5-merous. Fruit scarlet, 1.2–1.4 cm. long, 0.9–1.3 cm. wide, disc flat and scar-like.

UGANDA. Nile District: E. Madi, Zoka Forest, June 1933, *Eggeling* 1250!; Toro District: Bwamba Forest, 30 Sept. 1932, *A.S. Thomas* 752! & 5 Mar. 1939, *A.S. Thomas* 2808!
DISTR. U 1, 2; known only from Uganda (see Note)
HAB. Forest; 790–1220 m.
SYN. [*C. excelsa* sensu Thomas in Emp. Journ. Exp. Agric. 12: 8 (1944), *non* A. Chev.]
NOTE. Although A.S. Thomas (cit. supr.) distinguished between populations from Bwamba and Zoka forests, I have been unable to on the basis of herbarium specimens. For the most part *C. liberica* sens. lat. occurs at rather low altitudes, but some specimens (including *C. dewevrei* De Wild. & Dur. race *C. ituriensis* A. Chev.) have been recorded from the Ruwenzori Mts in Zaire from 900–1300 m. Forma *bwambensis* does seem to be distinct from these specimens, but it is harder to separate from certain specimens from the Maridi (Meridi) area of southern Sudan. The Sudanese specimens show more variation as a group and although some closely resemble forma *bwambensis*, I hesitate to include them here at the present time.

2. C. fadenii *Bridson* in K.B. 36: 827, fig. 1 (1982). Type: Kenya, Teita [Taita] Hills, *Faden et al.* 71/56 (K, holo.!, EA, iso.!)

Small tree 5–15 m. tall; young branches rather thick, glabrous, covered with brownish bark. Leaf-blades broadly elliptic to round, 6–10.8 cm. long, 4–7 cm. wide, rounded to obtuse at apex, acute to obtuse at base, moderately shiny, with lateral and tertiary nerves raised on both faces; domatia glabrous to sparsely hairy; petiole 0.6–1 cm. long; stipules triangular-ovate, 3–4 mm. long, obtuse or sometimes acute. Flowers 5-merous, 1–2(–3) per axil or rarely cauliflorous, borne singly; inflorescence-stalks up to 3 mm. long, obscured by bracteoles; bracteoles 2, the lower one usually larger than the upper one and somewhat lobed. Calyx-limb reduced to a rim. Corolla-tube 3 mm. long, 3 mm. wide at throat; lobes 9 mm. long, 3 mm. wide, rounded. Fruit 1.6–2.2 cm. long, 0.8–1 cm. wide, slightly narrowed toward base; disc not markedly prominent; inflorescence-stalks lengthening to 6–9 mm. Seeds light brown, 1 cm. long, 0.7 cm. wide. Fig. 122.

KENYA. Teita District: Teita Hills, Mbololo Hill, Mraru Ridge, 12 Sept. 1970, *Faden et al.* 70/559! & 2 Jan. 1971, *Faden et al.* 71/56! & 28 May 1972, *Faden* 72/269!
DISTR. K 7; endemic to Teita Hills
HAB. Mist or montane evergreen forest; 1440–1750 m.
SYN. *Coffea sp.* sensu Berthaud, Guillaumet, Le Pierres & Lourd in Café Cacao Thé 24: 104 (1980)

3. C. mongensis *Bridson* in K.B. 36: 829, fig. 2 (1982). Type: Tanzania, Lushoto District, Monga, *Peter* 18367 (K, holo.!, B, iso.!)

Glabrous shrub 3 m. tall; young branches covered with maroon bark which flakes and becomes paler with age. Leaf-blades obovate or sometimes broadly elliptic, 4.6–12.5 cm. long, 1.8–6.6 cm. wide, obtuse or occasionally rounded at apex, acute to cuneate or rarely obtuse at base, coriaceous, slightly shiny above; lateral and tertiary nerves raised on both faces; domatia glabrous or rarely ciliate; petiole 0.4–1 cm. long; stipules triangular-ovate, 2–3 mm. long, obtuse. Flowers 5-merous, 1–3 per axil, borne singly or in fascicles; inflorescence-branches up to 5 mm. long, not totally obscured by bracteoles; peduncle up to 1 mm. long; upper bracteoles up to 1.5 mm. long, lower bracteoles slightly shorter, not lobed, puberulous. Calyx-limb reduced to a rim. Corolla-tube 3.25 mm. long, 2.5 mm. wide at throat; lobes narrowly oblong, 7 mm. long, 1.5–2 mm. wide, obtuse. Fruit 1.3–2 cm. long, 0.7–1 cm. wide, slightly narrowed towards base; disc surrounded by corky calyx-rim; inflorescence-stalks lengthening to 5–7 mm. Seeds brownish-olive, 1.2–1.3 cm. long, 0.6 cm. wide; testa loosely attached, with very fine striations.

TANZANIA. Lushoto District: Shume Forest Reserve, Aug. 1955, *Semsei* 2179! & Monga, 24 Nov. 1916, *Peter* 18367!; Ulanga District, Uzungwa Mts., Mwanihana Forest Reserve, above Sanje village, 7–8 June 1980, *Rodgers & Vollesen* 1051!
DISTR. T 3, 6; not known elsewhere
HAB. Forest; 1150–1400 m.

SYN. [*C. zanquebariae* sensu A. Chev., Caféiers du Globe 2, t.68 (1942) & 3: 218 (1947), pro parte, quoad *Zimmermann* in *Herb. Amani* 2511, *non* Lour.]

4. C. sp. A

Shrub 3 m. tall; young stems glabrous, covered with brown bark. Leaf-blades broadly elliptic, 7.5–13.5 cm. long, 3.5–6.5 cm. wide, subacuminate at apex, acute at base, moderately shiny on both faces, with lateral and tertiary nerves apparent on both faces; domatia sparsely hairy; petiole 0.6–1 cm. long; stipules triangular, obtuse. Flowers 1–2 per axil, not on a common peduncle; bracteoles 2, not lobed. Corolla and calyx not known. Fruit orange, up to 1.5 cm. long, 1 cm. wide, not prominent at apex, scarcely tapered towards base; inflorescence-stalks lengthening to 1.3–1.5 cm., the portion above the upper bracteoles being the longest.

TANZANIA. Morogoro District: Kimboza Forest Reserve, 31 Mar. 1983, *Mwasumbi, Rodgers & Hall* 12393! & 9–17 July 1983, *Rodgers, Hall & Mwasumbi* 2511!
DISTR. T 6; only known from above specimens
HAB. Lowland evergreen forest on limestone; 300–350 m.

5. C. canephora *Froehner* in N.B.G.B. 1: 237 (1897) & in E.J. 25: 269 (1898); De Wild. in Actes Congr. Intern. Bot., Paris: 233 (1900), Les Caféiers: 20 & 25 (1901), Miss. Laurent: 330 (1906) & Ann. Jard. Buitenz., suppl. 3, 1: 368 (1910); Cheney, Coffee, Monogr. Econ. Sp.: 72, t.26 & 27 (1925); A. Chev., Caféiers du Globe 1: 82, fig. 14 & 15 (1929); Portères, Ann. Agric. Afr. Trop. Occ. Fr. 1, 2: 235 t. 29 & 30 (1937); Lebrun in Mém Inst. Col. Belge, Sect. Sci. Nat. Méd. 8°, 11(3): 122, t. 11–14 (1941); Thomas in Emp. Journ. Exp. Agric. 12: 3 (1944); A. Chev., Caféiers du Globe 2, t. 28–39 (1942) & 3: 186 (1947); F.P.N.A.: 354 (1947); F.P.S. 2: 432 (1952); Cramer, Rev. Lit. Coff. Res. Indonesia: 120 (1957); Wellman, Coffee: 80 (1961); Haarer, Modern Coffee Prod.: 17 (1962); F.W.T.A., ed. 2, 2: 154 (1963). Type: Gabon, cultivated in Libreville by Klaine from seeds sent from near Fernand-Vaz by A. Le Roy, *Klaine* in *Pierre* 247 (P, holo)

Tree or bush 3.5–8.5 m. tall; young branches glabrous, covered with light brown shiny bark. Leaf-blades oblong-elliptic, broadly elliptic or occasionally obovate to lanceolate, 12–31.5(–40) cm. long, 4.5–12(–22) cm. wide, distinctly acuminate at apex, obtuse to cuneate or sometimes rounded at base, papery to thinly coriaceous, shiny above, margin sometimes undulate in some cultivated forms; lateral nerves (8–)11–15(–17) pairs, main ones sometimes puberulent beneath; domatia rather inconspicuous, pubescent or absent; petiole 0.5–1.8(–2.1) cm. long; stipules deltate, 0.6–1.8 cm. long, obtuse or acute but always exceeded by a mucro. Flowers 5–6(–7)-merous, 8–30(–48) per axil, borne in 1–4(–7) congested fascicles; peduncle up to 3 (–7 in fruit) mm. long; inflorescence-branches and pedicels compressed in flowering stage so that calyces do not or scarcely exceed the surrounding bracteole; bracteoles stipule-like, with well-developed subfoliaceous lobes, up to 2 cm. long. Calyx-tube 1–2 mm. long; limb reduced to a rim. Corolla-tube 0.5–1.6 cm. long, 2–3 mm. wide at throat; lobes oblong, 0.8–1.9 cm. long, 1.8–5.5 mm. wide, obtuse to rounded. Fruit red, sometimes faintly striped, broadly oblong-ellipsoid, 0.9–1.7 cm. long, 0.6–1.2 cm. wide; disc small, cylindrical, somewhat prominent. Seeds fawn, 7–13 mm. long, 5.5–9.5 mm. wide.

UGANDA. W. Nile District: E. Madi, Zoka Forest, Jan. 1952, *Leggat* 66!; Toro District: Fort Portal, Itwara Forest, 29 Jan. 1945, *Greenway & Eggeling* 7052!; Mbale District: Kami [Khami], at boundary of W. Bugwe Forest Reserve, 20 Apr. 1951, *G.H.S. Wood* 176!

TANZANIA. Bukoba District: Bwanyai [Bwangai], Mar. 1939, *Doughty* 20!
DISTR. U 1–3; T 1; West Africa, Cameroun, Gabon, Zaire, Sudan and Angola, also widely cultivated
HAB. Forest; 700–1400 m.

SYN. [*C. arabica* sensu Hiern in Trans. Linn. Soc., Bot., sér 2, 1: 171 (1876), in F.T.A. 3: 181 (1877) &
Cat. Welw. Afr. Pl. 1: 488 (1898), pro parte, quoad *Welwitsch* 3183, *non* L.]
[*C. liberica* sensu Hiern in Trans. Linn. Soc., Bot., sér 2, 1: 171 (1876), in F.T.A. 3: 181 (1877) &
Cat. Welw. Afr. Pl. 1: 488 (1898), pro parte, quoad *Welwitsch* 3181 & 3182, non Hiern sensu
stricto]
C. arabica L. var. *stuhlmannii* Froehner in E.J. 25: 263 (1898); De Wild. in Actes Congr. Intern.
Bot., Paris: 232 (1900), Les Caféiers: 35 (1901) & in Ann. Jard. Buitenz., suppl. 3, 1: 364 (1910).
Types: Tanzania, Bukoba District, semi-cultivated in a banana plantation, *Stuhlmann* 3774 &
partially wild, *Stuhlmann* 1450 (B, syn.!, K, drawings!)
C. laurentii De Wild. in Actes Congr. Intern. Bot., Paris: 234 (1900), Les Caféiers: 28 (1901), Miss.
Laurent: 328 (1906) & in Ann. Jard. Buitenz., suppl. 3, 1: 373 (1910); Lebrun in Mém. Inst. Roy.
Col. Belge, Sect. Sci. Nat. Méd. 8°, 11(3): 136 (1941). Type: Zaire, cultivated at Lusambo,
Laurent (BR, holo.)
C. robusta Linden, Cat. Pl. Nouv. Hort. Colon.: 11 & 64 (1900); & in La Semaine Horticole 4: 472
(1900); De Wild., Miss Laurent: 328 (1906) & in Ann. Jard. Bot. Buitenz., suppl. 3, 1: 378 (1910);
Cheney, Coffee, Monogr. Econ. Sp.: 90, t. 36 (1925); F.W.T.A. 2: 96 (1931); U.O.P.Z.: 204
(1949); I.T.U, ed. 2: 340 (1952); Cramer, Rev. Lit. Coff. Res. Indonesia: 12 & 113 (1957). Type:
plants raised from seed collected in ?Zaire by *Luja*, and distributed by Linden (type not found,
see note)
C. welwitschii De Wild., Les Caféiers: 19 (1901) & in Ann. Jard. Buitenz., suppl. 3, 1: 381 (1910).
Type: Angola, *Welwitsch* 3183 (P, holo., BM!, COI, K!, iso.)
C. canephora Froehner var. *hiernii* De Wild., Les Caféiers: 20 (1901) & in Ann. Jard. Buitenz.,
suppl. 3, 1: 268 (1901). Type: Angola, *Welwitsch* 3182 (P, holo., BM, K, iso.!)
C. canephora Froehner var. *hinaultii* De Wild., Les Caféiers: 21 (1901), Miss. Laurent: 339, t. 100
(1906) & in Ann. Jard. Buitenz., suppl. 3, 1: 368 (1901); A. Chev., Caféiers du Globe 2, t. 30–32
(1942) & 3: 192 (1947). Type: Gabon, Nyo R., *Klaine* 1634 (P, holo.)
C. canephora Froehner var. *kouilouensis* De Wild., Les Caféiers: 21 (1901), Miss. Laurent: 334, t.
101 (1906) & in Ann. Jard. Buitenz. suppl. 3, 1: 368 (1910); Cheney, Coffee, Monogr. Econ Sp.:
75, t. 28 (1925); Lebrun in Mém. Inst. Roy. Col. Belge, Sec. Sci. Nat. Méd. 8, 11(3): 138 (1941).
Types: cultivated in Gabon on the Kouilou R. near Loango, *Klaine* 1928a (P, syn.) & 1928b (P,
syn., K, isosyn.!)
C. canephora Froehner var. *muniensis* De Wild., Les Caféiers: 23 (1901) & in Ann. Jard. Buitenz.,
suppl. 3, 1: 368 (1910). Type: cultivated in Gabon, in the Muni region on the Noya R., *Chalot* 61
(P, holo., K, iso.!)
C. canephora Froehner var. *oligoneura* De Wild., Les Caféiers: 23 (1901) & in Ann. Jard. Buitenz.,
suppl. 3, 1: 368 (1910). Type: cultivated in Gabon, Kouilou R., *Chalot* 3–54 (P, holo.)
C. canephora Froehner var. *trillesii* De Wild., Les Caféiers: 24 (1901) & in Ann. Jard. Buitenz.,
suppl. 3, 1: 369 (1910). Type: Gabon, Como R., *Trilles* 77 (P, holo.)
C. canephora Froehner var. *wildemanii* De Wild., Les Caféiers: 25 (1901) & in Ann. Jard. Buitenz.,
suppl. 3, 1: 369 (1910); Lebrun in Mém. Inst. Roy. Col. Belge, Sect. Sci. Nat. Méd. 8°, 11(3): 138
(1941). Type: Zaire, *Dewèvre* 9876 (BR, holo.)
C. canephora Froehner var. *opaca* De Wild. in Agric. Prat. Pays Chauds (Bull. Jard. Colon. Nogent
sur Marne) 4: 117, fig. (1904) & in Ann. Jard. Buitenz., suppl. 3, 1: 369 (1910). Type: cultivated
in Paris from material from Gabon, Kouillou region, supplied by *Chalot* (P, holo.)
C. maclaudii A. Chev. in Comptes Rendus Acad. Sci., Paris 140: 1474 (1905) & Expl. Bot. Afr. Occ.
Fr.: 336 (1920); F.W.T.A. 2: 96 (1931). Types: Guinée, Haut Konkouré, Labe, *Chevalier* 12330 &
Mont Bilima, *Chevalier* 12332 bis (P, syn., K, isosyn!)
C. canephora Froehner forma *sankuruensis* De Wild., Miss. Laurent: 330, fig. 52–53, t. 77 (1906);
Lebrun in Mém. Inst. Roy. Col. Belge, Sect. Sci. Nat. Méd. Mem. 8°, 11(3): 139 (1941). Types:
Zaire cultivated at Lusambo and wild from Ibaka Forest, *Laurent* (BR, syn.)
C. canephora Froehner var. *crassifolia* De Wild., Miss. Laurent: 333, t. 76 (1906) & in Ann. Jard.
Buitenz., suppl. 3, 1: 368 (1910); Lebrun in Mém. Inst. Roy. Col. Belge, Sect. Sci. Nat. Méd.
Mém. 8°, 11(3): 139 (1941); A. Chev., Caféiers du Globe 3: 195 (1947). Type: Zaire, Lusambo,
Laurent (BR, holo.)
C. bukobensis Zimmermann in Der Pflanzer 4: 326 (1908); T.T.C.L.: 492 (1949). Type as for *C.
arabica* var. *stuhlmannii*
C. canephora Froehner var. *sankuruensis* (De Wild.) De Wild. in Ann. Jard. Buitenz., suppl. 3, 1:
369 (1910)
C. ugandae Cramer in Meded. Dept. Landb. Ned. Ind. 11: 680 (1913). Type: cultivated in Java at
Bogor [Buitenzorg]
C. quillou Wester in Philipp. Agric. Rev. 9: 121 & 219, t. 2 (1916), *nomen* (see note)
C. canephora Froehner var. *laurentii* (De Wild.) A. Chev., Caféiers du Globe 2, t. 29 (1942) & 3: 190
(1947)
C. canephora Froehner var. *oka* A. Chev., Caféiers du Globe 2, t. 33 (1942) & 3: 193 (1947). Types:
Cameroun, Doumé, *Hedin* 268 & Makak Forest Reserve, *Jacques-Félix* 2300 & Guinée,
Bambaradou, *Barthe* 1 (P, syn.)
C. canephora Froehner var. *maclaudii* (A. Chev.) A. Chev., Caféiers du Globe 2, t. 34 (1942) & 3:
194 (1947)

C. canephora Froehner var. *stuhlmannii* (Froehner) A. Chev., Caféiers du Globe 2, t. 35 (1942) & 3: 193 (1947)

C. canephora Froehner var. *ugandae* (Cramer) A. Chev., Caféiers du Globe 2, t. 36 (1942) & 3: 194 (1947)

C. canephora Froehner var. *welwitschii* (De Wild.) A. Chev., Caféiers du Globe 2, t. 37 (1942) & 3: 195 (1947)

C. canephora Froehner var. *gossweileri* A. Chev., Caféiers du Globe 2, pl. 38 (1942) & 3: 191 (1947). Types: Angola, Capir, Cuanza Sul, Amboim near Carloanga-Cuva, *Gossweiler* 9957 & 9957b (COI, syn., BM, K, isosyn.!)

C. canephora Froehner subvar. *robusta* (Linden) A. Chev., Caféiers du Globe 3: 191 (1947)

C. canephora Froehner var. *nganda* Haarer, Modern Coffee Prod.: 19 & 21 (1962), *nom. superfl.*, based on var. *kouilouensis* (given as *C. kouilouensis*)

NOTE. This species is self-sterile and is very variable. None of the varieties cited above can be distinguished readily on the characters present on herbarium specimens. In the Flora area, in localities where this species is thought to be indigenous it is not always possible to decide if a plant is naturalised or truly wild (see Jervis in Tanganyika Notes & Records 8: 47 (1939) and A.S. Thomas in Emp. Journ. Exp. Agric. 12: 4 (1944)).

'Robusta Coffee' is the form most commonly referred to by growers who use this name in preference to the botanically correct *C. canephora*. *C. robusta* is scarcely distinguishable from *C. laurentii* and would seem no more than a broad-leaved form of *C. canephora*. Robbrecht (in B.J.B.B. 46: 402 (1976)) states that there were apparently no herbarium specimens deposited by Linden, however, Chevalier claims that the illustration of *C. canephora* var. *laurentii* (*cit. supr.* 2: 30, t. 29 (1942)) was based on a specimen raised from Linden's seeds ("cotype de "*C. robusta*" Linden, plante provenant de grains vendue par la firme Linden"). A seedling preserved at K bears the following information "seedling of *Coffea robusta* Linden, received from Linden and cultivated at Kew". Additional fertile sheets at Kew, cultivated at the St. Clair Experimental Station Trinidad (no. 6809) from material sent from Kew, probably from the original Linden material, have been annotated "lectotype of *C. robusta*", they are accompanied by notes and sketches, but in an unknown hand and are undated. Further research is required before neotypification is attempted; since the seeds were widely distributed several herbarium specimens could have been preserved.

Both Thomas (*l.c.*: 3 (1944)) and Haarer (*l.c.*: 19 & 21 (1962)) distinguish, in Uganda, a form which grows as a spreading or dome-shaped bush (rather than an upright tree) which they refer to as 'nganda'; Purseglove, Tropical Crops, Dicotyledons: 483 (1974) also maintains these two forms. Thomas says that such spreading forms are called '*C. quillou*' in Indonesia and Haarer cites *C. kouilouensis* Pierre ex De Wild. as a synonym of his var. *nganda*. This is very confusing since Chevalier (Caféiers du Globe 3: 189 (1947)) sinks var. *kouilouensis* into his var. *typica*. Many growers recognise one or more forms that have variously been referred to as "Kouillouensis, Kouilou, Quillou, Quillouensis or even Conillon Coffee". The name var. *kwilensis* has also been used, but this probably should be restricted to plants of *C. canephora* introduced into Indonesia from Kwilu near Limbé (Victoria) in Cameroun. Cramer is of the opinion that the cultivar Petit Indenie may also belong to the var. *kouilouensis* group. For more detailed discussion on this group of species see Cramer, Rev. Lit. Coff. Res. Indonesia: 119 (1957). From the taxonomic point of view the names var. *nganda* and var. *kouilouensis* are both best avoided for the cultivated forms of spreading habit. The protologue of var. *kouilouensis* gives insufficient detail of habit to tell if it could correspond to the growers' Quillou or not. I find no reason to doubt Chevalier's inclusion of var. *kouilouensis* in his var. *typica*, the types of both coming from approximately the same locality. Thomas states that the upright tree is referred to as *C. ugandae* by the Indonesians, while Cramer *l.c.*: 121, says that *C. ugandae* is probably identical with *C. arabica* var. *stuhlmannii* (*C. bukobensis*). However, some specimens from Amani cultivated as *C. bukobensis* are annotated as "shrubs to 10 ft. high with flat bushy crowns".

It is probably best to avoid the use of botanical varieties altogether as many clones are available to growers and usually referred to alphanumerically. Moens, in Bull. Agric. Congo 52 (6): 1171–1216 (1961) (Différenciation Morphologique des clones d'arbres-mères du Caféiers Robusta) gives detailed descriptions of some such clones.

Some forms of *C. canephora* approach *C. arabica* and may be indistinguishable from them on morphological grounds in some instances, see note after *C. arabica*, p. 713.

6. C. arabica *L.*, Sp. Pl.: 172 (1753); Sims in Bot. Mag. 32, t. 1303 (1810); DC., Prodr. 4: 499 (1830); A. Rich., Tent. Fl. Abyss. 1: 349 (1848); Hiern in Trans. Linn. Soc., Bot., ser. 2, 1: 170 (1876), pro parte, & in F.T.A. 3: 180 (1877), pro parte; P.O.A. B: 246 & C: 387 (1895); Froehner in N.B.G.B. 1: 233 (1897) & in E.J. 25: 261 (1898); De Wild. in Actes Congr. Intern. Bot., Paris: 231 (1900), Les Caféiers: 35 (1901), Miss. Laurent: 344, t. 67–70 (1906) & in Ann. Jard. Buitenz., suppl. 3, 1: 360 (1910); Cheney, Coffee, Monogr. Econ. Sp.: 48, t. 13–18 (1925); A. Chev., Caféiers du Globe 1: 71 (1929); Lebrun in Mém. Inst. Roy. Col. Belge, Sect. Sci. Nat. Méd. Mém. 8°, 11(3): 114 (1941); A.S. Thomas in Emp. Journ. Exp. Agric. 10: 207–212 (1942); A. Chev., Caféiers du Globe 2, pl. 17–24 (1942) & 3: 196 (1947); 442 (1949); F.P.S.: 2: 432 (1952); Cramer, Rev. Lit. Coff. Res. Indonesia: 89 (1957); K.T.S.: 436 (1961); Wellman, Coffee: 28 (1961); F.F.N.R.: 405 (1962); Haarer, Modern

Coffee. Prod.: 13 (1962); F.W.T.A., ed. 2, 2: 156 (1963); Fl. Rwanda 3: 154, fig. 47.1 (1985).
Type: cultivated in Holland, *Hort. Cliff.* (BM, holo.!)

Bush or tree 2–7 m. tall; young branches glabrous, covered with light brown shiny bark.
Leaf-blades elliptic to broadly elliptic or oblong-elliptic (or lanceolate or round in some
cultivated forms), 7–18 cm. long, 3–7.5 cm. wide, distinctly acuminate at apex, acute to
obtuse at base, thinly coriaceous to coriaceous, with 7–10 pairs of main lateral nerves,
shiny above, margin sometimes undulate; domatia rather inconspicuous, glabrous,
occasionally absent from some leaves; stipules triangular, 4–8 mm. long, acute, usually
exceeded by a mucro. Flowers (4–)5(–6)-merous, 2–20 per axil, borne in 1–3(–15)
fascicles; peduncle 0.5–2(–3) mm. long in flower (2–4 mm. long in fruit); pedicels 1–2(–3)
mm. long (calyces usually well clear of surrounding bracteole); bracteoles frequently with
subfoliaceous lobes up to 6 mm. long. Calyx-tube 1–2 mm. long; limb reduced to a rim or
sometimes irregularly toothed; teeth up to 1(–2) mm. long. Corolla-tube (0.5–)0.9–1.1 cm.
long, 2–3 mm. wide at throat; lobes oblong, 0.9–1.6 cm. long, (2–)4–6 mm. wide, rounded.
Fruit red (yellow or purple in cultivated forms), oblong-ellipsoid or sometimes ±
subglobose, 1–2 cm. long, 0.9–1.1 cm. wide; pedicel lengthening to 4–8 mm. Seeds fawn or
greenish fawn, 9–1.2(–1.5) cm. long, 6–7 mm. wide.

KENYA. Northern Frontier Province: Mt. Marsabit, July 1958, *J. Adamson* 58/1!, July 1942, *Bally* 1859!
& Aug. 1968, *Faden* 68/487!
DISTR. K 1; Sudan, Ethiopia; widely cultivated throughout the tropics
HAB. Forest; 1370–1525 m.

SYN. *C. laurifolia* Salisb., Prodr. Stirp. Hort. Chapel Allert.: 62 (1796), *nom. superfl.* for *C. arabica*, *non*
K. Kunth
C. vulgaris Moench., Meth. Pl. Hort. Marb.: 504 (1794), *nom. superfl.* for *C. arabica*
C. myrtifolia Roxb., Hort. Bengal: 15 (1815), *nom. nud.*
C. moka Heynh., Nom. Bot. 2: 153 (1846), *nom. superfl.* for *C. arabica*, *non* var. *mokka* Cramer

NOTE. *C. arabica* is an allotetraploid (2n = 44) and generally breeds true from seed. Between 40 and
50 subspecific taxa of *C. arabica* have been mentioned in the literature. Taxonomically these are
more appropriately referred to as cultivars. Modern opinion (e.g. Haarer, Modern Coffee
Production: 15 (1962)) favours the idea that these have been derived from two varieties of *C.
arabica* — namely var. *arabica* (including var. *typica* of Cramer and presumably var. *abyssinica* A.
Chev.) and var. *bourbon* Choussy (presumably including var. *culta* A. Chev.). Var. *arabica* is
presumed to have originated from the native Ethiopian *C. arabica* which was subsequently
distributed into cultivation from the Yemen, and var. *bourbon* arose as a spontaneous mutant
cultivated by the French on Réunion. The differences between these two varieties are impossible
to ascertain from herbarium specimens; var. *arabica* is said to have the young leaves bronze-tipped
and the fruit-bearing branches pendulous, while var. *bourbon* has the young leaves green and the
fruit-bearing branches bent down only at the tips. Sylvain (in Turrialba 5(1–2): 37–53 (1955)) stated
that in his opinion plants corresponding to both var. *arabica* and var. *bourbon* can be found in the
wild Ethiopian populations and Meyer (Economic Botany 19(2): 136–151 (1965), has noted
considerable variation and outcrossing in populations of *Coffea* in the rain-forest districts of
Ethiopia.
A helpful account of some of the infraspecific taxa of *C. arabica* can be found in Cramer, Review
of Literature of Coffee Research in Indonesia: 95–100 (1957), where the varieties are usefully
grouped under headings such as "small leaved varieties, varieties with one divergent characteristic
and varieties of divergent habit"; in Krug, Mendes & Carvalho, 'Taxonomia de *Coffea arabica* L.',
Inst. Agr. Est., Campinas, Bol. Tec. 62: (1939), where the varieties are very well illustrated, and a
succinct review by Purseglove, Tropical Crops, Dicotyledons: 461–463 (1974). (See note on p. 703
for narrow-leaved varieties.)
Although the distinction between *C. arabica* and *C. canephora* usually presents no difficulties,
some forms of *C. arabica* are morphologically indistinguishable from some forms of *C. canephora*.
Apart from the difference in chromosome number (2n = 44 for *C. arabica* and 2n = 22 for *C.
canephora*) it has been shown that stomatal counts per unit area can be used to distinguish them
(see J.A. Williams in Turrialba 22(3): 263 (1972)).

7. C. eugenioides *S. Moore* in J.B. 45: 43 (1907); Bullock in K.B. 1933: 81 (1933); De Wild.
in Ann. Jard. Buitenz., suppl. 3, 1: 370 (1910); Lebrun in Mém. Inst. Roy. Col. Belge, Sect.
Sci. Nat. Méd. Mém. 8°, 11(3): 83, t. III/6 (1941); Thomas in Emp. Journ. Exp. Agric. 12: 1
(1944); A. Chev., Caféiers du Globe 2: t. 71–73 (1942) & 3: 215 (1947), pro parte, excl.
specim. ex Malawi; F.P.S. 2: 432 (1952); Cramer, Rev. Lit. Coff. Res. Indonesia: 138 (1957);
K.T.S.: 437 (1961); Wellman, Coffee: 53 (1961); Haarer, Modern Coffee Prod.: 28 (1962);
Bridson in K.B. 36: 831, fig. 3 A–F (1982); Troupin, Fl. Pl. Lign. Rwanda: 556, fig. 187.1
(1982); Fl. Rwanda 3: 154, fig. 48.1 (1985). Type: Uganda, Toro District, near Mpanga,
Bagshawe 1076 (BM, holo.!)

Bush or small tree 1–4.5 m. tall; young branches glabrous or occasionally with a puberulous line on both sides, covered with moderately shiny light brown to buff bark. Leaf-blades elliptic, 2–12 cm. long, 0.8–5.6 cm. wide, acuminate at apex, acute at base, papery to subcoriaceous, dull to slightly shiny above, lateral and tertiary nerves raised on both faces; domatia absent or present, glabrous; petiole 2–7 mm. long; stipules triangular, 1.5–2.5 mm. long, bearing an aristate tip, 0.5–3.5 mm. long. Flowers 5-merous, usually 1–2 single-flowered inflorescences per axil (inflorescence-stalks up to 6 mm. long) or sometimes with fascicles of up to 5 flowers; peduncle up to 2.5 mm. long, usually with an additional common segment terminating in pedicels 1–1.5 mm. long; cupular bracteoles usually 2, the lower often with subfoliaceous lobes; small free, scale-like bracteoles usually present on the pedicels. Calyx-limb reduced to a rim. Corolla-tube 5.5–10 mm. long, 2–3 mm. wide at throat; lobes oblong-lanceolate, 5–12 mm. long, 2–4 mm. wide, acute. Fruit red, 8–10.5 mm. long, 6–8 mm. wide; inflorescence-stalk lengthening to 1.5–5 mm. Seeds yellowish or greenish fawn, 6–8 mm. long, 3–5 mm. wide.

UGANDA. W. Nile District: Zeio, *Eggeling* 1907!; Ankole District: Kalinzu Forest Reserve, Feb. 1953, *Osmaston* 2837!; Mengo District: Kasa Forest near Mityana, 8 Feb. 1950, *Dawkins* 507!
KENYA. Nandi District: N. bank of Yala R., 1–2 km. below bridge, 19 Apr. 1965, *Gillett* 16729!; Embu District: E. Mt. Kenya, 8 km. from Embu, *Doughty* 82a!; Masai District: Masai Reserve, southern Uaso Nyiro [Guaso Nyero] R., 6 Jan. 1920, *Dowson* 725!
TANZANIA. Bukoba District: Minziro Forest, Aug. 1952, *Procter* 65! & July 1950, *Watkins* 463 in *F.H* 3202!; Mpanda District: Kungwe-Mahali Peninsula, Musenabantu, 13 Aug. 1959, *Harley* 9332!
DISTR. U 1–4; K 3–6; T 1, 4, also cultivated in U 4, K 4 and T 2; Zaire, Rwanda, Sudan
HAB. Forest; 1050–2100 m.

SYN. *C. arabica* L. var. *intermedia* Froehner in E.J. 25: 264 (1897), pro parte, excl. *Whyte* s.n. Type: Kenya, Elgeyo [Ligaijo], *Fischer* 326 (B, holo.†); illustration in A. Chev., Caféiers du Globe 2, t. 71 (1942) based on type specimen.
    C. nandiensis Dowson in Ann. Appl. Biol. 8: 88 (1921); Bullock in K.B. 9: 401 (1930), *nomen*
    C. becquetii A. Chev. in Rev. Bot. Appliq. 14: 354 (1934). Type: Rwanda, Rubou–Rulindo–Kigali, *van Hoonacker* in *Becquet* 639 (BR, holo., K, iso.!)
    C. intermedia (Froehner) A. Chev. in Rev. Bot. Appliq. 19: 397 (1939)

NOTE. In my opinion *C. eugenioides* var. *kivuensis* (Lebrun) A. Chev. is a distinct species; *C. kivuensis* Lebrun has not been recorded from the Flora area.

8. C. salvatrix *Swynnerton & Phillipson* in J.B. 74: 314, fig. (1936); A. Chev. in Rev. Bot. Appliq. 20: 535, fig. 6 (1940) & Caféiers du Globe 2, t. 70 (1942) & 3: 221 (1947); Wellman, Coffee: 53 (1961); Bridson in K.B. 36: 833, fig. 3 G–M (1982). Type: cultivated in Chirinda, Zimbabwe, from material brought from Mozambique, Mossurise District, Sitatonga Hills, *Swynnerton* (BM, holo.!)

Glabrous bush or small tree 2–5 m. tall; young branches covered with shiny buff, usually dark spotted bark, often with transverse cracks. Leaf-blades elliptic to broadly elliptic, 6–12.2 cm. long, 2.5–5.6 cm. wide, acuminate at apex, acute at base, papery to subcoriaceous, slightly to moderately shiny above; lateral and tertiary nerves raised on both faces; domatia sparsely pubescent to pubescent; petiole 0.5–1 cm. long; stipule-limbs triangular, 2–3 mm. long, acute or truncate, then with an arista not exceeding 1 mm. Flowers 5-merous in 1–2 single-flowered inflorescences per axil (total inflorescence-stalks 0.8–1 cm. long) or with fascicles of 2–7 flowers; peduncle ± 2(–7 in fruit) mm. long, often with an additional common segment present, terminating in pedicels 3–4 mm. long; bracteoles cupular, the lower ones sometimes with rudimentary lobes; small scale-like bracteoles present on pedicel. Calyx-limb not exceeding disc. Corolla-tube 0.6–1 cm. long, 4 mm. wide at throat; lobes oblong, 1–1.1 cm. long, 0.3–0.5 cm. wide, rounded. Fruit greenish with rose striations, becoming black when mature, 7–9 mm. long, 6–8 mm. wide; stipe lengthening to 9 mm. Seeds 6 mm. long, 4 mm. wide.

TANZANIA. Rungwe District: Kirambo [Kilambo], 2 Apr. 1913, *Stolz* 1966!
DISTR. T 7; Mozambique, Zimbabwe, ?Malawi
HAB. Not given, but elsewhere from mixed evergreen forest at ± 800–1675 m.

NOTE. The inclusion of the above fruiting specimen in *C. salvatrix* is somewhat tentative as the material is poor. The lax inflorescence-branches are typical of this species and none of the vegetative characters conflict, although the leaves on this sheet reach 14 cm. long, 6.3 cm. wide, which is a little in excess of that otherwise recorded for *C. salvatrix*.

9. C. pseudozanguebariae *Bridson* in K.B. 36: 835, fig. 4 A–G (1982). Type: Tanzania, Tanga District, Magunga Estate, *Faulkner* 1077 (K, holo.!)

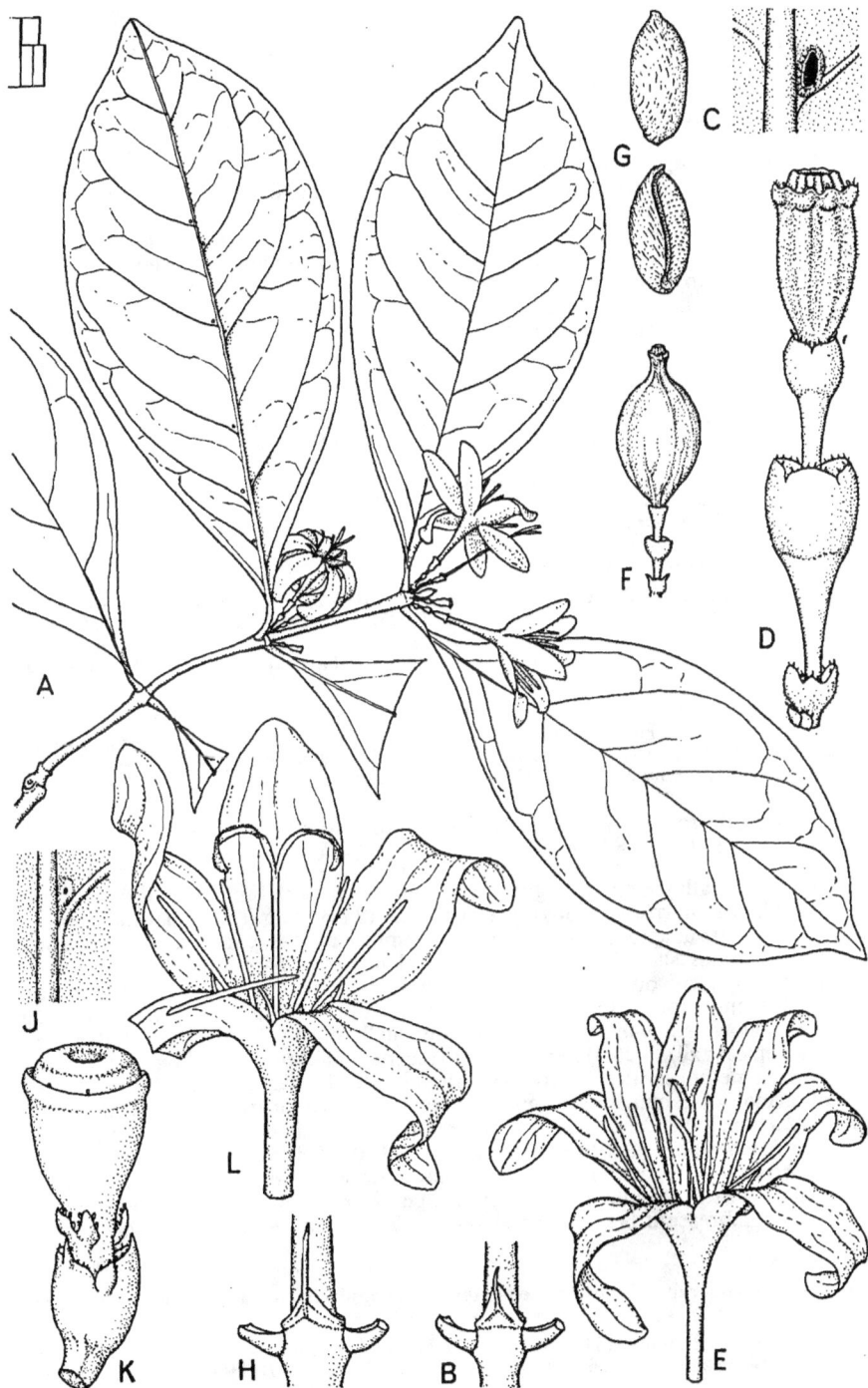

FIG. 123. *COFFEA PSEUDOZANGUEBARIAE* — **A**, flowering branch, × ⅔; **B**, stipule, × 3; **C**, domatium, × 6; **D**, single-flowered inflorescence, × 8; **E**, corolla, × 2; **F**, fruit, × 2; **G**, seed, 2 views, × 3. *C. SESSILIFLORA* subsp. *SESSILIFLORA* — **H**, stipule, × 3; **J**, domatium, × 6; **K**, single-flowered inflorescence, × 8; **L**, corolla, × 2. A, C, from *Greenway* 7557; B, from *Koritschoner* 1535; D, E, from *Faulkner* 1077; F, from *Peter* 39922, G, from *Tanner* 3379; H–L, from *Rawlins* 216. Drawn by Diane Bridson.

Bush or small tree 1.5–3.5 m. tall; very young branches glabrous to puberulous, covered with light rusty-brown to maroon-brown shiny bark, becoming pale greyish and dull with age. Leaf-blades elliptic to broadly elliptic, 3–4.8 cm. long, 1.5–7 cm. wide, acute to subacuminate or occasionally obtuse at apex, acute or sometimes slightly attenuated at base, subcoriaceous, dull above; lateral nerves apparent and tertiary nerves obscure to apparent on both faces; domatia well marked or less often small, ciliate to pubescent; petiole 2–6 mm. long; stipule-limbs triangular, shortly aristate, 2–4 mm. long altogether. Flowers (5–)6–8-merous, usually individually borne in the axils of the upper leaves (often including the apical axil), 1–2(–5) per axil; inflorescence-stalks 2–6 mm. long; upper bracteoles cupular or with free lobes; lower bracteoles ± 1–2 mm. long, cupular, sometimes slightly lobed; scale-like bracteoles present or sometimes absent. Calyx-limb much shorter than the disc. Corolla-tube 0.7–1.2 cm. long, 3–7 mm. wide at throat; lobes oblong to oblong-lanceolate, 1–1.4 cm. long, 5–7 mm. wide, rounded. Style-arms up to 6 mm. long, scarcely divergent. Fruit 8–11 mm. long, 4–5.5 mm. wide, distinctly beaked (constricted at apex into bottle-neck shape); stalk lengthening to 1 cm. Seeds mid-to light brown, 5–7.5 mm. long, 2.5–4 mm. wide; testa with distinct or fine striations. Fig. 123/A–G.

KENYA. Kwale District: Diani Forest, 29 Mar. 1973, *Kibuwa* 1225! & Shimba Hills, on road to Wireless Station, 23 Mar. 1968, *Magogo & Glover* 421!; Kilifi District Cha Shimba [Simba], 17 Nov. 1974, *Adams* 105!
TANZANIA. Lushoto District: Sigi near Lunguza, 24 Nov. 1945, *Greenway* 7557!; Pangani District: Mkwaja, Mkaramo [Nkaramo], 10 Jan. 1957, *Tanner* 3379!; Uzaramo District: Sinda I. near Dar es Salaam, 5 Jan. 1969, *Harris* 2700!; Zanzibar, Kituani, 4 Feb. 1933, *Vaughan* 2099!
DISTR. K 7; T 3, 6; Z; not known elsewhere
HAB. Forest and coastal bush; 0–800 m.

SYN. [*C. zanguebariae* sensu Busse in Tropenpflanzer 6: 143 (1902) & Rev. Cult. Colon. 11: 185 (1902); Cheney, Coffee, Monogr. Econ. Sp.: 68, t. 25 (1925), pro parte; A. Chev., Caféiers du Globe 1: 96 (1929), pro parte, quoad *Busse* s.n., & 2, t. 66 (1942) & 3: 218 (1947), pro parte, quoad *Last* s.n., *Sacleux* 1015; T.T.C.L.: 493 (1949), pro parte, quoad *Busse* s.n.; K.T.S.: 437 (1961), pro parte, quoad *Wakefield* s.n., *non* Lour.]

NOTE. The specimens from Lushoto District have the very young branches puberulous while the specimens seen from Pangani and Uzaramo Districts and Zanzibar have the young branches glabrous; most of the specimens from Kenya are glabrous, but one or two puberulous forms have also been noted. See note after *C. sessiliflora* var. *sessiliflora*.

10. C. sessiliflora *Bridson* in K.B. 41: 307 (1986). Type: Kenya, Kilifi District, Rabai, *Verdcourt* 2402 (K, holo.!, EA, iso.)

Shrub 1–3 m. tall; young stems glabrous or pubescent, with buff, fawn or light brown bark. Leaf-blades narrowly to broadly elliptic, 4.5–12 cm. long, 1.2–6 cm. wide, acute to shortly acuminate at apex, acute or sometimes cuneate at base, papery to subcoriaceous, dull or sometimes slightly shiny above; lateral nerves apparent and tertiary nerves obscure to apparent on both faces; domatia small, glabrous to sparsely pubescent; petioles 2–6 mm. long; stipule-limbs triangular, distinctly aristate, 3–5.5 mm. long altogether. Flowers 5–6-merous, usually individually borne in upper leaf-axils (often including apical axil), 1–2(–3) per axil; inflorescence-stalks not exceeding 2 mm.; upper bracteoles sometimes with free lobes, but often hidden inside lower cupular bracteoles; lower bracteoles 1–2 mm. long, often producing spathulate or sometimes linear-aristate lobes 1–2.5 mm. long. Calyx-limb not exceeding disc. Corolla-tube 1–1.4 cm. long, 5 mm. wide at top; lobes oblong-obovate, 1.3–1.6 cm. long, 0.6–0.8 cm. wide. Style-arms up to 8 mm. long, often divergent. Fruit broadly ellipsoid or oblong-ellipsoid, 0.8–1.2 cm. wide, apex either constricted into a bottle-neck shape or not; stalk lengthening to 2–4 mm., glabrous or glabrescent. Seeds greenish fawn, 5–7 mm. long, 4mm. wide, finely striated.

subsp. **sessiliflora**

Young stems and fruit glabrous; bracteoles sometimes producing linear to spathulate lobes. Fig. 123/H–L.

KENYA. Kwale District: Shimba Hills, Lango ya Mwagandi [Longomwagandi] Forest, 6 Mar. 1968, *Magogo & Glover* 209!; Lamu District: Gongoni, Oct. 1937, *Dale* in F.D. 3817! & Utwani, Oct.–Nov. 1956, *Rawlins* 216!
DISTR. K 7; not known elsewhere
HAB. Forest; 0–450 m.

SYN. [*C. zanguebariae* sensu K.T.S.: 437 (1961), pro parte, quoad *Dale* 3817 & *Rawlins* 216, *non* Lour.]
C. sp. A sensu Bridson in K.B. 36: 386, fig. 4 H–L (1982); Hamon, Anthony & Le Pierres in Bull. Mus. Hist. Nat. Paris, sér. 4, 6, B, Adansonia 2: 207–223 (1984)

NOTE. This taxon is often sympatric with *C. pseudozanguebariae* and apart from the morphological differences noted in the key, Hamon et al. (cit. supr.) have observed differing patterns of presence and absence of enzymes (esterases and malate dehydrogenases) and differing caffeine content in the two species, *C. pseudozanguebariae* containing little or none). Further they were able to demonstrate a slight difference in flowering periods between the two species, though they comment that under certain conditions flowering could perhaps overlap allowing hybridisation and introgression. Two of their collections from the Shimba Hills appear to be intermediate. Field observations of both *C. sessiliflora* and *C. pseudozanguebariae* would be of great interest.

subsp. **mwasumbii** *Bridson* in K.B. 41: 308 (1986). Type: Tanzania, Uzaramo District, Pugu Hills Forest Reserve, *Mwasumbi* 12493 (K, holo.!, DSM, iso.)

Young stems pubescent; fruit glabrescent; bracteoles producing linear-aristate lobes.

TANZANIA. Uzaramo District: Pugu Hills Forest Reserve, 3 June 1983, *Mwasumbi* 12493! & 9 July 1982, *Hawthorne* 1100! & 27 Aug. 1982, *Hawthorne* 1566!
DISTR. **T** 6; known only from the Pugu Hills
HAB. Forested hillslopes; 200 m.

11. C. sp. B; *Bridson* in K.B.: 838, fig. 5 A–E (1982).

Young branches covered with glabrous, shiny reddish-brown bark. Leaf-blades elliptic to broadly elliptic, 4–9.5 cm. long, 2.2–5 cm. wide, acute to acuminate at apex, acute at base, scarcely shiny above, with 5–7 main pairs of lateral nerves, lateral and tertiary nerves raised on both faces; domatia present, pubescent; petioles 4–5 mm. long; stipules triangular, 2–4 mm. long, apiculate. Flowers 6–7-merous, 1–7 per axil, borne in 1–2 fascicles, often with additional single-flowered inflorescences; peduncles not exceeding 2 mm.; inflorescence-stalks 1–2 mm. long in fascicled inflorescences and 3–5 mm. long in single-flowered inflorescences; bracteoles truncate, the lower tending to be lobed, 1–1.5 mm. long. Calyx-limb shorter than disc. Corolla-tube 6 mm. long, 3 mm. wide at throat; lobes oblong, 9 mm. long, 5 mm. wide, rounded. Fruit not known.

TANZANIA. Lushoto/Tanga Districts: E. Usambara Mts., Ngua, 28 Dec. 1932, *Greenway* 3306!
HAB. Rain-forest; 1025 m.
DISTR. **T** 3; known only from the above specimen.

12. C. sp. C.; *Bridson* in K.B. 36: 838, fig. 5 F, H (1982)

Young branches glabrous, covered with mid-brown bark. Leaf-blades elliptic to narrowly obovate, 6.5–11.8 cm. long, 2.3–4.5 cm. wide, distinctly acuminate at apex, acute to cuneate at base, subcoriaceous, slightly shiny on both faces; lateral nerves apparent but tertiary nerves rather obscure on both faces; domatia absent or obscure; petiole 0.5–1 cm. long; stipules triangular, up to 3 mm. long, acute or shortly apiculate. Flowers 1–2 per axil, borne singly; bracteoles 2, ± 1 mm. long, lower one with short linear lobes. Calyx and corolla not known. Fruit brown-green, 1.1–1.2 cm. long, 0.7 cm. wide; fruiting stalk ± 4 mm. long. Seed ± 9 mm. long (not fully developed).

TANZANIA. Iringa District: Mufindi Lulando [Lulanda] Forest, 17 Feb. 1979, *Cribb, Grey Wilson & Mwasumbi* 11478!
HAB. Forest; 1600 m.
DISTR. **T** 7; known only from the above specimen.*

13. C. sp. D; *Bridson* in K.B. 36: 838, fig. 5 J–N (1982)

Glabrous shrub ± 2–4 m. tall; young branches covered with ± maroon shiny bark. Leaf-blades elliptic, 7.3–13 cm. long, 3–6 cm. wide, shortly acuminate at apex, acute at base, subcoriaceous, shiny above; tertiary nerves rather more prominent above than beneath; domatia pubescent; petioles 6–8 mm. long; stipules triangular, 2–3 mm. long, apiculate. Flowers 5-merous, 1–3 per node, individually borne; inflorescence-stalks obscured by bracteoles, 2–3 mm. long; bracteoles usually 2, upper one ± truncate, lower one shortly lobed. Calyx-limb repand, shorter than disc. Corolla-tube 1–1.1 cm. long, ± 3 mm. wide at throat; lobes oblong-lanceolate, 1.1–1.6 cm. long, 0.3–0.4 cm. wide, acute. Fruit not known.

TANZANIA. Lindi District: Lake Lutamba, 10 Dec. 1934, *Schlieben* 5716!

* New collections reported by J. Lovett, but not yet seen.

HAB. Deciduous woodland; ± 240 m.
DISTR. T 8; known only from the above specimen.

14. C. sp. E; *Bridson* in K.B. 36: 840, fig. 6 A–F (1982)

Small tree 2 m. tall; young branches glabrous, covered with buff bark. Leaf-blades elliptic to broadly elliptic, 4–8.8 cm. long, 2–4.5 cm. wide, acute or sometimes obtuse at apex, obtuse or sometimes acute at base, chartaceous, dull to slightly shiny above; lateral nerves apparent, but tertiary nerves somewhat obscure on both faces; domatia large, glabrous; petiole 2–3 mm. long; stipule-limbs triangular, distinctly aristate, 3–5 mm. long altogether. Flowers 5-merous, usually individually borne in the upper leaf-axils (often including the apical axil), 1 per axil; inflorescence-stalks very reduced; bracteoles stipule-like, producing linear lobes up to 3 mm. long. Calyx-limb equalling disc. Corolla-tube 7–9 mm. long, 3.5 mm. wide at throat; lobes oblong, 8–11 mm. long, 3 mm. wide. Fruit not known.

TANZANIA. Morogoro District: Kitulanghalo [Kitulangolo] Forest Reserve, near Mikese village, 25 Apr. 1972, *T. & S. Pocs & Mwanjabe* 6559c!
HAB. Dry evergreen forest; 500–600 m.
DISTR. T 6; known only from the above specimen.

15. C. sp. F; Bridson in K.B. 36: 841, fig. 8 E, F (1982)

Evergreen tree up to 10 m. tall, much branched; young branches glabrous, covered with fawn rather shiny bark, older branches maroon-brown. Leaf-blades elliptic to broadly elliptic, 5–9 cm. long, 2.5–5.2 cm. wide, acute to shortly acuminate at apex, acute at base, subcoriaceous, dull; midrib prominent above; lateral and tertiary nerves apparent to obscure on both faces; domatia small, with few hairs; petiole 3–5 mm. long, rather stout; stipules truncate-triangular to triangular, 1–2 mm. long, shortly apiculate. Flowers 5-merous, individually borne, 1–2 per axil; inflorescence-stalks reduced, up to 1 mm. long; bracteoles 1–2; upper one ± 1.5 mm. long, bearing linear to slightly spathulate lobes, up to 4 mm. long. Calyx-tube up to 1.5 mm. long, ribbed; limb ± 1 mm. long, ± truncate or with small irregular teeth, distinctly exceeding the disc. Corolla-tube 4 mm. long, 3 mm. wide at throat; lobes 5 mm. long, 3–4 mm. wide, rounded. Fruit not known, but possibly ribbed.

TANZANIA. Rufiji District: Mafia, Chungaruma, 2 Oct. 1937, *Greenway* 5366!
DISTR. T 6; known only from the above specimen
HAB. Forest; 10 m.

NOTE. It is possible that this species could also produce flowers precociously at a different season in much the same way as *C. racemosa* Lour., a species from Mozambique, Zimbabwe and Natal.

16. C. sp. G; *Bridson* in K.B. 36: 841, fig. 8 A–D (1982)

Shrub 2 m. high, somewhat scandent; young branches covered with dull maroon bark; puberulous. Leaf-blades elliptic, 4.5–7 cm. long, 1.5–2.7 cm. wide, acuminate at apex, obtuse to acute at base, crisped at margin, papery to subcoriaceous, moderately shiny above; domatia with cavities partly defined, pubescent; petiole 1.5–3 mm. long; stipules with a triangular base, 1.5–2.5 mm. long, with an aristate tip up to 5 mm. long. Flowers individually borne, 1–2 per axil; inflorescence-branches reduced; bracteoles 1–2, with linear or subulate lobes 4–8 mm. long. Calyx-limb equalling disc, 5-toothed (on evidence from fruit). Corolla not known. Fruit 1.2–1.3 cm. long, 0.7 cm. wide, not stipitate. Seeds dull greenish, 9 mm. long, 4 mm. wide.

TANZANIA. Kilosa District: Ukaguru Mts., Mamiwa Forest Reserve, valley above Uponela road, 1 km. N. of Mandege Forest Station, 8 Aug. 1972, *Mabberley* 1417!
DISTR. T 6; known only from the above specimen
HAB. Valley forest; 1600 m.

NOTE. This species somewhat resembles *C. mufindiensis* Bridson but can easily be distinguished by its well-spaced acuminate leaves and the lobed bracteoles.

17. C. sp. H; *Bridson* in K.B. 36: 841, fig. 6 M–Q (1982)

Shrub; young branches covered with brown bark, puberulous. Leaf-blades broadly elliptic, 8–12.2 cm. long, 3.3–6.2 cm. wide, acute to shortly acuminate at apex, acute to cuneate at base, papery, moderately shiny above, perhaps crisped when young; domatia

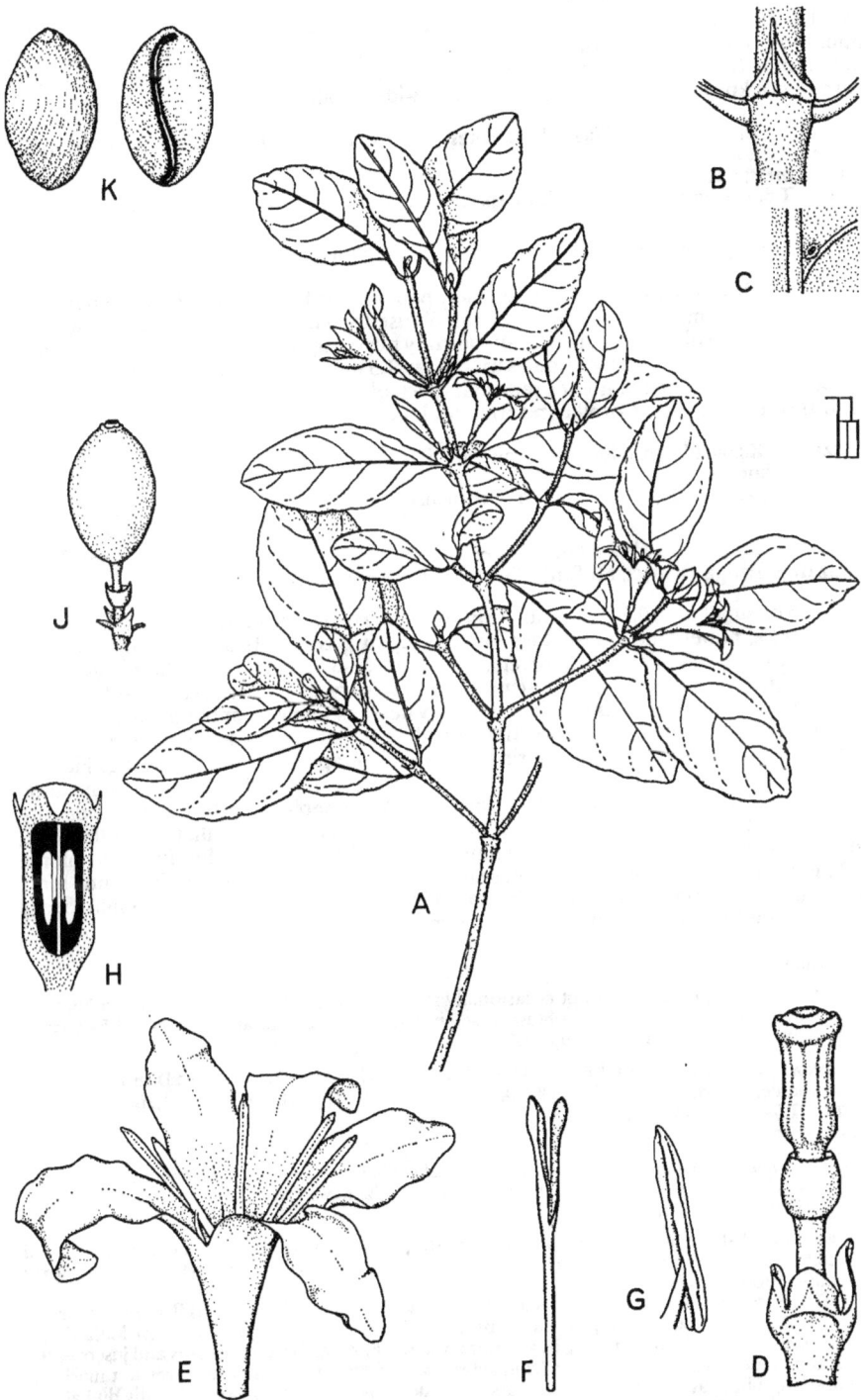

FIG. 124. *COFFEA MUFINDIENSIS* subsp. *MUFINDIENSIS* — **A**, flowering branch, × ⅔; **B**, stipule, × 3; **C**, domatium, × 6; **D**, single-flowered inflorescence, × 8; **E**, corolla, × 4; **F**, style and stigma, × 4; **G**, stamen, × 6; **H**, longitudinal section through ovary, × 10; **J**, fruit, × 2; **K**, seed, 2 views, × 3. A–C, from *Carmichael* 22; D–H, from *Director of Agriculture* s.n.; J, K, from *Greenway* 5793. Drawn by Diane Bridson.

small, pubescent or sometimes absent; petiole 5–7 mm. long; stipules triangular, 1.5–2.5 mm. long, acute or very shortly aristate. Flowers 1 per axil, inflorescence-stalks clearly apparent in fruit.; upper bracteole cupular, lower bracteole very slightly lobed. Calyx and corolla not known. Fruit 1 cm. long, 0.75 cm. wide, stipitate.

TANZANIA. Ulanga District: Magombera Forest Reserve, 20 May 1979, *Rodgers!* & 23 May 1979, *Rodgers, Homewood & Hall!*
HAB. Forest; ± 250 m.
DISTR. T 6; known only from the above specimens.

18. C. sp. I; *Bridson* in K.B. 36: 841, fig. 8 E, F (1982)

Shrub with young stems maroon, sparsely puberulous. Leaf-blades elliptic to narrowly obovate, 4–9.5 cm. long, 2.4–3.5 cm. wide, obtuse to acuminate at apex, acute at base, papery, slightly shiny above; lateral and tertiary nerves apparent on both faces; domatia absent; petiole 5–6 mm. long; stipules triangular, 1.5 mm. long, acute, caducous. Flowers 1 per axil; inflorescence-branches reduced; bracteoles 2, lower one slightly lobed. Calyx-limb shorter than disc. Corolla and fruit not known.

TANZANIA. Kigoma District: Kasakati, June 1965, *Suzuki* 104!
HAB. Not known
DISTR. T 4; known only from the above specimen.

19. C. mufindiensis *Bridson* in K.B. 36: 842, fig. 7 (1982) & in K.B. 41: 309 (1986). Type: Tanzania, Iringa District, Mufindi, *Entom. Lab. Morogoro* (K, holo.!)

Shrub or small tree, 0.5–4.5 m. tall; very young branches covered with light or greyish-brown bark, pubescent. Leaf-blades drying greyish green or brownish, elliptic or less often narrowly elliptic, 2–6.3(–8) cm. long, 0.8–3(–3.5) cm. wide, obtuse, acute or subacuminate at apex, obtuse to acute at base, papery to subcoriaceous, dull to shiny above, with lateral nerves apparent and tertiary nerves somewhat apparent on both faces, margins often crisped or undulate; domatia glabrescent to pubescent; petiole 1–5 mm. long; stipules deltoid at base, tapering to a subulate apex, 2–7(–8) mm. long. Flowers 5-merous, 1–2 per axil borne individually; inflorescence-stalks up to 3 mm. long, sometimes obscured by bracteoles; bracteoles 2, upper one ± truncate, lower one sometimes shortly lobed. Calyx-limb shorter than disc, sometimes ciliate. Corolla white, sometimes reddish in bud; tube 2.5–6 mm. long, 3.5–4 mm. wide at throat; lobes oblong, 5–11 mm. long, 2.5–3.5 mm. wide, obtuse to rounded. Fruit orange to red, 8–10 mm. long, 5.5–7 mm. wide, glabrous to sparsely puberulous; stalk lengthening to 2–8 mm. Seeds fawn to light brown, 5–7 mm. long, 4–5.5 mm. wide.

subsp. mufindiensis

Leaves crowded, usually dull but occasionally shiny above, papery, tending to be undulate or crisped at margins when dry; apex obtuse to acute or rarely subacuminate; calyx-tube 1.5–2.5 mm. long, glabrous; limb never ciliate. Fig. 124.

TANZANIA. Mpwapwa District: Kiboriani Mts., 3 Oct. 1938, *Greenway* 5793!; Ulanga District: Uzungwa Mts., Sanje, 10 Oct. 1984, *D. Thomas* 3819!; Iringa District: Imagi Mt., 15 Dec. 1961, *Richards* 15684!
DISTR. T 5–7; not known elsewhere
HAB. Forest; 1600–2100 m.

SYN. *C. nufindiensis* A. Chev., Caféiers du Globe 3: 217 (1947) *nom. non rite publ.* based on Tanzania, N. Mpwapwa, *Hornby* 734 (K) & Iringa District, Uzungwa [Nzungma] Forest, *Herb. Amani* H9/34 (BR, EA, K)

NOTE. I decided that it would be more appropriate to select one of the specimens from Mufindi which was originally annotated by Hutchinson as *C. mufindiensis* for the type of this species, than to readopt those cited by Chevalier.
    There are three additional subspecies of *C. mufindiensis*, subsp. *lundaziensis* Bridson, occurring in Northern Zambia, subsp. *pawekiana* (Bridson) Bridson occurring in Northern Malawi and subsp. *australis* Bridson occurring in southern Malawi, E. Zimbabwe mountains and just over the Mozambique border. These differ from subsp. *mufindiensis* in having the leaves not markedly crowded, subcoriaceous or less often papery, moderately shiny or only occasionally dull above, acute to subacuminate at the apex and less distinctly undulate at margin; also the calyx-tube is 1–1.5 mm. long, usually puberulous and the limb frequently ciliate. A table showing subspecific differences in *C. mufindiensis* is given by Bridson in K.B. 41: 310 (1986). The following specimen: —Njombe District, Mdando Forest Reserve, *Gillett* 17865, seems intermediate between subsp. *mufindiensis* and subsp. *lundaziensis*, the leaves being well spaced and the calyx-limb ciliate.

R.M. Davies in *Herb. Amani* H12/31 & H35/30 notes in the data that "this wild coffee can be found all over the Mufindi District and on a stream near Musekera Estate, Tukuyu District (Rungwe)". No Rungwe records are known; collections from this area could indicate that the character disjunction between subsp. *mufindiensis* and subsp. *lundaziensis* is not sufficient for their continued recognition. See note for *C. klaurathii* after *C. zanguebariae.*

20. C. zanguebariae *Lour.,* Fl. Cochinch.: 145 (1790); Roem. & Schultes, Syst. Veg. 5: 197 (1819); DC., Prodr. 4: 500 (1830); Hiern in Trans. Linn. Soc., Bot. 1: 172 (1876) & in F.T.A. 3: 182 (1877); K. Schum. in P.O.A. B: 247 (1895); Froehner in N.B.G.B. 1: 234 (1897) & in E.J. 25: 274 (1898); De Wild. in Actes Congr. Intern. Bot., Paris: 235 (1900), Les Caféiers: 42 (1901) & in Ann. Jard. Buitenz., suppl. 3, 1: 381 (1910); Cheney, Coffee Monogr. Econ. Sp.: 68 (1925), pro parte; A. Chev., Caféiers du Globe 1: 96 (1929) pro parte: Merr., Comment. Lour. Fl. Cochinch. in Trans. Amer. Phil. Soc. n.s. 25, 2: 369 (1935); A. Chev., in Rev. Bot. Appliq. 20: 532 (1940), Caféiers du Globe 2, t. 67 (1942) & 3: 281 (1947), pro parte; T.T.C.L.: 493 (1949), pro parte; Wellman, Coffee: 53 (1961); Haarer, Modern Coffee Prod: 26 (1962); Vollesen in Op. Bot. 59: 68 (1980); Bridson in K.B. 36: 847, fig. 9 (1982). Types: Tanzania/Mozambique (Zanzibar Coast) and cultivated in Mozambique Is., *Loureiro* (syntypes unknown)

Shrub or tree, 3–5 m. tall, glabrous; young branches covered with light rusty-brown or occasionally creamish coloured bark. Leaves usually restricted to new growth; blades broadly elliptic, sometimes round or rarely elliptic, 5.4–13.8 cm. long, 2.2–7.8 cm. wide, obtuse to acute or slightly acuminate at apex, acute to obtuse at base, papery, dull or rarely somewhat shiny above, nerves raised on both faces, often undulate at margin with a marginal nerve; domatia pubescent with cavity partly defined; petioles 3–6 mm. long; stipules triangular, 2–3 mm. long, acute or more often shortly apiculate. Flowers precocious, 7–8-merous, borne on older stems, usually at nodes from which the leaves have been shed, 2–5 per axil, often borne in fascicled inflorescences; peduncle not exceeding 2 mm. long; inflorescence-stalks compact in flowering stage; bracteoles usually 2, slightly lobed; scale-like bracteoles present on the pedicels. Calyx-tube somewhat ribbed; limb equalling disc, somewhat dentate. Corolla-tube 6–10 mm. long, 2–3 mm. wide at throat; lobes ± oblong, 7–12 mm. long, 3–4 mm. wide, rounded. Fruit pale green turning blackish, 0.9–1.9 cm. long, 0.6–0.9 cm. wide, ribbed, occasionally beaked, minutely puberulous; pedicel lengthening to 3 mm. long. Seeds fawnish, 7–10 mm. long, 6 mm. wide.

TANZANIA. Kilwa District: Selous Game Reserve, Kingupira, 13 Nov. 1975, *Vollesen* in *M.R.C.* 2982! & 21 Nov. 1975, *Vollesen* in *M.R.C.* 3023! & Balenje, 10 Feb. 1971, *Rodgers* 1220!
HAB. Riverine thicket; 125 m.
DISTR. T 8; Mozambique, ? cultivated in Madagascar (*fide* Lanessen, P.C. Ut. Col. Fr.: 44 (1886) & Cheney, *l.c.*)

SYN. *Amajoua africana* Spreng., Syst., Veg. 2: 126 (1825), nom. *superfl.*, based on *C. zanguebariae*
    Hexepta axillaris Raf., Sylva Tel.: 164 (1838), nom. *superfl.*, based on *C. zanguebariae*
    Coffea ibo Froehner in N.B.G.B. 1: 231 & 234 (1897) & in E.J. 25: 272 (1898); Lecomte, Le Café: 36 (1899); De Wild. in Actes Congr. Intern. Bot., Paris: 231 (1900), Les Caféiers: 39 (1901) & in Ann. Jard. Buitenz., suppl. 3, 1: 371 (1910); Cheney, Coffee, Monor. Econ. Sp.: 28 (1925); A. Chev., Caféiers du Globe 1: 97 (1929); Haarer, Modern Coffee Prod.: 26 (1962). Type: Mozambique, Inhambane, collected by Governor of Mozambique, communicated by *Henriques* (B, holo.!, COI, iso.)
    C. schumanniana Busse in Tropenpflanzer 6: 142, t. (1902) & in Rev. Cult. Colon. 11: 184 (1902); De Wild. in Ann. Jard. Buitenz., suppl. 3, 1: 379 (1910); A. Chev., Caféiers du Globe 1: 98 (1929) & 3: 214 (1947); T.T.C.L.: 493 (1949); Wellman Coffee: 53 (1961); Haarer, Modern Coffee Prod.: 26 (1962). Type: Tanzania, Masasi District, lower Rovuma valley near Mbangala R., *Busse* 1077 (B, holo. †)
    C. zanzibarensis R.M. Grey in Rep. Harvard Bot. Gard. Cienfuegos, Cuba 1900–1926: 31 (1927), nomen
    [*C. racemosa* sensu A. Chev., Caféiers du Globe 2, pl. 62 (1942) & 3: 219 (1947), pro parte, quoad type of *C. ibo, non* Lour.]
    [*C. arabica* L. var. *mokka* sensu A. Chev., Caféiers du Globe 3: 202 (1947), pro parte, quoad *Busse* 1077, *non* Cramer]

NOTE. In the absence of the syntypes two taxa have been confused under the epithet *C. zanguebariae;* this species and *C. pseudozanguebariae.* Loureiro described the flowers of *C. zanguebariae* as being "polynatus" and having a 6–7-lobed corolla; this species has 2–5 flowers per axil but the corolla is 7–8-lobed (in the 3 flowering specimens known to me), while *C. pseudozanguebariae* has 1–2(–3) flowers per axil and the corolla is (5–)6–8-lobed. Furthermore Loureiro's description of the fruit agrees with this species rather than with the small-seeded, distinctly beaked fruit of *C. pseudozanguebariae,* which would seem less likely to have been cultivated as mentioned by Loureiro or so extensively used as suggested by Cheney. The fact that no gatherings of *C. pseudozanguebariae* are known south of latitude 8°S further substantiates the choice of this taxon as being the correct *C. zanguebariae*, since this corresponds with the related species, *C. racemosa*, also based on a *Loureiro* type. According to Haarer this species seems to have had some limited commercial use under the names "Ibo Coffee" or "Zanzibar Coffee".

This species is closest to *C. racemosa* Lour. which differs in its smaller, coriaceous leaves. No gatherings of *C. racemosa* are known from the flora area, (although Chevalier cites *Braun* 1443 (B †) as being cultivated at Amani). It has been recorded from Ibo and Mozambique Islands and possibly could occur in southern Tanzania, near the coast. Two specimens from Mozambique (*Gomes e Sousa* 1684 & 1942) are somewhat intermediate between *C. zanguebariae* and *C. racemosa*.

Chevalier (Caféiers du Globe 3: 219 (1947)) included the nomen *C. klaurathii* De Wild. (in Ann. Jard. Buitenz., suppl. 3, 1: 372 (1910)) under the synonymy of *C. racemosa*. Since the specimen on which *C. klaurathii* was based (*Klaurath* s.n. B †) came from Iringa, this association is highly unlikely. Only *C. mufindiensis* and *C. sp. C* have been recorded from Iringa District and although De Wildeman's description more or less complies with *C. mufindiensis*, Chevalier's illustration (Caféiers du Globe 2, pl. 64 (1942)) is less convincing especially with regard to the ribbed fruit.

21. C. sp. J

Shrub 1–3 m. high; young branches puberulous, covered with greyish bark. Leaf-blades narrowly obovate to obovate, 3–5.5 cm. long, 1.2–2.5 cm. wide, acute at apex, cuneate at base, becoming coriaceous when mature, shiny above, midrib and lateral nerves prominent above, less conspicuous beneath; domatia with well-defined cavity, pubescent; petiole 1.5–3 mm. long, pubescent at base above; stipules triangular, 2–3 mm. long, apiculate, pubescent. Flowers possibly precocious, 1 per axil, often borne at apex of lateral branches; inflorescence-branches presumably reduced; bracteoles usually 2, with subulate to linear lobes 3–6 mm. long. Corolla not known. Fruit 1.1 cm. long, 0.7 cm. wide (perhaps not fully mature), 10-ribbed; calyx-limb persistent, 1–1.5 mm. long, dentate, exceeding the disc. Seed light green, 8 mm. long, 4 mm. wide.

TANZANIA. Kilwa District: Selous Game Reserve, Nungu thicket, 7 Nov. 1970, *Ludanga* in *M.R.C.* 1169! & 18 Jan. 1977, *Vollesen* in *M.R.C.* 4344!
HAB. *Brachystegia microphylla* thicket; ± 700 m.
DISTR. T 8: not known elsewhere

SYN. *C. sp. nov. aff. C. racemosa* Lour. sensu Vollesen in Op. Bot. 59: 68 (1980)
    C. sp. K sensu Bridson in K.B. 36: 852, fig. 10 N–S (1982)

NOTE. With regard to vegetative characters this species bears a strong resemblance to *C. racemosa* Lour., but the lobed bracteoles and well-developed calyx-limb render it clearly distinct.

22. C. rhamnifolia (*Chiov.*) *Bridson* in K.B. 38: 320 (1983). Type: Somalia, Fra Bur Eibi to Saheroi, *Paoli* 1163 (FT, holo.!)

Small shrub 1.2–3 m. tall, with numerous short branches; very young branches pubescent, older branches covered with ash-grey bark. Leaves appearing after flowers, narrowly obovate, 0.9–5 cm. long, 0.7–2.8 cm. wide, obtuse to rounded at apex, sometimes apiculate, cuneate at base, glabrous save for the midrib which is puberulous above; marginal nerve apparent; domatia absent; petiole 1–2 mm. long; stipule-limbs triangular, ± 2 mm. long, gradually tapering to an acute or acuminate apex. Flowers precocious, (5–)6–7-merous, scented, terminal on short lateral spurs, solitary (or ?paired, see note); pedicel ± 1 mm. long, immediately subtended by a pair of young foliage leaves which are in turn subtended by 1–2 pairs of light rusty-brown, narrowly ovate scarious bracts 2–5 mm. long. Calyx-tube glabrous to puberulous, 1.5 mm. long; limb ± 0.25 mm. long, ± subequalling the disc, truncate. Corolla-tube 0.8–1.2 cm. long, 2.5–7.5 mm. wide at throat; lobes oblong to oblong-elliptic, 1.2–2.3 cm. long, 0.3–0.6 cm. wide, acute. Fruit 8–9 mm. long, 7 mm. wide; stipe extending to 2–3 mm. Seeds fawnish brown, 7–9 mm. long, 2.5–4 mm. wide; surface with very close, fine reticulations.

KENYA. Tana River District: Galole–Garissa road, Nov. 1964, *Makin* in *E.A.H.* 13041! & 13048! & 21 Dec. 1964, *Gillett* 16505!
DISTR. K 7; Somalia
HAB. *Acacia-Commiphora* bushland on red sandy loam; 95 m.

SYN. *Paolia jasminoides* Chiov. in Result. Sci. Miss. Stef.-Paoli, Coll. Bot. 1: 93 (1916); Tennant in K.B. 22: 436, t. 2, fig. 1 (1968). Types: Somalia, between Erenlei and Berdale, *Paoli* 931 bis (FT, syn.!) & between Baghei Gadudu and Audinle, *Paoli* 984 (FT, syn.!), *non Coffea jasminoides* Hiern (1876)
    Plectronia rhamnifolia Chiov. in Result. Sci. Miss. Stef. Paoli, Coll. Bot. 1: 95 (1916) & in Fl. Somala 2: 243, fig. 143 (1932)

Canthium rhamnifolium (Chiov.) Cufod., E.P.A.: 1010 (1965), *nom non rite publ.*
Coffea paolia Bridson in K.B. 34: 376 (1979) & in K.B. 36: 852, fig. 11 (1982). Type as for *Paolia jasminoides*

NOTE. This species links *Coffea* sensu stricto to *Psilanthus*. The flower type of *C. rhamnifolia* is typical of *Coffea*, with both anthers and stigma well exserted, while the terminal position of the flowers and presence of scarious bracts is frequently met in *Psilanthus*. Leroy (in Ass. Sc. Intern. Café, 9ᵉ Coloque, Londres: 475 (1980)) placed *C. rhamnifolia* in *Coffea* subgen. *Baracoffea* (Leroy) Leroy together with three Madagascan species having the corolla-tube distinctly longer than the lobes. The illustration of this species in K.B. 22: 436 (1968) is somewhat erroneous, especially with regard to the inflorescence; no 2- or 3-flowered inflorescences have been noted.

86. PSILANTHUS

Hook.f. in G.P. 2: 115 (1873); Hiern in F.T.A. 3: 185 (1877)

Shrubs or small trees. Leaves opposite, petiolate, papery to subcoriaceous; domatia usually present, hairy; stipules shortly sheathing, with a truncate to triangular limb usually bearing a linear to subulate lobe, often caducous. Flowers appearing precociously or not, ♂, (4–)5-merous, terminal, less often axillary or both terminal and axillary, sessile to shortly pedicellate; bracts and bracteoles absent or glumaceous, flower sometimes subtended by young leaves above the bracteoles. Calyx-tube ovate to campanulate; limb subequalling the disc, 5-lobed or irregularly toothed, and margin beset with colleters. Corolla white; tube cylindrical but widened at apex, equalling to over twice as long as the lobes, glabrous or sparsely pubescent outside; lobes contorted in bud, spreading. Anthers set at mouth of corolla, sessile, linear, shortly apiculate at apex, scarcely hastate at base, usually attached above to well above the mid-point, with the apical part exserted or sometimes entirely included, occasionally tending to adhere towards base. Disc annular. Ovary 2-locular; placentas small, attached to middle or somewhat above middle of septum; style cylindrical, slender, glabrous, usually short, reaching less than ⅓ of the way up the corolla-tube; stigma included, bifid; arms linear to narrowly oblong or ?lanceolate, sometimes not separating. Fruit a drupe, ellipsoid or broader than long and distinctly 2-lobed, containing 2, 1-seeded cartilaginous pyrenes; calyx-lobes greatly or not at all accrescent. Seed oblong-ellipsoid, grooved on inner face; endosperm pale in colour, horny, asymmetrically folded from groove, embryo somewhat curved; testa thin, smooth at low magnifications.

Subgen. **Afrocoffea** *(P.Moens) Bridson* in K.B. 42: 454 (1987)

Coffea sect. *Paracoffea* Miq., Fl. Ind. Batavae: 308 (1856)

Coffea subgen. *Afrocoffea* P.Moens in B.J.B.B. 32: 131 (1962)

Cofeanthus A. Chev., Caféiers du Globe 3: 226 (1947), invalid name based on *Psilanthus ebracteolatus* Hiern

Paracoffea (Miq.) Leroy in Journ. Agric. Trop. Bot. Appl. 14: 276 & 606 (1967), excl. subgen. *Insularoparacoffea*, comb. non rite publ.

Psilanthus subgen. *Paracoffea* (Miq.) Leroy in Ass. Sci. Intern. Café, 9ᵉ Colloque, Londres: 475 (1980)

A subgenus of ± 18 species in Tropical Africa, India, Malesia and northernmost Australia (Torres Strait Is.). Three species occur in western tropical Africa, with three additional rather poorly known species from the Flora area.

Subgen. *Psilanthus* is restricted to W. Africa and the Zaire Basin. It contains *P. mannii* Hook.f. and possibly some other poorly known taxa and is easily distinguished from subgen. *Afrocoffea* by the large accrescent calyx-lobes.

Corolla-lobes distinctly acuminate; flowers appearing with
    mature leaves, both terminal and axillary; leaves acuminate
    and usually apiculate at apex . . . . *1. P. semseii*
Corolla-lobes acute or obtuse; flowers appearing precociously
    or with immature leaves, always terminal on short lateral
    branches:
    Corolla 4-lobed, tube 3.9 cm. long; immature leaves elliptic *2. P. sp. A*
    Corolla ?5-lobed, tube 1.3–3.2 cm. long; leaves obovate,
      obtuse to rounded at apex . . . . *3. P. leroyi*

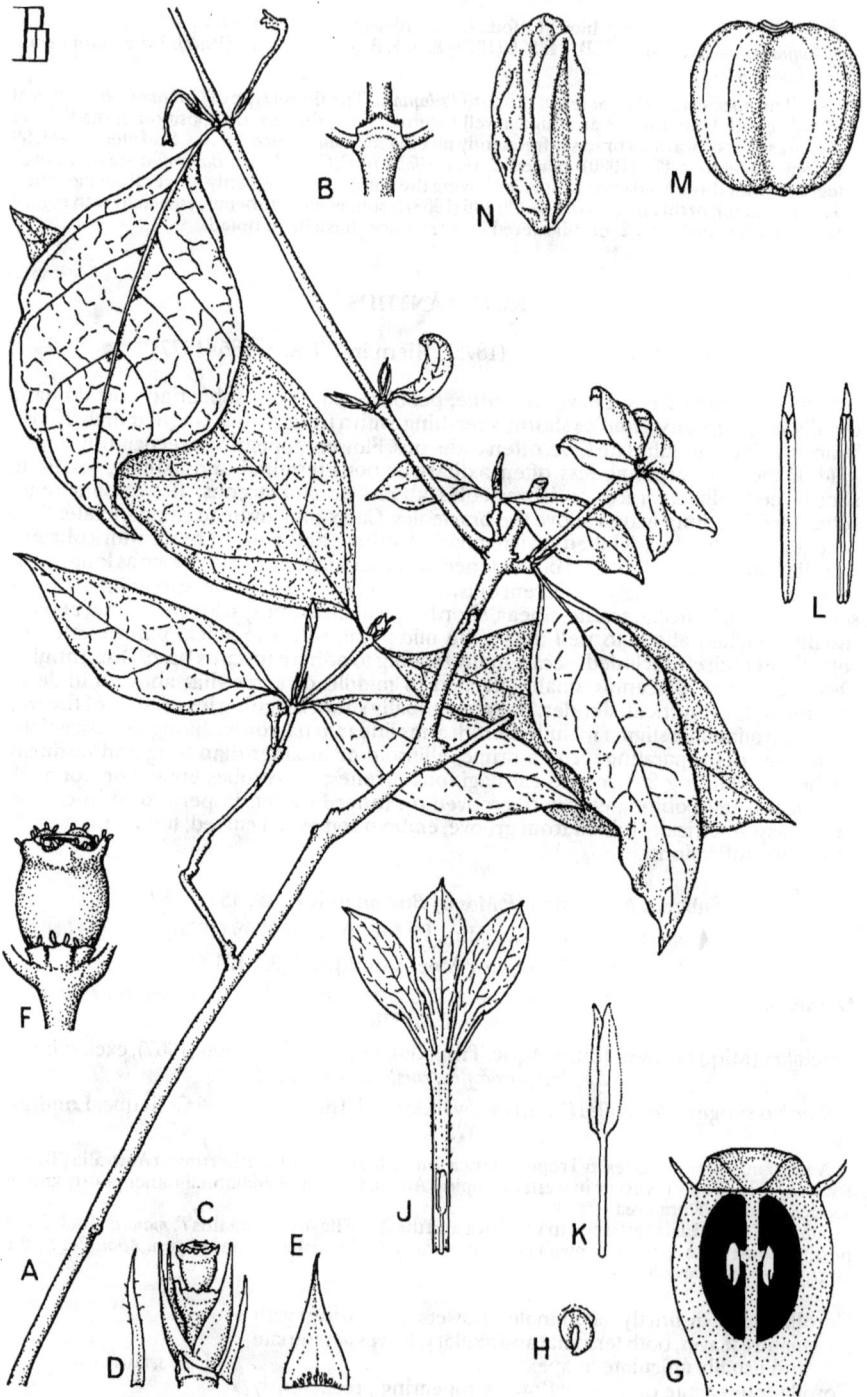

FIG. 125. *PSILANTHUS SEMSEII* — **A**, flowering branch, × ⅓; **B**, stipule, × 3; **C**, inflorescence, × 3; **D**, lower scarious bract, × 3; **E**, upper scarious bract, × 3; **F**, calyx with bracteolar sheath, × 10; **G**, longitudinal section through ovary, × 20; **H**, ovules, from front, × 20; **J**, half corolla, × 1; **K**, style and stigma, × 3; **L**, stamen, 2 views, × 4; **M**, fruit, × 3; **N**, malformed seed, × 4. A, C–L, from *Semsei* 3400; B, M, N, from *Rodgers et al.* s.n. Drawn by Diane Bridson.

1. P. semseii *Bridson* in K.B. 36: 584, fig. 12 (1982). Type: Tanzania, Ulanga District, Magombera Forest, *Semsei* 3400 (K, holo.!, EA, iso.)

Shrub 0.95–2.5 m. tall; young branches covered with reddish-brown bark, glabrous. Leaf-blades elliptic to broadly elliptic, 4.2–10.2 cm. long, 1.9–4.4 cm. wide, acuminate and usually apiculate at apex, cuneate at base, thinly papery to papery, drying yellowish or bluish green, not discolourous, midrib and lateral nerves whitish beneath, entirely glabrous; hair-lined domatia present; petiole 2–3 mm. long; stipules truncate to shortly triangular, 1–2 mm. long, apiculate. Flowers sweet-scented, both terminal and axillary, solitary, sessile, immediately subtended by a membranous stipule-like bracteole beset with colleters on the margin and often bearing 2 foliate appendages up to 3 mm. long, which are in turn subtended by 2 pairs of triangular glumaceous bracteoles, up to 5 mm. long, beset with colleters towards base inside. Calyx-tube ovate, ± 1 mm. long, glabrous; limb ± 0.25 mm. long, almost equalling disc, divided to about halfway into shortly triangular lobes, each bearing several colleters along the margin. Corolla-tube 1.3–2.8 cm. long, 2–3 mm. wide at apex, sparsely pubescent outside, glabrous inside; lobes narrowly ovate to ovate, 1.5–1.8 cm. long, 6–7 mm. wide, distinctly acuminate. Anthers attached near the apex with only tips exserted. Style ± 5 mm. long; stigma-lobes lanceolate, 5 mm. long, very thinly membranous, positioned well below the anthers. Fruit bilobed, 7 mm. long, 8 mm. wide, sessile. Seed (imperfectly developed) oblong-ellipsoid, distinctly grooved on ventral face. Fig. 125.

TANZANIA. Ulanga District: near Mbingu, Ndagara, 30 Sept. 1952, *Carmichael* 130! & Magombera Forest Reserve, 20 May 1979, *Rodgers, Homewood & Hall!* & 8 Nov. 1961, *Semsei* 3400!
DISTR. T 6; not known elsewhere
HAB. Forest; 250 m.

SYN. [*Paracoffea bengalensis* sensu Leroy in Comptes Rendus Acad. Sci., Paris, sér. D, 291: 595 (1980), *non* (Roem. & Schultes) Leroy (1967)]
[*P. sp. A* sensu Leroy in Comptes Rendus Acad. Sci. Paris, sér. D, 291: 595 (1980), *nomen* & in Bull. Mus. Hist. Nat. Paris sér. 4, 3, sect. B, Adansonia 3: 252, t. 1 (1981)]
Rubiaceae gen. nov. aff. Coffea sensu Vollesen in Op. Bot. 59: 73 (1980)

NOTE. It is difficult to decide if there are 2 small collateral ovules attached near the centre of the septum or one ovule with placental tissue; more material is required for clarification of this point.

2. P. sp. A; *Bridson* in K.B. 36: 855, fig. 13 A–F (1982).

Small glabrous bush to 1 m. high; young branches slender, covered with very thin buff bark with reddish flecks, peeling; older branches dull buff. Only immature leaves known; blades drying blackish, ± elliptic, membranous, glabrous; stipules brown when dry, triangular 1.5–2 mm. long. Flowers terminal on very short lateral branches up to 6 mm. long, sessile, subtended by a membranous stipule-like collar up to 0.75 mm. long, with a shallow irregularly lobed margin, sometimes bearing 2 immature leaves which are in turn subtended by 1–2 pairs of glumaceous bracteoles; bracteoles red-brown when dry, linear to linear-triangular, 3–5 mm. long, caducous. Calyx-tube ± 0.75 mm. long; limb-tube ± 0.5 mm. long, bearing unequal lobes; lobes 4–5, up to 3 mm. long, bearing several colleters along margin. Corolla glabrous; tube 3.9 cm. long, 3.5 mm. wide at top, 1.5 mm. wide at base; lobes 2.3 cm. long, 6.5 mm. wide, acute. Anthers 4–5, 1 cm. long, attached above a point ⅔ from the base, with upper portion exserted and twisted, sometimes with basal portions tending to adhere to each other. Style 0.7 mm. long; stigma 3 mm. long, rather thickened, narrowly elliptic-linear, with arms not separating. Fruit not known.

TANZANIA. Ulanga District: Masagati [Massagati], Nyama [Njame] R., 27 Nov. 1931, *Schlieben* 1487!
DISTR. T 6; known only from the above specimen
HAB. Deciduous woodland; ± 500 m.

SYN. *Psilanthus sp. B* sensu Leroy in Comptes Rendus Acad. Sci, Paris, sér. D, 291: 595 (1980) & in Bull. Mus. Hist. Nat., Paris, sér. 4 , 3, sect. B, Adansonia 3: 254, t. 2 (1981)

NOTE. Although I think it very likely that this species is correctly placed in this genus (on account of the bracteole type), I would not consider formally describing it in the absence of seeds as the chief generic distinction between *Argocoffeopsis* (including *Argocoffea*) (from West Africa, Zaire Basin and Zambia) and *Psilanthus* lies here (the seeds of *Argocoffeopsis* being plano-convex to hemispherical without a groove). The style of *Psilanthus* is usually much shorter (as in this species) than in *Argocoffeopsis*. The 4-lobed corolla of this species is not common in *Psilanthus*, but as flowers with 4–5 calyx-lobes and 4–5 anthers have been noted on the above specimen it is possible that 5-lobed corollas also occur. Furthermore, the stigma is perhaps not typical, but this would seem insufficient reason for excluding it from *Psilanthus* at present.

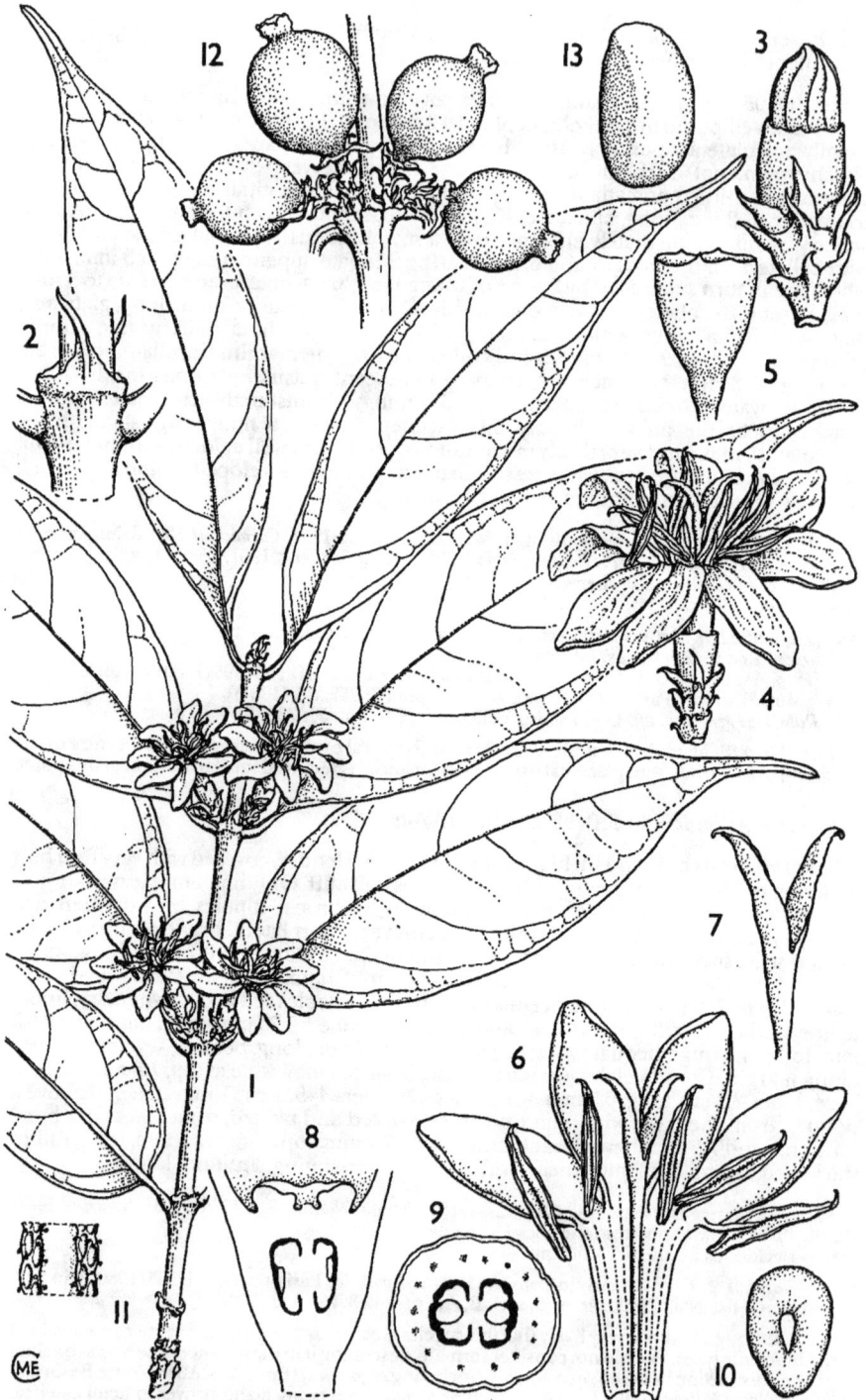

FIG. 126. *CALYCOSIPHONIA SPATHICALYX* — 1, flowering branch, × ⅔; 2, stipules, × 3; 3, flower-bud, with bracteoles, × 4; 4, flower with bracteoles, × 2; 5, calyx, × 4; 6, half corolla, × 3; 7, stigma, × 4; 8, longitudinal section of ovary, × 12; 9, transverse section of ovary, × 12; 10, placenta with ovule, × 14; 11, detail of anther, × 10; 12, fruits, × 1; 13, seed, lateral view, × 2. 1, 2, from *de Witte* 4201; 3–7, from *Gutzwiller* 1843; 8–10, from *Lebrun* 1937; 11, from *Gossweiler* 14020; 12, 13, from *Semsei* 1458. Drawn by Mrs M.E. Church, with 11 by Sally Dawson.

3. P. leroyi *Bridson* in K.B. 36: 857, fig. 13 G–J (1982). Type: Uganda, Karamoja District, Mt. Zulia, *J. Wilson* 911 (EA, holo.!, K, iso.!)

Glabrous shrub to 2–6 m.; young branches light brown or red, tending to flake when older. Leaves restricted to short spurs off the main branches; blades (possibly not fully mature) obovate, 2.5–5.2 cm. long, 1.6–3.3 cm. wide, obtuse to rounded at apex, attenuate at base, papery, dull on both faces; domatia absent or occasionally a few hairs present in the nerve-axils; petiole 3–5 mm. long; stipules truncate, up to 1 mm. long. Flowers appearing precociously or partly so, terminal on short lateral branches, subsessile or with pedicels not exceeding 1 mm., subtended by a pair of leaves. Calyx-tube ± ovate, ± 2 mm. long, glabrous or minutely puberulous; limb up to 0.25 mm. long, equalling disc, divided into 5 shallow lobes bearing colleters along the margin. Corolla-tube 1.3–3.2 cm. long, 2 mm. wide at throat, sparsely covered with short adpressed hairs outside; lobes oblong-elliptic or ovate, 7–11 mm. long, up to 8 mm. wide, obtuse to rounded. Anthers not exserted. Style very short, 0.25–4 mm. long; stigma-lobes linear, divergent, up to 2 mm. long. Fruit oblong-ellipsoid, 7 mm. long, 6 mm. wide, minutely puberulous; pedicel lengthening to 3 mm. Immature seed oblong-ellipsoid, 6 mm. long, grooved.

UGANDA. Karamoja District: near Mt. Zulia, Karapedo R., Mar. 1960, *J. Wilson* 911!
DISTR. U 1; Ethiopia
HAB. River bank; 1220 m.

SYN. *Paracoffea sp. C*; Leroy in Bull. Mus. Hist. Nat. Paris, sér. 4, sect. B, Adansonia 3: 256, t. 3 (1981).

87. CALYCOSIPHONIA

Robbrecht in B.J.B.B. 51: 370 (1981)

Shrubs or small trees. Leaves opposite, petiolate, papery to subcoriaceous, glabrous; domatia absent or present in axils of lateral nerves but rather inconspicuous and sparsely ciliate; stipules shortly sheathing, with triangular limbs bearing a subulate to linear caducous lobe, pubescent and beset with colleters inside. Flowers ⚥, 7–8-merous, axillary with 1–4 flowers borne in each axil, stalked; bracts 2–3, cupular or saucer-shaped (referred to as calyculi), usually totally obscuring the pedicel, pubescent and beset with numerous colleters inside, often with linear or subfoliaceous lobes. Calyx-tube campanulate, obscured by the calyculi; limb truncate or with rudimentary teeth. Corolla white or ? yellow, glabrous outside; tube cylindrical below but widening above, slightly shorter than or slightly longer than the lobes; lobes contorted in bud, 7–8, spreading. Stamens 7–8, attached at the mouth of corolla-tube, exserted, erect; filaments, about ⅓ the length of the anther; anthers narrowly linear-lanceolate, with thecae transversely divided into cells, apiculate at apex, scarcely hastate at base, attached about ⅓ of their length from the base. Disc annular. Ovary 2-locular; placentas small, fleshy, attached near the middle of the septum; ovules 1–2, deeply immersed in the placenta; style cylindrical, slender, glabrous, exceeding the corolla-tube; stigma exserted, 2-lobed. Fruit rather fleshy, 1–2-seeded; calyx-limb persistent. Seeds dark in colour, ± hemispherical; small depression present on apex or on curved face near apex; testa apparently absent, surface smooth to somewhat rugulose at low power magnifications; endosperm dark purplish in colour (preserved material), entire; embryo straight, with radicle inferior.

A tropical African genus with 2 species.

C. spathicalyx (*K.Schum.*) *Robbrecht* in B.J.B.B. 51: 373 (1981) & in Distr. Pl. Afr. 22, map 757 (1981). Type: Cameroun, Yaoundé, *Zenker & Staudt* 79 (B, holo.†, K, iso.!)

Small tree or straggling shrub with arching branches 2–8.5 m. tall; young branches glabrous or puberulous; older branches covered with thin light brown shiny bark. Leaves commonly drying brownish black, glabrous; blades oblong-elliptic, elliptic or sometimes oblanceolate to narrowly obovate, 6.5–22 cm. long, 2.3–8 cm. wide, acute or rounded then abruptly acuminate at apex with the acumen 0.8–4 cm. long and 1–4.5 mm. wide, acute to obtuse or sometimes cuneate at base; tertiary nerves reticulate, raised on both faces; petioles 0.5–1.3 cm. long; stipule-limbs 1.5–3 mm. long, light brown when mature; lobe subulate to linear, 2–5 mm. long. Flowers 1–3 per axil; inflorescence-stalks 2–4(–7 in fruit) mm. long; calyculi (2–)3, 1–5 mm. long, usually the upper one the biggest and often provided with either linear lobes, 1–5 mm. long, or subfoliaceous lobes, 0.5–1.7 cm. long and 2.5–4 mm. wide. Calyx-tube 1.5–2 mm. long; limb 2.5–7 mm. long, characteristically split down one side, ± truncate or shortly and unequally toothed, glabrous and without colleters inside. Corolla-tube 0.7–1.4 cm. long, 3–4.5 mm. wide at top, 0.75–1 mm. wide at base, pubescent inside; lobes oblong, oblong-lanceolate or occasionally narrowly ovate, 7–1.7 cm. long, 3–6 mm. wide, acute to acuminate. Anthers dark in colour. Stigma exserted; lobes ± linear or sometimes slightly spathulate. Fruit green-yellow, ellipsoid, 0.9–1.3 cm. long, 0.4–1.2 cm. wide, (1–)2-seeded. Seeds blackish, usually hemispherical, but occasionally biconvex (in case of 1-seeded fruit), 0.6–1.1 cm. across, sometimes with an indistinct ovate ridge which becomes more sharply defined at the narrow (apical) end. Fig. 126.

UGANDA. W. Nile District: E. Madi, Zoka Forest, Jan. 1952, *Leggat* 74!; Toro District: Bwamba Forest, 2 Feb. 1945, *Greenway & Eggeling* 7066! & 20 Jan. 1932, *Hazel* 149!
KENYA. Kwale District: Shimba Hills, Lango ya [Longo] Mwagandi, 22 May 1968, *Magogo & Glover* 1100!
TANZANIA.   Morogoro District: Turiani, Manyangu Forest, Nov. 1953, *Semsei* 1458!; Ulanga District: Magombera Forest Reserve, 24 May 1976, *Vollesen* in *M.R.C.* 3683!; Ulanga/Njombe District: Ruhudji, 1931, *Schlieben* 1474!
DISTR. U 1, 2; K 7; T 3, 6, ?7; Ivory Coast, Ghana, Nigeria, Central African Republic, Cameroun, Zaire, Gabon, Sudan, Malawi and Angola
HAB. Forest; 250–915 m.

SYN. *Coffea spathicalyx* K.Schum. in E.J. 23: 587 (1897); Froehner in N.B.G.B. 1: 232 (1897) & in E.J. 25: 266 (1898); De Wild., Les Caféiers: 41 (1901) & in Ann. Jard. Buitenz, suppl. 3: 379 (1910); Pellegrin, Fl. Mayombe 3: 25 (1938); Thomas in Emp. Journ. Exp. Agric. 12: 9 (1944); F.P.S. 2: 432 (1952); Hepper in F.W.T.A., ed. 2, 2: 156 (1963)
      Calycosiphonia spathicalyx (K.Schum.) Lebrun in Mem. Inst. Roy. Col. Belge, Sect. Sci. Nat. Méd. Mem. 8°, 11(3) (Rech. Morph. & Syst. Caf. Congo): 68, Pl. 7 (1941); Leroy in Journ. Agric. Trop. Bot. Appl. 8: 710 (1961), *nom. invalid.*
      Tricalysia spathicalyx (K.Schum.) A.Chev., Caféiers du Globe 2: 36, t. 144 (1942) & 3: 240 (1947), *nom. invalid.*

NOTE. The disjunction in the distribution of this species would seem to suggest that the coastal element could represent a separate subspecies, but I have found little morphological evidence to support this on the limited material available.

88. BELONOPHORA*

Hook.f. in G.P. 2: 109 (1873) & Ic. Pl. 12, t. 1127 (1873); Keay in B.J.B.B. 28: 297 (1958)

Diplosporopsis Wernham, Cat. Talb. Nig. Pl.: 47 (1913)

Trees or shrubs with much the appearance of *Coffea*. Leaves paired; stipules elliptic, lanceolate or triangular, often tapering-subulate, deciduous. Flower sessile in fascicles on both sides of successive nodes. Calyx bracteolate at the base; tube campanulate; limb with 5 rounded imbricate lobes. Corolla salver-shaped, the tube slender, somewhat narrowed at the base, adpressed pubescent outside, glabrous at the throat; lobes 5, narrowly oblong, contorted to the right as seen in plan view, adpressed puberulous. Anthers sessile, included in the corolla-tube, linear. Disc cushion-shaped. Ovary 2-locular, each locule with a pendulous placenta bearing 2 collateral ovules inserted on its inner face; style included, divided into 2 papillose subulate branches. Fruits globose or ellipsoid-globose, 2-locular, with leathery wall, marked at the apex with evident remains of the disc and calyx-lobes. Seeds 2–4 per fruit, pendulous; albumen horny; testa apparently absent; embryo with radicle superior.

A small genus of about half a dozen species mainly in western Africa.

B. hypoglauca (*Hiern*) *A.Chev.* in Rev. Bot. Appliq. 19: 398 (1939); Keay, F.W.T.A., ed. 2, 2: 158 (1963). Types: Angola, Pungo Andongo, near Catete, *Welwitsch* 3174 (LISU, syn., BM, K, isosyn.!) & between Catete and Pedra Pungo, *Welwitsch* 3175 (LISU, syn., BM, K, isosyn.!) & Catete, R. Tangue, *Welwitsch* 3176 (LISU, syn., BM, isosyn.)

Large shrub or small tree (1.5–)3–12 m. tall with much the habit of a coffee; young stems dark, shiny and wrinkled when dry, later covered with pale corky bark. Leaf-blades oblong or oblanceolate-oblong to elliptic-obovate, 6.5–20(–27) cm. long, 1.7–9(–12.5) cm.

* This genus was omitted in error from the key in part 1.

FIG. 127. *BELONOPHORA HYPOGLAUCA* — 1, flowering branch, × ⅔; 2, stipules, × 2; 3, flower, × 4; 4, calyx with bracts, × 12; 5, corolla opened out, × 6; 6, style, × 12; 7, longitudinal section of ovary, × 12; 8, placenta with ovules, 2 views, × 30; 9, fruits, × 1; 10, seed, × 3. 1–8, from *Eggeling* 1510; 9, from *Dawe* 538; 10, from *Mutimushi* 3277. Drawn by Mrs M.E. Church.

venation obscure above; petiole 0.5–1.3 cm. long; stipules subulate, attenuate from a triangular base, 1–1.8 cm. long. Flowers scented, in dense fascicles up to 1.5 cm. long, 2 cm. wide, 3–10-flowered, mostly in the axils of fallen leaves; bracteoles ovate, 1 mm. long, 1 mm. wide, ciliate. Calyx-tube 2 mm. long; lobes elliptic, rounded, ± 1–2 mm. long and wide, usually ciliate. Corolla white; tube 0.9–1.7(–1.9) cm. long; lobes 3–9(–11) mm. long, 1–4(–4.5) mm. wide, finely adpressed pubescent outside. Anthers 4.5–5.5 mm. long, their tips 1–4(–6) mm. below the throat. Fruits red, 1.2–1.3(–1.7) cm. long, 1–1.5 cm. wide; stipe ± 4 mm. long. Fig. 127.

UGANDA. W. Nile District: W. Madi, Meturu, Amua, Mar. 1935, *Eggeling* 1791!; Bunyoro District: Budongo Forest, Feb. 1935, *Eggeling* 1613!; Mengo District; Mawokota, Feb. 1905, *E. Brown* 159! TANZANIA. Buha District: Mkalinzi, Kavura Hill, Nov. 1956, *Procter* 561!
DISTR. U 1, 2, 4; T 4; Sierra Leone to N. Nigeria, Cameroun, ?Zaire, Central African Republic, Sudan (Equatoria), Zambia and Angola
HAB. Evergreen forest, woodland, particularly near streams; 1050–1560 m.

SYN. *Coffea hypoglauca* Hiern in Trans. Linn. Soc., ser. 2, 1: 173 (1876), F.T.A. 3: 183 (1877) & Cat. Afr. Pl. Welw. 1: 490 (1898)
Belonophora glomerata M.B.Moss in K.B. 1929: 195 (1929); F.P.S. 2: 425 (1952). Type: Sudan, Imatong Mts., Laboni Forest, *Chipp* 46 (K, holo.!)

NOTE. The single specimen from T 4 seen has flowers twice the size of those from Uganda, but considerable variation is found in W. Africa; nevertheless the distribution of large- and small-flowered variants might reward further study.

89. HEINSENIA

K.Schum. in E.J. 23: 453 (1897)

Aulacocalyx Hook.f. subgen. *Heinsenia* (K.Schum.) Verdc. in K.B. 36: 514 (1981)

Trees or shrubs with branching similar to that of many *Rothmannia* species and characteristic evergreen foliage. Leaves opposite, petiolate, thinly coriaceous, usually with ± transversely parallel venation; domatia absent or present and pubescent; stipules triangular, acute or acuminate, soon falling. Inflorescences 1–several or occasionally ± many-flowered on very short terminal branchlets subtended by a single leaf but often lateral and pseudoaxillary owing to the sympodial growth of the stem; bracts small. Calyx-tube ovoid to tubular; limb shortly cylindrical, with 5 triangular to lanceolate teeth. Corolla ± pubescent to densely adpressed silky outside, narrowly campanulate; corolla-lobes 5, ovate to lanceolate. Anthers linear, elongate, apiculate, sessile, usually medifixed, included or slightly to completely exserted; pollen grains simple. Ovary 2-locular; each locule with a pendulous placenta carrying 2–10 embedded ovules; style filiform or elongate-clavate, glabrous or rarely pubescent, apical part sometimes 10-ribbed, and divisible into lobes up to 4–5 mm. long bearing the stigmatic surfaces; the whole style is divisible almost to the base if pulled gently. Fruit fleshy, 1–several-seeded, subglobose, crowned by the persistent calyx-limb. Seeds subglobose with abundant endosperm, apparently devoid of a testa; embryo with radicle superior.

A monotypic genus restricted to tropical Africa, closely related to *Aulacocalyx* Hook.f.

I reluctantly followed Petit (B.J.B.B. 32: 183 (1962)) in merging the two genera as did Hallé (Fl. Gabon 17, Rubiacées 2: 154 (1970)) but willingly conform with Robbrecht and Puff's decision (E.J. 108: 124 (1986)) to maintain *Heinsenia*. It has a distinctly different corolla shape although in other characters, particularly the nature of the placenta and ovules (in both genera the placentas were originally mistaken for ovules), also the distinctive branching and leaf-venation, it does stand very close to *Aulacocalyx*. See additional note, p. 736.

H. diervilleoides *K.Schum.* in E.J. 23: 454 (1897); T.T.C.L.: 500 (1949); I.T.U., ed. 2: 346 (1952); Brenan in Mem. N.Y. Bot. Gard. 8: 452 (1954); Garcia in Mem. Junta Invest. Ultram., sér. Bot., 4: 40 (1958) & Mem. Junta Invest. Ultram. sér. II, 6: 26 (1959); K.T.S.: 445 (1961); Robbrecht & Puff in E.J. 108: (1986). Type: Tanzania, E. Usambara Mts., Derema [Nderema], *Heinsen* 23 (B, holo.†, BR, K, iso.!)

Shrub or small understorey tree 3–12(–15) m. tall, with rather bushy crown and striate pale grey-brown bark; stems pale, glabrous or the young parts slightly pubescent, the bark ± fissuring longitudinally but not peeling off. Leaf-blades often red when young or with

FIG. 128. *HEINSENIA DIERVILLEOIDES* subsp. *DIERVILLEOIDES* — 1, flowering branch, × ⅔; 2, stipules, × 4; 3, flower-buds, × 3; 4, flower, × 3; 5, half corolla, × 3; 6, stigma, × 5; 7, longitudinal section of ovary, × 8; 8, transverse section of ovary, × 10, 9, placenta with ovule, 2 views, × 30; 10, young fruit, × 2; 11, mature fruits, × 1; 12, seed, × 3. 1–5, 7–9, from *Verdcourt* 108; 6, 11, from *Drummond & Hemsley* 1363; 10, from *Maas Geesteranus* 5278; 12, from *Gardner* 18566. Drawn by Mrs M.E. Church.

the nerves red beneath, oblong to oblanceolate or oblong-elliptic or much narrower and lanceolate to oblanceolate, 5–16.5 cm. long, 1–7.7 cm. wide, acuminate at the apex, usually distinctly so with acute or attenuate point, rarely only shortly, cuneate at the base, slightly coriaceous, glabrous or very sparsely to densely adpressed bristly on the nerves beneath; petiole often red when young, 4–10 mm. long; stipules 4–7 mm. long, with thickened median part drawn out to a subulate point. Inflorescence elements (1–)2–6(or more)-flowered; pedicels 2–5(-6) mm. long (the longest fruiting measurements) pubescent. Calyx-tube 1–1.5 mm. long; limb 2–4 mm. long, the lobes with a median thickened part protruding as subulate apices from the thinner sinuses, or sometimes the limb splitting irregularly as well. Corolla mostly white spotted with pink inside, but apparently often pure white or pure pink, narrowly campanulate, but strongly narrowed at the extreme base; tube (0.7–)0.9–1.2 cm. long, sparsely to densely adpressed pubescent outside; lobes ovate, (3.5–)5–8.5 mm. long, 4–6 mm. wide, obtuse. Anthers included or exserted for ⅓ of their length. Style white, glabrous, slightly 10-ribbed. Fruit greenish purple, subglobose, 1.1–1.3 cm. diameter, slightly pubescent particularly near the persistent calyx and the pedicel. Seeds ± ovoid or half-globose, 7 mm. long, 4.5 mm. wide, rugulose.

subsp. **diervilleoides**

Leaf-blades oblong to oblong-oblanceolate or lanceolate to oblanceolate but not elliptic, up to 15 × 7.5 cm., long-acuminate at the apex. Inflorescences mostly but not always several-flowered. Subulate part of calyx-lobes shorter, 0.25–1.5 mm. long. Corolla not drying black. Fig. 128.

UGANDA. Ankole District: Kalinzu Forest, Aug. 1936, *Eggeling* 3175!; Mbale District: Mt. Elgon, Sipi, Feb. 1940, *St. Clair-Thompson in Eggeling* 3947!; Mengo District: Mabira Forest, *E. Brown* 429!
KENYA. Northern Frontier Province: Marsabit, July 1958, *T. Adamson* 11A!; Nandi District: Kaimosi, Sirwa Farm, Nov. 1971, *Tweedie* 4163!; Kwale District: Shimba Hills, Lango ya Mwagandi [Longo Mwagondi], 16 Mar. 1968, *Magogo & Glover* 309!
TANZANIA. Mbulu District: Great North Road, Pienaars Heights, 5 May 1962, *Polhill & Paulo* 2339!; Lushoto District: just below Amani, 12 Mar. 1950, *Verdcourt* 108!; Songea District: Liwiri [Luwiro] -Kiteza Forest Reserve, 18 Oct. 1956, *Semsei* 2525!
DISTR. U 2–4; K 1 (Marsabit), 3–5, 7; T 2, 3, 6–8; E. Zaire, Rwanda, Sudan, Mozambique, Malawi, Zimbabwe
HAB. Rain-forest and other moist evergreen forest; 375–2400 m.

SYN. *Heinsenia lujae* De Wild., Pl. Nov. Hort. Thenensis 1: 9, t. 3 (1904); K.T.S.: 445 (1961) adnot. Type: Mozambique, Morrumbala, *Luja* 357 (BR, holo.) non *Aulococalyx lujae* De Wild. (1915)
  H. sylvestris S.Moore in J.L.S. 40: 85, t. 4 (1911). Types: Zimbabwe, Chirinda Forest, *Swynnerton* 576 & 689 (BM, syn., K, isosyn.!)
  H. brownii S.Moore in J.L.S. 40: 86 (1911). Type: Uganda, Mengo District, Mabira Forest, *E. Brown* 429 (BM, holo.!)
  H. infundibuliformis Petit in B.J.B.B. 31: 3 (1961). Type: Zaire, Masisi, Mbati, *Léonard* 2672 (BR, holo., BM, K, iso.!)
  Aulococalyx diervilleoides (K.Schum.) Petit in B.J.B.B. 32: 184 (1962); Verdc. in K.B. 36: 514 (1981); Fl. Pl. Lign. Rwanda: 542, fig. 181/2 (1982) & Fl. Rwanda 3: 140, fig. 41/2 (1985)
NOTE. Moore (J.L.S. 40: 87 (1911)) keys out 4 species of *Heinsenia* on characters of corolla indumentum, pedicel length, calyx-lobe length and width of the basal part of the corolla-tube but these characters vary too much to allow separate species to be maintained; he also shows the style of *H. sylvestris* divided for most of its length into lobes.
  Several field notes state flowers pink or flowers white and it is not certain if the spots were not mentioned in these cases. Probably they can be absent.

subsp. **mufindiensis** (*Verdc.*) *Verdc.* comb. nov. Type: Tanzania, Iringa District, W. Mufindi, *Brenan, Greenway & Gilchrist* 8261 (K, holo.!)

Leaf-blades distinctly more shortly acuminate, elliptic, up to 7 × 3 cm. Inflorescences 1–2-flowered. Subulate part of calyx-lobes longer, up to 2 mm. long. Corolla drying black.

TANZANIA. Iringa District: W. Mufindi, Nyumbanitu, 1 Nov. 1947, *Brenan, Greenway & Gilchrist* 8261!
DISTR. T 7; not known elsewhere
HAB. Evergreen forest; 1500–2000 m.

SYN. *Aulococalyx diervilleoides* (K.Schum.) Petit subsp. *mufindiensis* Verdc. in K.B. 36: 515 (1981)
NOTE. The material looks more distinctive than the description suggests but on the basis of sparse material subspecific rank is all I care to give it; *Procter* 3370! (Mufindi Scarp Forest Reserve, Oct. 1966) has distinctly long acuminate leaves and is more or less intermediate.

Tribe 21. CREMASPOREAE

One genus only   . . . .   **90. Cremaspora**

90. CREMASPORA

Benth. in Hook., Niger Fl.: 412 (1849)

Shrubs, small trees or sometimes lianes or with subscandent branches; stems and foliage almost glabrous to densely hairy; branches appearing supra-axillary and often supported by rounded leaves very different from the main foliage. Leaves opposite, shortly petiolate, mostly thinly coriaceous, mostly oblong-elliptic; stipules keeled, acuminate, soon deciduous. Flowers sweetly scented, rather small in sessile mostly dense axillary clusters; bracts and bracteoles triangular-acuminate, not forming cupules. Calyx-tube ovoid; limb campanulate with 5 triangular to lanceolate teeth. Corolla white, cylindrical or very narrowly funnel-shaped, glabrescent or pubescent outside, glabrous to pilose at the throat; lobes 5, narrowly oblong, ± obtuse or acute, strictly contorted, pubescent outside. Stamens 5, inserted at the mouth of the tube; filaments very short; anthers buff, dorsifixed, linear, apiculate, ± bifid at the base, fully exserted. Disc shallowly bowl-shaped, becoming more cylindrical in fruit. Ovary 2-locular, with a single ovule in each locule, pendulous from the apex; style green, filiform, exserted, hairy save at extreme base, the apical 1 mm. divisible into 2 lobes which do not spread and bear the stigmatic surface on their inner faces. Fruit ovoid, indehiscent, leathery-walled, 2-locular, (1–)2-seeded. Seeds half-ovoid, compressed, rugulose; testa characteristically finely transversely wrinkled-striate; albumen horny, not ruminate; embryo small with inferior radicle; placenta elongated, lying along whole side of seed.

A genus of 3 or perhaps 4 species occurring in tropical Africa, Comoro Is. and Madagascar; one a rare W. African species and one very variable and widespread.

Most of the taxa described in this genus from Madagascar have been correctly transferred to *Polysphaeria* by Cavaco. Others described from W. Africa belong in *Coffea* and *Tricalysia*. All these genera have a rather similar facies and can be readily confused if the floral details are not carefully observed. *Cremaspora triflora* subsp. *confluens* has been reported a host of a coffee borer. Hiern, Bentham and myself have all reported the radicle as superior.

Venation not markedly raised and closely reticulate when dry
    (very widespread) . . . .   *1. C. triflora*
Venation markedly raised and closely reticulate when dry (T 6,
    Uluguru Mts.) . . . .   *2. C. sp. A*

1. **C. triflora** (*Thonn.*) *K.Schum.* in E. & P. Pf. 4(4): 88 (1891) & in P.O.A. C: 383 (1895); A.Chev., Caféiers du Globe 3: 255 (1947); T.T.C.L.: 493 (1949); F.P.S. 2: 433 (1952); F.F.N.R.: 405 (1962); Keay in F.W.T.A., ed. 2, 2: 148 (1963); Hepper, W. Afr. Herb. Isert & Thonning: 106 (1976); Verdc. in K.B. 35: 131 (1980); Fl. Pl. Lign. Rwanda: 556, fig. 187/2 (1982) & Fl. Rwanda 3: 156, fig. 48/2 (1985). Type: Ghana, near Asiama, *Thonning* 299 (C, holo., C, P-JU, S, iso.)

Shrub or small tree (1–)1.8–9 m. tall, usually much branched, the branches elongate, arched or sometimes subscandent or sometimes a true liane; stems glabrescent to densely hairy or tomentose with grey to rusty hairs. Leaf-blades oblong, oblong-elliptic, obovate-oblong or lanceolate to oblanceolate, (2–)4–18 cm. long, (0.5–)2–9 cm. wide, acuminate at the apex, cuneate to broadly rounded at the base, often drying an olive- or grey-green, glabrous above, almost glabrous to pubescent or hairy beneath, particulaly on the nerves; venation not markedly raised nor closely reticulate; leaves subtending lateral branches mostly rounded-reniform, rounded to mucronulate at the apex, up to ± 3.5–4 cm. in diameter; stipules triangular to triangular-lanceolate, 5–7 mm. long, glabrescent to densely pubescent. Flowers in 2 opposite clusters at each node, the combined inflorescence 2–3 cm. wide. Calyx-lobes 0.5–2 mm. long. Corolla white or yellowish; tube 3–5.5(–10) mm. long; lobes 3–7 mm. long. Style 0.9–1.6 cm. long. Fruit red, 0.7–1.5 cm. long, 5–7(–12) mm. wide, glabrescent to quite densely hairy, crowned with the persistent calyx and sometimes drawn out into a beak beneath it rendering the fruit ± narrowly urceolate. Seeds 5.5–7 mm. long, 3.2–5.5 mm. wide, 2–3.5 mm. thick.

subsp. **triflora**; Verdc. in K.B. 35: 131 (1980)

Leaf-blades often smaller, usually ± 8–12 × 3.5–5 cm. but very variable. Calyx-lobes triangular, usually shorter, 0.5–1.5 mm. long. Fruits not drawn out into a beak. Fig. 129/12–14.

UGANDA. W. Nile District: Koboko [Kobboko], Mar. 1955, *Eggeling* 1839!; Bunyoro District: Budongo Forest, Siba area, Sept. 1935, *Eggeling* 2195!; Mengo District: Kirerema, Nov. 1913, *Dummer* 404!.

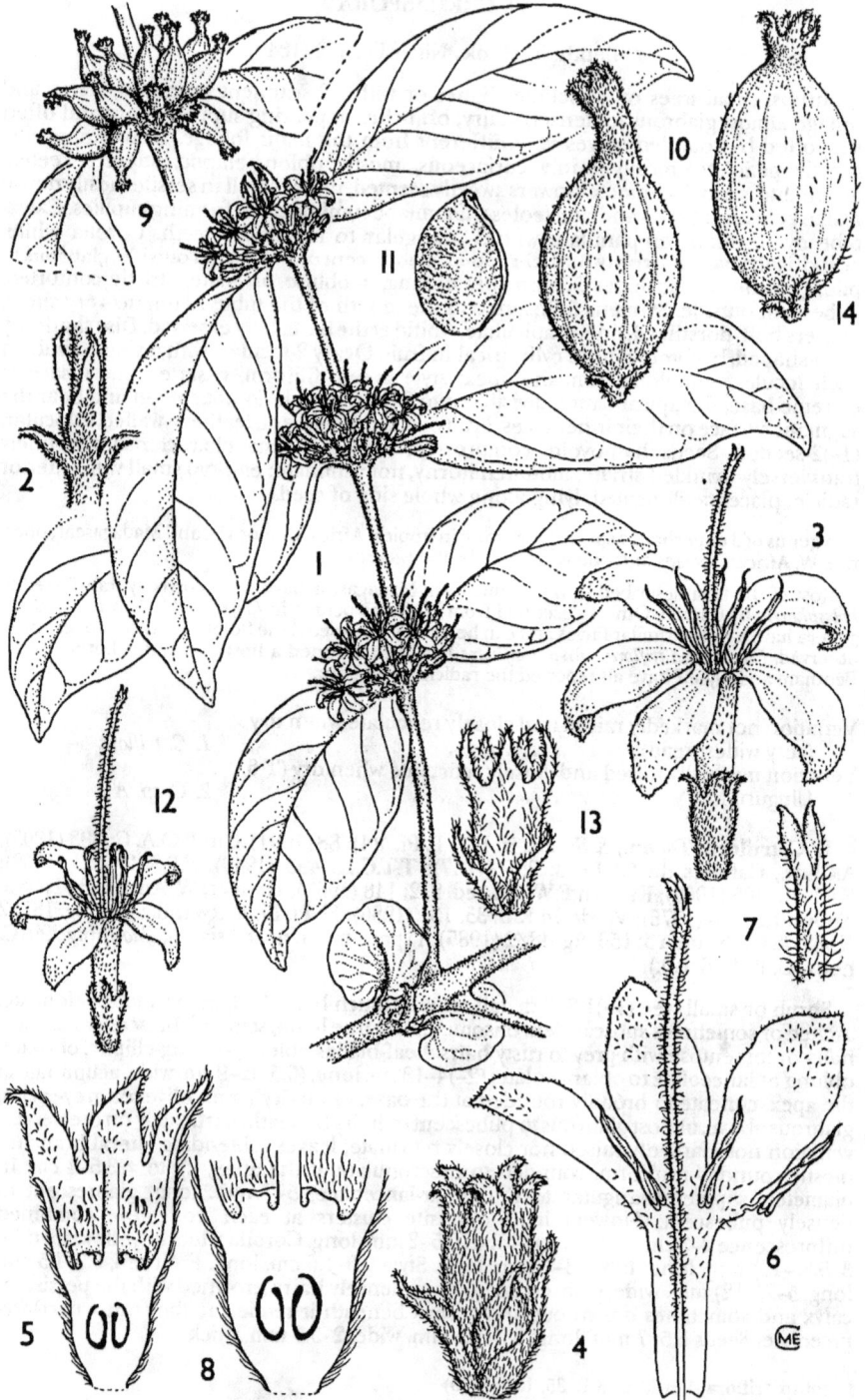

FIG. 129. *CREMASPORA TRIFLORA* subsp. *CONFLUENS* — 1, flowering branch, × ½; 2, stipules, × 2; 3, flower, × 3; 4, calyx and bracts, × 6; 5, longitudinal section of calyx, × 10; 6, half corolla, × 4; 7, stigma, × 10; 8, longitudinal section of ovary, × 16; 9, part of fruiting branch, × 1; 10, fruit, × 4; 11, seed, × 3. Subsp. *TRIFLORA* — 12, flower, × 3; 13, calyx, with bracts, × 8; 14, fruit, × 4. 1–8, from *Faulkner* 1052; 9–11, from *Peter* 55953; 12, from *Dummer* 404; 13, 14, from *Chandler* 1919. Drawn by Mrs M.E. Church.

KENYA. Teita District: Mwatate R. valley, 5.8 km. Mwatate-Wundanyi, 29 Aug. 1970, *Faden et al.* 70/444!

TANZANIA. Bukoba District: Nshamba, Oct. 1935, *Gillman* 570!; Lushoto District: E. Usambara Mts. Sangerawe, 21 Nov. 1947, *Brenan & Greenway* 8360!; Mpanda District: Mahali Mts., Mokoloka, 19 Sept. 1958, *Newbould & Jefford* 2457!; Iringa District: N. Gologolo Mts., 13 Sept. 1970, *Thulin & Mhoro* 963!

DISTR. U 1, 2, 4; **K** 7; **T** 1–4, 6, 7; W. Africa from Cape Verde Is. and Senegal to Cameroun, Gabon, Zaire, Burundi, Sudan, Mozambique, Malawi, Zimbabwe and Angola

HAB. Evergreen forest, thicket; 885–1950 m.

SYN. *Psychotria triflora* Thonn. in Schumach., Beskr. Guin. Pl.: 108 (1827)
Coffea microcarpa DC., Prodr. 4: 499 (1830), *nom. illegit. non* Ruíz & Pavón. Type: Senegal, Cap Rouge, Casamance, *Leprieur & Perrottet* (P, holo.)
C. *hirsuta* G.Don, Gen. Hist. Gard. 3: 581 (1834), as '*hirsutus*'. Type: Sierra Leone, *Don* (BM, holo.!)
Cremaspora africana Benth. in Hook., Niger Fl.: 412 (1849); Hiern in F.T.A. 3: 126 (1877) & Cat. Afr. Pl. Welw. 1: 471 (1898), *nom. superfl.* Type as for *Coffea hirsuta*
C. *bocandeana* Webb in Hook., Journ. Bot. & Kew Gard. Misc. 2: 371 (1850). Type: Cape Verde Is., *Bocandé* (?FT, holo.)
C. *heterophylla* F.Didr. in Kjoeb. Vid. Meddel: 187 (1854), *nom. superfl.* Type as for *Coffea hirsuta*
C. *microcarpa* (DC.) Baill. in Bull. Soc. Linn. Paris 1: 206 (1879), *nom. illegit.* (mentions *Psychotria triflora* in syn.)
?C. *heterophylla* K.Schum. in P.O.A. C: 383 (1895), *nom. illegit., non* F.Didr. Types: Malawi, *Buchanan* 27 (B, syn.†, BM, K, isosyn.!) & 969 (B, syn.†, K, isosyn.!)
?C. *coffeoides* Hemsley in K.B. 1896: 18 (1896). Type: Malawi, N. side of the Ruo, 1895, *Johnston* (K, holo.!)

NOTE. This subspecies is very variable and a number of weakly defined races could be separated. In Uganda and Tanzania (**T** 1, 4) the plants are mostly scandent with spreading hairy stems and small hairy leaves and longer stipules similar to W. African material. In the forests of the Usambara and Uluguru Mts. the plant is rarely scandent and the leaves are distinctly more glabrescent or quite glabrous beneath, the stipules distinctly shorter and more triangular and the stems more sparsely adpressed pubescent. Specimens from the Shagayu Forest, W. Usambaras, are distinctive. *Wallace* 933 from W. Usambaras bears the information 'very tall tree'. Some specimens from the Sudan and Zambia have particularly large fruits, 1.5 × 1.2 cm.; the type of C. *coffeoides* has very small glabrescent fruits 7 × 5.5 mm. and the *Thulin & Mhoro* specimen cited above is similar; C. *heterophylla* described from flowering material is probably the same. Without further material the status of these fruit variations is not clear. Faden says of his 70/444 cited above "fruits turning dark blue"; this needs confirmation and is at variance with all other collectors.

subsp. **confluens** (*K.Schum.*) *Verdc.* in K.B. 35: 132 (1980). Type: Zanzibar, Mkokotoni [Kokotoni], *Stuhlmann* 607 (B, syn.†) & Tanzania, Uzaramo District, Tambani, *Stuhlmann* 6144 (B, syn.†)

Leaf-blades often larger, up to 17 × 7.5 cm. Calyx-lobes triangular-lanceolate, usually longer, 1.5–2 mm. long. Fruits shortly but distinctly drawn out into a beak at the apex beneath the persistent calyx-limb. Fig. 129/1–11.

KENYA. Kwale District: Mwachi, Sept. 1936, *Dale* in F.D. 3559!; Kilifi District: Mida, Sept. 1929, *R.M. Graham* in F.D. 2072!; Lamu District: Utwani Forest, Dec. 1956, *Rawlins* 280!

TANZANIA. Moshi District: 8 km. S. of Moshi, by R. Njoro, 3 Nov. 1955, *Milne-Redhead & Taylor* 7213!; Tanga District: Magunga, 11 Oct. 1952, *Faulkner* 1052!; Morogoro District: Kimboza Forest Reserve, July 1952, *Semsei* 832!; Zanzibar I., Massazine, 8 Sept. 1959, *Faulkner* 2345!

DISTR. **K** 7; **T** 2, 3, 6–8; **Z**; **P**; ?Malawi, also cultivated in India last century

HAB. Evergreen forest, fringing forest and bush, thicket; 0–700(–1225) m.

SYN. C. *confluens* K.Schum. in P.O.A. C: 383 (1895); T.T.C.L.: 493 (1949)
[C. *triflora* sensu K.T.S.: 437 (1961), *non* (Thonn.) K.Schum. senso stricto]

NOTE. The boundaries between this subspecies and the typical one in the Usambara and Uluguru Mts. are apparently quite sharp; subsp. *confluens* occurs commonly up to 540 m. and subsp. *triflora* descends to just below 900 m. Elsewhere subsp. *confluens* reaches 1225 m. in **K** 7 (Teita Hills, *Beentje et al.* 1173), 900 m. in Moshi District and Mahenge and 700 m. in **T** 7. Subsp. *comorensis* (Baill.) Verdc. is very similar in fruit but all specimens seen have uniformly narrower leaves and although similar leaves can be found in some specimens of subsp. *confluens* (e.g. *Peter* 55953, E. Usambaras, Pandeni to Longusa, 12 Dec. 1916) I have kept the Comoro populations separate. Some specimens from the Flora Zambesiaca area also have beaked fruits, e.g. Malawi, Karonga, *Scott Elliot* 8405, but the separation of the two subspecies in East Africa is so satisfactory that I am certain it is valid; these intermediates from other areas certainly rule out considering the two specifically distinct. Only one puzzling specimen from the Flora area has been seen — *Thulin & Mhoro* 2957 (Tanzania, Kilosa District, Ukaguru Mts., Matandu Mt. 3–5 km. WNW of Mandege Forest Station, in montane forest at 1700–1900 m., 5 June 1978). This has the foliage and adpressed indumentum of the Usambara form of subsp. *triflora* and comes from a similar habitat and altitude but the fruits (not mature) show a definite short beak. Further material may indicate this is a distinct subspecies.

2. C. sp. A

Shrub to ± 2 m., with light bark; older stems ± quadrangular and grooved, the bark longitudinally fissured to reveal reddish under-bark; young stems with short adpressed pubescence. Leaf-blades oblong-elliptic, 3–10 cm. long, 2–5 cm. wide, shortly and ± obtusely acuminate at the apex, cuneate at the base, subcoriaceous, the venation raised and closely reticulate on both surfaces, midrib narrowly channelled above, raised beneath and with adpressed whitish hairs but leaf-surface otherwise glabrous; petiole ± 5 mm. long, pubescent; stipules narrowly triangular, 7 mm. long, 3 mm. wide, keeled. Flowers not known; persistent calyx on young fruit with limb-tube 1.5 mm. long, with narrowly triangular lobes 1.2 mm. long, densely adpressed pubescent. Undeveloped fruit ± 7 mm. long, similarly pubescent.

TANZANIA. Morogoro District: N. Uluguru Mts., Lupanga Peak, 1 Aug. 1981, *Lovett* 209! & 210!
DISTR. T 6; not known elsewhere
HAB. Evergreen forest; 1900 m.

NOTE. It does not seem possible to assimilate this into *C. triflora*, but further flowering and fruiting material is needed to describe it.

New names validated by D.M. Bridson and B. Verdcourt

Galiniera saxifraga *(Hochst.) Bridson*, p. 696

Heinsenia diervilleoides *K.Schum.* subsp. **mufindiensis** *(Verdc.) Verdc.*, p. 732

Additional genus

Since this part went to press the following species, originally described from Zaire and also known from N. & W. Zambia, has been collected from T 4, Ufipa District, Zambian border, Kalambo Falls, 22 June 1986, *J. & J. Lovett* 846!

Aulacocalyx laxiflora Petit in B.J.B.B. 31: 4 (1961)

Shrub or small tree 2–8 m. tall with ± glabrous elliptic-oblong leaves to 15 × 6 cm. Inflorescences 3–10-flowered, the corolla white with adpressed pubescent cylindrical tube 0.5–2 cm. long, otherwise very similar to *Heinsenia diervilleoides* (see p. 730).

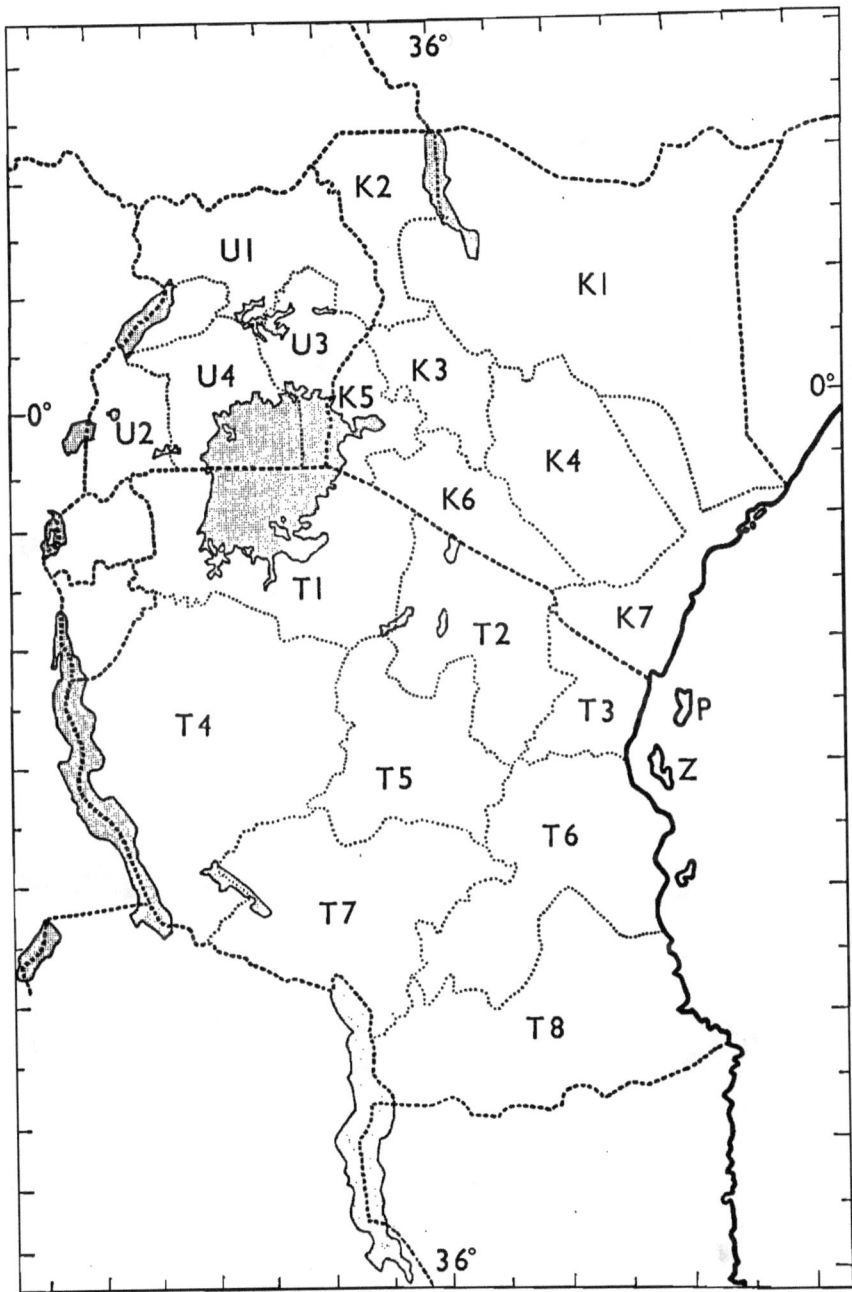

For Product Safety Concerns and Information please contact our EU
representative GPSR@taylorandfrancis.com
Taylor & Francis Verlag GmbH, Kaufingerstraße 24, 80331 München, Germany

9 789061 913375